Handbook of

NATURALLY OCCURRING COMPOUNDS

Volume II
Terpenes

Handbook of NATURALLY OCCURRING COMPOUNDS

Volume II
Terpenes

T. K. DEVON

PFIZER MEDICAL RESEARCH LABORATORIES
GROTON, CONNECTICUT

A. I. SCOTT

STERLING CHEMISTRY LABORATORY
YALE UNIVERSITY, NEW HAVEN, CONNECTICUT

Academic Press New York and London 1972

ACADEMIC PRESS, INC.
111 Fifth Avenue, New York, New York 10003

United Kingdom Edition published by
ACADEMIC PRESS, INC. (LONDON) LTD.
24/28 Oval Road, London NW1 7DD

LIBRARY OF CONGRESS CATALOG CARD NUMBER: 76-187258

PRINTED IN THE UNITED STATES OF AMERICA

CONTENTS

PREFACE

With the advent of spectroscopy as a tool in the structural elucidation of natural products there has been a rapid increase in our knowledge of the structures and variety of naturally occurring compounds. It has consequently become an increasingly difficult task to maintain current awareness in the natural products field, and the acquisition of retrospective data is a problem only partly alleviated at present by literature reviews, monographs, and compendia. It was in an attempt to pool the chemical and biochemical data of natural products that the Card Index File of Naturally Occurring Compounds was conceived and initiated at the University of Sussex in 1966. This project, the first essential step in the construction of a more comprehensive system, undertook to search for, list, and classify all reported naturally occurring compounds whose structures had been determined (with the exclusion of polymeric and macromolecular compounds). The interest so often expressed in this file led, in 1970, to the reorganization of part of the data into the publishable format now presented.

It is somewhat inevitable, when attempting so wide-ranging a project, that the selection criteria should be controversial; thus the decision to include steroidal aglycones as "natural products" is likely to be frowned upon by the purists. Others, however, may consider the absence of antibiotic degradation products a severe limitation. Similarly, the lack of inclusion, at this stage, of spectroscopic and botanical data may appear limiting; however, an attempt to be so comprehensive so rapidly would have greatly delayed the availability of the present material. It was felt that the present format and material could well stand alone in satisfying particular needs in this field; the provision of a literature reference for each compound should assist in the further acquisition of data.

No pretense of complete comprehensiveness is being claimed for this handbook, nor will it likely be found to be error-free; however, considerable care has been taken to abstract the literature as deeply and thoroughly as our resources have permitted. A certain degree of editing of material for structural and stereochemical correlations has been undertaken and has hopefully removed the bulk of errors and ambiguities from the handbook. It would, of course, be greatly appreciated if diligent users who spot errors or omissions would forward them to the authors so that subsequent yearbooks can set the records straight.

We would like to indicate our debt to the numerous authors of earlier compendia, monographs, and reviews from whose works the original Card Index File of Naturally Occurring Compounds gained much of its foundation. For the facilities enabling the construction of the file our appreciation is tendered to the Chemical Laboratory, University of Sussex. At Yale University the encouraging assistance of Drs. Phil Bays, Paul Reichardt, and Jim Sweeny was gratefully received, as also was that of John Harrison, librarian of the

Kline Science Library. The final preparation of material for publication was ably performed by Mrs. Diane Devon, assisted by Miss Sue Wilson, Miss Margaret Hufnagel, and Mrs. Pamela McDaniel, for all of whose efforts we are most appreciative. Finally, assistance from members of the Pfizer Electronic Data Processing Department in the preparation of computer generated indexes is gratefully acknowledged.

T. K. Devon
A. I. Scott

GUIDE TO HANDBOOK USAGE

Contents

This Handbook contains at present most of the known naturally occurring compounds to which structures have been assigned.

Limitations

Excluded from the Handbook are polymeric compounds, such as proteins and polysaccharides, synthetic derivatives of natural products, and degradation products (artifacts).

Data

Each structure is stored in the Handbook with its name, molecular formula, molecular weight, optical rotation (α_D), melting point, literature reference (usually the latest), and classification number.

DATA FORMAT

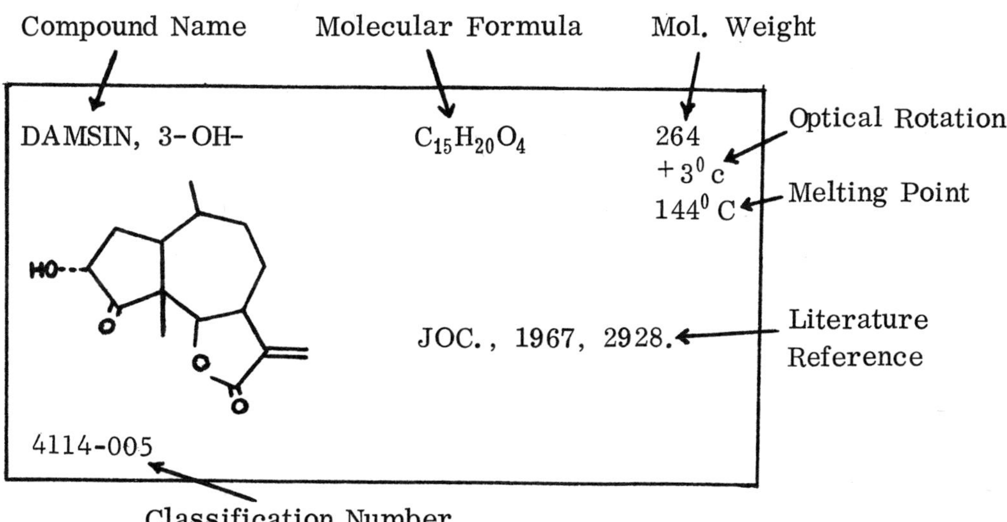

Indices

Compounds can be retrieved by name using the ALPHABETICAL INDEX or by structural type using the STRUCTURAL CLASSIFICATION GUIDES at the beginning of each class. MOLECULAR WEIGHT AND MOLECULAR FORMULA INDICES are also provided at the end of the book.

Code Numbers

CLASSIFICATION CODE NUMBERS (at the top of each page) are used to specify structural types, and the key to these is held within the individual Structural Guides. The first two digits represent the major class and the second two the subclass. Each individual compound also possesses a COMPOUND SEQUENCE NUMBER which, in combination with the Classification Code Number, supplies a unique address for that compound. The compounds are stored in the Handbook in ascending sequence, new compounds being simply inserted at the end of its appropriate section.

Classification Guide

The Catalogue of each primary class of compounds will be preceded by its corresponding Classification Guide which consists of:

Introduction
Biogenetic Chart
Main Skeleton Key
Less Common Skeletons Index

Main Volumes

The wide range of structural types of naturally occurring compounds have been classified and then collected into three groups, each of which will be issued as a separate volume:

Volume I	Acetogenins, Shikimates, and Carbohydrates
Volume II	Terpenes
Volume III	Alkaloids and Related Nitrogenous Compounds

Supplements

Each April an annual supplement for the preceding year will be published for each volume which will contain new compounds, structure changes in earlier reported compounds, and any additional data and indices. At suitable intervals these supplements will be cumulated and merged into the main volumes.

ABBREVIATIONS

Nonstandard Journal Abbreviations

Aust. J. C.	Australian Journal of Chemistry
Can. J. C.	Canadian Journal of Chemistry
Helv.,	Helvetica Chimica Acta
Ind. J. C.	Indian Journal of Chemistry
JACS	Journal of the American Chemical Society
JCS	Journal of the Chemical Society
JOC	Journal of Organic Chemistry
J. Ind. C. S.	Journal of the Indian Chemical Society

Symbols Used for Solvents (optical rotations)

a	Acetone	m	Methanol
b	Benzene	n	HCl
c	Chloroform	p	Pyridine
d	Dioxan	r	Diethyl ether
e	Ethanol	t	Carbon tetrachloride
h	Hexane	w	Water

Other Abbreviations

ac	Acetate	i	Iso
Ac	Acetyl	Me	Methyl
Amorph	Amorphous	m.e.	Methyl ester
Ang	Angeloyl	n	Normal
A/C	Absolute configuration	Ph	Phenyl
Bu^t	Butyl	pic	Picrate
Bz	Benzoyl	Pr	Propyl
Caff	Caffeoyl	Sen	Senecioyl
Cinn	Cinnamoyl	Tig	Tigloyl
Coum	Coumaroyl	Val	Valeroyl
Et	Ethyl	Van	Vanilloyl
		Ver	Veratroyl

INTRODUCTION

At the last count there were some 11,000 structurally determined naturally occurring compounds in our files: of these approximately 4,000 are *primarily* terpenoid in their biogenetic origin and have been assembled within their biogenetic/structural classes into this Volume. It should be noted that nitrogenous terpenoids (e.g. steroidal alkaloids) are considered here as primarily terpenoid in origin and therefore included in this Volume rather than Volume III (Alkaloids). In the few cases where ambiguity exists in classification (e.g. complex isopentenylation of phenolic compounds) then the compounds in question will be doubly classified.

The terpenes have been classically identified through the recognition of an "isoprene" pattern in their carbon skeletons. It is the number of these significant C_5 units in a compound that has given rise to a simple primary classification system. The organization of the isoprenyl carbon skeleton within each primary class then gives rise to the various secondary classes. These are a little more arbitrary, being selected skeletons which represent several naturally occurring compounds. Thus the terpenes have been classified primarily on their carbon number (C_{10}, C_{15}, etc.) and then on their carbon skeleton. Each terpene category has been assigned a two-digit code number (see below) and the carbon skeletons used in subclassifying these categories have been given a further two-digit code number (found at the beginning of each primary section). The skeletons used for this classification are those most commonly recurring; less common skeletons have been assigned secondary code numbers in the range 90-99 (individual sections vary slightly in their treatment of these skeletons).

The biogenetic/structural nature of the classification schemes employed is illustrated at the beginning of each primary class by a chart interrelating the skeleton types found in that category. These charts represent, for the most part, hypothetical biogenetic relationships and as such merely provide a plausible framework upon which biogenetic ideas and structures can be placed. It was felt that the inclusion of these charts might assist in providing a rapid overall picture of each class and, occasionally, in clarifying the classification of particularly unusual skeletons.

Summary of Terpenoid Classification Schemes

PCCN	Primary Class	SCCN	Secondary Class
40	Monoterpenes	01-15 99	Main Skeletons Less Common Skeletons
41	Sesquiterpenes	01-30 90 91 92 93 94 98	Main Skeletons Less Common Skeletons/ Acyclic Monocyclic Bicyclic Tricyclic Tetracyclic N-heterocyclic
42	Diterpenes	01-20 91 92 93 99	Main Skeletons Less Common Skeletons/ Unusually Cyclised Rearranged Degraded Misc.
43	Sesterterpenes	01-03	Main Skeletons
44	Triterpenes	01-29 91 92 93 94 99	Main Skeletons Less Common Skeletons/ Unusually Cyclised Rearranged Degraded Ring A contracted Misc.
45	Steroids	01-19 91 92 99	Main Skeletons Less Common Skeletons/ Rearranged Degraded Misc.
46	Carotenoids	01-06 99	Main Skeletons Less Common Skeletons
47	Polyprenoids	01	Not subclassified
49	Miscellaneous compounds of terpenoid origin		Subclassified according to number of C atoms

PCCN = Primary Code Classification Number
SCCN = Secondary Code Classification Number

40-- THE MONOTERPENES

This collection of 380 compounds has been placed into 15 well defined structural/biogenetic categories (see Main Skeleton Key) and one miscellaneous section (4099) containing 15 less common skeletons. Thus nearly 400 compounds have been indexed by 30 skeletons. The biogenetic interrelationships of these skeletons are suggested on the chart following.

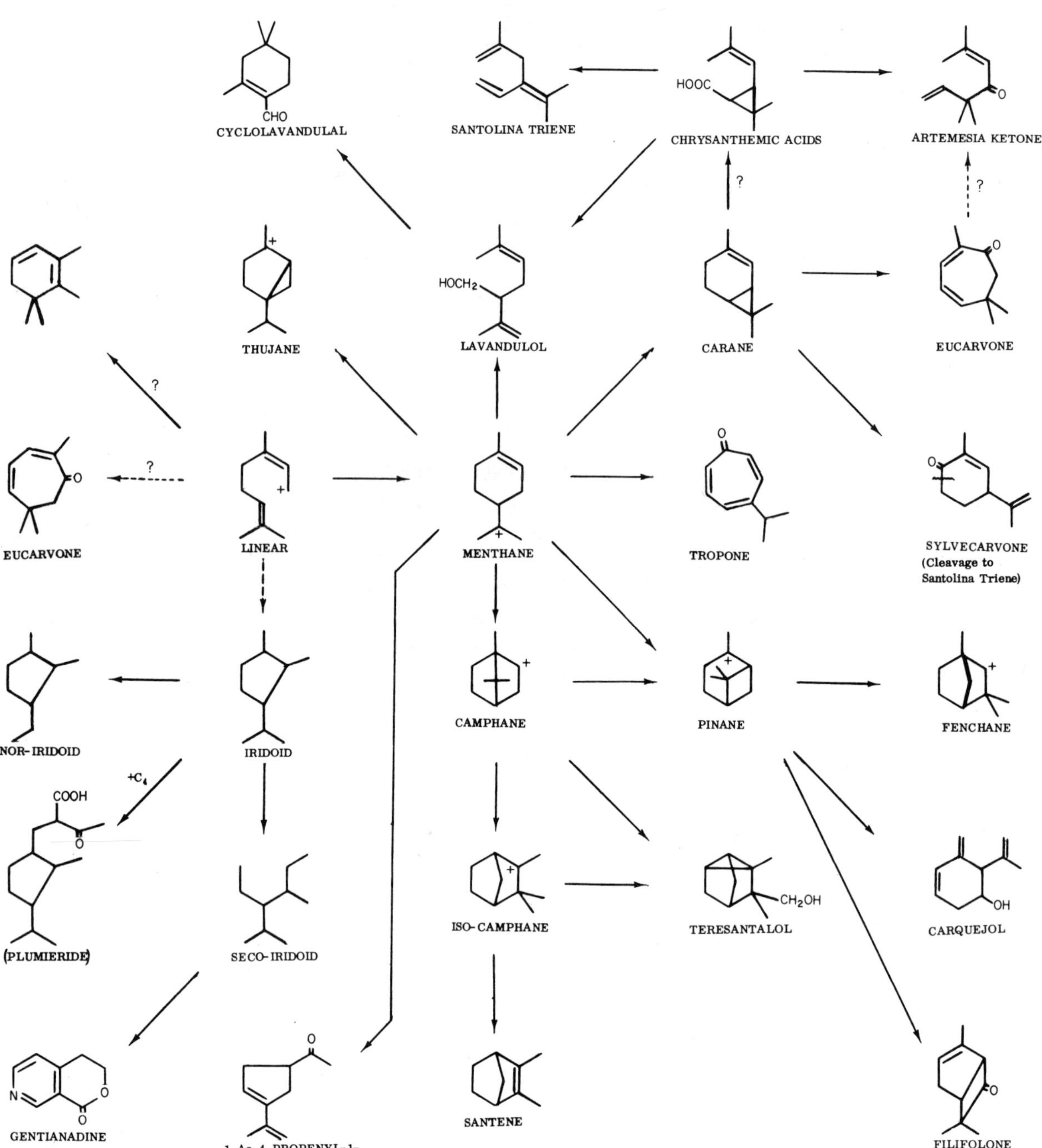

CHO
CYCLOLAVANDULAL
SANTOLINA TRIENE
HOOC
CHRYSANTHEMIC ACIDS
O
ARTEMESIA KETONE
?
?
+
HOCH₂
THUJANE
LAVANDULOL
CARANE
O
EUCARVONE
?
O
?
+
EUCARVONE
LINEAR
MENTHANE
O
TROPONE
O
SYLVECARVONE
(Cleavage to
Santolina Triene)
+
CAMPHANE
PINANE
+
FENCHANE
NOR-IRIDOID
IRIDOID
+C₄
COOH
O
(PLUMIERIDE)
SECO-IRIDOID
+
ISO-CAMPHANE
CH₂OH
TERESANTALOL
OH
CARQUEJOL
O
N
O
O
GENTIANADINE
1-Ac-4-PROPENYL-1-
CYCLOPENTENE
SANTENE
O
FILIFOLONE

40--	Monoterpenes - Main Skeleton Key

01 Linear	06 Thujane	11 Tropone
02 Menthane	07 Fenchane	12 Eucarvone
03 Pinane	08 Iridoid	13 Skytanthus Alks
04 Camphane	09 Seco-Iridoid	14 Gentiana Alks
05 Carane	10 Cyclocitral	15 Plumierides

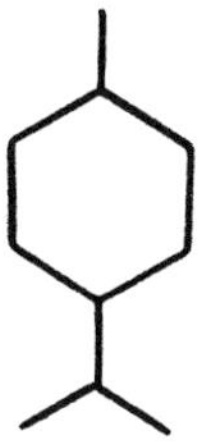

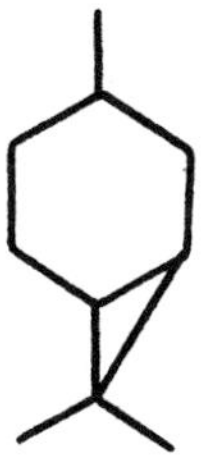

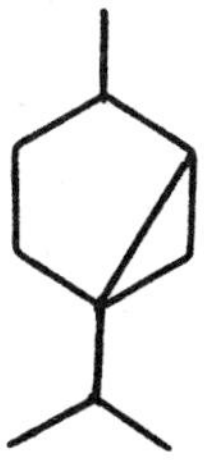

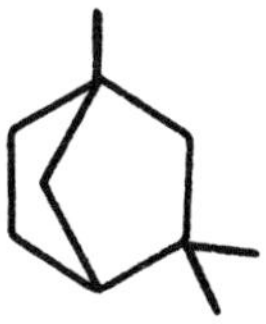

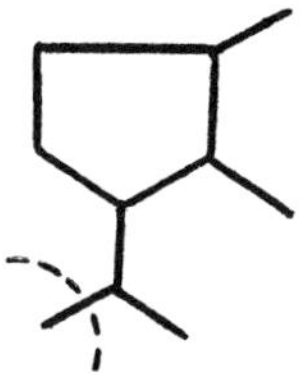

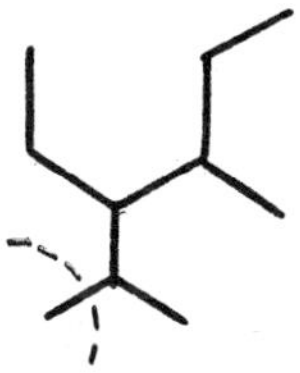

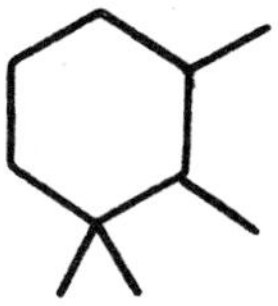

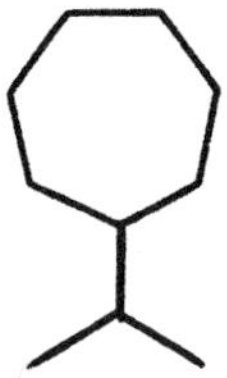

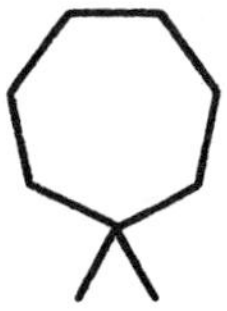

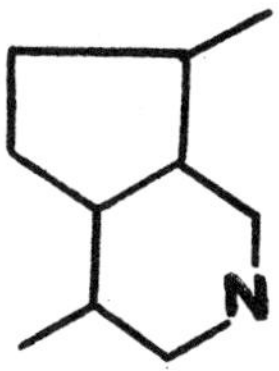

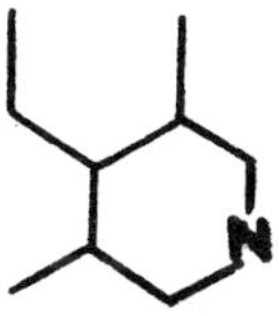

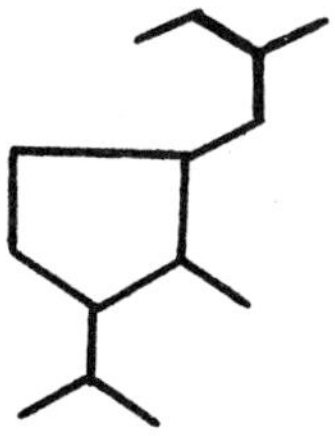

4099 Monoterpenes - Less Common Skeletons

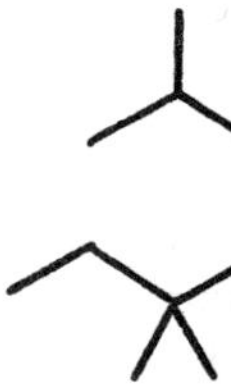

006 007 008 011	009	010 012 027
004 005	026	013 014
018	019	029 030 031 032

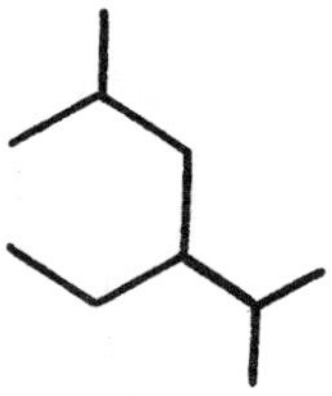

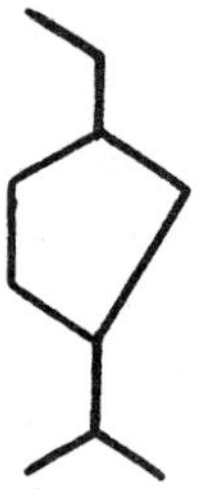

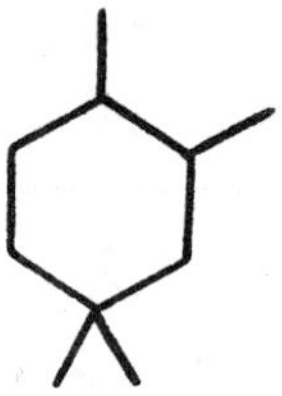

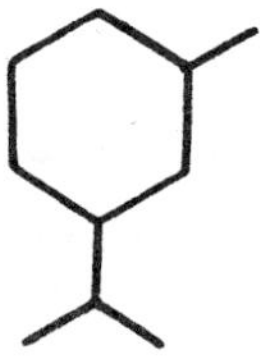

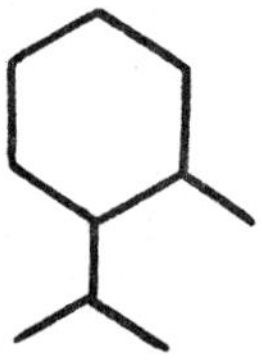

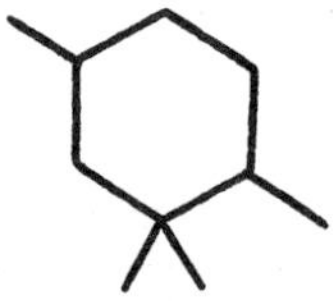

4099	Monoterpenes - Less Common Skeletons

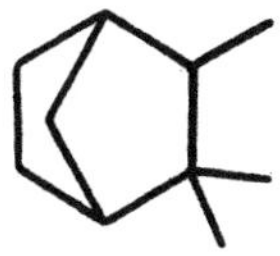 003 021 022	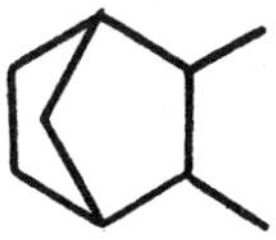 001	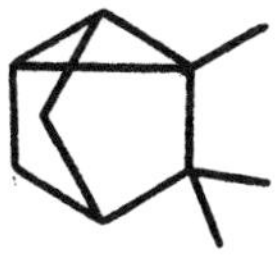 015 016
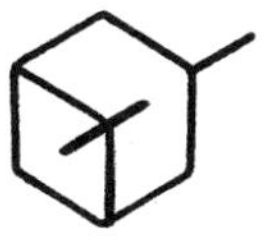 020	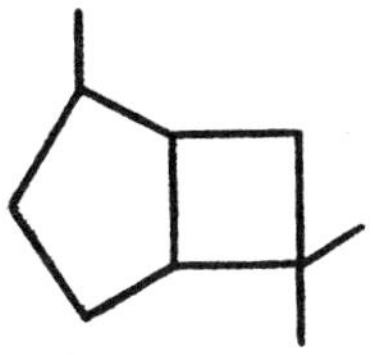 024 025	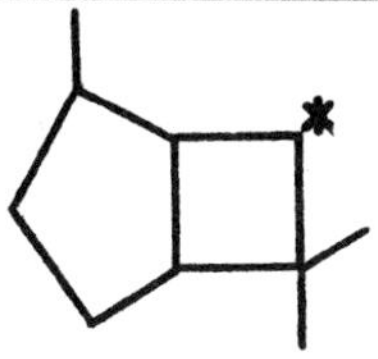 017 * attached to an orcinol ring
 023		

OCIMENE — $C_{10}H_{16}$ — 136 — OI — Oil

JACS 1952 2688

4001-001

MYRCENE — $C_{10}H_{16}$ — 136 — OI — Oil

JOC 1942 397

4001-002

HYMENTHERENE — $C_{10}H_{16}$ — 136 — -5^0 — Oil

J. Ind. Chem. Soc. 1952 (29) 23

4001-003

COSMENE — $C_{10}H_{14}$ — 134 — OI — Oil

JCS 1954 4006

4001-004

CITRONELLOL (+)-β- — $C_{10}H_{20}O$ — 156 — $+2^0$ — Oil

CH_2OH

JACS 1951 2385

4001-005

CITRONELLOL (-)-β- — $C_{10}H_{20}O$ — 156 — -4.3^0 — Oil

CH_2OH

JACS 1951 2385

4001-006

CITRONELLOL GLUCOSIDE — $C_{16}H_{30}O_6$ — 318

CH_2-O-Glucosyl

Phytochem. 1969 1339

4001-007

GERANIOL — $C_{10}H_{18}O$ — 154 — OI — Oil

CH_2OH

JCS 1955 2830

4001-008

GERANYL ACETATE — $C_{12}H_{20}O_2$ — 196 — OI — Oil

CH_2OAc

Ind. Soap J., April, 1945.

4001-009

GERANIOL-β-D-GLUCOSIDE — $C_{16}H_{28}O_6$ — 316

CH_2O—Glu

Phytochem. 1969 1341

4001-010

NEROL | $C_{10}H_{18}O$ | 154 OI Oil

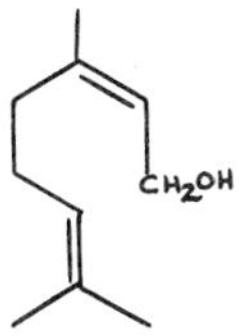

Bull. Chem. Soc. Jap. 1964 (37) 158

4001-011

TAGETOL | $C_{10}H_{18}O$ | 154 -17^{0}e Oil

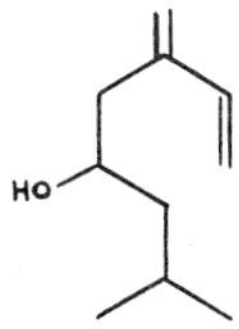

Tet 1966 1929

4001-016

NEROL GLUCOSIDE | $C_{16}H_{28}O_6$ | 316

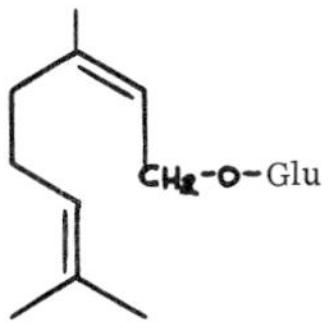

Phytochem 1969 1339

4001-012

MYRCENOL | $C_{10}H_{16}O$ | 152 Oil

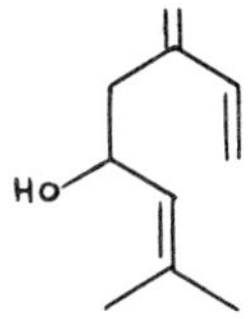

Science 1966 (154) 509

4001-017

LINALOOL, d- | $C_{10}H_{18}O$ | 154 $+19^{0}$ Oil

Tet. 1962 (18) 37

4001-013

TAGETONE, trans- | $C_{10}H_{16}O$ | 152 OI Oil

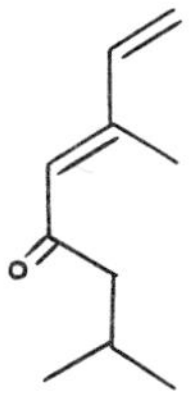

JCS 1963 2535

4001-018

LINALOOL, l- | $C_{10}H_{18}O$ | 154 -20^{0} Oil

Tet. 1962 (18) 37

4001-014

TAGETONE, cis- | $C_{10}H_{16}O$ | 152 OI Oil

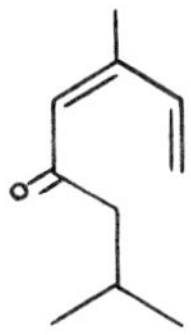

JCS 1963 2535

4001-019

BUPLEUROL | $C_{10}H_{20}O$ | 156 OI Oil

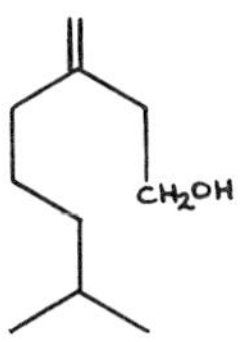

Chem. Zentr., I 1926, 1304.

4001-015

MYRCENOL, iso- | $C_{10}H_{16}O$ | 152 OI Oil

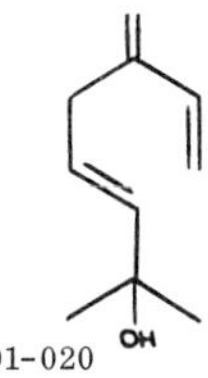

Tet. 1966 1929

4001-020

MYRCENONE $C_{10}H_{14}O$ 150 OI Oil Helv. 1948 (31) 29 4001-021	CITRAL, cis- $C_{10}H_{16}O$ 152 OI Oil CHO Compt. Rendu. 1923, (177) 669 4001-026
OCIMENONE $C_{10}H_{14}O$ 150 OI Oil Helv. 1948 (31) 29 4001-022	CITRAL, trans- $C_{10}H_{16}O$ 152 OI Oil CHO Compt. Rendu. 1923(177) 669 4001-027
OCT-7-EN-4-ONE, 2, 6-di-Me- $C_{10}H_{18}O$ 154 +1.5^{0} Oil J. Ind. Chem. Soc., 1964 (41) 752 4001-023	CITRONELLIC ACID $C_{10}H_{18}O_2$ 170 -6.6^{0} Oil COOH Chem. Abstr., 1933 (27) 1495 4001-028
CITRONELLAL, D- $C_{10}H_{18}O$ 154 +12.5^{0} Oil CHO Tet. Lett. 1959 (3) 1 4001-024	PERILLAKETONE $C_{10}H_{14}O_2$ 166 OI Oil Chem. Ind. 1962 1618 4001-029
CITRONELLAL, L- $C_{10}H_{18}O$ 154 - 14^{0} Oil CHO JCS 1950 3457 4001-025	BATATIC ACID $C_{10}H_{12}O_4$ 196 +17.5^{0} 89^{0}C COOH Chem. Ind. 1954 1427 4001-030

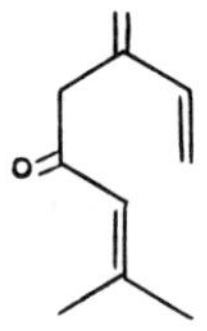

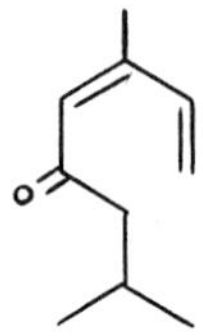

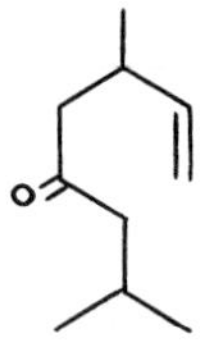

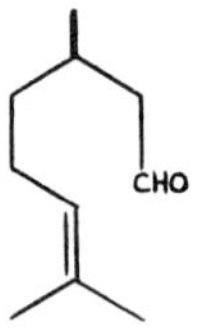

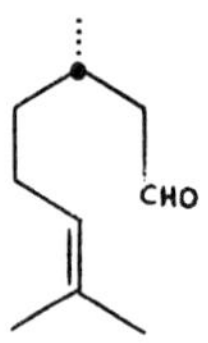

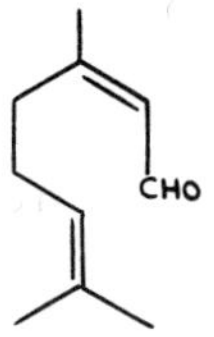

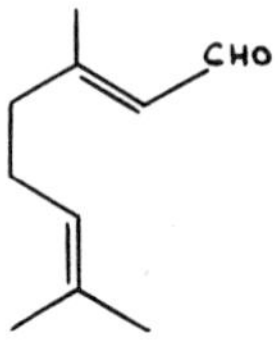

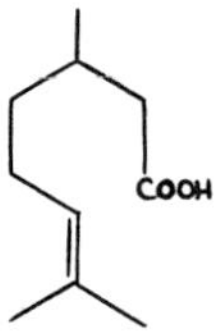

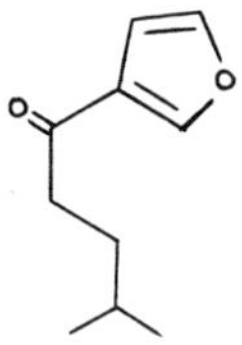

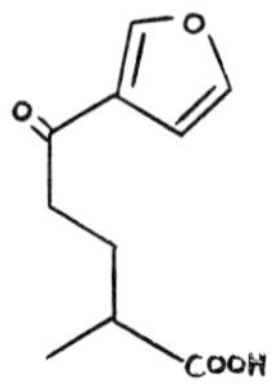

Name	Formula	MW / mp / state	Reference	No.
PERILLENE	$C_{10}H_{14}O$	150 OI Oil	Ber., 1936(62)2459	4001-031
ROSEFURAN	$C_{10}H_{14}O$	150 OI Oil	JACS 1968 1227	4001-036
CLAUSENANE, α-	$C_{10}H_{12}O$	148 OI Oil	J. Sci. Ind. Res. 1948(7B) 11	4001-032
IPOMEANINE (COOMe)	$C_9H_{10}O_3$	166 OI Oil	Chem. Ind. 1954 902	4001-037
EGOMAKETONE	$C_{10}H_{12}O_2$	164 OI Oil	Chem. Ind. 1962 1618	4001-033
LINALOOL EPOXIDE (OH)	$C_{10}H_{18}O_2$	170 -6° Oil	Tet. Lett.1962 769	4001-038
ELSHOLTZIONE	$C_{10}H_{14}O_2$	166 OI Oil	Helv. 1931 (14) 1277	4001-034
GERANIC ACID (COOH)	$C_{10}H_{16}O_2$	168 OI Oil	Nature 1962 (194) 704	4001-039
ELSHOLTZIONE, β-dehydro- (Naginata ketone)	$C_{10}H_{12}O_2$	164 OI Oil	Chem. Ind. 1960 236	4001-035
NEROLIC ACID (COOH)	$C_{10}H_{16}O_2$	168 OI Oil	Nature 1962 (194) 704	4001-040

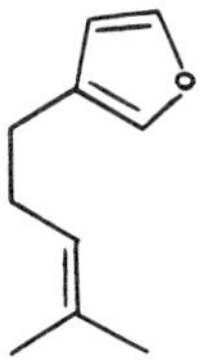

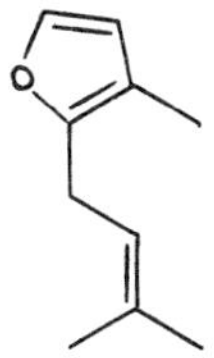

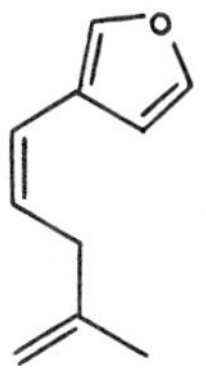

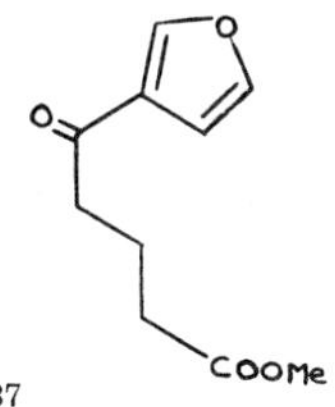

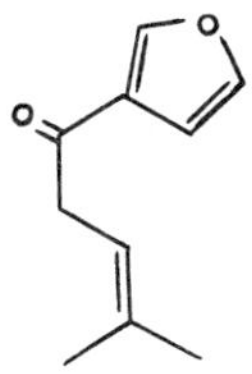

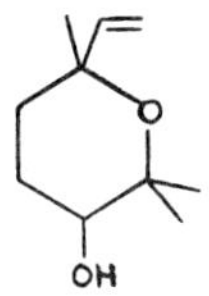

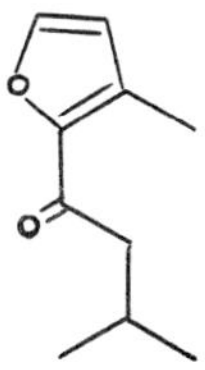

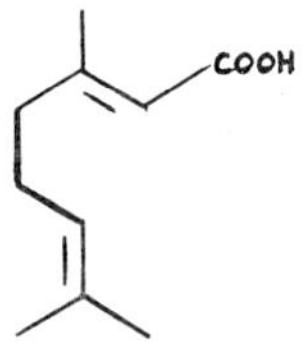

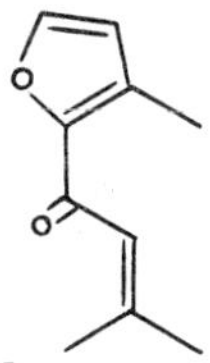

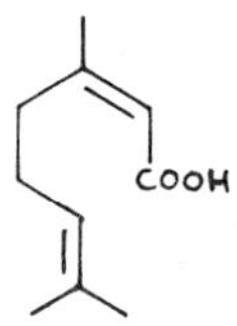

deleted 4001-041	LAVENDER PYRAN, Dehydro- $C_{10}H_{18}O$ 154 - 18^{0} Oil Bull. Soc. Chim. Fr., 1961, 645. 4001-046
LINALOYL OXIDE, DL- $C_{10}H_{18}O$ 154 Rac. Oil Helv., 1966, 2055. 4001-042	LAVENDER PYRAN, Dehydro-epi- $C_{10}H_{18}O$ 154 -15^{0} Oil Bull. Soc. Chim. Fr., 1961, 645. 4001-047
OCIMENOYL OXIDE, DL- $C_{10}H_{18}O$ 154 Rac. Oil Helv., 1966, 2055. 4001-043	EGOMAKETONE, iso- $C_{10}H_{12}O_2$ 164 OI Oil Aust. J. C., 1966 (19) 891. 4001-048
LAVENDER PYRAN $C_{10}H_{20}O$ 156 +5^{0} Oil Bull. So. Chem. Fr., 1961, 645. 4001-044	ELSHOLTZIDIOL $C_{10}H_{16}O_3$ 184 59^{0}C Experimentia, 1970 (26) 817. 4001-049
LAVENDER PYRAN, Epi- $C_{10}H_{20}O$ 156 +6^{0} Oil Bull. Soc. Chem. Fr., 1961, 645. 4001-045	LILAC ALCOHOL A $C_{10}H_{18}O_2$ 170 +15^{0}c Bull. Chem. Soc. Jap., 1970, 3319. 4001-050

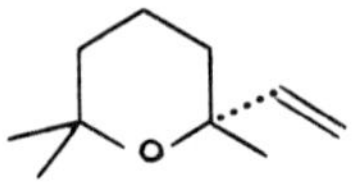

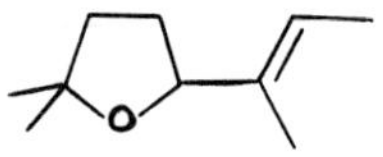

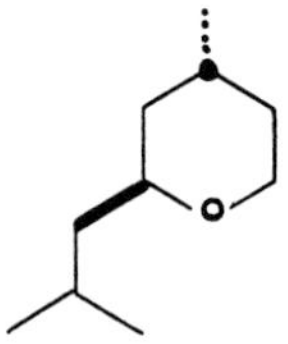

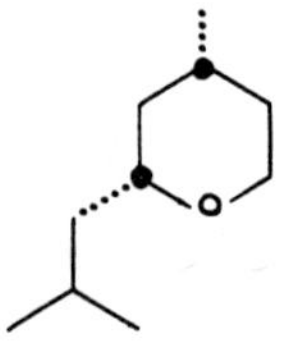

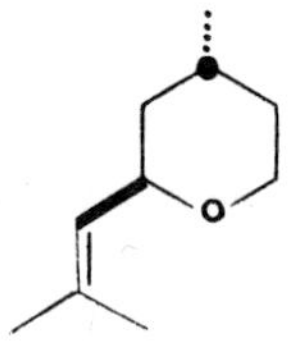

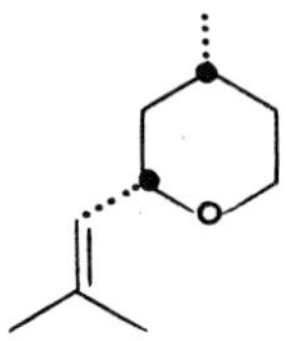

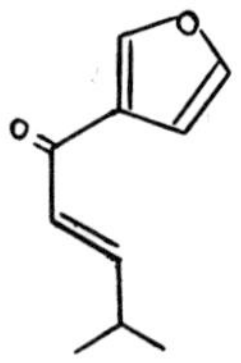

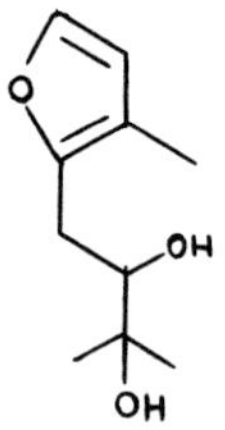

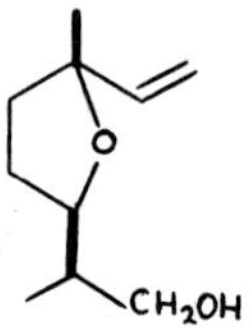

LILAC ALCOHOL B

$C_{10}H_{18}O_2$

170
-2^0c

Stereo-isomer of
4001-050

Bull. Chem. Soc. Jap., 1970, 3319.

4001-051

Compound	Formula	Data	Reference	No.
LIMONENE, D-	$C_{10}H_{16}$	136 +127^0 Oil	JCS 1937 152	4002-001
PHELLANDRENE, D-β-	$C_{10}H_{16}$	136 +65^0 Oil	Ann. 1905, 2	4002-006
LIMONENE, L-	$C_{10}H_{16}$	136 -123^0 Oil	Ber. 1930 1724	4002-002
PHELLANDRENE, L-β-	$C_{10}H_{16}$	136 -74^0 Oil	JCS., 1947, 1039.	4002-007
LIMONENE, iso-	$C_{10}H_{16}$	136 -140^0 Oil	JACS 1920 1204	4002-003
TERPINENE, α-	$C_{10}H_{16}$	136 OI Oil	JACS 1935 586	4002-008
PHELLANDRENE, D-α-	$C_{10}H_{16}$	136 +115^0 Oil	JCS 1930 2781	4002-004
TERPINENE, γ -	$C_{10}H_{16}$	136 OI Oil	JOC 1942 397	4002-009
PHELLANDRENE, L-α-	$C_{10}H_{16}$	136 -177^0 Oil	JCS 1923 1657	4002-005
TERPINOLENE	$C_{10}H_{16}$	136 OI Oil	JOC 1942 399	4002-010

MENTHA-1, 3-8-TRIENE, p- $C_{10}H_{14}$ 134
OI
Oil

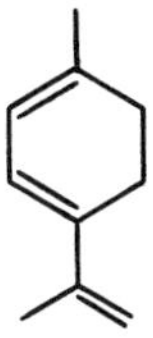

Bull. Soc. Chim. Fr., 1967, 4679.

4002-011

CYMENE, p- $C_{10}H_{14}$ 134
OI
Oil

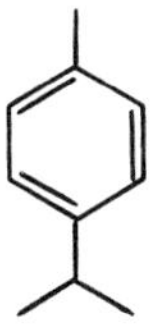

JACS., 1924, 686.

4002-012

CYMENENE, p- $C_{10}H_{12}$ 132
OI
Oil

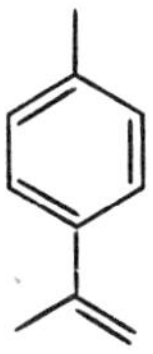

JCS., 1942, 188.

4002-013

CARVEOL, D-dihydro- $C_{10}H_{18}O$ 154
$+34^0$
Oil

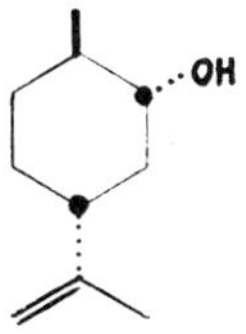

Ber. 1950 (83) 193

4002-014

CARVEOL, L-iso-dihydro- $C_{10}H_{18}O$ 154
-28^0 e
39^0C

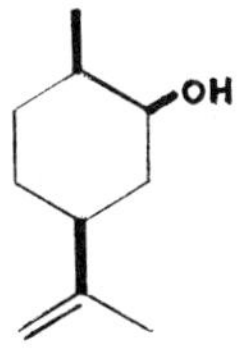

Ber. 1950 (83) 193

4002-015

CARVEOL, D-neo-dihydro- $C_{10}H_{18}O$ 154

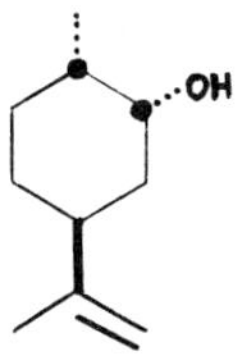

Ber. 1950 (83) 193

4002-016

CARVEOL, L-neo-dihydro- $C_{10}H_{18}O$ 154
-33^0
215^0C

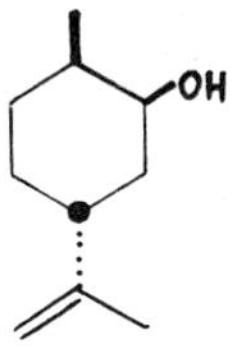

Ber. 1950 (83) 193

4002-017

TERPINEOL, D-α- $C_{10}H_{18}O$ 154
$+95^0$
39^0C

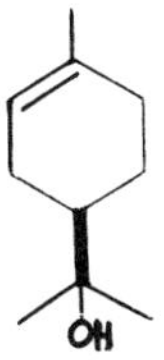

Ann. 1949 (564) 109

4002-018

TERPINEOL, L-α- $C_{10}H_{18}O$ 154
-106^0r
38^0C

Ann. 1949 (564) 109

4002-019

TERPINEOL, DL-α- $C_{10}H_{18}O$ 154
Rac.
36^0C

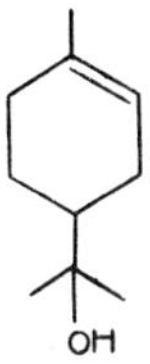

Chem. Abstr., 1965, (62) 13513

4002-020

Compound	Formula	Data	Reference	No.
TERPINEOL, dihydro-α-	$C_{10}H_{20}O$	156 35°C	Ber. 1927 (60) 1372	4002-021
PIPERITOL, d-	$C_{10}H_{18}O$	154 +46° Oil	JCS 1930 2770	4002-026
TERPINEOL, γ-	$C_{10}H_{18}O$	154 70°C	Ind. For. Rec., 1923 (10) 1	4002-022
PIPERITOL, l-	$C_{10}H_{18}O$	154 -35° Oil	JCS 1930 2770	4002-027
CARVEOL	$C_{10}H_{16}O$	152	JCS 1934 233	4002-023
PULEGOL, iso-	$C_{10}H_{18}O$	154 Oil	Chem. Abstr., 1932 (26) 4679	4001-028
MENTHOL, L-	$C_{10}H_{20}O$	156 -50° 43°C	JCS., 1934, 1779.	4002-024
MENTHA-1(7), 8-DIEN-2-OL, (+)-cis-	$C_{10}H_{16}O$	152 +94° Oil	Bull. Soc. Chim. Fr. 1960,37	4002-029
MENTHOL, D-neo-	$C_{10}H_{20}O$	156 +20° Oil	JCS., 1934, 1779.	4002-025
MENTHA-2, 8-DIEN-1-OL, (+)-cis-	$C_{10}H_{16}O$	152 +180° Oil	Bull. Soc. Chim. Fr. 1960 37	4002-030

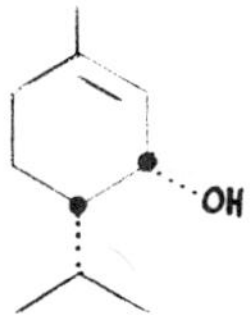

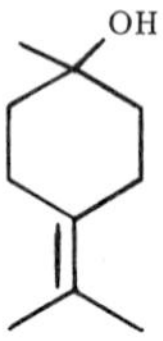

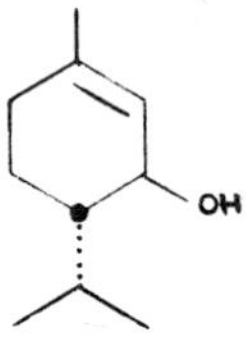

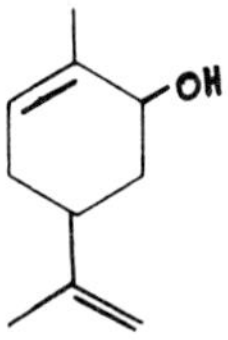

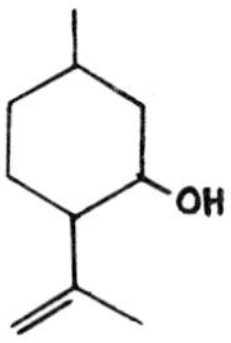

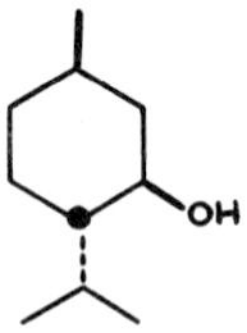

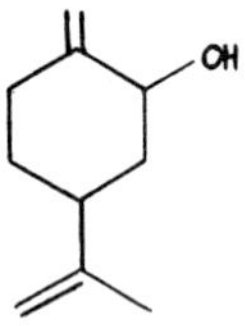

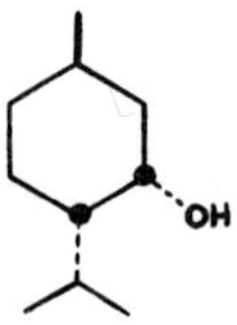

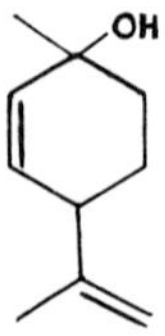

No.	Name	Formula	Data	Reference
4002-031	MENTHA-2, 8-DIEN-1-OL, (+)-trans-	$C_{10}H_{16}O$	152 $+64^0$ Oil	Bull. Soc. Chim. Fr. 1960 37
4002-032	SOBREROL, (+)-cis-	$C_{10}H_{18}O_2$	170 $+15^0$ 108^0C	Ber. 1953 (86) 1437
4002-033	SOBREROL, (+)-trans-	$C_{10}H_{18}O_2$	170 $+150^0$ c 149^0C	Ber. 1953 (86) 1437
4002-034	PERILLYL ALCOHOL, d-	$C_{10}H_{16}O$	152 $+12^0$ Oil	Helv. 1945 1220
4002-035	PERILLYL ALCOHOL, l-	$C_{10}H_{16}O$	152 -18^0 Oil	Helv. 1945 1220
4002-036	UROTERPENOL	$C_{10}H_{18}O_2$	170 Oil	Biochem. J. 1966 (101) 727
4002-037	MULLILAM DIOL	$C_{10}H_{18}O_3$	186 $\pm 0^0$ e 171^0C	Tet. 1967 2495
4002-038	MENTHANDIOL, (-)-2, 5-trans-	$C_{10}H_{20}O_2$	172 -13^0 m 144^0C	Tet. Lett. 1967 3357
4002-039	ARTEMISOL	$C_{10}H_{18}O$	154 Oil	JOC 1941 612
4002-040	TERPIN, cis-	$C_{10}H_{20}O_2$	172 104^0C	Chem. Abstr. 1942 (36) 1441

TERPIN, trans-

$C_{10}H_{20}O_2$ 172

157°C

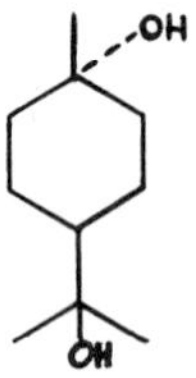

Chem. Abstr. 1942 (36) 1441

4002-041

CARVACROL, para-OMe-

$C_{11}H_{16}O_2$ 180 OI Oil

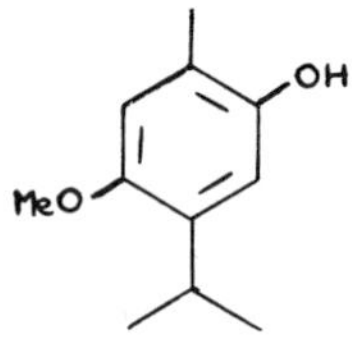

JOC 1955 443

4002-046

TERPINEN-4-OL, D-

$C_{10}H_{18}O$ 154 +25° Oil

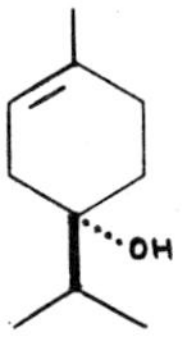

Helv. 1965 10

4002-042

THYMOQUINOL

$C_{10}H_{14}O_2$ 166 OI 140°C

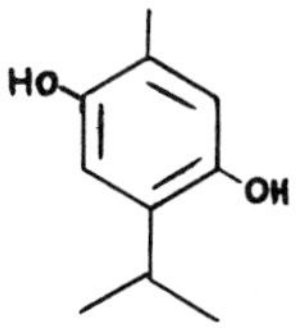

JOC 1955 82

4002-047

TERPINEN-4-OL, DL-

$C_{10}H_{18}O$ 154 Rac. 212-4°C

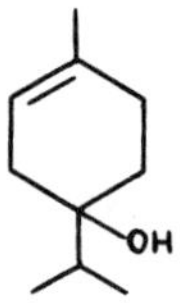

Chem. Abstr. 1934 3524

4002-043

THYMOQUINOL, di-Me-Ether-

$C_{12}H_{18}O_2$ 194 OI Oil

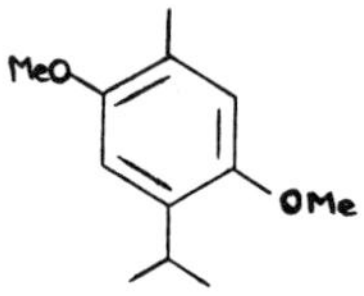

4002-048

CARVACROL

$C_{10}H_{14}O$ 150 OI Oil

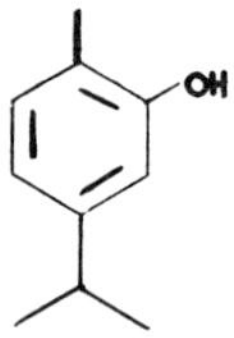

Planta Med., 1965, (13) 56.

4002-044

THYMOL

$C_{10}H_{14}O$ 150 OI 51°C

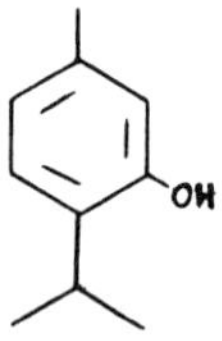

Arch. Pharm., 1966, (299) 468.

4002-049

CARVACROL METHYL ETHER

$C_{11}H_{16}O$ 164 OI Oil

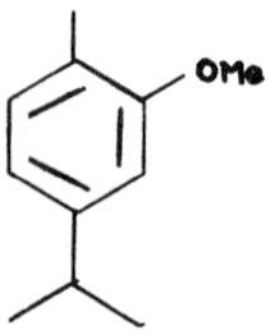

Chem. Abstr. 1954 8756

4002-045

THYMOL METHYL ETHER

$C_{11}H_{16}O$ 164 OI Oil

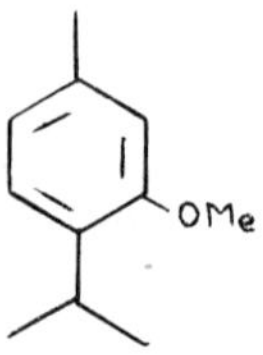

Chem. Abstr. 1947 3585

4002-050

Compound	Formula	Data	Reference	No.
THYMOL, para-OMe-	$C_{11}H_{16}O_2$	180 OI Oil	JOC 1955 443	4002-051
CINEOL, 1, 8-	$C_{10}H_{18}O$	154 Oil	Tet. Lett. 1959 (3) 1	4002-056
CUMINYL ALCOHOL	$C_{10}H_{14}O$	150 OI Oil	Chem. Abstr. 1951 (45) 8207	4002-052
PINOL, (+)-dihydro-	$C_{10}H_{18}O$	154 + 104^0 Oil	Chem. Ber. 1950 193	4002-057
LIBOCEDROL	$C_{22}H_{30}O_4$	358 OI 87^oC	JOC 1955 788	4002-053
PULEGONE, D-	$C_{10}H_{16}O$	152 $+28^0$ Oil	Chem. Abstr., 1952 1622	4002-058
ASCARIDOL	$C_{10}H_{16}O_2$	168 Oil	JACS 1949 1133	4002-054
PULEGONE, D-iso-	$C_{10}H_{16}O$	152 $+34^0$ Oil	Chem. Abstr., 1954 3932	4002-059
CINEOL, 1, 4-	$C_{10}H_{18}O$	154 Oil	Chem. Abstr. 1950 10266	4002-055
SANTOLINENONE, α-	$C_{10}H_{16}O$	152 Oil	Gazz., 1915, 167.	4002-060

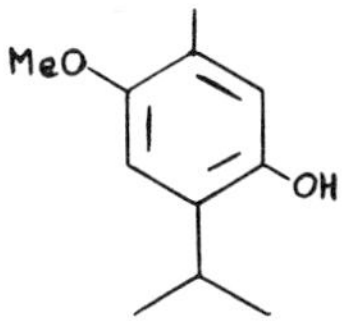

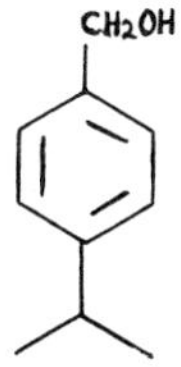

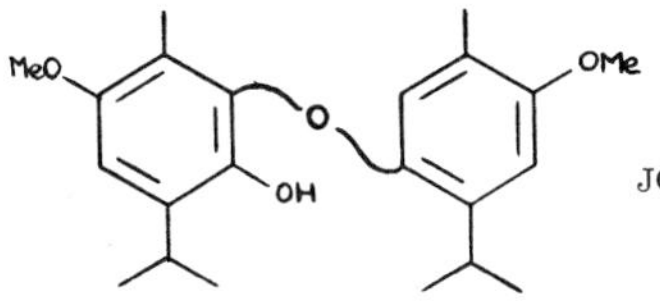

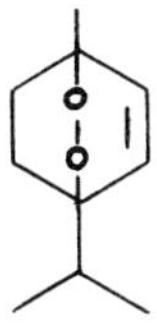

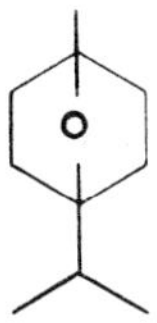

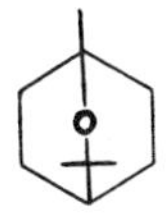

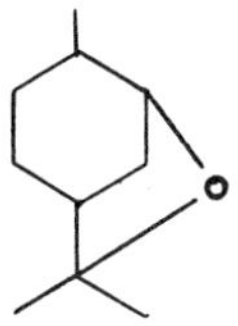

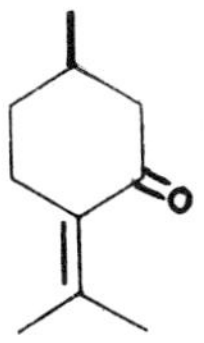

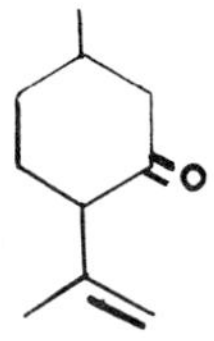

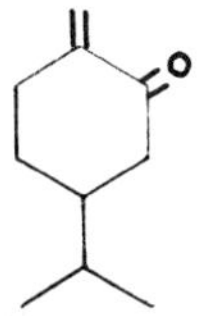

Name	Formula	Data	Reference	No.
MENTHONE, d-	$C_{10}H_{18}O$	154 + 28^0 205^{0}C	Chem. Abstr. 1955 3478	4002-061
PIPERITONE, l-	$C_{10}H_{16}O$	152 -68^0 Oil	JCS 1921 1644	4002-066
MENTHONE, l-	$C_{10}H_{18}O$	154 -29^{0}c Oil	Chem. Abstr. 1955 3478	4002-062
PIPERITENONE	$C_{10}H_{14}O$	150 Oil	Helv. 1943 162	4002-067
MENTHONE, l-iso-	$C_{10}H_{18}O$	154 -94^0 Oil	Chem. Abstr. 1954 9625	4002-063
PIPERITENONE-OXIDE	$C_{10}H_{14}O_2$	166 +151^0 26^{0}C	JACS 1956 5022	4002-068
MENTHONE, d-iso-	$C_{10}H_{18}O$	154 +93^0 Oil	Chem. Abstr. 1954 9625	4002-064
PIPERITONE-OXIDE, l-	$C_{10}H_{16}O_2$	168 -177^{0}e 15^{0}C	JACS 1956 3792	4002-069
PIPERITONE, d-	$C_{10}H_{16}O$	152 +62^0 Oil	JCS 1921 1644	4002-065
PIPERITENONE, iso-	$C_{10}H_{14}O$	150 Oil	Helv. 1942 732	4002-070

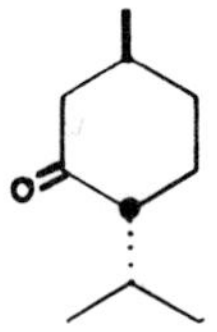

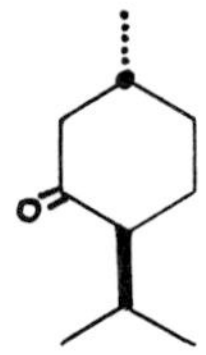

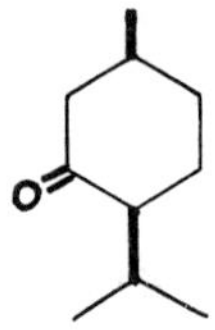

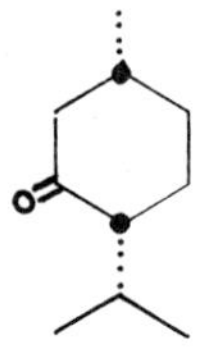

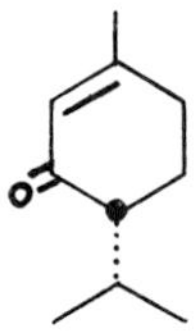

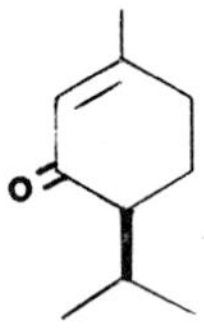

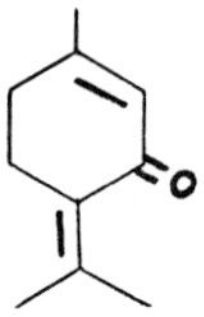

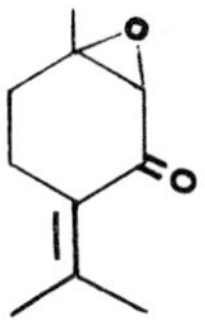

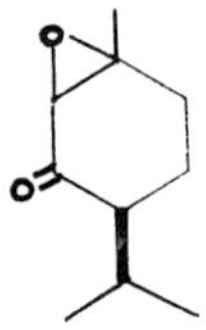

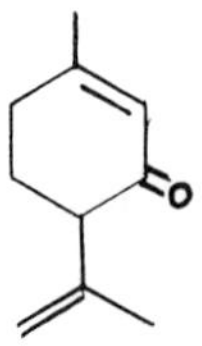

Compound	Formula	Data	Reference	No.
CARVONE, d-	$C_{10}H_{14}O$	150 +69° Oil	Chem. Abstr. 1953 6610	4002-071
LIPPIONE	$C_{10}H_{14}O_2$	166 Oil	JACS 1956 5022	4002-076
CARVONE, L-	$C_{10}H_{14}O$	150 -62° Oil	Phytochemistry, 1966 , 823.	4002-072
DIOSPHENOLENE	$C_{10}H_{14}O_2$	166 OI 75°C	Helv. 1966 (49) 2012	4002-077
CARVONE, dihydro-l-	$C_{10}H_{16}O$	152 -19° Oil	Chem. Abstr. 1954 11731	4002-073
BUCCOCAMPHOR	$C_{10}H_{16}O_2$	168 84°C	Ber. 1906 (39) 1160	4002-078
CARVOMENTHONE, l-	$C_{10}H_{18}O$	154 -25° Oil	JCS 1922 876	4002-074
PHELLANDRAL, d-	$C_{10}H_{16}O$	152 +123°	JCS 1937 1448	4002-079
CARVOTACETONE, d-	$C_{10}H_{16}O$	152 +60° Oil	JCS 1922 876	4002-075
PHELLANDRAL, l-	$C_{10}H_{16}O$	152 -139° Oil	JACS 1953 4851	4002-080

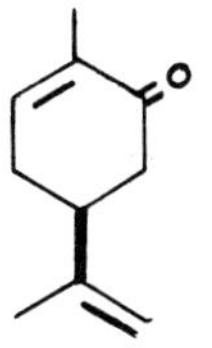

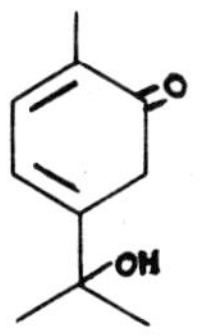

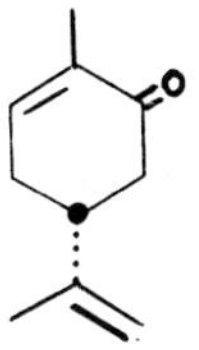

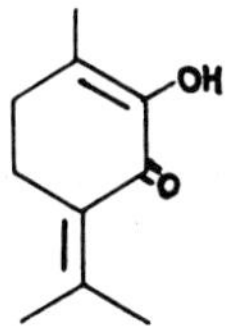

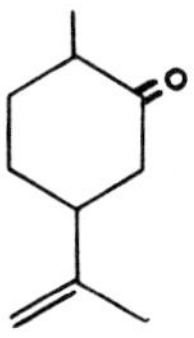

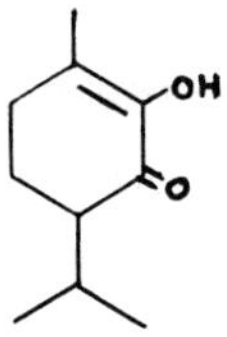

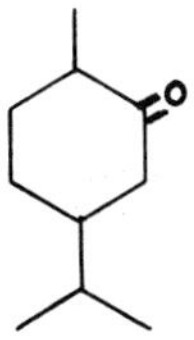

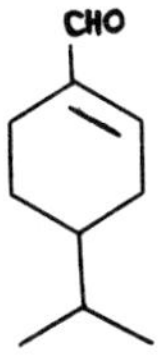

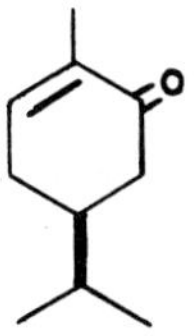

Name	Formula	Data	Reference	No.
PERILLA-ALDEHYDE, d-	$C_{10}H_{14}O$	150 + 136° Oil	Chem. Abstr. 1942 6754	4002-081
PERILLA-ALDEHYDE, l-	$C_{10}H_{14}O$	150 -146° Oil	Ber., 1911 (44) 53 & 460	4002-082
CUMINALDEHYDE	$C_{10}H_{12}O$	148 OI Oil	JCS 1922 266	4002-083
OLEUROPEIC ACID	$C_{10}H_{16}O_3$	184 -117°c 163°C	Gazz. 1965(95) 1279	4002-084
OLEUROPIC ACID SUCROSIDE	$C_{22}H_{36}O_{13}$	508 - 4°w 200 °C	Tet. Lett. 1966 5673	4002-085
PHELLANDRINIC ACID	$C_{10}H_{16}O_2$	168 144° C	JCS 1940 808	4002-086
BURSERA LACTONE	$C_{10}H_{14}O_3$	182 +61° 191°C	JCS 1964 4254	4002-087
DIPENTENE dl - Limonene	$C_{10}H_{16}$	136 Rac. Oil	Ber. 1930 1724	4002-088
TERPINENE, β-	$C_{10}H_{16}$	136 OI Oil	Chem. Abstr. 1950 (44) 4637	4002-089
LIMONENE 1, 2-epoxide	$C_{10}H_{16}O$	152 Oil	Bull. Soc. Chim. Fr. 1960 37	4002-090

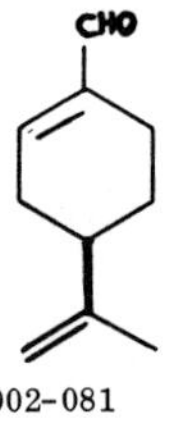

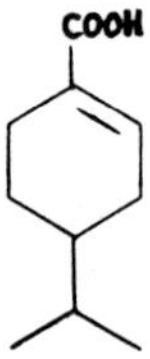

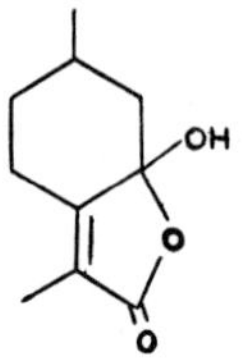

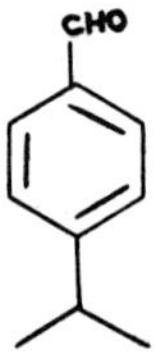

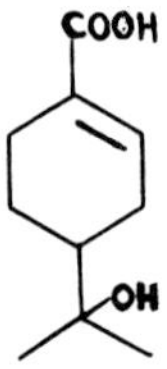

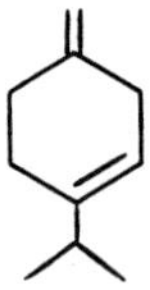

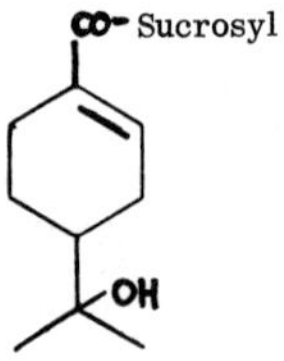

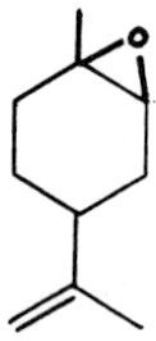

Structure labels: CHO; CHO; CHO; COOH, OH; CO- Sucrosyl, OH; COOH; OH, O, O; O

YABUNIKKEOL , Cis-
$C_{10}H_{16}O$ 152 $+41^0$ 34^0C
Bull. Chem. Soc. Jap., 1970, 1599.
4002-091

PEPERIC ACID
$C_{10}H_{14}O_3$ 182 188^0C
JACS , 1950, 399.
4002-096

YABINIKKEOL , Trans-
$C_{10}H_{16}O$ 152 -18^0 Oil
Bull. Chem. Soc. Jap., 1970 , 1599.
4002-092

MACROPONE
$C_{10}H_{12}O_2$ 164 OI Oil
Chem. Abstr. 1954 12019
4002-097

MENTHA-2, 8-DIEN-1-OL
$C_{10}H_{16}O$ 152 Oil
Phytochemistry , 1969 , 1671 .
4002-093

ANTHEMOL
$C_{10}H_{16}O$ 152 OI Oil
Ann. , 1879 (195) 92.
4002-098

EVODONE
$C_{10}H_{12}O_2$ 164 -60^0 74^0C
Aust. J. C., 1956, 241.
4002-094

PULEGOL, D-iso-
$C_{10}H_{18}O$ 154 $+34^0$ Oil
Chem. Abstr. 1953 10179
4002-099

MENTHOFURAN
$C_{10}H_{14}O$ 150 $+93^0$ Oil
JACS , 1950 , 5313.
4002-095

MENTHYL ACETATE, 1, 2-epoxy-
$C_{12}H_{20}O_3$ 212
Pharm. Weekbl., 1970 (105) 733.
4002-100

HELENIUM LACTOL $C_{10}H_{10}O_2$ 162

75°C

Tet. Lett., 1969, 4703.

4002-101

$C_{18}H_{24}O_5$ 320

Oil

* epoxide

R_1 = R_2 = i-butyroyl

Ber., 1969 (102) 864.

4002-105

SHISOOL, cis- $C_{15}H_{28}O$ 224

Oil

Bull. Chem. Soc. Jap., 1970. 2637.

4002-102

$C_{19}H_{26}O_5$ 334

Oil

* epoxide

R_1 = α-Me-Butyroyl

R_2 = i-Butyroyl

Ber., 1969 (102) 864.

4002-106

SHISOOL, trans- $C_{15}H_{28}O$ 224

Oil

Bull. Chem. Soc. Jap., 1970, 2637.

4002-103

BRUCEOL, deoxy- $C_{19}H_{20}O_4$ 312

Chem. Comm., 1968, 368.

4002-107

$C_{18}H_{26}O_4$ 306

OI

Oil

Ber., 1969 (102) 864.

R_1 = R_2 = i-Butyroyl

4002-104

PINENE, d-α-	$C_{10}H_{16}$	136 $+51^0$ Oil	PINOCAMPHEOL, Trans-L-	$C_{10}H_{18}O$	154 -55^0e 68^0 C
	JACS 1931 1030			JACS 1937 2509	
4003-001			4003-006		
PINENE, l-α-	$C_{10}H_{16}$	136 -51^0 Oil	PINOCAMPHEOL, Cis-L-	$C_{10}H_{18}O$	154 -36^0b 57^0C
	JACS 1931 1030			Chem. Abstr., 1960 (54)23203.	
4003-002			4003-007		
PINENE, d-β-	$C_{10}H_{16}$	136 $+21^0$ Oil	MYRTENOL, d-	$C_{10}H_{16}O$	152 $+46^0$ Oil
	Perf. Ess. Oil Rec. 1959 (50) 823			Chem. Abstr. 1943 4064	
4003-003			4003-008		
PINENE, l-β-	$C_{10}H_{16}$	136 -21^0 Oil	MYRTENOL, l-	$C_{10}H_{16}O$	152 -46^0 Oil
	Chem. Abstr. 1952 (46) 6333			Chem. Abstr. 1943 (37) 4716	
4003-004			4003-009		
PINOCARVEOL, l-	$C_{10}H_{16}O$	152 -62^0 6^0 C	VERBENOL, cis-	$C_{10}H_{16}O$	152 $+65^0$ Oil
	JOC 1955 1003			Science 1966 (154) 509	
4003-005			4003-010		

Name	Formula	Data	Name	Formula	Data
VERBENOL, trans-	$C_{10}H_{16}O$	152 +168^0 Oil	CHRYSANTHENONE	$C_{10}H_{14}O$	150 -88^0 Oil
4003-011	J. Insect Physiol. 1969 (15) 363		4003-016	Chem. Ind. 1958 293	
BENIHIOL	$C_{10}H_{18}O$	154 +23^0 Oil	MYRTENAL, d-	$C_{10}H_{14}O$	150 + 13^0 Oil
4003-012	Chem. Abstr. 1943 4064		4003-017	Chem. Abstr. 1950 9121	
VERBENONE	$C_{10}H_{14}O$	150 + 249^0 10^{0}C	MYRTENOIC ACID	$C_{10}H_{14}O_2$	166 +51^0 e 54^{0}C
4003-013	Perf. Ess. Oil Rec. 1959 (50) 823		4003-018	Chem. Abstr. 1947 3448	
PINOCARVONE, l-	$C_{10}H_{14}O$	150 -69^0 Oil	MYRTENOIC ACID, dihydro-	$C_{10}H_{16}O_2$	168 Oil
4003-014	Chem. Abstr. 1954 2329		4003-019	Chem. Abstr. 1947 3448	
PINOCAMPHONE	$C_{10}H_{16}O$	152 -18^0 Oil	PINENE, DL- β-	$C_{10}H_{16}$	136 Rac. Oil
4003-015	Chem. Abstr. 1949 2977		4003-020	J. Prakt. Chem., 1921, 41.	

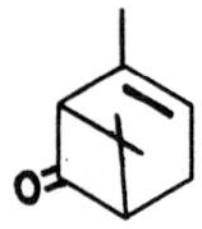

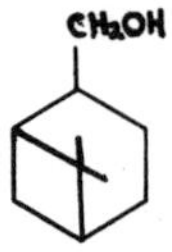

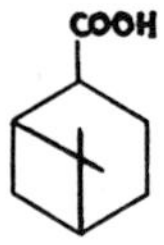

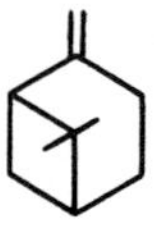

PINENE, DL- α- $C_{10}H_{16}$ 136 Rac. Oil JACS , 1931, 1030 4003-021	PAEONIFLORIN $C_{23}H_{28}O_{11}$ 480 -13⁰m Glu.O, OH, Benz. O Tet. 1969 1825 4003-023
MYRTENOL ISOVALERATE $C_{15}H_{24}O_2$ 236 Oil CH_2O -i-val Chem. Listy, 1958, (52) 1784. 4003-022	ARITASONE $C_{20}H_{28}O_2$ 300 -119⁰ 106⁰C Chem. Abstr. 1956 5579 4003-024

BORNEOL, d- $C_{10}H_{18}O$ 154 + 36^{0}e 2 08^{0}C Can. J. C., 1964, 1057 4004-001	CAMPHOR, l- $C_{10}H_{16}O$ 152 -40^{0}e 178^{0}C Chem. Comm., 1968, 1553. 4004-005
BORNEOL, l- $C_{10}H_{18}O$ 154 -36^{0}e 204^{0}C Coll. Czech. Chem. Comm. 1966, 1113. 4004-002	SANTENONE $C_9H_{14}O$ 138 -5^{0}e 50^{0}C Schimmel's Rep., 1910, Oct., p. 98. 4004-006
SANTENONE ALCOHOL $C_9H_{16}O$ 140 60^{0}C Chem. Abstr., 1937 (31) 6644 4004-003	CAMPHOR, (±)- $C_{10}H_{16}O$ 152^0 Rac. 179^{0}C Chem. Abstr. 1962 (57) 13902 4004-007
CAMPHOR, d- $C_{10}H_{16}O$ 152 +40^{0}c 178^{0}C Chem. Comm. 1968, 1553. 4004-004	

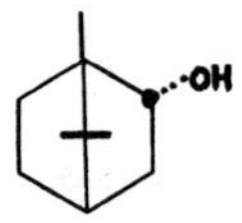

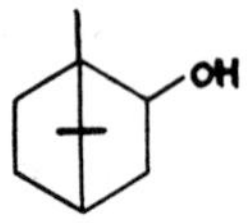

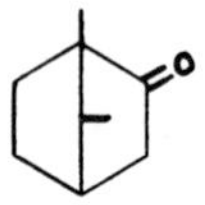

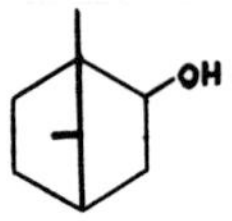

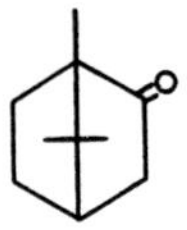

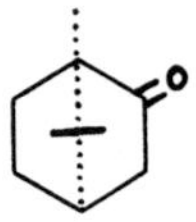

CARENE, l-3- $C_{10}H_{16}$ 136 -5^0 Oil J. Ind. Inst. Sci. 1926 (9A) 137 4005-001	CHAMIC ACID $C_{10}H_{14}O_2$ 166 $+258^0c$ Oil COOH Acta Chem. Scand. 1956 1381 Arkiv. Chemie 1964 123 4005-004
CARENE, d-3- $C_{10}H_{16}$ 136 $+7^0$ Oil Ann. 1924 268 4005-002	CHAMINIC ACID $C_{10}H_{14}O_2$ 166 $+6^0m$ 105^0C COOH Acta Chem. Scand. 1956 1381 Arkiv. Chemie 1964 123 4005-005
CARENE, d-4- $C_{10}H_{16}$ 136 $+62^0$ Oil JCS 1929 909 4005-003	

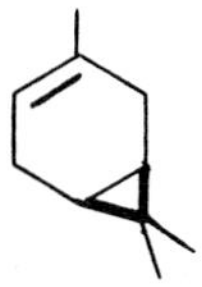

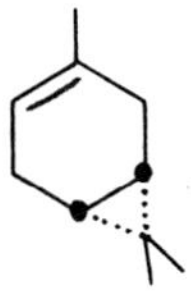

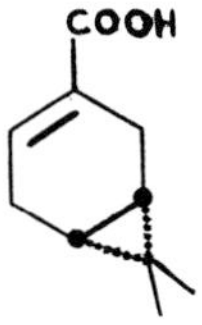

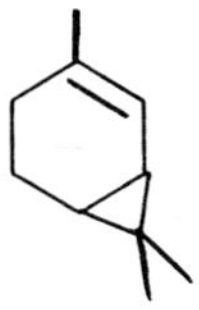

Name	Formula	MW / Rotation / State	Reference	No.
THUJANE	$C_{10}H_{18}$	138 +62⁰ Oil	Ber. 1937 935	4006-001
THUJANOL, trans-4-	$C_{10}H_{18}O$	154 + 32⁰	Chem. Abstr. 1969 (70) 54818	4006-006
SABINENE, d-	$C_{10}H_{16}$	136 + 89⁰ Oil	JACS 1935 336	4006-002
THUJYL ALCOHOL (Natural Alcohol is a mixture)	$C_{10}H_{18}O$	154 Oil	Arkiv. Kemi 1964 (22) 137	4006-007
SABINENE, l-	$C_{10}H_{16}$	136 -46⁰ Oil	JACS 1935 336	4006-003
THUJONE, α-	$C_{10}H_{16}O$	152 -20⁰ Oil	Arkiv. Kemi 1964 (22) 137	4006-008
THUJENE, d-α-	$C_{10}H_{16}$	136 + 39⁰ Oil	JCS 1912 471	4006-004
THUJONE, β -	$C_{10}H_{16}O$	152 +76⁰ Oil	Arkiv. Kemi 1964 (22) 137	4006-009
SABINOL	$C_{10}H_{16}O$	152 +8⁰ Oil	JCS 1939 1040	4006-005
UMBELLULONE	$C_{10}H_{14}O$	150 -36⁰ Oil	JOC 1966 684	4006-010

SALVENE $C_{10}H_{18}$ 138 Oil Chem. Abstr., 1956 (50) 17341 4006-011	SABINENE, DL- $C_{10}H_{16}$ 136 Rac. Oil Perf. Ess. Oil Rec., 1961 (52) 643 4006-012

FENCHOL $C_{10}H_{18}O$ 154

39^0C

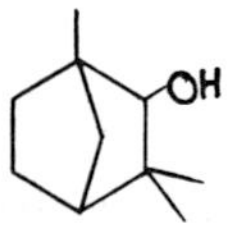

JCS 1925 1472

4007-001

FENCHYL-p-COUMARATE $C_{19}H_{24}O_3$ 300

215^0C

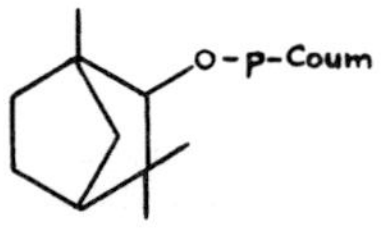

O-p-Coum

Phytochem. 1968 148

4007-003

FENCHONE $C_{10}H_{16}O$ 152

$+72^0$

4^0C

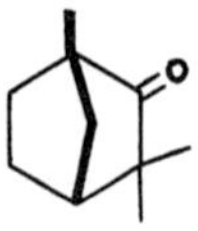

Chem. Abstr. 1956 9693

4007-002

OSMANE $C_{10}H_{20}$ 140 OI Oil Yak. Zashi, 1957 (77) 566 4008-001	IRIDODIOL, all-cis- $C_{10}H_{20}O_2$ 172 CH₂OH CH₂OH Bull. Chem. Soc. Jap. 1964 (37) 1888 4008-006
IRIDODIOL, α- $C_{10}H_{20}O_2$ 172 80°C Bull. Chem. Soc. Jap. 1964 (37) 1888 4008-002	IRIDODIOL, dehydro- $C_{10}H_{18}O_2$ 170 Tet. Lett. 1968 5325 4008-007
IRIDODIOL, β- $C_{10}H_{20}O_2$ 172 Bull. Chem. Soc. Jap. 1964 (37) 1888 4008-003	IRIDODIAL $C_{10}H_{16}O_2$ 168 Oil Aust. J. Chem. 1965 1989 4008-008
IRIDODIOL, γ- $C_{10}H_{20}O_2$ 172 Bull. Chem. Soc. Jap. 1964 (37) 1888 4008-004	DOLICHODIAL $C_{10}H_{14}O_2$ 166 -26°b Oil Proc. Chem. Soc. 1962 380 4008-009
IRIDODIOL, δ- $C_{10}H_{20}O_2$ 172 Bull. Chem. Soc. Jap. 1964 (37) 1888 4008-005	ANISOMORPHAL $C_{10}H_{14}O_2$ 166 Oil Tet. Lett. 1962 (1) 29 4008-010

ETHER, HOP — $C_{10}H_{16}O$ — 152 — Oil

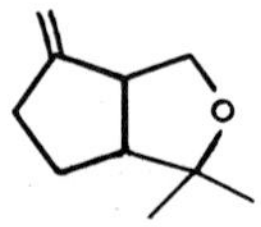

Tet. Lett. 1968 1645

4008-011

IRIDOMYRMECIN, iso- — $C_{10}H_{16}O_2$ — 168 — -62^0c — 59^0C

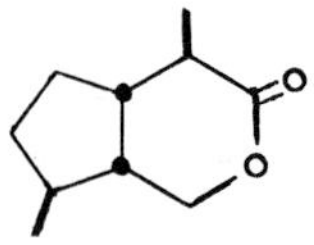

Tet. 1959 (6) 217

4008-016

MATATABIETHER — $C_{10}H_{16}O$ — 152 — -150^0 — Oil

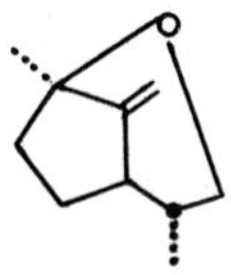

Tet. Lett. 1968 5321

4008-012

NEPETALACTONE, cis-trans- — $C_{10}H_{14}O_2$ — 166 — $+11^0c$ — Oil

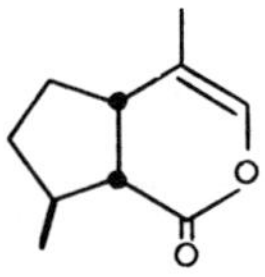

Experimentia 1963 (19) 564

4008-017

MATATABIOL, Neo- — $C_{10}H_{18}O_2$ — 170 — $+21^0$ — Oil

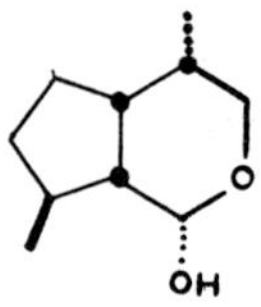

Tet. Lett. 1968 5325

4008-013

NEPTETALACTONE, trans-cis- — $C_{10}H_{14}O_2$ — 166 — $+22^0c$ — Oil

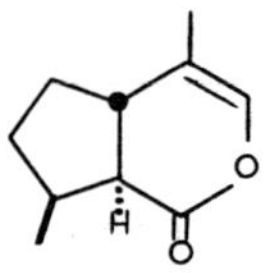

Experimentia 1963 (19) 564

4008-018

MATATABIOL, iso-neo- — $C_{10}H_{18}O_2$ — 170

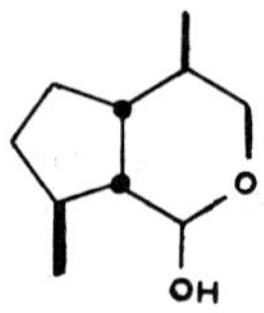

Tet. Lett. 1968 5325

4008-014

NEPETALACTONE, dihydro- — $C_{10}H_{16}O_2$ — 168 — $+72^0$ — Oil

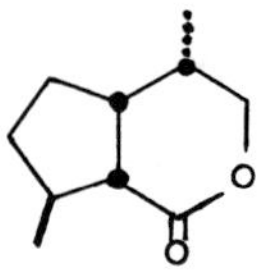

Tet. Lett. 1965 4097

4008-019

IRIDOMYRMECIN — $C_{10}H_{16}O_2$ — 168 — -210^0e — 61^0C

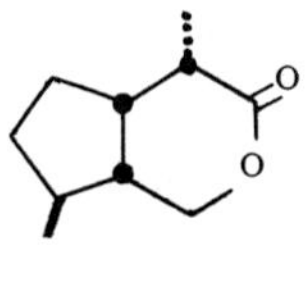

Tet. 1965 1247

4008-015

NEPETALACTONE, Iso-dihydro — $C_{10}H_{16}O_2$ — 168 — $+2.7^0$ — Oil

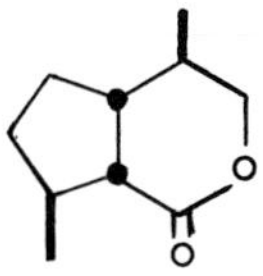

Tet. Lett. 1965 4079

4008-020

Compound	Formula	Data	Reference	No.
NEPETALACTONE, neo-	$C_{10}H_{14}O_2$	166, Oil	Tet. Lett 1965 4097	4008-021
VALTRATE, homo-dihydro- (R_1 = Isoval. R_2 = Isocapryl. R_3 = Ac.)	$C_{23}H_{34}O_8$	438, -72^0m, 50^0C	Tet. 1968 313	4008-026
GENIPIC ACID	$C_9H_{12}O_4$	184, -105^0e, amorph	Tet. 1964 1781	4008-022
VALTRATE, deoxy-dihydro- (R_1 = Isoval., R_2 = Isoval.)	$C_{22}H_{32}O_7$	408, 69^0C	Tet. 1968 313	4008-027
VALTRATE (R_1= Ac, R_2= Isoval., R_3= Isoval.)	$C_{22}H_{30}O_8$	422, $+172^0$m, Oil	Tet. 1968 313	4008-023
VALTRATE, homo-deoxy-dihydro- (R_1 = Isoval., R_2 = Isocapryl)	$C_{23}H_{34}O_7$	422	Tet. 1968 313	4008-028
VALTRATE, acetyl- (R_1 = Ac; R_2 =) R_3 =) Isoval., 3-OAc-Isoval.)	$C_{24}H_{34}O_{10}$	482, $+163^0$m, 84^0C	Tet. 1968 313	4008-024
LOGANIN	$C_{17}H_{26}O_{10}$	390, -83^0, 223^0C	JCS 1969 721	4008-029
VALTRATE, dihydro- (R_1 = Isoval., R_2 =Isoval., R_3 = Ac.)	$C_{22}H_{32}O_8$	424, -81^0m, 64^0C	Tet. 1968 313	4008-025
LOGANIN, deoxy	$C_{17}H_{26}O_9$	374	Chem. Comm. 1970 826	4008-030

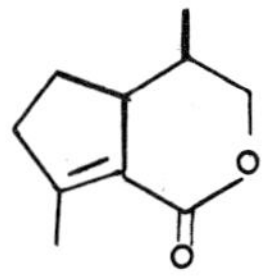

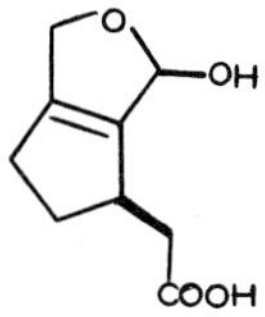

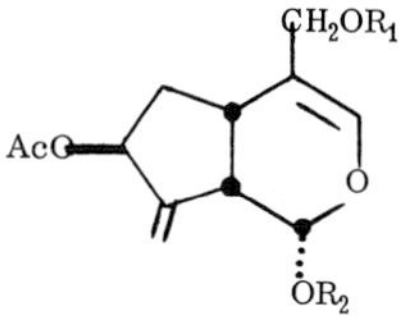

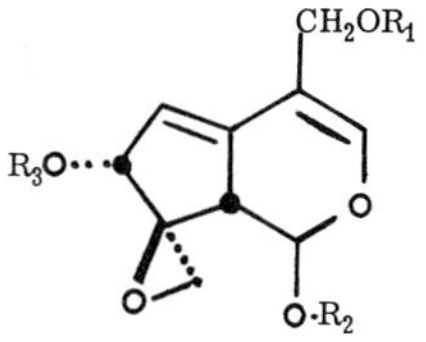

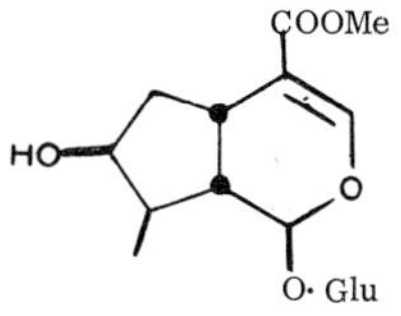

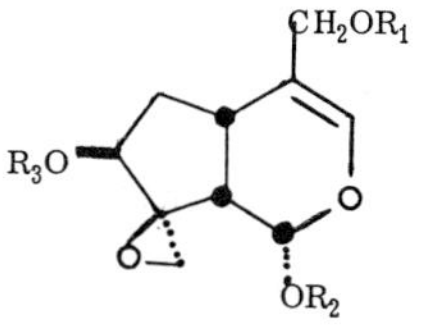

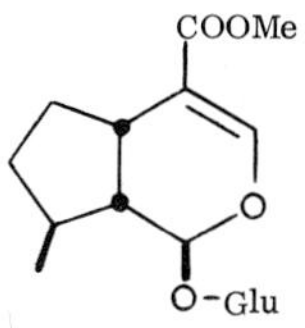

No.	Name	Formula	MW	$[\alpha]_D$	m.p.	Notes	Reference
4008-031	LOGANIC ACID	$C_{16}H_{24}O_{10}$	376		amorph	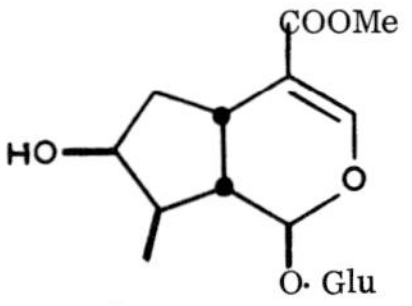	Chem. Comm. 1968 138
4008-032	MONOTROPEIN	$C_{16}H_{22}O_{11}$	390	-131^0w	162^0C		Chem. Pharm. Bull. 1968 1018
4008-033	SCANDOSIDE	$C_{16}H_{22}O_{11}$	390	-53^0w	141^0C		Tet. Lett. 1968 683
4008-034	DAPHYLLOSIDE	$C_{19}H_{26}O_{12}$	446	$+20^0$w	96^0C		Yakugaki Zasshi 1966 (86) 943
4008-035	THEVIRIDOSIDE	$C_{17}H_{24}O_{11}$	404	-23^0w			Helv. 1969 (52) 478
4008-036	GENIPIN	$C_{11}H_{14}O_5$	226	$+135^0$m	120^0C	R = H	JOC 1961 1192
4008-037	GENIPIN-1-GENTIOBIOSIDE	$C_{23}H_{34}O_{15}$	550	0^0	228^0C	R = Gentiobiosyl	Chem. Pharm. Bull. 1970 1066
4008-038	GENIPINIC ACID	$C_{11}H_{14}O_6$	242	-126^0e	amorph	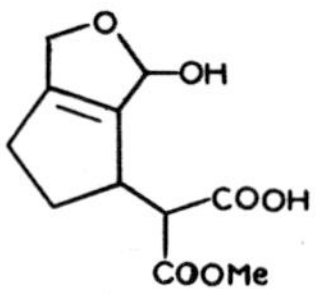	Tet. 1964 1781
4008-039	LAMIOL	$C_{16}H_{26}O_{10}$	378	-153^0			Tet. 1967 4709
4008-040	LAMIOSIDE	$C_{18}H_{28}O_{11}$	420	-133^0m			Tet. 1967 4709

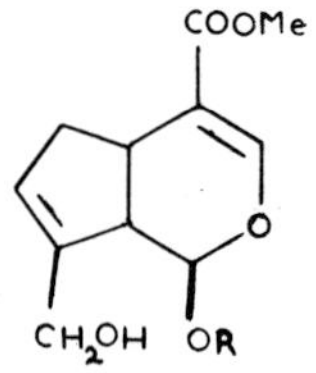

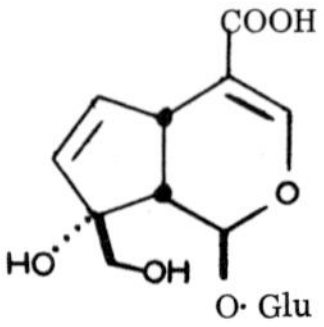

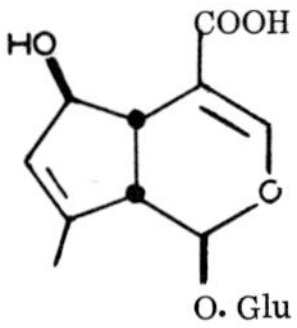

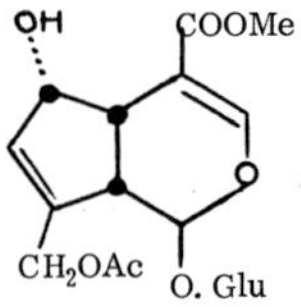

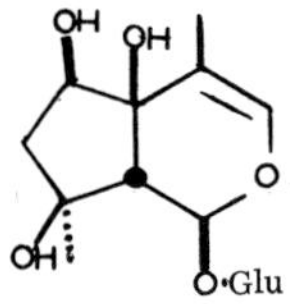

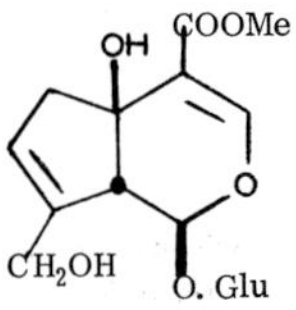

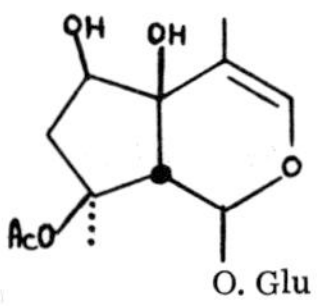

VERBENALOL $C_{11}H_{14}O_5$ 226 -20°e 134°C COOMe, O, OH Tet. Lett. 1968 2471 4008-041	PAEDEROSIDE $C_{18}H_{22}O_{10}S$ 430 -196°m 122°C Ac·SCH$_2$, O-Glu Tet. Lett. 1968 683 4008-046
VERBENALIN $C_{17}H_{24}O_{10}$ 388 -173°w 183°C COOMe, O. Glu Tet. 1962 (18) 1049 4008-042	AUCUBIN $C_{15}H_{22}O_9$ 346 -162°w 182°C HO, CH$_2$OH, O-Glu JCS 1961 5194 4008-047
ASPERULOSIDE $C_{18}H_{22}O_{11}$ 414 -200°w 132°C CH$_2$OAc, O-Glu JCS 1954 4182 4008-043	PROCUMBIDE $C_{15}H_{24}O_{11}$ 380 -81°e 212°C OH, OH, HO, HO, O. Glu Ann. Chem. 1968 (712) 138 4008-048
ASPERULOSIDE, des-Ac- $C_{16}H_{20}O_{10}$ 372 -88°e CH$_2$OH, O-Glu Tet. Let. 1968 683 4008-044	MELITTOSIDE, mono- $C_{15}H_{22}O_{10}$ 362 -180°w OH, OH, OH, O. Glu Ric. Sci. 1967 (37) 840 4008-049
PAEDEROSIDIC ACID $C_{18}H_{24}O_{11}S$ 448 +28°m HO, COOH, Ac·S·CH$_2$, O-Glu Tet. Let. 1968 683 4008-045	MELITTOSIDE $C_{21}H_{32}O_{15}$ 524 -29°w 168°C OH, O. Glu, OH, O. Glu Gazz. 1967 (97) 1209 4008-050

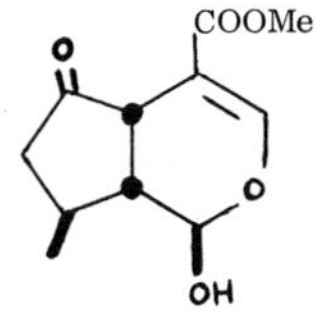

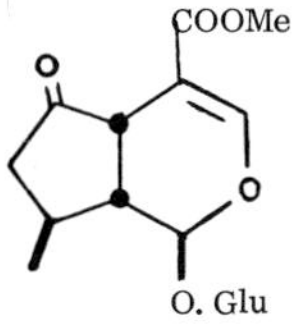

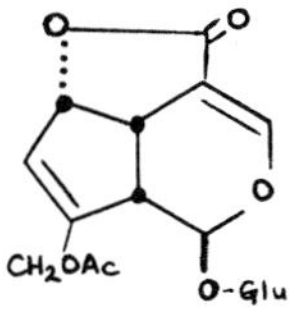

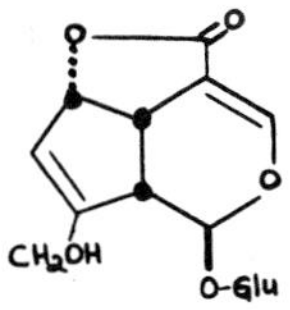

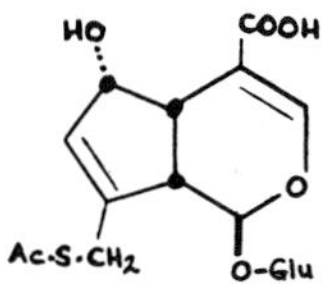

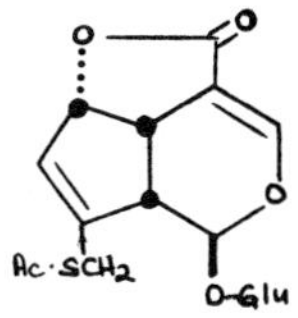

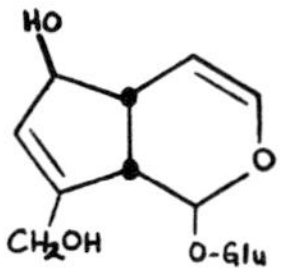

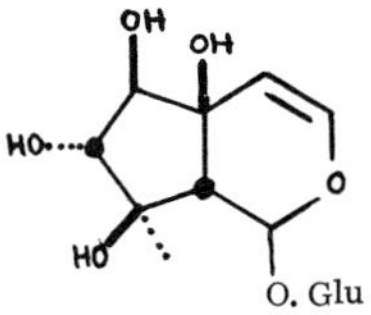

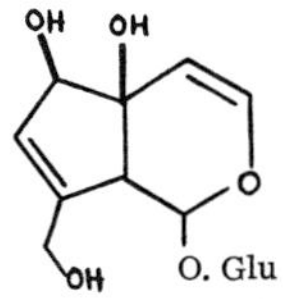

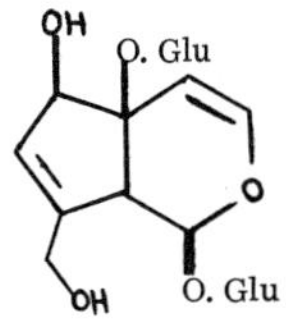

ANTIRRINOSIDE $C_{15}H_{22}O_{10}$ 362 -78°d OH OH O O. Glu Gazz. 1968 (98) 177 4008-051	GLOBULARIN $C_{24}H_{28}O_{11}$ 492 -73°e OH O O.Cin O.Glu Chem. Abstr. 1967 (66) 55420 4008-056
AGNUSIDE $C_{22}H_{26}O_{11}$ 466 -91°e 145°C OH O O-Glu O p-OH-benzoyl Arch. Pharm. 1960 (293) 556 4008-052	PICROSIDE I $C_{24}H_{28}O_{11}$ 492 -82°m OH O CH_2OH O.(6-Cin-Glu) Tet. Lett. 1969 2837 4008-057
CATALPOL $C_{15}H_{22}O_{10}$ 362 -122°e 208°C OH O CH_2OH O. Glu Tet. Lett. 1962 321 4008-053	UNEDOSIDE $C_{14}H_{20}O_9$ 332 -112° 233°C OH O O. Glu Tet. Lett. 1966 1245 4008-058
CATALPOL, 6-Me-ether $C_{16}H_{24}O_{10}$ 376 -122°m 237°C OMe O CH_2OH O. Glu Biochem. J. 1965 (95) 1 4008-054	HARPAGIDE $C_{15}H_{24}O_{10}$ 364 -154°e OH OH HO O O. Glu Tet. Lett. 1965 3439 4008-059
CATALPOSIDE $C_{22}H_{26}O_{12}$ 482 -184°m 216°C R = p-OH-benzoyl OR O CH_2OH O. Glu JOC 1966 500 4008-055	MARPAGIDE, 8-OAc- $C_{17}H_{26}O_{11}$ 406 -132°m 155°C OH OH AcO O O. Glu Tet. Lett. 1965 3437 4008-060

HARPAGOSIDE $C_{24}H_{30}O_{11}$ 494
-43°m

OH OH
O
O
Cin
O.Glu

Tet. Lett. 1965 3437

4008-061

$C_{11}H_{16}O_5$ 228
+21°c
104°C

COOMe
HO
O
O

Aust. J. C., 1969, 1283.

4008-065

MONOTROPEIN, bis-desoxy-dihydro- $C_{15}H_{24}O_9$ 360
-90°e
157°c

COOH
O
O.Glu

Phytochem. 1970
641

4008-062

ANTIRRIDE $C_{15}H_{22}O_8$ 330
-124°
84°C

HO
O
O·Glu

Gazz., 1969 (99) 807.

4008-066

SYRINGOPICROSIDE $C_{24}H_{30}O_{11}$ 494

CO·O
OH
O
O
O·Glu

Tet. 1970 2365

4008-063

GALIRIDOSIDE $C_{15}H_{22}O_9$ 346
-78°w
190°C

OH
O
O
O·Glu

Helv., 1970, 2010.

4008-067

JASMINIM $C_{26}H_{38}O_{12}$ 542
-245°e
160°C

O
CO
O
O·Glu
CH₂OH
O
O

Tet., 1970, 4561.

4008-064

LOGANIN, seco-

$C_{17}H_{24}O_{10}$ 388
-96^{0}m
amorph

JCS, 1969, 1187.

4009-001

GENTIOPICROSIDE

$C_{16}H_{18}O_9$ 354
-196^{0}w
122^{0}C

Tet. Lett., 1968, 4429.

4009-002

MORRONISIDE

$C_{17}H_{26}O_{11}$ 406
-72^{0}e

Tet. Lett., 1969, 2725.

4009-003

KINGISIDE

$C_{17}H_{24}O_{11}$ 404
-91^{0}e

Tet. Lett., 1969, 2725.

4009-004

MENTHIAFOLIN

$C_{26}H_{36}O_{12}$ 540
-68^{0}m
186^{0}C

Chem. Comm., 1968, 1278.

4009-005

FOLIAMENTHIN

$C_{26}H_{36}O_{12}$ 540
-63^{0}m
195^{0}C

Chem. Comm., 1968, 1276.

4009-006

DIHYDRO-FOLIAMENTHIN

$C_{26}H_{38}O_{12}$ 542

✱ Dihydro

Chem. Comm., 1968, 1278.

4009-007

ELENOLIDE

$C_{10}H_{12}O_6$ 228

Aglycone

4009-008

OLEUROPEIN

$C_{25}H_{32}O_{13}$ 540

Chem. Abstr., 1966 (64) 8124.

4009-009

SWERTIAMARIN

$C_{16}H_{22}O_{10}$ 374

Bull. Soc. Chim. Fr., 1964, 403.

4009-010

AMAROSWERIN cf. Swertiamarin - biphenylacyl on 3 position of glucosyl group $C_{29}H_{30}O_{14}$ 602 -13^0m Amorph OH OH OH Tet. Lett., 1968, 4921. 4009-011	AMAROGENTIN cf. Sweroside - biphenylacyl on 3 position of glucosyl group $C_{29}H_{30}O_{13}$ 586 -117^0m 229^0C Tet. Lett., 1968, 4919. 4009-013
SWEROSIDE $C_{16}H_{22}O_9$ 358 -236^0w O O O O·Glu Tet. Lett., 1966, 5229. 4009-012	

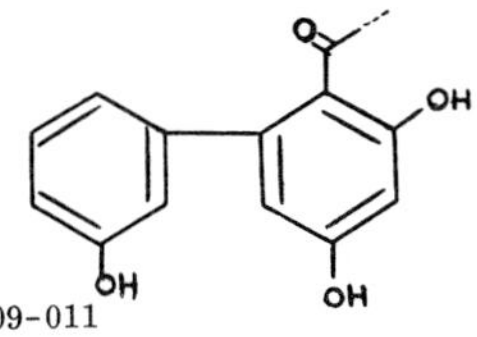

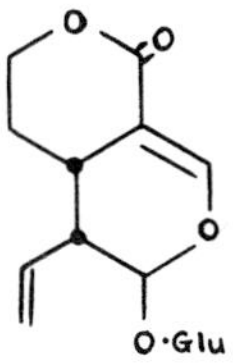

SAFRANAL

$C_{10}H_{14}O$ 150
OI
Oil

CHO

Helv. 1955 (38) 1863

4010-001

PYRONENE, α-

$C_{10}H_{16}$ 136

Oil

Compt. rendu, 1935, (201) 219.

4010-004

CYCLOCITRAL, 4-OH-β-

$C_{10}H_{16}O_2$ 168
-87°e
Oil

HO
CHO

Ber. 1941 (74) 219

4010-002

PYRONENE, β-

$C_{10}H_{16}$ 136
OI
Oil

Compt. rendu, 1935, (201) 219.

4010-005

PICROCROCIN

$C_{16}H_{26}O_7$ 330

156°C

Glu.
O
CHO

Ber. 1941 (74) 219

4010-003

KARAHANA ETHER

$C_{10}H_{16}O$ 152

Oil

O

Tet. Lett. 1968 1645

4010-006

THUJAPLICIN, α- $C_{10}H_{12}O_2$ 164 OI 34°C Prog. Org. Chem. 1952 (1) 22 4011-001	DOLABRINOL, α- $C_{10}H_{10}O_3$ 178 OI Oil JOC., 1962, 3368. 4011-006
THUJAPLICIN, β- $C_{10}H_{12}O_2$ 164 OI 52°C Prog. Org. Chem. 1952 (1) 22 4011-002	DOLABRIN, β- $C_{10}H_{10}O_2$ 162 OI 59°C Chem. Ind. 1967 1070 4011-007
THUJAPLICIN, γ- $C_{10}H_{12}O_2$ 164 OI 80°C Prog. Org. Chem. 1952 (1) 22 4011-003	PYGMAEIN $C_{11}H_{14}O_3$ 194 OI 40°C JOC., 1959, 1584. 4011-008
NEZUKONE $C_{10}H_{12}O$ 148 OI Oil Agr. Biol. Chem. 1968 249 4011-004	UTAHIN $C_{20}H_{20}O_5$ 340 OI 313°C Chem. Comm., 1968 233 4011-009
TROPOLONE, 5-Et- $C_9H_{10}O_2$ 150 OI 81°C Exp. 1966 (22) 141 4011-005	THUJAPLICINOL, α- $C_{10}H_{12}O_3$ 180 OI JOC., 1961, 173 4011-010

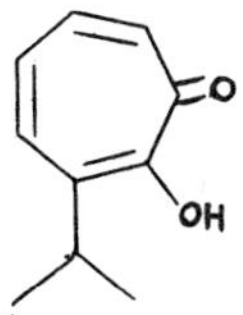

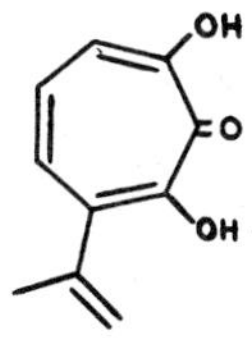

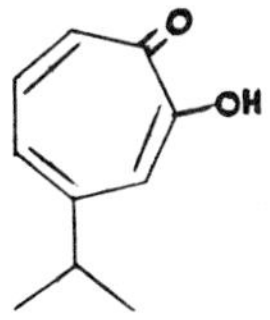

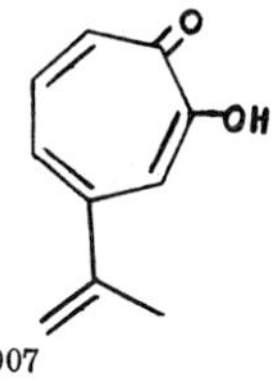

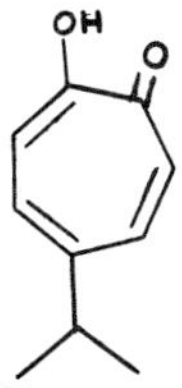

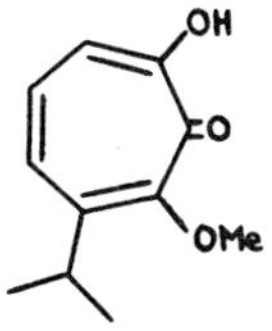

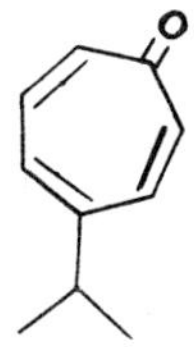

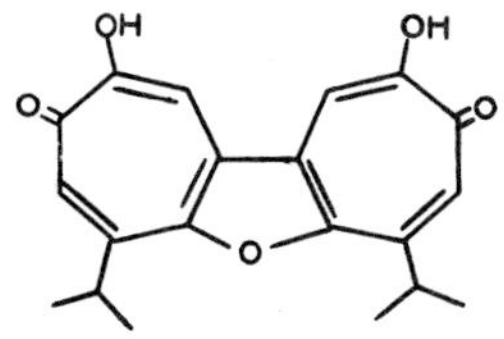

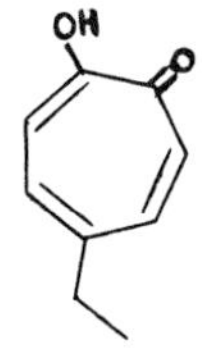

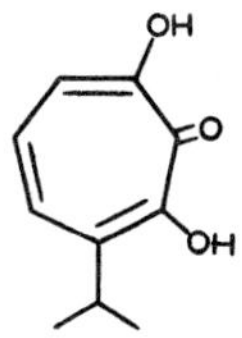

THUJAPLICINOL, β-

$C_{10}H_{12}O_3$ 180
OI

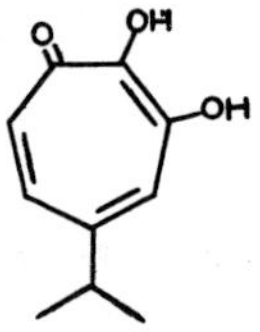

Chem. Abstr., 1968, (67) 114357.

4011-011

CHAMAECIN

Oil

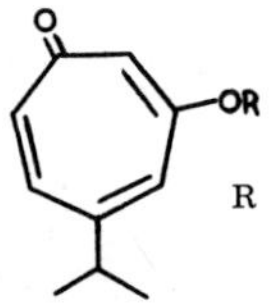

R = unknown

Chem. Abstr., 1956 (50) 17341.

4011-013

PYGMAEIN, Iso-

$C_{11}H_{14}O_3$ 194
OI
112°C

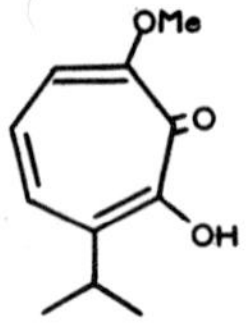

JOC, 1962, 3368.

4011-012

EUCARVONE $C_{10}H_{14}O$ 150 OI Oil JOC., 1961, 1609. 4012-001	THUJAIC ACID $C_{10}H_{12}O_2$ 164 OI 88°C COOH Chem. Abstr. 1953 9281 4012-003
KARAHANAENONE $C_{10}H_{16}O$ 152 Oil Tet. Lett. 1968 1645 4012-002	

SKYTANTHINE, α-
$C_{11}H_{21}N$ 167
+79°
p: 120°C

JACS, 1967, 2476.

4013-001

SKYTANTHINE-I, OH-
$C_{11}H_{21}NO$ 183
+38°
95°C

Tet., 1967, 3147.

4013-006

SKYTANTHINE, β-
$C_{11}H_{21}N$ 167
+16°
p: 135°C

Tet., 1962, 183.

4013-002

SKYTANTHINE-II, OH-
$C_{11}H_{21}NO$ 183
-38°m
120°C

Tet., 1967, 3147.

4013-007

SKYTANTHINE, δ-
$C_{11}H_{21}N$ 167
+9°
p: 139°C

Tet., 1962, 183

4013-003

SKYTANTHINE, Nor-N-Me-
$C_{10}H_{19}N$ 153
+35°c
Oil

Tet., 1969, 1523.

4013-008

SKYTANTHINE, Oxy-
$C_{11}H_{21}NO$ 183
94°C

Tet., 1969, 1523.

4013-004

SKYTANTHINE, Dehydro-
$C_{11}H_{19}N$ 165
Oil

Chem. Ind., 1963, 984.

4013-009

SKYTANTHINE, Iso-oxy-
$C_{11}H_{21}NO$ 183
91°C

Tet., 1969, 1523.

4013-005

TECOSTANINE
$C_{11}H_{21}NO$ 183
0°m
85°C

Ann. Pharm. Fr., 1963 (21) 699.

4013-010

TECOMANINE — $C_{11}H_{17}NO$ — 179, -175^0c, Oil

Tet., 1969, 1523.

4013-011

RW 47 — $C_9H_{11}NO$ — 149, $+27^0c$, 131^0C

Aust. J. C., 1967, 2505.

4013-012

VENOTERPENE — $C_9H_{11}NO$ — 149, 130^0C

Tet. Lett., 1968, 2763.

4013-013

TECOSTIDINE — $C_{10}H_{13}NO$ — 163, -4^0c, p: 112^0C

Bull. Soc. Chim. Fr., 1961, 2901.

4013-014

ACTINIDINE — $C_{10}H_{13}N$ — 147, -7^0c, p: 143^0C

Tet. Lett., 1968, 5325.

4013-015

ACTINIDINE, 4-nor- — $C_9H_{11}N$ — 133, $+3^0c$, Oil

Tet., 1969, 1523.

4013-016

BOSCHNIAKINE — $C_{10}H_{11}NO$ — 161, $+21^0c$, Oil

Tet., 1969, 1523.

4013-017

BOSCHNIAKINIC ACID — $C_{10}H_{11}NO_2$ — 177, +, 220^0C

Tet., 1967, 4635.

4013-018

ACTINIDINE, N-(p-OH-phenethyl)- — $C_{18}H_{22}NO$ — 268, Cl $+51^0m$, Cl 202^0C

R = Me

Acta Chem. Scand., 1967, 53.

4013-019

TECOSTIDINE, N-(p-OH-phenethyl)- — $C_{18}H_{22}NO_2^+$ — 284, 223^0C

R = CH_2OH

Acta Chem. Scand., 1967, 52.

4013-020

SKYTANTHINE-N-OXIDE, β- $C_{11}H_{21}NO$ 183
0^0m
220^0C

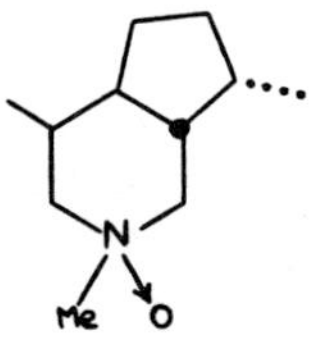

Me O

Chem. Ind., 1969, 1631.

4013-021

GENTIANINE $C_{10}H_9O_2N$ 175 OI 82^0C Bull. Chem. Soc. Jap., 1959 (32) 1155. 4014-001	GENTIOFLAVINE $C_{10}H_{11}NO_3$ 193 219^0C Tet., 1968, 1323. 4014-006
GENTIANADINE $C_8H_7NO_2$ 149 OI 77^0C Chem. Abstr., 1967 (67) 117007. 4014-002	GENTIANAMINE $C_{11}H_{11}O_3N$ 205 150^0C Chem. Abstr., 1967 (67) 117007. 4014-007
GENTIALUTINE $C_9H_{11}ON$ 149 OI 130^0C Chem. Abstr., 1965 (62) 13507. 4014-003	JASMININE $C_{11}H_{12}N_2O_3$ 220 -38^0c 175^0C Aust. J. C., 1968, 1321. 4014-008
GENTIATIBETINE $C_9H_{11}NO_2$ 165 162^0C Chem. Abstr., 1967 (67) 117012. 4014-004	FONTAPHILLINE $C_{16}H_{15}NO_3$ 269 OI Chem. Abstr., 1967 (67) 99972. 4014-009
OLIVERIDIN $C_{10}H_{13}O_2N$ 179 Khim. Prir. Soed., 1969, 608. 4014-005	BAKANKOSIDE $C_{16}H_{23}NO_8$ 357 Tet. Lett., 1961, 544. 4014-010

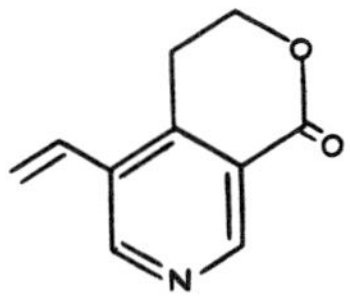

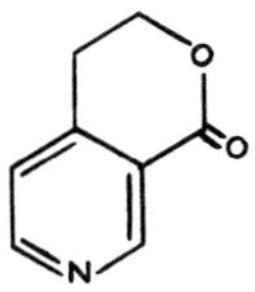

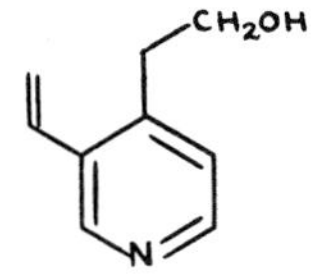

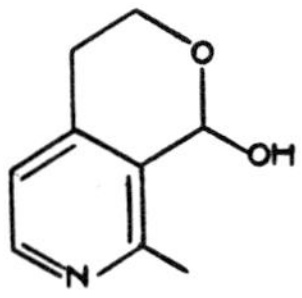

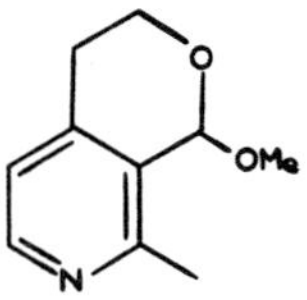

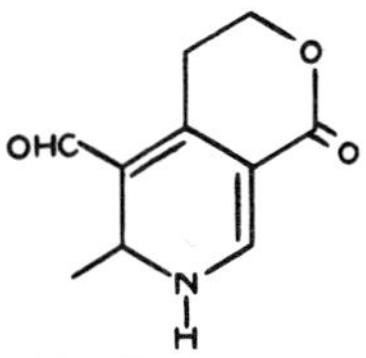

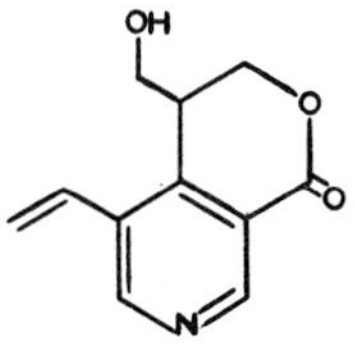

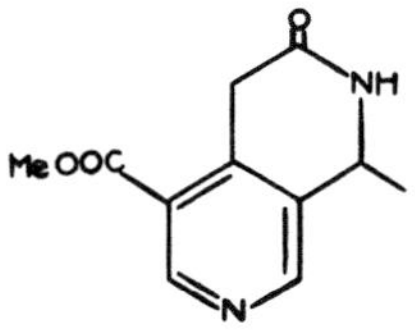

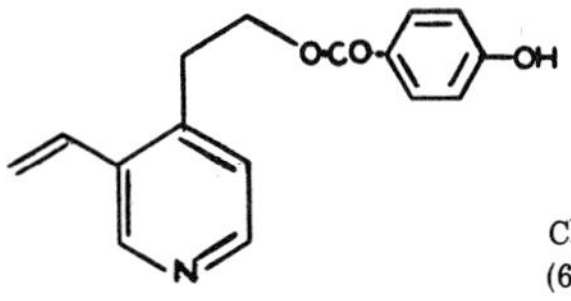

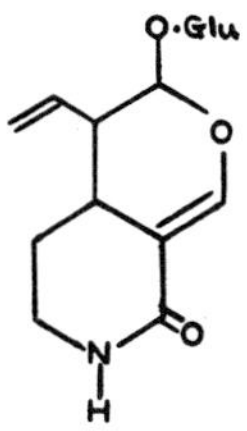

PLUMIERIDE $C_{21}H_{26}O_{12}$ 470 -114°w 225°C Experimentia, 1964, 250. COOMe O O·Glu OH O 4015-001	PLUMERICIN, dihydro- $C_{15}H_{16}O_6$ 292 + 257°c 150°C Helv., 1961 (44) 1447. COOMe O O O O Et 4015-004
FULVOPLUMIERIN $C_{14}H_{12}O_4$ 244 OI 152°C Helv., 1953, 1468. COOMe O O 4015-002	PLUMERICIN, iso- $C_{15}H_{14}O_6$ 290 +216°c 201°C Helv., 1961 (44) 1447. COOMe O O O O 4015-005
PLUMERICIN $C_{15}H_{14}O_6$ 290 + 197°c 212°C Helv. 1961 (44) 1447. COOMe O O O O 4015-003	PLUMERICIN ACID, dihydro- $C_{14}H_{14}O_6$ 278 190°C Helv., 1961 (44) 1447. COOH O O O O 4015-006

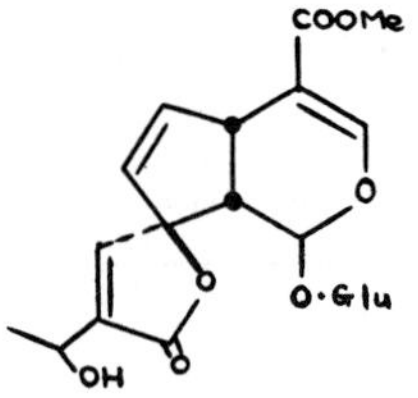

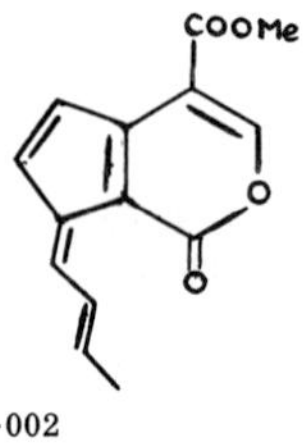

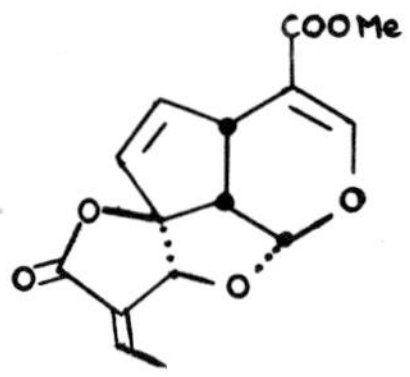

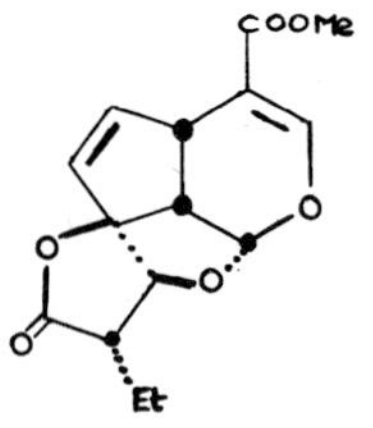

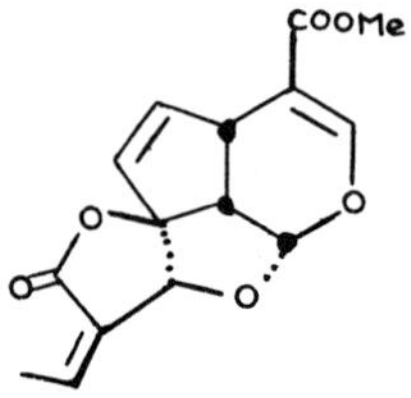

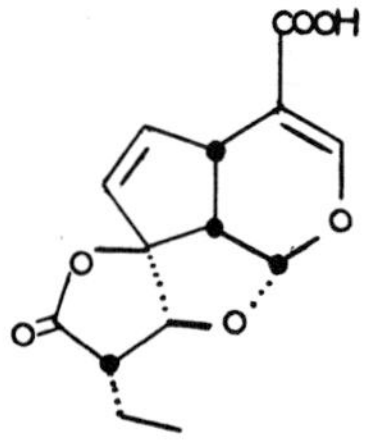

Name	Formula	Data	Reference	No.
SANTENE	C_9H_{14}	122 ± 0° Oil	Ann. 1931 (486) 209	4099-001
ARTEMISIA KETONE	$C_{10}H_{16}O$	152 OI Oil	Helv. 1936, 646.	4099-006
SANTOLINENONE, β- (?)	$C_{10}H_{16}O$	152 Oil	Gazz., 1916 (46) 251.	4099-002
ARTEMISIA KETONE, Iso-	$C_{10}H_{16}O$	152 OI Oil	J. Pharm. Soc. Jap., 1920 (464) 837.	4099-007
CAMPHENE, D-	$C_{10}H_{16}$	136 +104° 51°C	Ann. 1941 186	4099-003
ARTEMISIA ALCOHOL	$C_{10}H_{18}O$	154	Yak. Zasshi, 1957, (77) 1307.	4099-008
CHRYSANTHEMIC MONOACID (ex pyrethrins)	$C_{10}H_{16}O_2$	168 +20° 18°C	Biochim. Biophys. Acta, 1962, 312.	4099-004
LAVANDULOL	$C_{10}H_{18}O$	154 - 10° Oil	Helv., 1947, 1453.	4099-009
CHRYSANTHEMIC DIACID (ex pyrethrins)	$C_{11}H_{16}O_4$	212 +104°t Oil	Biochim. Biophys. Acta 1962, 312.	4099-005
LYRATOL	$C_{10}H_{16}O$	152 +62° Oil	Tet., 1969, 3217.	4099-010

Structure labels: 4099-002: O; 4099-004: HOOC; 4099-005: MeOOC, HOOC; 4099-006: O; 4099-007: O; 4099-008: OH; 4099-009: $HOCH_2$; 4099-010: $HOCH_2$

YOMOGI ALCOHOL A OH 4099-011	$C_{10}H_{18}O$ Experimentia, 1970, (20) 8.	154 OI Oil	TERESANTALIC ACID COOH 4099-016	$C_{10}H_{14}O_2$ J. Ind. Chem. Soc., 1944 (21) 377.	166 -77^0 b 157^0 C
SANTOLINA TRIENE 4099-012	$C_{10}H_{16}$ Tet. Lett., 1964, 3775.	136 OI Oil	CANNABICYCLOL OH C_5H_{11} O 4099-017	$C_{21}H_{30}O_2$ Chem. Comm. 1968 894	314 146^0C
LAVANDULAL, Cyclo- CHO 4099-013	$C_{10}H_{16}O$ Tet. Lett., 1967, 2645.	152 OI Oil	SYLVECARVONE O 4099-018	$C_{10}H_{14}O$ Current Sci. 1966 235	150 Oil
LAVANDULIC ACID, Cyclo- COOH 4099-014	$C_{10}H_{16}O_2$ Chem. Ind., 1967, 1256.	168 OI 110^0C	CARQUEJOL Acetate AcO 4099-019	$C_{12}H_{16}O_2$ Helv. 1967 963	192 Oil
TERESANTALOL OH 4099-015	$C_{10}H_{16}O$ J. Ind. Chem. Soc., 1944 (21) 271.	152 $+12^0$e 113^0C	ORTHODENE 4099-020	$C_{10}H_{16}$ Ber. 1938 1591	136 $+33^0$ Oil

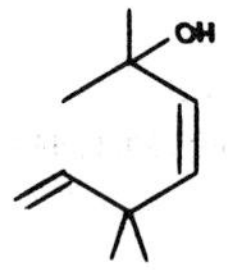

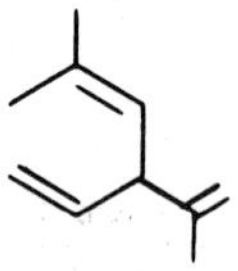

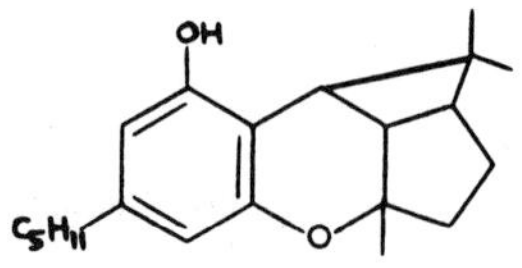

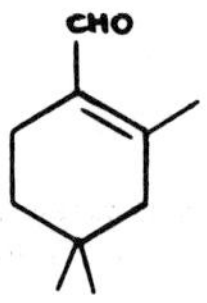

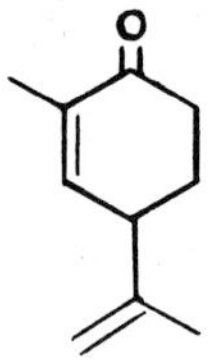

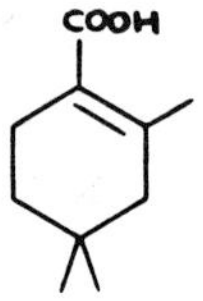

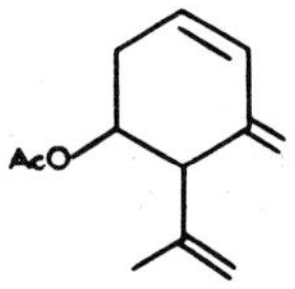

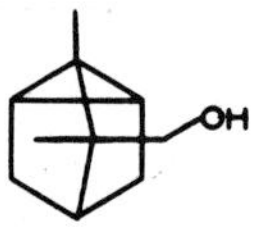

CAMPHENE, L-	$C_{10}H_{16}$	136 -94^0 52^0C
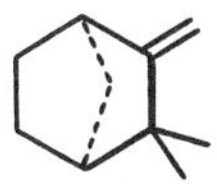	Ber., 1892, 147; 162.	
4099-021		
CYCLOPENTENE, 1-Ac- 4-Isopropenyl-1-	$C_{10}H_{14}O$	150 0^0 Oil
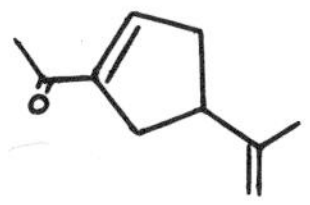	JACS 1960 636	
4099-026		
CAMPHENE, dl-	$C_{10}H_{16}$	136 Racemate 50^0C
	Ann. Chim. 1919 (12) 334	
4099-022		
CHRYSANTHEMUM LACTONE	$C_{10}H_{14}O_2$	166 OI 64^0C
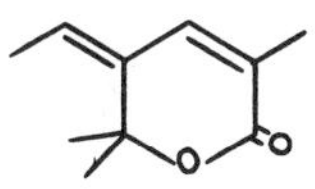	Tet. Lett., 1969 2413.	
4099-027		
CYCLOBUTANE, 1-(2-OH-ethyl)- 2-isopropenyl-1-Me-	$C_{10}H_{18}O$	154
CH_2OH	JACS., 1970, 425.	
4099-023		
CHAKSINE	$C_{11}H_{20}O_2N_3{}^+$	226 SO_4:$+61^0w$ .. 318^0C
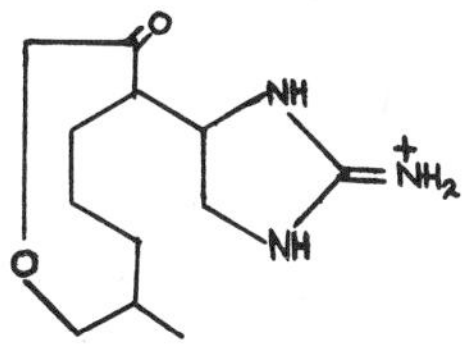 NH, $\overset{+}{N}H_2$, NH	JACS 1958 1521	
4099-028		
FILIFOLONE, D-	$C_{10}H_{14}O$	150 $+307^O$ Oil
	Chem. Comm. 1967 1037	
4099-024		
R = H CH_2OR CHO	$C_{15}H_{20}O_4$	264
	Tet. Lett., 1970, 1453.	
4099-029		
FILIFOLONE, L-	$C_{10}H_{14}O$	150 $-270^O c$ Oil
H, H	Chem. Comm. 1967 1037	
4099-025		
R = Ac	$C_{17}H_{22}O_5$	306
	Tet. Lett., 1970, 1453.	
4099-030		

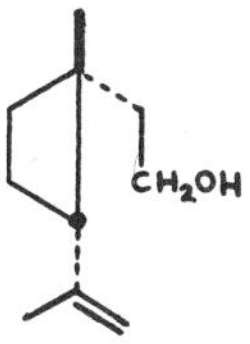

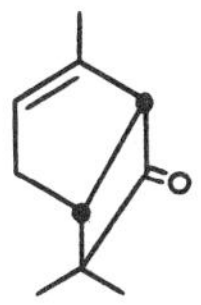

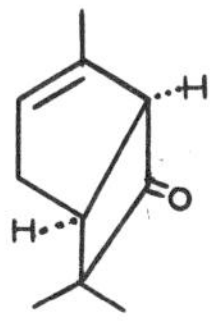

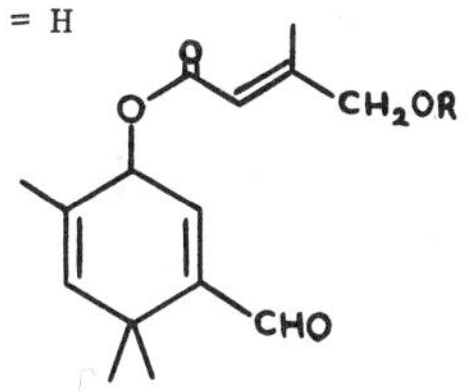

$C_{15}H_{20}O_3$ 248

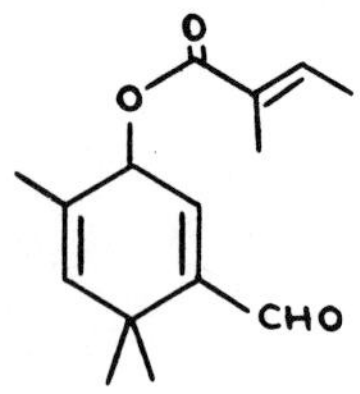

CHO

4099-031

$C_{17}H_{22}O_5$ 306

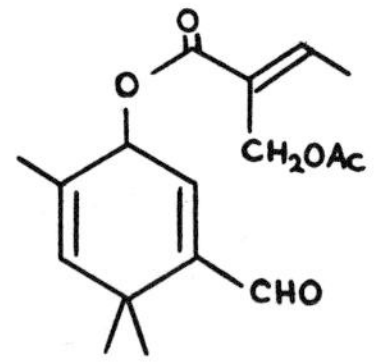

CH_2OAc

CHO

Tet. Lett., 1970, 1453.

4099-032

41-- THE SESQUITERPENES

This is the largest of the terpenoid classes, there being almost 1000 sesquiterpenes recorded here. From this large group 30 main structural types emerge (see Main Skeleton Key) to represent over 700 of the compounds, the remainder being described by some 70 less common skeletons. It is apparent that this large number of less common skeletons is too unwieldy to be placed into one category as was possible with the monoterpenes (viz. 4099). Thus the subdivision of this group into mono, bi, tri, and tetracyclic skeletons (4191, 4192, 4193, and 4194 respectively). The nitrogenous sesquiterpenes have, for convenience, been collected into one class–category 4198.

Proposed biogenetic interrelationships between the sesquiterpene skeletons are displayed on the following chart. Two acyclic farnesol precursors have been utilized here in generating the wide variety of sesquiterpene skeletons, but there are alternative hypothetical schemes employing three geometric isomers which can be found elsewhere.

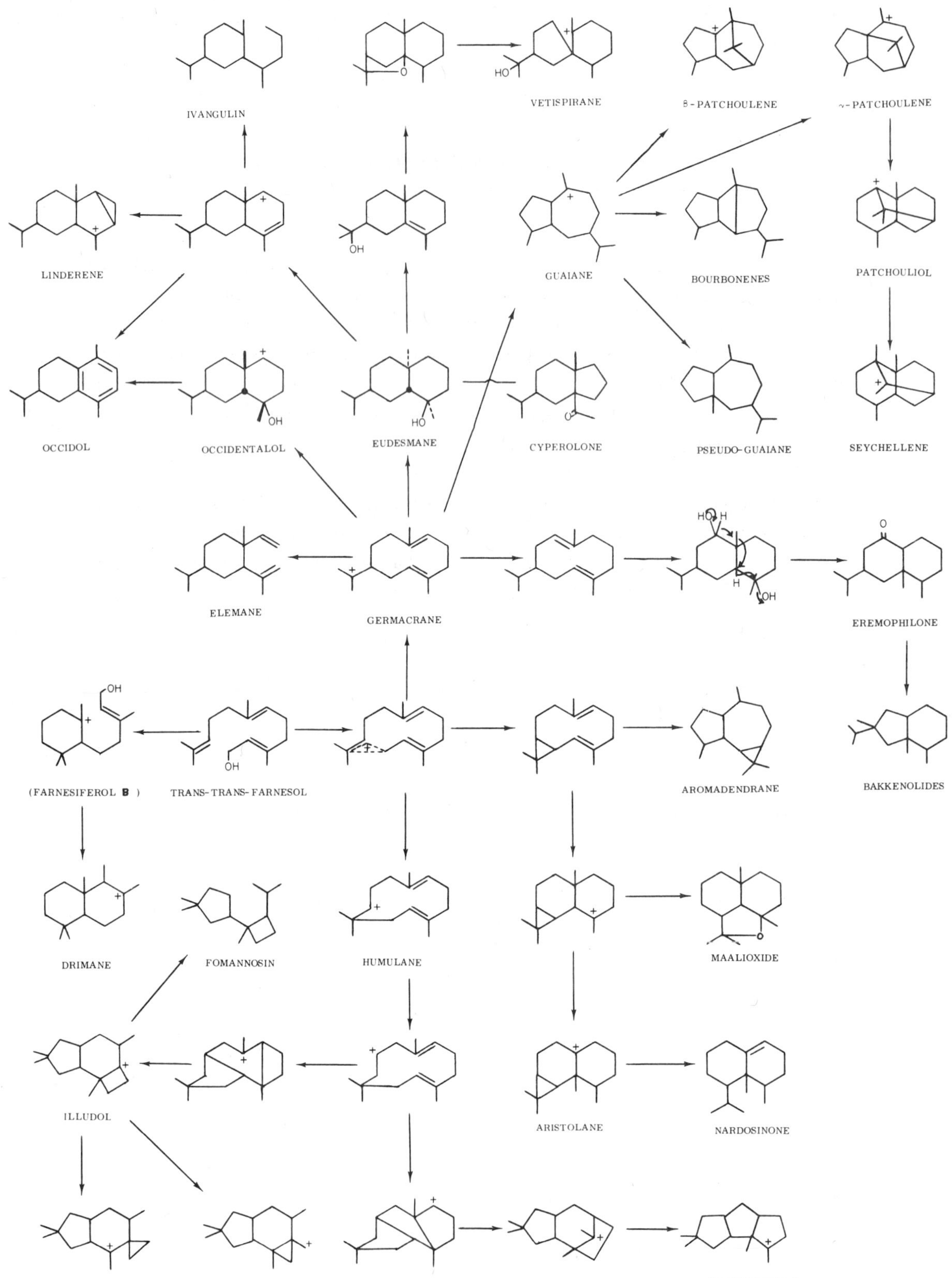
IVANGULIN
VETISPIRANE
β-PATCHOULENE
α-PATCHOULENE
LINDERENE
GUAIANE
BOURBONENES
PATCHOULIOL
OCCIDOL
OCCIDENTALOL
EUDESMANE
CYPEROLONE
PSEUDO-GUAIANE
SEYCHELLENE
ELEMANE
GERMACRANE
EREMOPHILONE
(FARNESIFEROL B)
TRANS-TRANS-FARNESOL
AROMADENDRANE
BAKKENOLIDES
DRIMANE
FOMANNOSIN
HUMULANE
MAALIOXIDE
ILLUDOL
ARISTOLANE
NARDOSINONE
ILLUDINS
MARASMIC ACID
HIRSUTIC ACID C

SESQUITERPENES

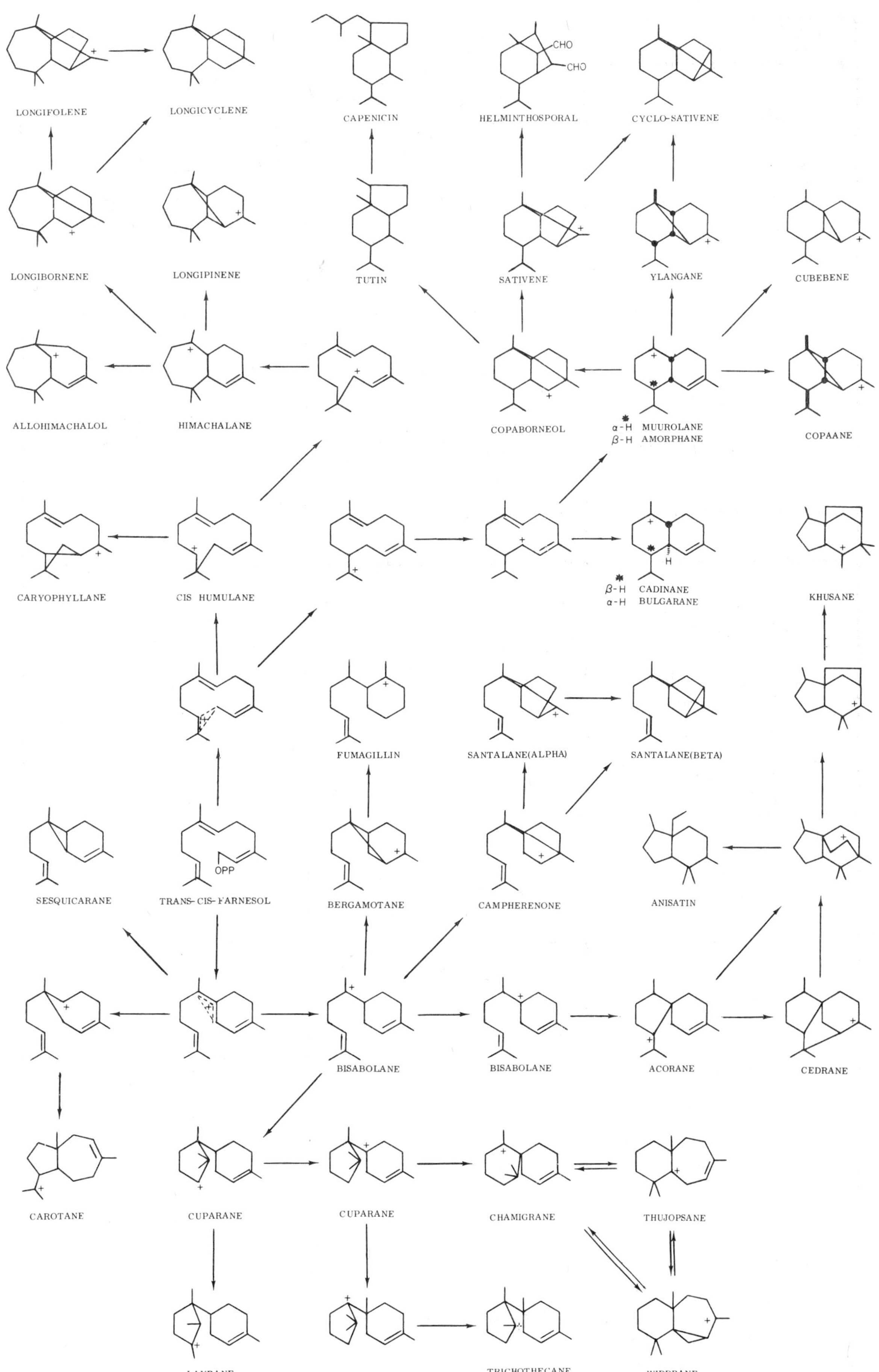

CHO
CHO
LONGIFOLENE
LONGICYCLENE
CAPENICIN
HELMINTHOSPORAL
CYCLO-SATIVENE
LONGIBORNENE
LONGIPINENE
TUTIN
SATIVENE
YLANGANE
CUBEBENE
ALLOHIMACHALOL
HIMACHALANE
COPABORNEOL
α-H MUUROLANE
β-H AMORPHANE
COPAANE
H
CARYOPHYLLANE
CIS HUMULANE
β-H CADINANE
α-H BULGARANE
KHUSANE
FUMAGILLIN
SANTALANE(ALPHA)
SANTALANE(BETA)
OPP
SESQUICARANE
TRANS-CIS-FARNESOL
BERGAMOTANE
CAMPHERENONE
ANISATIN
BISABOLANE
BISABOLANE
ACORANE
CEDRANE
CAROTANE
CUPARANE
CUPARANE
CHAMIGRANE
THUJOPSANE
LAURANE
TRICHOTHECANE
WIDDRANE

41--	Sesquiterpenes-Main Skeleton Key

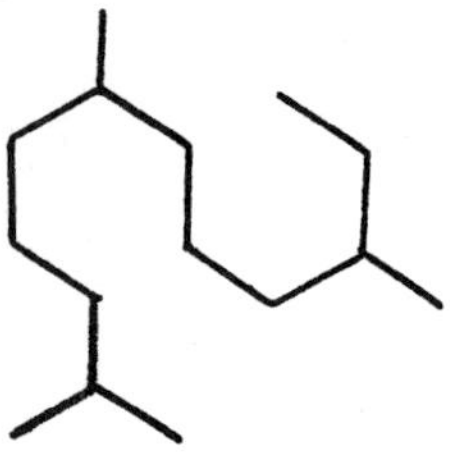

01 Linear	06 Bisabolane	11 Laurane
02 Germacrane	07 Cadinane	12 Acorane
03 Humulane	08 Carotane	13 Widdrane
04 Guaiane	09 Himachalane	14 Pseudo-guaiane
05 Cedrane	10 Trichothecane	15 Cuparane

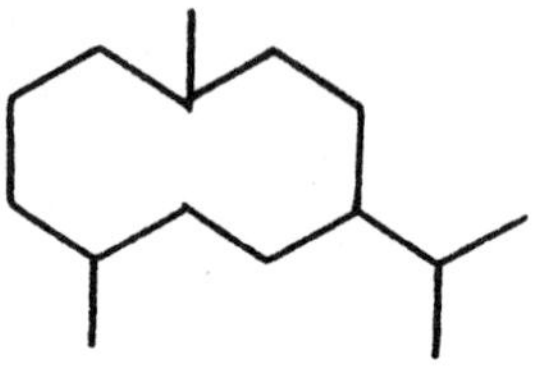

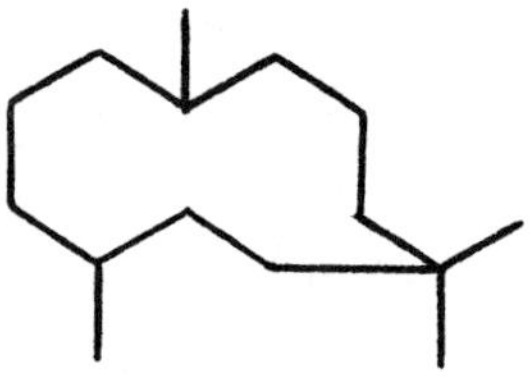

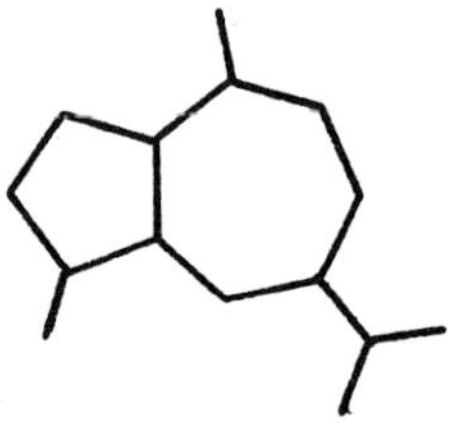

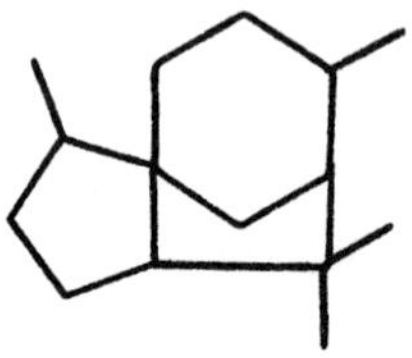

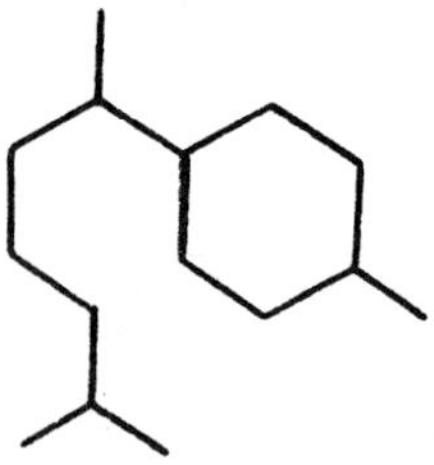

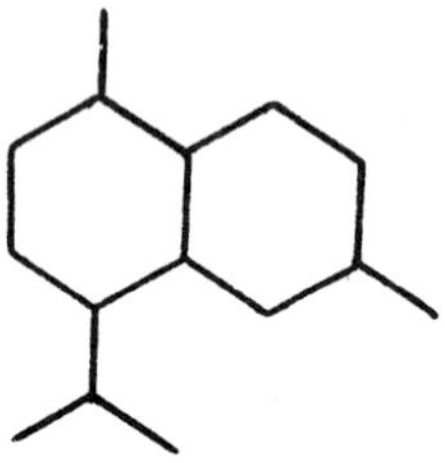

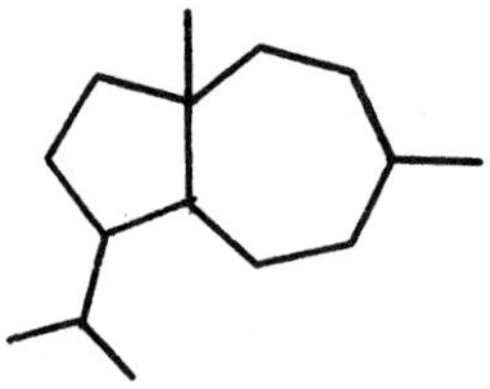

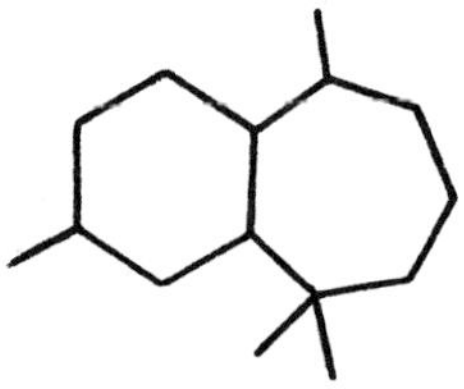

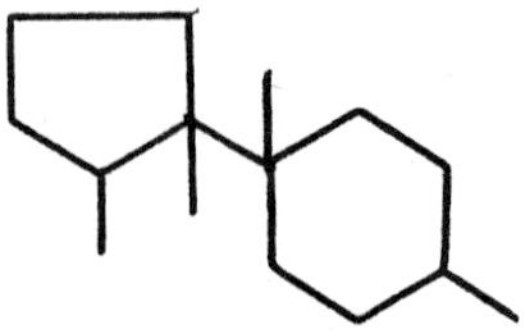

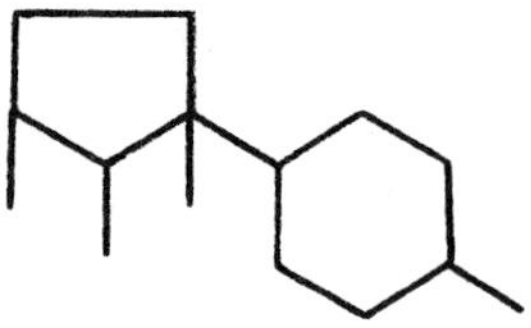

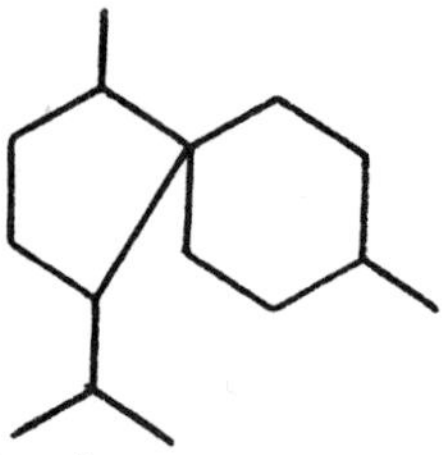

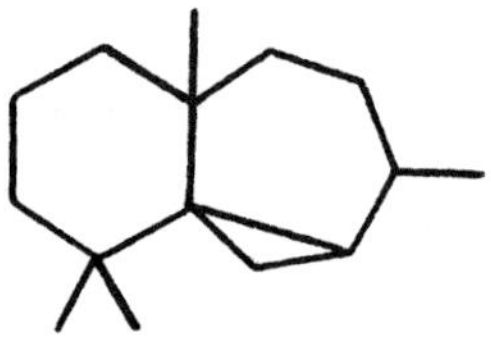

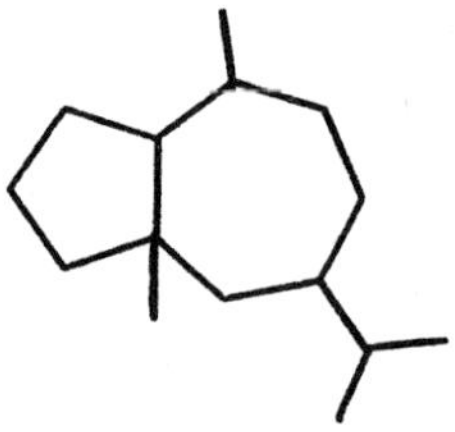

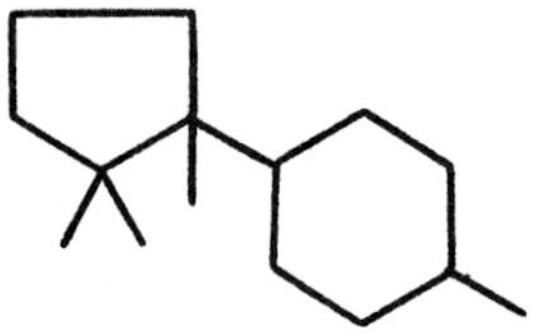

41--	Sesquiterpenes- Main Skeleton Key

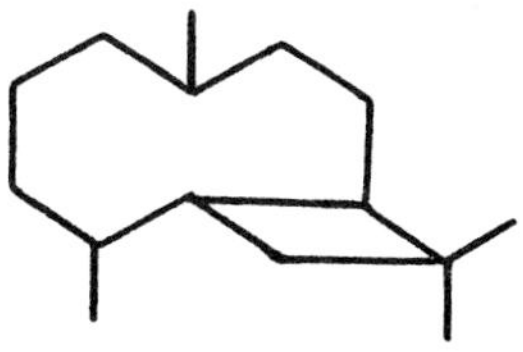 16 Caryophyllane	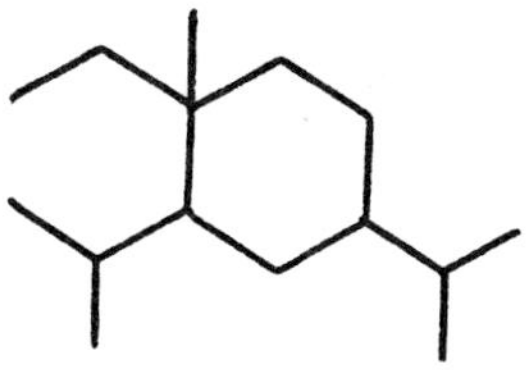 21 Elemane	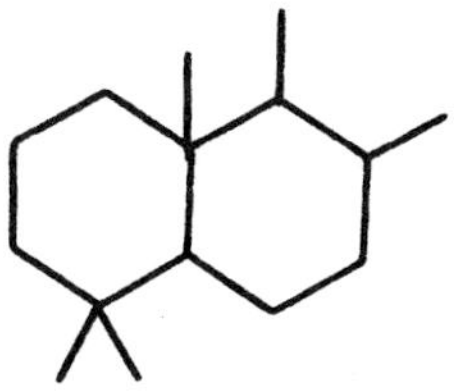 26 Drimane
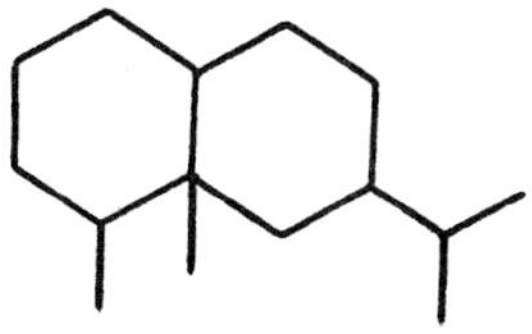 17 Eremophilane	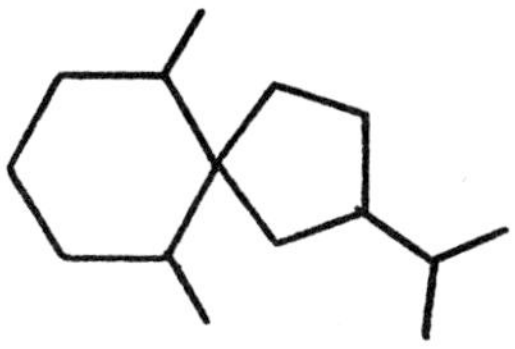 22 Vetispirane	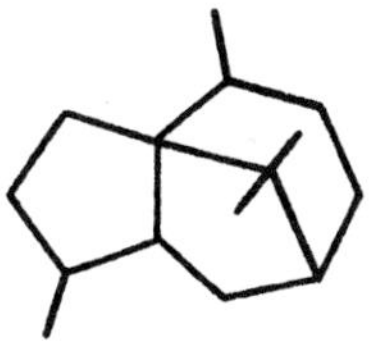 27 Patchoulane
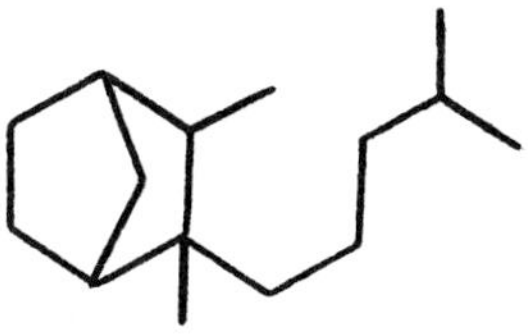 18 Santalane	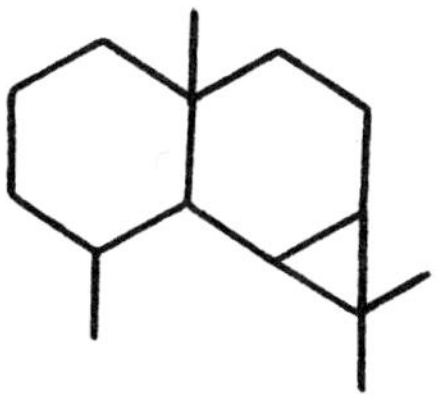 23 Maaliane	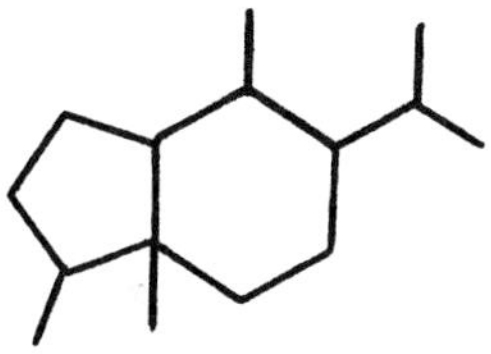 28 Tutin group
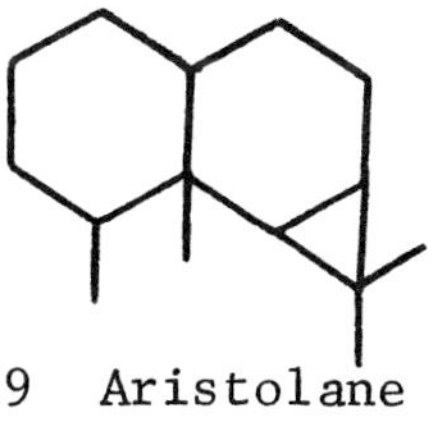 19 Aristolane	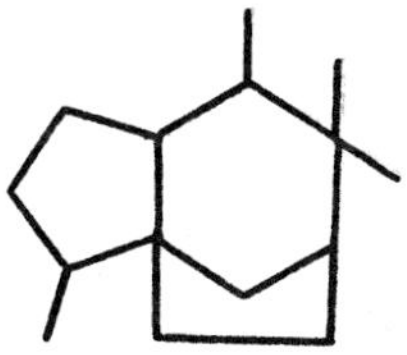 24 Khusane	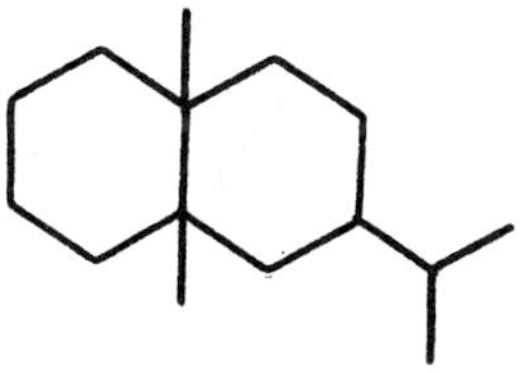 29 Valerane
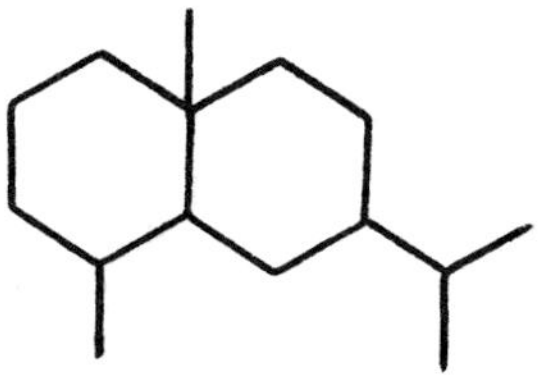 20 Eudesmane	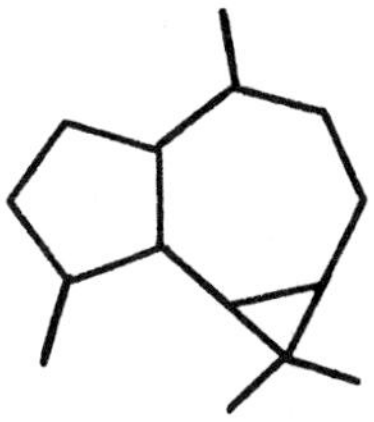 25 Aromadendrane	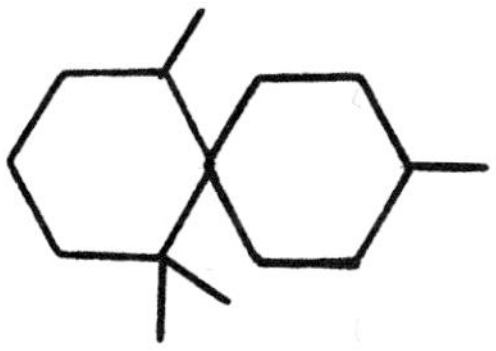 30 Chamigrane

4190	Sesquiterpenes-Less Common Skeletons-Acyclic

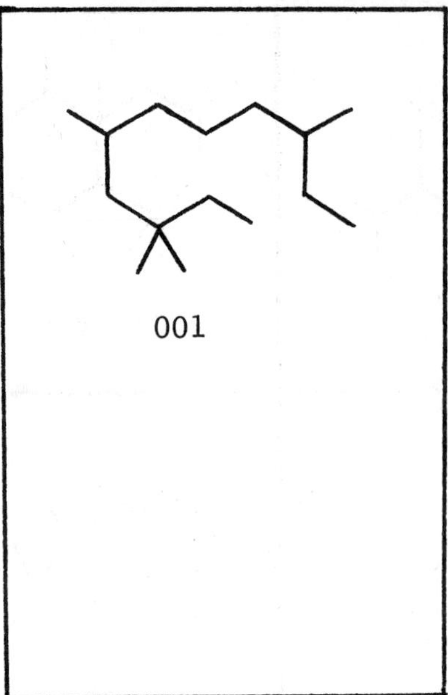

4191	Sesquiterpenes - Less Common Skeletons - Monocyclic

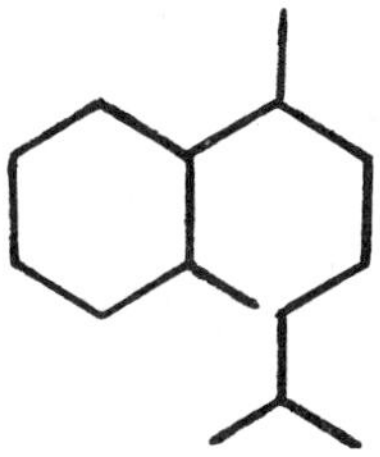 001 002	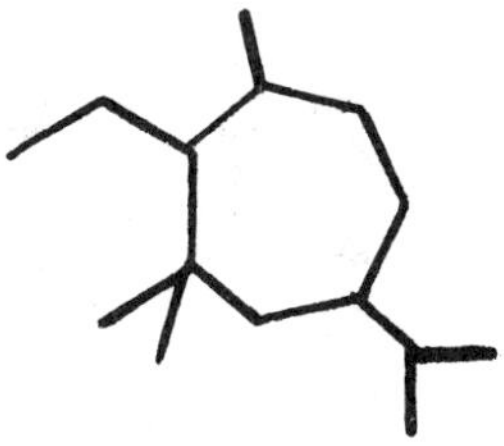 003 004 005 006 007 008 009	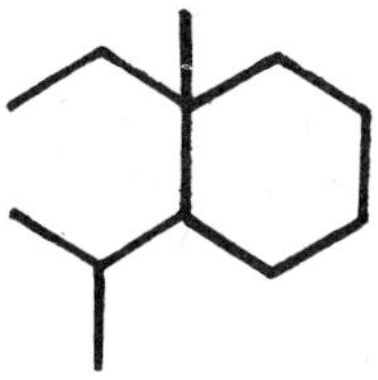 043
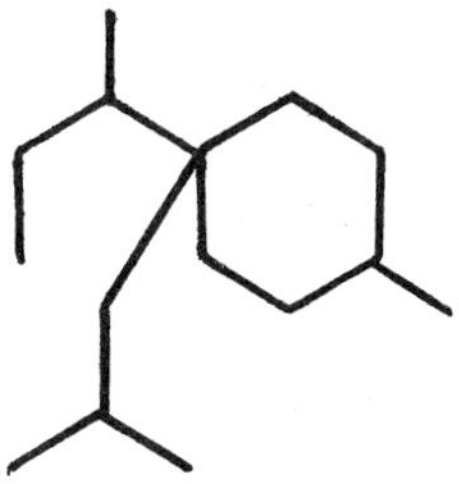 040	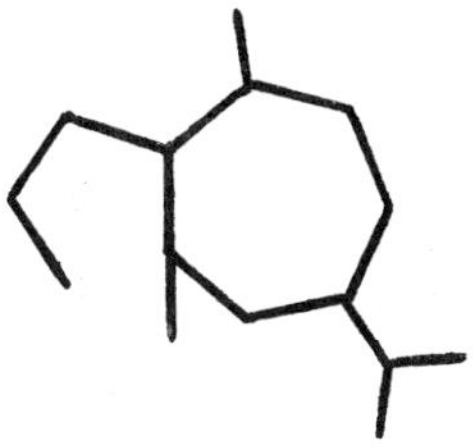 010 011 012 046	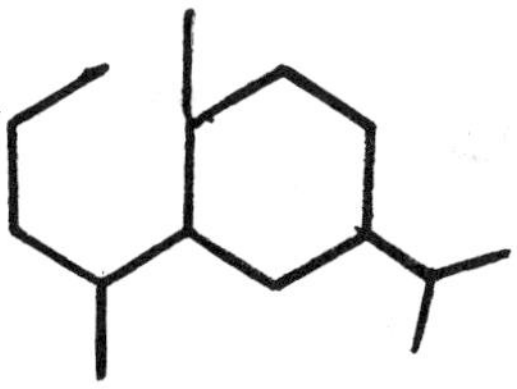 022
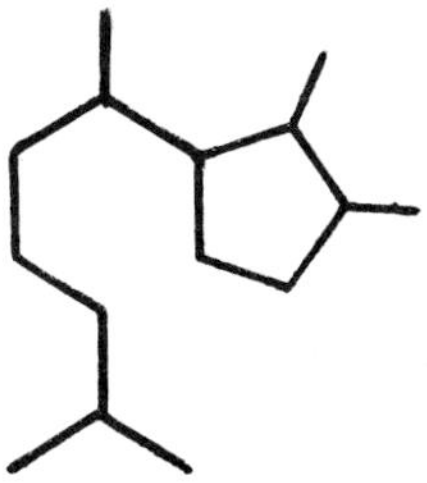 042	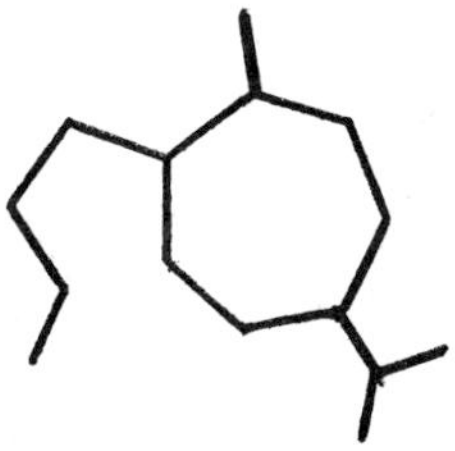 013 020 014 021 015 016 017 018 019	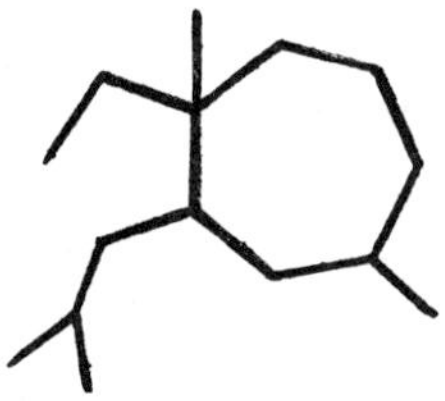 024 025 026

4191	Sesquiterpenes-Less Common Skeletons-Monocyclic

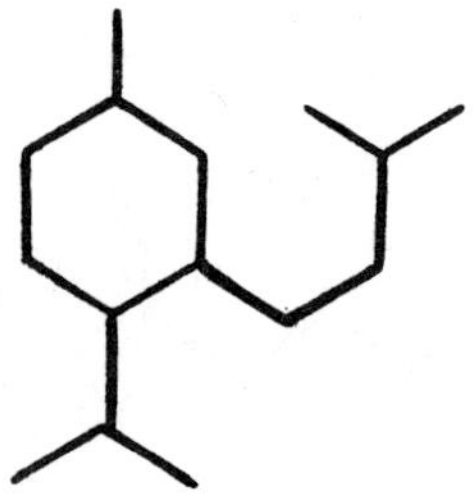

031	027 028 029 030	037
032 033	045	038 039 040 041 047
034 035 036	023	

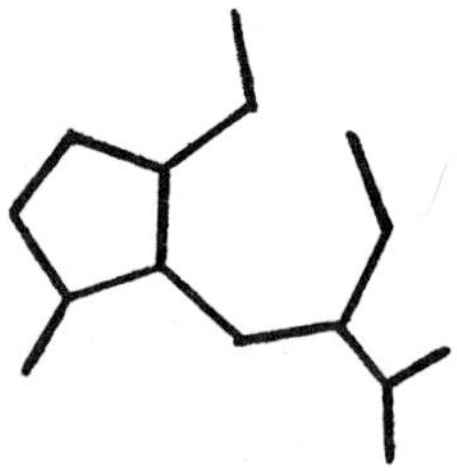

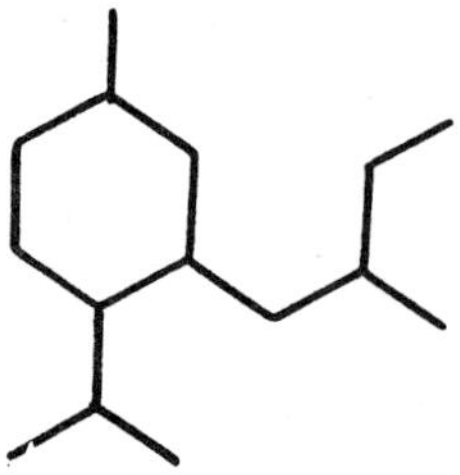

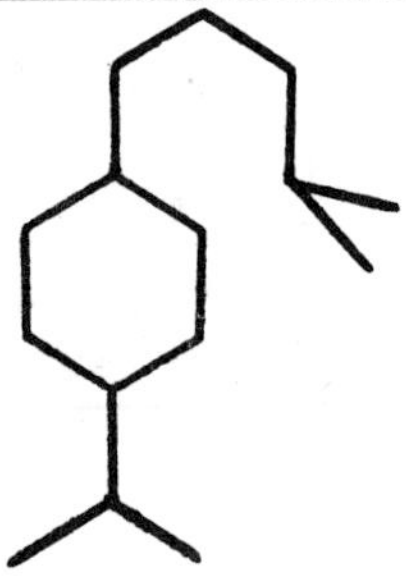

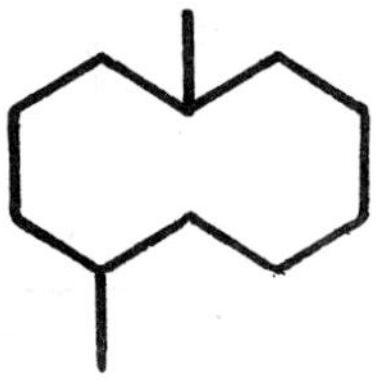

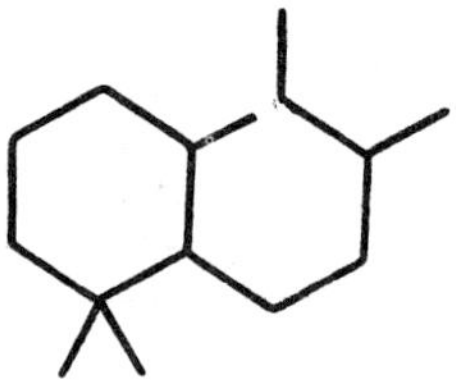

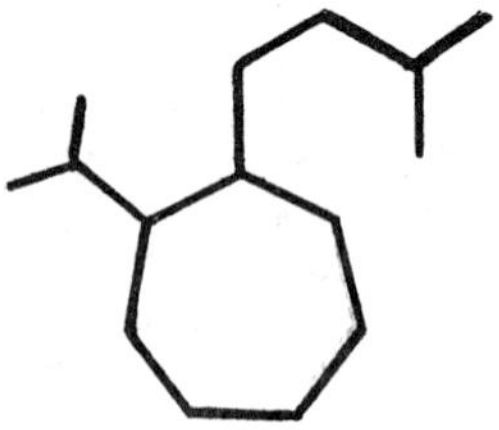

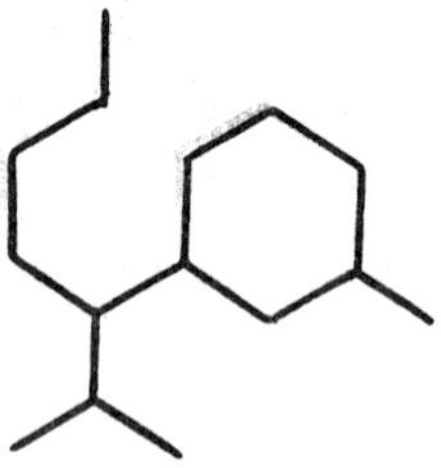

4192	Sesquiterpenes-Less Common Skeletons-Bicyclic

011	012	076
002 003 004	013	064 079
005 006	020	065 066 067

4192 Sesquiterpenes-Less Common Skeletons-Bicyclic

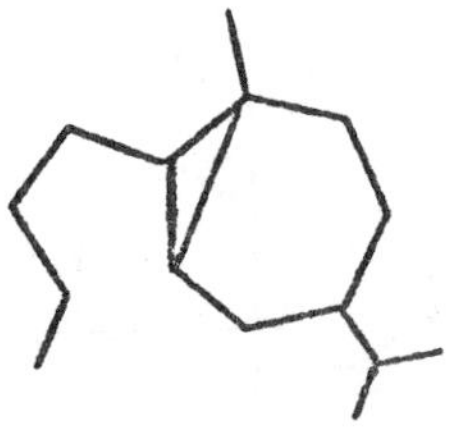

010	057 058	049 050
059 060 061 062 063	073	021 022
017 046	016	027 029

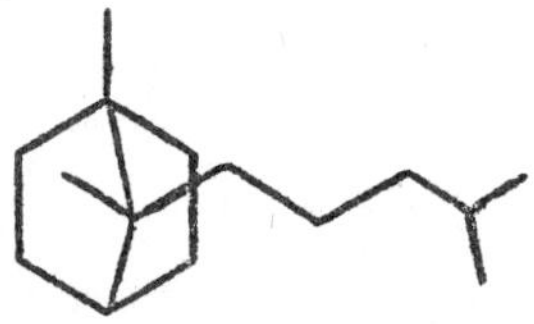

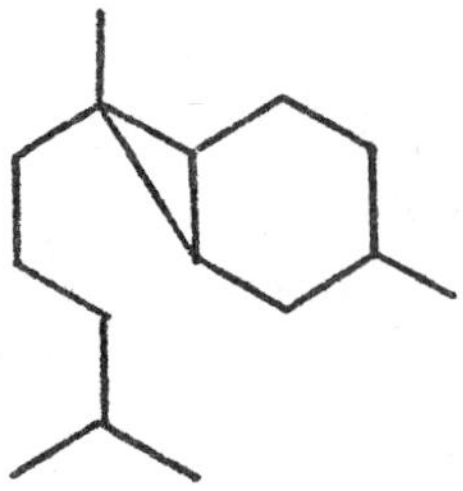

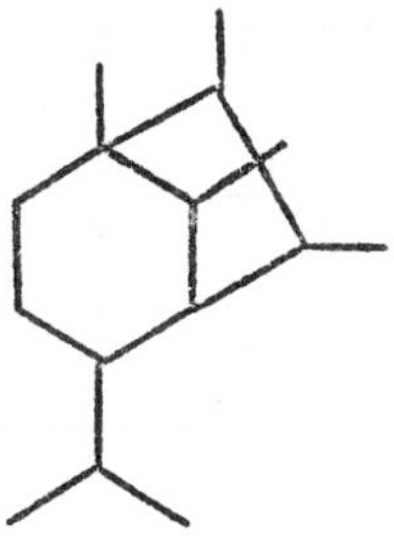

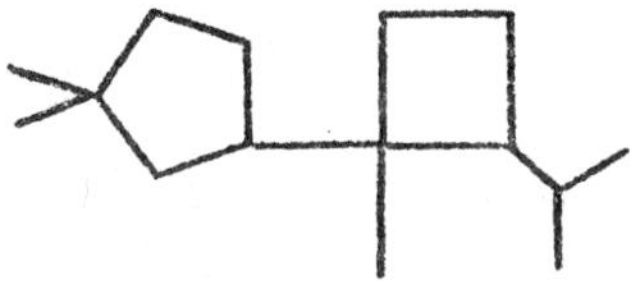

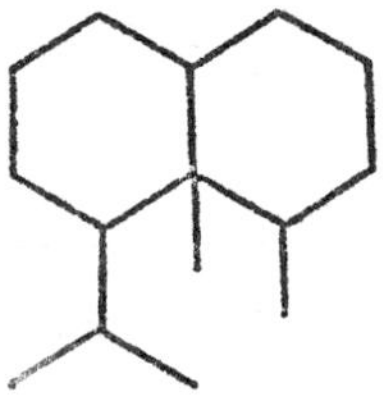

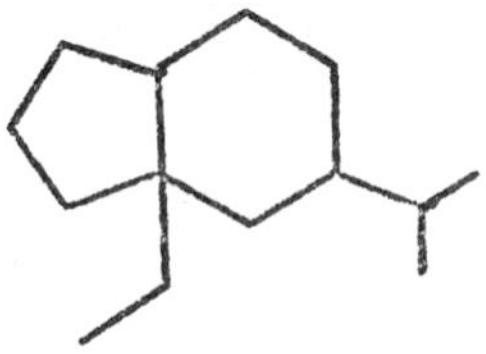

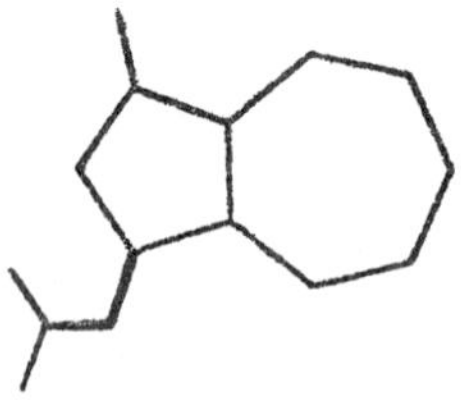

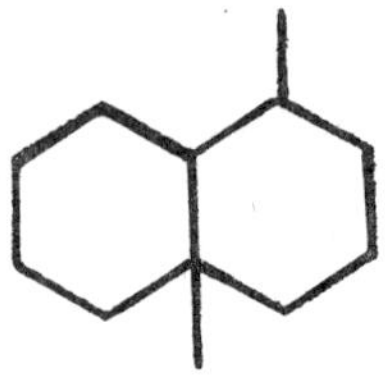

4192	Sesquiterpenes-Less Common Skeletons-Bicyclic

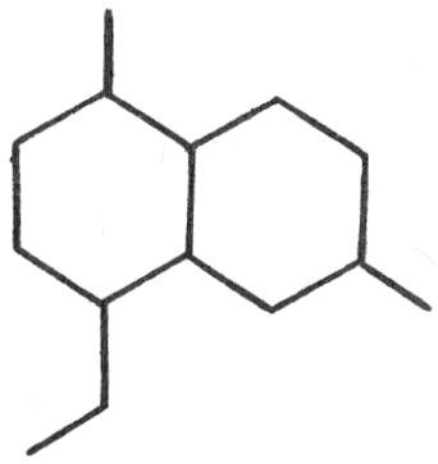 023 024	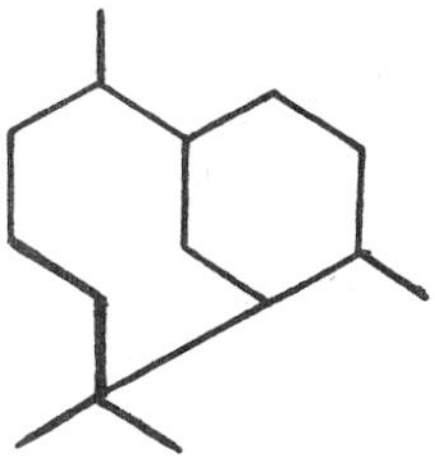 051 052	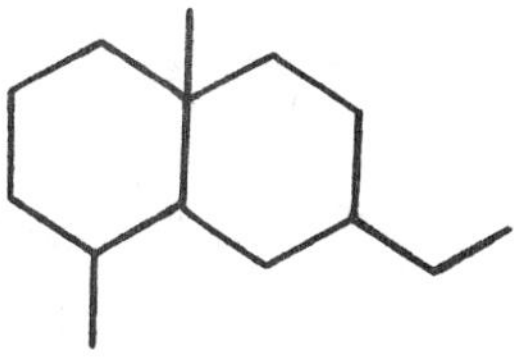 036 037 038 039 040 041 042 043
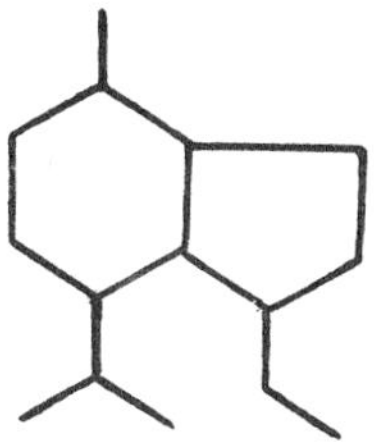 077	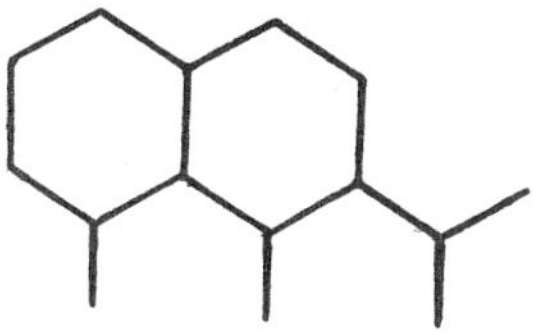 030 031 033 035	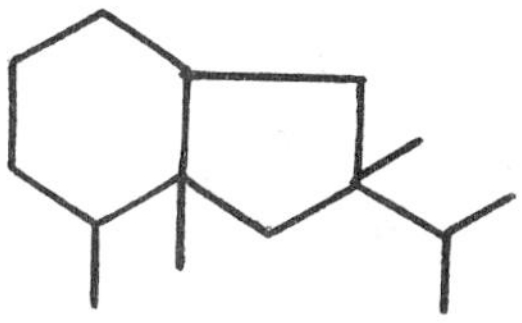 056 055 054 053
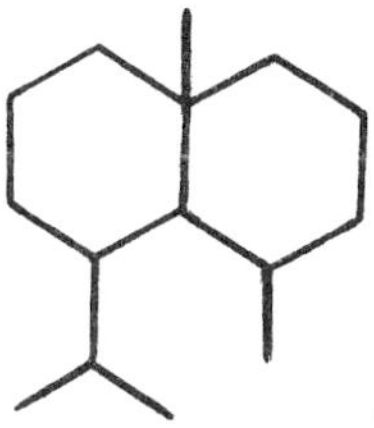 025 026	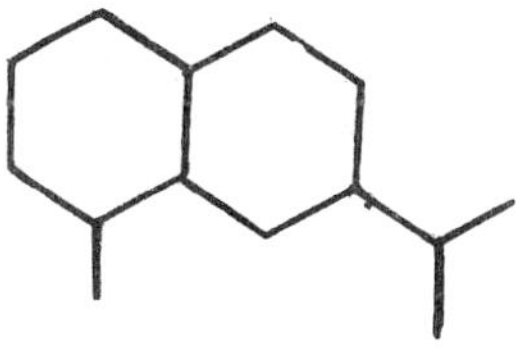 034 032	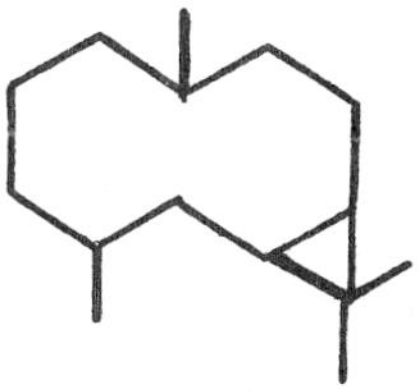 072

4192	Sesquiterpenes-Less Common Skeletons- Bicyclic

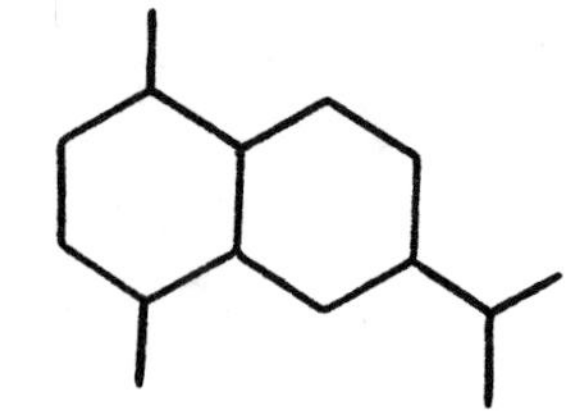 028 078	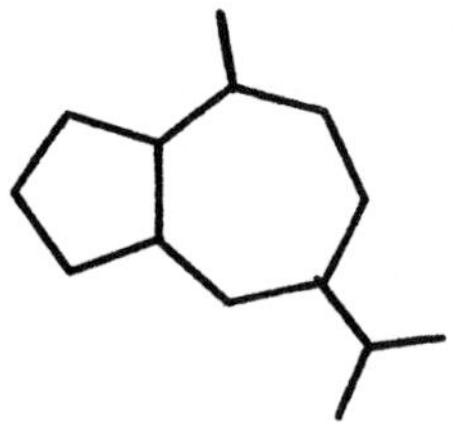 001 008 009	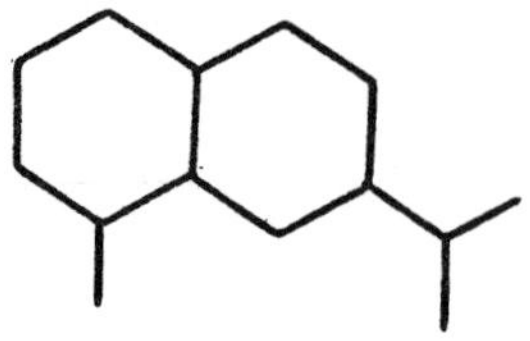 044
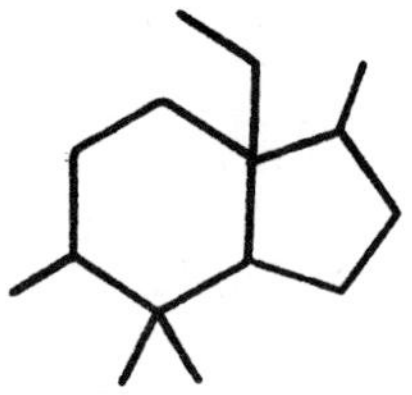 074 075	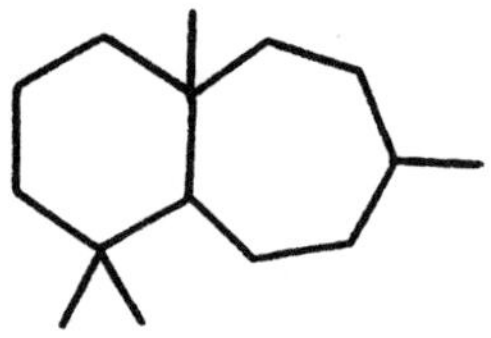 018 019	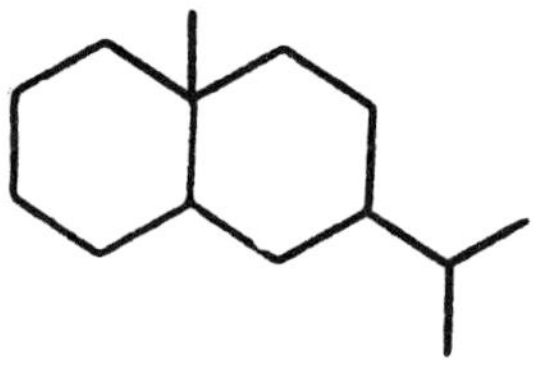 045
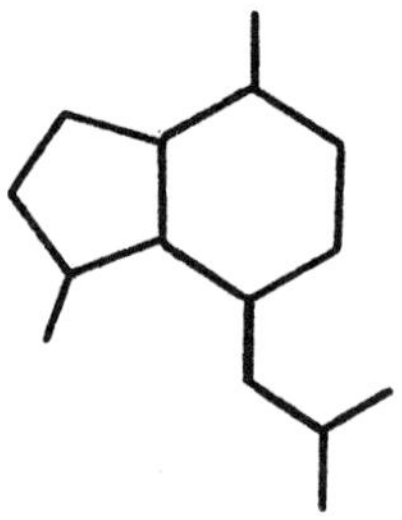 014 015	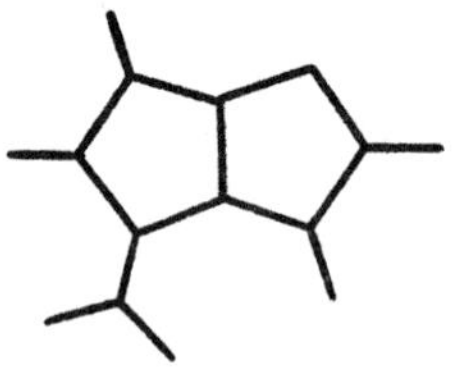 068	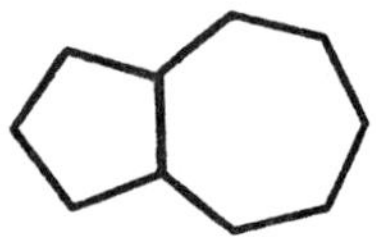 007

4192	Sesquiterpenes-Less Common Skeletons-Bicyclic

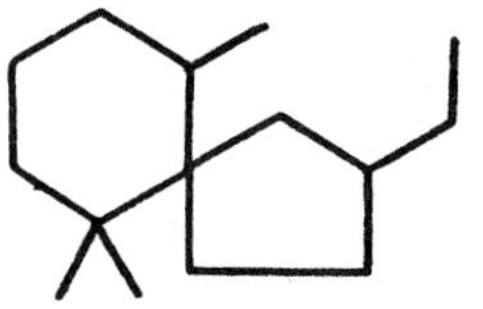 071	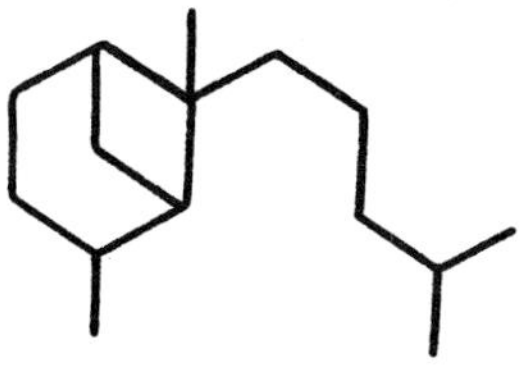 047 048	

4193	Sesquiterpenes - Less Common Skeletons - Tricyclic

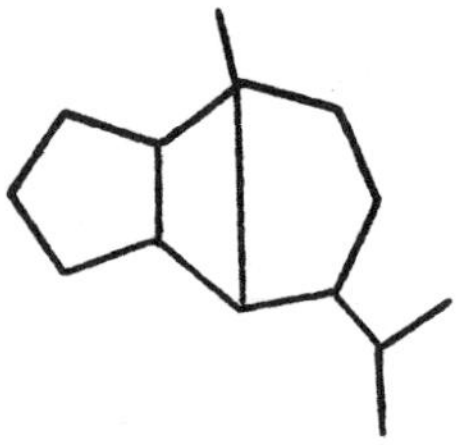

009	030	040
001 002 003 004	031	036
033 032	044	039 038 037

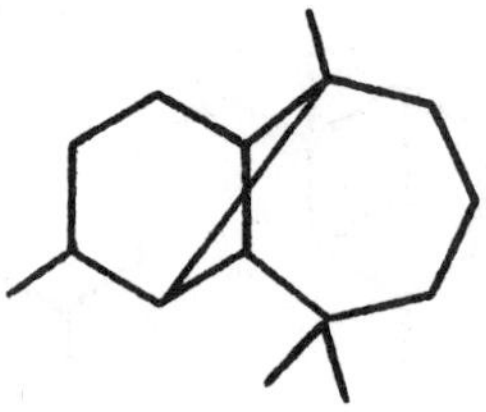

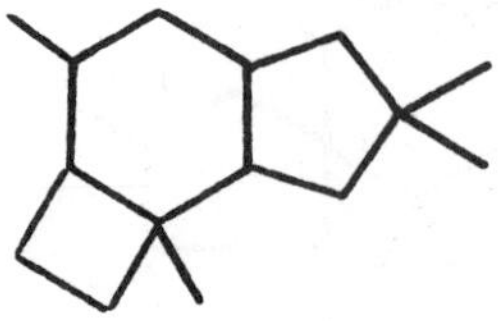

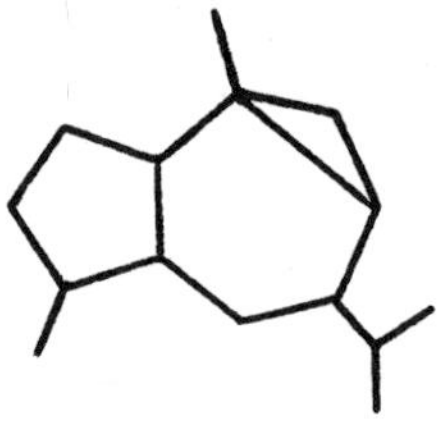

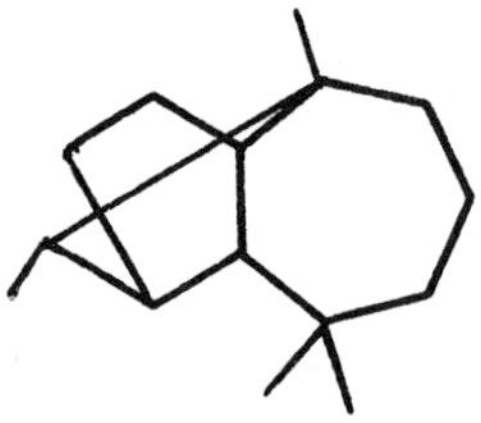

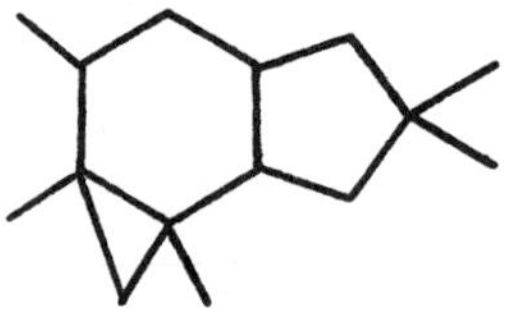

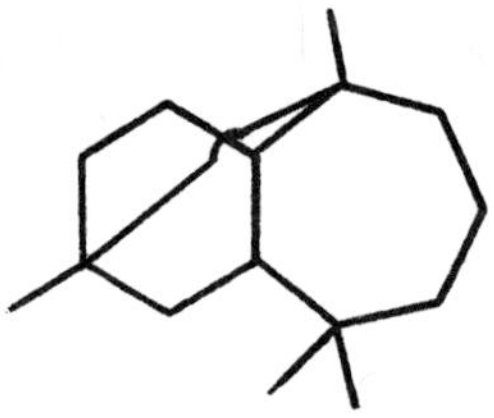

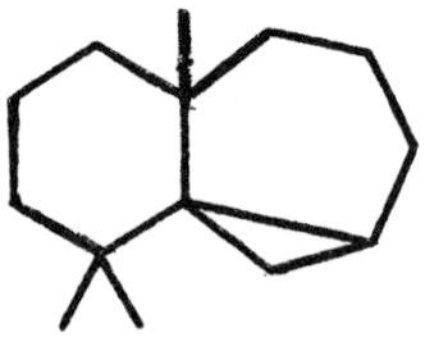

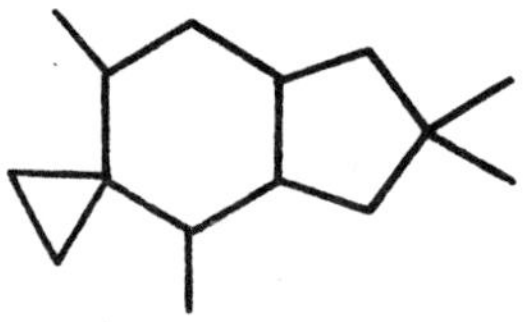

4193	Sesquiterpenes-Less Common Skeletons-Tricyclic

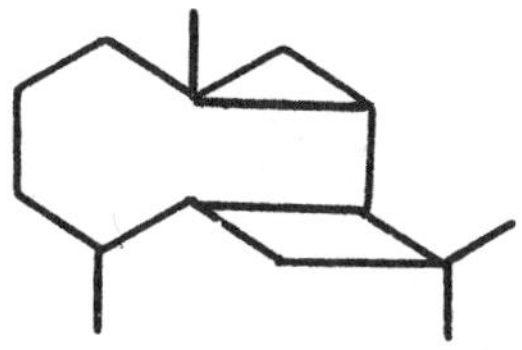

045	019 020 021	012 013 014 015 016 017
022 023 024 025 026 027 028	029	006
010	018	007 008

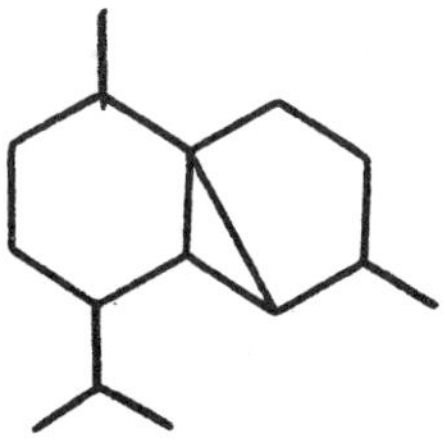

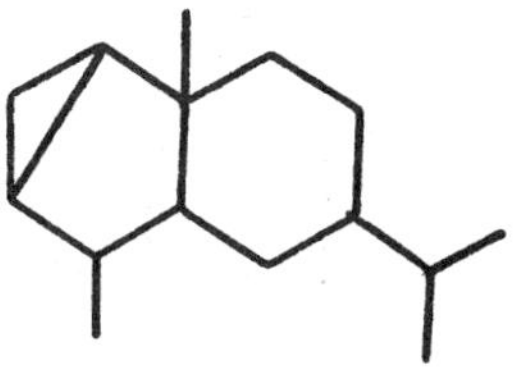

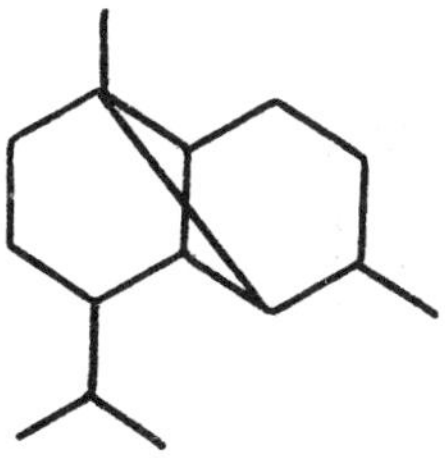

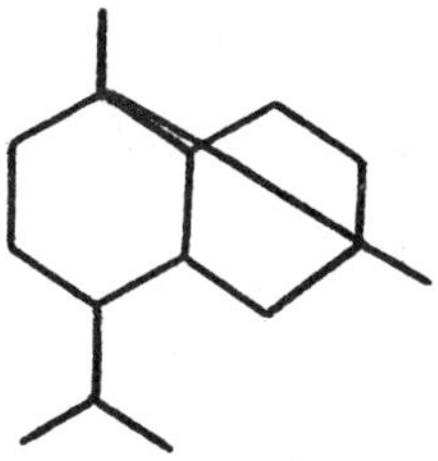

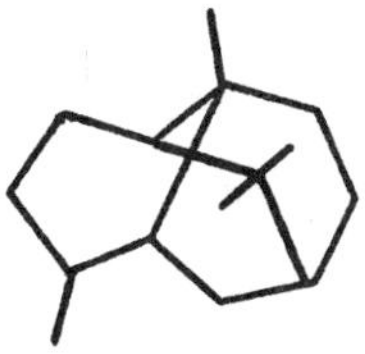

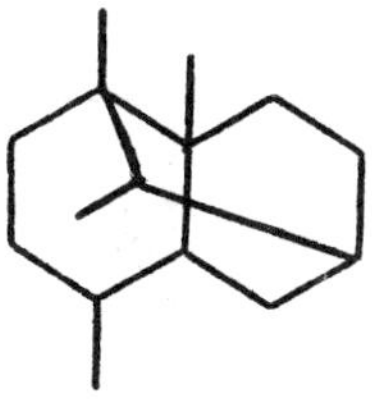

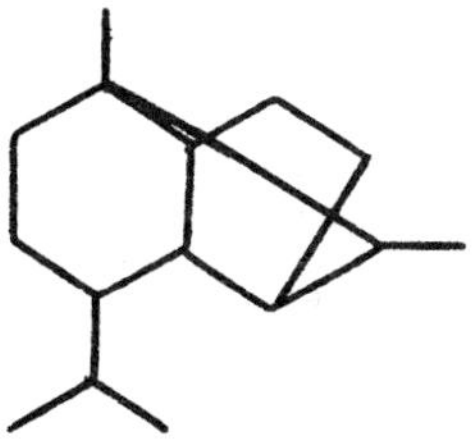

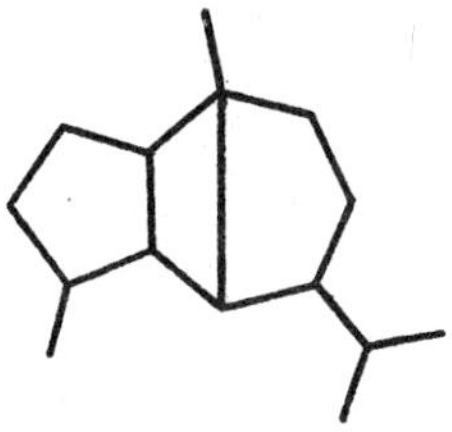

4193	Sesquiterpenes-Less Common Skeletons-Tricyclic

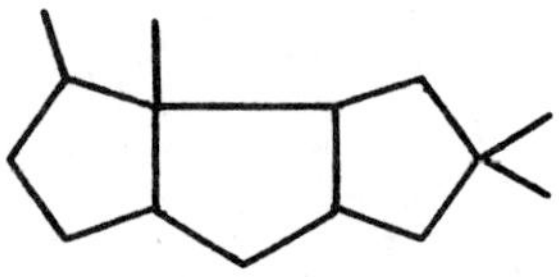

041 042 043 052	011	046
005	048	047
051	049	034 035

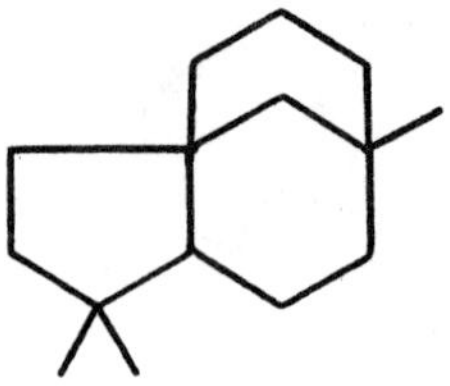

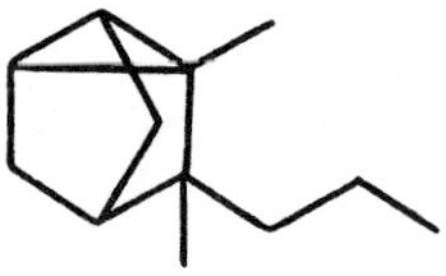

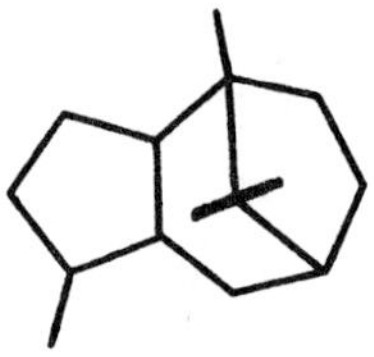

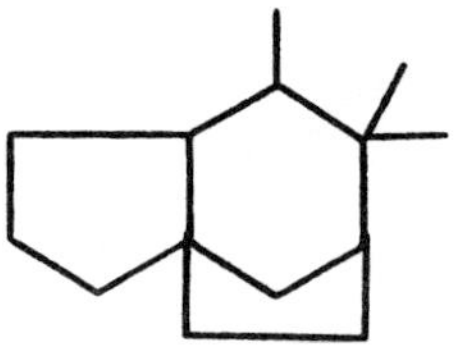

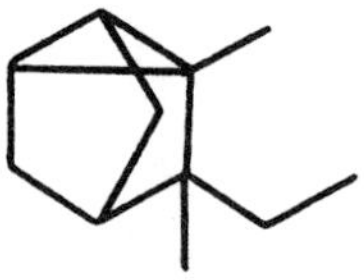

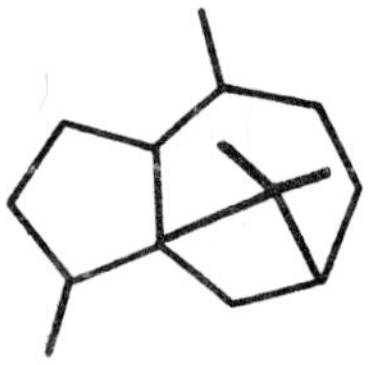

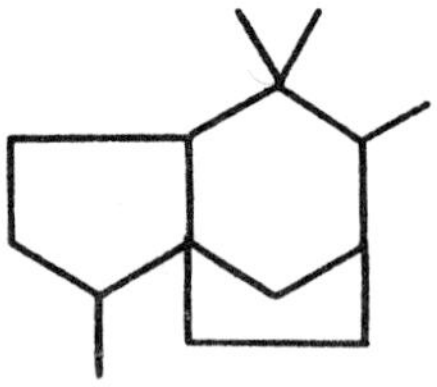

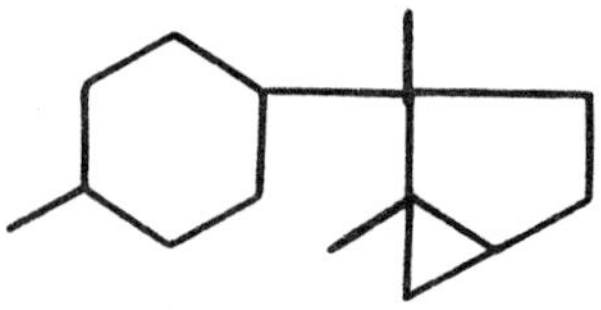

4194	Sesquiterpenes-Less Common Skeletons-Tetracyclic

 007	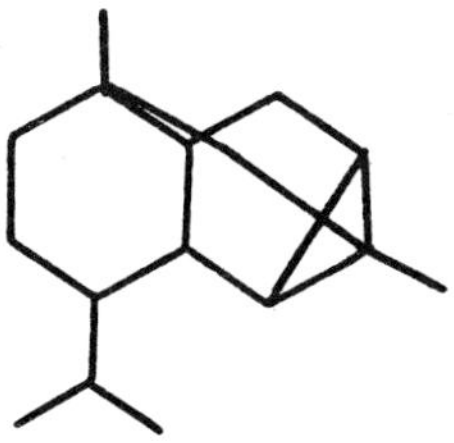 001 002 003 004 005	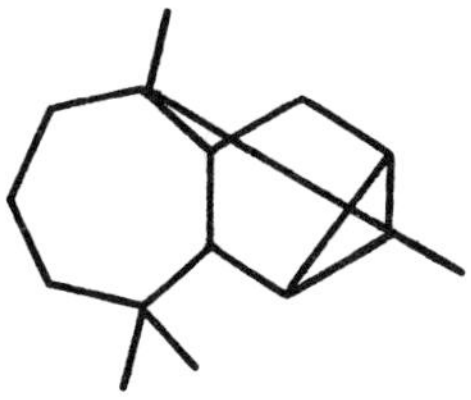 006

4198	Sesquiterpenes-Less Common Skeletons-N-heterocyclic

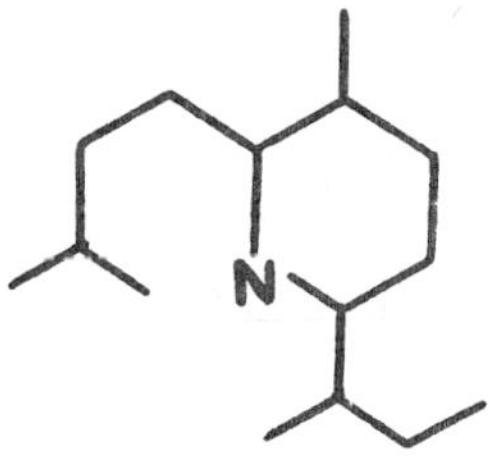

N 001 004 002 005 003	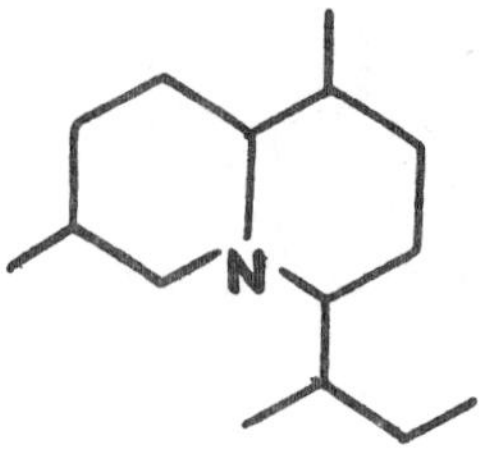 N 006 011 * 007 012 * 008 013 * 009 014 * 010 * dimeric	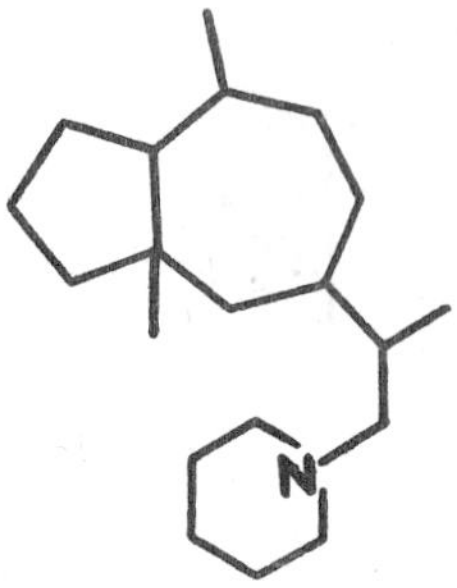 N 015
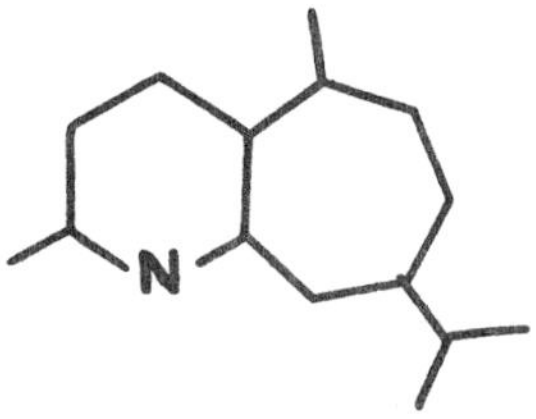 N 016	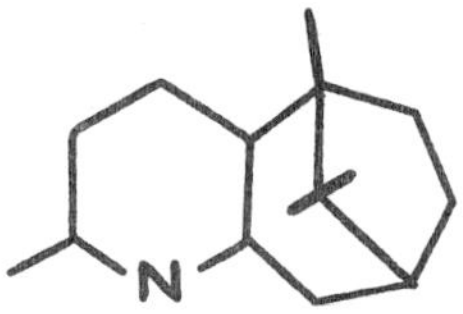 N 017	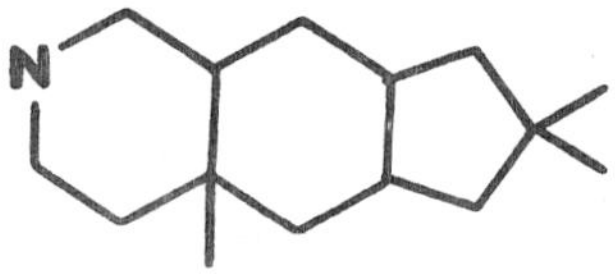 N 018
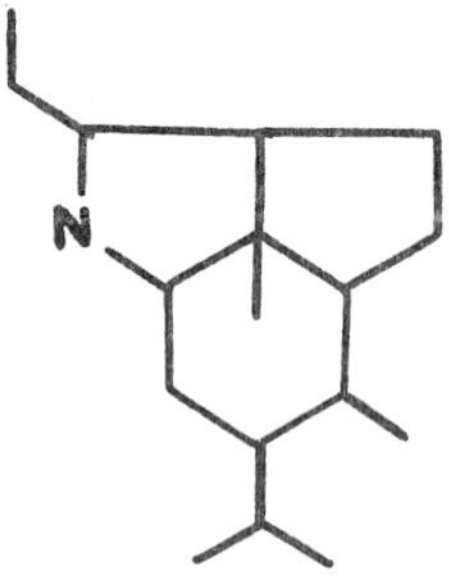 N 019	Me_2N 020	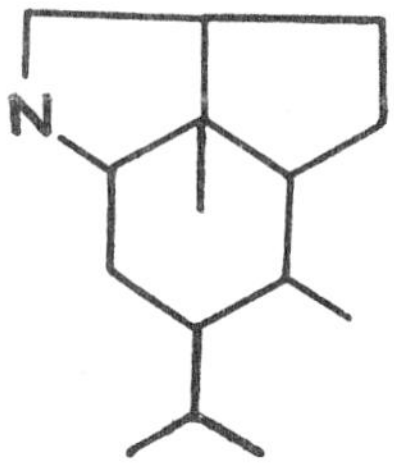 N 021 022 023 024 025

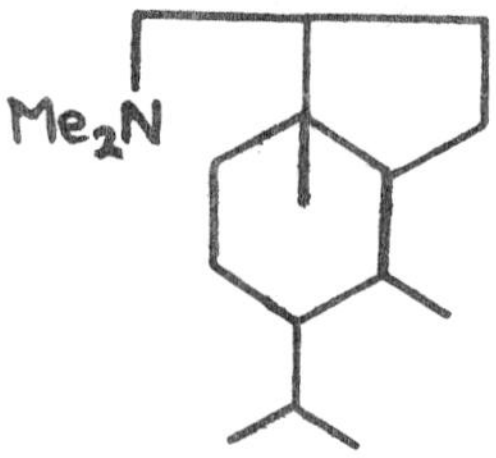

FARNESENE, α- | $C_{15}H_{24}$ | 204 | OI | Oil

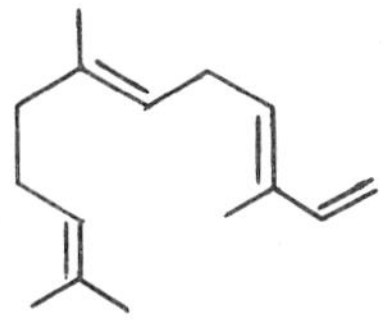

Aust. J. C. 1969 197

4101-001

FARNESENE, β- | $C_{15}H_{24}$ | 204 | OI | Oil

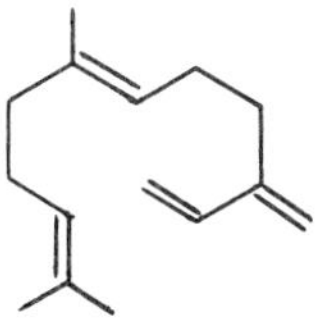

Acta Chem. Scand. 1952 (6) 883

4101-002

FARNESOL | $C_{15}H_{26}O$ | 222 | OI | Oil

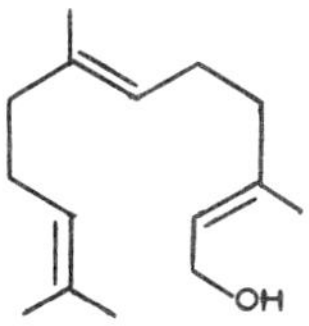

JCS 1966 2144

4101-003

NEROLIDOL | $C_{15}H_{26}O$ | 222 | $+14^0$ | Oil

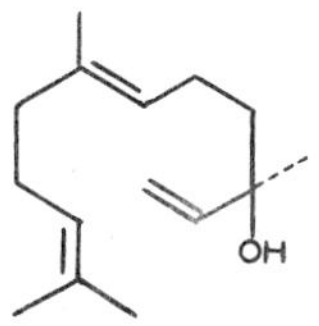

Coll. Czech. Chem. Comm. 1962 (27) 1726

4101-004

CAPARRAPIDIOL | $C_{15}H_{28}O_2$ | 240 | $+8^0$c | Oil

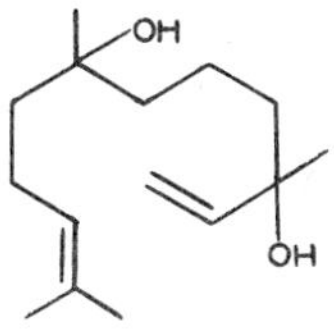

Tet. Lett. 1966 3731

4101-005

CAPARRATRIOL | $C_{15}H_{30}O_3$ | 258 | $+3^0$c | 97^0C

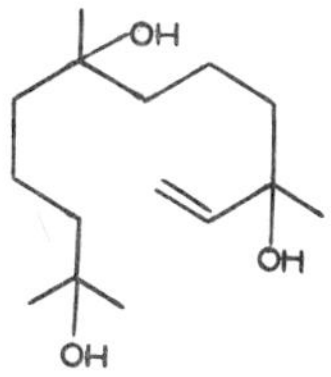

Tet. Lett. 1966 3731

4101-006

FARNESAL | $C_{15}H_{24}O$ | 220 | OI | Oil

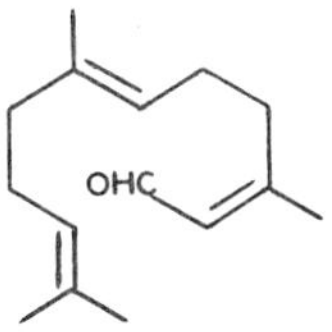

Tet. Lett. 1967 3893

4101-007

SINENSAL | $C_{15}H_{22}O$ | 218 | OI | Oil

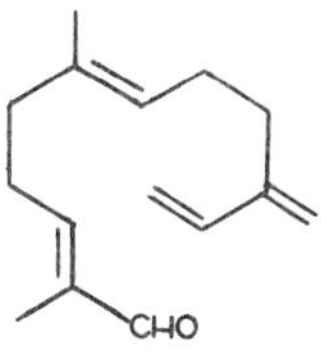

JOC 1965 1690

4101-008

DAVANONE | $C_{15}H_{24}O_2$ | 236 | $+78^0$ | Oil

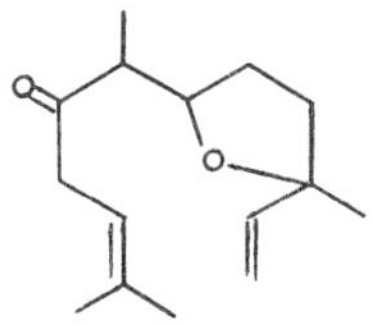

Chem. Abstr. 1969 (70) 11836

4101-009

TORREYOL, neo- | $C_{15}H_{22}O_2$ | 234 | OI | Oil

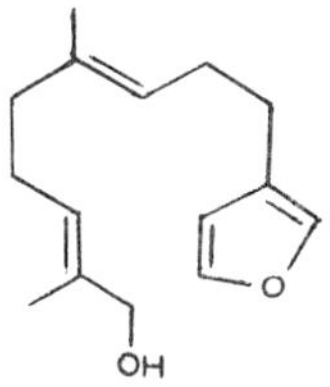

Bull. Chem. Soc. Jap. 1965 381

4101-010

TORREYAL $C_{15}H_{20}O_2$ 232 +2° neat Oil Tet. Let. 1963 1171 4101-011	FURAN, 3-(4', 8'-diMe-6-keto-nonyl)- $C_{15}H_{24}O_2$ 236 Oil Chem. Abstr., 1963 (59) 1565. 4101-016
IPOMEAMARONE $C_{15}H_{22}O_3$ 250 +85° Arch. Biochem. Biophys. 1962 (99) 52 4101-012	CECROPIA JUVENILE HORMONE $C_{18}H_{30}O_3$ 294 Oil JACS., 1970, 737. 4101-017
NGAIONE, (-)- $C_{15}H_{22}O_3$ 250 -27° neat Oil Aust. J. C. 1970 (23) 107 4101-013	JUVENATE, 12-homo-cis- $C_{17}H_{28}O_3$ 280 Proc. Nat. Acad. Sci. (U.S.A.), 1968 (60) 853. 4101-018
DENDROLASINE $C_{15}H_{22}O$ 218 OI Oil Tet. 1957 156, 177 4101-014	MYOPORONE $C_{15}H_{22}O_3$ 250 Oil Chem. & Ind. 1957 491 Tet. 1958 68 4101-019
FREELINGYNE $C_{15}H_{12}O_3$ 240 OI 164°C Tet. Lett. 1966 (18) 1939 4101-015	$C_{20}H_{30}O_3$ 318 Tet. Lett., 1969 5109. 4101-020

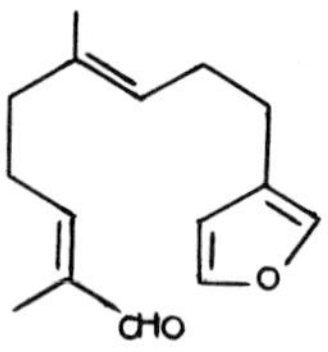

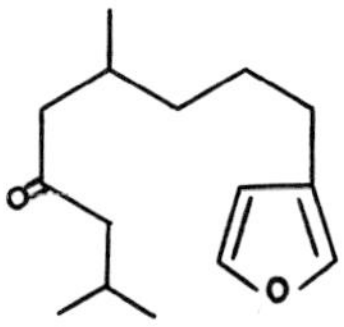

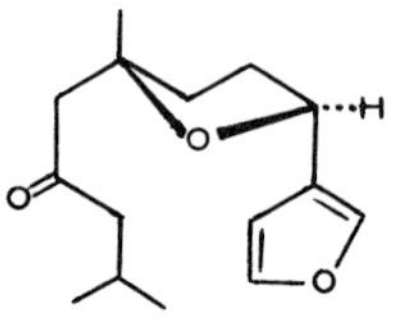

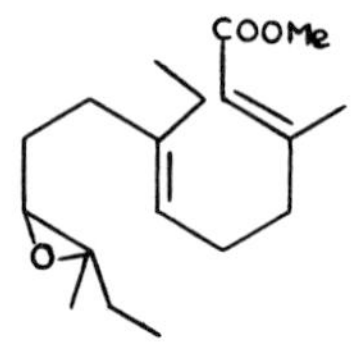

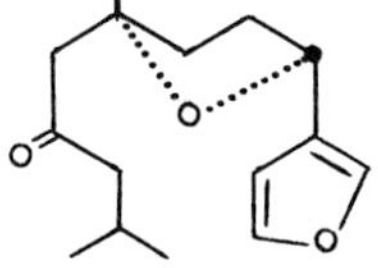

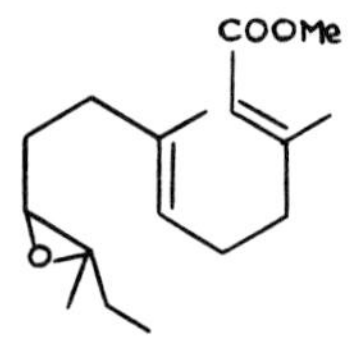

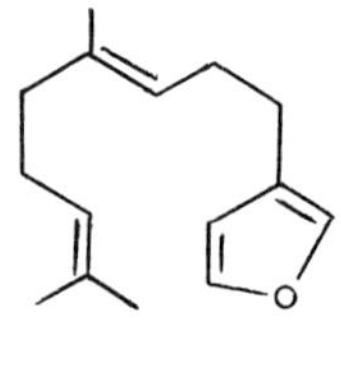

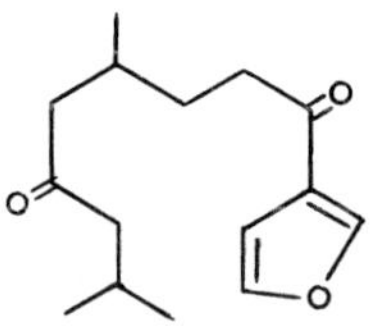

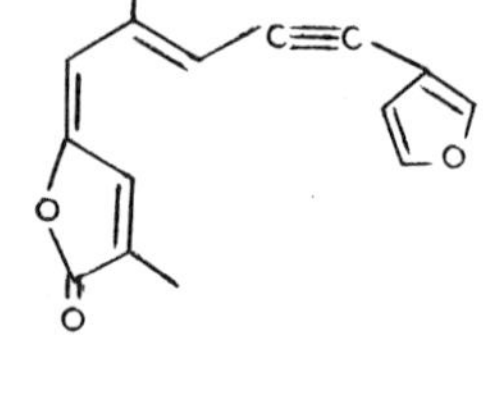

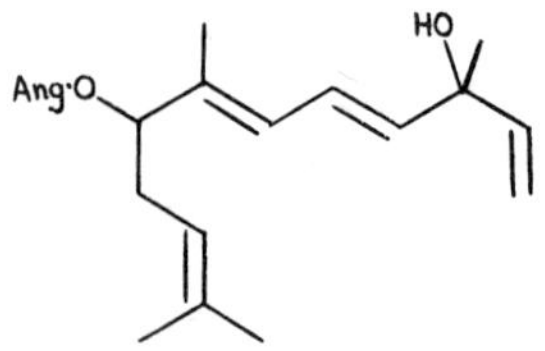

DODECA-2, 6, 10-TRIENE, 10-cis-2, 6, 10-triMe- $C_{15}H_{26}$ 206 OI Oil Aust. J. C., 1970, 2337. 4101-021	FARNESENE, c, t-allo- $C_{15}H_{24}$ 204 OI Oil Bull. Chem. Soc. Jap., 1969 (42) 3615. 4101-023
DODECA-2, 6, 10-TRIENE, 10-trans-2, 6, 10-triMe- $C_{15}H_{26}$ 206 OI Oil Aust. J. C., 1970, 2337. 4101-022	FARNESENE, t, t-allo- $C_{15}H_{24}$ 204 OI Oil Bull. Chem. Soc. Jap., 1969 (42) 3615. 4101-024

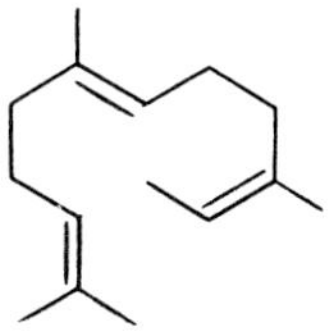

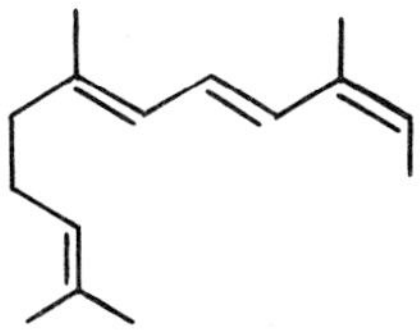

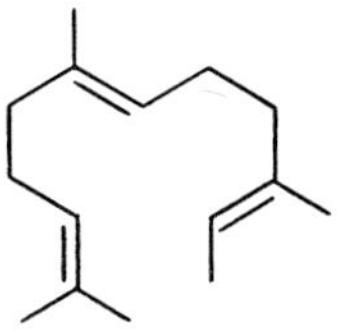

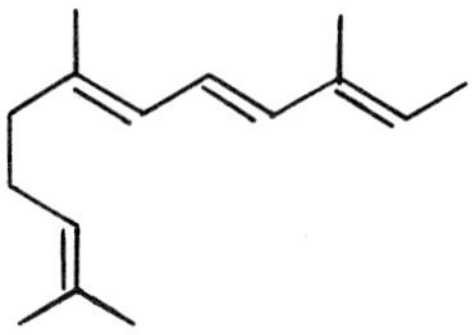

GERMACRENE A — $C_{15}H_{24}$ — 204 — -3^0t — Oil

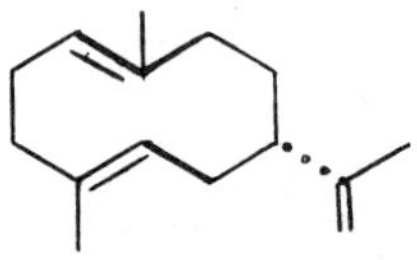

Tet. Lett. 1970 497

4102-001

SHIROMODIOL, mono-acetate- — $C_{17}H_{28}O_4$ — 296 — -45^0c — 80^0C

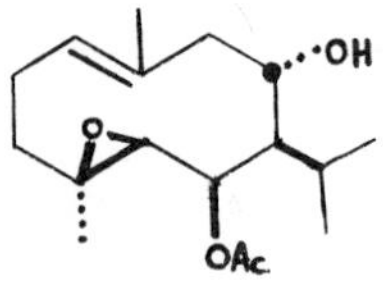

Chem. Comm. 1970 128

4102-006

GERMACRENE B — $C_{15}H_{24}$ — 204 — OI — Oil

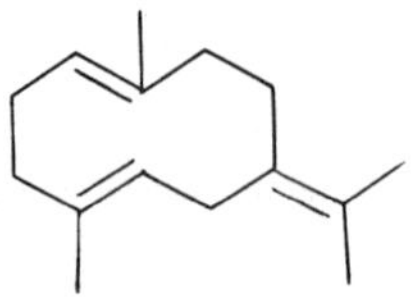

Tet. Lett. 1969 3097

4102-002

SHIROMODIOL, di-acetate- — $C_{19}H_{30}O_5$ — 338 — -62^0c — 112^0C

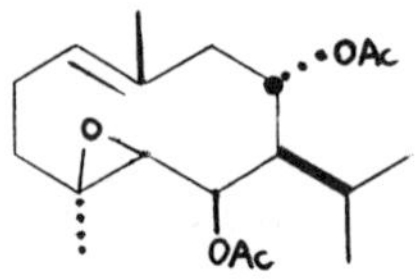

Tet. Lett. 1968 4673

4102-007

GERMACRENE C — $C_{15}H_{24}$ — 204 — 0^0

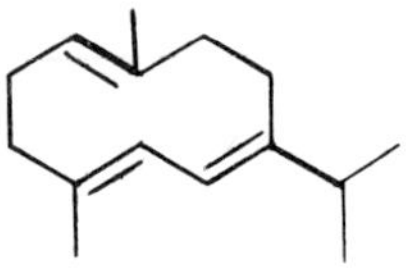

Tet. Lett. 1969 1799

4102-003

deleted

4102-008

GERMACRENE D — $C_{15}H_{24}$ — 204 — -240^0

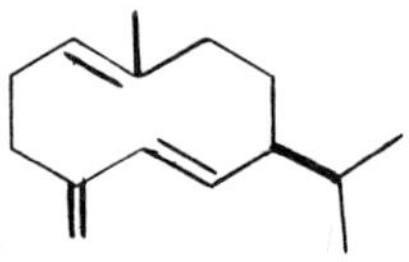

Tet. Lett. 1969 3097

4102-004

GERMACRONE — $C_{15}H_{22}O$ — 218 — OI — Oil

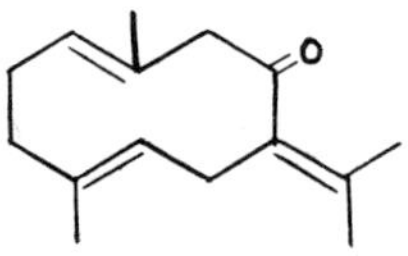

JACS 1960 (82) 5749

4102-009

HEDYCARYOL — $C_{15}H_{26}O$ — 222 — $+30^0$c — Oil

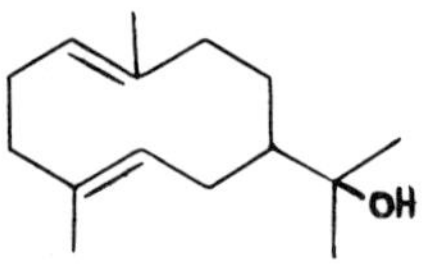

Chem. Comm. 1968 1229

4102-005

CALAMENDIOL, pre-iso- — $C_{15}H_{24}O$ — 220 — Oil

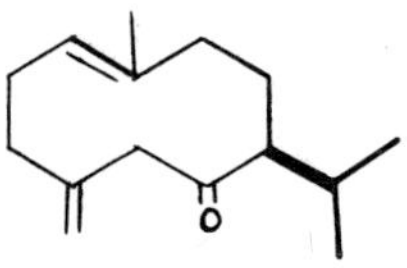

Tet. Lett. 1970 855

4102-010

CURDIONE $C_{15}H_{24}O_2$ 236 +26° 61°C Chem. Pharm. Bull. 1967 1390 4102-011	ZEDERONE $C_{15}H_{18}O_3$ 246 +266° 154°C Chem. Pharm. Bull. 1966 550 4102-016
ARISTOLACTONE $C_{15}H_{20}O_2$ 232 +156°e 110°C Tet. Lett. 1964 2391 4102-012	SERICENIC ACID $C_{15}H_{18}O_3$ 246 0° 203°C COOH Tet. Lett. 1968 4957 4102-017
FURANODIENE $C_{15}H_{20}O$ 216 OI 44°C Tet. Lett. 1968 931 4102-013	SERICENINE $C_{16}H_{20}O_3$ 260 0° 115°C COOCH3 Tet. Lett. 1968 4957 4102-018
FURANODIENONE $C_{15}H_{18}O_2$ 230 OI 90°C Chem. Comm. 1969 662 4102-014	LINDERALACTONE $C_{15}H_{16}O_3$ 244 +102° 140°C Chem. Comm. 1968 637 4102-019
FURANODIENONE, iso- $C_{15}H_{18}O_2$ 230 70°C Chem. Comm. 1969 662 4102-015	LINDERANE $C_{15}H_{16}O_4$ 260 +179° 190°C Chem. Comm. 1968 637 JCS 1969 1491 4102-020

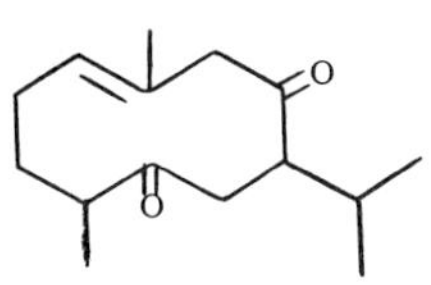

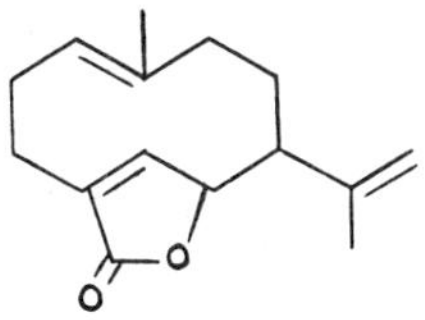

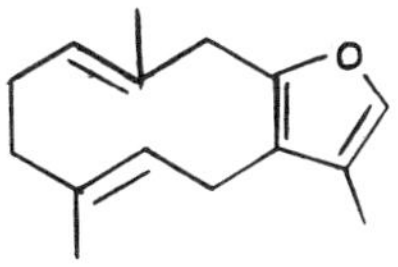

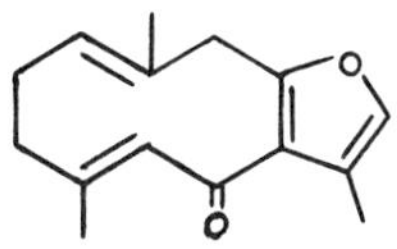

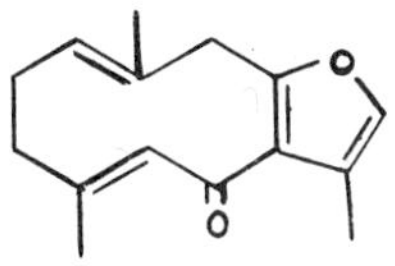

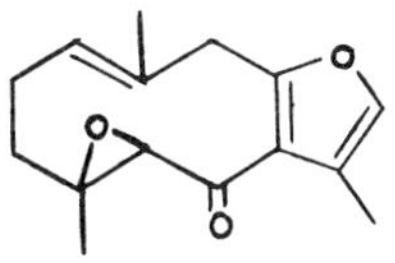

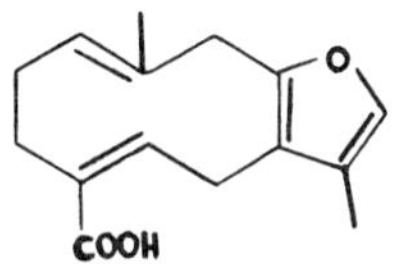

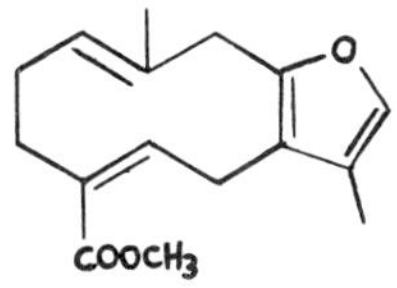

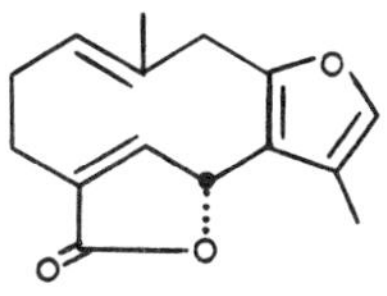

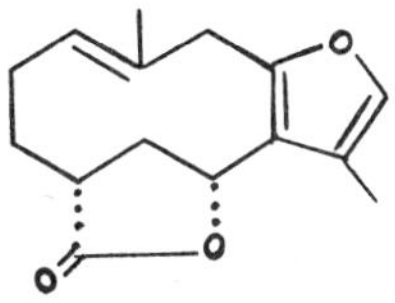

LINDERANE, Neo-

$C_{15}H_{16}O_4$ 260 $+ 32^0$ 170^0C

Chem. Comm. 1968 1168
Tet. 1967 (23) 261

4102-021

LINDERANE, pseudo-neo-

$C_{15}H_{16}O_4$ 260 $+90^0$ 201^0C

Chem. Comm. 1968 941, 1168

4102-022

LINDERADINE

$C_{15}H_{16}O_5$ 276 -69^0 131^0C

Chem. Comm. 1968 940, 1168

4102-023

LITSEALACTONE

$C_{17}H_{18}O_5$ 302 $+58^0$ 158^0C

Chem. Comm. 1968 940

4102-024

ZEYLANINE

$C_{17}H_{18}O_5$ 302 $+271^0$ 175^0C

Tet. 1967 (23) 261

4102-025

ZEYLANANE

$C_{17}H_{18}O_6$ 318 $+231^0$ 151^0C

* epoxide

Chem. Comm., 1968, 940.

4102-026

ZEYLANICINE

$C_{17}H_{18}O_6$ 318 -153^0 235^0C

Tet. 1967 (23) 261

4102-027

ZEYLANIDINE

$C_{17}H_{18}O_7$ 334 -174^0 220^0C

* epoxide

Tet. 1967, 261.

4102-028

LITSEACULANE

$C_{17}H_{18}O_6$ 318 $+76^0$ 146^0C

Chem. Comm. 1968 940

4102-029

SALONITOLIDE

$C_{15}H_{22}O_4$ 266

Coll. Czeck. 1968 2239; 1965 2863

4102-030

SCABIOLIDE	$C_{21}H_{28}O_8$	408 + 101°c 120°C	MIKANOLIDE, dihydro-	$C_{15}H_{16}O_6$	292 +91°d 242°C
4102-031	Coll. Czech. Chem. Comm., 1968, 2239.		* dihydro 4102-036	Tet. Lett., 1967, 3111. JOC., 1970, 1453.	
UVEDALIN COOMe OAc 4102-032	$C_{23}H_{28}O_9$ JOC 1970 2605	448 +13°c 132°C	MIKANOLIDE, desoxy- 4102-037	$C_{15}H_{16}O_5$ JOC 1970 1453	276 + 99° 199°C
POLYDALIN COOMe OAc OH 4102-033	$C_{23}H_{28}O_{10}$ JOC 1970 2605	464 +8°c 182°C	SCANDENOLIDE AcO 4102-038	$C_{17}H_{18}O_7$ JOC 1970 1453	334 + 62° 232°C
ISABELIN 4102-034	$C_{15}H_{16}O_4$ Tet. 1969 4762 Chem. Comm. 1968 1679	260 -57°c 170°C	SCANDENOLIDE, dihydro- AcO 4102-039	$C_{17}H_{20}O_7$ JOC 1970 1453	336 + 83° 279°C
MIKANOLIDE * 4102-035	$C_{15}H_{14}O_6$ Tet. Lett. 1967 3111	290 +53°d 232°C	ELEPHANTOPIN 4102-040	$C_{19}H_{20}O_7$ JACS 1966 3674	360 -398° 263°C

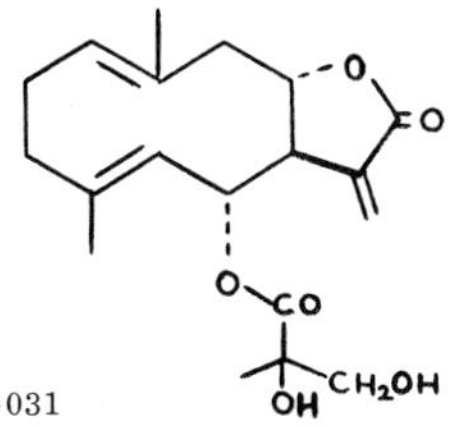

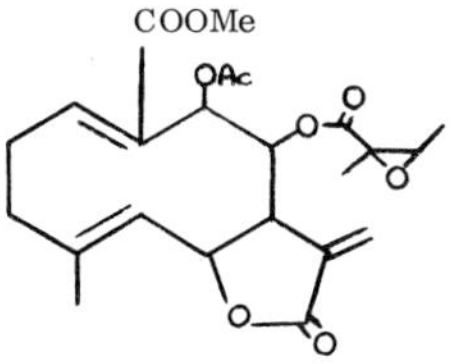

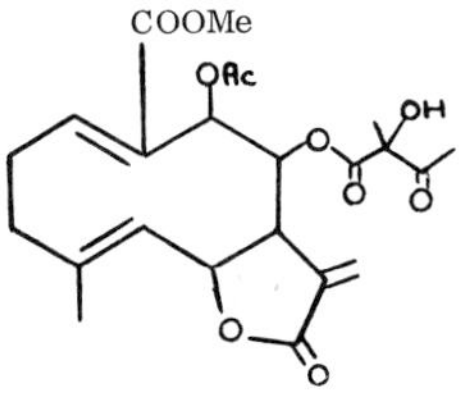

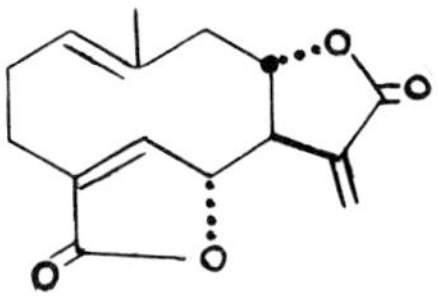

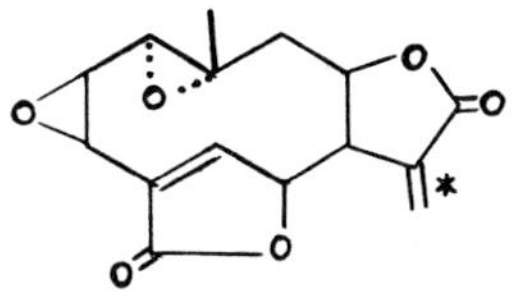

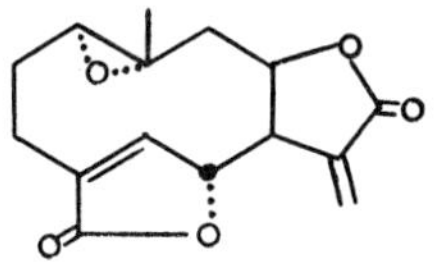

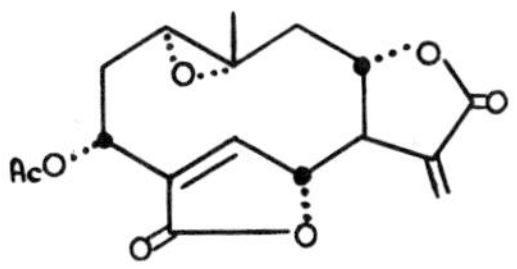

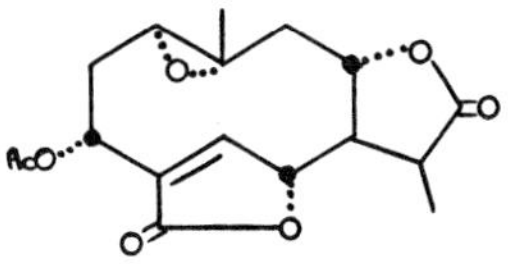

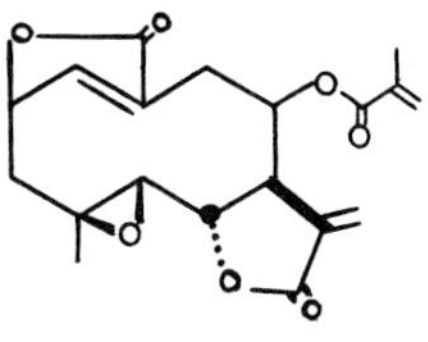

ELEPHANTIN $C_{20}H_{22}O_7$ 374 -380° 243°C Chem. Comm. 1970 128 JACS 1966 3674 4102-041	COSTUNOLIDE, OH- $C_{15}H_{20}O_3$ 248 R = OH Coll. Czech. Chem. Comm., 1963, 1618. 4102-046
ZEXBREVIN $C_{19}H_{22}O_6$ 346 + 41°c 218°C Tet. 1970 1657 4102-042	EUPATOLIDE $C_{15}H_{20}O_3$ 248 185°C Coll. Czech. Chem. Comm. 1963 1715 4102-047
ARTEMISIIFOLIN $C_{15}H_{20}O_4$ 264 +55°m 129°C CH₂OH OH Chem. Comm., 1970, 148. Phytochem., 1970, 199. 4102-043	TULIPINOLIDE $C_{17}H_{22}O_4$ 290 +260° b 181°C OAc JOC 1970 1928 4402-048
COSTUNOLIDE $C_{15}H_{20}O_2$ 232 +128°c 106°C R = H Coll. Czech. Chem. Comm., 1968, 2239. 4102-044	TULIPINOLIDE, epi- $C_{17}H_{22}O_4$ 290 +76°c 91°C OAc JOC 1970 1928 4102-049
COSTUNOLIDE, dihydro- $C_{15}H_{22}O_2$ 234 +113°c 77°C R = H * dihydro Tet., 1963, 1061. 4102-045	BALACHANOLIDE $C_{15}H_{22}O_3$ 250 + 183°c 154°C OH Coll. Czech. Chem. Comm. 1961 2612 4102-050

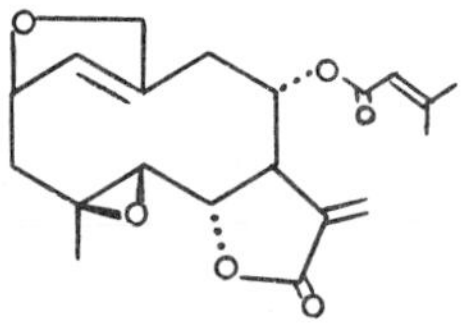

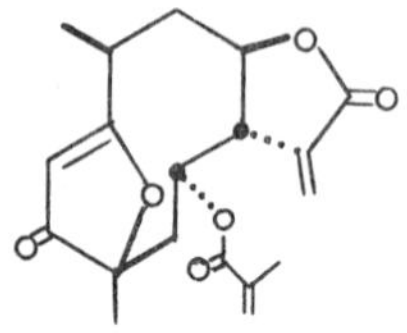

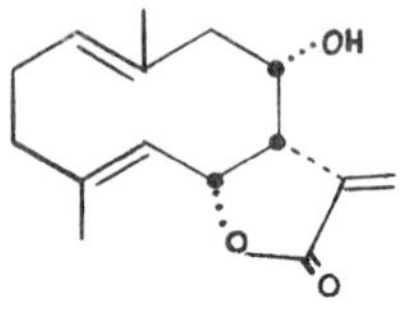

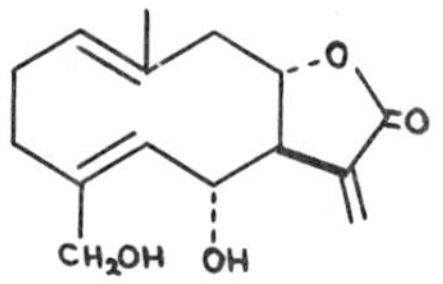

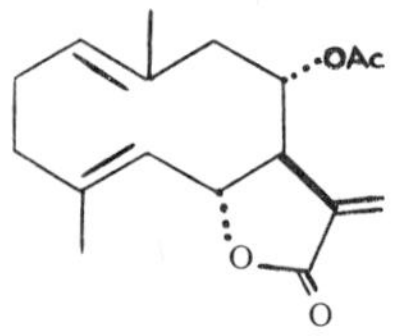

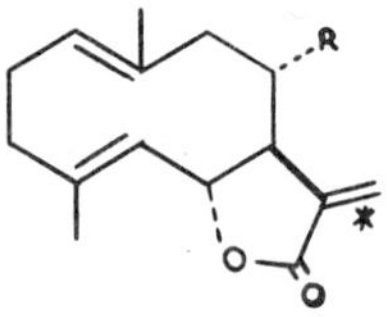

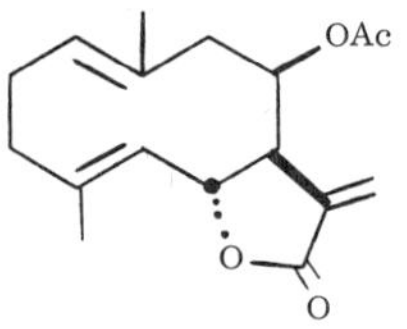

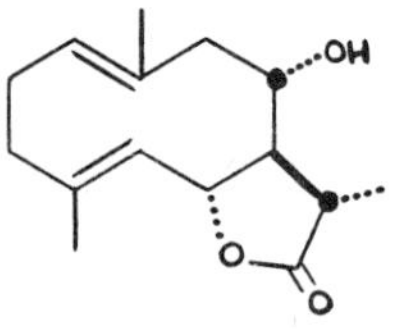

BALCHANOLIDE, iso- $C_{15}H_{22}O_3$ 250 $+122^0$c 133^0C

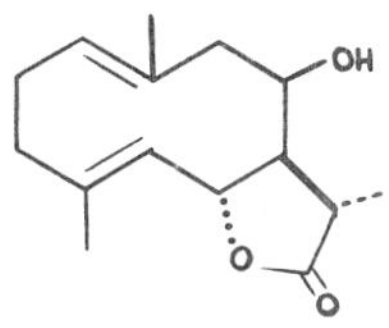

Coll. Czech. Chem. Comm. 1963 1715

4102-051

BALCHANOLIDE, OH- $C_{15}H_{22}O_4$ 266 $+105^0$e 163^0C

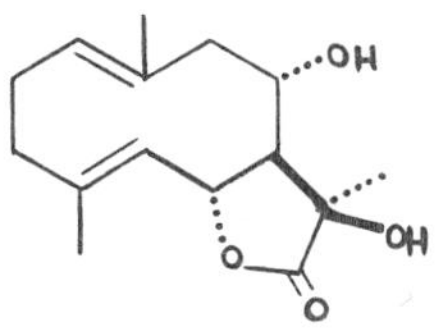

Coll. Czech. Chem. Comm. 1961 1826

4102-052

BALCHANOLIDE, Ac- $C_{17}H_{24}O_4$ 292 $+102^0$c 125^0C

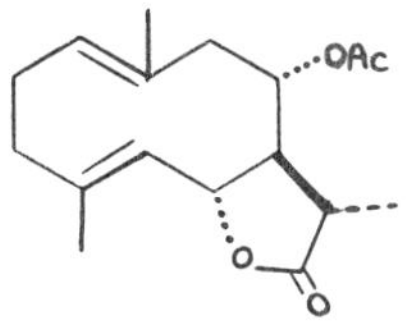

4102-053

PARTHENOLIDE $C_{15}H_{20}O_3$ 248 -78^0c 115^0C

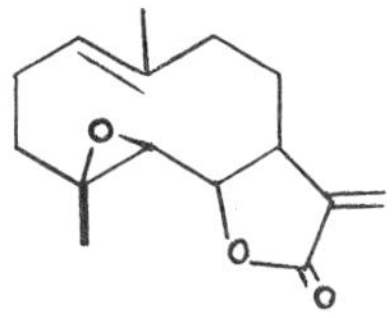

Tet. 1965 1509

4102-054

PELENOLIDE, OH- $C_{15}H_{24}O_3$ 252 112^0C

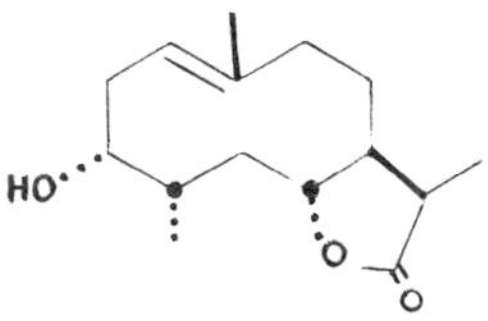

Coll. Czech. Chem. Comm. 1966 1967

4102-055

TAMAULIPIN-A $C_{15}H_{20}O_3$ 248 $+171^0$ 160^0C

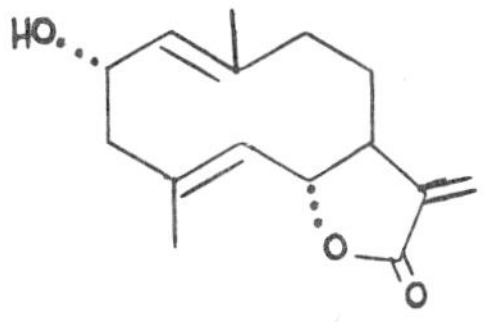

Tet. 1968 4093

4102-056

TAMAULIPIN-B $C_{15}H_{20}O_3$ 248 $+99^0$ 141^0C

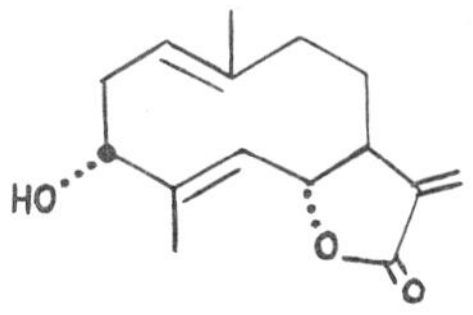

Chem. Comm. 1967 1235

4102-057

SALONITENOLIDE $C_{15}H_{20}O_4$ 264 $+199^0$ 136^0C

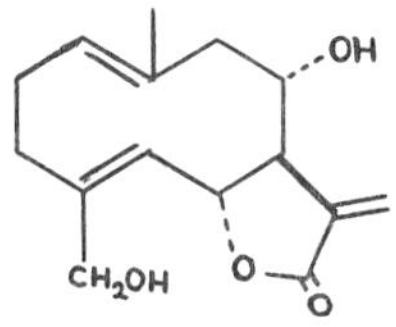

Coll. Czech. Chem. Comm., 1968, 2239; 1970, 148.

4102-058

ALBIOCOLIDE $C_{15}H_{20}O_4$ 264 $+73^0$m 104^0C

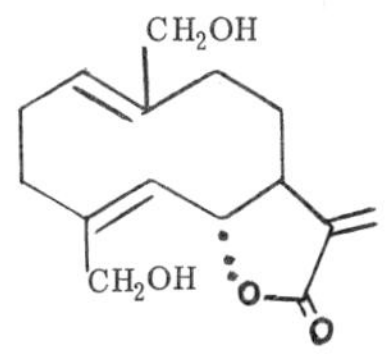

Coll. Czech. Chem. Comm. 1967 3934; 1968 2239

4102-059

NOBILIN $C_{20}H_{28}O_5$ 344 0^0c 177^0C

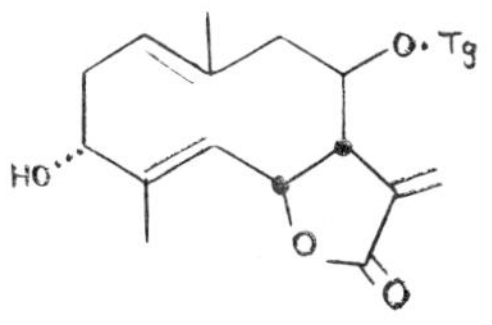

Tet. Lett., 1970, 5017.

4102-060

HELIANGINE $C_{20}H_{26}O_6$ 362 -110°c 228°C

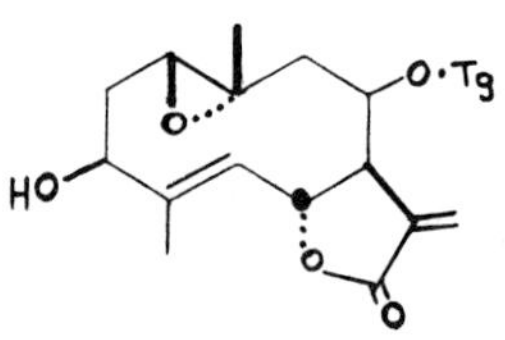

Tet. 1966 (22) 3173

4102-061

JURINEOLIDE $C_{20}H_{26}O_7$ 378 +135°m 171°C

Coll. Czech. Chem. Comm. 1969 229

4102-066

ARCTIOPICRIN $C_{19}H_{26}O_6$ 350 +133°e 116°C

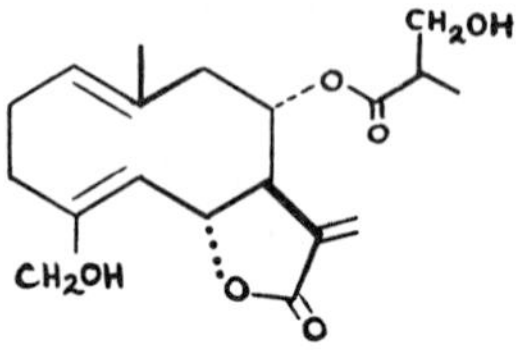

Coll. Czech. Chem. Comm. 1968 2239; 1970 148

4102-062

PELENOLIDE "a", keto- $C_{15}H_{22}O_3$ 250

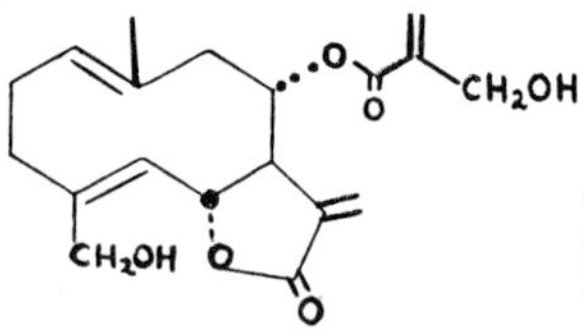

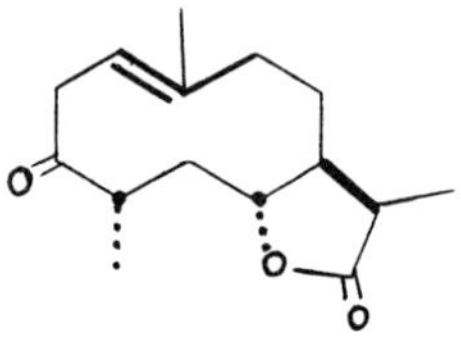

Coll. Czech. Chem. Comm. 1967 3917; 1956 1494

4102-067

CNICIN $C_{20}H_{26}O_7$ 378 +158 330°C

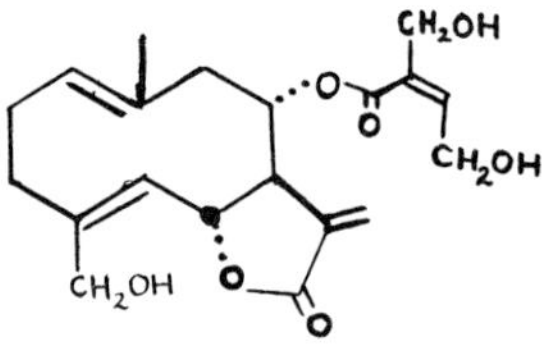

Tet. Lett. 1969 2931

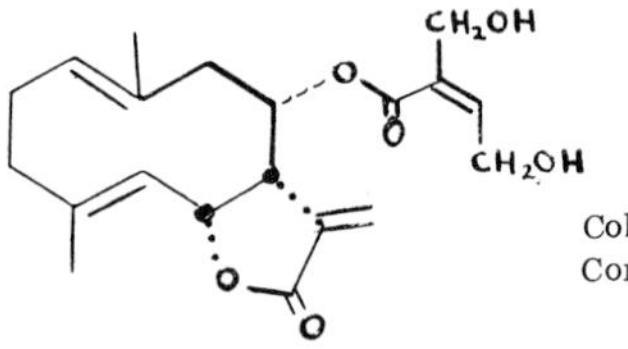

4102-063

PELENOLIDE "b", keto- $C_{15}H_{22}O_3$ 250

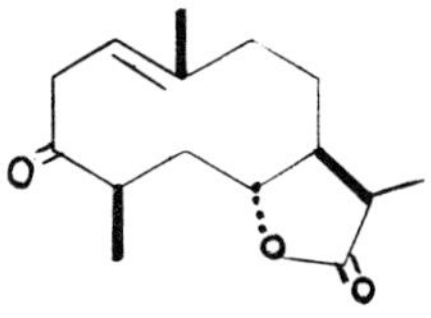

Coll. Czech. Chem. Comm. 1956 1967

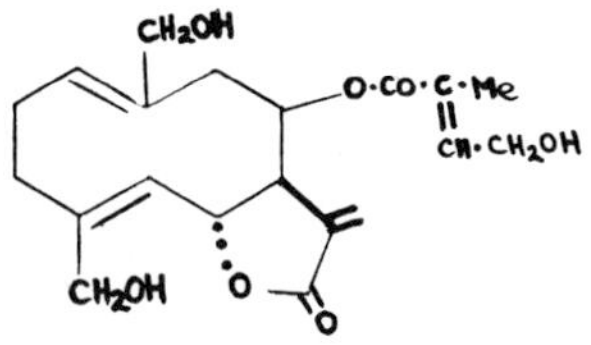

4102-068

ONOPORDOPICRIN $C_{19}H_{24}O_6$ 348 +166°c 55°C

Coll. Czech. Chem. Comm. 1968 1730; 2239; 1970 148

4102-064

UROSPERMAL A $C_{15}H_{18}O_5$ 278 -2° m 164°C

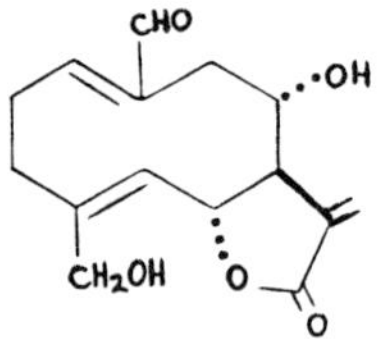

Chem. Comm. 1970 435

4102-069

EUPATORIOPICRINE $C_{20}H_{26}O_6$ 362 + 95°c 159°C

Coll. Czech. Chem. Comm. 1962 2654

4102-065

UROSPERMAL B $C_{15}H_{18}O_5$ 278 192°C

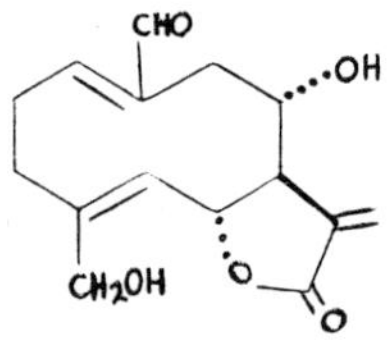

Chem. Comm. 1970 435

4102-070

VERNOLIDE	$C_{19}H_{22}O_7$	362 +230°c 182°C
	Tet. Lett. 1967 1333	
4102-071		
LANUGINOLIDE	$C_{17}H_{24}O_5$	308 -57°c 185°C
	Chem. Comm., 1970, 1534.	
4102-076		
SHIROMOOL	$C_{15}H_{26}O_2$	238
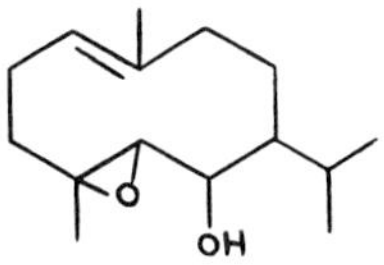	Agr. Biol. Chem., 1970 (34) 946.	
4102-072		
CHAMISSONIN	$C_{15}H_{20}O_4$	264 -20°e 125°C
	Tet. Lett., 1969, 3161.	
4102-077		
CHIHUAHUIN	$C_{17}H_{22}O_5$	306
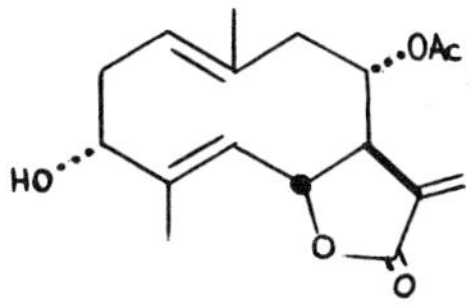	Chem. Comm., 1970, 148.	
4102-073		
LINDERACTONE, neo-	$C_{15}H_{16}O_3$	244 +100° 117°C
	JCS., 1969, 2786.	
4102-078		
ELEPHANTOPIN, deoxy-	$C_{19}H_{20}O_6$	344 200°C
	Tet. Lett., 1970. 2863.	
4102-074		
VERNOLIDE, hydroxy-	$C_{19}H_{22}O_8$	378 +135°d 150°C
	Compt. Rend., 1969 (268) 82.	
4102-079		
PARTHENOLIDE, dihydro-	$C_{15}H_{22}O_3$	250 -68°c 138°C
	Chem. Comm., 1970, 1534.	
4102-075		
ARTEMORIN	$C_{15}H_{20}O_3$	248 116°C
	Phytochem., 1970, 2377.	
4102-080		

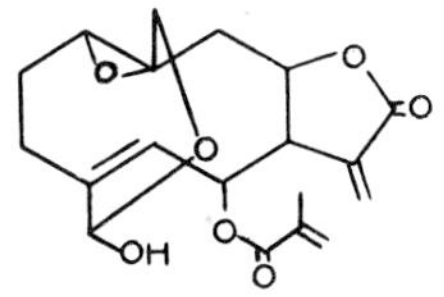

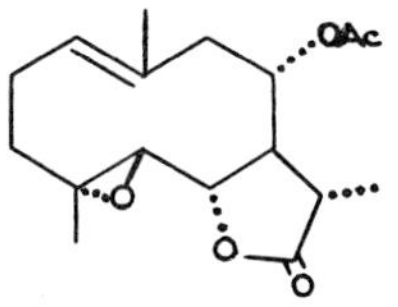

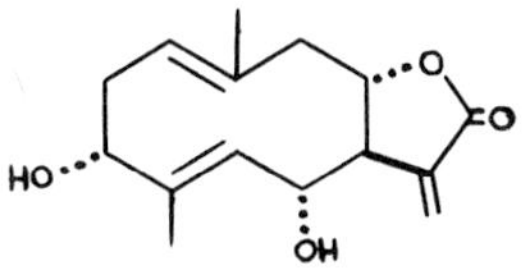

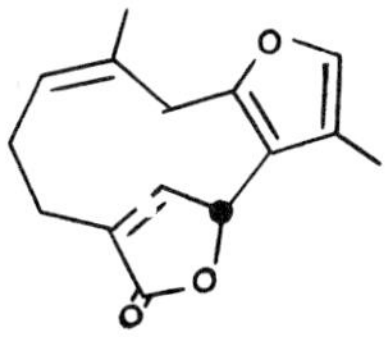

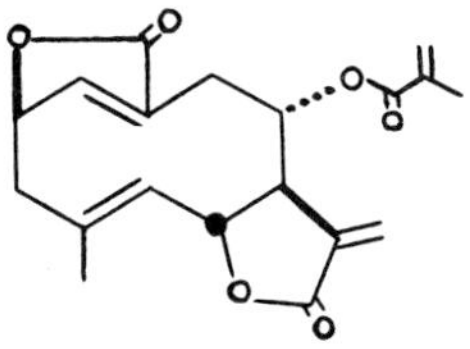

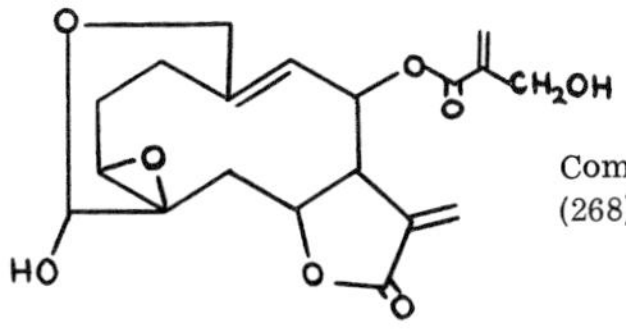

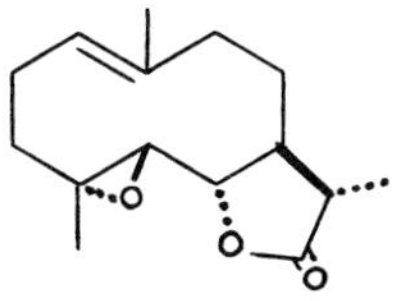

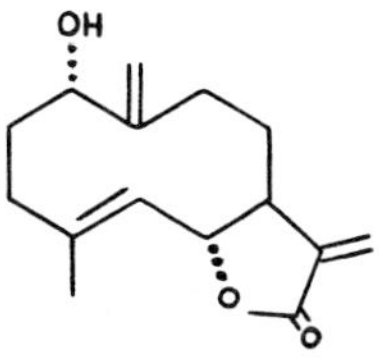

RIDENTIN $C_{15}H_{20}O_4$ 264 Phytochem., 1969. 2009. 4102-081	VERLOTORIN, anhydro- $C_{15}H_{18}O_3$ 246 124°C Phytochem., 1970, 2377. 4102-083
VERLOTORIN $C_{15}H_{20}O_4$ 264 Phytochem., 1970, 2377. 4102-082	

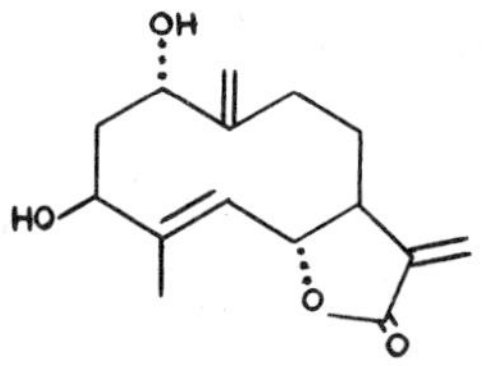

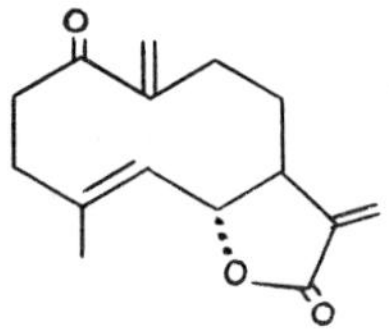

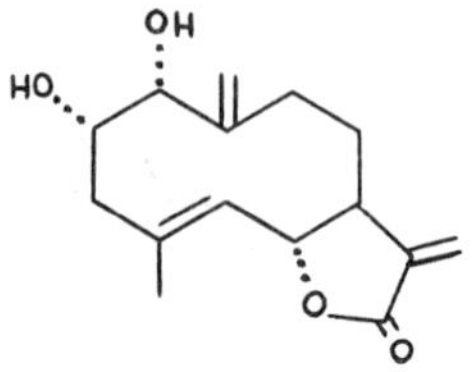

Name	Formula	Data	Reference	No.
HUMULENE, α-	$C_{15}H_{24}$	204 OI Oil	Tet. 1968 4113	4103-001
HUMULENE, α- di-epoxide	$C_{15}H_{24}O_2$	236 0°c 105°C	Tet. 1962, 575	4103-006
HUMULENE, β-	$C_{15}H_{24}$	204 OI Oil	Coll. Czech. Chem. Comm. 1961 1832	4103-002
ZERUMBONE	$C_{15}H_{22}O$	218 OI 66°C	Tet. 1968 4113	4103-007
HUMULENOL-II, (+)-	$C_{15}H_{24}O$	220 + 12° 111°C	Tet. 1968 4115	4103-003
CAUCALOL di-acetate	$C_{19}H_{30}O_5$	338 122°C	Tet. Lett. 1966 623	4103-008
HUMULENE EPOXIDE-I, (-)-	$C_{15}H_{24}O$	220 -21° 106°C	Tet. 1968 4115	4103-004
HUMULOL	$C_{15}H_{26}O$	222 -1°m Oil	Bull. Chem. Soc. Jap., 1969, 2405.	4103-009
HUMULENE EPOXIDE-II, (-)-	$C_{15}H_{24}O$	220 -31° 113°C	Tet. 1968 4115	4103-005
HUMULADIENOL	$C_{15}H_{24}O$	220 +5°c Oil	Bull. Chem. Soc. Jap., 1969, 2088.	4103-010

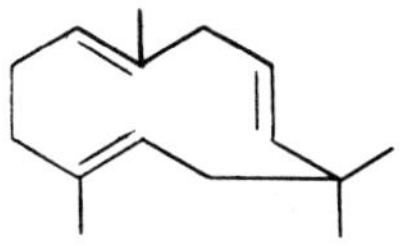

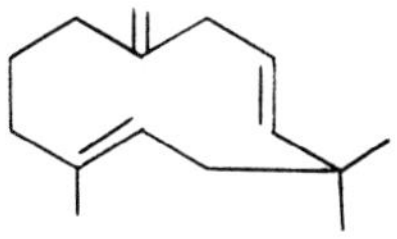

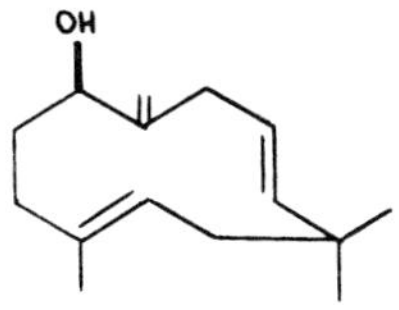

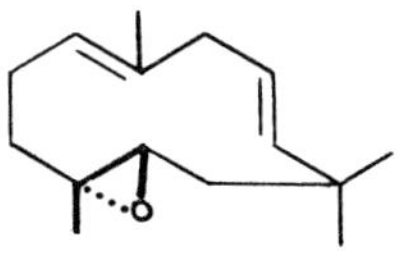

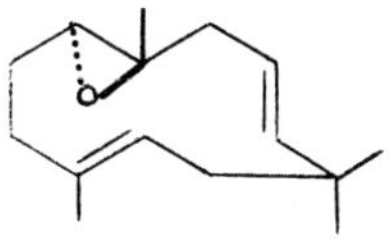

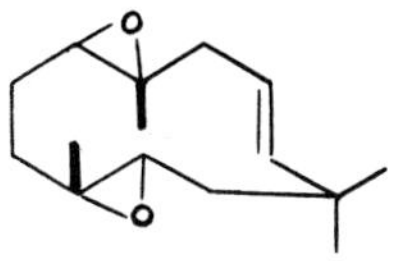

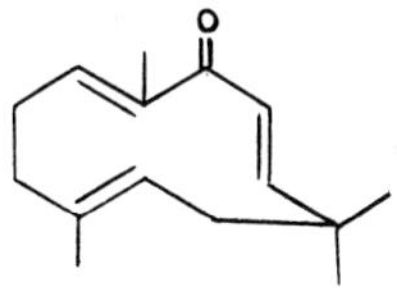

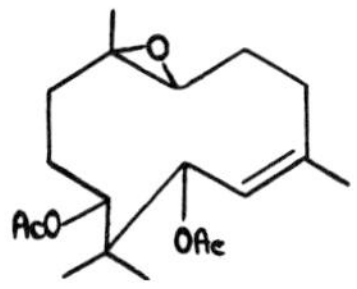

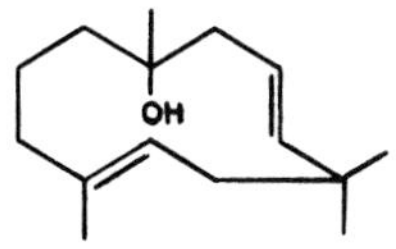

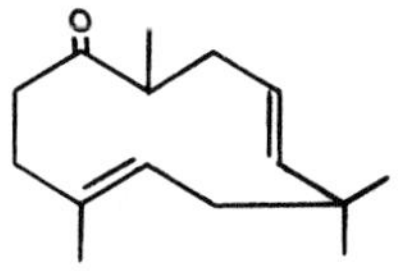

HUMULENE EPOXIDE III	$C_{15}H_{24}O$	220 +2^0 Oil
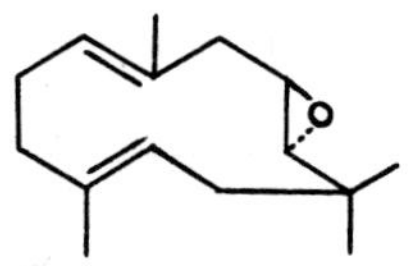	Tet., 1971, 635.	
4103-011		

GUAIENE, α- $C_{15}H_{24}$ 204 -65°d JACS 1962 1307 4104-001	GUAIA-3, 7-DIENE $C_{15}H_{24}$ 204 -59° Oil Coll. Czech. Chem. Comm. 1964 1042 4104-006
GUAIENE, ε- $C_{15}H_{24}$ 204 -24° Oil Chem. Ind. 1962 1715 4104-002	GUAIAZULENE $C_{15}H_{18}$ 198 OI 32°C Chem. Abstr. 1956,7405 4104-007
BULNESENE, α- $C_{15}H_{24}$ 204 0° Oil JACS 1962 1307 4104-003	LACTARAZULENE $C_{15}H_{16}$ 196 OI Oil Coll. Czech. Chem. Comm., 1970 1300 Chem. Listy 1953 (47) 1856 Artefact ? 4104-008
BULNESENE, β- $C_{15}H_{24}$ 204 JACS 1962 1307 4104-004	LACTAROFULVENE $C_{15}H_{16}$ 196 OI Oil Tet. 1968, 2081 4104-009
CHIGADMARENE, α- $C_{15}H_{24}$ 204 -150° Chem. Abstr. 1953 8698 4104-005	APITONENE-1 $C_{15}H_{26}$ 206 ? Chem. Abstr. 1968 (68) 114770 4104-010

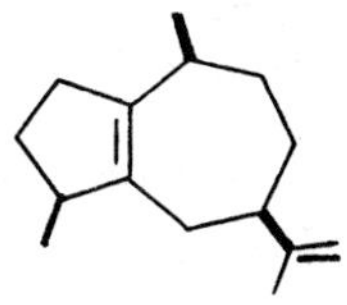

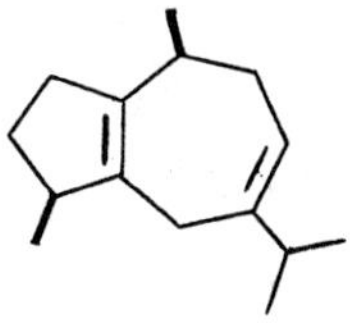

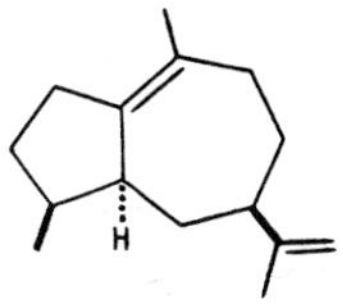

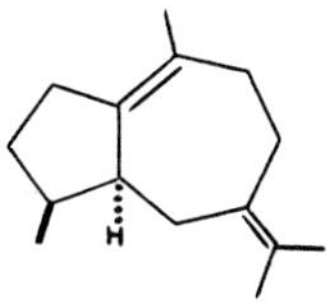

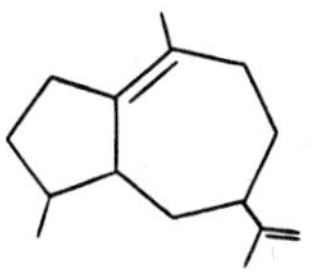

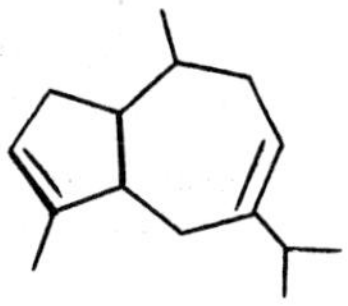

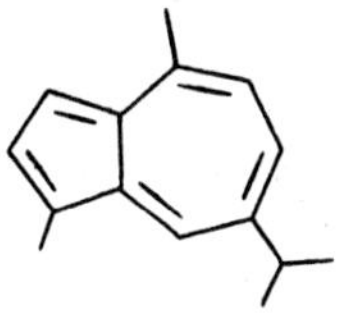

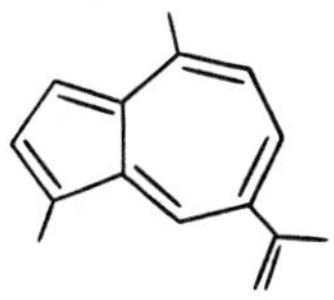

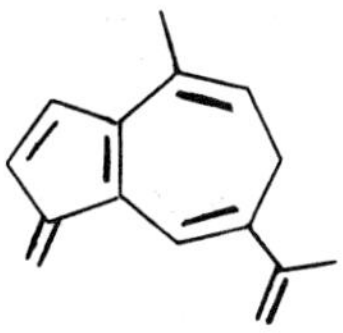

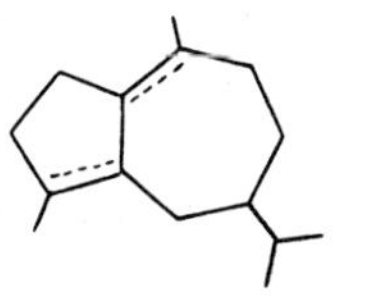

LACTAROVIOLIN	$C_{15}H_{14}O$	210 OI 53°C	NARDOL	$C_{15}H_{26}O$	222 -10°c Oil
Artefact ?	Coll. Czech. Chem. Comm., 1970 1300 Chem. Ind. 1954 1202			Tet. Lett. 1966 1035	
4104-011			4104-016		
CHAMAZULENE CARBOXYLIC ACID	$C_{15}H_{16}O_2$	228	BULNESOL	$C_{15}H_{26}O$	222 +4° e 68°C
	Natwiss. 1954 257			JACS 1962 1307	
4104-012			4104-017		
GUAIENE, (+)-epoxy-	$C_{15}H_{24}O$	220 +11°c Oil	POGOSTOL	$C_{15}H_{26}O$	222 -20°c Oil
	Tet. Lett. 1967 4661			Chem. Pharm. Bull. 1968 1608	
4104-013			4104-018		
GERMACROL	$C_{15}H_{22}O$	218 56°C	PARTHENIOL	$C_{15}H_{24}O$	220 +117°c 128°C
	Ann. 1952 (576) 116			JACS 1948 2075 Chem. Ind., 1962 1424	
4104-014			4104-019		
GUAIOL	$C_{15}H_{26}O$	222 -42°c 91°C	GUAI-1, 3, 5, 9, 11-PENTAENE, 14-OH	$C_{15}H_{18}O$	214 OI
	Chem. Abstr. 1955 6549			Coll. Czech. Chem. Comm. 1970 1300.	
4104-015			4104-020		

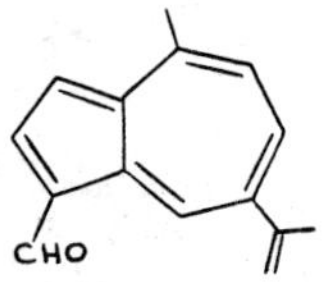

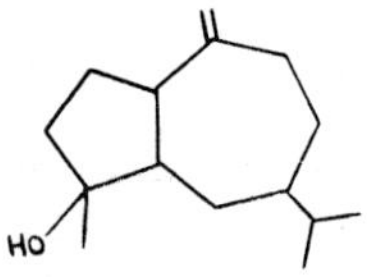

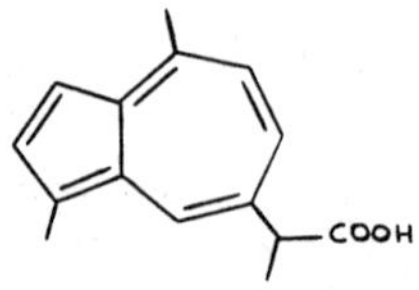

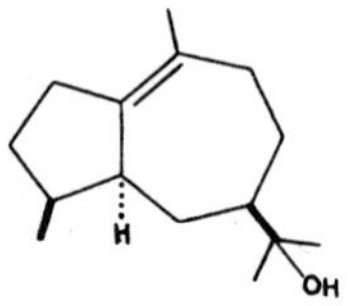

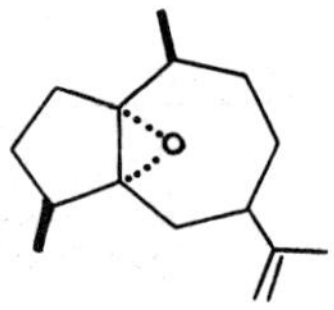

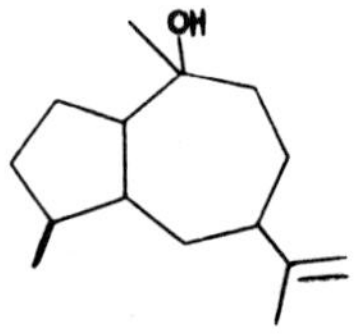

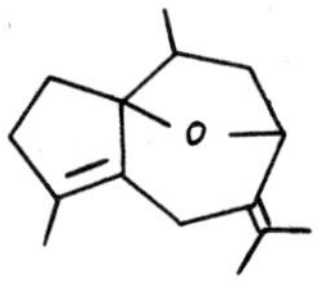

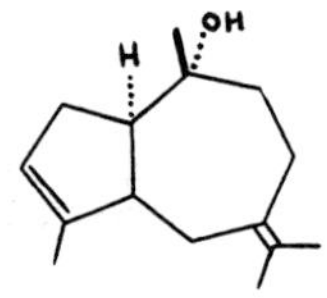

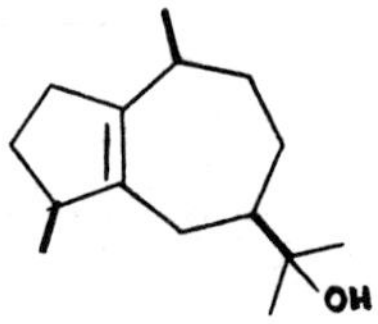

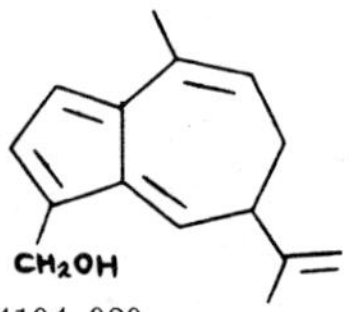

GUAI-1, 3, 5, 9, 11-PENTAENE, 14-STEARYLOXY-

$C_{33}H_{52}O_2$ 480
OI

Coll. Czech. Chem. Comm., 1970, 1300

4104-021

ROTUNDONE, (-)-

$C_{15}H_{22}O$ 218
-92.7^{0}c

Tet. Lett. 1967 461.

4104-022

TORILIN

$C_{22}H_{32}O_5$ 376
-45^0 e
77^{0}C

Tet. 1969 4751

4104-023

CURCUMENOL

$C_{15}H_{22}O_2$ 234
+141^{0}c
Oil

Chem. Pharm. Bull. 1968, 1605

4104-024

CURCUMENOL

$C_{15}H_{22}O_2$ 234
+397^0
119^{0}C

Chem. Pharm. Bull. 1968 39.

4104-025

GALBANOL

$C_{15}H_{26}O$ 222
+2.8^0
Oil

Ang. Chem. Int. Edn. 1963 (2) 695

4104-026

GUAIOXIDE

$C_{15}H_{26}O$ 222
-39^{0}c
Oil

Chem. Comm. 1968 649.

4104-027

LIGULOXIDE

$C_{15}H_{26}O$ 222
-53^0
36^{0}C

Chem. Comm. 1968, 1534.

4104-028

LIGULOXIDOL

$C_{15}H_{26}O_2$ 238
-37^0
Oil

Tet., 1970, 2911

Chem. Comm. 1968 1534

4104-029

LIGULOXIDOL ACETATE

$C_{17}H_{28}O_3$ 280
-52^0
85^{0}C

Chem. Comm. 1968 106

4104-030

Name	Formula	Data	Reference	No.
KESSANE	$C_{15}H_{26}O$	222 -7.2°c Oil	Chem. Pharm. Bull., 1963 (11) 547	4104-031
GAILLARDIN	$C_{17}H_{22}O_5$	306 -15°c 200°C	Tet. Lett. 1969 973.	4104-036
KESSANOL	$C_{15}H_{26}O_2$	238 +18°c 118°C	Chem. Pharm. Bull. 1967 324	4104-032
IVALIN, pseudo-	$C_{15}H_{20}O_3$	248 -145°c 122°C	JOC 1965 118	4104-037
KESSANOL, 8-epi-	$C_{15}H_{26}O_2$	238	Chem. Pharm. Bull. 1967, 324.	4104-033
IVALIN, dihydro-pseudo-	$C_{15}H_{22}O_3$	250	JOC 1965 118	4104-038
KESSYL ALCOHOL, α-	$C_{15}H_{26}O_2$	238 -38°c 85°C	Tet. 1967 (23) 553	4104-034
HELENIUM LACTONE	$C_{15}H_{20}O_3$	248 -88°c 112°C	Chem. Pharm. Bull. 1968 (8) 1601	4104-039
KESSYL GLYCOL	$C_{15}H_{26}O_3$	254 -19°c 119°C	Tet. 1967 (23) 553	4104-035
VIRGINOLIDE	$C_{15}H_{16}O_3$	244 +4° c 134°C	JOC 1967 507	4104-040

GEIGERINE — $C_{15}H_{20}O_4$ — 264; -64^0c; 191^0C

Proc. Chem. Soc. 1960 279.

4104-041

HELENALIN, iso- — $C_{15}H_{18}O_4$ — 262; 262^0C

Chem. Listy 1957 51 1521

4104-042

MEXICANIN H — $C_{15}H_{18}O_4$ — 262; -45^0; 150^0C

Tet. Lett. 1966 1029

4104-043

MOKKU-LACTONE — $C_{15}H_{20}O_2$ — 232; $+18^0$; 36^0C

Chem. Abstr. 1967 95219c

4104-044

COSTUS LACTONE, dehydro- — $C_{15}H_{18}O_2$ — 230; -13^0; 60^0C

Tet. 1965 3575

4104-045

ZALUZANIN C — $C_{15}H_{18}O_3$ — 246; $+37^0$; 95^0C

Tet., 1967, 3903.

4104-046

ZALUZANIN D — $C_{17}H_{20}O_4$ — 288; $\pm 0^0$; 104^0C

Tet., 1967, 3903.

4104-047

PYRETHROSIN — $C_{17}H_{22}O_5$ — 306; -31^0c; 200^0C

Chem. Abstr. 1967 (66) 65658

4104-048

SIEVERSININ — $C_{15}H_{20}O_3$ — 248; $+57^0$ c; 142^0C

Chem. Abstr. 1967 32803

4104-049

GLOBICINE — $C_{17}H_{22}O_5$ — 306; $+66^0c$; 149^0C

Coll. Czech. Chem. Comm. 1963 1202

4104-050

CUMAMBRIN A — $C_{17}H_{22}O_5$ — 306; +90°; 178°C Tet. 1968 5627. 4104-051	ARTABSIN — $C_{15}H_{20}O_3$ — 248; -49°c; 134°C Tet. Lett. 1968 3855 4104-056
CUMAMBRIN B — $C_{15}H_{20}O_4$ — 264; +81°; 87°C Tet. 1968 5627. 4104-052	ARBORESCIN — $C_{15}H_{20}O_3$ — 248; +63°c; 145°C Tet. Lett. 1963 1127 4104-057
LIGUSTRIN — $C_{15}H_{18}O_3$ — 246; +56°; 136°C Tet. 1968 6987 4104-053	ESTAFIATINE — $C_{15}H_{18}O_3$ — 246; -10°c; 105°C Tet. 1963 1101 4104-058
MATRICIN — $C_{17}H_{22}O_5$ — 306; -131°c; 159°C J. Pharm. Sci. 1967 650 4104-054	SOLSTITIALIN — $C_{15}H_{20}O_5$ — 280; 206°C Acta Cryst. 1970 (26B) 554 4104-059
CHAMAZULENOGEN, pro- — $C_{15}H_{20}O_3$ — 248; -49°; 134°C Chem. Ind. 1956 492 4104-055	EUPAROTIN — $C_{20}H_{24}O_7$ — 376; -124°e; 200°C JACS 1967 466 4104-060

EUPAROTIN ACETATE $C_{22}H_{26}O_8$ 418 -191°e 156°C

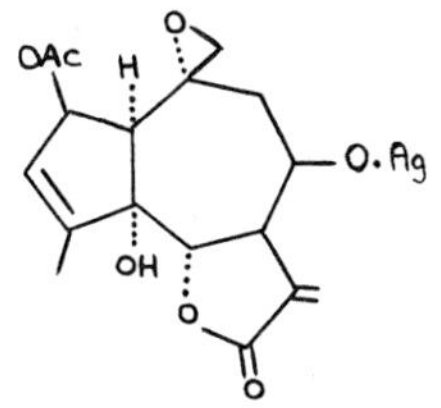

JACS 1967 465

4104-061

EUPACHLORIN ACETATE $C_{22}H_{27}O_8Cl$ 454 -192°m 162°C

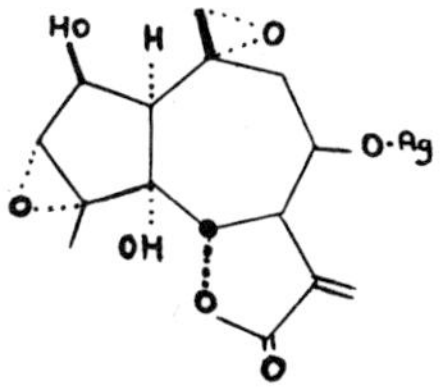

* acetyl

Tet. Lett., 1968, 3517.

4104-066

EUPATOROXIN $C_{20}H_{24}O_8$ 392 -98°m 199°C

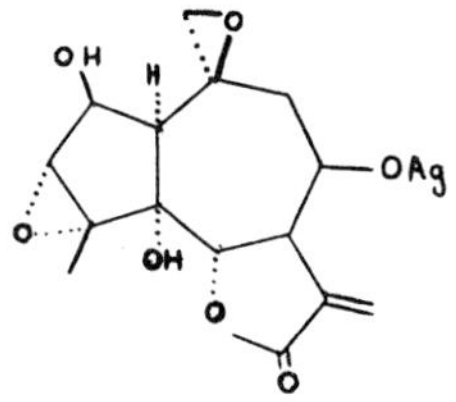

Tet. Lett. 1968 3517

4104-062

EUPACHLOROXIN $C_{20}H_{25}O_8Cl$ 428

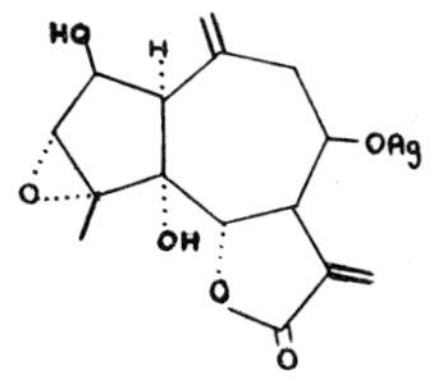

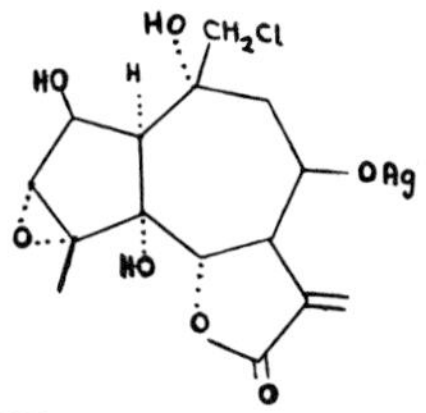

Tet. Lett. 1968 3517

4104-067

EUPATOROXIN, 10-epi- $C_{20}H_{24}O_8$ 392 -109°m 231°C

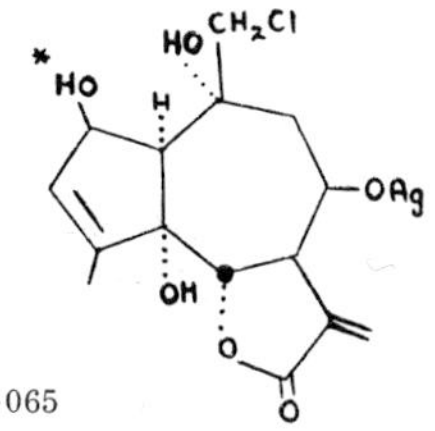

Tet. Lett. 1968 3517.

4104-063

CYNAROPICRINE $C_{19}H_{22}O_6$

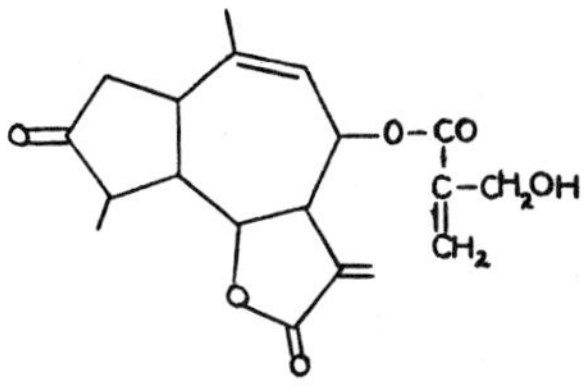

Coll. Czech. Chem. Comm. 1960 2777

4104-068

EUPATUNDIN $C_{20}H_{24}O_7$ 376 -80°e 188°C

Tet. Lett. 1968 3517

4104-064

ARBIGLOVIN $C_{15}H_{18}O_3$ 246 +199° 202°C

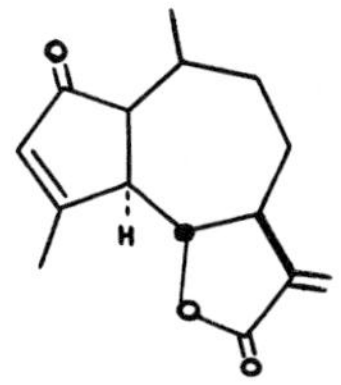

JACS., 1967, 5511.

4104-069

EUPACHLORIN $C_{20}H_{25}O_7Cl$ 412 -110°e 220°C

Tet. Lett. 1968 3517

4104-065

CARPESIA LACTONE $C_{15}H_{20}O_3$ 248 Oil

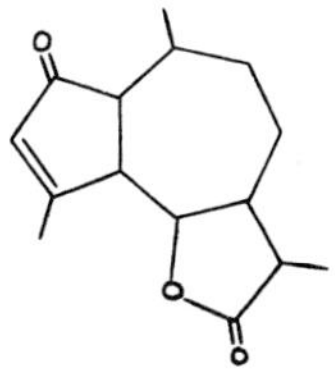

Chem. Abstr. 1956 891; 1680

4104-070

JACQUINELIN $C_{15}H_{18}O_4$ 262 $+27^0$ 76^0C

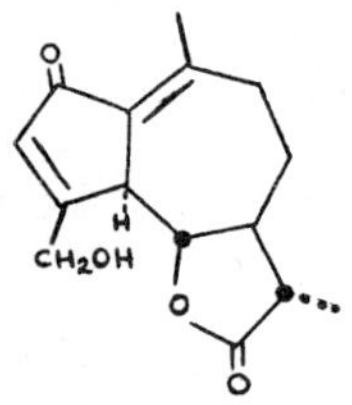

JCS 1966 1298
JACS., 1967, 5511.

4104-071

ACHILLINE, OAc- $C_{17}H_{20}O_5$ 304 $+116^0c$ 194^0C

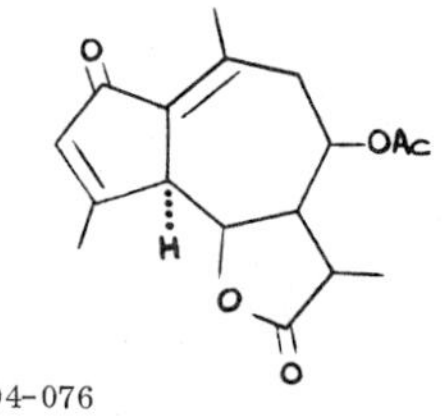

Tet. Lett. 1963 137

4104-076

LACTUCIN $C_{15}H_{16}O_5$ 276 $+78^0p$ 223^0C

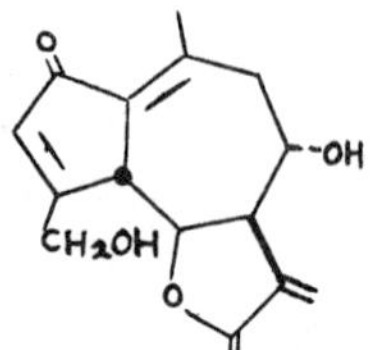

JCS, 1958, 936
Tet. 1963 1389

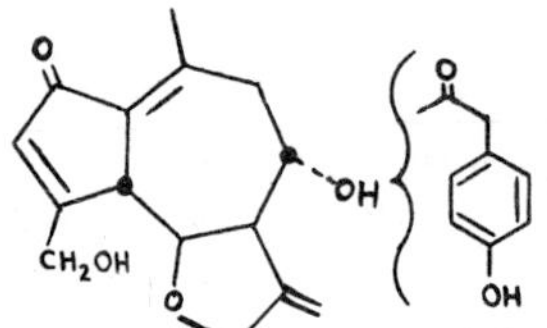

(?)

4104-072

MATRICARIN $C_{17}H_{20}O_5$ 304 $+23^0c$ 194^0C

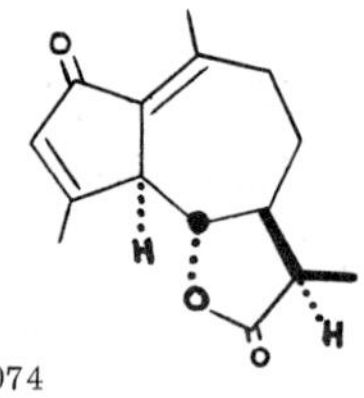

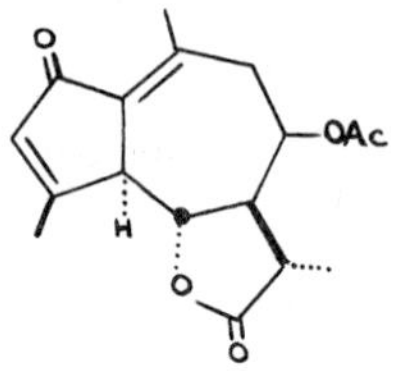

Helv. 1967 1961
Tet., 1969 2099

4104-077

LACTUCOPICRIN $C_{23}H_{22}O_7$ 410 $+67^0p$ 117^0C

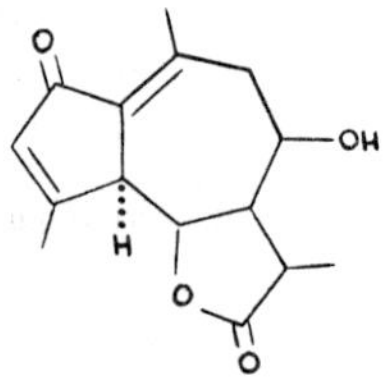

JCS 1958 963

(?)

4104-073

MATRICARIN, des-Ac- $C_{15}H_{18}O_4$ 262 $+10^0m$ 131^0C

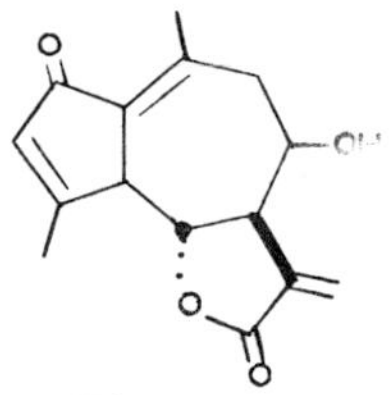

Chem. Abstr.
1964 (60) 5561

4104-078

ACHILLINE $C_{15}H_{18}O_3$ 246 $+160^0c$ 144^0C

Tet. 1969 2117

4104-074

MATRICARIN, des-OAc $C_{15}H_{18}O_3$ 246 $+58^0c$ 202^0C

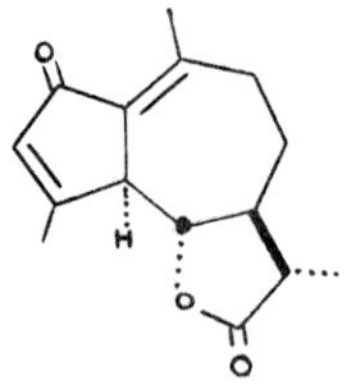

Tet. 1969 2099

4104-079

ACHILLINE, 8-OH- $C_{15}H_{18}O_4$ 262 $+110^0c$ 161^0C

Tet. Lett. 1963 137.

4104-075

GROSSHEMIN $C_{15}H_{18}O_4$ 262 $+125^0$ 201^0C

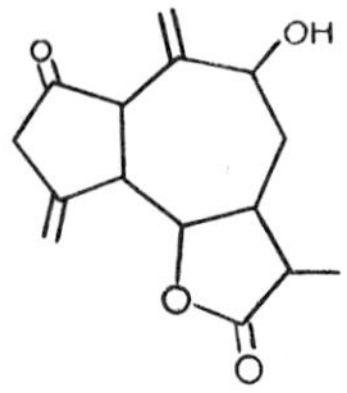

Chem. Abstr. 1965
(63) 636

4104-080

Name	Formula	Data	Reference	No.
SAURINE	$C_{15}H_{18}O_5$	278 317⁰C	Chem. Abstr. 1967 28919	4104-081
AMBERBOIN	$C_{15}H_{20}O_4$	264 +164⁰ 145⁰C	Chem. Abstr. 1969 (70) 29087	4104-082
CURCUMOL	$C_{15}H_{24}O_2$	236 -41⁰ 142⁰C	Chem. Pharm. Bull. 1966 1241	4104-083
BADKHYZIN (Angeloyl)	$C_{20}H_{24}$	344 -68⁰n 140⁰C	Chem. Abstr. 1966 2144g	4104-084
CALOCEPHALIN	$C_{17}H_{22}O_5$	306 +28⁰c 146⁰C	Aust. J. C. 1966 143	4104-085
FERULIN	$C_{15}H_{16}O_3$	244 178⁰C	Khim. Prir. Soedin., 1970 (6) 13.	4104-086
BAHIA I (R = H)	$C_{15}H_{18}O_4$	262 +4⁰c 210⁰C	Can. J. C., 1969 2849.	4104-087
BAHIA II	$C_{20}H_{24}O_7$	376 +48⁰c 134⁰C	Can. J. C., 1969, 2849.	4104-088
GURJUNENE, γ -	$C_{15}H_{24}$	204 +147⁰ Oil	Tet., 1969, 1786.	4104-089
MONTANOLIDE	$C_{22}H_{30}O_7$	406 -72⁰ 133⁰C	Coll. Czech. Chem. Comm., 1970, 3296.	4104-090

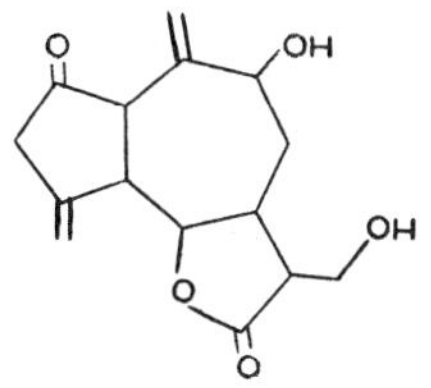

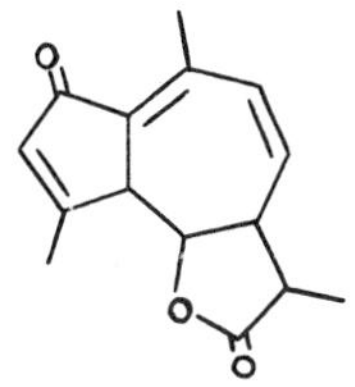

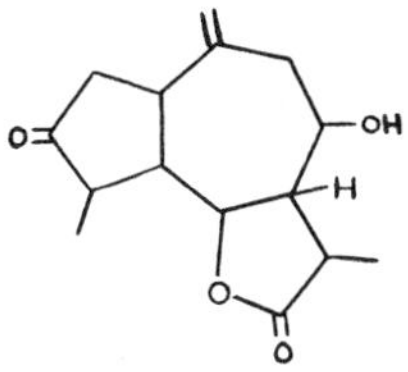

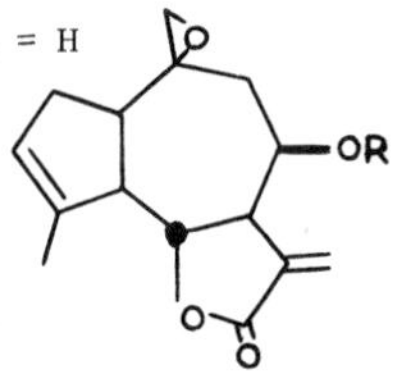

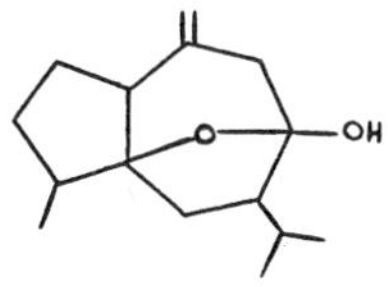

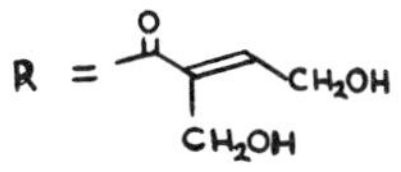

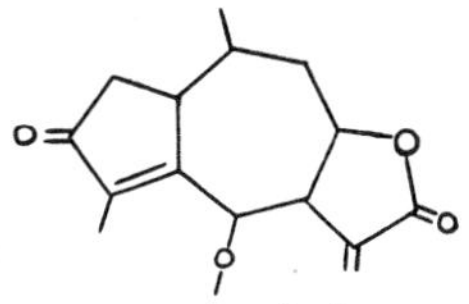

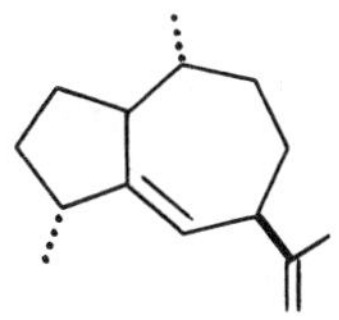

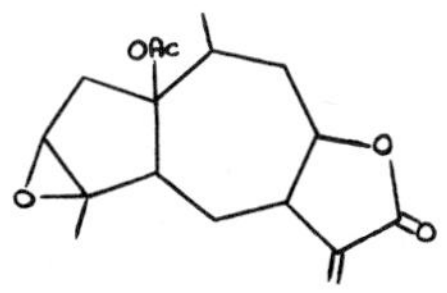

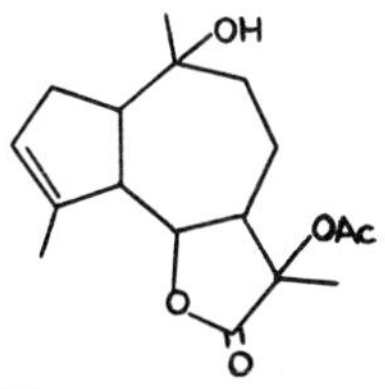

CHRYSARTEMIN A	$C_{15}H_{18}O_5$	278 +51° 250°C
4104-091	Phytochem., 1970, 1615.	

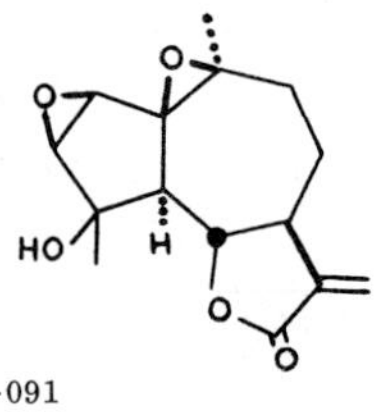

CHRYSARTEMIN B	$C_{15}H_{18}O_5$	278 +37° 262°C
4104-092	Phytochem., 1970, 1615.	

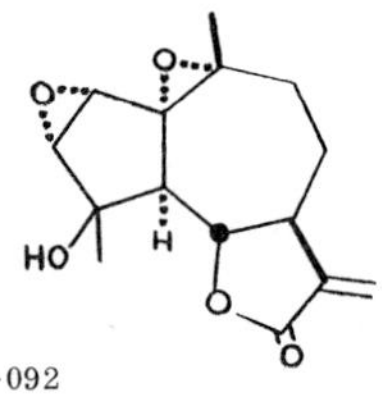

CEDRENE, α- | $C_{15}H_{24}$ | 204 | -92^{0} | Oil

JACS, 1955, 1072.

4105-001

CEDRENE, β- | $C_{15}H_{24}$ | 204 | $+11^{0}$ | Oil

4105-002

CEDROL | $C_{15}H_{26}O$ | 222 | $+9^{0}$ | $85^{0}C$

JACS, 1955, 1072.

4105-003

BIOTOL, α- | $C_{15}H_{24}O$ | 220 | -27^{0} | $78^{0}C$

Tet. Lett., 1968, 843.

4105-004

BIOTOL, β- | $C_{15}H_{24}O$ | 220 | $+44^{0}$ | $84^{0}C$

Tet. Lett., 1968, 843.

4105-005

CEDREN-13-OL, 8- | $C_{15}H_{24}O$ | 220

Tet., 1968, 3399.

4105-006

CEDRANEDIOL, 8S, 14- | $C_{15}H_{26}O_2$ | 238 | $+1^{0}$ | $147^{0}C$

Tet., 1968, 3399.

4105-007

CEDRANE-8, 9-DIOL | $C_{15}H_{26}O_2$ | 238

Chem. Abstr., 1968 (69) 80107.

4105-008

PIPITZOL, α- | $C_{15}H_{20}O_3$ | 248 | $+192^{0}$ | $146^{0}C$

Tet. Lett., 1965, 1577.

4105-009

PIPITZOL, β- | $C_{15}H_{20}O_3$ | 248 | -172^{0} | $131^{0}C$

Tet. Lett., 1965, 1577.

4105-010

CEDRANOLIDE, 8, 14- $C_{15}H_{22}O_2$ 234; -66^0; Oil

Tet., 1968, 3399.

4105-011

CEDRANOXIDE, 8, 14- $C_{15}H_{24}O$ 220

Tet., 1968, 3399.

4105-012

CEDROLIC ACID $C_{15}H_{24}O_3$ 252; -47^0c; 184^0C

OH

COOH

Acta Chem. Scand., 1961, 945.

4105-013

JALARIC ACID A $C_{15}H_{20}O_5$ 280; $+37^0e$; 179^0C

CHO HO H

COOH

CH_2OH

Tet. Lett., 1963, 513.

4105-014

SHELLOLIC ACID $C_{15}H_{20}O_6$ 296; $+18^0e$; 206^0C

HOOC H HO H

COOH

CH_2OH

Tet., 1962, 552.

4105-015

SHELLOLIC ACID, epi- $C_{15}H_{20}O_6$ 296; $+49^0e$; 233^0C

COOH OH

COOH

CH_2OH

Tet., 1962, 1321.

4105-016

LACCIJALARIC ACID $C_{15}H_{20}O_4$ 264; 165^0C

CHO OH

COOH

Tet., 1969, 3855.

4105-017

LACCISHELLOLIC ACID $C_{15}H_{20}O_5$ 280; 197^0C

COOH OH

COOH

Tet., 1969, 3855.

4105-018

LACCISHELLOLIC ACID, epi- $C_{15}H_{20}O_5$ 280; $+43^0$; 236^0C

COOH OH

COOH

Tet., 1969, 3855.

4105-019

LAKSHOLIC ACID $C_{15}H_{22}O_5$ 282; 182^0C

CH_2OH OH

COOH

CH_2OH

Tet., 1969, 3845.

4105-020

LAKSHOLIC ACID, epi- $C_{15}H_{22}O_5$ 282

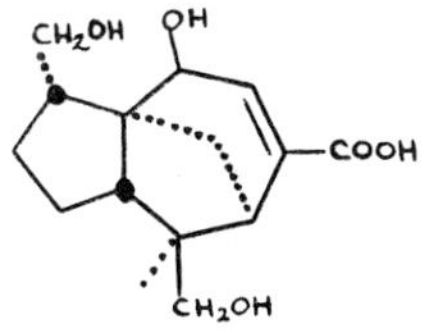

Tet., 1969, 3845.

4105-021

BISABOLENE, β-d- $C_{15}H_{24}$ 204 +52° Oil

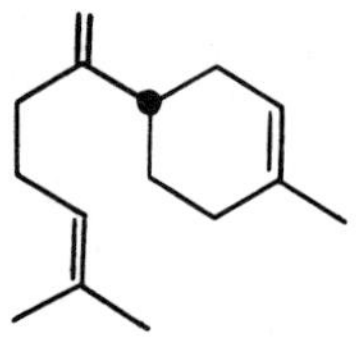

Coll. Czech. Chem. Comm. 1961 (26) 1362

4106-001

BISABOLENE, β- - $C_{15}H_{24}$ 204 -84° Oil

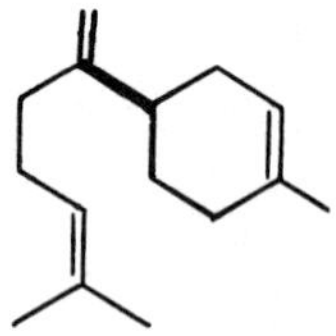

Chem. Abstr. 1953 (47) 8698

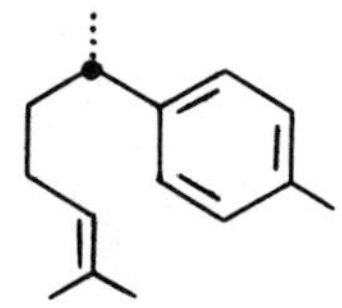

4106-002

BISABOLENE, iso- $C_{15}H_{24}$ 204 -47° c Oil

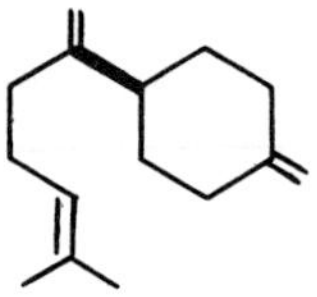

Tet. 1962 (18) 1165

4106-003

ZINGIBERENE, α- $C_{15}H_{24}$ 204 -120° Oil

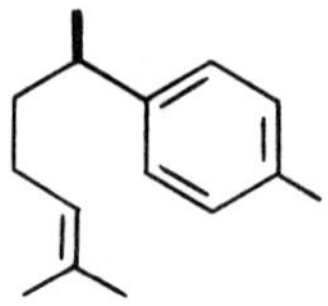

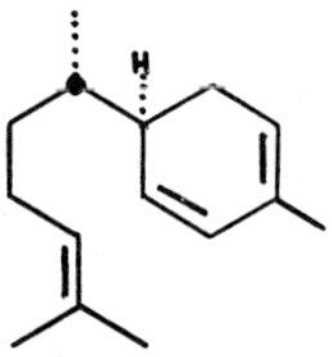

Helv. 1956 (33) 171

4106-004

ZINGIBERENE, β- $C_{15}H_{24}$ 204 -81° Oil

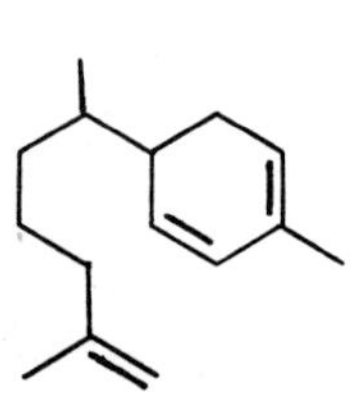

Can. J. C. 1964 (42) 2610

4106-005

CURCUMENE, α-d- $C_{15}H_{22}$ 202 +38° c Oil

Tet. 1965 2593

4106-006

CURCUMENE, l-α- $C_{15}H_{22}$ 202 -34° Oil

Tet. 1968 4113

4106-007

CURCUMENE, d-β- $C_{15}H_{24}$ 204 +27° Oil

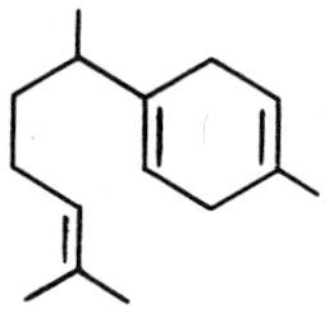

Bull. Soc. Chim. Fr. 1951 987

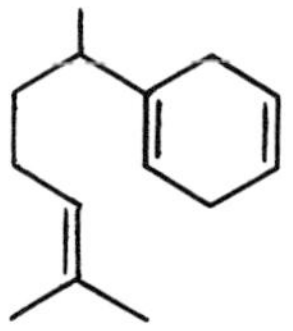

4106-008

CURCUMENE, l-β- $C_{15}H_{24}$ 204 -48° Oil

JCS 1939 1504

4106-009

CURCUMENE, γ- $C_{15}H_{24}$ 204 +32° Oil

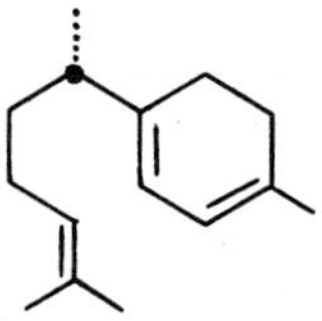

JCS 1949 838

4106-010

BISABOLOL, α-l- $C_{15}H_{26}O$ 222 -56° Oil OH Perf. Ess. Oil Rec. 1958 (49) 711 4106-011	LANCEOL $C_{15}H_{24}O$ 220 -69° Oil CH_2OH JCS 1951 1888 4106-016
ANYMOL $C_{15}H_{26}O_6$ 222 -63° Oil OH Austral. J. C. 1953 166 4106-012	BISABOLOL, oxide- $C_{15}H_{26}O_2$ 238 99°C OH O Coll. Czech. Chem. Comm. 1953 116 4106-017
BISABOLOL, α-d- $C_{15}H_{26}O$ 222 +52° Oil stereochemistry ? Coll. Czech. Chem. Comm. 1953 122 4106-013	TURMERONE $C_{15}H_{22}O$ 218 +82° Oil O Chem. Abstr. 1952 (46) 6333 4106-018
BISABOLOL, β - $C_{15}H_{26}O$ 222 +23° c Oil OH JOC 1968 909 4106-014	TURMERONE, (+)- $C_{15}H_{20}O$ 216 +68° Oil O Tet. 1964 2921 4106-019
NUCIFEROL $C_{15}H_{22}O$ 218 +41° Oil CH_2OH Bull. C. Soc. Jap. 1965 381 4106-015	ATLANTONE, α- $C_{15}H_{22}O$ 218 OI Oil O Helv. 1934 (17) 372 4106-020

ATLANTONE, γ- $C_{15}H_{22}O$ 218 OI Oil Helv. 1934 (17) 372 4106-021	TODOMATSUIC ACID $C_{15}H_{24}O_3$ 252 +86⁰ 58⁰ C COOH Bull. Chem. Soc. Jap. 1961 (34) 741 4106-026
CRYPTOMERIONE $C_{15}H_{22}O$ 218 -38⁰ Bull. Chem. Soc. Jap. 1964 (37) 1029 4106-022	JUVABIONE $C_{16}H_{26}O_3$ 266 + 80⁰c Oil COOMe Tet. Lett. 1967 1053; Chem. Comm. 1969 715 4106-027
CRYPTOMERONE $C_{15}H_{24}O_3$ 252 OH OH Tet. Lett. 1969,3185 4106-023	JUVABIONE, dehydro- $C_{16}H_{24}O_3$ 264 +102⁰ c Oil Chem. Comm. 1968 1057 4106-028
PEREZONE $C_{15}H_{20}O_3$ 248 - 17⁰ e 105⁰ C HO Chem. Comm. 1965 354 4106-024	BISABOLANGELONE $C_{15}H_{20}O_3$ 248 149⁰C OH Tet. Lett. 1966 3541 4106-029
NUCIFERAL $C_{15}H_{20}O$ 216 + 62⁰ c Oil CHO Bull. C. Soc. Jap. 1965 381 4106-025	PANICULIDE A $C_{15}H_{20}O_4$ 264 121⁰C R = Me HO R Chem. Comm., 1968, 1493. 4106-030

PANICULIDE B — $C_{15}H_{20}O_5$ — 280; + 4°m; 146°C R = CH_2OH Chem. Comm., 1968, 1493. 4106-031	BILOBANONE — $C_{15}H_{20}O_2$ — 232; +6.7°; Oil Chem. Comm. 1967 678; Chem. Abstr. 1968 (69) 59424 4106-034
PANICULIDE C — $C_{15}H_{18}O_5$ — 278; +37°m; Oil Chem. Comm., 1968, 1493. 4106-032	ANGELIKOREANOL — $C_{15}H_{20}O_3$ — 248; +132°c; 158° C Tet. Lett., 1970, 4381. Cf. 029 (?) 4106-035
SESQUIPHELLANDRENE, β- — $C_{15}H_{24}$ — 204; -4°; Oil Aust. J. C., 1966, 283. 4106-033	

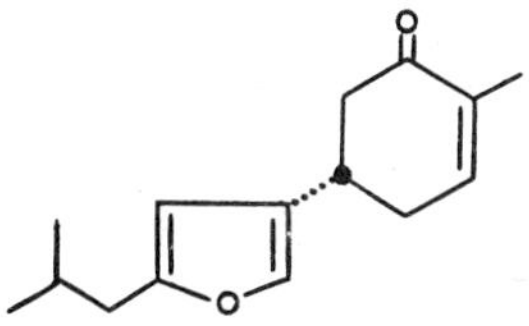

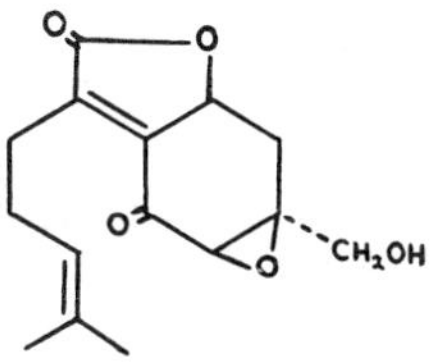

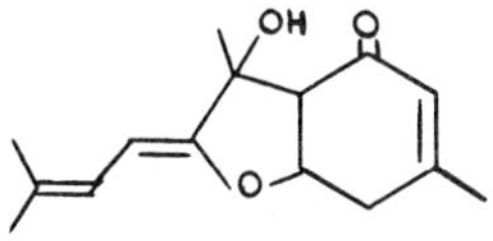

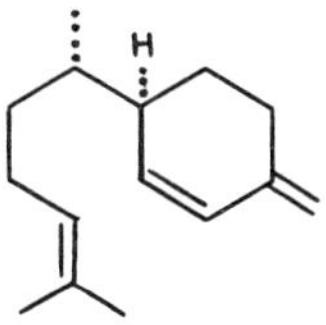

Name	Formula	Data	Reference	No.
CADINENE, d-β-	$C_{15}H_{24}$	204 +70°	J. App. Chem. 1952 (2) 187	4107-001
CADINENE, γ 2-	$C_{15}H_{24}$	204 -40° c Oil	Tet., 1963,241.	4107-006
CADINENE, l-β-	$C_{15}H_{24}$	204 Oil	Coll. Czech. Chem. Comm. 1958 2181	4107-002
CADINENE, tetrahydro-γ-	$C_{15}H_{28}$	208 -31° Oil	Perf. Ess. Oil Rec., 1961 (52) 233.	4107-007
CADINENE, d-γ-	$C_{15}H_{24}$	204 +148° Oil	Tet. 1968 4113	4107-003
CADINENE, d-(+)-δ-	$C_{15}H_{24}$	204 +92° c Oil	Tet. Lett. 1968 519; 1913	4107-008
CADINENE, γ1-	$C_{15}H_{24}$	204 -20°	Bull. Chem. Soc. Jap. 1964 1053	4107-004
DYSOXYLONENE	$C_{15}H_{24}$	204 0° Oil	Tet. Lett. 1968 519.	4107-009
BULGARENE, (-)-ε-	$C_{15}H_{24}$	204 -17° Oil	Coll. Czech. Chem. Comm. 1967 808; 822	4107-005
CADINENE , omega-	$C_{15}H_{24}$	204 +92° Oil	Coll. Czech. Chem. Comm. 1967 808; 1968 822 Tet. Lett. 1968 519	4107-010

Name	Formula	Data	Reference	No.
CADINA-9, 11-DIENE	$C_{15}H_{24}$	204 $+68^0$ Oil	Chem. Abstr. 1966 15431	4107-011
SESQUIBENIHEN	$C_{15}H_{24}$	204 $+36^0$ Oil	Chem. Abstr. 1947 (41) 3448	4107-016
METROSIDERENE	$C_{15}H_{24}$	204 -2^0 Oil	JCS 1954 1179	4107-012
EUPATENE, D-	$C_{15}H_{24}$	204 $+19^0$ Oil	Perf. Ess. Oil Rec., 1956 (47) 315.	4107-017
CADINA-1, 9(15)-DIENE	$C_{15}H_{24}$	204 Oil	Chem. Abstr. 1955 (49) 7197	4107-013
EUPATENE, L -	$C_{15}H_{24}$	204 -6^0 Oil	Perf. Ess. Oil Rec., 1956 (47) 315.	4107-018
CADINENE, 4(14)-	$C_{15}H_{26}$	206 -33^0 Oil	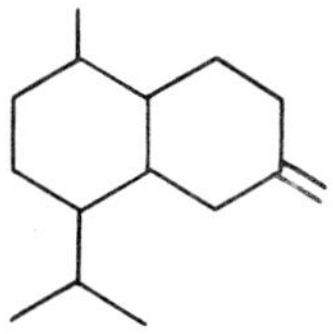	4107-014
deleted				4107-019
CADINA-4, 11-DIENE	$C_{15}H_{24}$	204 -31^0 Oil	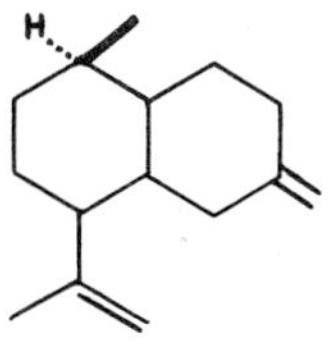	4107-015
MUUROLENE, (-)-α	$C_{15}H_{24}$	204 -52.5^0 Oil	Coll. Czeck. Chem. Comm. 1966 3373	4107-020

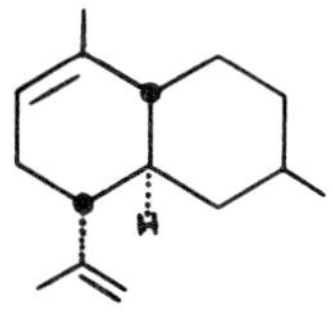

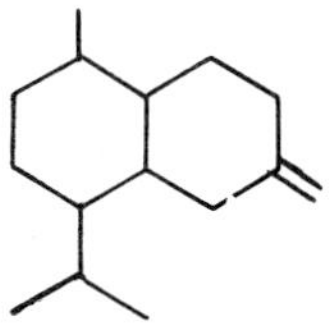

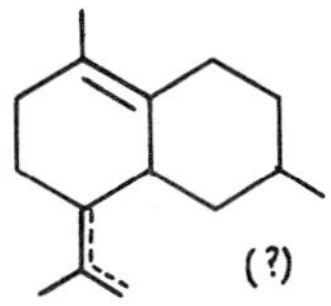

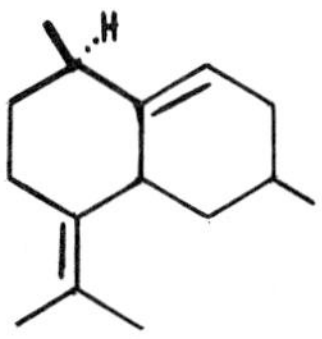

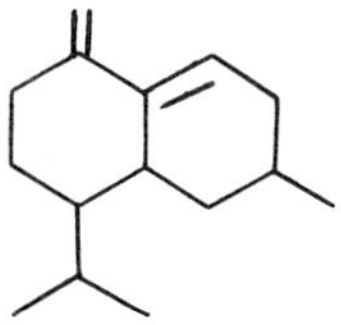

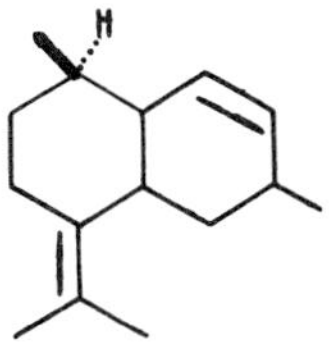

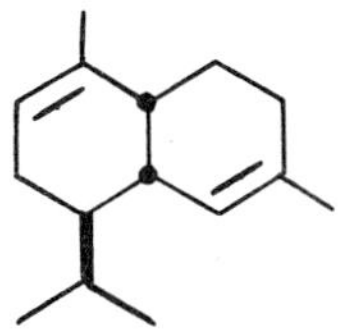

(?)

MUUROLENE, (+)-γ- — $C_{15}H_{24}$ — 204 — $+1.5^0$ — Oil

Coll. Czech. Chem. Comm. 1966 3373

4107-021

MUUROLENE, ε- — $C_{15}H_{24}$ — 204 — $+57^0$ — Oil

Acta Chem. Scand. 1964 (18) 572

4107-022

AMORPHENE, (-)-γ- — $C_{15}H_{24}$ — 204 — -75^0 — Oil

Coll. Czech. Chem. Comm. 1966 2025

4107-023

CALAMENENE — $C_{15}H_{22}$ — 202 — -68^0c — Oil

Bull. Chem. Soc. Jap. 1964 1029

4107-024

CALACORENE, α- — $C_{15}H_{20}$ — 200 — $+52^o$ — Oil

Tet. 1958 246

4107-025

CALACORENE, β- — $C_{25}H_{20}$ — 200 — $+50^0$ — Oil

Coll. Czech. Chem. Comm. 1953 512

4107-026

CADALENE — $C_{15}H_{18}$ — 198 — OI — Oil

Coll. Czech. Chem. Comm. 1953 500

4107-027

CYTHOCOL — $C_{15}H_{24}O$ — 220 — 0^0c — Oil

Perf. Ess. OilRec., 1961 (52) 558.

OH

4107-028

deleted

4107-029

VETICADINOL — $C_{15}H_{26}O$ — 222 — Impure

Bull. Soc. Chim. Belges., 1961 (70) 5.

OH

4107-030

CADINOL, α-

$C_{15}H_{26}O$ 222
-47°e
75°C

Coll. Czech. Chem. Comm. 1958 (23) 1297

4107-031

CALAMEONE

$C_{15}H_{26}O_2$ 238
-10°e
168°C

Chem. Comm., 1970, 1323.

4107-036

"CADINOL, T-"

$C_{15}H_{26}O$ 222
-5°c
Oil

Chem. Comm. 1967 565

4107-032

CALAMENDIOL, iso-

$C_{15}H_{26}O_2$ 238
73° C

Tet. Lett., 1969 3729.

4107-037

MURROLOL, T-

$C_{15}H_{26}O$ 222
-103°c
79°C

Tet. Lett. 1968 1915

4107-033

CUBENOL

$C_{15}H_{26}O$ 222

Tet. Lett. 1967 2073

4107-038

MUUROL-3-ENE-2β, 9α-DIOL

$C_{15}H_{26}O_2$ 238
-48°e
206°C

4107-034

CUBENOL, epi-

$C_{15}H_{26}O$ 222
-101°c

Tet. Lett. 1967 2073

4107-039

MUUROL-3-ENE-2α, 9α-DIOL

$C_{15}H_{26}O_2$ 238
-111°m
169°C

Tet. Lett. 1969 2375

4107-035

CADIN-8-EN-10-OL

$C_{15}H_{26}O$ 222
+5° m
75° C

Chem. Pharm. Bull., 1968, 903.

4107-040

ARTICULOL — $C_{15}H_{24}O$ — 220

Tet. 1967 2169

4107-041

KHUSINOL — $C_{15}H_{24}O$ — 220
$+124^{0}$c
87^{0}C

Tet. 1963 233

4107-042

KHUSINOL, epi- — $C_{15}H_{24}O$ — 220
-87^{0}c
Oil

Tet. 1965 3197

4107-043

KHUSINOLOXIDE — $C_{15}H_{24}O_2$ — 236
-25^{0}
113^{0}

Tet. 1967 (23) 1267

4107-044

CADINOL, γ- — $C_{15}H_{26}O$ — 222
$+100^{0}$
136^{0}C

Chem. Abstr. 1955 (49) 6549

4107-045

CADINOL, d-δ- — $C_{15}H_{26}O$ — 222
$+118^{0}$
140^{0}C

Acta Chem. Scand. 1966 2893

4107-046

TORREYOL, (-)- — $C_{15}H_{26}O$ — 222
-106^{0}c
140^{0}C

Acta Chem. Scand. 1966 2893

4107-047

KHUSOL — $C_{15}H_{24}O$ — 220
-137^{0}c
101^{0}C

Tet. 1963 1073

4107-048

CYMBOPOL — $C_{15}H_{26}O$ — 222
-8.4^{0}
Oil

Chem. Abstr. 1953 (47) 1340; J. Chem. Soc. Jap. 1932 636

4107-049

CADALENE, 7-OH- — $C_{15}H_{18}O$ — 214
OI
88^{0}C

Chem. Abstr. 1968 (68) 114765

4107-050

Name	Formula	Data	Reference	No.
CHILOSCYPHONE	$C_{15}H_{22}O$	218 -46^0d 122^0C	Tet. Lett. 1969 1599; 1970 1289	4107-051
MANSONONE A	$C_{15}H_{20}O_2$	232 $+680^0$c 118^0C	Tet. Lett. 1966, 2767; Tet. Lett. 1965 4857	4107-056
CALACORONE	$C_{15}H_{24}O$	220 $+60^0$ Oil	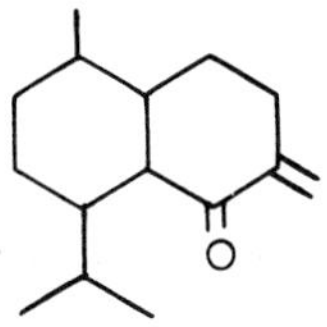	4107-052
MANSONONE B	$C_{15}H_{20}O_3$	248 68^0C	Tet. Lett. 1965 4857	4107-057
ARTICULONE	$C_{15}H_{22}O$	218 -27^0 c Impure	Tet. 1967 2169	4107-053
MANSONONE C	$C_{15}H_{16}O_2$	228 135^0C	Tet. Lett. 1966 2767 Tet. Lett. 1965 4857	4107-058
CADINOL, 2-KETO-α-	$C_{15}H_{24}O$	236 -123^0c 111^0C	Tet. Lett. 1968 3881	4107-054
MANSONONE D	$C_{13}H_{14}O_3$	242 174^0C	Tet. Lett. 1965 4857	4107-059
MUUROL-3-ENE-9α-OL-2-ONE	$C_{15}H_{24}O_2$	236 -104^0e 126^0C	Tet. Lett. 1969 2375	4107-055
MANSONONE E	$C_{15}H_{14}O_3$	242 $+82^0$c 145^0C	Tet. Lett. 1966 2767	4107-060

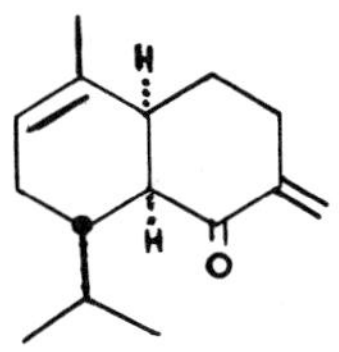

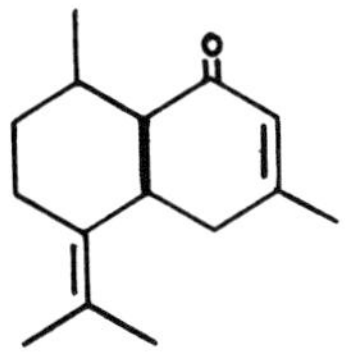

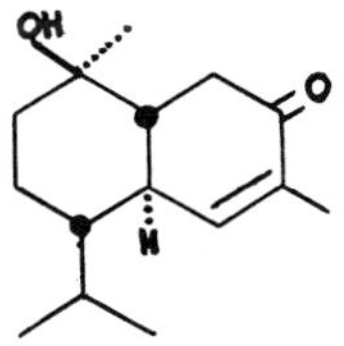

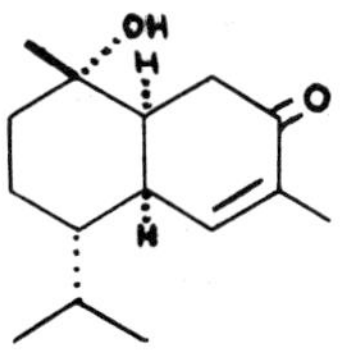

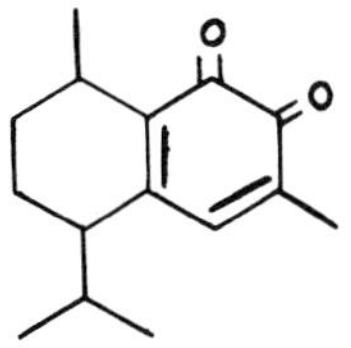

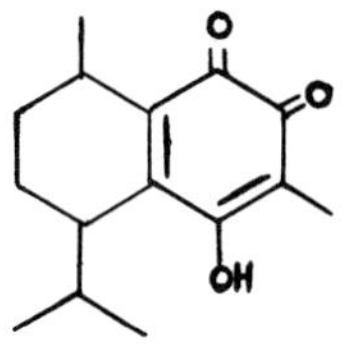

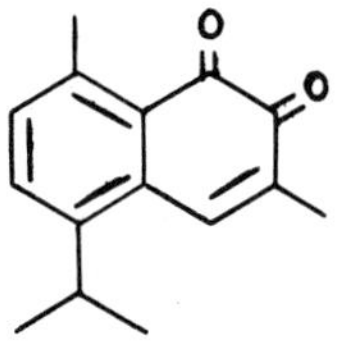

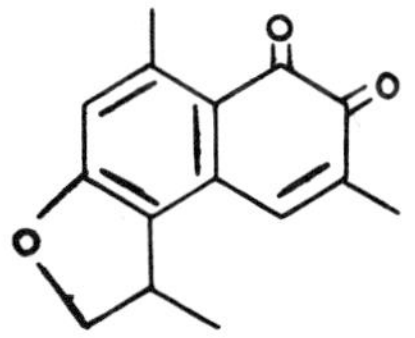

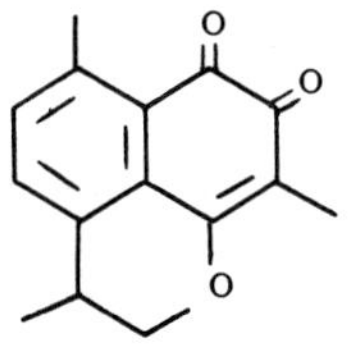

MANSONONE F	$C_{15}H_{12}O_3$	240 OI 215°C	Tet. Lett. 1965 4857	4107-061
MANSONONE G	$C_{15}H_{16}O_3$	244 OI 212°C	Tet. Lett. 1966 2767	4107-062
MANSONONE H	$C_{15}H_{14}O_4$	258 OI >320°C	Tet. Lett. 1966 2767	4107-063
MANSONONE I	$C_{15}H_{14}O_4$	258 214°C	Chem. Abstr. 1967 (67) 90610	4107-064
MANSONONE L	$C_{15}H_{12}O_4$	256 OI	Tet. Lett. 1969 3583	4107-065
CADALENAL, 1, 2, 3, 4-tetrahydro-7-OH-	$C_{15}H_{20}O_2$	232	Chem. Abstr. 1968 (68) 114765	4107-066
CADALENAL, 7-OH-	$C_{15}H_{16}O_2$	228 OI	Chem. Abstr. 1968 (68) 114765	4107-067
CADALENAL, 1-OMe-7-OH-	$C_{16}H_{18}O_3$	258 OI 138°C	Chem. Abstr. 1968 (68) 114765	4107-068
ZIZANENE	$C_{15}H_{24}$	204 +120° Oil	Tet. Lett., 1970, 4651.	4107-069
CADINAN-3, 4, 9-TRIOL	$C_{15}H_{28}O_3$	256	Chem. Abstr., 1970 (73) 77418.	4107-070

GOSSYPOL	$C_{30}H_{30}O_8$	518 OI 184^{0}C
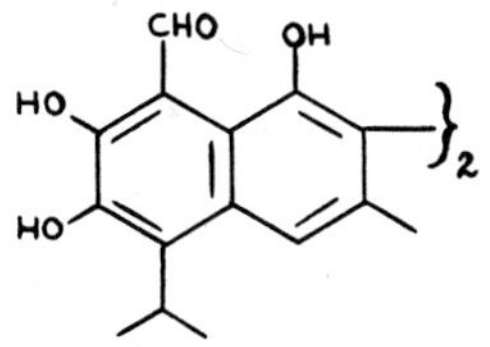	JACS., 1947 1268.	
4107-071		

CURZERENONE, Pyro-	$C_{15}H_{16}O$	212 OI 77^{0}C
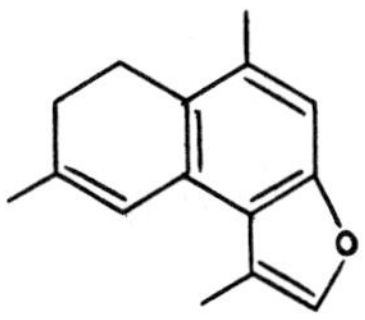	Tet. Lett., 1968, 4417.	
4107-073		

CALAMENENE, 7-OH-	$C_{15}H_{22}O$	218 -30^{0}c
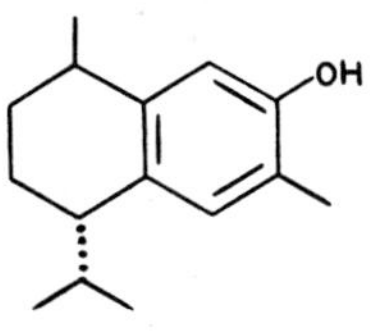	Chem. Ind., 1969, 922.	
4107-072		

CAROTOL $C_{15}H_{26}O$ 222 +2.5° Oil Bull. Soc. Chim. Fr. 1967 2059 4108-001	LASERPITINE $C_{25}H_{38}O_7$ 450 +121°m 117°C Chem. Abstr. 1967 65653 4108-004
DAUCOL $C_{15}H_{26}O_2$ 238 -17°e 115°C Bull. Soc. Chim. Fr. 1964 2021 4108-002	LASERPITINE, iso- $C_{25}H_{38}O_7$ 450 -36°m 159°C Chem. Abstr. 1967 82276 4108-005
LASEROL $C_{15}H_{26}O_5$ 286 +177°m 191°C Chem. Abstr. 1967 65653 4108-003	

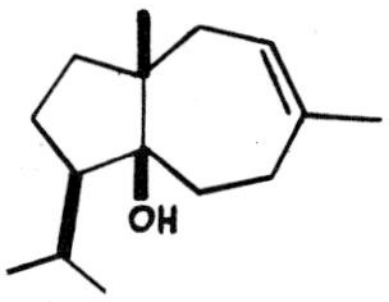

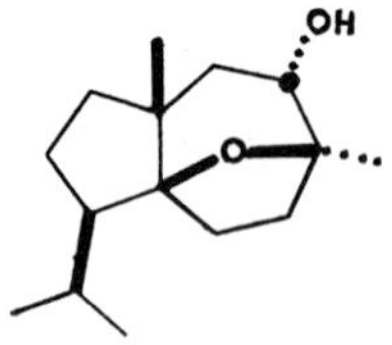

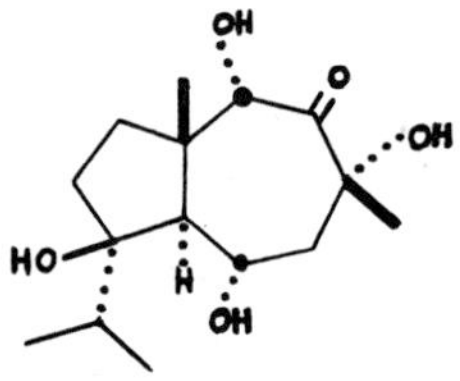

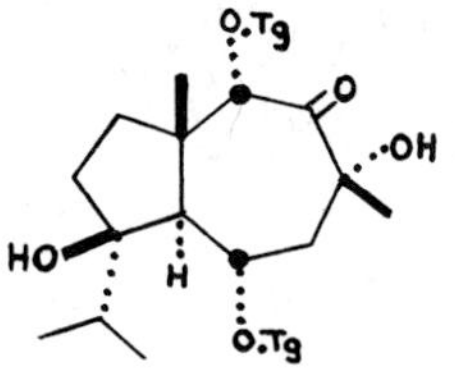

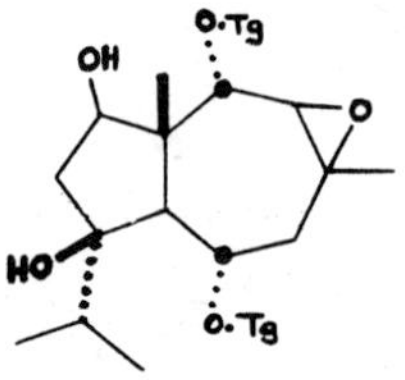

HIMACHALENE, α- $C_{15}H_{24}$ 204 -198°c Oil Tet. 1968 3861 4109-001	HIMACHALOL, (+)- $C_{15}H_{26}O$ 222 +73°c 68°C Tet. 1968 3861 4109-003
HIMACHALENE, β- $C_{15}H_{24}$ 204 +226° Oil Tet. 1968 4109-002	

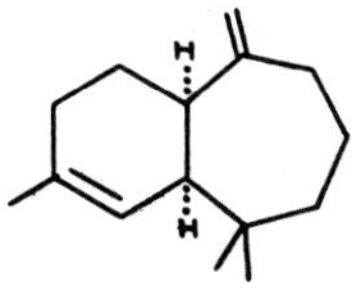

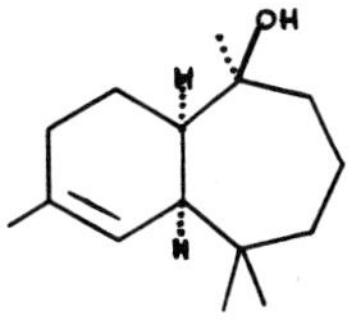

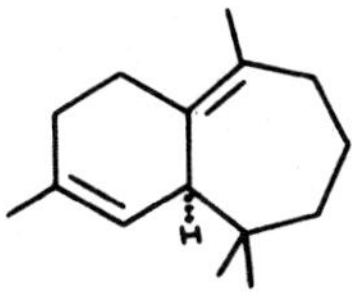

TRICHODIENE

$C_{15}H_{24}$ 204 $+21^0$c Oil

Tet. Lett., 1970, 2671.

4110-001

TRICHODIOL

$C_{15}H_{24}O_3$ 252 $+52^0$c 82^0C

Tet. Lett., 1970, 1177.

4110-002

RORIDIN C

$C_{15}H_{22}O_3$ 250 -33^0c 118^0C

Acta Chem. Scand., 1966 (20) 1044.

4110-003

TRICHODERMIN

$C_{17}H_{24}O_4$ 292

Acta Chem. Scand., 1966 (20) 1044.

4110-004

TRICHOTHECIN

$C_{19}H_{24}O_5$ 332 $+44^0$ 118^0C

JCS, 1959, 1105.

4110-005

VERRUCAROL

$C_{15}H_{22}O_4$ 266 -39^0c 156^0C

JOC, 1965, 726.

4110-006

SCIRPENOL, diOAc-

$C_{19}H_{26}O_7$ 366

JCS, 1966, 116.

4110-007

CROTOCIN

$C_{19}H_{24}O_5$ 332

Tet. Lett., 1967 (17) 1665.

4110-008

SCIRP-9-EN-3α-OL, 8α-isovaleroyloxy-4β, 15-diacetoxy-

$C_{24}H_{34}O_9$ 466 15^0e 151^0C

Tet., 1968, 3329.

4110-009

SCIRP-9-ENE, 4, 8, 15-triOAc-3α, 7α-diOH-

$C_{21}H_{28}O_{10}$ 440 187^0C

JCS, 1970, 378.

4110-010

NIVALENOL 4110-011	$C_{15}H_{20}O_7$ 312 222°C JCS, 1970, 375.	VERRUCARIN J cf. Verrucarin A 4110-016	$C_{27}H_{32}O_8$ 484 Helv., 1965, 1669.
FUSARENONE 4110-012	$C_{17}H_{22}O_8$ 354 +58°m 92°C JCS, 1970, 375.	RORIDIN A 4110-017	$C_{29}H_{40}O_9$ 532 130°c 200°C Helv., 1966, 2527.
VERRUCARIN A 4110-013	$C_{27}H_{34}O_9$ 502 JCS, 1966, 1395.	RORIDIN D cf. Roridin A 4110-018	$C_{29}H_{38}O_9$ 530 +29°c 234°C Helv., 1966, 2547.
VERRUCARIN A, 2'-dehydro- cf. Verrucarin A 4110-014	$C_{27}H_{32}O_9$ 500 +118°c 237°C Helv., 1966, 2594.	FUSARENONE, Ac- 4110-019	$C_{19}H_{24}O_9$ 396 136°C Chem. Comm., 1969, 1266.
VERRUCARIN B cf. Verrucarin A 4110-015	$C_{27}H_{32}O_9$ 500 +94°c 330°C Nature, 1961, 4798.		

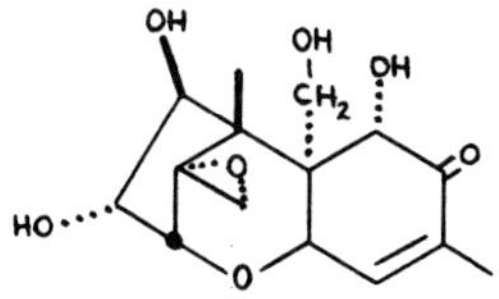

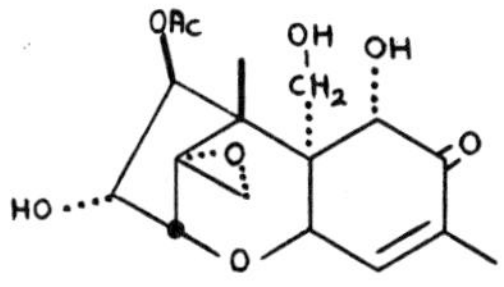

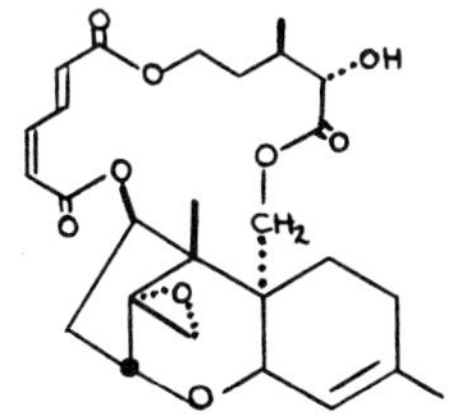

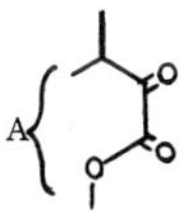

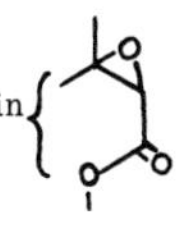

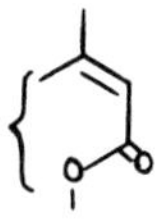

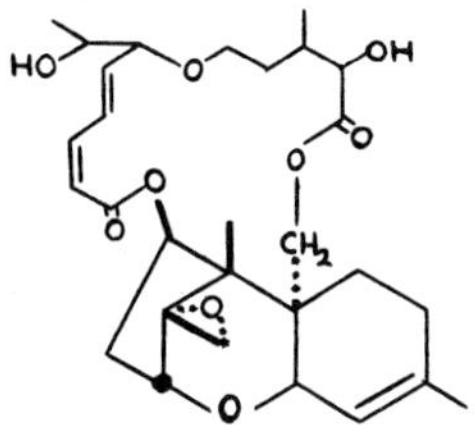

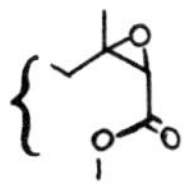

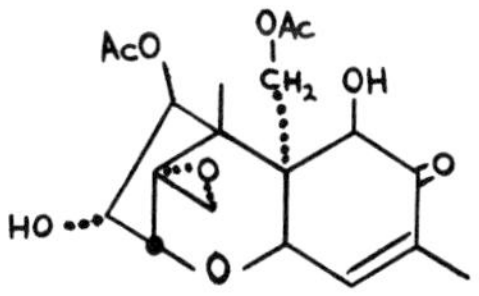

LAURENE $C_{15}H_{20}$ 200 +49°c Oil Tet. Lett., 1967, 3187. 4111-001	APLYSIN, de-Br- $C_{15}H_{20}O$ 216 Tet., 1963, 1485. 4111-004
LAURENISOL $C_{15}H_{19}OBr$ 294 +86°c Tet. Lett., 1969, 1343. 4111-002	APLYSINOL $C_{15}H_{19}O_2Br$ 310 -56°c 160°C Tet., 1963, 1485. 4111-005
APLYSIN $C_{15}H_{19}OBr$ 294 -85°c 85°C Chem. Comm., 1967, 271. 4111-003	LAURINTEROL, iso- $C_{15}H_{19}OBr$ 294 Oil Tet., 1970, 3271. 4111-006

ALASKENE, α- — $C_{15}H_{24}$ — 204 — Oil

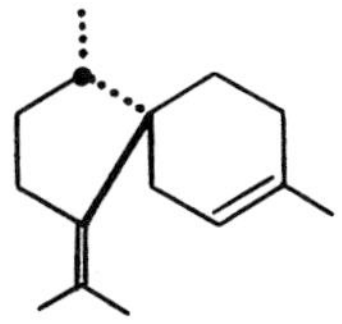

Tet. Lett. 1970 2277

4112-001

ALASKENE, β- — $C_{15}H_{24}$ — 204 — Oil

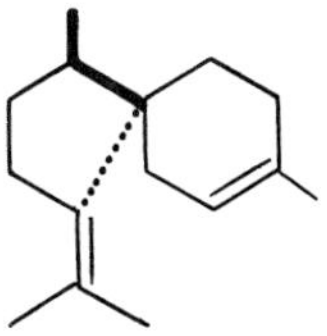

Tet. Lett. 1970 2277

4112-002

ACORENOL — $C_{15}H_{26}O$ — 222 — -36^0

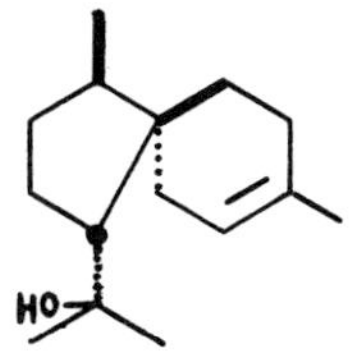

Tet. Lett. 1970 143

4112-003

ACORENOL, β- — $C_{15}H_{26}O$ — 222 — ±O

Tet. Lett. 1970 1371

4112-004

ACORADIENE, β- — $C_{15}H_{24}$ — 204 — $+23^0$

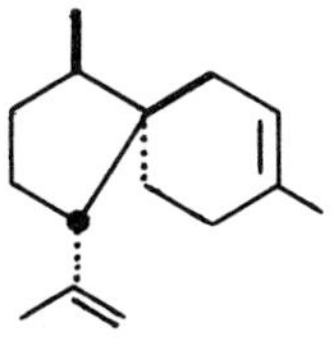

Tet. Lett. 1970 1371

4112-005

ACORDADIENE, γ- — $C_{15}H_{24}$ — 204 — -66^0

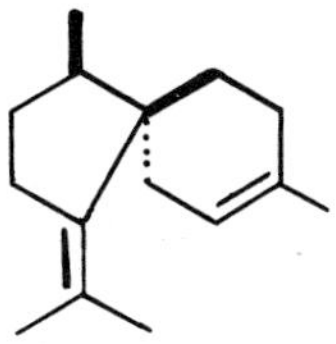

Tet. Lett. 1970 1371

4112-006

ACORADIENE, δ- — $C_{15}H_{24}$ — 204 — $+15^0$

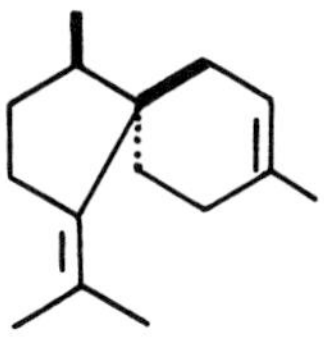

Tet. Lett. 1970 1371

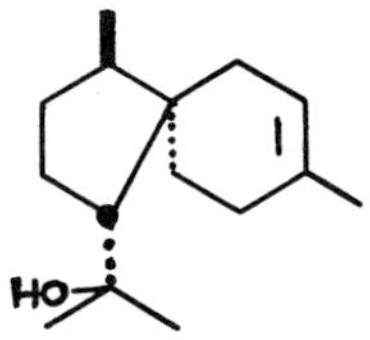

4112-007

ACORONE — $C_{15}H_{24}O_2$ — 236 — $+144^0$e — 100^0C

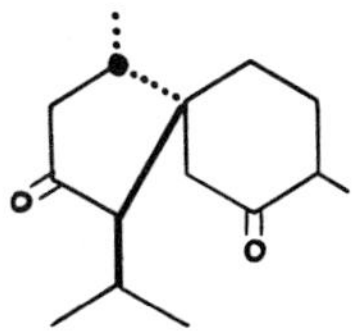

JCS 1966 579

4112-008

ACORONE, Iso- — $C_{15}H_{24}O_2$ — 236 — -91^0e — 98^0C

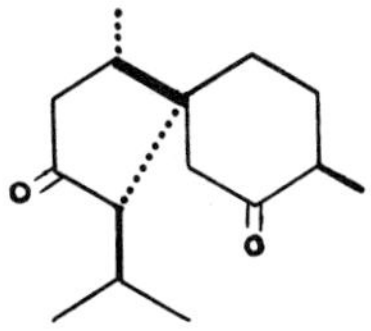

Chem. Comm. 1965 276

4112-009

ACORONE, crypto- — $C_{15}H_{24}O_2$ — 236 — $+98^0$e — 108^0C

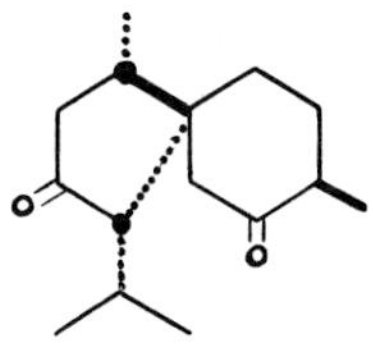

Coll. Czech. 1964 539

4112-010

ACORENONE $C_{15}H_{24}O$ 220 -22° oil Coll. Czech. Chem. Comm. 1961 (26) 3183 4112-011	ACORADIENE $C_{15}H_{24}$ 204 Oil Tet. Lett., 1970, 143 4112-013
ACORENONE B $C_{15}H_{24}O$ 220 Chem. Comm. 1968 1135 4112-012	

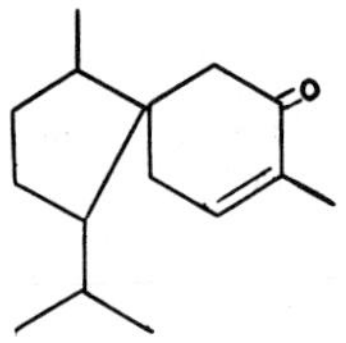

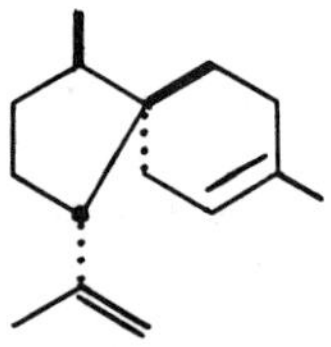

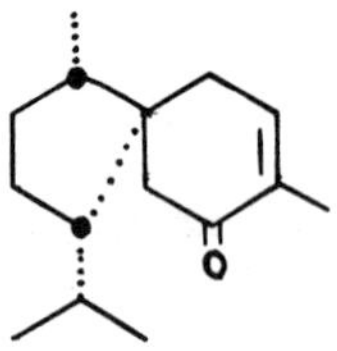

THUJOPSENE

$C_{15}H_{24}$ 204
-80
Oil

Acta Chem. Scand. 1961 592 1676; 1963 738

4113-001

WIDDRENIC ACID

$C_{15}H_{22}O_2$ 234
-86°c
170°C

COOH

JOC 1965 3947

4113-003

THUJOPSADIENE

$C_{15}H_{22}$ 202
Oil

Tet. Let. 1969 3095

4113-002

DAMSINIC ACID $C_{15}H_{22}O_3$ 250 112°C JOC 1970 486 4114-001	CONFERTIFLORIN $C_{17}H_{22}O_5$ 306 +25° 145°C Tet. 1967 2529 4114-006
AMBROSIN $C_{15}H_{18}O_3$ 246 -155°e 146°C Tet. Lett. 1961 82 4114-002	CONFERTIFLORIN, des-Ac- $C_{15}H_{20}O_4$ 264 +17° 203°C Tet. 1967 2529 4114-007
FRANSERIN $C_{15}H_{22}O_4$ 266 0° 226°C Can. J. C. 1968 1535 4114-003	HYSTERIN $C_{17}H_{24}O_5$ 308 -80°c 168°C JOC 1966 675 4114-008
DAMSINE $C_{15}H_{20}O_3$ 248 -72°c 110°C Coll. Czech. Chem. Comm. 1963 2257 4114-004	AMBROSIN, Neo- $C_{15}H_{18}O_3$ 246 126°C Phytochem. 1967 1563 4114-009
DAMSIN, 3-OH- $C_{15}H_{20}O_4$ 264 +3°(C) 144°C JOC., 1967, 2928. 4114-005	AMBROSIOL $C_{15}H_{22}O_4$ 266 -111°c 116°C JOC 1966 681 4114-010

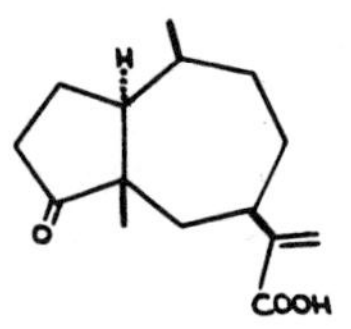

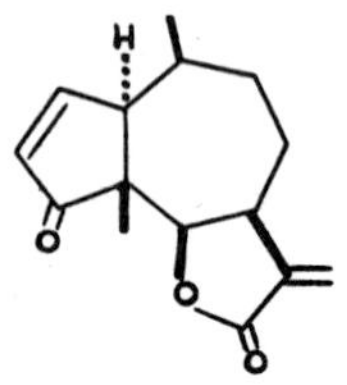

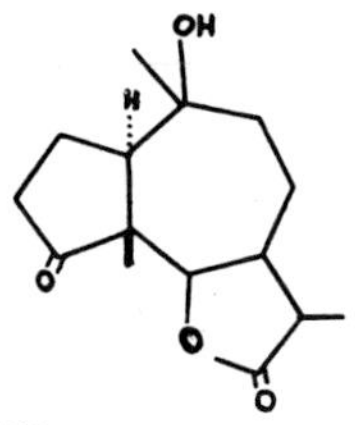

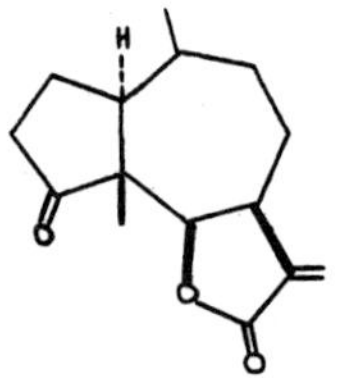

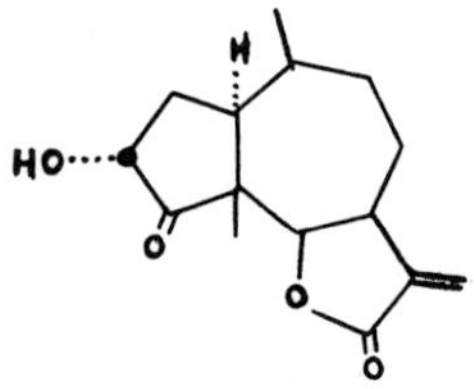

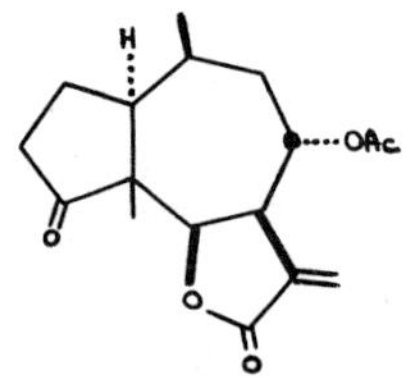

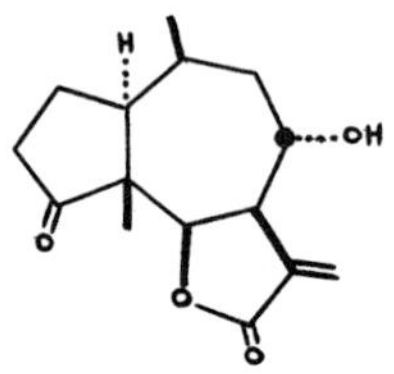

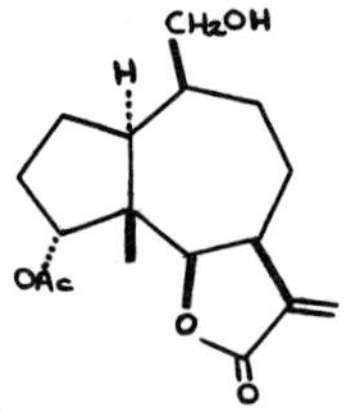

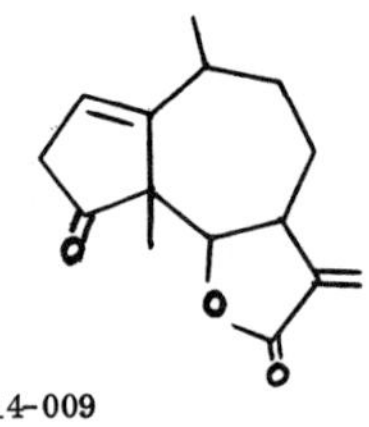

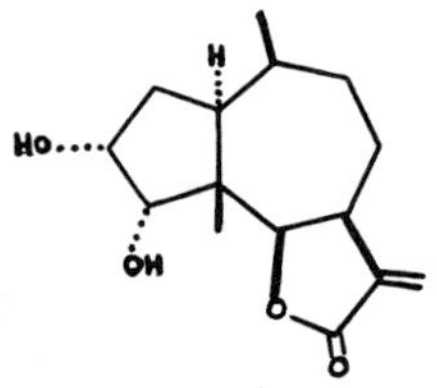

PARTHENIN — $C_{15}H_{18}O_4$ — 262, +7°c, 165°C

Tet., 1966, 3279.

4114-011

CORONOPILIN — $C_{15}H_{20}O_4$ — 264, -30° e, 178°C

JOC 1964 2553

4114-012

CONCHOSIN A — $C_{15}H_{18}O_5$ — 278, -30°e, 151°C

Tet. 1970 2775

4114-013

CONCHOSIN B — $C_{17}H_{20}O_6$ — 320, 144°C

Tet. 1970 2775

4114-014

TETRANEURIN A — $C_{17}H_{22}O_6$ — 322, + 4°, 187°C

Tet. 1969 805

4114-015

TETRANEURIN B — $C_{17}H_{22}O_6$ — 322, -45°, 195°C

Tet. 1970 2167

4114-016

TETRANEURIN C — $C_{19}H_{26}O_7$ — 366, -109°, 145°C

Tet. 1970 2167

4114-017

TETRANEURIN D — $C_{17}H_{24}O_6$ — 324, -73°, 204°C

Tet. 1970 2167

4114-018

CONFERTIN — $C_{15}H_{20}O_3$ — 248, 146°C

Can. J. C. 1968 1535

4114-019

GAILLARDILIN — $C_{17}H_{22}O_6$ — 322, -2°, 198°C

Tet. 1966 693

4114-020

AMARILIN — $C_{15}H_{20}O_4$ — 264 — $+5^{0}c$ — $196^{0}C$

Tet. 1965 1711

4114-021

MEXICANIN C — $C_{15}H_{20}O_4$ — 264 — $-80^{0}c$ — $252^{0}C$

Tet. 1963 1359

4114-026

MEXICANIN A — $C_{15}H_{18}O_4$ — 262 — $-27^{0}c$ — $139^{0}C$

Tet. 1966 (22) 1709

4114-022

HELENALIN — $C_{15}H_{18}O_4$ — 262 — $-102^{0}e$ — $170^{0}C$

JACS 1956 3860

4114-027

LINIFOLINE B — $C_{17}H_{20}O_5$ — 304 — $150^{0}C$

JOC 1962 4043

4114-023

MEXICANIN-I — $C_{15}H_{18}O_4$ — 262 — $+43^{0}c$ — $258^{0}C$

Tet. 1964 79

4114-028

AROMATIN — $C_{15}H_{18}O_3$ — 246 — -6^{0} — $160^{0}C$

Tet. 1966 (22) 3279

4114-024

LINIFOLINE A — $C_{17}H_{20}O_5$ — 304 — $+33^{0}c$ — $202^{0}C$

JOC 1968 2780

4114-029

AROMATICIN — $C_{15}H_{18}O_3$ — 246 — $+18^{0}$ — $233^{0}C$

Tet. 1965 1711

4114-025

BIGELOVIN — $C_{17}H_{20}O_5$ — 304 — $+46^{0}e$ — $190^{0}C$

Tet. 1966 (22) 3279

4114-030

BALDUILIN A — $C_{17}H_{20}O_5$ — 304 — $+57^0c$ — 232^0C

Tet. 1966 (22) 3279

4114-031

FASTIGILIN C — $C_{20}H_{24}O_6$ — -86^0 — 198^0C

Tet. 1966 1907

4114-036

TENULIN, iso- — $C_{17}H_{22}O_5$ — 306 — $+6^0e$ — 160^0C

Tet. 1966 (22) 1709

4114-032

FASTIGILIN A — $C_{20}H_{26}O_6$ — 362 — -82^0 — 176^0C

Tet. 1966 (22) 1907

4114-037

BREVILIN A — $C_{20}H_{26}O_5$ — 346 — 116^0C

JOC 1968 2780
JACS 1959 1481

4114-033

FASTIGILIN B — $C_{20}H_{26}O_6$ — 362 — 260^0C

Tet. 1966 1907

4114-038

THURBERILIN — $C_{20}H_{26}O_5$ — $+20^0$ e — 162^0C

Tet. 1966 (22) 3279

4114-034

ODORATIN — $C_{15}H_{22}O_4$ — 266 — $+71^0$ — 166^0C

Can. J. C. 1968 1539

4114-039

TENULIN, des-Ac-iso- — $C_{15}H_{20}O_4$ — 264 — 246^0C

JOC 1964 1549

4114-035

CUMANIN — $C_{15}H_{22}O_4$ — 266 — $+161^0$ — 120^0C

Tet. 1966 1499

4114-040

GEIGERININE $C_{15}H_{22}O_4$ 266 -11^0e 202^0C

JCS 1963 4989

4114-041

PERUVININ $C_{15}H_{20}O_4$ 264 $+34^0$ 170^0C

Tet. 1967 (23) 529

4114-042

PULCHELLIN $C_{15}H_{22}O_4$ 266 -36^0c 166^0C

Tet. Lett. 1970 131

4114-043

STEVIN $C_{17}H_{24}O_5$ 308 $+161^0$ 185^0C

Tet. 1967 4265

4114-044

ALTERNILIN $C_{17}H_{24}O_6$ 324 $+11^0$ 194^0C

JOC 1968 2780

4114-045

FLEXUOSIN A $C_{17}H_{24}O_6$ 324 $+12^0$c 220^0C

Tet. 1964 979

4114-046

FLEXUOSIN B $C_{20}H_{28}O_6$ 364 $+44^0$c 117^0C

Tet. 1964 979

4114-047

SPATHULIN $C_{19}H_{26}O_8$ 382 $+17^0$ 262^0C

JOC 1967 1042

4114-048

PAUCIN $C_{23}H_{32}O_{10}$ 468 $+52^0$ 178^0C

Tet. Lett. 1969 515

R= ac-glucosyl

4114-049

PERUVINE $C_{15}H_{20}O_4$ 264 $+155^0$c 192^0C

Tet. 1966 1723

4114-050

TETRANEURIN E — $C_{17}H_{24}O_6$ — 324 — -70^0m — 200^0C

R = H

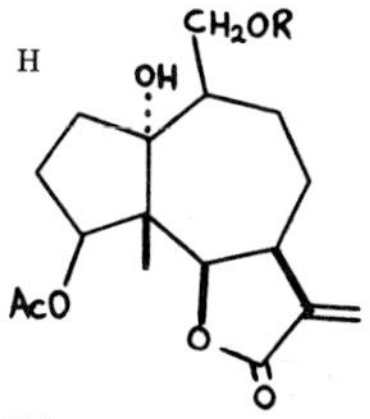

JOC., 1970, 2888.

4114-051

TETRANEURIN F — $C_{19}H_{26}O_7$ — 366 — -47^0m — 136^0C

R = Ac

JOC., 1970, 2888.

4114-052

OPODIN, 11, 13-dehydro– — $C_{20}H_{24}O_4$ — 328 — 114^0C

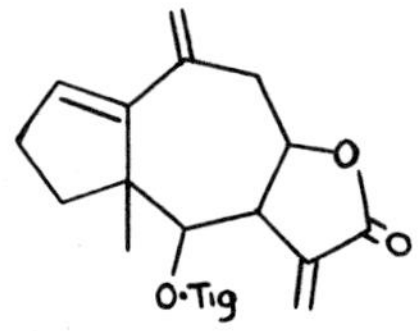

Khim. Prir. Soedin., 1969 (5) 490.

4114-053

LINEARIFOLIN A — $C_{20}H_{24}O_6$ — 360 — -90^0c — 188^0C

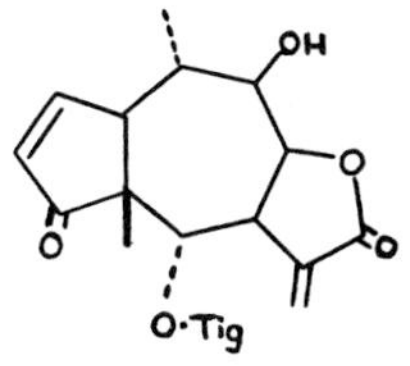

JOC., 1970, 4117.

4114-054

LINEARIFOLIN B — $C_{20}H_{24}O_6$ — 360 — -104^0c — 215^0C

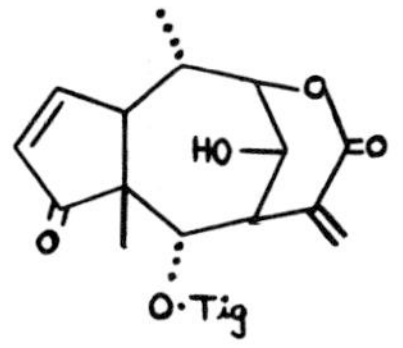

JOC., 1970, 4117.

4114-055

AUTUMNOLIDE — $C_{15}H_{20}O_5$ — 280 — $+21^0$c — 189^0C

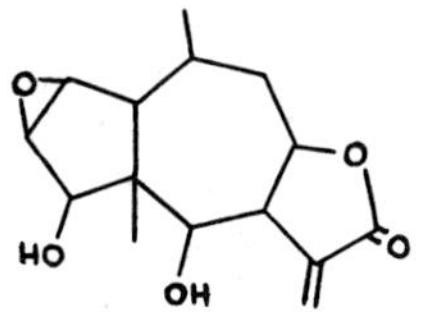

JOC., 1969, 2915.

4114-056

CORONOPILIN, dehydro– — $C_{15}H_{18}O_3$ — 246 — -60^0c — 121^0C

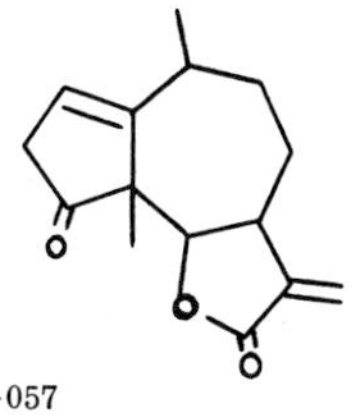

Khim. Prir. Soedin., 1969 (5) 487.

4114-057

IVOXANTHINE — $C_{15}H_{20}O_4$ — 264 — $+47^0$c — 165^0C

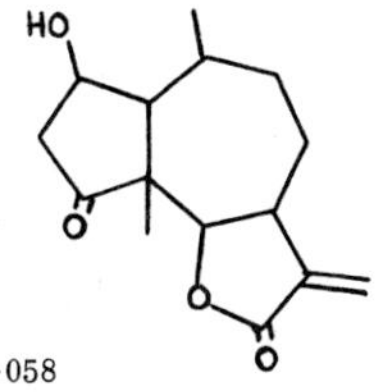

Khim. Prir. Soedin., 1969 (5) 487.

4114-058

CUPARENENE, α-	$C_{15}H_{24}$	204 +64⁰ Oil
	JOC 1966 315	
4115-001		
CUPARENOL, β-	$C_{15}H_{22}O$	218 +64⁰ Oil
HO	Tet. Lett. 1968 843	
4115-006		
CUPARENENE, β-	$C_{15}H_{24}$	204 Oil
	JOC 1966 315	
4115-002		
CUPARENOL, γ-	$C_{15}H_{22}O$	218 +92⁰ Oil
CH_2OH	Tet. Lett. 1968 843	
4115-007		
CUPARENE	$C_{15}H_{22}$	202 +65⁰ Oil
	Tet. 1958 (4) 361	
4115-003		
CUPARENONE, α-	$C_{15}H_{20}O$	216 +177⁰c 52⁰C
O	Tet. Lett. 1967 3187	
4115-008		
CUPARENOL, α-	$C_{15}H_{22}O$	218 +12⁰ 73⁰C
HO	Tet. Lett., 1968 843	
4115-004		
CUPARENONE, β-	$C_{15}H_2O$	216 +48⁰c Oil
O	Tet. Lett. 1964 73	
4115-009		
CUPARENOL, α-iso-	$C_{15}H_{22}O$	218 +81⁰ 78⁰C
HO	Tet. Lett., 1968 843	
4115-005		
CUPARENIC ACID	$C_{15}H_{20}O_2$	232 +63⁰c 159⁰C
COOH	Tet. 1958 (4) 361	
4115-010		

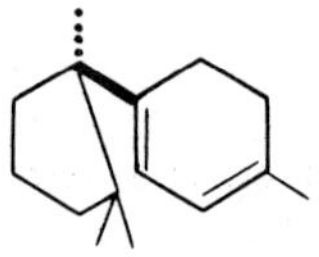

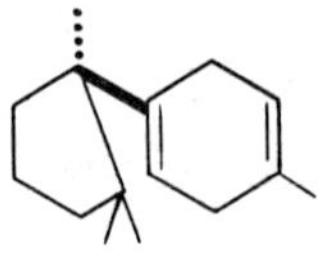

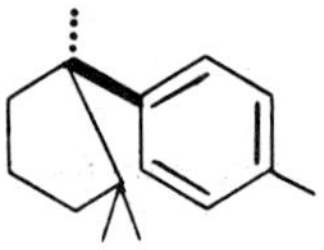

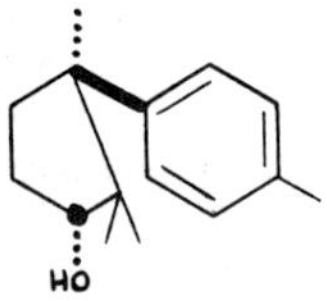

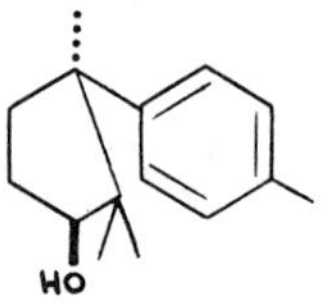

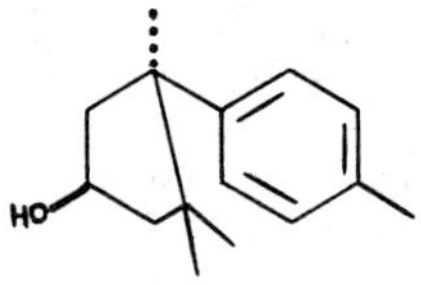

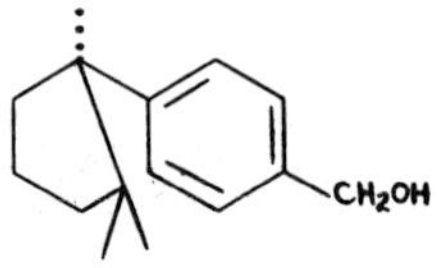

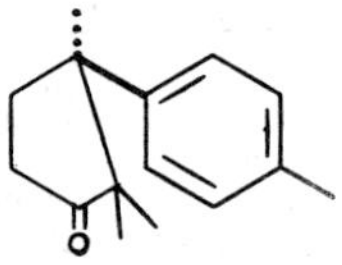

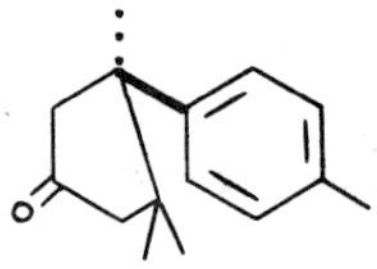

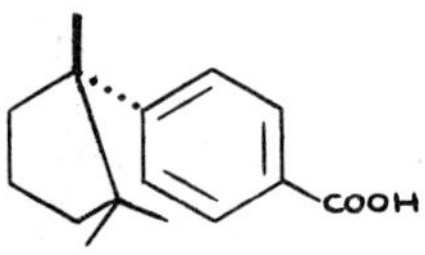

HELICOBASIDINE $C_{15}H_{20}O_4$ 264 -123^0c 191^0C Chem. Comm. 1970 926 4115-011	HELICOBASIDIN, deoxy- $C_{15}H_{20}O_3$ 248 -186^0c 195^0C Chem. Pharm. Bull., 1967 380 4115-012

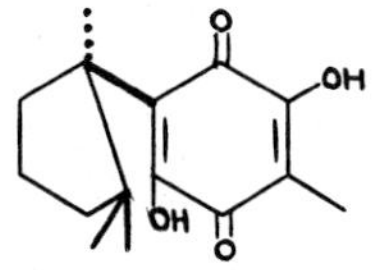

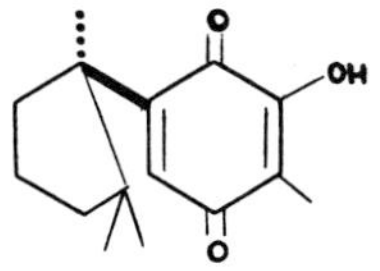

CARYOPHYLLENE	$C_{15}H_{24}$	204 -14^0c Oil
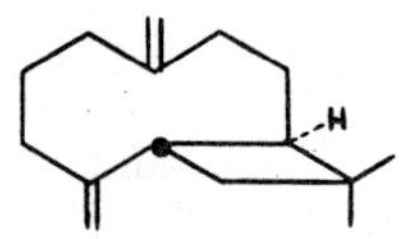	Tet., 1968, 4113.	
4116-001		
CARYOPHYLLENE, D-β-, (+)-	$C_{15}H_{24}$	204 $+10^0$ Oil
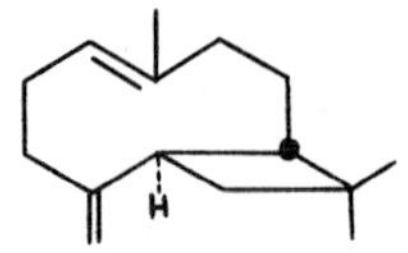	Chem. Abstr., 1963 (59) 7313g.	
4116-002		
CARYOPHYLLENE, L-β-	$C_{15}H_{24}$	204 -9^0 Oil
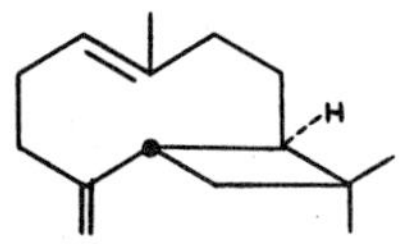	Tet., 1968, 4113.	
4116-003		
CARYOPHYLLENE, iso-	$C_{15}H_{24}$	204 $+28^0$ Oil
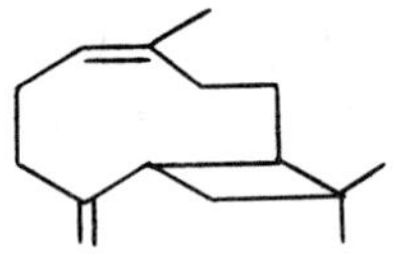	JACS, 1964, 485.	
4116-004		
CARYOPHYLLENE OXIDE	$C_{15}H_{24}O$	220 -70^0c 64^0C
	Tet., 1968, 4113.	
4116-005		
MULTIJUGENOL, α-	$C_{15}H_{24}O$	220 $+115^0$ Oil
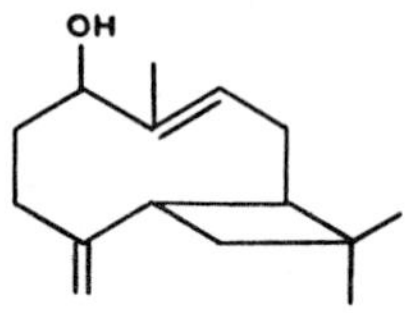	Tet. Lett., 1971, 658.	
4116-006		
BETULENOL, α-	$C_{15}H_{24}O$	220 -20^0 Oil
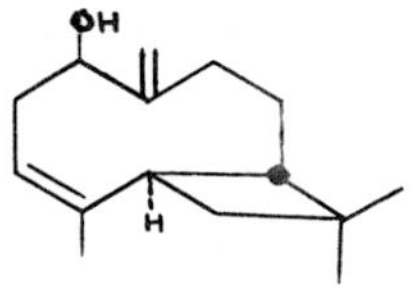	Coll. Czech. Chem. Comm., 1959 3780	
4116-007		
BETULENOL, β-	$C_{15}H_{24}O$	220 -36^0 Oil
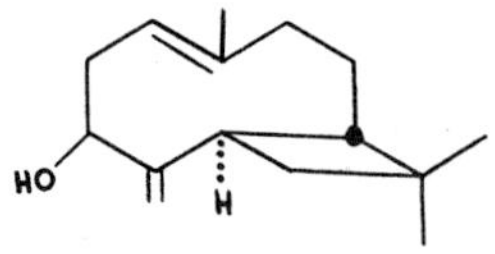	Coll. Czech. Chem. Comm., 1959 3780	
4116-008		

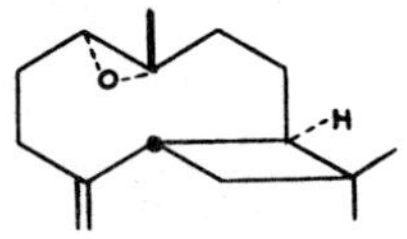

Name	Formula	Data	Reference	No.
EREMOPHILENE	$C_{15}H_{24}$	204 -142° Oil	Tet. Lett. 1968 3315	4117-001
VALENCENE	$C_{15}H_{24}$	204	Tet. Lett. 1965 4779	4117-002
NOOTKATENE	$C_{15}H_{22}$	202 -177°c Oil	Tet. Lett. 1965 4779	4117-003
VETIVENE, γ-	$C_{15}H_{22}$	202 Oil	Tet. Lett. 1970 1760	4117-004
VETIVENOL, bicyclo-	$C_{15}H_{24}O$	220 +122° Oil	Chem. Pharm. Bull. 1968 2449	4117-005
VALERIANOL	$C_{15}H_{26}O$	222 +134°c Oil	Chem. Pharm. Bull. 1968 833	4117-006
EREMOLIGENOL	$C_{15}H_{26}O$	222 -93°c Oil	JCS 1966 1545	4117-007
VALENCENIC ACID, iso-	$C_{15}H_{22}O_2$	234 136°C	Tet. Lett. 1968 6099	4117-008
EREMOPHILONE	$C_{15}H_{22}O$	218 -172°e 42°C	JACS 1960 6354	4117-009
EREMOPHILA-1, 11-DIEN-9-ONE, 7α(H)-	$C_{15}H_{22}O$	218 44°C	Tet. Lett. 1969 307	4117-010

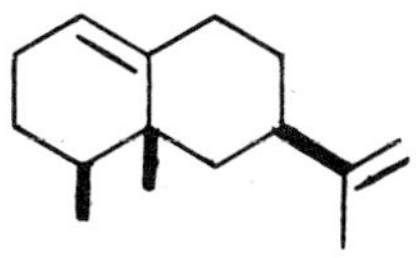

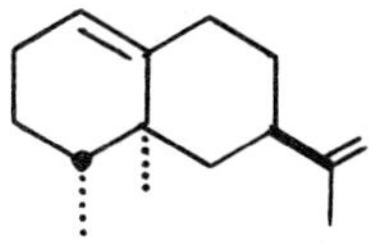

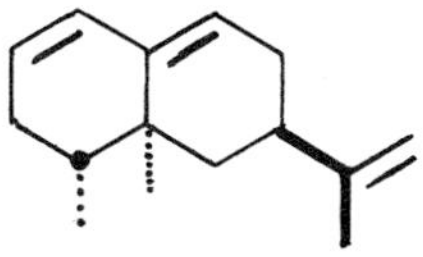

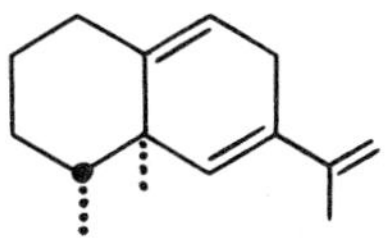

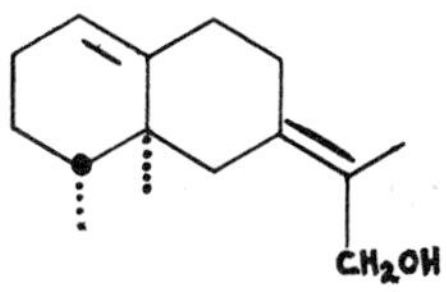

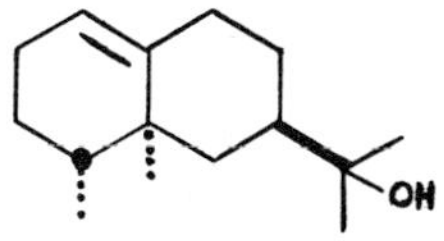

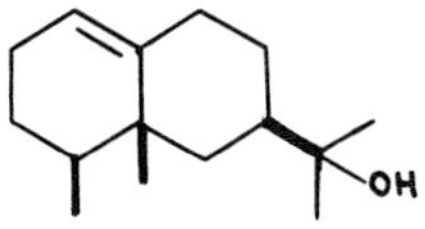

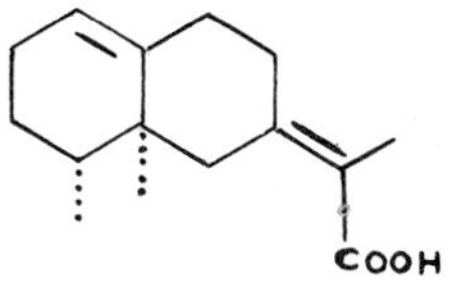

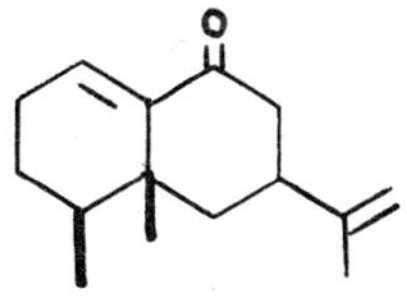

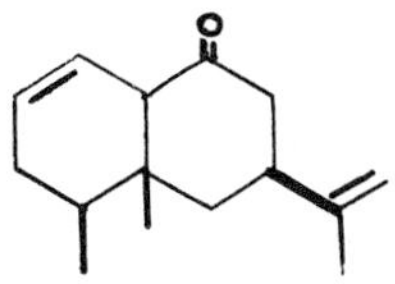

FUKINONE $C_{15}H_{24}O$ 220 +68° Oil Tet. 1968 5871 4117-011	NARDOSTACHONE $C_{15}H_{20}O$ 216 +209°c Oil Tet. Lett., 1970 413 4117-016
EREMOPHILONE, allo- $C_{15}H_{22}O$ 218 Chem. Ind. 1966 2170 4117-012	EREMOPHILONE, oxy- $C_{15}H_{22}O_2$ 234 +141°c 66°C JACS 1960 6354 4117-017
NOOTKATONE $C_{15}H_{22}O$ 218 +196°c 36°C Tet. Lett. 1965 4779 4117-013	EREMOPHILONE, oxy-dihydro- $C_{15}H_{24}O_2$ 236 +91°c 102°C Ind. J. C. 1963 171 4117-018
NOOTKATONE, 7-epi- $C_{15}H_{24}O$ 220 Phytochem. 1970 145 4117-014	EREMOPHILA-10, 11-DIEN-9-ONE, 8α-OH- $C_{15}H_{22}O_2$ 234 -34°m Oil Aust. J. C. 1966 303 4117-019
VETIVONE, α- $C_{15}H_{22}O$ 218 +238°e 51°C Chem. Comm. 1967 (2) 89 4117-015	EREMOPHILA-1, 11-DIEN-9-ONE, 8α-OH- $C_{15}H_{22}O_2$ 234 +59° 103°C Aust. J. C. 1966 303 4117-020

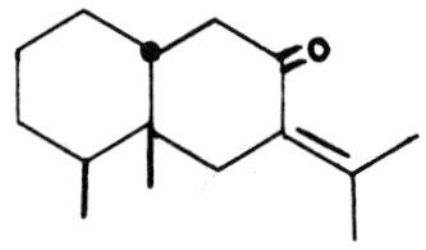

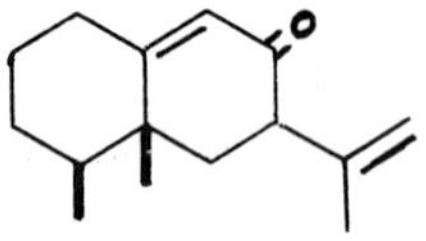

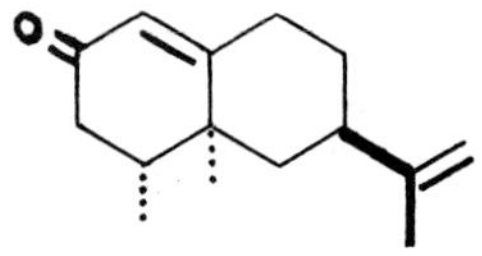

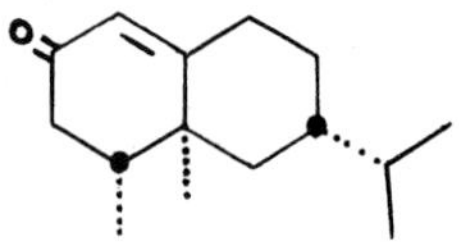

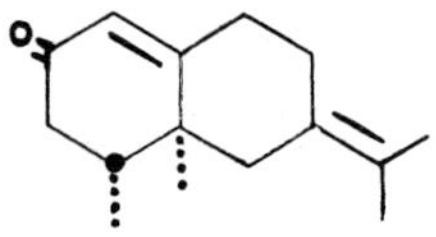

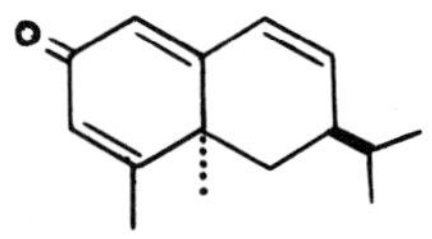

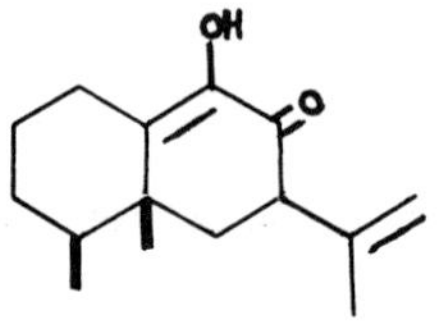

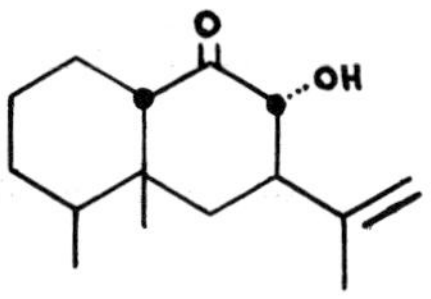

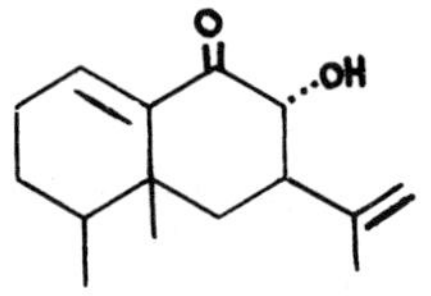

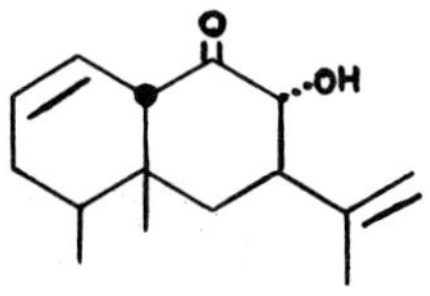

PETASOL, iso-

$C_{15}H_{22}O_2$ 234 115^{0}c 125^{0}C

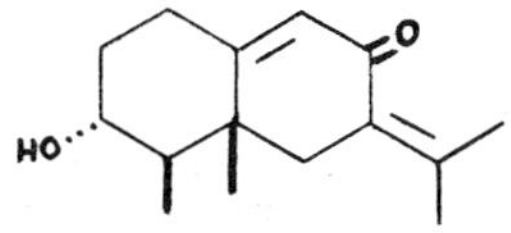

JACS 1960 4337

4117-021

PETASINE

$C_{20}H_{28}O_3$ 316 $+49^{0}$c 72^{0}C

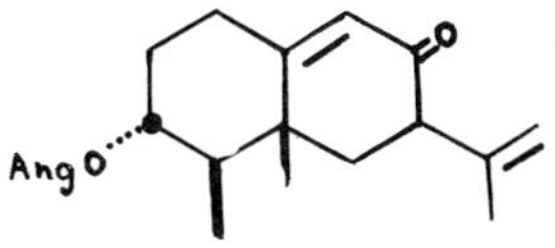

JACS 1960 4337

4117-022

PETASINE, iso-

$C_{20}H_{28}O_3$ 316 $+28^{0}$c 96^{0}C

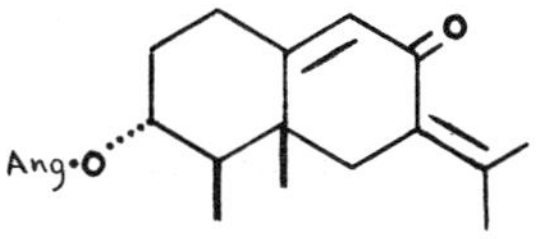

Chem. Comm. 1966 701

4117-023

PETASITIN

$C_{20}H_{28}O_4$ 332 -26^{0}

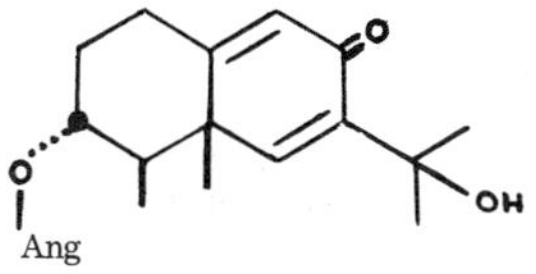

Tet. Lett. 1968 629

4117-024

WARBURGDIONE

$C_{15}H_{18}O_{12}$ 230 $+25^{0}$c 127^{0}C

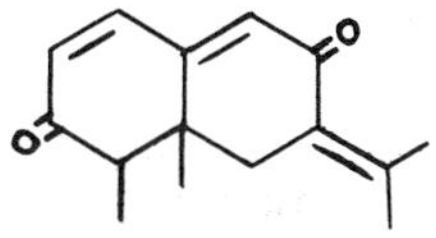

Chem. Comm. 1966 (19) 701

4117-025

EREMOPHILANE, furano-

$C_{15}H_{22}O$ 218 -8^{0} Oil

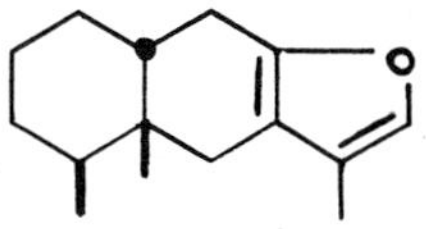

Coll. Czech. Chem. Comm. 1964 2182

4117-026

EREMOPHILANE, 9-OH-furo-

$C_{15}H_{22}O_2$ 234 $+14^{0}$m Oil

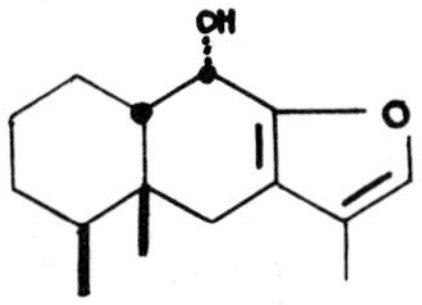

Coll. Czech. Chem. Comm. 1969 336

4117-027

PETASALBINE, L-

$C_{15}H_{22}O_2$ 234 -12^{0}c 80^{0}C

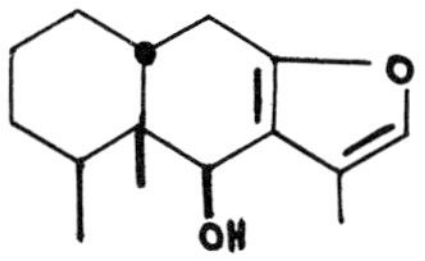

Tet. 1965 2605

4117-028

ALBOPETASINE

$C_{20}H_{28}O_3$ 316 0^{0}c 106^{0}C

O-Angeloyl

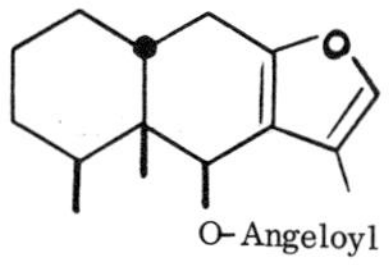

Coll. Czech. Chem. Comm., 1962 1400

4117-029

EREMOPHILANDIOL, furano-

$C_{15}H_{22}O_3$ 250 -24^{0}c 179^{0}C

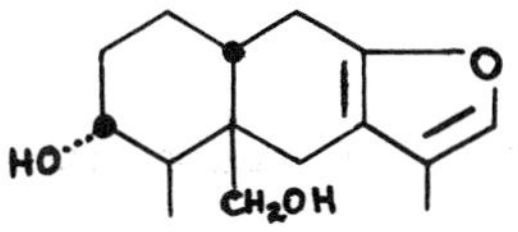

Coll. Czech. Chem. Comm. 1962 1394

4117-030

PETASIN, furano- $C_{20}H_{28}O_4$ 332 -30°c 115°C Tet. Lett. 1961 697 4117-031	EREMOPHILONE, furano- $C_{15}H_{20}O_2$ 232 0°c 150°C Coll. Czech. Chem. Comm. 1964 2189 4117-036
PETASOL, furano- $C_{15}H_{22}O_3$ 250 154°C Coll. Czech. Chem. Comm. 1964 1922 4117-032	LIGULARONE $C_{15}H_{20}O_2$ 232 -58° 65° C Tet. 2605 1965 4117-037
deleted 4117-033	LIGULARENONE, furano - $C_{15}H_{18}O_2$ 230 +3° 95°C Bull. Soc. Chim. Fr. 1968 1047 4117-038
EURYOPSOL $C_{15}H_{22}O_4$ 266 174°C Tet. 1969 5227 4117-034	LIGULARENONE, 10- β - H- $C_{15}H_{18}O_2$ 230 Chem. Abstr. 1967 99986 4117-039
KABLICIN $C_{25}H_{34}O_6$ 430 -18° 72°C Tet. Lett. 1968 1401 4117-035	WARBURGIN $C_{16}H_{16}O_4$ 272 +120°c 160°C Chem. Comm. 1966 393 4117-040

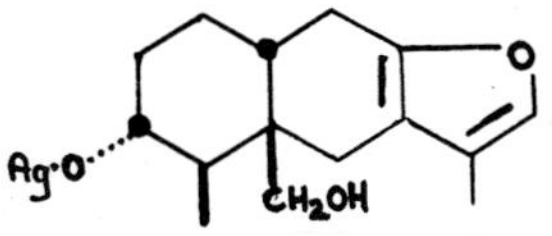

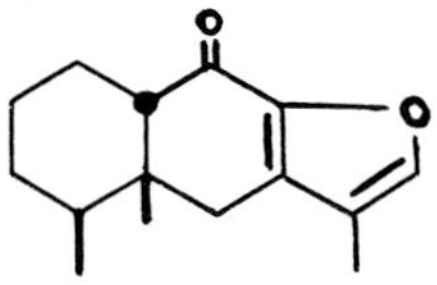

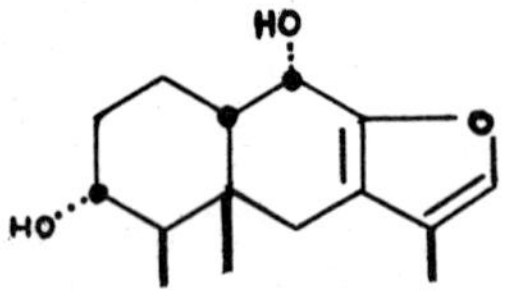

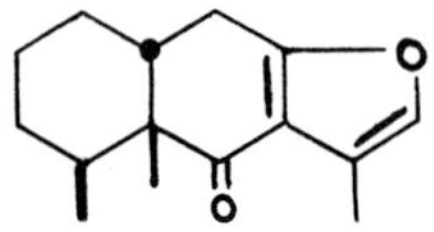

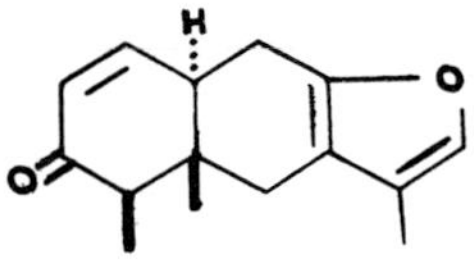

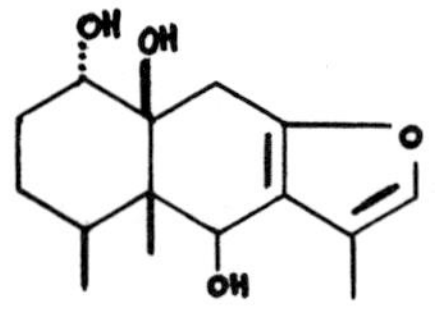

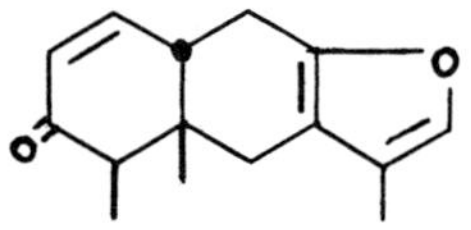

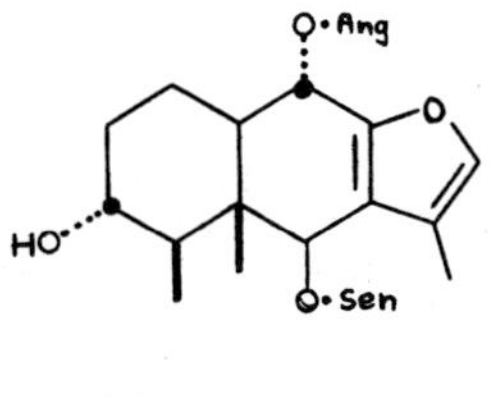

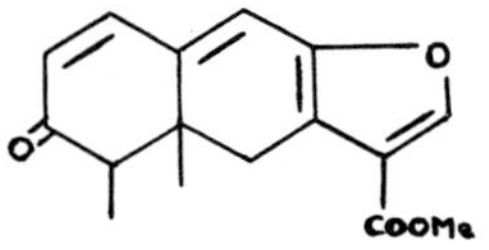

EURYOPSONOL $C_{15}H_{20}O_3$ 248 -36°c 230°C Tet. 1967 2431 HO 4117-041	LIGULARENOLIDE $C_{15}H_{18}O_2$ 230 +33°m 135°C Tet. Lett. 1968 3739 4117-046
DECOMPOSTIN $C_{17}H_{20}O_4$ 288 +60° 184°C Coll. Czech. Chem. Comm. 1969 2792 OAc 4117-042	EREMOPHILENOLIDE $C_{15}H_{22}O_2$ 234 +17°c 125°C Tet. 1963 1101 4117-047
ADENOSTYLONE $C_{19}H_{26}O_3$ 302 -69° 77°C Coll. Czech. Chem. Comm. 1969 1739; 2792 4117-043	deleted 4117-048
ADENOSTYLONE, iso- $C_{19}H_{26}O_3$ 302 0° 93°C Coll. Czech. Chem. Comm. 1969 1739; 2792 4117-044	EREMOPHILENOLIDE, 6-OH- $C_{15}H_{22}O_3$ 250 Coll. Czech. Chem. Comm. 1964 2189 OH 4117-049
ADENOSTYLONE, neo- $C_{20}H_{24}O_4$ 328 -86° 107°C Coll. Czech. Chem. Comm. 1969 2792 Angeloyl 4117-045	EREMOPHILAN-14β, 6α-OLIDE, furano- $C_{15}H_{18}O_3$ 246 -47°d 137°C Chem. Comm. 1969 551 4117-50

PETASITOLIDE A

$C_{20}H_{28}O_4$ 332 +48°c 147°C

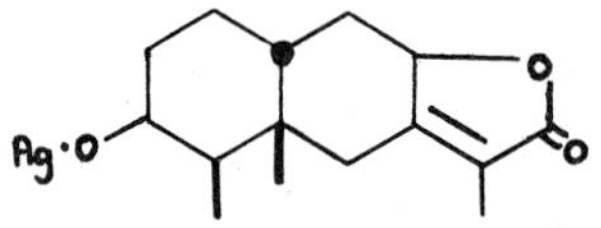

Coll. Czech. Chem. Comm. 1962 1394

4117-051

PETASITOLIDE B, S-

$C_{19}H_{26}O_4S$ 350 -33° 200°C

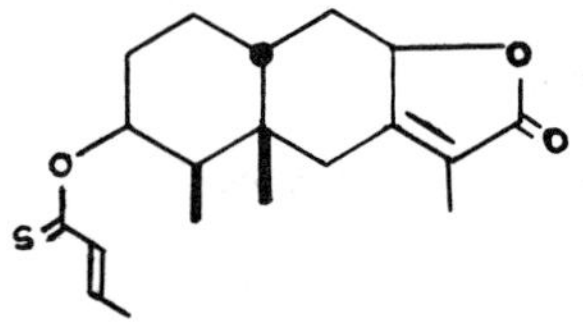

Coll. Czech. Chem. Comm. 1962 1394

4117-054

PETASITOLIDE B

$C_{20}H_{28}O_4$ 332 +32°c 146°C

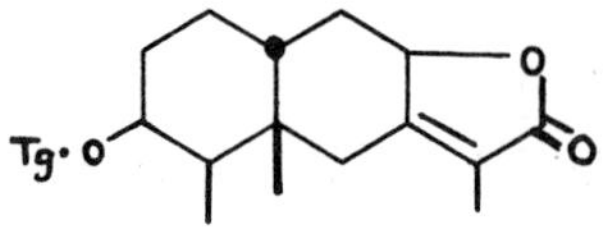

Coll. Czech. Chem. Comm. 1962 1394

4117-052

EREMOPHILANE, di-OMe-dihydro-furo-

$C_{17}H_{28}O_3$ 280 +66° 103°C

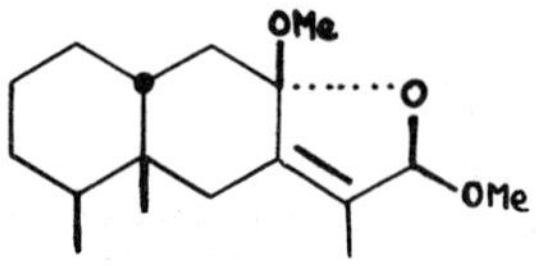

Coll. Czech. Chem. Comm. 1966 371

4117-055

PETASITOLIDE A, S-

$C_{19}H_{26}O_4S$ 350 +48°c 202°C

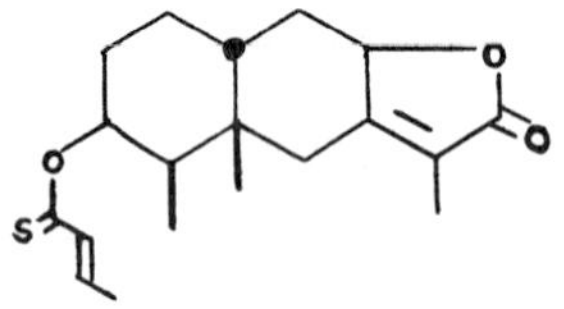

Coll. Czech. Chem. Comm. 1962 1393

4117-053

VETIVENENE, β-

$C_{15}H_{22}$ 202 -160° Oil

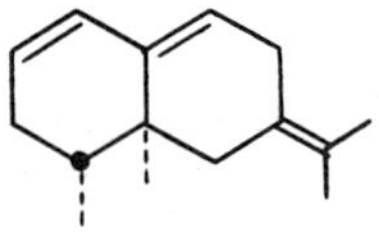

Tet. Lett., 1970, 1759.

4117-056

SANTALENE, α-	$C_{15}H_{24}$	204 +15° Oil	Tet. Lett., 1963, 2123.
4118-001			
SANTALENE, β-	$C_{15}H_{24}$	204 -93° Oil	Helv., 1935, 355.
4118-002			
SANTALENE, epi-β-	$C_{15}H_{24}$	204 Oil	Tet. Lett., 1963, 1949.
4118-003			
SANTALOL, α-	$C_{15}H_{24}O$	220 +17° Oil	Tet. Lett., 1963, 2123.
4118-004			
SANTALOL, β-	$C_{15}H_{24}O$	220 -91° Oil	Tet. Lett., 1970, 41.
4118-005			
SANTALAL, (-)-α-	$C_{15}H_{22}O$	218 -3°c Oil	Tet. Lett., 1968, 1533.
4118-006			
SANTALIC ACID, β-	$C_{15}H_{22}O_2$	234 Oil	J. Ind. C.S., 1944 (21) 337.
4118-007			

CALARENOL	$C_{15}H_{24}O$	220 + 48°c Oil	ARISTOLENE, L-	$C_{15}H_{24}$	204 -99° Oil
	Tet., 1967, 1997.			Chem. Comm., 1968, 1070.	
4119-001			4119-005		
DEBILONE	$C_{15}H_{22}O_2$	234 135°C	FERULENE, α-	$C_{15}H_{24}$	204 +68° Oil
	Coll. Czech. Chem. Comm., 1970, 745.			Tet. Lett., 1965, 3017.	
4119-002			4119-006		
ARISTOLEN-2-ONE, 1(10)-	$C_{15}H_{22}O$	218	GURJUNENE, β-	$C_{15}H_{24}$	204 +44° Oil
	Chem. Abstr., 1969 (70) 47620.			Tet., 1964, 963.	
4119-003			4119-007		
ARISTOLA-1(10), 8-DIEN-2-ONE	$C_{15}H_{20}O$	216	ARISTOLONE	$C_{15}H_{24}O$	220 -339° 101°C
	Chem. Abstr., 47620 (70) 1969			Tet. Lett., 1962, 827.	
4119-004			4119-008		

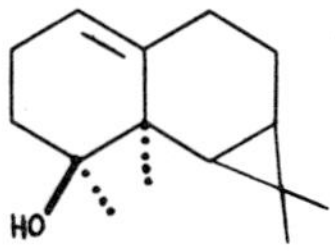

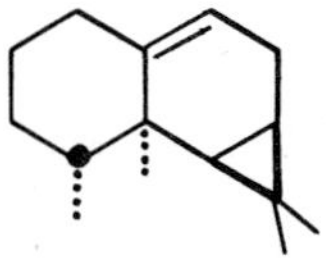

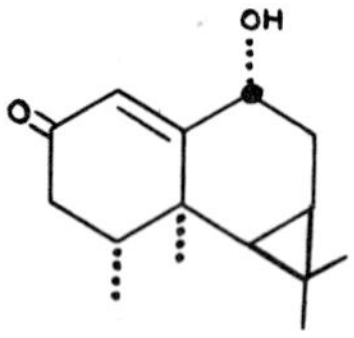

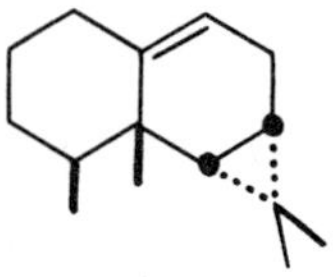

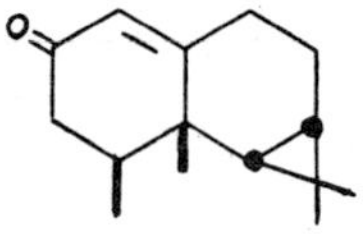

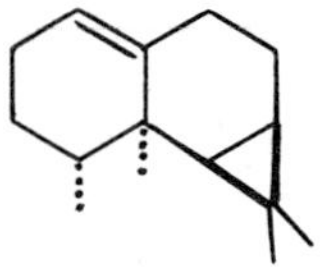

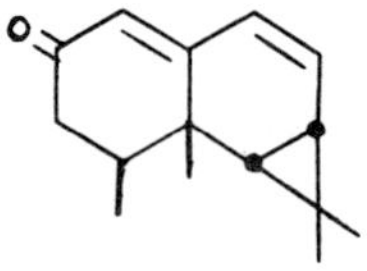

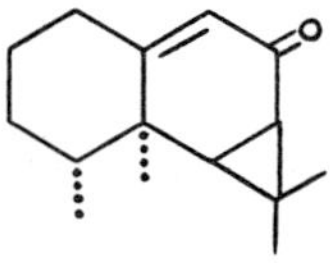

SELINENE, α- — $C_{15}H_{24}$ — 204, $+63^0$, Oil

Tet., 1960 (8) 181

4120-001

SELINENE, β- — $C_{15}H_{24}$ — 204, $+38^0$, Oil

Tet. 1960 (8) 181

4120-002

SELINENE, γ- — $C_{15}H_{24}$ — 204, $+3^0c$, Oil

Chem. Ind., 1951 312

4120-003

SELINA-4, 11-DIENE — $C_{15}H_{24}$ — 204, $+32^0$ m, Oil

Chem. Pharm. Bull., 1967 903

4120-004

VETISELINENE — $C_{15}H_{24}$ — 204

Tet. Lett. 1970 1759

4120-005

SIBIRENE — $C_{15}H_{24}$ — 204, $+120^0$, Oil

Chem. Abstr., 1967 18771

4120-006

EUDESMOL, α- — $C_{15}H_{26}O$ — 222, $+33^0c$, 81^0C

Coll. Czech. Chem. Comm. 1961 2045

4120-007

EUDESMOL, β- — $C_{15}H_{26}O$ — 222, $+68^0c$, 82^0C

JACS 1955 3646

4120-008

EUDESMOL, γ - — $C_{15}H_{26}O$ — 222, $+67^0$, 74^0C

Chem. Ind. 1962 1759

4120-009

OCCIDENTALOL — $C_{15}H_{24}O$ — 220, $+363^0c$, 98^0C

JOC 1969 736

4120-010

VERBESINOL, α-

$C_{15}H_{26}O$ 222
-150^0
135^0C

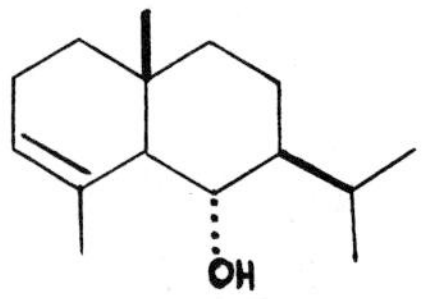

JACS 1961 1511

4120-011

JUNENOL, d(+)-

$C_{15}H_{26}O$ 222
$+52^0c$
60^0C

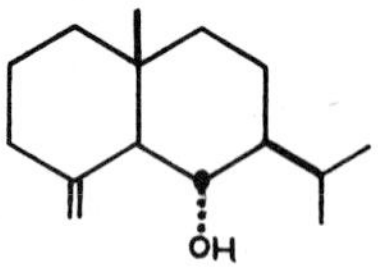

Tet. 1964 2596.

4120-012

JUNENOL, L-(-)-

$C_{15}H_{26}O$ 222
-57^0e
65^0C

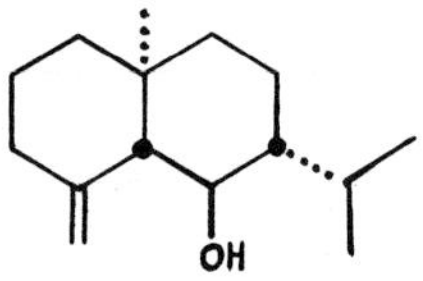

Tet. 1962 969

4120-013

COSTOL, α-

$C_{15}H_{24}O$ 220
$+14^0c$
Oil

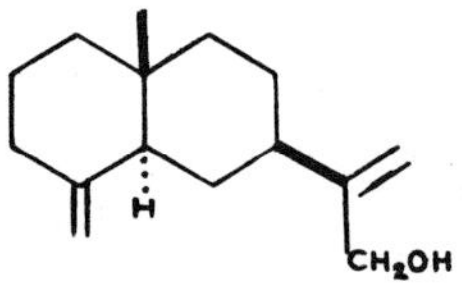

Tet. 1967 1993

4120-014

VETISELINENOL

$C_{15}H_{24}O$ 220
-18^0c

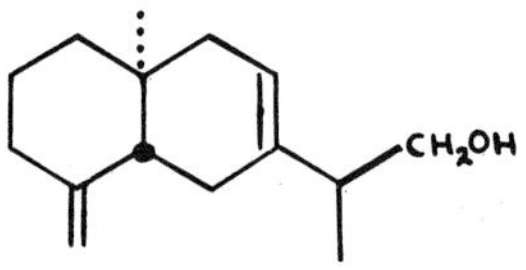

Tet. Lett., 1970 231

4120-015

AGAROL

$C_{15}H_{26}O$ 222
-22^0c
Oil

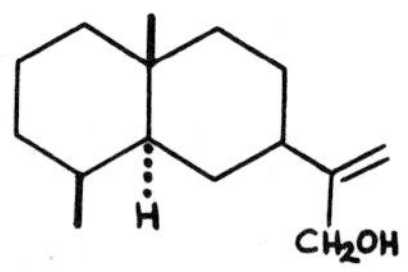

Tet. Lett. 1959 (13) 7

4120-016

INTERMEDEOL

$C_{15}H_{26}O$ 222
$+11^0e$
48^0C

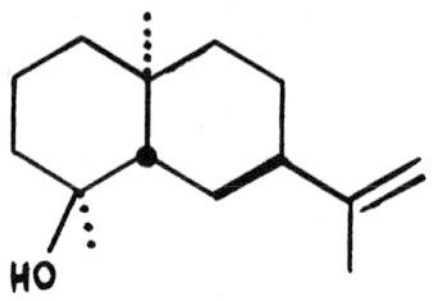

Tet. Lett., 1968
3223

4120-017

INTERMEDEOL, neo-

$C_{15}H_{26}O$ 222
$+8^0e$
Oil

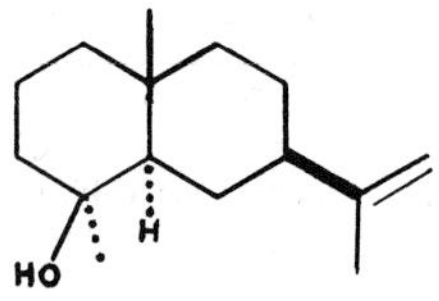

Chem. Ind. 1964 194

4120-018

SELIN-11-EN-4α-OL

$C_{15}H_{26}O$ 222
-18^0
94^0C

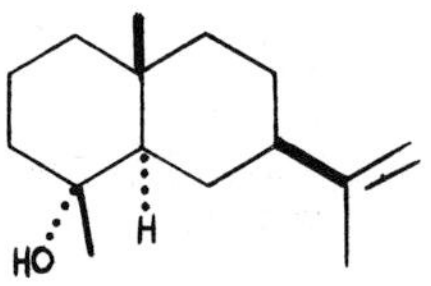

Tet. Lett., 1968

4120-019

CAMPHOR, (+)-Juniper-

$C_{15}H_{26}O$ 222
0^0
166^0C

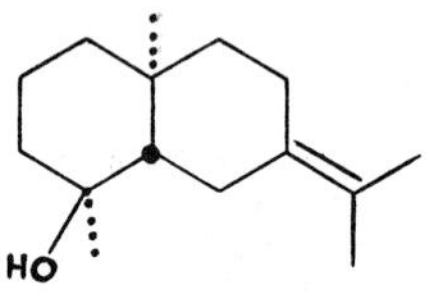

Tet. Lett., 1968 3223

4120-020

CYPEROL $C_{15}H_{24}O$ 220 +131° 112°C Chem. Pharm. Bull., 1967 1929 4120-021	CANARONE $C_{15}H_{24}O$ 220 +34°c Oil (?) Tet. 1965 3197 4120-026
CYPEROL, iso- $C_{15}H_{24}O$ 220 -5° Oil Chem. Pharm. Bull., 1967 1929 4120-022	ROTUNOL, α- $C_{15}H_{22}O_2$ 234 88°C Tet. Lett., 1969 2741 4120-027
PTEROCARPOL $C_{15}H_{26}O_2$ 238 +39°c 105°C Tet., 1968 6231 4120-023	ROTUNOL, β- $C_{15}H_{22}O_2$ 234 118°C Tet. Lett., 1969 2741 4120-028
CRYPTOMERIDIOL $C_{15}H_{28}O_2$ 240 -22°c 137°C Chem. Pharm. Bull. 1961 619 4120-024	CYPERONE, D-α- $C_{15}H_{22}O$ 218 +138° Oil JCS 1955 528; 2423 4120-029
OPLODIOL $C_{15}H_{26}O_2$ 238 - 52° 106°C JCS 1967 423 4120-025	CYPERONE, β- $C_{15}H_{22}O$ 218 +239° Oil JCS 1955 525 4120-030

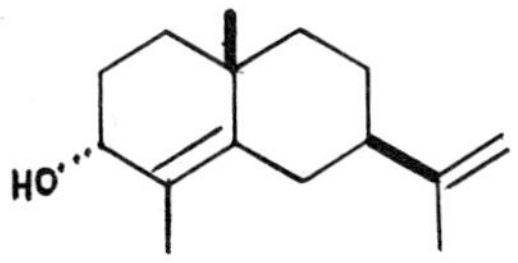

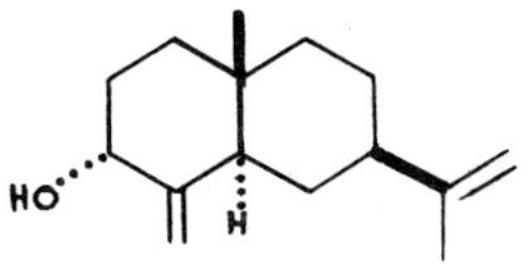

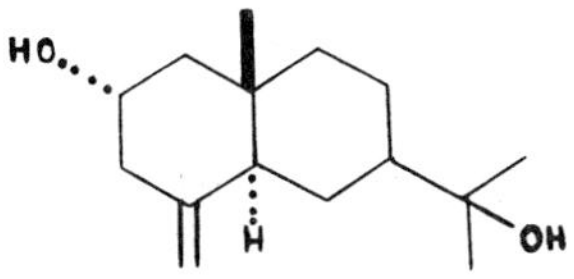

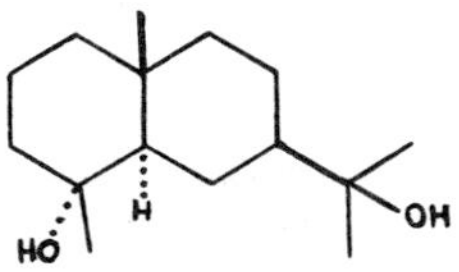

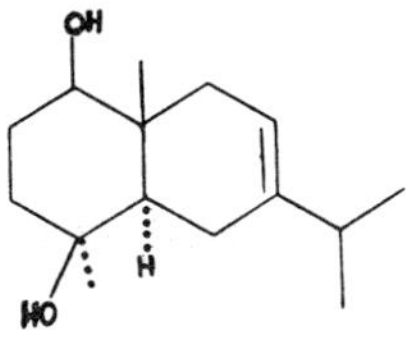

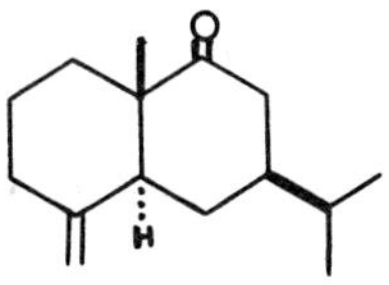

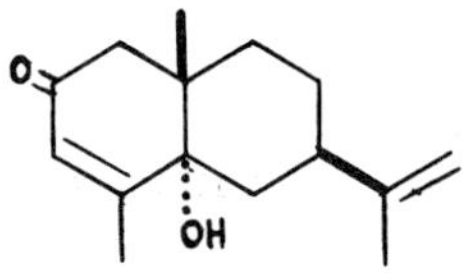

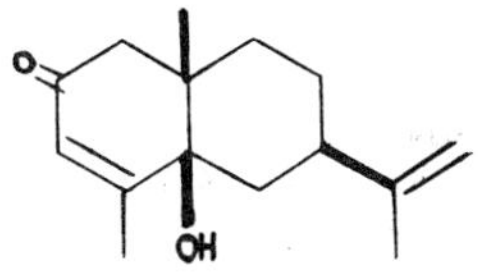

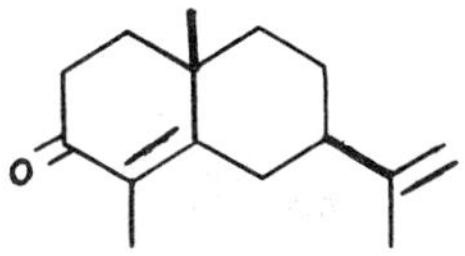

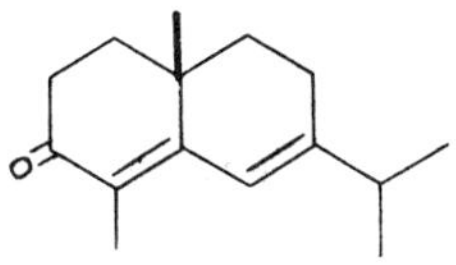

CARISSONE $C_{15}H_{24}O_2$ 236 +137°c 78°C JCS 1955 3027 4120-031	AGAROFURAN, dihydro- $C_{15}H_{26}O$ 222 -77°c Oil Tet. 1963 1079 4120-036
COSTIC ACID $C_{15}H_{22}O_2$ 234 +23°c 88°C JOC 1966 1632 4120-032	AGAROFURAN, 4-OH-dihydro- $C_{15}H_{26}O_2$ 238 -76°c 130°C Tet. 1963 1519 4120-037
ILICIC ACID $C_{15}H_{24}O_3$ 252 - 35° 177°C JOC 1966 1632 4120-033	AGAROFURAN, 3, 4-diOH-dihydro- $C_{15}H_{26}O_3$ 254 -41°c 176°C Tet. 1963 1519 4210-038
AGAROFURAN, α- $C_{15}H_{24}O$ 220 +37°c Oil Tet. 1963 1079 JACS., 1967, 5665. 4120-034	CURCOLONE $C_{15}H_{18}O_3$ 246 +14°c 140°C Chem. Pharm. Bull., 1968 827 4120-039
AGAROFURAN, β- $C_{15}H_{24}$ 220 -127°c Oil Tet. 1963 1079 4120-035	LINDESTRENE $C_{15}H_{18}O$ 214 - 64° Oil Tet. 1964 2655 4120-040

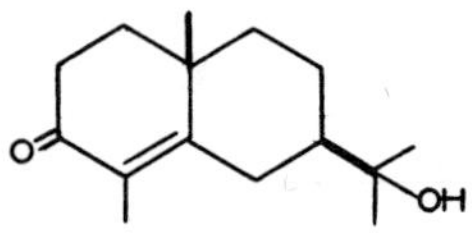

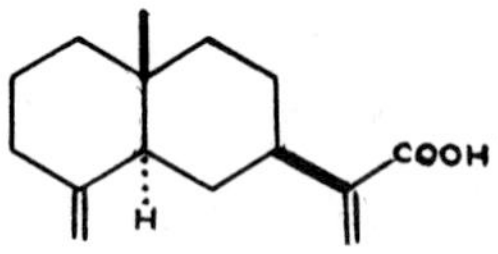

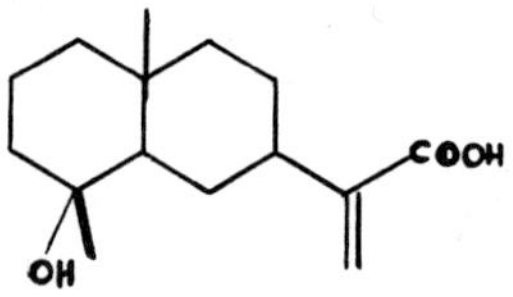

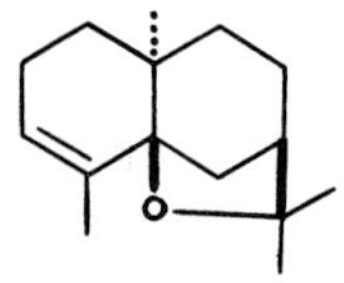

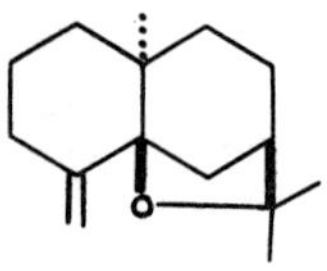

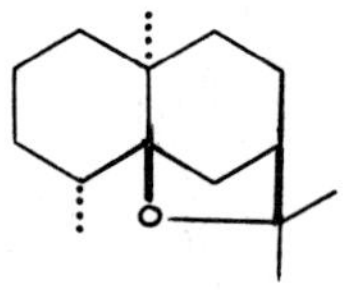

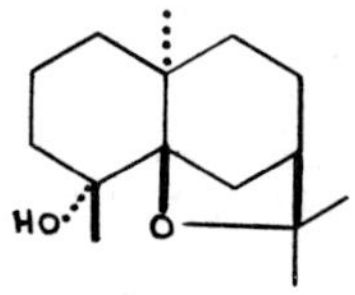

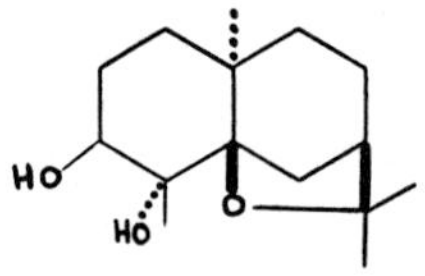

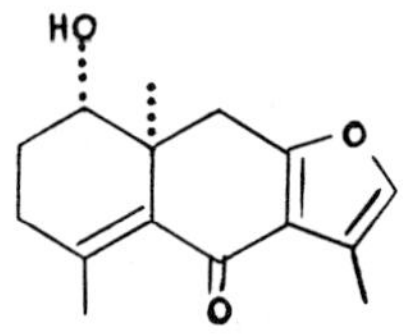

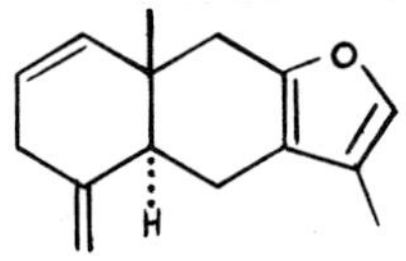

ATRACTYLONE	$C_{15}H_{20}O$	216 +40°c 38°C
Chem. Pharm. Bull. Jap. 1962 641		
4120-041		
ENCELIN	$C_{15}H_{16}O_3$	244 -16°c 195°C
JOC 1968 656		
4120-046		
ALANTOLACTONE	$C_{15}H_{20}O_2$	232 +175°c 75°C
JCS 1963 534		
4120-042		
FARINOSIN	$C_{15}H_{18}O_4$	262 -111°c 200°C
JOC 1968 3743		
4120-047		
ISOALANTOLACTONE	$C_{15}H_{20}O_2$	232 +172°c 115°C
JOC 1964 3727		
4120-043		
PINNATIFIDINE	$C_{15}H_{18}O_3$	246 +302°e 165°C
JOC 1962 4041		
4210-048		
ISOALANTOLACTONE , dihydro-	$C_{15}H_{22}O_2$	234 +22°e 173°C
JOC 1965 726		
4120-044		
YOMOGIN	$C_{15}H_{16}O_3$	244 -88°c 202°C
JOC 1966 2523		
4120-049		
LINDESTRENOLIDE	$C_{15}H_{18}O_2$	230 +189° Oil
JCS 1968 569		
4120-045		
LINDESTRENOLIDE, OH-	$C_{15}H_{18}O_3$	246 +210° 220°C
JCS 1968 569		
4120-050		

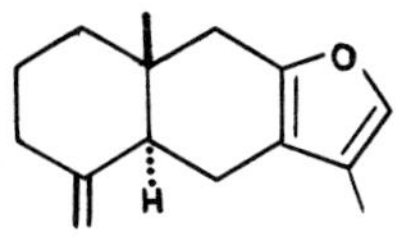

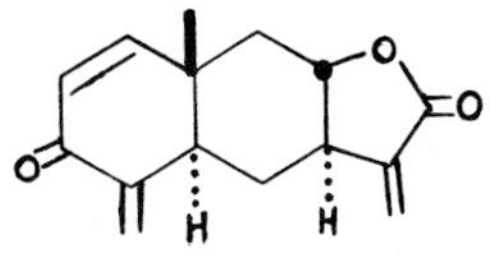

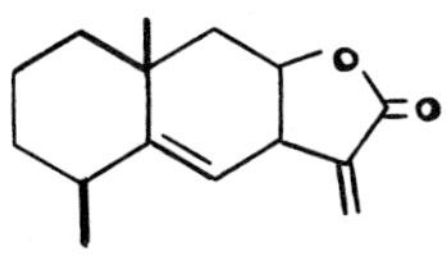

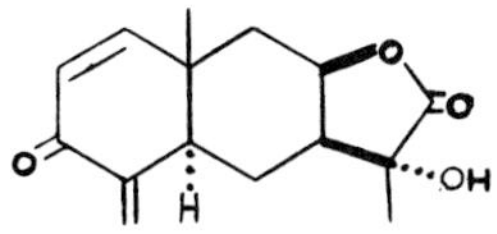

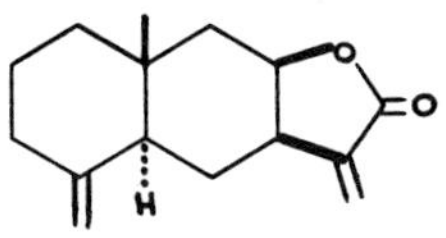

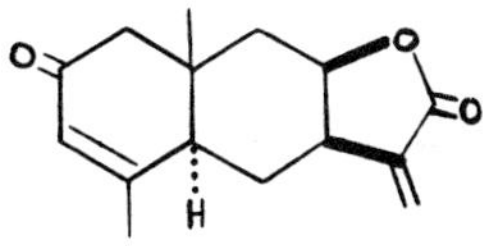

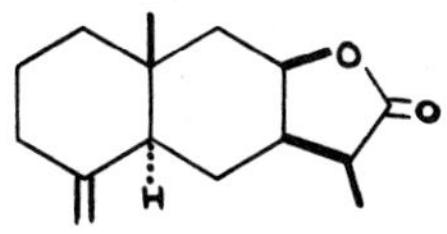

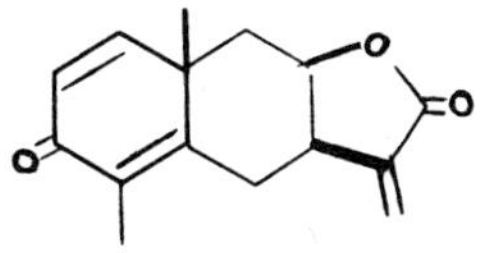

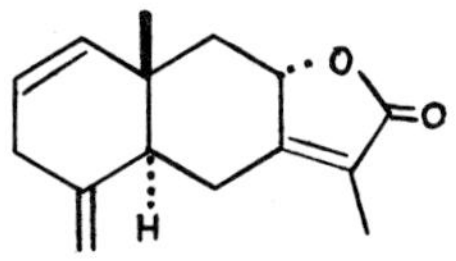

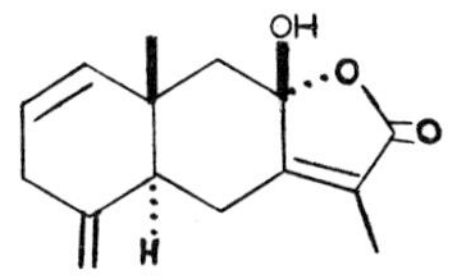

PULCHELLIN B

$C_{17}H_{22}O_5$

306
$+93^0$
216^0c

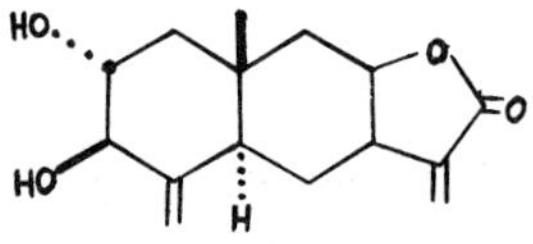

JOC 1970 627

4120-051

PULCHELLIN C

$C_{15}H_{20}O_4$

264
$+125^0$
200^0C

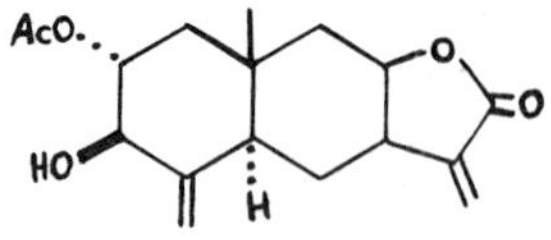

JOC 1970 627

4120-052

IVASPERIN

$C_{15}H_{20}O_4$

264
$+141^0$m
150^0C

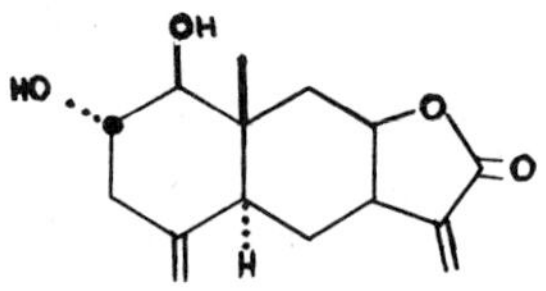

JOC 1964 1022

4120-053

MICROCEPHALIN

$C_{15}H_{22}O_4$

266
$+75^0$c
206^0C

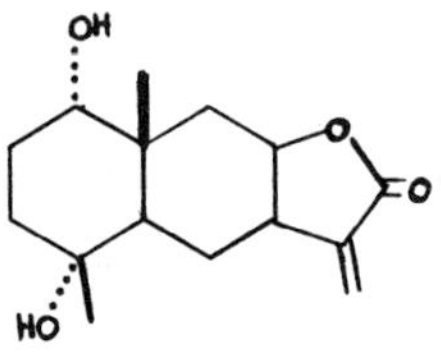

JOC 1964 1700

4120-054

PYRETHROSIN, β-cyclo-

$C_{17}H_{22}O_5$

306
$+58^0$m
166^0C

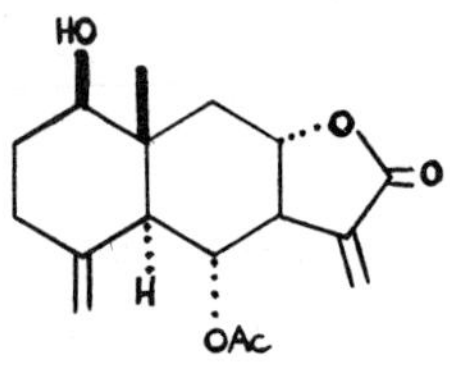

Can. J. C., 1969 1139

4120-055

IVALIN

$C_{15}H_{20}O_3$

248
$+142^0$c
131^0C

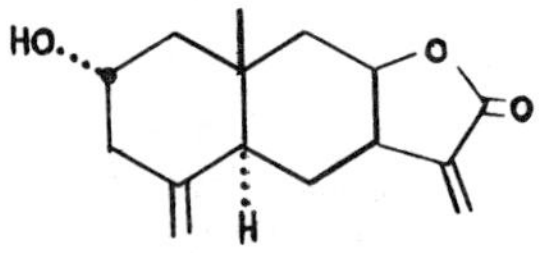

JOC 1962 905

4120-056

ASPERILIN

$C_{15}H_{20}O_3$

248
$+150^0$c
151^0C

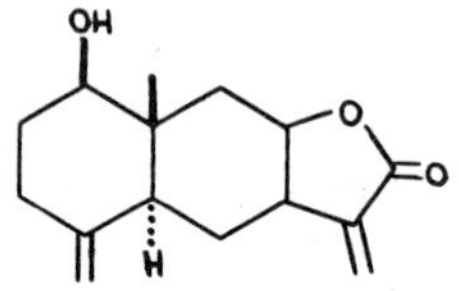

JOC 1964 1022

4120 057

IVANGUSTIN

$C_{15}H_{20}O_3$

248
$+85^0$
121^0C

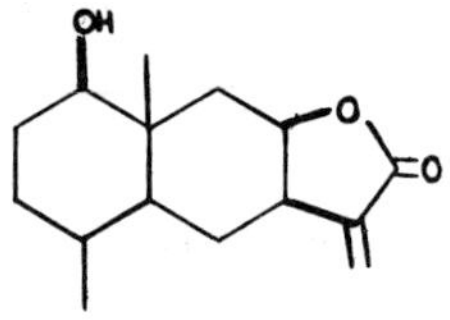

JOC 1967 3658

4120-058

TELEKINE

$C_{15}H_{20}O_3$

248
$+234^0$c
160^0C

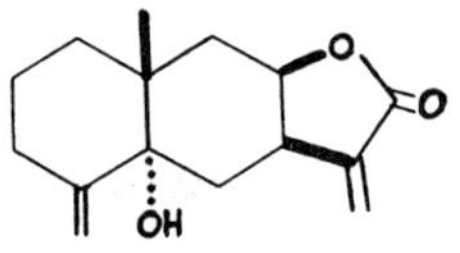

Coll. Czech. Chem. Comm., 1961 1350

4120-059

TELEKINE, iso-

$C_{15}H_{20}O_3$

248
$+100^0$c
144^0C

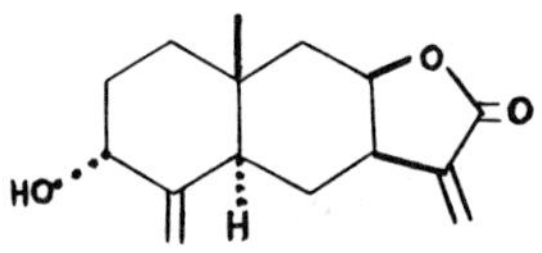

Coll. Czech. Chem. Comm., 1961 1350

4120-060

Compound	Formula	Data	Reference	No.
TELEKIN, 3-epi-iso-	$C_{15}H_{20}O_3$	248 +157° 176°C	JOC 1968 3743	4120-061
SANTONIN, β-	$C_{15}H_{18}O_3$	246 -137°c 217°C	JACS 1955 (77) 1044	4120-066
SANTAMARINE	$C_{15}H_{20}O_3$	248 +96° 135°C	Tet. 1965 1741	4120-062
SANTONIN, ψ -	$C_{15}H_{20}O_4$	264 -169°c 183°C	JCS 1963 534	4120-067
DOUGLANINE	$C_{15}H_{20}O_3$	248 +133°c 116°C	Tet. Lett. 1967 2159	4120-063
SANTONIN, desoxy-ψ -	$C_{15}H_{20}O_3$	248 -207°e 101°C	JCS 1963 534	4120-068
MIBULACTONE	$C_{15}H_{22}O_4$	266 +156° 228°C	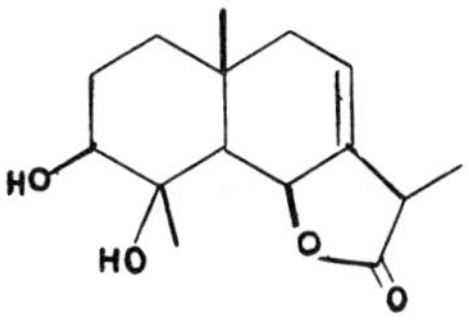	4120-064
FINITIN	$C_{15}H_{20}O_3$	248 -168°c 153°C	JCS 1963 534	4120-069
SANTONIN, α-	$C_{15}H_{18}O_3$	246 -173°e 170°C	Tet. Lett., 1963 275	4120-065
SANTONIN, 11-OH-	$C_{15}H_{18}O_4$	262 -134°e 286°C	Ber. 1937 (70) 812	4120-070

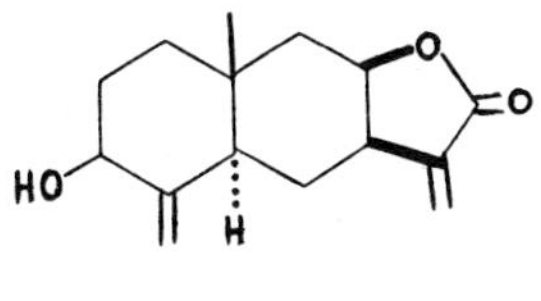

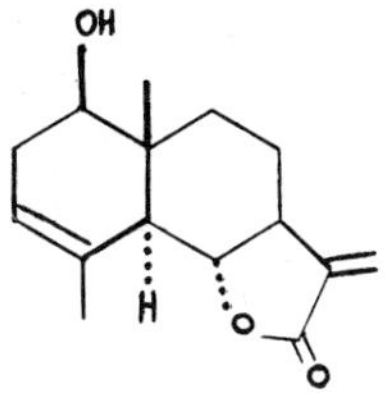

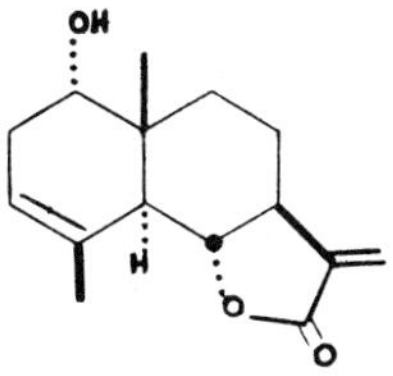

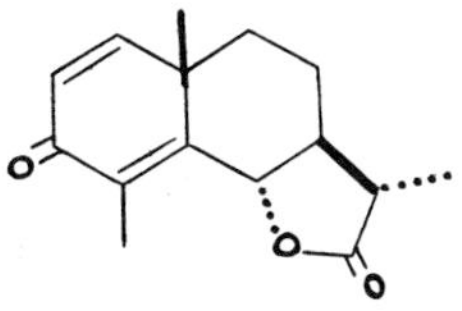

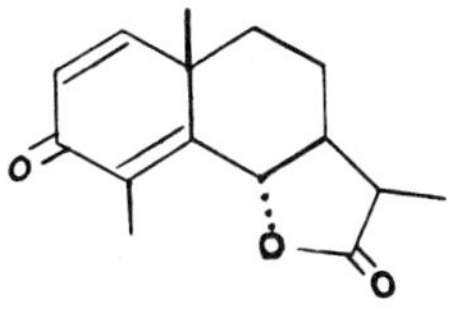

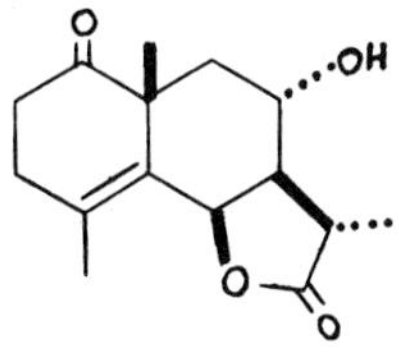

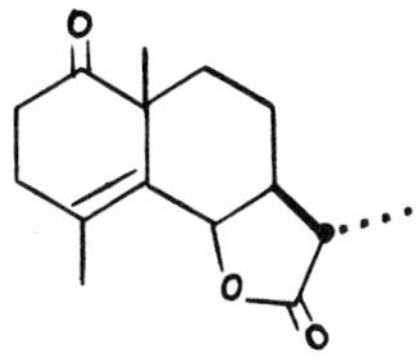

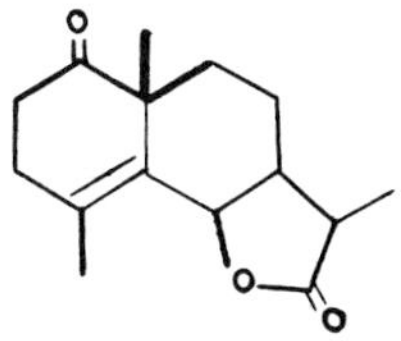

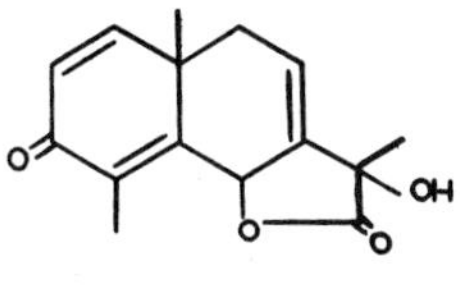

TUBERIFERINE $C_{15}H_{18}O_3$ 246 +9° 161°C Tet. Lett., 1967 3475 4120-071	SELINADIENE, 3,7(11)- $C_{15}H_{24}$ 204 Oil Chem. Ind., 1966, 1225. 4120-076
ARTEMISIN $C_{15}H_{18}O_4$ 262 -84° 203°C Proc. Chem. Soc. 1962 112 4120-072	DICTYOPTEROL $C_{15}H_{24}O$ 220 -31°t Oil Chem. Abstr., 1967, 55600. 4120-077
ARGLANINE $C_{15}H_{18}O_4$ 262 +111° 207°C Tet. Lett. 1967 (21) 2013 4120-073	DICTYOPTERONE $C_{15}H_{22}O$ 218 -12°t Chem. Abstr., 1967, 55600. 4120-078
VULGARIN $C_{15}H_{20}O_4$ 264 + 49° 175°C Tet., 1965 (21) 1231 4120-074	COSTAL, α- $C_{15}H_{22}O$ 218 +24° Oil Chem. Pharm. Bull. 1967,903 4120-079
ARTICALIN $C_{15}H_{20}O_4$ 264 +45°c 226°C Phytochem. 1969 1297 4120-075	COSTAL, β- $C_{15}H_{22}O$ 218 Tet. Lett., 1965, 3777. 4120-080

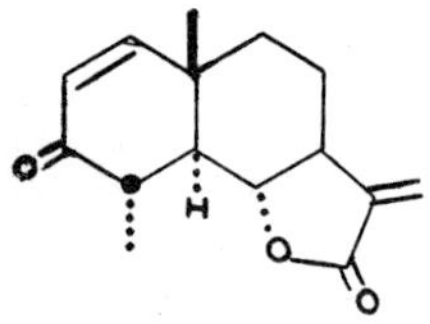

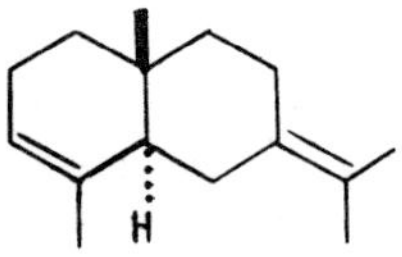

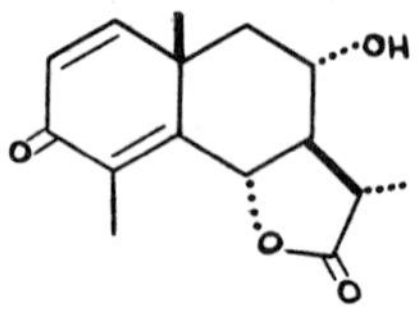

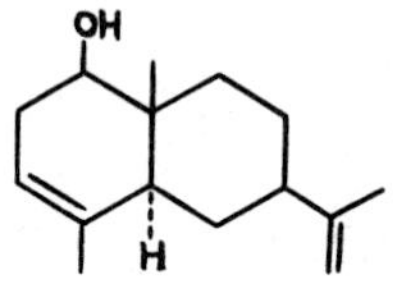

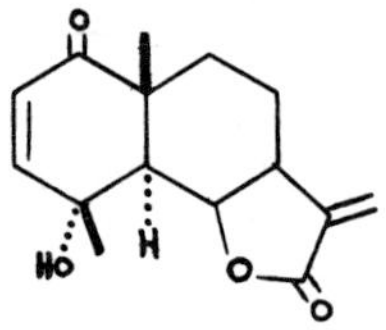

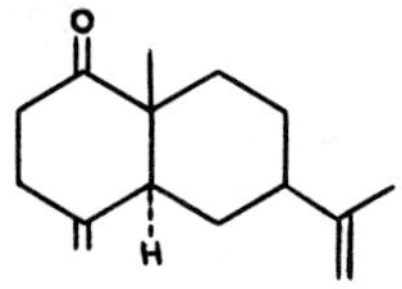

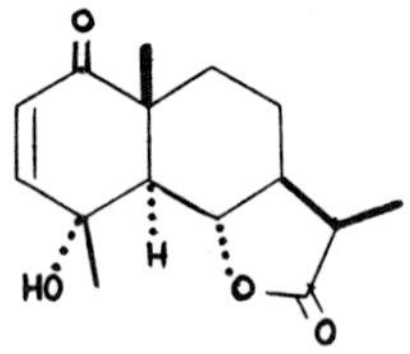

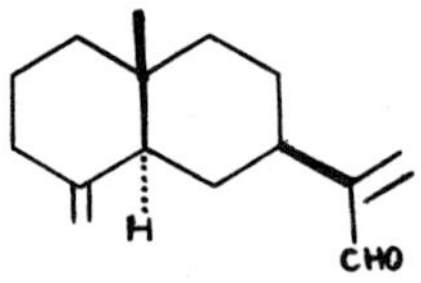

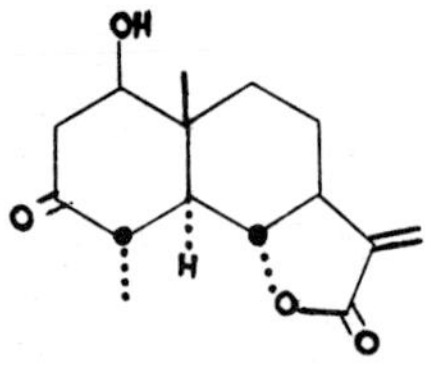

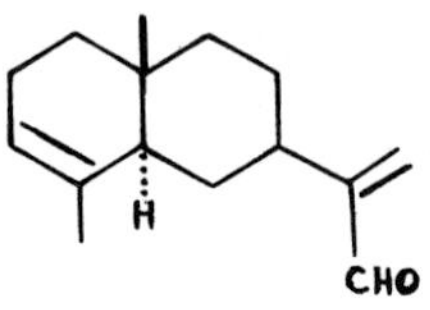

TUBIFERINE $C_{15}H_{18}O_3$ 246 $+9^0c$ 161^0C Chem. Abstr., 1968 (69) 27548. 4120-081	SELINA-4(14), 7(11)-DIENE $C_{15}H_{24}$ 204 Oil Chem. Ind., 1966, 1225. 4120-086
PULCHELLIN E $C_{17}H_{22}O_5$ 306 JOC, 1970, 627. 4120-082	LUDOVICIN-A $C_{15}H_{20}O_4$ 264 $+128^0c$ 215^0C Phytochemistry, 1970, 403. 4120-087
PULCHELLIN F $C_{20}H_{26}O_5$ 346 JOC, 1970, 627. 4120-083	LUDOVICIN-B $C_{15}H_{20}O_4$ 264 $+138^0c$ 152^0C Phytochemistry, 1970, 403. 4120-088
MAYTINE $C_{29}H_{37}NO_{12}$ 591 R = H JACS., 1970, 6667. 4120-084	LUDOVICIN-C $C_{15}H_{18}O_4$ 262 $+95^0c$ 194^0C Phytochemistry, 1970, 403. 4120-089
MAYTOLINE $C_{29}H_{37}NO_{13}$ 607 R = OH JACS., 1970, 6667. 4120-085	REYNOSIN $C_{15}H_{20}O_3$ 248 $+180^0c$ 146^0C Phytochem., 1970, 823. 4120-090

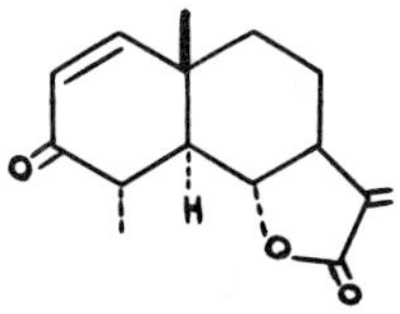

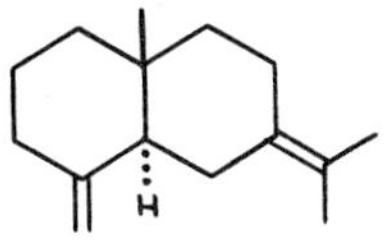

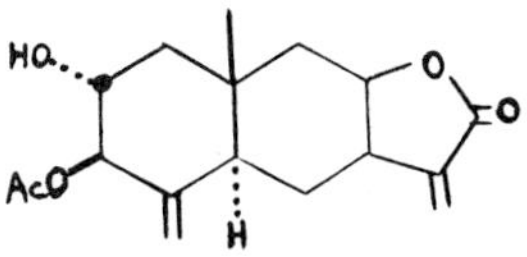

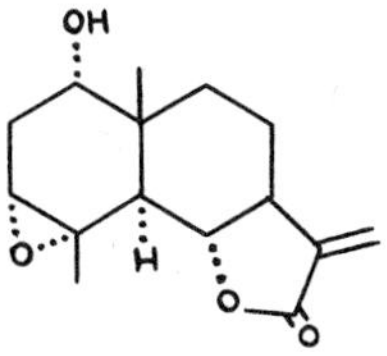

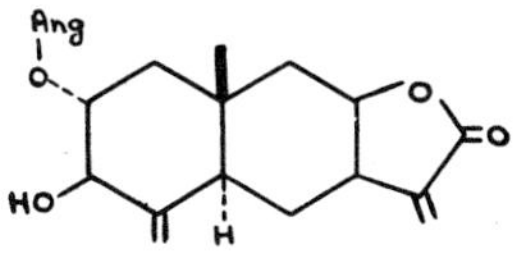

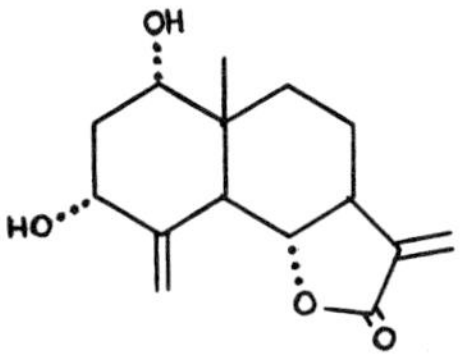

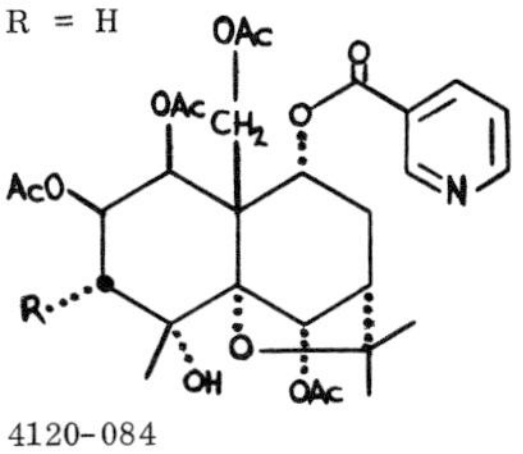

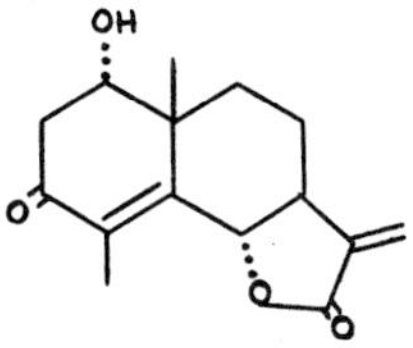

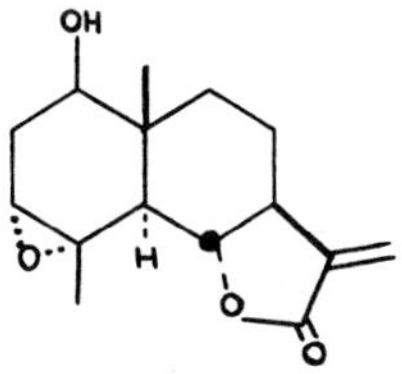

SANTAMARINE, epoxy- $C_{15}H_{20}O_4$ 264
+87°c
243°C

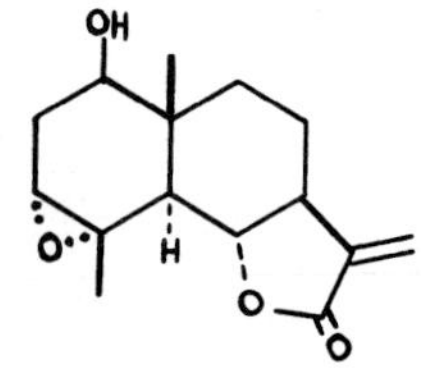

Phytochem., 1970, 823.

4120-091

ELEMENE, d-α- | $C_{15}H_{24}$ | 204 | +116°c | Oil

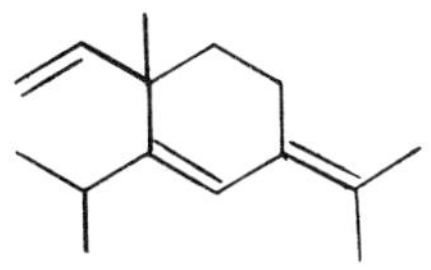

Tet. 1964 [18] 1509

4121-001

ELEMENOL, epi-δ- | $C_{15}H_{24}O$ | 220 | + 184°m

OH

Tet. Lett. 1968 2901

4121-006

ELEMENE, d-β- | $C_{15}H_{24}$ | 204 | +14° | Oil

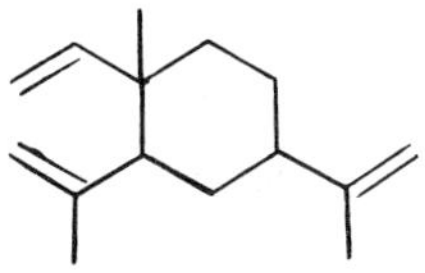

Bull. Chem. Soc. Jap. 1964 (37) 1053

4121-002

ELEMOL | $C_{15}H_{26}O$ | 222 | -10°b | 49°C

H OH

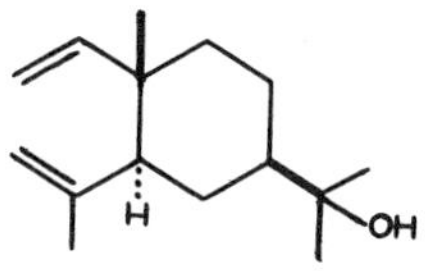

Chem. Abstr. 1955 (49) 14699

4121-007

ELEMENE, l-β- | $C_{15}H_{24}$ | 204 | -17° | Oil

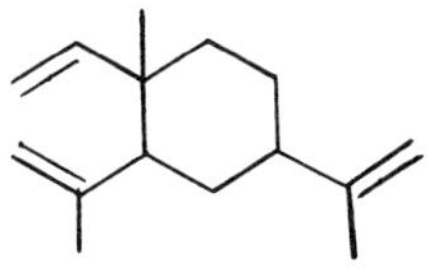

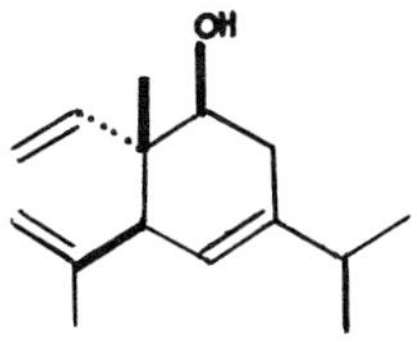

Chem. Abstr. 1956 (50) 3320

4121-003

ELEMENONE, α- | $C_{15}H_{22}O$ | 218 | -3°neat | Oil

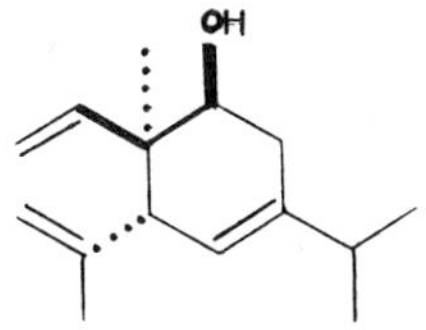

O

4121-008

ELEMENE, δ- | $C_{15}H_{24}$ | 204 | Oil

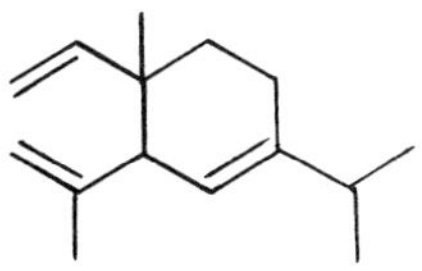

Aust. J. Chem. 1964 (17) 1270

4121-004

ELEMENONE, β- | $C_{15}H_{22}O$ | 218 | Oil

O

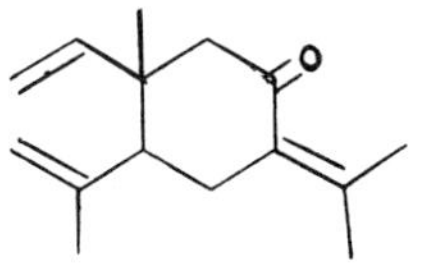

Coll. Czech. Chem. Comm. 1959 2371

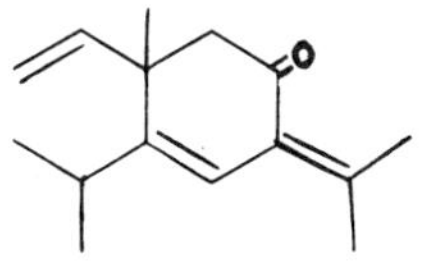

4121-009

ELEMENOL, δ- | $C_{15}H_{24}O$ | 220 | -13°m

OH

Tet. Lett. 1968 2901

4121-005

ELEMENONE, - | $C_{15}H_{22}O$ | 218 | Oil

O

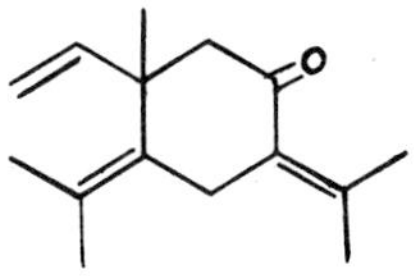

4121-010

SHYOBUNONE $C_{15}H_{24}O$ 220 Tet. Lett. 1968 5315 4121-011	GERMAFURENOLIDE, OH-iso- $C_{15}H_{20}O_3$ 248 +5^0 160^{0}C JCS 1968 569 4121-016
SHYOBUNONE, epi- $C_{15}H_{24}O$ 220 Tet. Lett. 1968 5315 4121-012	GERMAFURENOLIDE, iso- $C_{15}H_{20}O_2$ 232 +5^0 85^{0}C JCS 1968 569 4121-017
SHYOBUNONE, iso- $C_{15}H_{24}O$ 220 Tet. Lett. 1968 5315 4121-013	SERICEALACTONE $C_{16}H_{20}O_5$ 292 150^{0}C Tet. Let. 1968 2647 4121-018
ELEMENAL $C_{15}H_{22}O$ 218 -11^0 c Oil Tet. Lett. 1965 3777 4121-014	SERICEALACTONE, desoxy- $C_{16}H_{20}O_4$ 276 137^{0}C Tet. Lett. 1968 2647 4121-019
GERMACRENE, iso-furano- $C_{15}H_{20}O$ 216 0^0 Tet. Lett. 1968 2855 4121-015	CURZERENONE $C_{15}H_{18}O_2$ 230 0^0 Tet. Lett. 1968 931 4121-020

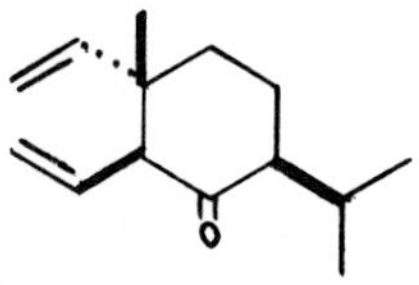

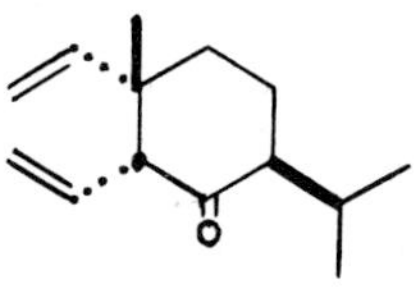

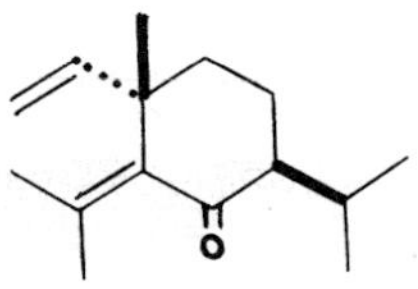

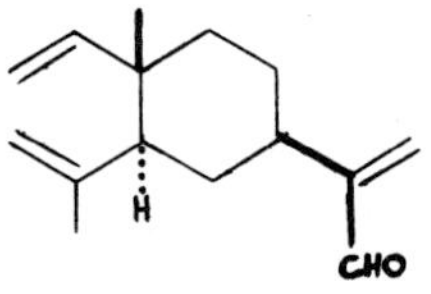

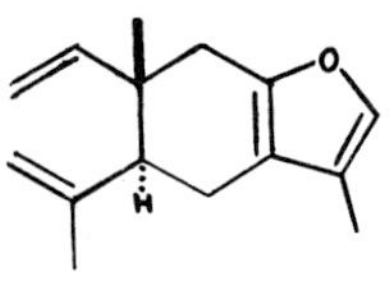

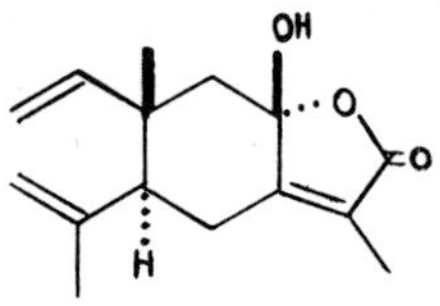

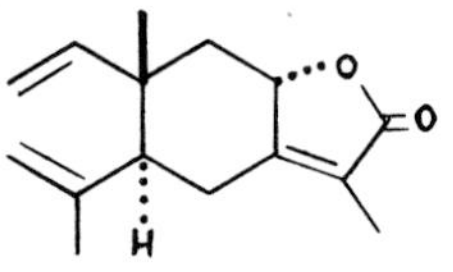

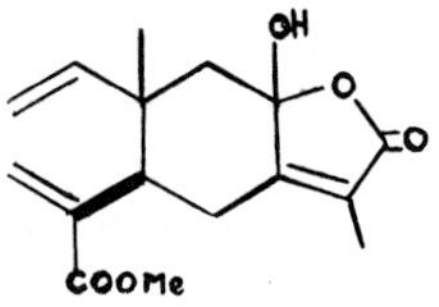

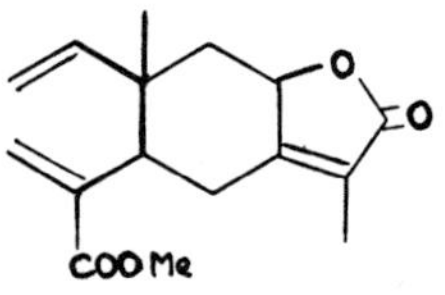

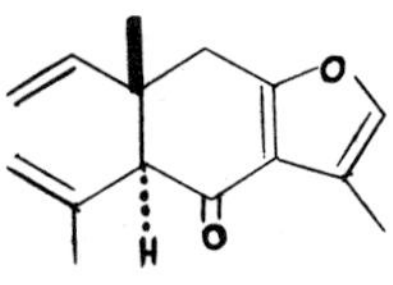

CURZERENONE, epi

$C_{15}H_{18}O_2$ 230 0°

Tet. Lett. 1968 931

4121-021

SERICENINE

$C_{16}H_{20}O_3$ 260 0° Oil

Tet. Lett. 1968 1999

4121-026

TEMISINE

$C_{15}H_{22}O_3$ 250 +70°c 228°C

Ber. 1941 (74B) 952
Chem. Abstr. 1949 (34) 5427

4121-022

MISCANDENIN

$C_{15}H_{14}O_5$ 274 -181° 233°C

JOC., 1970, 1453.

4121-027

SAUSSUREA LACTONE

$C_{15}H_{22}O_2$ 234 +66°c 148°C

Tet. 1963 1061

4121-023

OCCIDENOL

$C_{15}H_{24}O_2$ 236 -13°c Oil

Tet. Lett., 1970, 235.

4121-028

VERNOLEPIN

$C_{15}H_{16}O_5$ 276 +72° 181°C

JACS 1968 3596

4121-024

ZEDOARONE

$C_{15}H_{18}O_2$ 230 Oil

Chem. Abstr.
1968 (69) 77524

4121-029

LINDERALACTONE, iso-

$C_{15}H_{16}O_3$ 244 -225°d 120°C

JCS, 1969 1491

4121-025

VERNOMENIN

$C_{15}H_{16}O_5$ 276 -62° a amorph

JACS., 1968, 3596.

4121-030

VETISPIRENE, α- $C_{15}H_{22}$ 202 +220[0] Oil Tet. Lett., 1970, 1759. 4122-001	AGAROSPIROL $C_{15}H_{26}O$ 222 -5[0] Oil Tet., 1965, 115. 4122-005
VETISPIRENE, β- $C_{15}H_{22}$ 202 -90[0] Oil Tet. Lett., 1970, 1760. 4122-002	HINESOL $C_{15}H_{26}O$ 222 -48[0]c 60[0]C Coll. Czech. Chem. Comm., 1962, 1914. 4122-006
VETIVENENE, α-iso- $C_{15}H_{22}$ 202 -120[0] Oil JACS, 1967, 2750. 4122-003	VETIVONE, β- $C_{15}H_{22}O$ 218 -39[0]e 44[0]C JOC, 1970, 192. 4122-007
VETIVENENE, β-iso- $C_{15}H_{22}$ 202 -68[0] Oil Coll. Czech. Chem. Comm., 1960 (25) 2540. 4122-004	

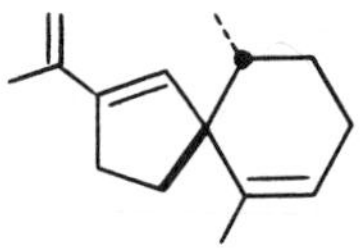

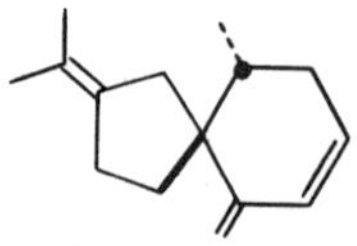

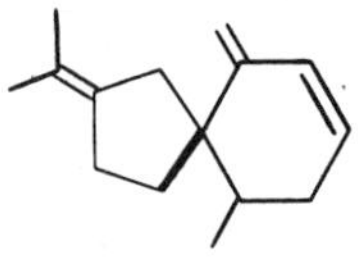

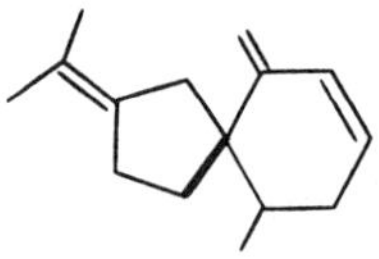

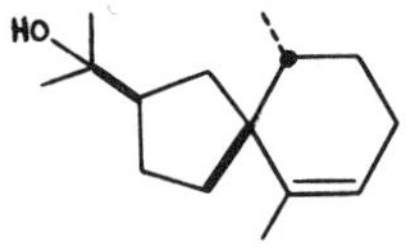

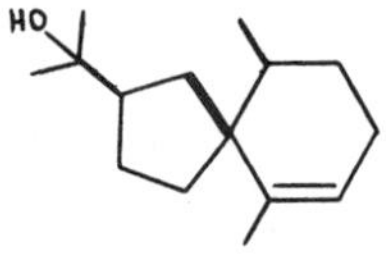

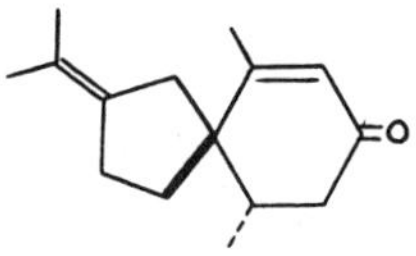

MAALIENE, (+)-α- $C_{15}H_{24}$ 204 +92° Oil Coll. Czech. Chem. Comm., 1967 (32) 808. 4123-001	MAALIENE, (+)-γ- $C_{15}H_{24}$ 204 +11° h Oil H Chem. Comm., 1968, 1070. 4123-003
MAALIENE, β- $C_{15}H_{24}$ 204 -136°c Oil Tet. Lett., 1962, 827. 4123-002	MAALIOL $C_{15}H_{26}O$ 222 +39°c 104°C OH Tet., 1964, 963. 4123-004

KHUSENE $C_{15}H_{24}$ 204 + 13° Oil Tet. Lett., 1968, 2497. 4124-001	ZIZANOIC ACID, epi- $C_{15}H_{22}O_2$ 234 110°C COOH Tet. Lett., 1968, 6098. 4124-005
KHUSENOL $C_{15}H_{24}O$ 220 + 30°c Oil OH Chem. Comm., 1969, 999. 4124-002	KHUSENIC ACID, iso- $C_{15}H_{22}O_2$ 234 COOH Tet. Lett., 1968, 2497. 4124-006
ZIZANOL $C_{15}H_{24}O$ 220 + 10° Oil HO Tet. Lett., 1970, 1755. 4124-003	TRICYCLOVETIVENOL, iso- $C_{15}H_{24}O$ 220 +29° Oil CH_2OH Perf. Ess. Oil Rec. 1969 60 314. 4124-007
KHUSENIC ACID $C_{15}H_{22}O_2$ 234 + 40° Oil COOH JOC, 1968, 1771. 4124-004	

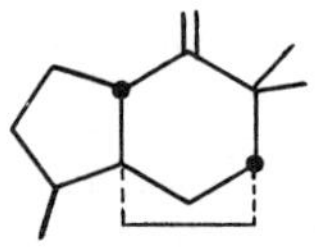

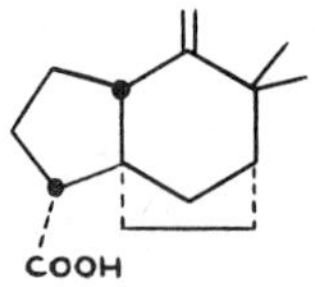

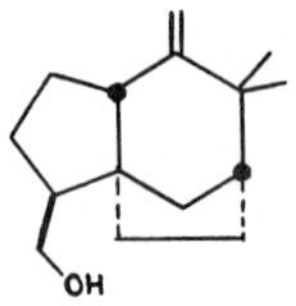

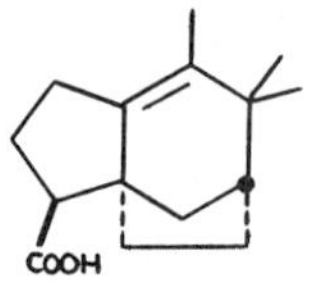

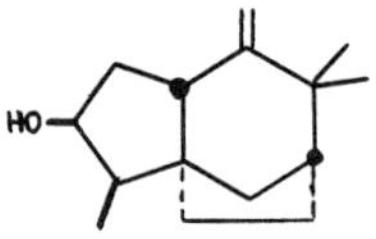

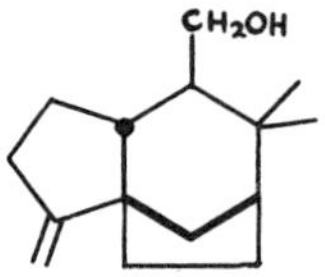

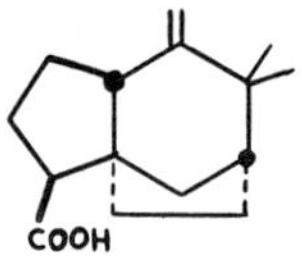

Name	Formula	Data	Reference	No.
AROMADENDRENE, D-	$C_{15}H_{24}$	204 + 25^0 Oil	JACS, 1966 (88) 4113.	4125-001
LEDOL, D-	$C_{15}H_{26}O$	222 + 5^{0}c 105^{0}C	Coll. Czech. Chem. Comm., 1960, 1837.	4125-006
AROMADENDRENE, Allo-	$C_{15}H_{24}$	204 -22^0 Oil	JACS, 1966 (88) 4113.	4125-002
PALUSTROL, L-	$C_{15}H_{26}O$	222 -17^0 Oil	Coll. Czech. Chem. Comm., 1961, 811.	4125-007
GURJUNENE, α-L-	$C_{15}H_{24}$	204 -180^0 Oil	Bull. Soc. Chim. Fr., 1963, 1950.	4125-003
SPATHULENOL, D-	$C_{15}H_{24}O$	220 + 56^0 Oil	Chem. Ind., 1963, 1245.	4125-008
GLOBULOL, L-	$C_{15}H_{26}O$	222 -43^{0}c 88^{0}C	Coll. Czech. Chem. Comm., 1960,1837.	4125-004
CYCLOCOLORENONE	$C_{15}H_{22}O$	218 -400^0 e Oil	JACS, 1966, 3403.	4125-009
VIRIDIFLOROL, D-	$C_{15}H_{26}O$	222 + 4^{0}c 74^{0}C	Coll. Czech. Chem. Comm., 1960, 1837.	4125-005

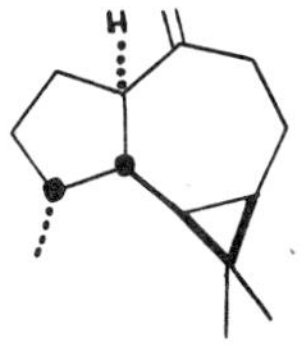

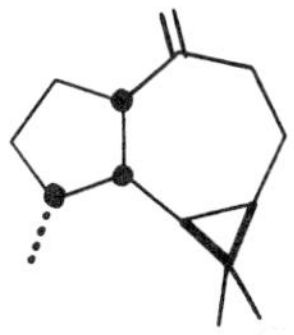

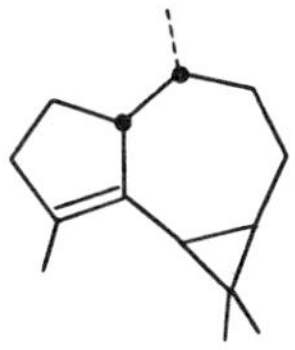

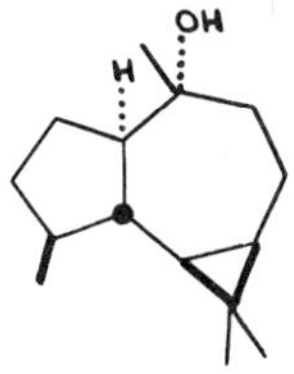

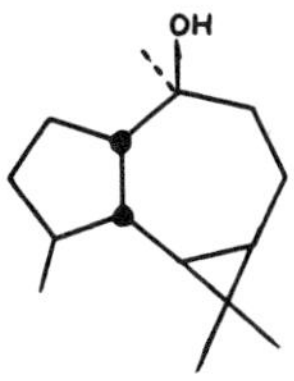

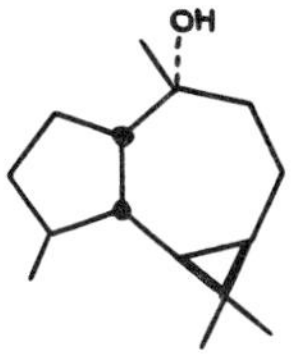

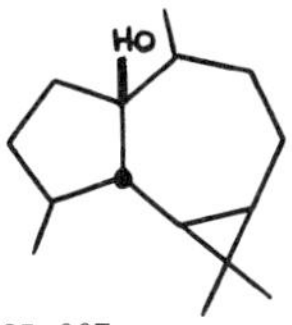

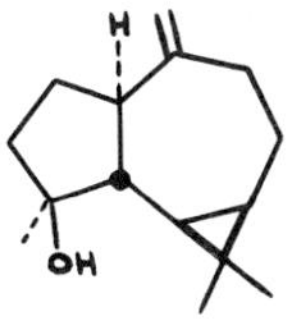

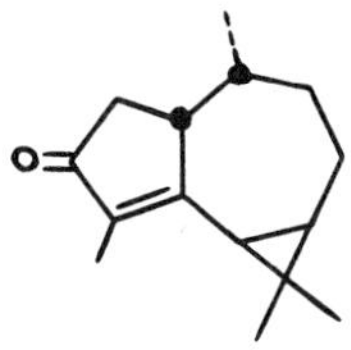

DRIMENOL — $C_{15}H_{26}O$ — 222 — -18^0 — 98^0C

Chem. Abstr., 1967 (67) 73695.

4126-001

FARNESIFEROL A — $C_{24}H_{30}O_4$ — 382 — -55^0c — 156^0C

Chem. Comm., 1966, 413.

4126-002

TADEONAL — $C_{15}H_{22}O_2$ — 234 — -126^0c — 53^0C

Aust. J. C., 1962, 389.

4126-003

deleted

4126-004

CINNAMODIAL — $C_{17}H_{24}O_5$ — 308 — -421^0 — 142^0C

Tet., 1969, 3896.

4126-005

CINNAMOLIDE — $C_{15}H_{22}O_2$ — 234 — -29^0 — 126^0C

Tet., 1969, 3896.

4126-006

CONFERTIFOLINE — $C_{15}H_{22}O_2$ — 234 — $+72^0c$ — 152^0C

JCS, 1960, 4685.

4126-007

BEMADIENOLIDE — $C_{15}H_{20}O_2$ — 232 — $+22^0$ — 125^0C

Tet., 1969, 3904.

4126-008

FUTRONOLIDE — $C_{15}H_{22}O_3$ — 250 — 216^0C

Tet., 1963, 635.

4126-009

BEMARIVOLIDE — $C_{17}H_{24}O_4$ — 292 — -258^0 — 138^0C

Tet., 1969, 3903.

4126-010

Name	Formula	MW / $[\alpha]_D$ / m.p.	Reference	No.
FRAGROLIDE	$C_{15}H_{20}O_3$	248 + 149^0 165^{0}C	Tet., 1969, 3903.	4126-011
CINNAMOSMOLIDE	$C_{17}H_{24}O_5$	308 -332^0 204^{0}C	Tet., 1969, 3896.	4126-012
IRESIN	$C_{15}H_{22}O_4$	266 +21^{0}c 142^{0}C	Tet., 1959 (7) 37.	4126-013
IRESIN, iso-	$C_{15}H_{22}O_4$	266	JACS, 1968, 5318.	4126-014
IRESONE, dihydro-	$C_{15}H_{22}O_4$	266 216^{0}C	Bull. Soc. Chim. Belges., 1958 (67) 632.	4126-015
DRIMENIN	$C_{15}H_{22}O_2$	234 -42^0 133^{0}C	JCS, 1960, 4685.	4126-016
DRIMENIN, iso-	$C_{15}H_{22}O_2$	234 + 87^{0}c 131^{0}C	JCS, 1960, 4685.	4126-017
UGANDENSOLIDE	$C_{17}H_{24}O_5$	308 + 23^{0}c 218^{0}C	Tet., 1969, 2887.	4126-018
VALDIVIOLIDE	$C_{15}H_{22}O_3$	250 + 111^{0}c 177^{0}C	Tet., 1963, 635.	4126-019
FUEGINE	$C_{15}H_{22}O_4$	266 +76^{0}c 171^{0}C	Tet., 1963, 635.	4126-020

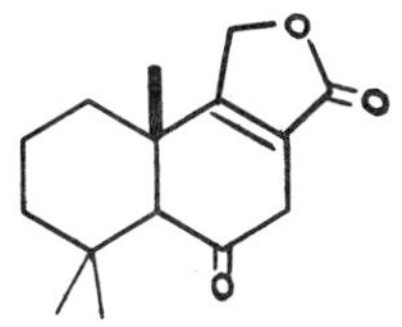

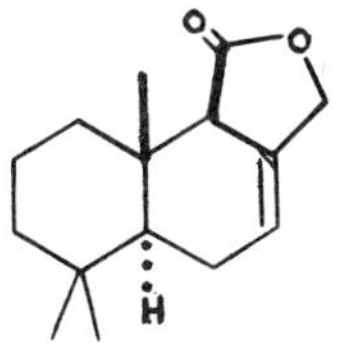

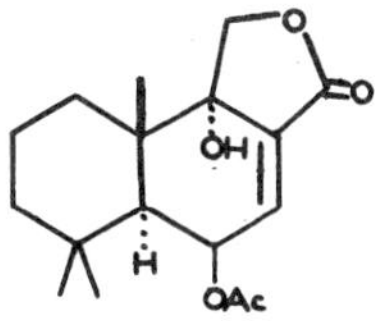

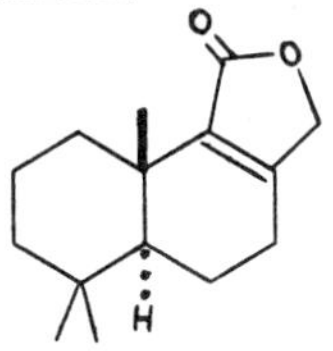

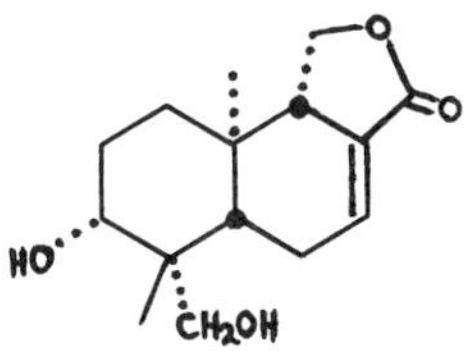

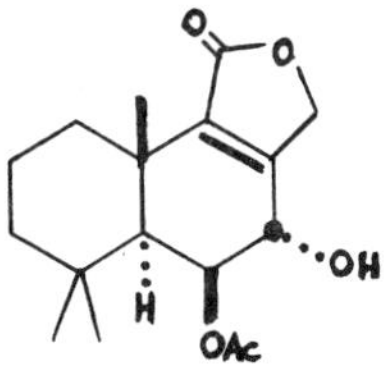

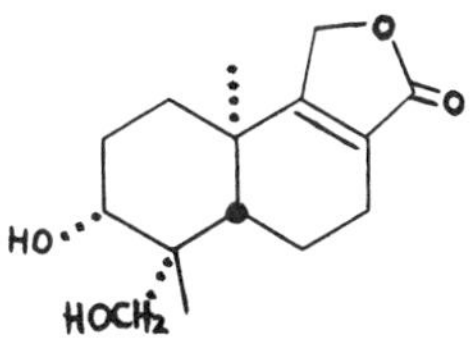

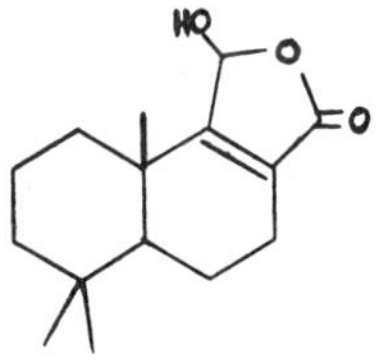

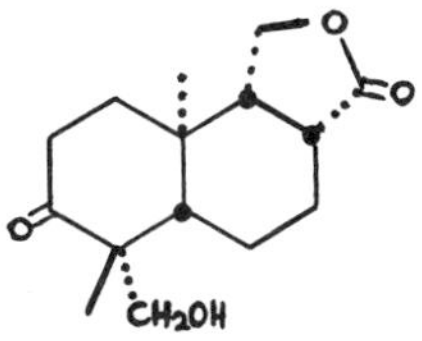

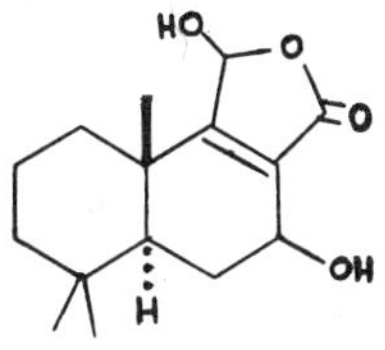

WINTERINE $C_{15}H_{20}O_3$ 248 +109°c 158°C 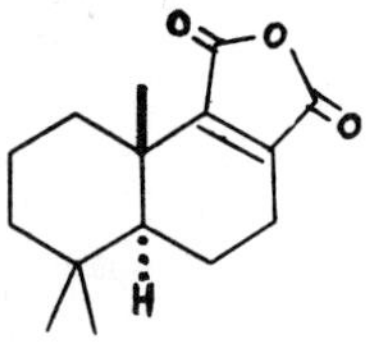 Tet., 1963, 635. 4126-021	TAURANIN $C_{22}H_{30}O_4$ 358 155°C 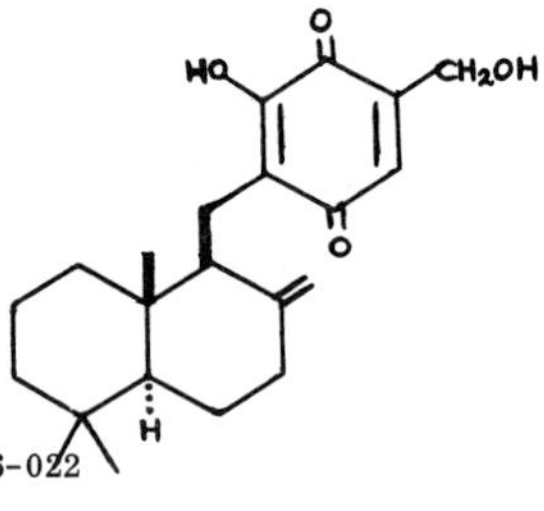 Tet. Lett., 1964, 1227. 4126-022

PATCHOULENE, α - $C_{15}H_{24}$ 204 +51⁰ Oil JACS 1961 927 4127-001	SUGETRIOL $C_{15}H_{24}O_3$ 252 +62° 221°C HO, OH, OH Chem. Pharm. Bull. 1967, 1433. 4127-006
PATCHOULENE, γ - $C_{15}H_{24}$ 204 Oil H JACS, 1961, 927. 4127-002	PATCHOULENONE $C_{15}H_{22}O$ 218 -97°c 52°C O Coll. Czech. Chem. Comm. 1964 1675 4127-007
CYPERENE $C_{15}H_{24}$ 204 H Chem. Pharm. Bull., 1968, 1900. 4127-003	PATCHOULENONE, iso- $C_{15}H_{22}O$ 218 +40° 48°C O Tet. Lett. 1965 4053 4127-008
PATCHOULENOL $C_{15}H_{24}O$ 220 -54°c 74°C OH, H Tet. Lett. 1967 2447 4127-004	CYPERENONE $C_{15}H_{22}O$ 218 O Tet. 1968 3891 4127-009
CYPERENOL $C_{15}H_{24}O$ 220 -12°c 94°C CH_2OH Tet. Lett. 1967 2447 4127-005	SUGEONOL ACETATE $C_{17}H_{24}O_3$ 276 +63° 94°C O, OAc Chem. Pharm. Bull. 1968 53 4127-010

TUTIN $C_{15}H_{18}O_6$ 294 +9° 213°C Tet. Lett., 1965, 4191. 4128-001	CORIAMYRTIN $C_{15}H_{18}O_5$ 278 +79° 230°C Chem. Pharm. Bull., 1967, 1697. 4128-005
HYENANCIN $C_{15}H_{18}O_7$ 310 +32° 232°C Tet. Lett., 1964, 371. 4128-002	PICROTOXININ $C_{15}H_{16}O_6$ 292 206°C Tet. Lett., 1960 (19) 21. 4128-006
HYENANCIN, iso- $C_{15}H_{20}O_7$ 312 -61°w 298°C Chem. Abstr., 1964 (61) 5697. 4128-003	PICROTIN $C_{15}H_{18}O_7$ 310 -70°e 225°C Tet. Lett., 1960 (19) 21. 4128-007
HYENANCIN METHYL ETHER, dihydro- $C_{16}H_{20}O_7$ 324 Chem. Abstr., 1968 (68) 3008. 4128-004	

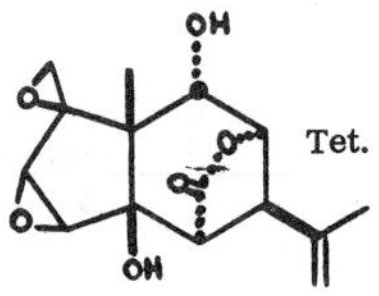

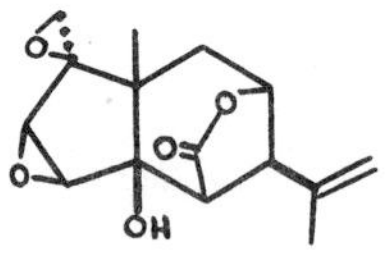

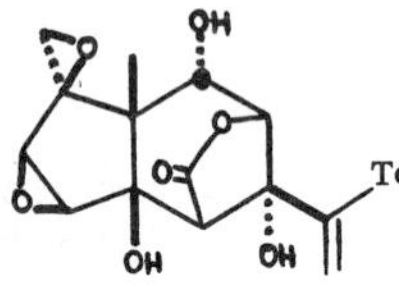

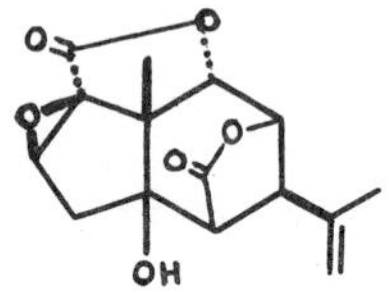

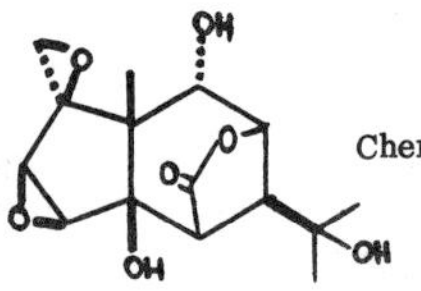

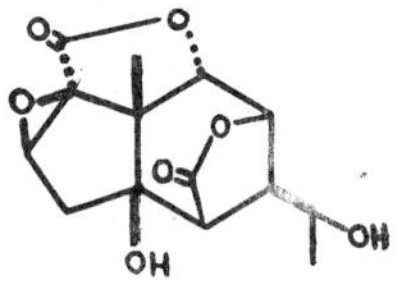

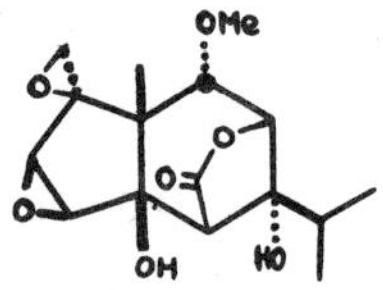

VALERANONE

$C_{15}H_{26}O$ 222 -55°c Oil

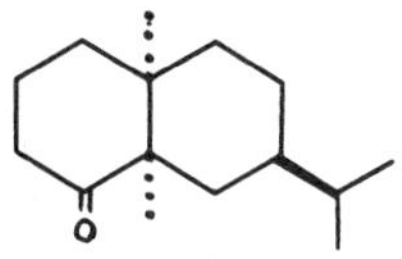

Chem. Pharm. Bull., 1965, 1408.

4129-001

CRYPTOFAURONOL

$C_{15}H_{26}O_2$ 238 -7° 90°C

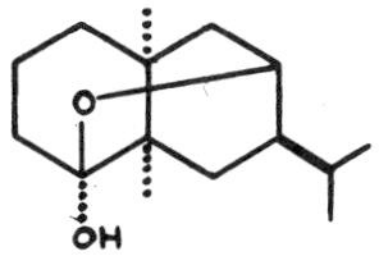

Chem. Pharm. Bull. (Tokyo), 1966 (14) 735.

4129-004

KANOKONOL

$C_{15}H_{26}O_2$ 238 -72° 52°C

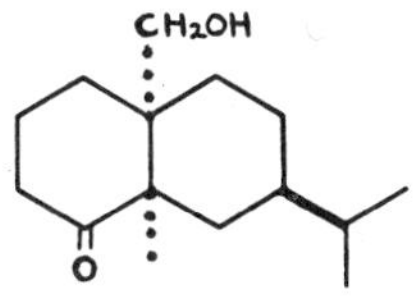

Chem. Pharm. Bull., 1965, 1417.

4129-002

FAURONYL ACETATE

$C_{17}H_{28}O_3$ 280 -78° 85°C

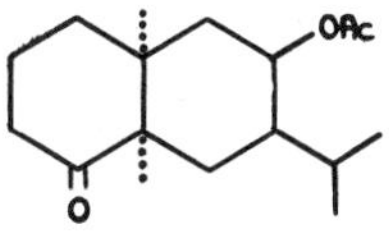

Chem. Pharm. Bull. (Tokyo), 1966 (14) 735.

4129-005

KANOKONOL, Acetate-

$C_{17}H_{28}O_3$ 280 -55°c Oil

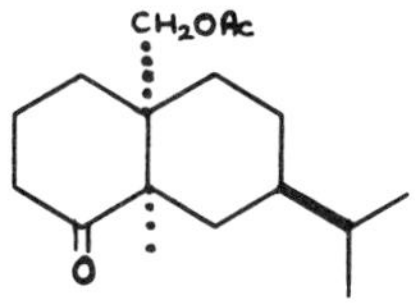

Tet., 1964, 1291.

4129-003

CHAMIGRENE, α- $C_{15}H_{24}$ 204 - 14⁰c Oil Tet. Lett., 1968, 2483. 4130-001	CHAMIGRENAL $C_{15}H_{22}O$ 218 -80⁰c Oil CHO Tet. Lett., 1968, 2483. 4130-003
CHAMIGRENE $C_{15}H_{24}$ 204 -53⁰c Oil Chem. Comm., 1967, 186. 4130-002	

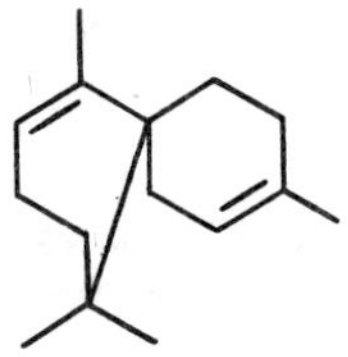

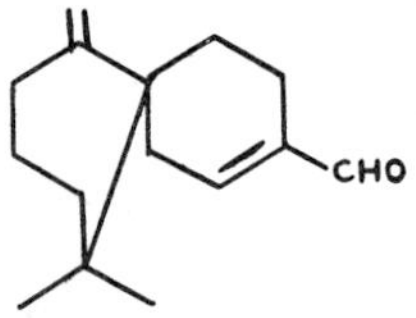

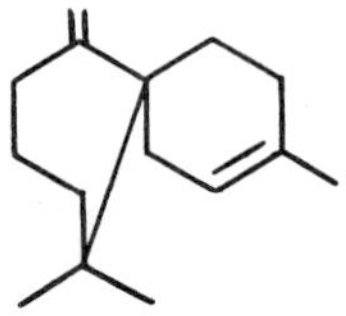

ARTEMONE	$C_{15}H_{24}O_2$	236 +41°c Oil
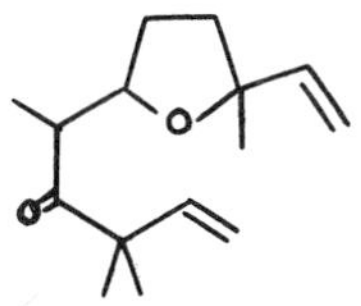	Tet. Lett., 1970, 5021.	
4190-001		

FUMAGILLIN $C_{26}H_{34}O_7$ 458
-27°m
190°C

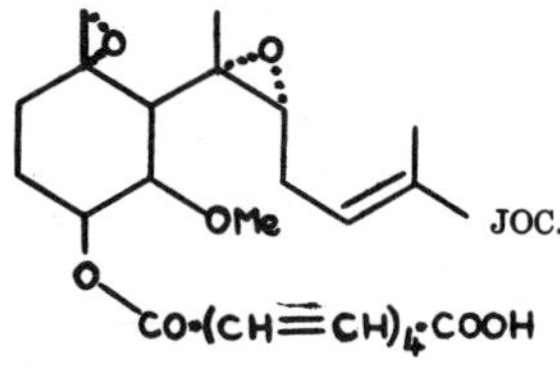

JOC., 1963, 928.

4191-001

OVALICIN $C_{16}H_{24}O_5$ 296
-75°
96°C

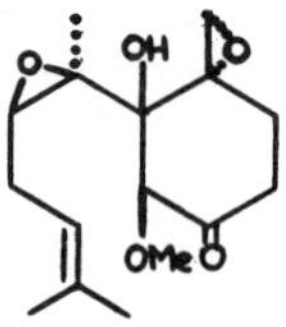

Agr. Biol. Chem., 1970 (34) 649.

4191-002

VERMEERIN $C_{15}H_{20}O_4$ 264
-58°
147°C

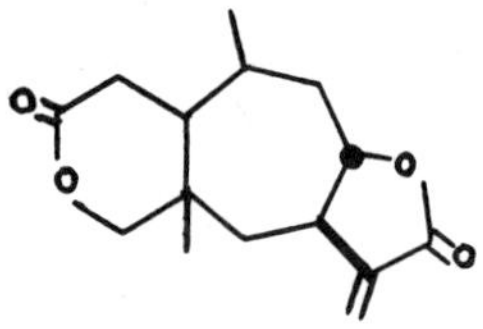

Tet., 1967, 4157.

4191-003

PSILOTROPIN $C_{15}H_{20}O_4$ 264

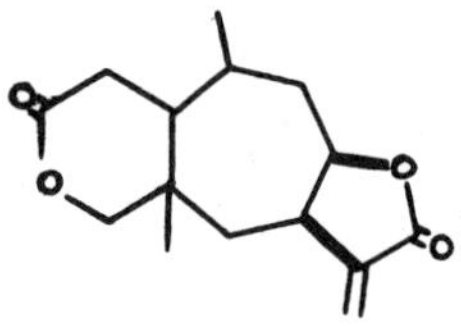

Phytochemistry, 1970, 59.

4191-004

THEMOIDIN $C_{15}H_{22}O_4$ 266
+62°c
220°C

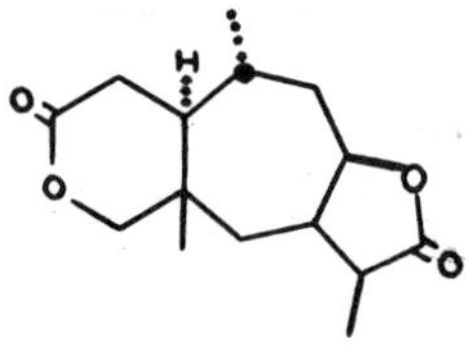

JOC., 1970, 2611.

4191-005

ANTHEMOIDIN $C_{15}H_{22}O_4$ 266
-115°c
220°C

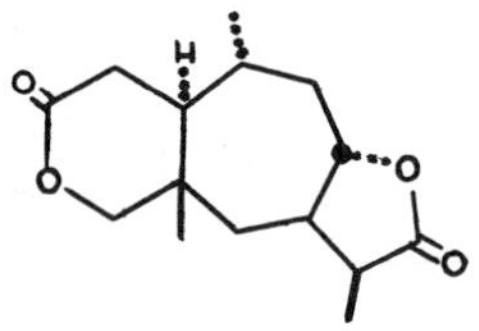

JOC., 1970, 2611.

4191-006

HYMENOLIDE $C_{17}H_{26}O_5$ 310
-49°c
137°C

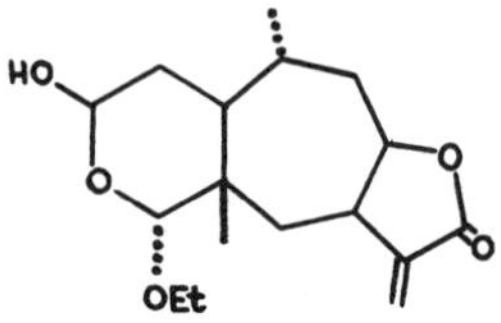

JOC., 1970, 2611.

4191-007

HYMENOXYNIN $C_{21}H_{34}O_9$ 430
-38°c
126°C

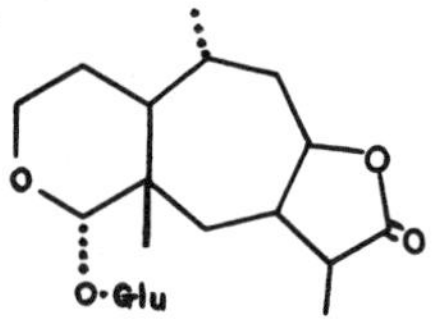

JOC., 1970, 2611.

4191-008

GREENEIN $C_{15}H_{20}O_4$ 264
+114°c
170°C

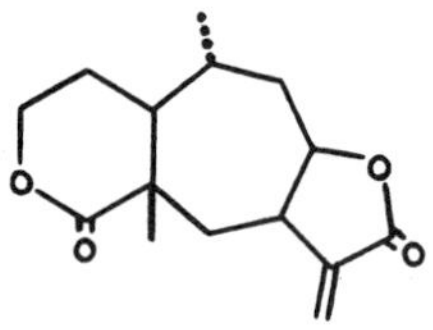

JOC., 1970, 2611.

4191-009

PSILOSTACHYIN $C_{15}H_{20}O_5$ 280
-125°c
215°C

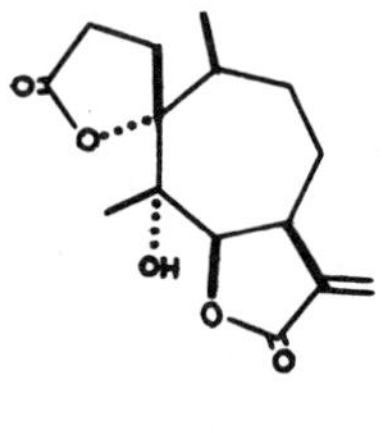

Tet., 1966, 1139.

4191-010

PSILOSTACHYIN B $C_{15}H_{18}O_4$ 262 -5°c 123°C Tet., 1966, 1943. 4191-011	XANTHANOL $C_{17}H_{24}O_5$ 308 -87°c 79°C R_1 = Ac R_2 = H OR1 OR2 JOC., 1969, 153. 4191-016
PSILOSTACHYIN C $C_{15}H_{20}O_4$ 264 -82°c 225°C Tet., 1966, 3279. 4191-012	XANTHANOL, iso- $C_{17}H_{24}O_5$ 308 +28°c 101°C R_1 = H R_2 = Ac JOC., 1969, 153. 4191-017
PARTHEMOLLIN $C_{15}H_{20}O_4$ 264 -130° 117°C OH JOC., 1970, 1110. 4191-013	XANTHININ $C_{17}H_{22}O_5$ 306 -53°c 124°C OAc JCS., 1965, 7009. 4191-018
IVALBIN $C_{15}H_{22}O_4$ 266 -45° 161°C OH OH JOC., 1967, 682. 4191-014	GAFRININ $C_{17}H_{24}O_5$ 308 -16° 110°C OH OAc Tet., 1968, 1687. 4191-019
XANTHATIN $C_{15}H_{18}O_3$ 246 JCS., 1965, 7009. 4191-015	GRIESENIN $C_{15}H_{16}O_4$ 260 +284°e 196°C * Tet., 1968, 6039. 4191-020

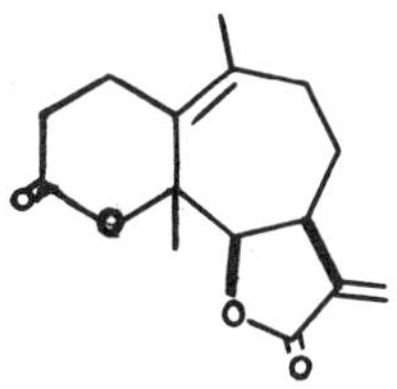

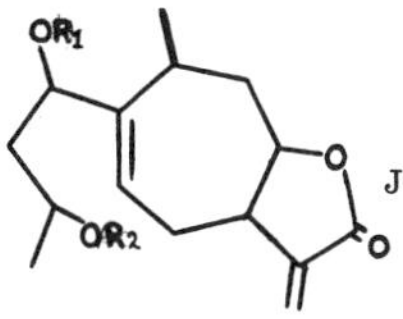

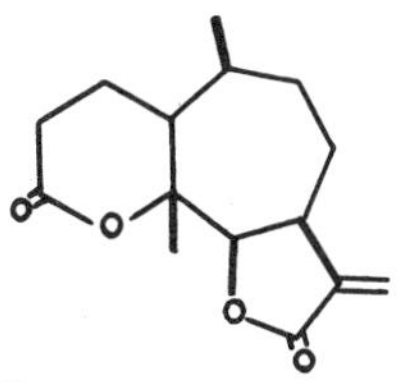

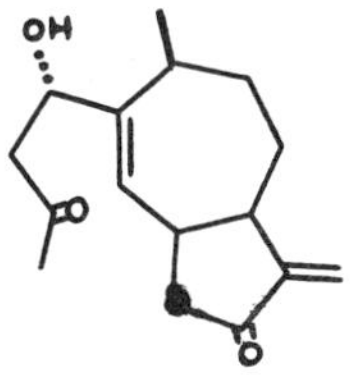

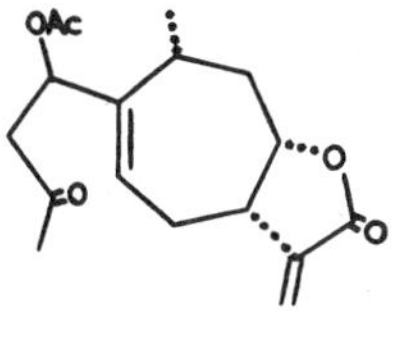

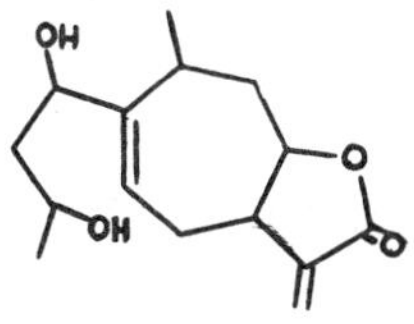

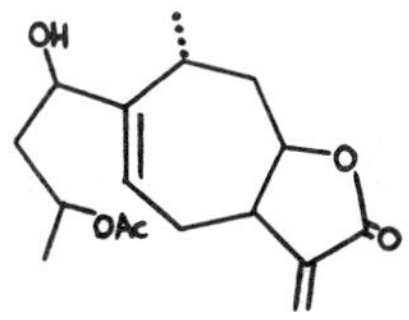

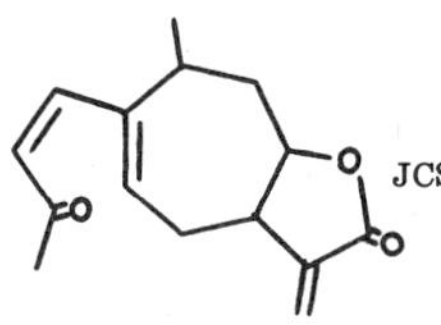

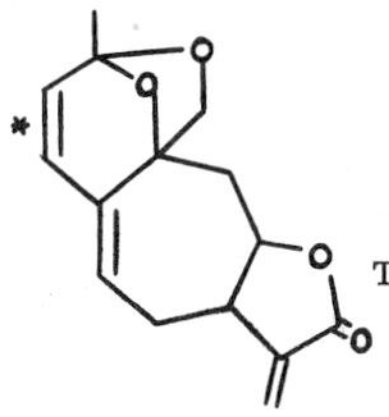

GRIESENIN, dihydro- $C_{15}H_{18}O_4$ 262 +92°e 140°C

*dihydro-

Tet., 1968, 6039.

4191-021

IVANGULIN $C_{16}H_{22}O_4$ 278 +109° 84°C

JOC., 1967, 3658.

4191-022

SESQUICHAMAENOL $C_{15}H_{22}O_2$ 234 0°m 110°C

Chem. Comm., 1970, 1538.

4191-023

ZERUMBONE, (±)-Ψ-photo- $C_{15}H_{22}O$ 218 OI Oil

Tet., 1968, 4113.

and optical antipode

4191-024

ZERUMBONE, (±)-dihydro-Ψ-photo- $C_{15}H_{24}O$ 220 OI Oil

Tet., 1968, 4113.

4191-025

ZERUMBONE, (±)-dihydro-Ψ-photo- $C_{15}H_{24}O$ 220 OI Oil

*epimer

Tet., 1968, 4113.

4191-026

FUROPELARGONE A $C_{15}H_{22}O_2$ 234 -105°c Oil

Tet., 1964, 1789.

4191-027

FUROPELARGONE B $C_{15}H_{22}O_2$ 234 +49°c Oil

Tet., 1964, 1789.

4191-028

FUROPELARGONE C $C_{15}H_{20}O_2$ 232 Oil

Tet., 1968, 2049.

4191-029

FUROPELARGONE D $C_{15}H_{20}O_2$ 232 Oil

Tet., 1968, 2049.

4191-030

CALACONE $C_{15}H_{24}O$ 220 +1° Oil

Coll. Czech. Chem. Comm., 1961 (26) 1021.

4191-031

PROCERIN $C_{15}H_{18}O_2$ 230 OI 71°C

OH O

Acta Chem. Scand., 1961, 645.

4191-036

FUROVENTALENE $C_{15}H_{18}O$ 214 OI

O

Tet. Lett., 1969, 3315.

4191-032

HUMBERTIOL $C_{15}H_{24}O$ 220 +15°c Oil

OH

Bull. Soc. Chim. Fr., 1970, 2401.

4191-037

MYRCENE, cyclo-isopropenyl- $C_{15}H_{24}$ 204 -11° Oil

Helv., 1931 (14) 1336.

4191-033

SICCANOCHROMENE A $C_{22}H_{30}O_2$ 326 +69°e

✱ O OH

Tet. Lett., 1968, 3643.

4191-038

NOOTKATIN $C_{15}H_{20}O_2$ 232 OI 95°C

OH O

Acta Chem. Scand., 1950, 1031.

4191-034

SICCANOCHROMENE B $C_{22}H_{30}O_3$ 342 +121°e amorph.

✱epoxide

Tet. Lett., 1968, 3643.

4191-039

NOOTKATINOL $C_{15}H_{22}O_3$ 250 OI 108°C

OH O OH

JOC., 1963, 2929.

4191-035

FARNESIFEROL B $C_{24}H_{30}O_4$ 382 +10°c

O• Coumarin

HO

Helv., 1959 (42) 2557.

4191-040

Name	Formula	Data	Reference	No.
FARNESIFEROL C	$C_{24}H_{30}O_4$	382 -30°c	Helv., 1959 (42) 2557.	4191-041
GEIJERENE, pre-	$C_{12}H_{18}$	162 Ol Oil	Aust. J. C., 1968, 2256.	4191-045
NEROLIDIOL, cyclo-	$C_{15}H_{28}O_2$	240 -20°c	Tet. Lett., 1970, 1293.	4191-042
CANABRIN	$C_{15}H_{20}O_5$	280 -134° 210°C	Phytochem., 1970, 1611.	4191-046
GEIJERENE	$C_{12}H_{18}$	162 Oil	Aust. J. C., 1964, 1270.	4191-043
FARNESOL, trans-γ-monocyclo-	$C_{15}H_{26}O$	222 +18°e Oil	Chem. Comm., 1971, 527.	4191-047
ACORIC ACID	$C_{15}H_{24}O_4$	268 +27°c 167°C	JCS., 1964, 2923.	4191-044

AZULENE, 4-methyl-7-isopropyl- $C_{14}H_{16}$ 184
OI
Oil

Chem. Abstr., 1970 (73) 59211.

4192-001

CHAMAZULENE $C_{14}H_{16}$ 184
OI
Oil

Coll. Czech. Chem. Comm. 1954 186

4192-002

CHAMAZULENE, 3,6-dihydro- $C_{14}H_{18}$ 186
OI
Oil

Tet., 1968, 2080.

4192-003

CHAMAZULENE, 5,6-dihydro- $C_{14}H_{18}$ 186
OI
Oil

Tet., 1968, 2080.

4192-004

AZULENE, 1,4-diMe- $C_{12}H_{12}$ 156
OI
Oil

Ber., 1966, 2669.

4192-005

AZULENE, 1-COOMe-4-Me- $C_{13}H_{12}O_2$ 200
OI
Oil

Ber., 1966, 2669.

4192-006

AZULENE $C_{10}H_8$ 128
OI
99^0C

Chem. Revs., 1952 (50) 127.

4192-007

MEXICANIN E $C_{14}H_{16}O_3$ 232
-47^0c
101^0C

Chem. Comm., 1966 (6) 151.

4192-008

MEXICANIN E, dihydro- $C_{14}H_{18}O_3$ 234
-188^0c
134^0C

Tet., 1965, 1711.

4192-009

CARABRONE $C_{15}H_{20}O_3$ 248
$+117^0$e
90^0C

Tet. Lett., 1969, 335.

4192-010

MEXICANIN D	$C_{15}H_{18}O_4$	262 +107°c 252°C
		JACS., 1963, 19.
4192-011		
CHANOOTINE	$C_{15}H_{18}O_3$	246 0°p 172°C
		Arkiv. Kemi, 1964 (22) 129.
4192-016		
TENULIN	$C_{17}H_{22}O_5$	306 -23°c 197°C
		JACS., 1962, 3857.
4192-012		
CYPEROLONE	$C_{15}H_{24}O_2$	236 +31° 42°C
		Chem. Pharm. Bull., 1966, 1439.
4192-017		
ZIERONE	$C_{15}H_{22}O$	218 -179°c Oil
		JCS., 1962, 1981.
4192-013		
WIDDRENE OXIDE	$C_{15}H_{26}O_2$	238 -9°c 154°C
		Acta Chem. Scand., 1962, 1553.
4192-018		
VALERENIC ACID	$C_{15}H_{22}O_2$	234 -120°e 141°C
		Coll. Czech. Chem. Comm., 1965, 553.
4192-014		
WIDDROL	$C_{15}H_{26}O$	222 +104°c 98°C
		JACS., 1967, 3232.
4192-019		
VALERENOLIC ACID	$C_{15}H_{22}O_3$	250 -94°e 172°C
		Coll. Czech. Chem. Comm., 1965, 553.
4192-015		
HIMACHALOL-Allo, (+)-	$C_{15}H_{26}O$	222 +37°c 85°C
		Tet., 1968, 3861.
4192-020		

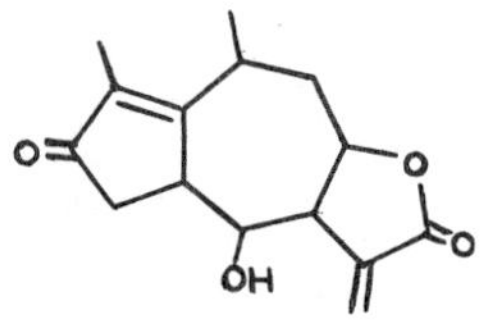

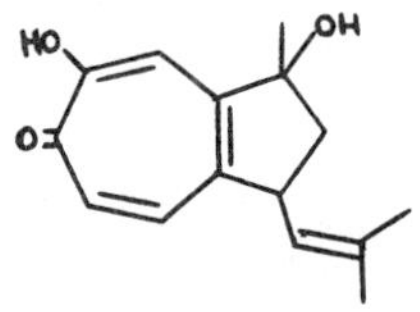

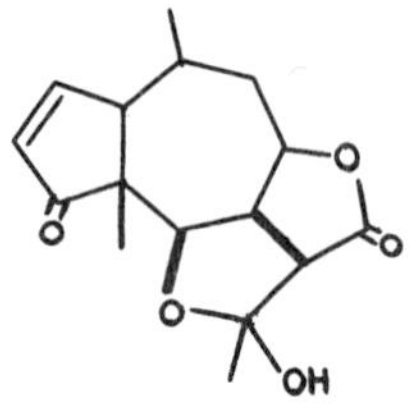

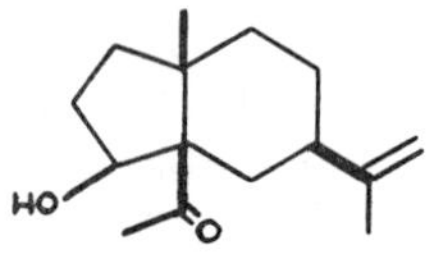

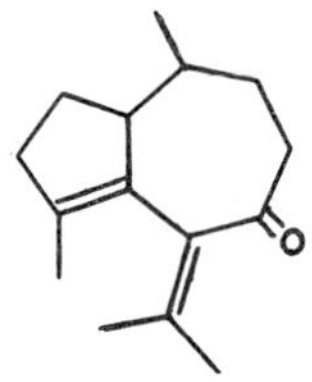

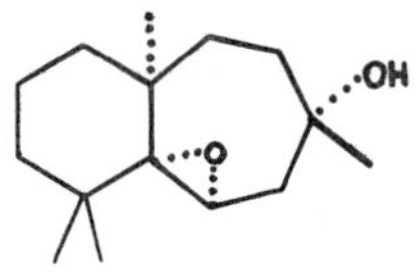

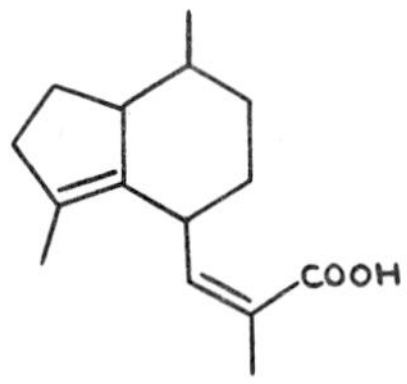

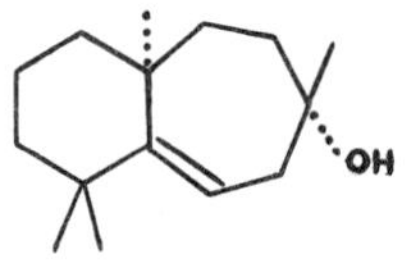

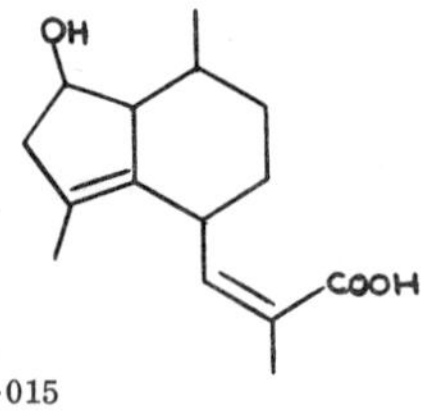

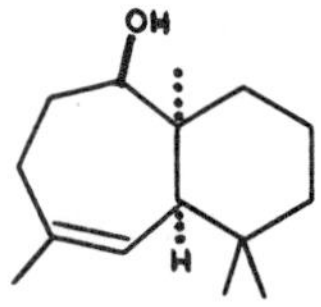

NARDOSINONE	$C_{15}H_{22}O_3$	250 109°C	Tet. Lett., 1968, 3615.	4192-021
GORGONENE-β-(+)-	$C_{15}H_{24}$	204 +14°	Chem. Comm., 1968, 1070.	4192-026
NARDOSINONE, iso-	$C_{15}H_{22}O_3$	250 +13.2°c	Ann., 1970 (733) 152.	4192-022
COGEIJERENE	$C_{12}H_{18}$	162 +0.7° Oil	Aust. J. C., 1964, 1270.	4192-027
KHUSILAL	$C_{14}H_{18}O$	202 -261° Oil	Tet., 1964, 2617.	4192-023
OCCIDOL	$C_{15}H_{22}O$	218 +77°c 70°C	Tet. 1964, 67.	4192-028
KHUSITONE	$C_{14}H_{20}O$	204 -135° Oil	Tet., 1964, 2633.	4192-024
GEOSMIN	$C_{12}H_{22}O$	182	Tet. Lett., 1968, 2971.	4192-029
MAALIOXIDE	$C_{15}H_{26}O$	222 +33° 65°C	Tet., 1964, 963.	4192-025
CACALOL	$C_{15}H_{18}O_2$	230 +10°c 93°C	JCS., 1969, 1184.	4192-030

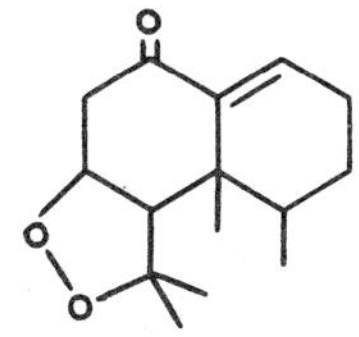

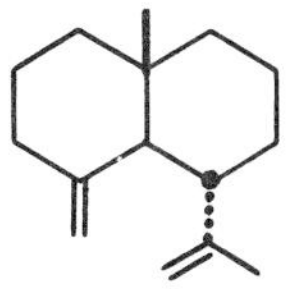

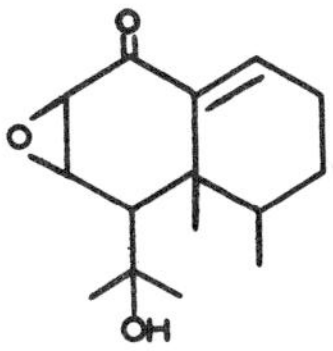

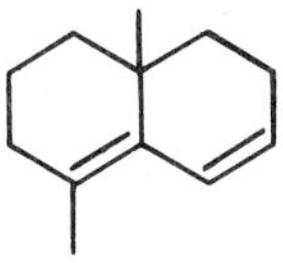

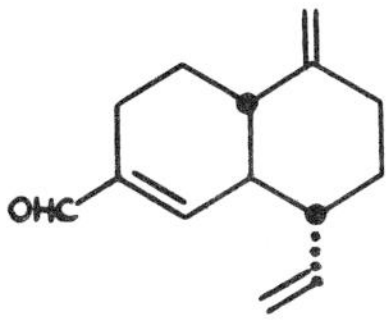

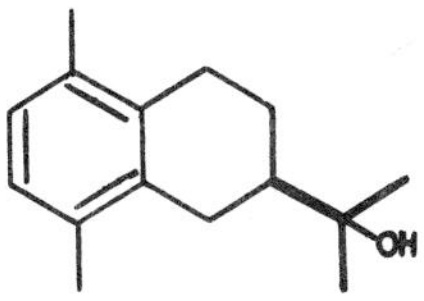

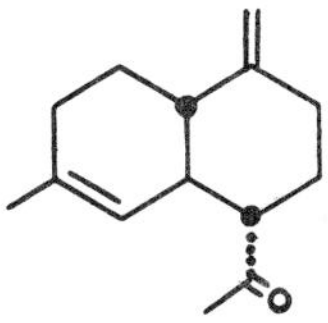

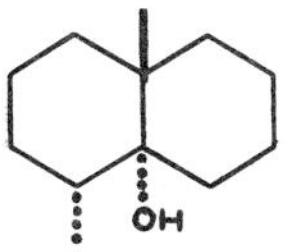

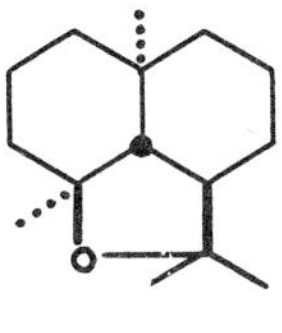

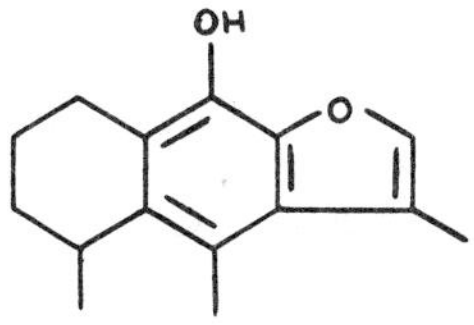

CACALONE	$C_{15}H_{16}O_3$	244 +84°c 120°C
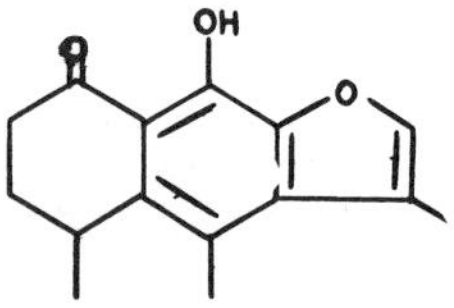	JCS., 1969, 1184.	
4192-031		
CHAMAECYNONE	$C_{14}H_{18}O$	202 -93°m 92°C
	Tet. Lett., 1966, 3663.	
4192-036		
MATURONE	$C_{14}H_{10}O_4$	242 OI 170°C
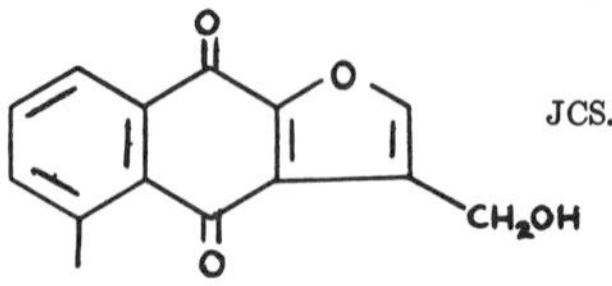	JCS., 1969, 1184.	
4192-032		
CHAMAECYNONE, iso-	$C_{14}H_{18}O$	202 Oil
4-epimer	Tet. Lett., 1966, 3663.	
4192-037		
MATURININ	$C_{16}H_{14}O_3$	254 OI 95°C
	JCS., 1969, 1184.	
4192-033		
CHAMAECYNONE, OH-iso-	$C_{14}H_{18}O_2$	218 -40°e 162°C
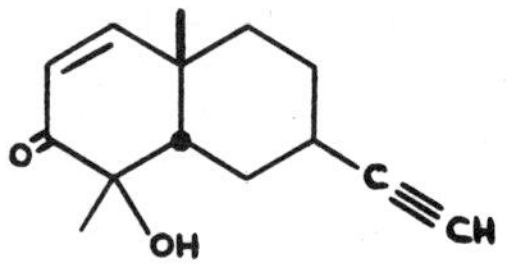	Tet. Lett., 1966, 3663.	
4192-038		
MATURINONE	$C_{14}H_{10}O_3$	226 OI 168°C
	JCS., 1969, 1184.	
4192-034		
CHAMAECYNONE, dihydro-iso-	$C_{14}H_{20}O$	204 Oil
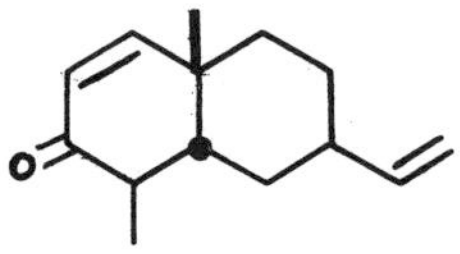	Tet. Lett., 1966, 3665.	
4192-039		
MATURIN	$C_{16}H_{14}O_4$	270 OI 120°C
	JCS., 1969, 1184.	
4192-035		
CHAMAECYNENOL	$C_{14}H_{20}O$	204 +9.3° Oil
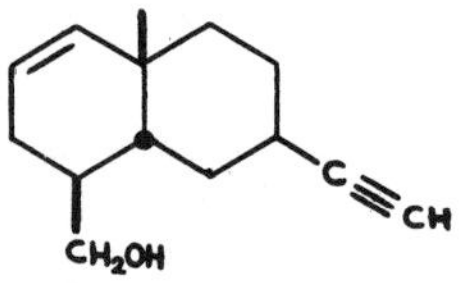	Chem. Ind., 1968, 1638.	
4192-040		

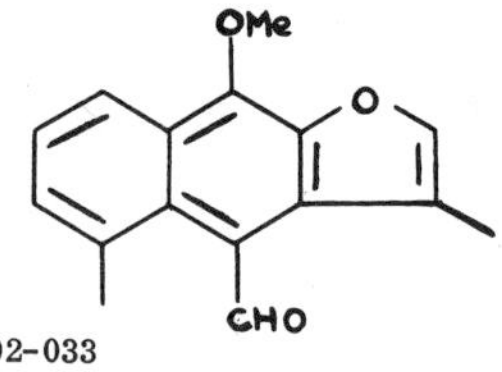

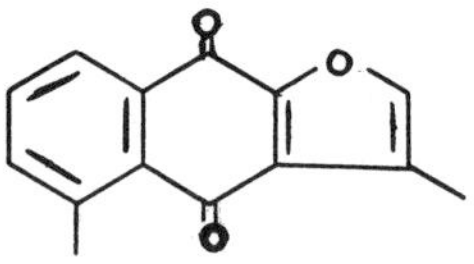

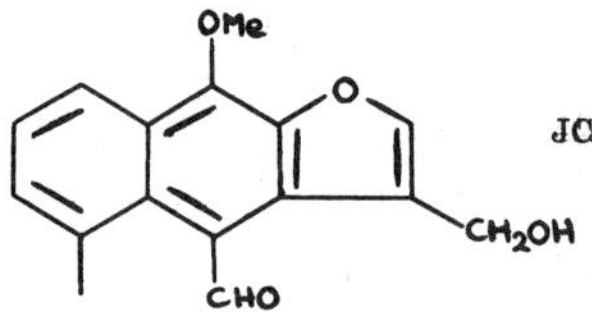

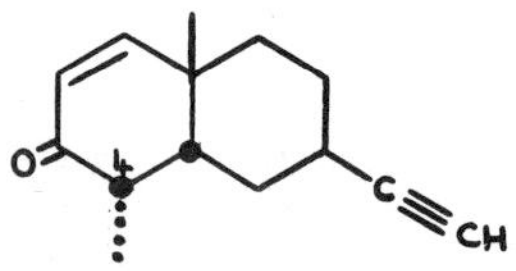

Compound	Formula	Data	Reference	No.
CHAMAECYNENOL ACETATE	$C_{16}H_{22}O_2$	246 Oil	Chem. Ind., 1968, 1638.	4192-041
FAURINONE	$C_{15}H_{26}O$	222 +22°c Oil	Chem. Pharm. Bull., 1968, 1779.	4192-046
CHAMAECYNENOL, dehydro-	$C_{14}H_{18}O$	202 -193°m Oil	Tet. Lett., 1968, 3639.	4192-042
BERGAMOTENE, α-	$C_{15}H_{24}$	204 -44°c Oil	Coll. Czech. Chem. Comm., 1950 (15) 373.	4192-047
CHAMAECYNENAL, dehydro-	$C_{14}H_{16}O$	200 Oil	Tet. Lett., 1968, 3639.	4192-043
BERGAMOTENE, β-	$C_{15}H_{24}$	204 +36°c Oil	Tet., 1966, 1917.	4192-048
RISHITIN	$C_{14}H_{22}O_2$	222 -29°e 66°C	JCS 1969 1073	4192-044
SIRENIN	$C_{15}H_{24}O_2$	236	JACS., 1968, 6434.	4192-049
AGAROFURAN, oxo-nor-	$C_{14}H_{22}O_2$	222 -119°c 56°C	Tet., 1963, 1519.	4192-045
SESQUICARENE	$C_{15}H_{24}$	204 -77°c Oil	Tet. Lett., 1968, 1253.	4192-050

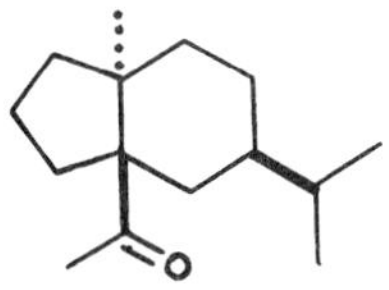

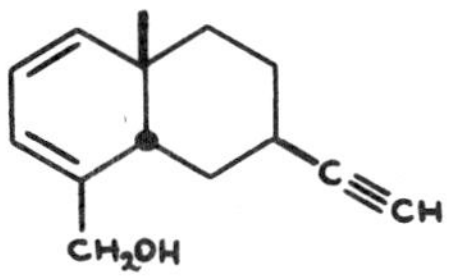

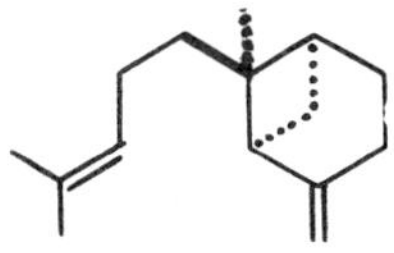

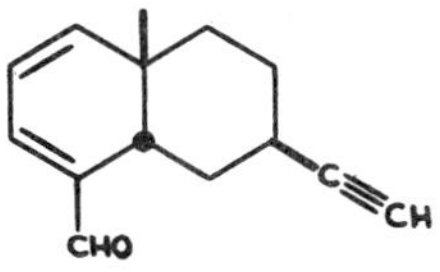

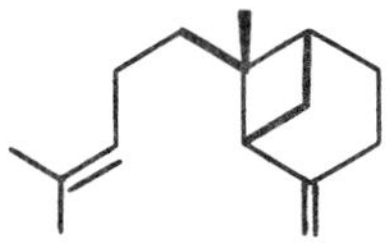

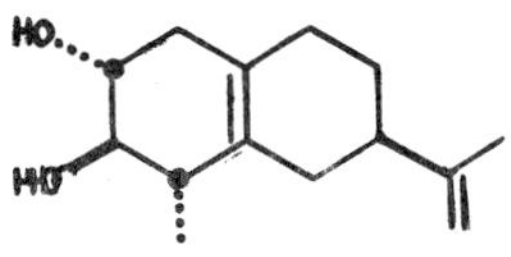

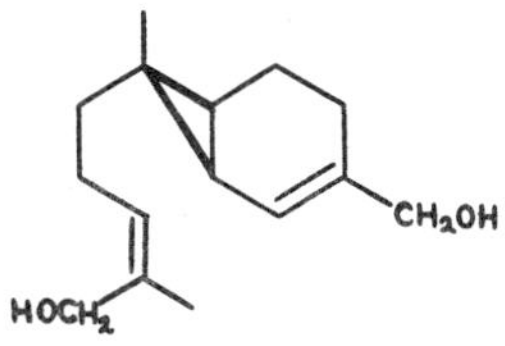

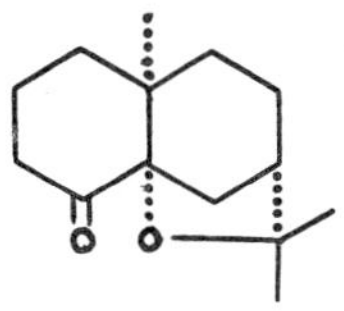

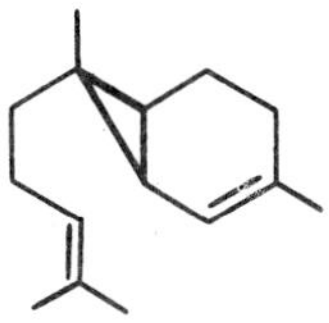

BAZZANENE $C_{15}H_{24}$ 204 +48° Oil Experimentia, 1969 (25) 1139. 4192-051	BAKKENOLIDE C $C_{20}H_{28}O_5$ 348 167°C O-Tig, OH Tet. Lett., 1968, 1993. 4192-056
BAZZANENOL $C_{15}H_{24}O$ 220 +19° HO Experimentia, 1970 (26) 347. 4192-052	CAMPHERENOL $C_{15}H_{26}O$ 222 -62° OH Tet. Lett., 1967, 5070. 4192-057
FUKINANOLIDE $C_{15}H_{22}O_2$ 234 +12°c 80.5°C Tet. Lett., 1968, 371. 4192-053	CAMPHERENONE $C_{15}H_{24}O$ 220 -33° Tet. Lett., 1967, 5069. 4192-058
FUKINOLIDE $C_{22}H_{30}O_6$ 390 -126°c 102°C OR, OAc, R=Ang Chem. Ind., 1968, 318. 4192-054	HELMINTHOSPORAL $C_{15}H_{22}O_2$ 234 -49°c 58°C CHO, CHO JACS., 1963, 3527. 4192-059
FUKINOLIDE-S $C_{21}H_{28}O_6S$ 408 -161°c 201°C R = S-Ang Chem. Ind., 1968, 318. 4192-055	HELMINTHSPOROL $C_{15}H_{24}O_2$ 236 -29°c 98°C CH₂OH, CHO Chem. Abstr., 1966 (65) 19273. 4192-060

HELMINTHOSPORAL, Pre- $C_{19}H_{32}O_3$ 308 OEt, O, OEt Can. J. C., 1965 (43) 1357. 4192-061	LAMBICIN $C_{18}H_{24}O_7$ 352 -17⁰c 181⁰C HO, O, HO, O–CO, OH, O Tet., 1969, 4337. 4192-066
HELMINTHOSPOROL, Pre- $C_{17}H_{28}O_2$ 264 O, OEt Can. J. C., 1965 (43) 1357. 4192-062	CAPENICIN $C_{20}H_{24}O_8$ 392 +77⁰d 244⁰C MeOOC, O, HO, O–CO Tet. Lett., 1966, 4819. 4192-067
HELMINTHOSPOROL, 9-OH-pre- $C_{15}H_{24}O_3$ 252 ac. -2⁰c ac. 65⁰C O, OH, OH JCS., 1970, 686. 4192-063	PENTALENOLACTONE $C_{15}H_{16}O_5$ 276 H, COOH, O, O, O Tet. Lett., 1970, 4901. 4192-068
ILLUDALIC ACID $C_{15}H_{16}O_5$ 276 0⁰e 200⁰C OH, OHC, HO, O, O JOC., 1969, 240. 4192-064	deleted 4192-069
PRETOXIN $C_{19}H_{24}O_8$ 380 +3⁰c 129⁰C O, O, HO, HO, O, OH, O Tet., 1969, 4337. 4192-065	deleted 4192-070

SPIROLAURENONE	$C_{15}H_{23}OBr$	298 -71°c Oil
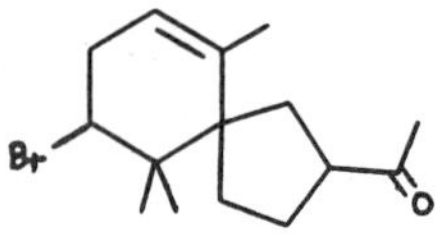	Tet. Lett., 1970, 4995.	
4192-071		
PINGUISONE	$C_{15}H_{20}O_2$	232 63°C
(?)	Coll. Czech. Chem. Comm., 1969, 582.	
4192-076		
GERMACRENE, bicyclo-	$C_{15}H_{24}$	204 +61°a
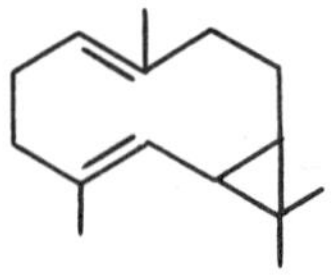	Tet. Lett., 1969, 3097.	
4192-072		
OPLOPANONE	$C_{15}H_{26}O_2$	238 -20°d 97°C
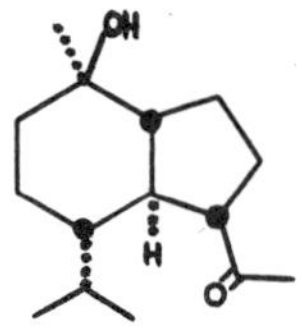	Tet., 1966, Suppl. 7, 219.	
4192-077		
FOMANNOSIN	$C_{15}H_{18}O_4$	262
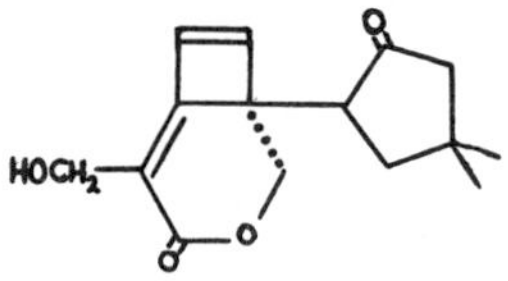	JACS., 1967, 1260.	
4192-073		
RISHITINOL	$C_{15}H_{22}O_2$	234 +47°c 128°C
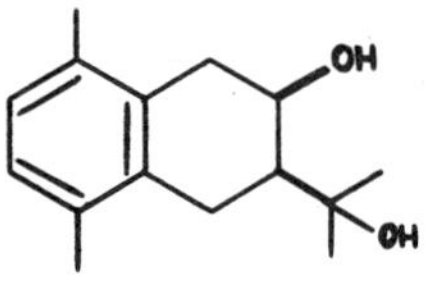	Tet. Lett., 1971, 83.	
4192-078		
ANISATIN	$C_{15}H_{20}O_8$	328 -27°d 228°C
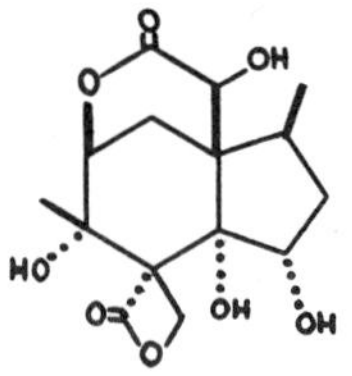	Tet., 1968, 1255.	
4192-074		
ILLUDOIC ACID	$C_{15}H_{18}O_5$	278 135°C
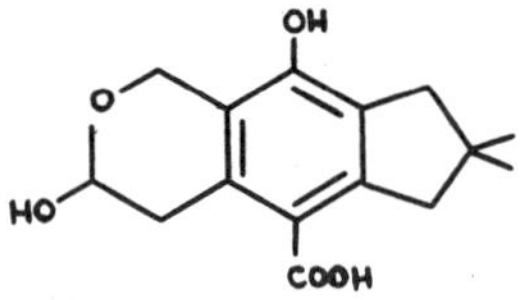	JOC., 1969, 240.	
4192-079		
ANISATIN, neo-	$C_{15}H_{20}O_7$	312 -25°d 238°C
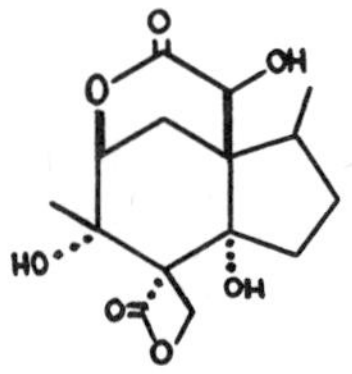	Tet. Lett., 1966, 4739.	
4192-075		

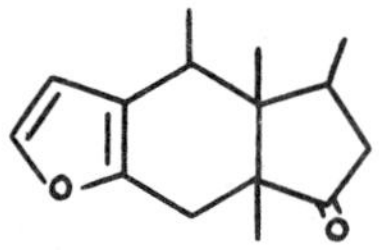

ZALUZANIN A — $C_{15}H_{20}O_4$ — 264; -10^0e; 265^0C

R = H

Tet., 1967, 29.

4193-001

PATCHOULIOL — $C_{15}H_{26}O$ — 222; -129c; 56^0C

Proc. Chem. Soc. 1963 383

4193-006

ZALUZANIN B — $C_{17}H_{22}O_5$ — 306; -12^0e; 224^0C

R = Ac

Tet., 1967, 29.

4193-002

BOURBONENE, α- — $C_{15}H_{24}$ — 204; $+25^0$; Oil

Tet., 1967, Suppl. 8, 53.

4193-007

IVAXILLARIN — $C_{15}H_{18}O_4$ — 262; -241^0m; 187^0C

JOC., 1966, 3232.

4193-003

BOURBONENE, β- — $C_{15}H_{24}$ — 204; -92^0; Oil

Tet., 1967, Suppl. 8, 53.

4193-008

AXIVALIN — $C_{17}H_{22}O_5$ — 306; -132^0m; 139^0C

JOC., 1966, 3232.

4193-004

BOURNONAN-1-ONE, 11-nor- — $C_{14}H_{22}O$ — 206; Oil

Bull. Soc. Chim. Fr., 1968, 2452.

4193-009

PATCHOULENE, β- — $C_{15}H_{24}$ — 204; -43^0c; Oil

JACS 1964 4438

4193-005

SEYCHELLENE — $C_{15}H_{24}$ — 204; -72^0c; Oil

Tet. Lett., 1968, 3849.

4193-010

CLOVANDIOL $C_{15}H_{26}O_2$ 238, $+2^0$, 152^0C Tet., 1971, 635. HO, OH 4193-011	LINDEROXIDE $C_{16}H_{20}O_2$ 244, -52^0c, Oil Tet., 1968, 625. O, CH_2OMe 4193-016
LINDENENE $C_{15}H_{18}O$ 214, -50^0c, Oil JCS., 1969 1920. O, H 4193-012	LINDEROXIDE, iso- $C_{16}H_{20}O_2$ 244, -238^0, Oil JCS 1967 631 O, CH_2OMe 4193-017
LINDENENONE $C_{15}H_{16}O_2$ 228, -333^0, 108^0C JCS., 1969, 2786. O, O 4193-013	SATIVENE $C_{15}H_{24}$ 204 JACS., 1965, 3275. 4193-018
LINDERENE $C_{15}H_{18}O_2$ 230, 144^0C Tet. 1964 2991 O, H, OH 4193-014	CUBEBOL $C_{15}H_{26}O$ 222, -62^0c, 64^0C Coll. Czech. Chem. Comm., 1960, 919. H, OH 4193-019
LINDERENE, Ac- $C_{17}H_{20}O_3$ 272, $+26^0e$, 80^0C Tet., 1964, 2991. 4193-015	CUBEBENE, α- $C_{15}H_{24}$ 204, -20^0c Tet. Lett., 1966, 6365. 4193-020

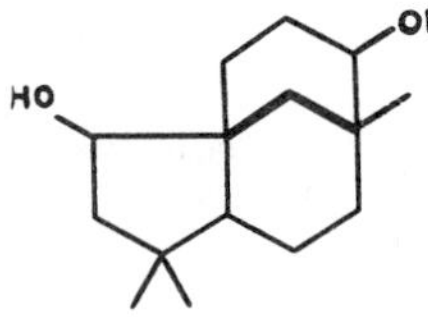

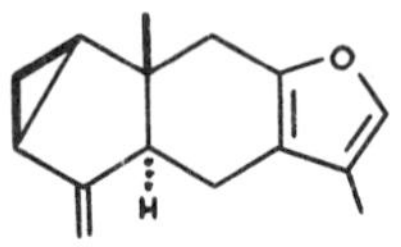

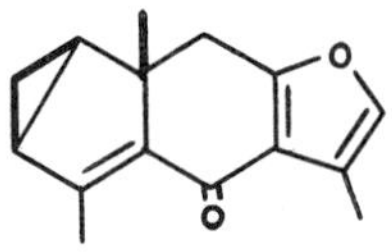

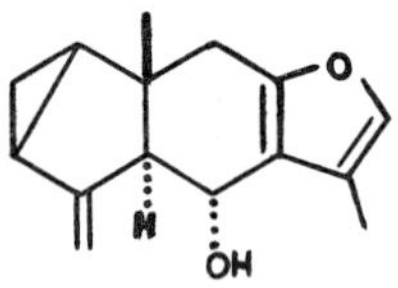

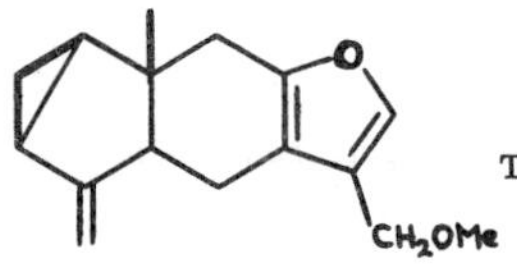

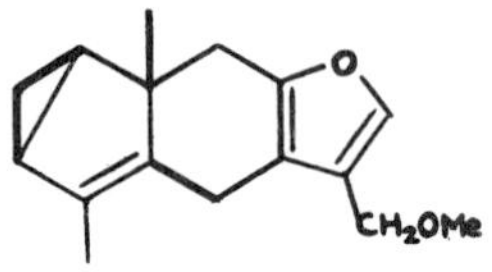

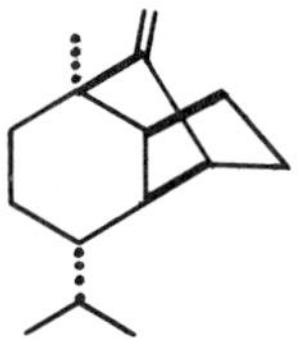

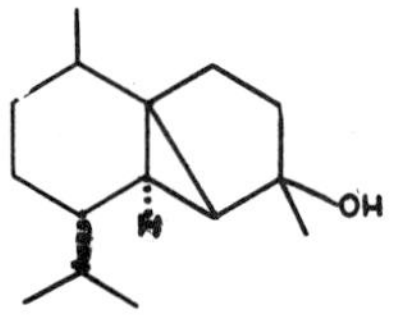

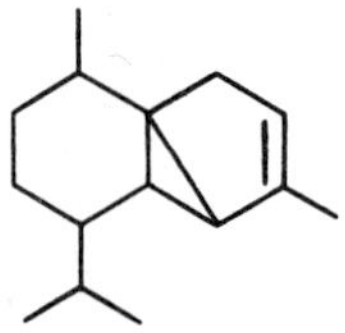

Name	Formula	Data	Reference	No.
CUBEBENE, β-	$C_{15}H_{24}$	204	Tet. Lett., 1966, 6365.	4193-021
YLANGENE, α-L-	$C_{15}H_{24}$	204 -23^0 Oil	Tet. Lett., 1969, 1601.	4193-026
MUSTAKONE	$C_{15}H_{22}O$	218 $+0.34^0$ Oil	Tet., 1965, 607.	4193-022
YLANGENE, β-	$C_{15}H_{24}$	204 Oil	JOC., 1964, 2100.	4193-027
COPADIENE, (+)-	$C_{15}H_{22}$	202 $+17^0c$ Oil	Tet. Lett., 1967, 4661.	4193-023
BRACHYLAENALONE A	$C_{15}H_{20}O_2$	232	Chem. Comm., 1969, 630.	4193-028
COPAENE	$C_{15}H_{24}$	204 -6.5^0 Oil	Tet. Lett., 1969, 1601.	4193-024
COPABORNEOL	$C_{15}H_{26}O$	222 $+27^0$ 47^0C	Acta Chem. Scand., 1967, 585.	4193-029
YLANGENE, α-D-	$C_{15}H_{24}$	204 $+51^0$ Oil	Tet. Lett., 1969, 1601.	4193-025
LONGIPINENE, α-	$C_{15}H_{24}$	204 $+37^0c$ Oil	Tet., 1960, 237.	4193-030

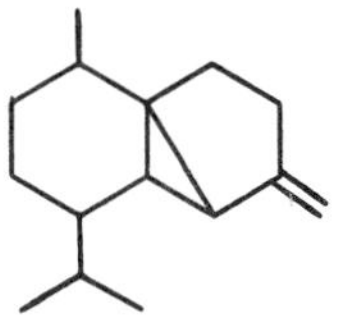

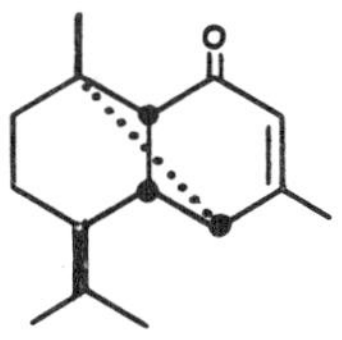

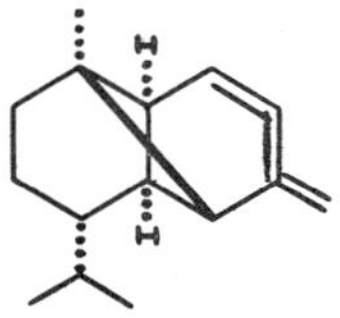

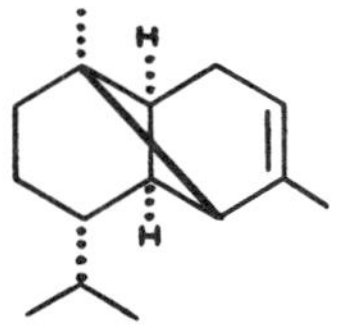

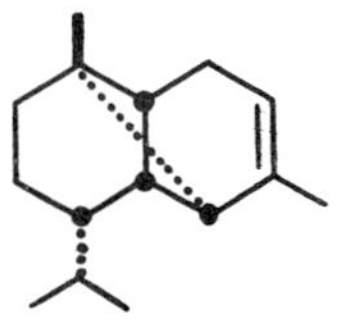

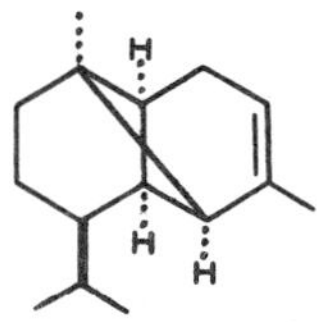

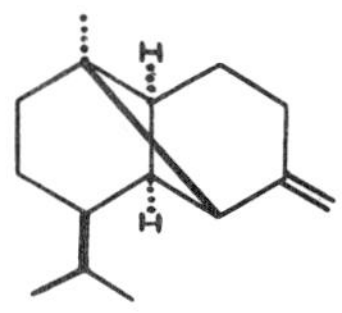

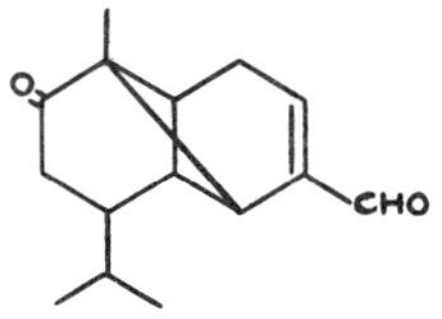

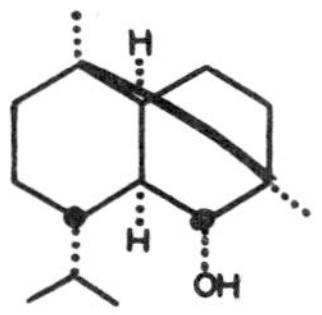

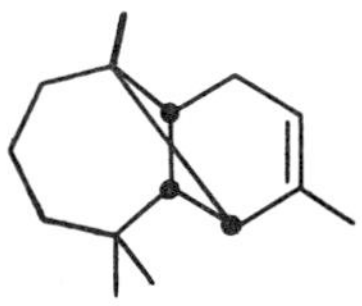

Compound	Formula	Data	Reference	No.
LONGIFOLENE, D-	$C_{15}H_{24}$	204 +42⁰c Oil	JACS., 1964, 438.	4193-031
LONGIBORNEOL	$C_{15}H_{26}O$	222 +25⁰e 109⁰C	Tet. Lett., 1964, 3761.	4193-032
CULMORIN	$C_{15}H_{26}O_2$	238	JCS., 1968, 148.	4193-033
LAURINTEROL R = Br	$C_{15}H_{19}OBr$	294 +13⁰c 54⁰C	Chem. Comm., 1967, 271.	4193-034
LAURINTEROL, debromo- R = H	$C_{15}H_{20}O$	216 -12⁰c	Tet. Lett., 1966 (17) 1836.	4193-035
MARASMIC ACID	$C_{15}H_{18}O_4$	262 +182⁰ 173⁰C	JACS., 1966, 2838.	4193-036
ILLUDIN S	$C_{15}H_{20}O_4$	264 125⁰C	Chem. Comm., 1970, 310.	4193-037
ILLUDIN S, dihydro-	$C_{15}H_{22}O_4$	266 amorph.	Tet. Lett., 1969, 3965.	4193-038
ILLUDIN M	$C_{15}H_{20}O_5$	280	JACS., 1963, 831.	4193-039
ILLUDOL	$C_{15}H_{24}O_3$	252 -116⁰e 131⁰C	JACS., 1967, 4562.	4193-040

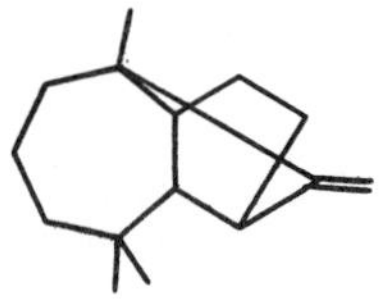

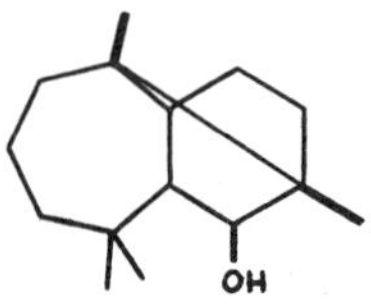

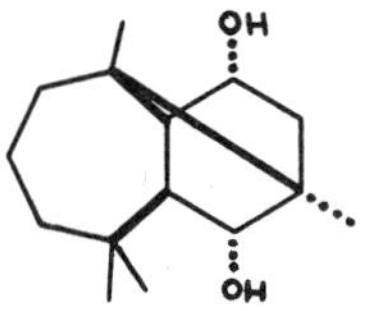

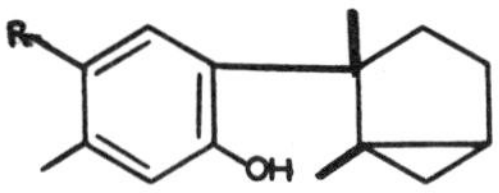

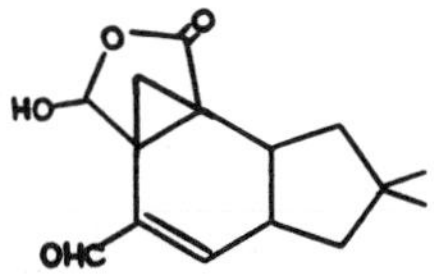

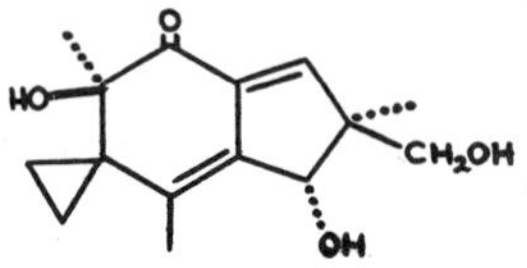

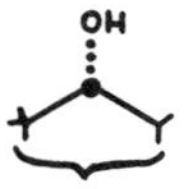

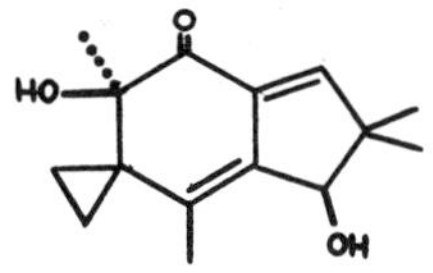

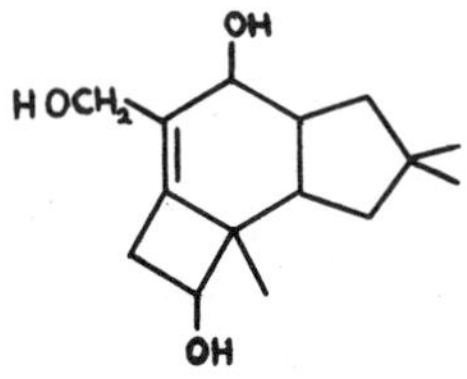

CORIOLIN $C_{15}H_{20}O_5$ 280 175°C A/C R = H Tet. Lett., 1971, 1955 4193-041	EKASANTALAL, tricyclo- $C_{12}H_{18}O$ 178 Oil CHO Tet. Lett., 1970, 38. 4193-046
CORIOLIN C $C_{23}H_{34}O_7$ 422 R = CO•CHOH•$(CH_2)_5$•Me Tet. Lett., 1970, 1637. Tet. Lett., 1971, 1955. 4193-042	ECSANTALAL, nor- $C_{11}H_{16}O$ 164 Oil CHO Ber., 1910 (43) 1893. 4193-047
CORIOLIN B $C_{23}H_{36}O_6$ 408 216°C Tet. Lett., 1971, 1955. R=CO. $(CH_2)_6$. Me 4193-043	KHUSIMONE $C_{14}H_{20}O$ 204 - 78°C Chem. Abstr., 1971 (74) 23019. 4193-048
MAYURONE $C_{14}H_{20}O$ 204 +253°c 70°C Tet. Lett., 1965, 3773. 4193-044	ZIZAENE, pre- $C_{15}H_{24}$ 204 Chem. Ind., 1971, 62. 4193-049
HUMULADIOL, tricyclo- $C_{15}H_{26}O_2$ 238 Chem. Comm., 1970, 117. 4193-045	deleted 4193-050

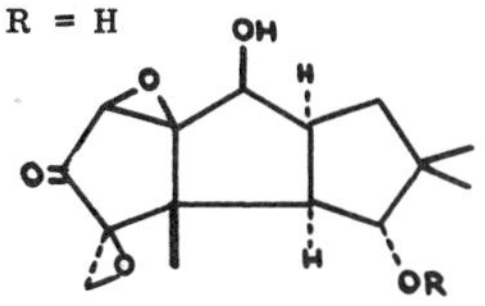

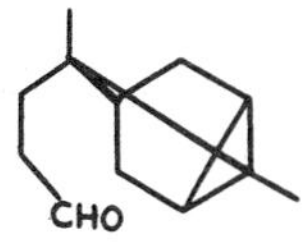

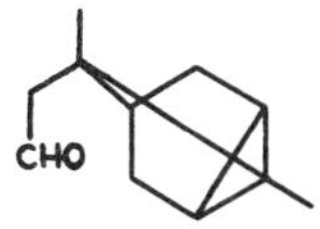

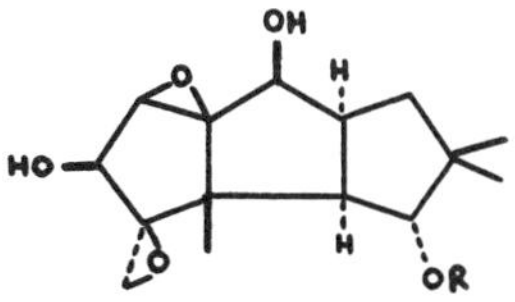

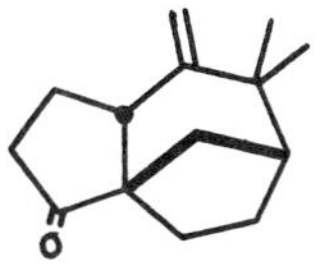

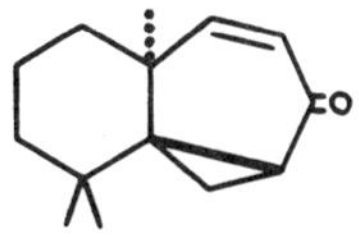

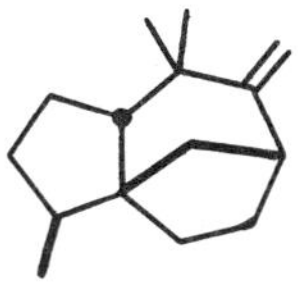

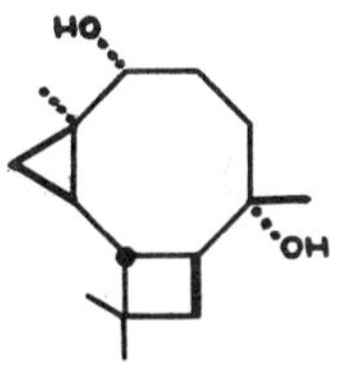

KHUSIMENE	$C_{15}H_{24}$	204	HIRSUTIC ACID C	$C_{15}H_{20}O_4$	264 +115° 182°C
CH₂OH	Chem. Abstr., 1968 (69) 10554.		HO COOH O	Chem. Comm., 1965, 310.	
4193-051			4193-052		

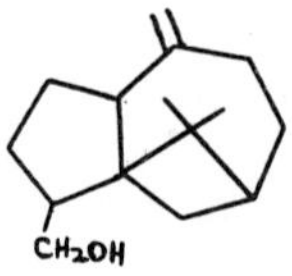

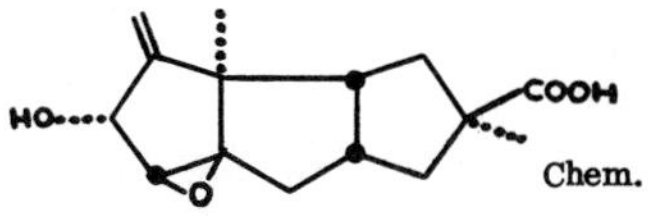

COPACAMPHENIC ACID, cyclo-	$C_{15}H_{22}O_2$	234 -15^0c 152^0C
R = COOH	Tet. Lett., 1969, 3170.	
4194-001		

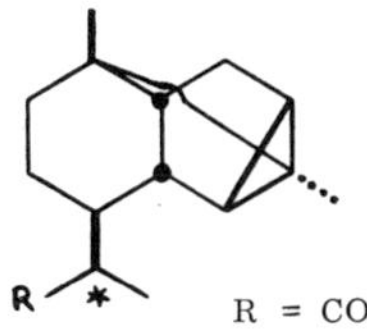

CYCLOSATIVENE	$C_{15}H_{24}$	204 $+94^0c$ Oil
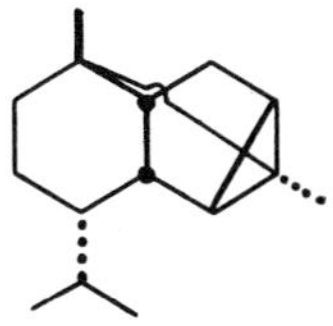	Tet. Lett., 1968, 3833.	
4194-005		

COPACAMPHENIC ACID, epi-cyclo-		234 $+78^0c$ 168^0C
* epi R = COOH	Tet. Lett., 1969, 3170.	
4194-002		

LONGICYCLENE	$C_{15}H_{24}$	204 $+34^0$ Oil
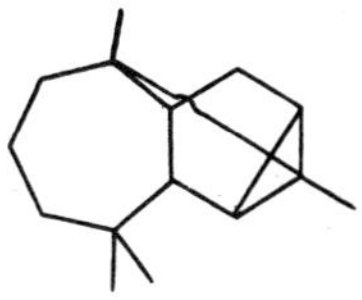	Tet., 1968, 4099.	
4194-006		

COPACAMPHENOL, cyclo-	$C_{15}H_{24}O$	220
R = CH_2OH	Tet. Lett., 1970, 231.	
4194-003		

ISHWARONE	$C_{15}H_{22}O$	218 $+23^0c$ 57^0C
	Tet., 1970, 2371.	
4194-007		

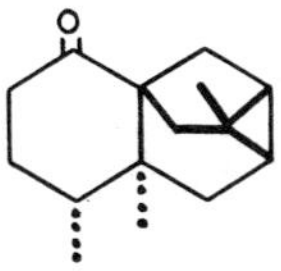

COPACAMPHENOL, epi-cyclo-	$C_{15}H_{24}O$	220
* epi R = CH_2OH	Tet. Lett., 1970, 231.	
4194-004		

NUPHARAMINE $C_{15}H_{25}NO_2$ 251 -35^{0}c J. Pharm. Soc. Jap., 1959, 729, 734. 4198-001	NUPHARIDINE, desoxy- $C_{15}H_{23}NO$ 233 -112^{0}c 21^{0}C Bull. Soc. Chim. Fr., 1959 (32) 892. 4198-006
NUPHARAMINE, anhydro- $C_{15}H_{23}NO$ 233 Oil R = CH_3 Chem. Abstr., 1968 (68) 29907. 4198-002	NUPHARIDINE, 7-epi-deoxy- $C_{15}H_{23}NO$ 233 -40^{0}e 257^{0}C Phytochem., 1970. 659. 4198-007
NUPHAMINE, (-)- $C_{15}H_{23}NO_2$ 249 -60^{0}c Oil R = CH_2OH Chem. Pharm. Bull., 1965, 1247. 4198-003	NUPHARIDINE $C_{15}H_{23}NO_2$ 249 +15^{0}w 221^{0}C N-oxide Bull. Soc. Chim. Fr., 1959 (32) 892. 4198-008
NUPHENINE $C_{15}H_{23}NO$ 233 -23^{0}m R = Me Tet. Lett., 1965, 4229. 4198-004	NUPHARIDINE, dehydro-desoxy- $C_{15}H_{21}NO$ 231 Chem. Pharm. Bull., 1965, 907. 4198-009
NUPHAMINE, 3-epi- $C_{15}H_{23}NO_2$ -47^{0}e R = CH_2OH Phytochem., 1970, 1851. 4198-005	CASTORAMINE $C_{15}H_{23}NO_2$ 249 -31^0 65^{0}C Tet. Lett., 1959 (12) 1. 4198-010

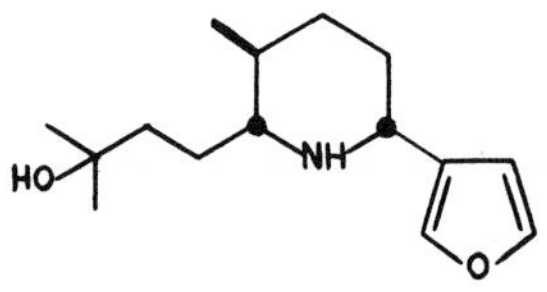

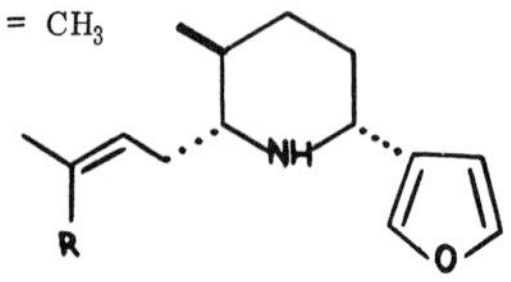

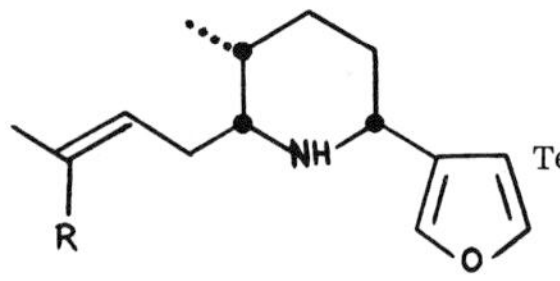

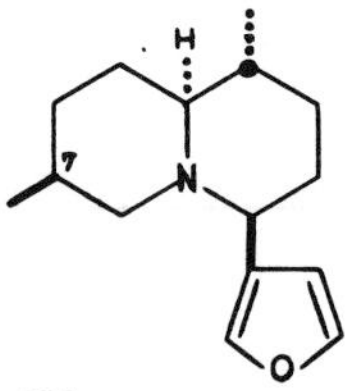

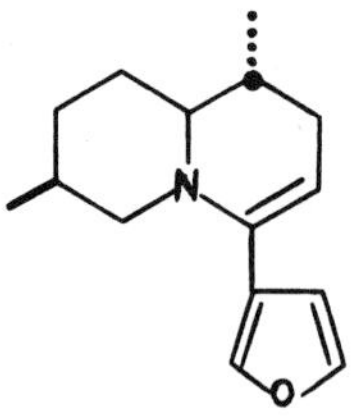

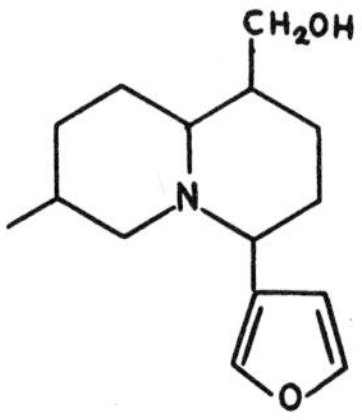

NUPHARIDIN, thio-bi- $C_{30}H_{42}N_2O_2S$ 494 Tet. Lett., 1964, 927. 4198-011	GUAI-PYRIDINE, epi- $C_{15}H_{21}N$ 215 -34^0e Oil JACS., 1966, 3109. 4198-016
NUPHARIDINE, neo-thio-bi- $C_{30}H_{42}N_2O_2S$ R = H Tet. Lett., 1965, 4149. 4198-012	PATCHOULI-PYRIDINE $C_{15}H_{21}N$ 215 -31^0e 25^0C JACS., 1966, 3109. 4198-017
NUPHLUTINE A, 6, 6-di-oh-thio- $C_{30}H_{42}N_2O_4S$ $+45^0$ R = OH Tet. Lett., 1970, 4477. 4198-013	ILLUDININE $C_{16}H_{17}NO_3$ 271 228^0C JOC., 1969, 240. 4198-018
NUPHLUTINE B, 6, 6'-diOH-thio- $C_{30}H_{42}N_2O_4S$ -69^0 stereoisomer Tet. Lett., 1970, 4477. 4198-014	DENDRINE $C_{19}H_{29}NO_4$ 335 -114^0c 191^0C Acta Chem. Scand., 1970, 1108. 4198-019
PULCHELLIDINE $C_{20}H_{33}NO_4$ 351 -22^0e 185^0C Tet. Lett., 1970, 131; 135. 4198-015	NOBILINE $C_{17}H_{27}O_3N$ 293 88^0C Tet. Lett., 1964, 79. 4198-020

DENDRAMINE $C_{16}H_{25}O_3N$ 279 -27°c 187°C Chem. Pharm. Bull. (Tokyo), 1966, 676. 4198-021	DENDROXINE $C_{17}H_{25}O_3N$ 291 -30°e 115°C Chem. Pharm. Bull. (Tokyo), 1966, 672. 4198-024
DENDROBINE $C_{16}H_{25}O_2N$ 263 Chem. Pharm. Bull. (Jap.), 1966, 1058. 4198-022	DENDROXINE, 6-OH- $C_{17}H_{25}O_4N$ 307 Chem. Pharm. Bull. (Tokyo), 1966, 676. 4198-025
DENDROBINE, 2-OH- $C_{16}H_{25}NO_3$ 279 -45°c 104°C Acta Chem. Scand., 1970, 1210. 4198-023	

42-- THE DITERPENES

The 650 compounds presently in this section of the terpene family can be placed into 20 main classes (see Main Skeleton Key) accounting for 560 of the compounds. The remaining 90 compounds with less common skeletons have been collected into four subclasses; 4291 embraces those skeletons whose biogenetic origin derives from some unusual cyclization of the linear C_{20} precursor—unusual, that is, in the sense of rarity rather than by virtue of some unexpected mode of cyclization. Subclass 4292 contains skeletons related to the main classes by rearrangement. Similarly subclass 4293 features skeletons derived from the main classes through degradation of the carbon skeleton. The final class 4299 acts as a receptacle for diterpenoid skeletons which cannot be suitably placed elsewhere—e.g. compounds derived by both degradation and rearrangement. The unique skeletons from these latter classes are displayed under their appropriate headings (see Less Common Sekeletons).

A notable feature of diterpene biogenesis is the natural occurrence of two stereochemical series—normal and antipodal (enantio- or ent-). The classification provided has not attempted to differentiate these series within a given carbon skeleton; it should also be noted in particular that the kaurance, ent-kaurane and phyllocladane skeletons are together in class 4206.

Proposed biogenetic interrelationships between the diterpene skeletons are outlined on the chart following.

TERPENES

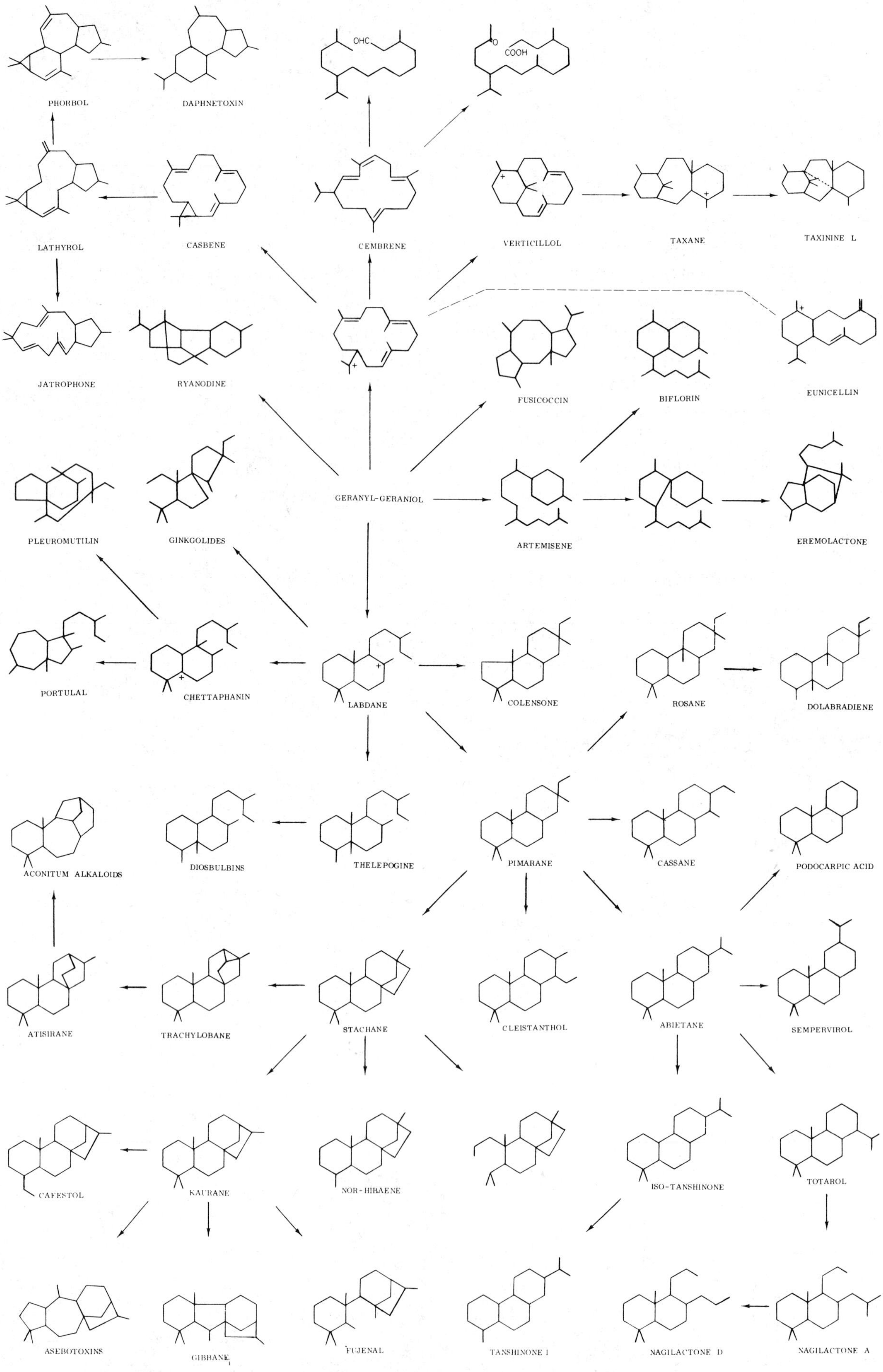

PHORBOL
DAPHNETOXIN
OHC
COOH
LATHYROL
CASBENE
CEMBRENE
VERTICILLOL
TAXANE
TAXININE L
JATROPHONE
RYANODINE
FUSICOCCIN
BIFLORIN
EUNICELLIN
PLEUROMUTILIN
GINKGOLIDES
GERANYL-GERANIOL
ARTEMISENE
EREMOLACTONE
PORTULAL
CHETTAPHANIN
LABDANE
COLENSONE
ROSANE
DOLABRADIENE
ACONITUM ALKALOIDS
DIOSBULBINS
THELEPOGINE
PIMARANE
CASSANE
PODOCARPIC ACID
ATISIRANE
TRACHYLOBANE
STACHANE
CLEISTANTHOL
ABIETANE
SEMPERVIROL
CAFESTOL
KAURANE
NOR-HIBAENE
ISO-TANSHINONE
TOTAROL
ASEBOTOXINS
GIBBANE
FUJENAL
TANSHINONE I
NAGILACTONE D
NAGILACTONE A

42--	Diterpenes-Main Skeleton Key

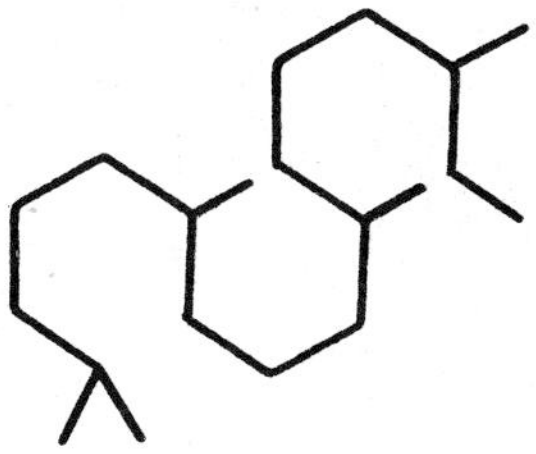

01 Linear	05 Totarol group
02 Labdane	06 Kaurane (see also 14)
03 Pimarane	07 Cassane
04 Abietane	08 Rosane

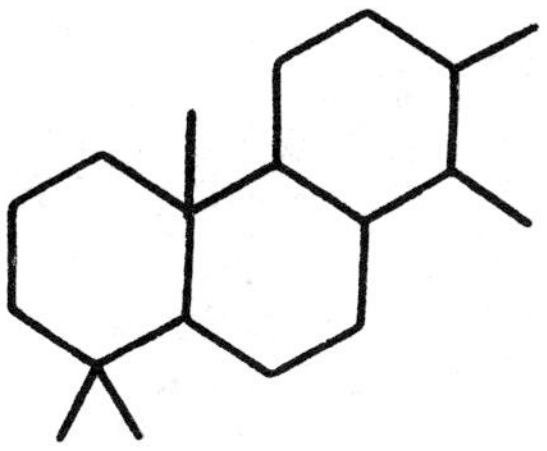

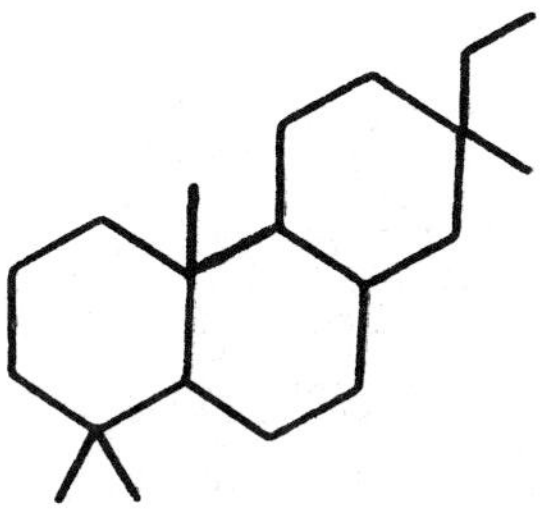

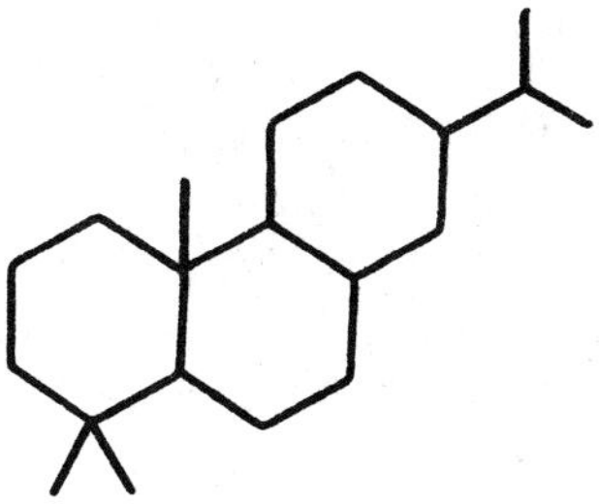

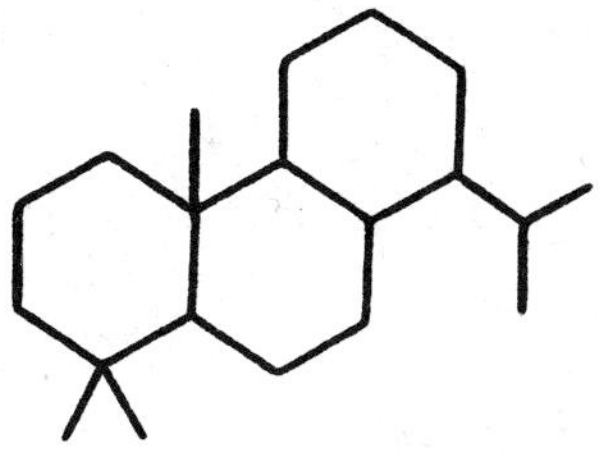

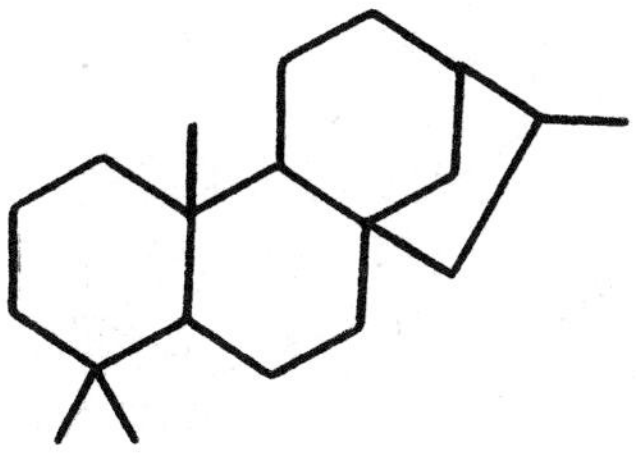

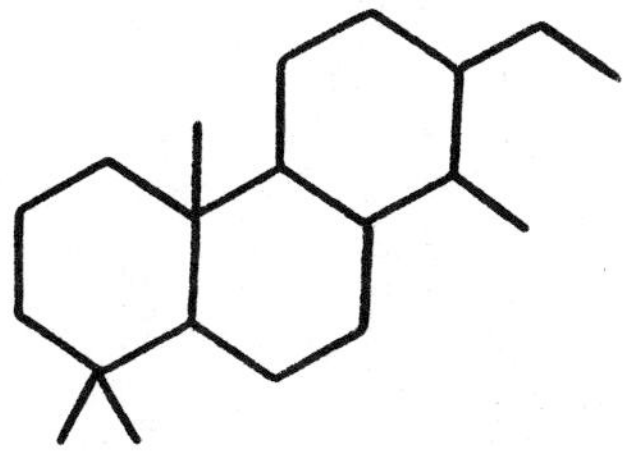

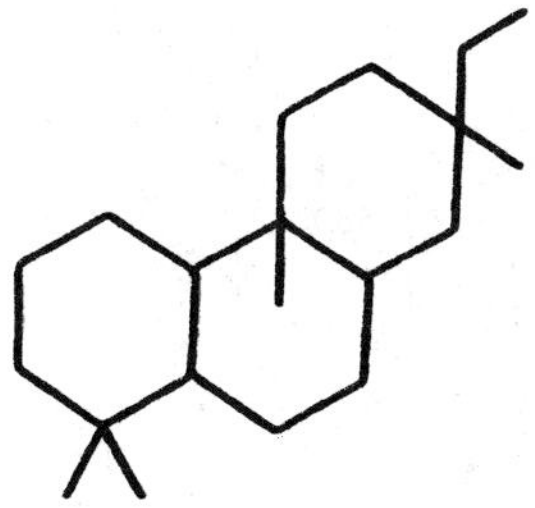

42__	Diterpenes- Main Skeleton Key

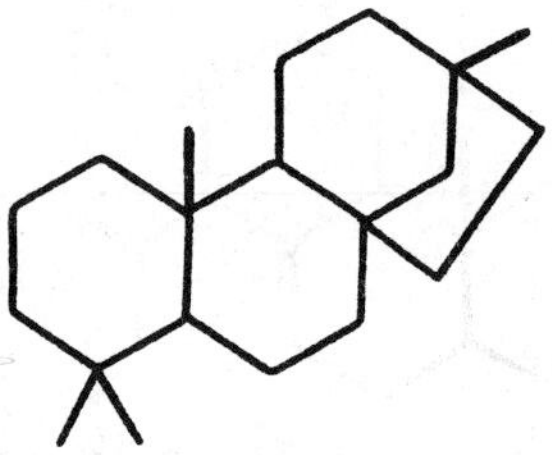

09 Stachane	13 Aconitum Alkaloids 14 Kaurane Alkaloids
10 Gibbane	15 Taxane
11 Trachylobane	16 Cembranc
12 Atisane	17 Kolevane

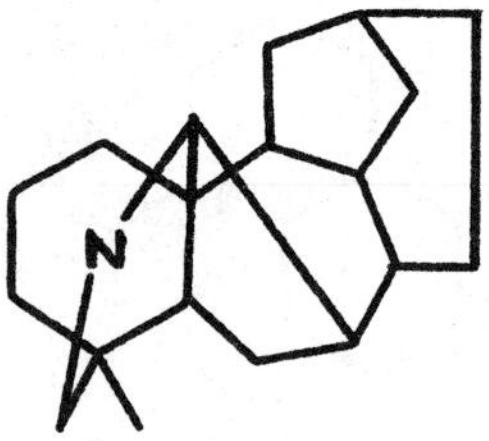

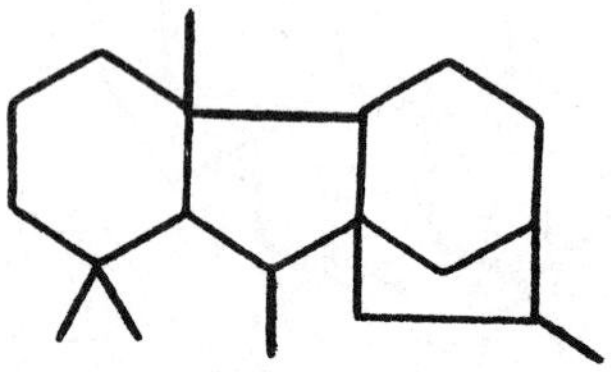

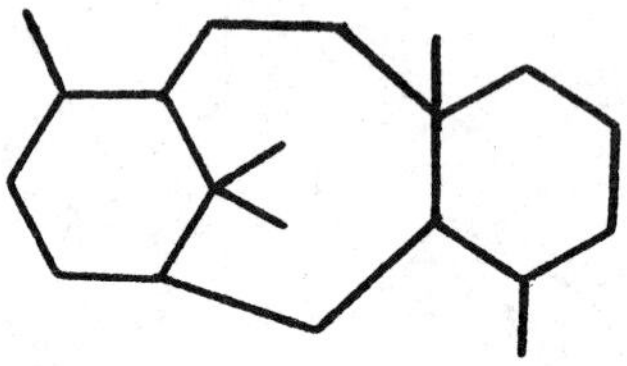

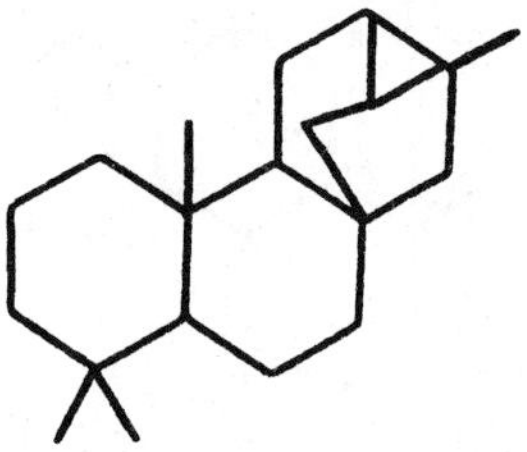

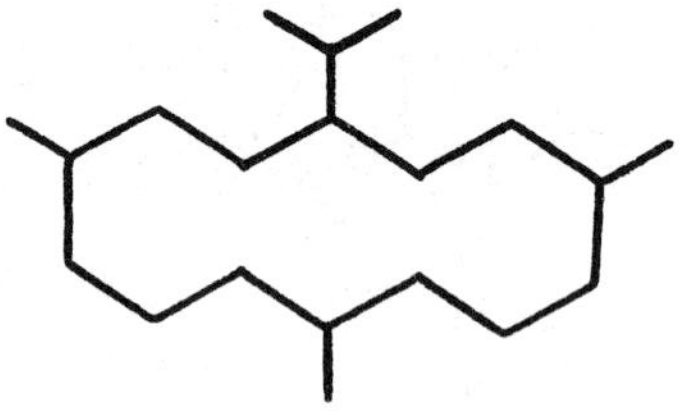

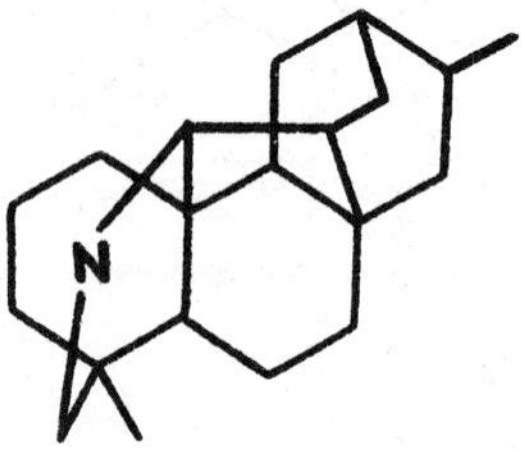

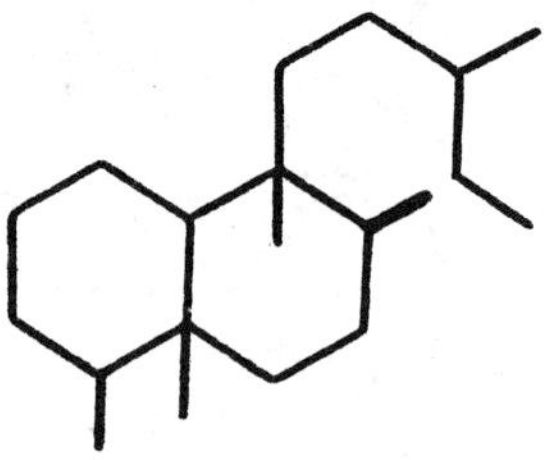

42--	Diterpenes-Main Skeleton Key

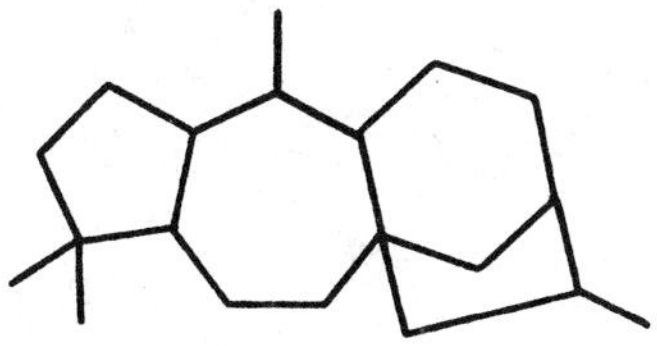

018 Ericacane	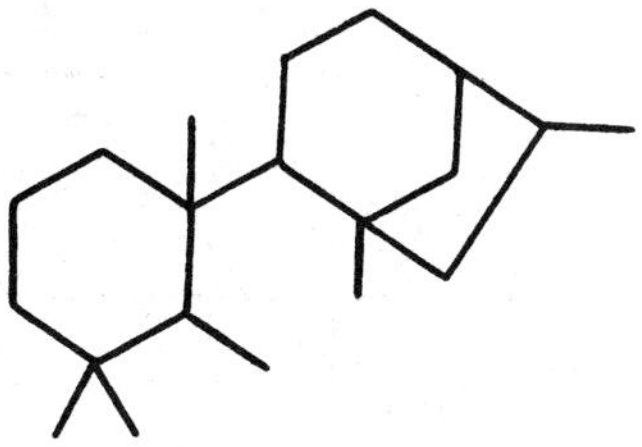
019 Fujinane	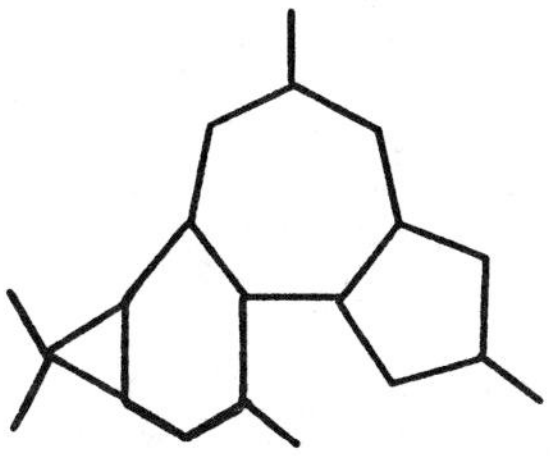
020 Tigliane	

4291	Diterpenes-Less Common Skeletons-Unusually Cyclised

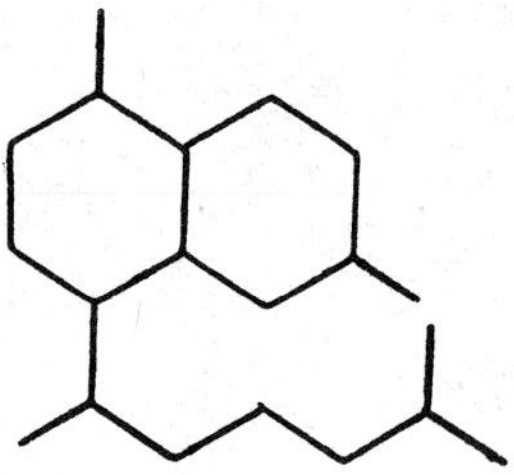

001	006
002 003 004 005	007 008
021	009

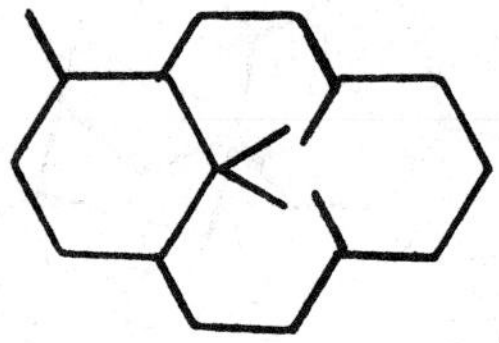

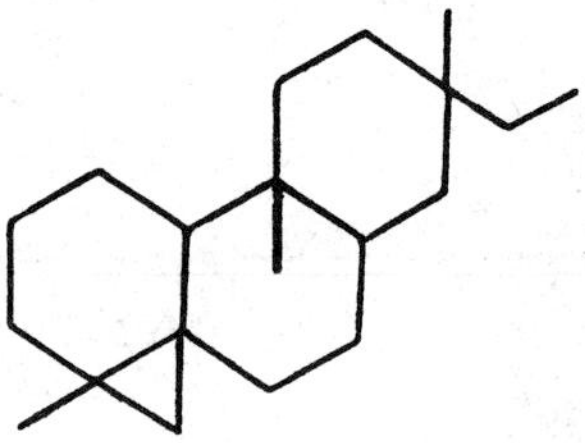

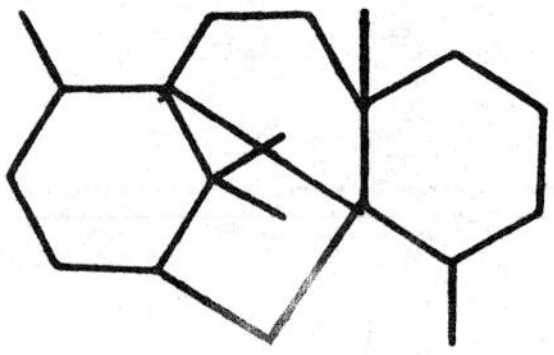

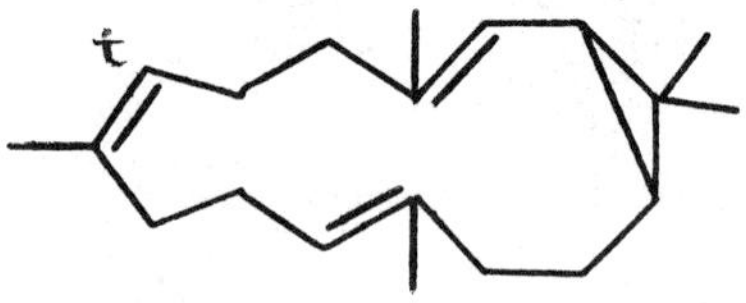

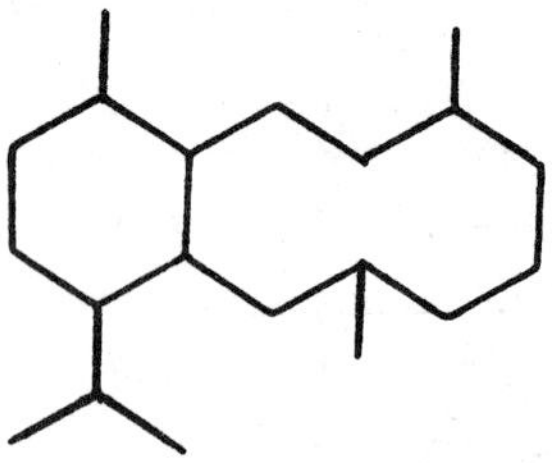

429]	Diterpenes-LessCommon Skeletons-Unusually Cyclised

010

011

012
013
014

015

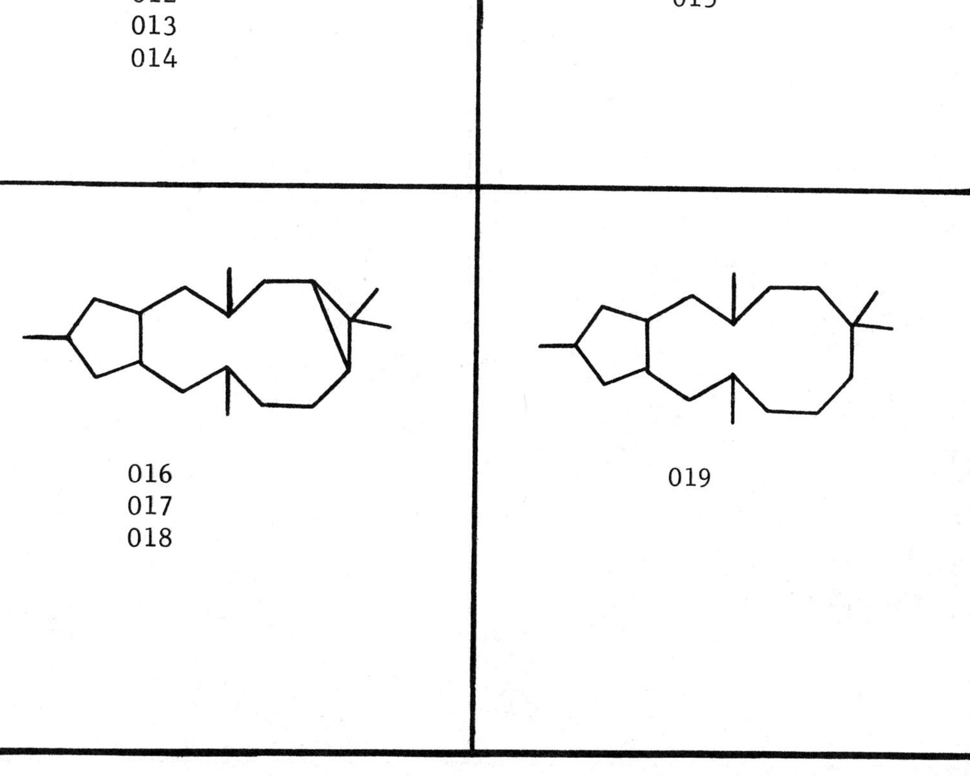

4291	Diterpenes-Less Common Skeletons-Unusually Cyclised

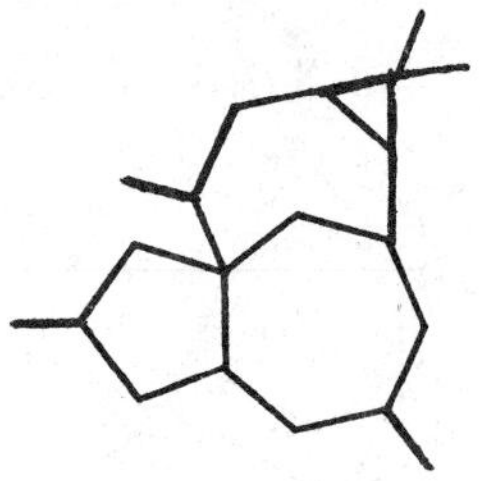

020

4292	Diterpenes-Less Common Skeletons-Rearranged

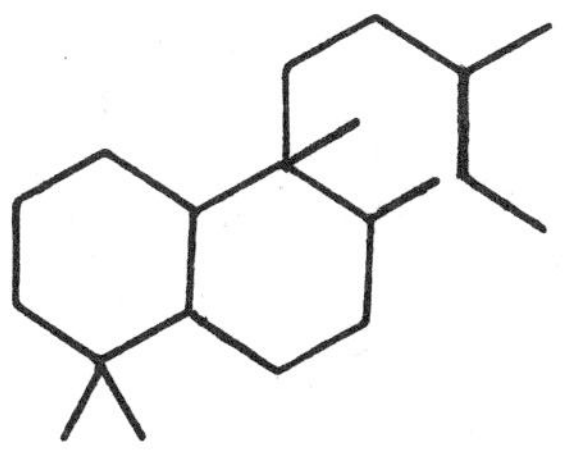

001

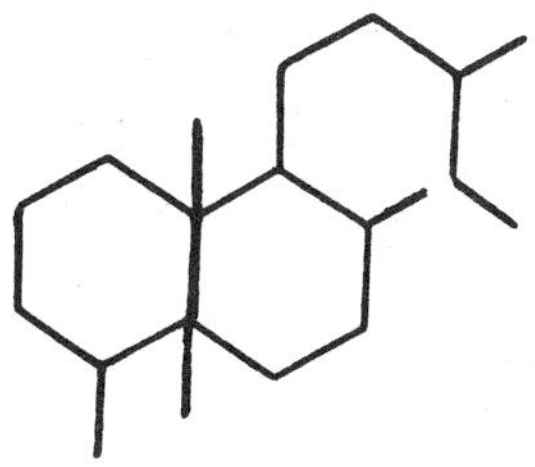

006

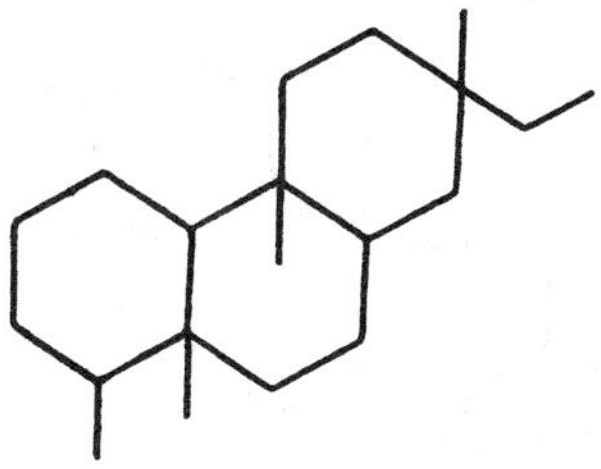

002
003
004

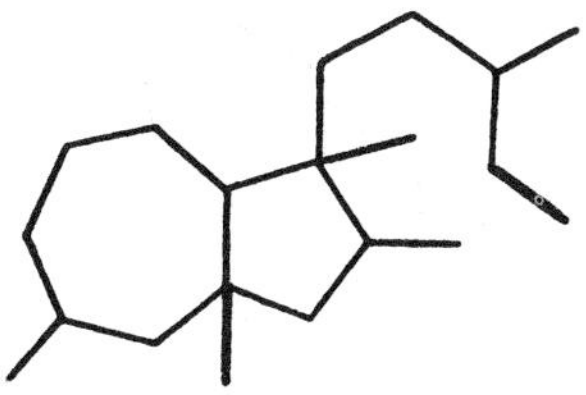

007

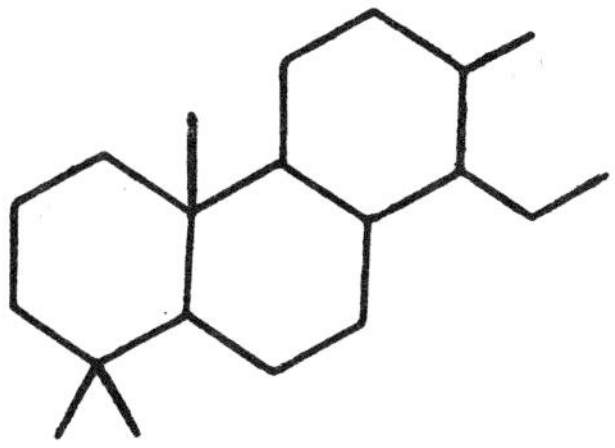

005

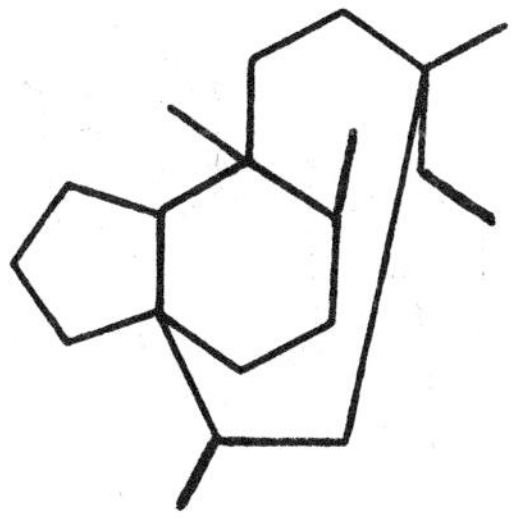

008

4292	Diterpenes-Less Common Skeletons-Rearranged

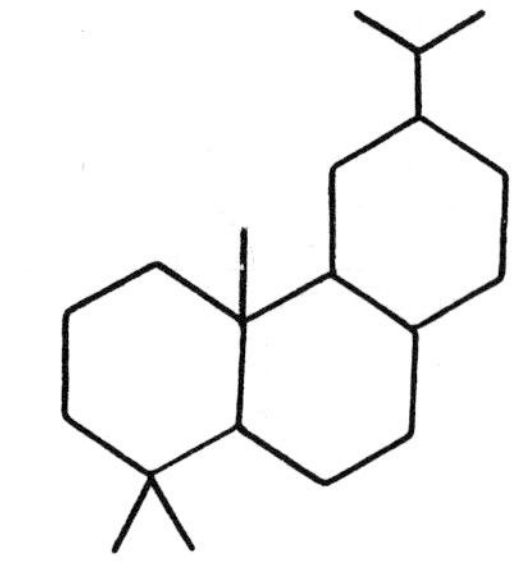

009	013 014
010 011	012
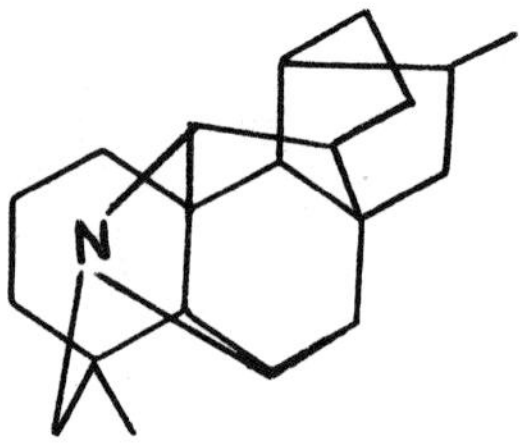 N 015	

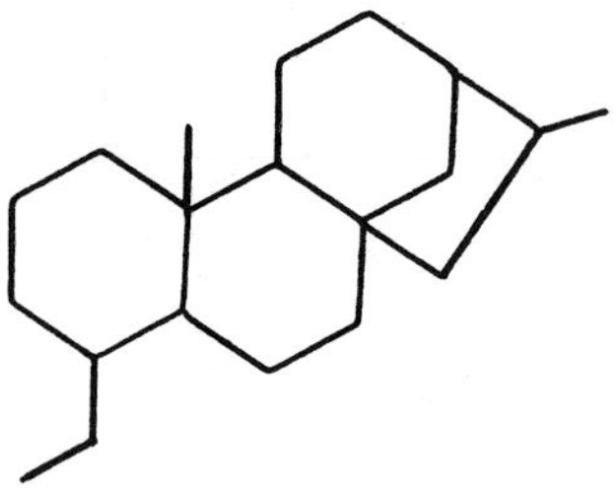

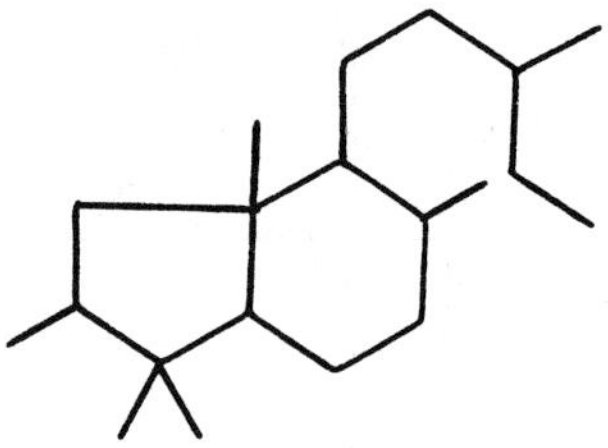

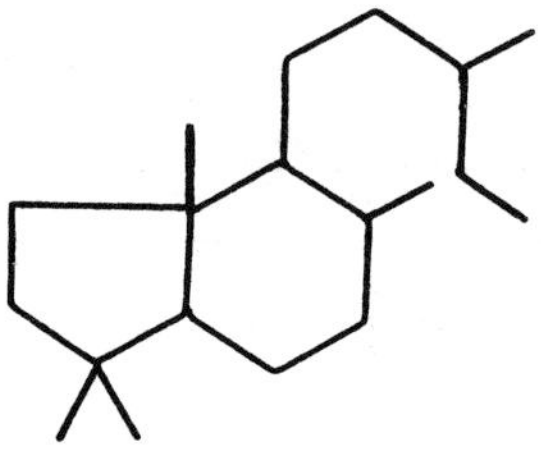

4293	Diterpenes-Less Common Skeletons-Degraded

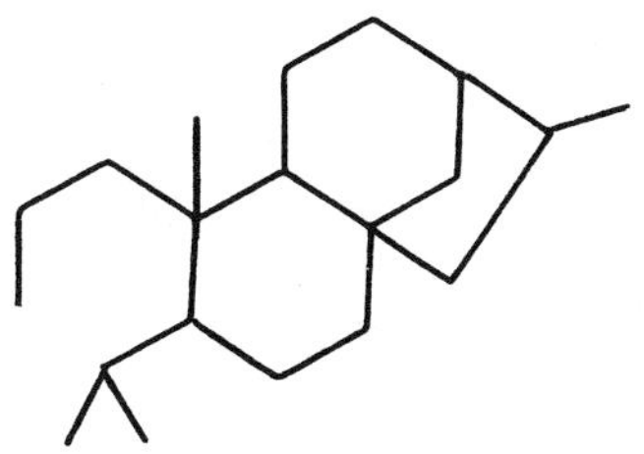

001	003
002	047
048	004

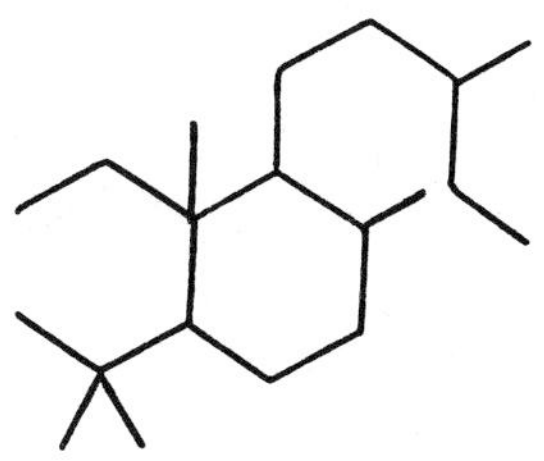

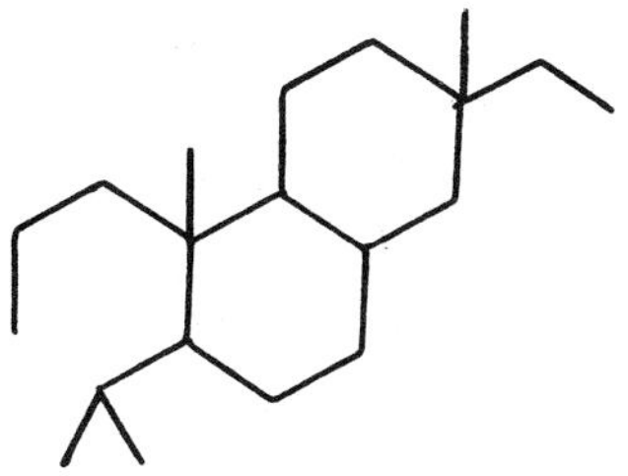

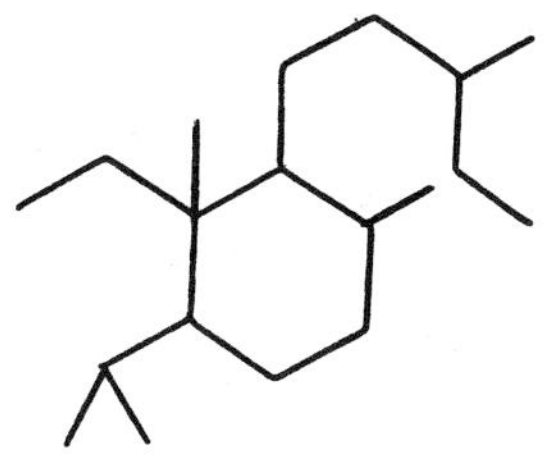

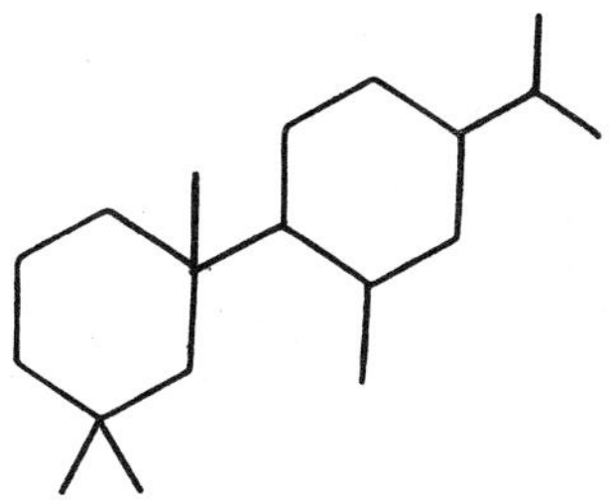

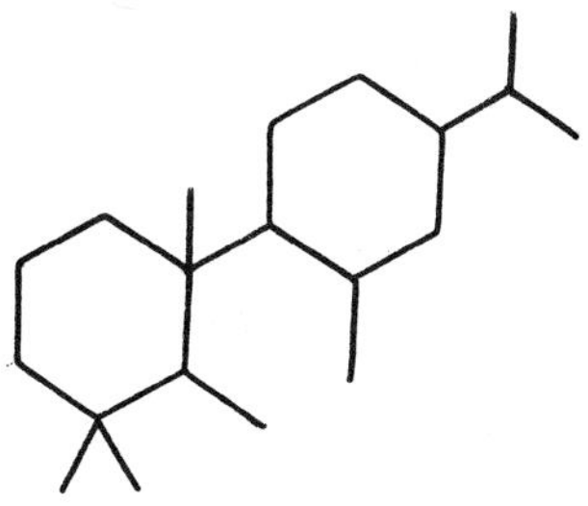

4293	Diterpenes-Less Common Skeletons-Degraded

N 005 006 007 008 009	011 010
033 034 035 036	041 042 043 044
032	031

4293	Diterpenes-Less Common Skeletons-Degraded

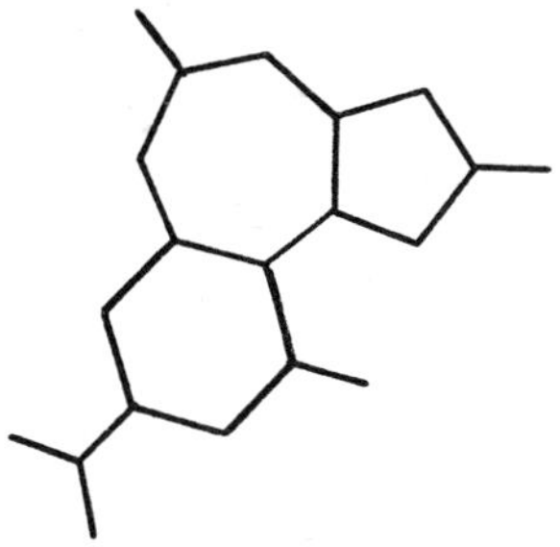

038 039	040
013	014 015
012	037

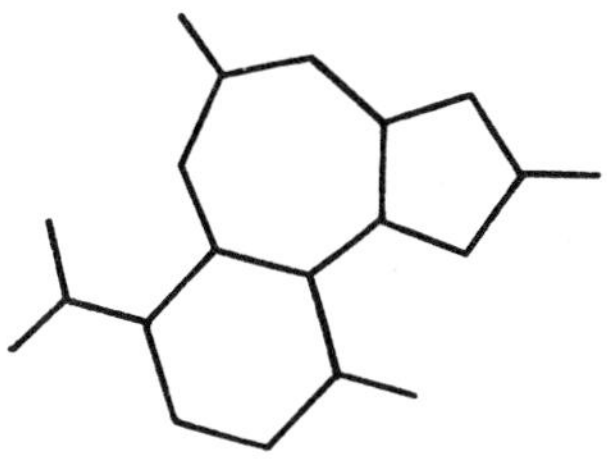

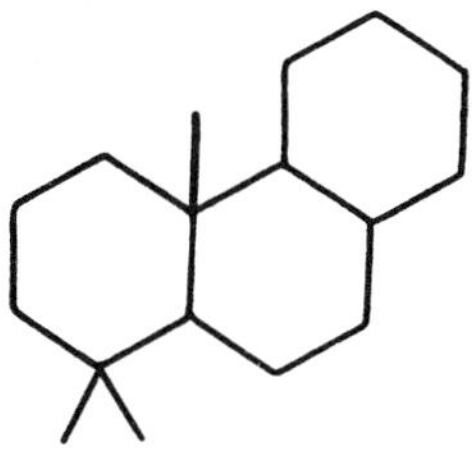

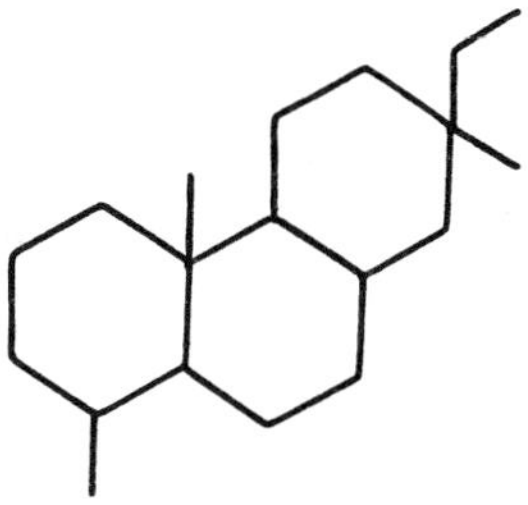

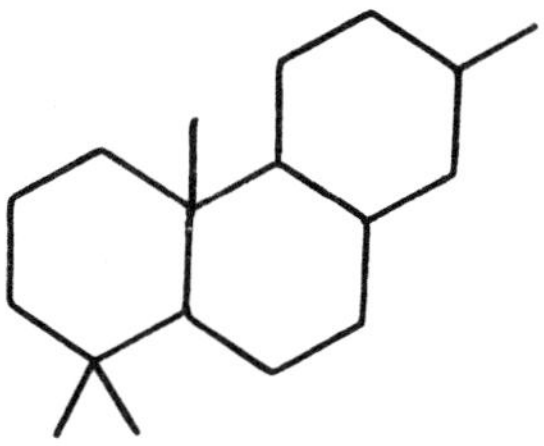

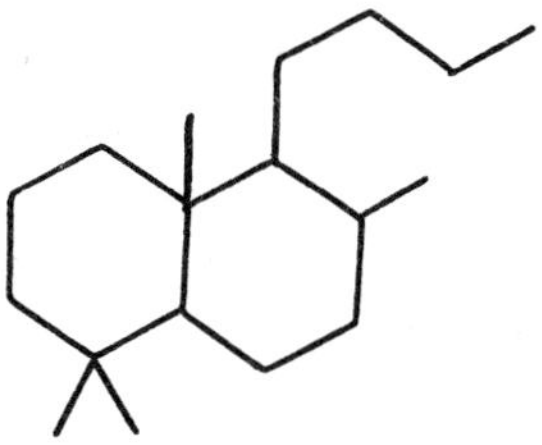

4293	Diterpenes-Less Common Skeletons-Degraded

028 029	045 046
016 017 018 019 020 021 022 023	026 030
024 025	

4299	Diterpenes — Less Common Skeletons - Misc.

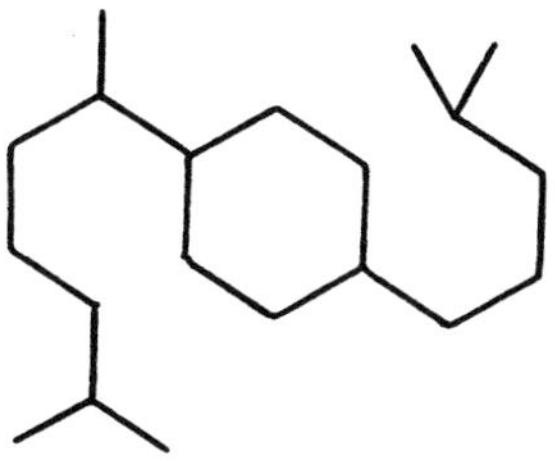

001	004
002 003	005 006 007

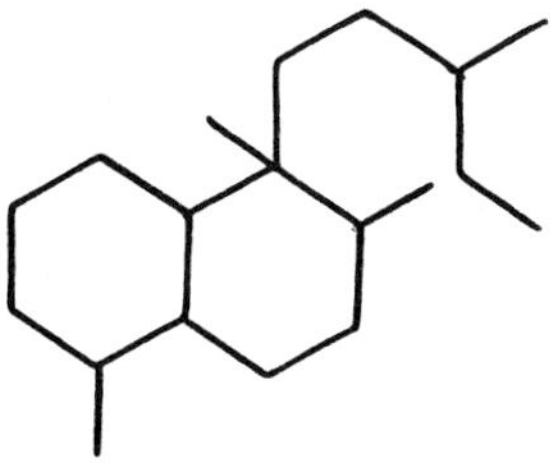

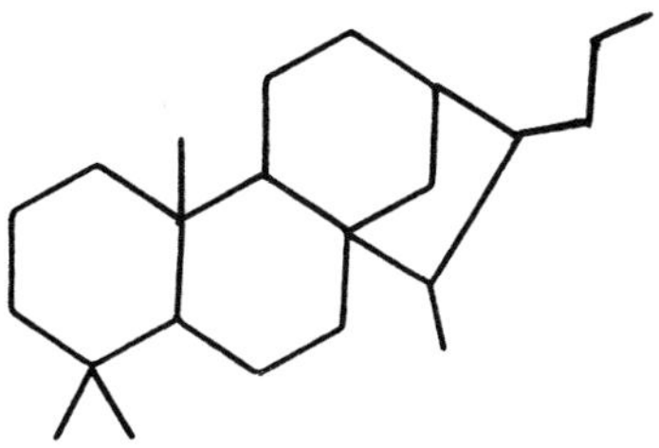

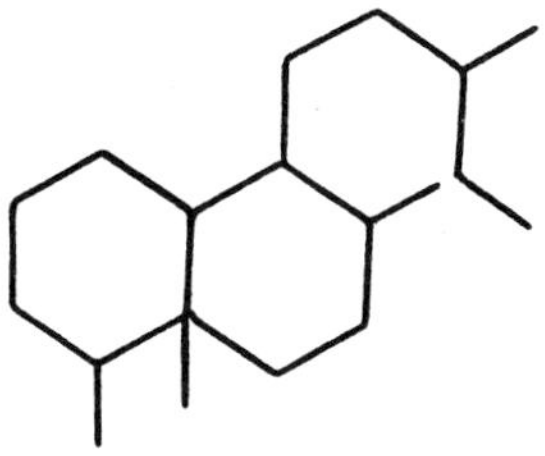

GERANYLGERANIOL $C_{20}H_{34}O$ 290
OI
Oil

CH₂OH

Tet. Lett., 1967 (2) 189.

4201-001

PHYTADIENE, 1,3-cis- $C_{20}H_{38}$ 278

Science, 1965 (147) 1148.

4201-006

GERANYL-LINALOOL, 3R- $C_{20}H_{34}O$ 290
-10.1^{0}c
Oil

OH

Acta Chem. Scand., 1967 (21) 825.

4201-002

PHYTADIENE, 1,3-trans- $C_{20}H_{38}$ 278

Science, 1965 (147) 1148.

4201-007

GERANYL-LINALOOL, 3S- $C_{20}H_{34}O$ 290
+14.5^{0}
Oil

Bull. Soc. Chim. Fr., 1958, 1128.

4201-003

PHYTADIENE, 2,4- $C_{20}H_{38}$ 278

Science, 1965 (147) 1148.

4201-008

PHYTOL $C_{20}H_{40}O$ 296

Oil

CH₂OH

JCS., 1966, 2144.

4201-004

PHYTANIC ACID $C_{20}H_{40}O_2$ 312

Oil

COOH

Nature, 1966 (210) 841.

4201-009

PHYTADIENE, neo- $C_{20}H_{38}$ 278

Oil

Perf. Ess. Oil Rec., 1962 (53) 685.

4201-005

SCLARENE $C_{20}H_{32}$ 272 Oil Aust. J. C., 1967 (20) 157. 4202-001	COMMUNOL, cis- $C_{20}H_{32}O$ 288 HOCH$_2$ Acta Chem. Scand., 1966 (20) 1074. 4202-006
BIFORMENE $C_{20}H_{32}$ 272 + 12^0 Oil JCS, 1961, 2187. 4202-002	ELLIOTINOL $C_{20}H_{32}O$ 288 + 14^0 15^{0}C CH$_2$OH JOC, 1965 (30) 429. 4202-007
LABDA-8(20)-13-DIEN-15-OL $C_{20}H_{34}O$ 290 CH$_2$OH Chem. Abstr., 1967 (67) 32822. 4202-003	OZOL $C_{20}H_{32}O$ 288 Oil CH$_2$OH JCS, 1968, 1063. 4202-008
ABIENOL $C_{20}H_{34}O$ 290 + 20^0 64^{0}C OH Zhur. Ob. Khim., 1962 (32) 656. 4202-004	MANOOL, (+)- $C_{20}H_{34}O$ 290 + 30^{0}e 53^{0}C OH Nature, 1952, 1018. 4202-009
ABIENOL, iso- $C_{20}H_{34}O$ 290 OH Chem. Abstr., 1966 (65) 13772. 4202-005	MANOOL, 13-epi- $C_{20}H_{34}O$ 290 + 51^0 138^{0}C OH Tet. Lett., 1965, 2633. 4202-010

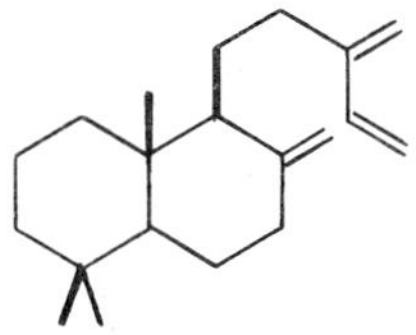

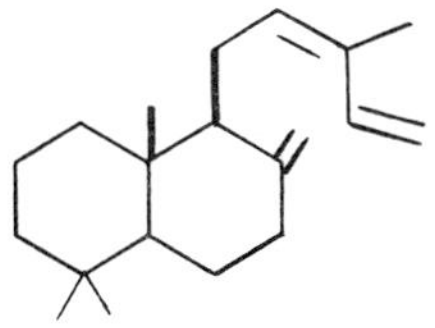

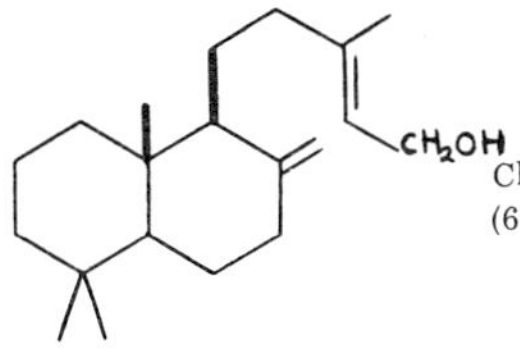

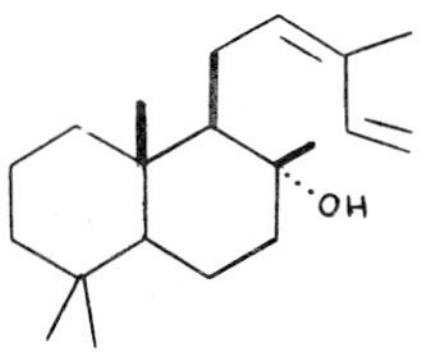

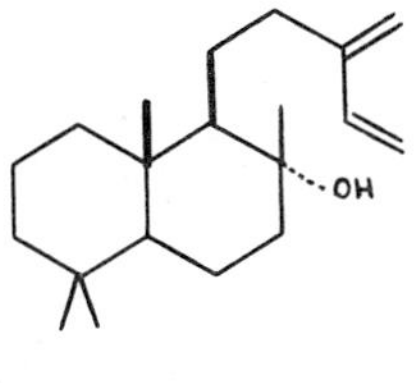

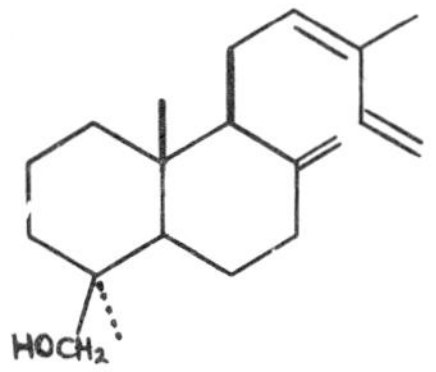

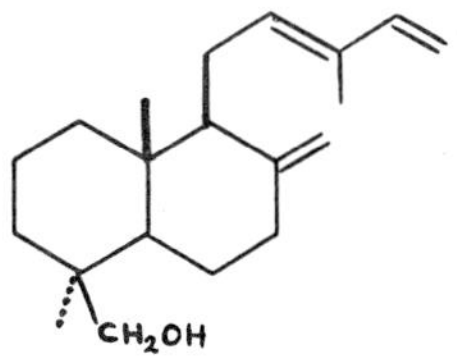

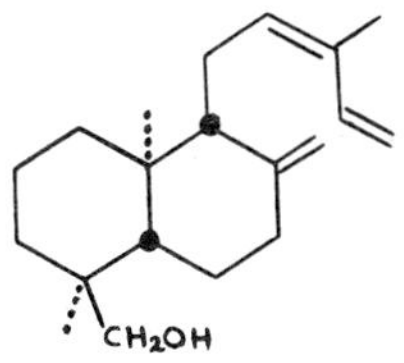

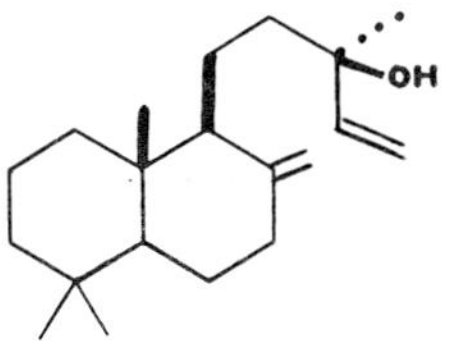

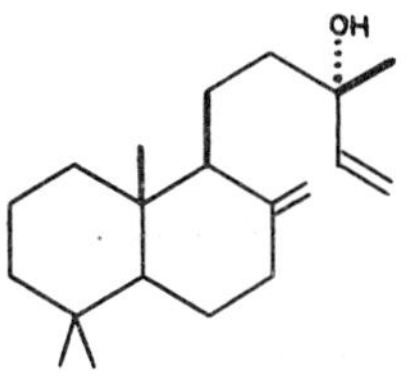

MANOOL, enantio-13-epi- $C_{20}H_{34}O$ 290 Tet., Supp. 8, 1966, 203. 4202-011	LARIXYL ACETATE $C_{22}H_{36}O_3$ 348 +67^0 82^{0}C Tet. Lett., 1965, 3523. 4202-016
MANOOL, ent-18-OH-13-epi- $C_{20}H_{34}O_2$ 306 -59^0 150^{0}C Tet., 1967, Supp. 8, Pt. I, 203. 4202-012	LABD-8(20)-EN-15, 18-DIOL, enantio- $C_{20}H_{34}O_2$ 306 - 36^0 Tet., Supp. 8, Pt. I, 203. 4202-017
TORULOSOL $C_{20}H_{34}O_2$ 306 +31^{0}c 111^{0}C Acta Chem. Scand., 1961, 1303. 4202-013	EPERUANE-8, 15-DIOL $C_{20}H_{38}O_2$ 310 +4^0 76^{0}C Tet., 1965, 1175. 4202-018
TORULOSOL, epi- $C_{20}H_{34}O_2$ 306 +43^0 114^{0}C Chem. Abstr., 1967 (67) 32822. 4202-014	LABDANE-8, 15-DIOL $C_{20}H_{38}O_2$ 310 - 10^{0}c 84^{0}C JCS, 1962, 4705. 4202-019
LARIXOL $C_{20}H_{34}O_2$ 306 +58^0 101^{0}C Tet. Lett., 1967 (3) 219. 4202-015	LABD-13-EN-8, 15-DIOL, enantio- $C_{20}H_{36}O_2$ 308 +1^0 132^{0}C Aust. J. C., 1965, 1441. 4202-020

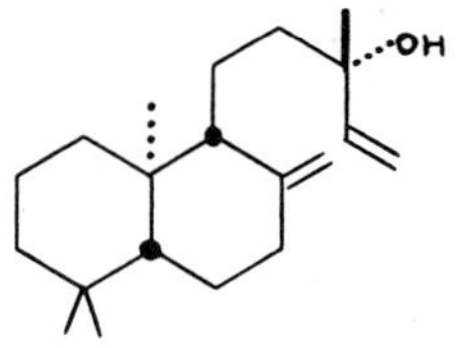

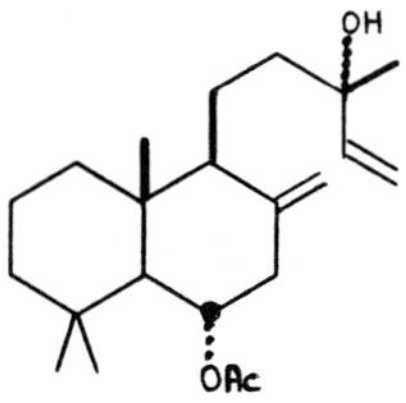

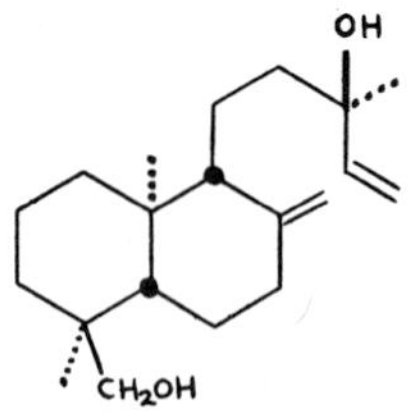

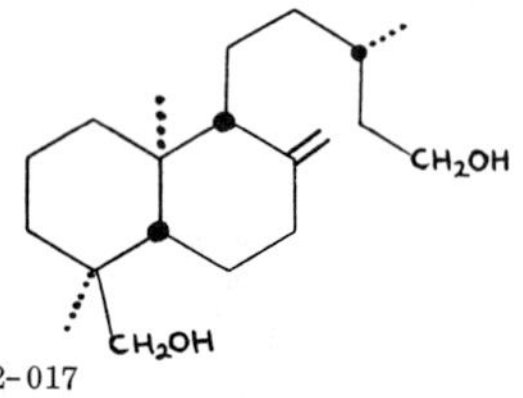

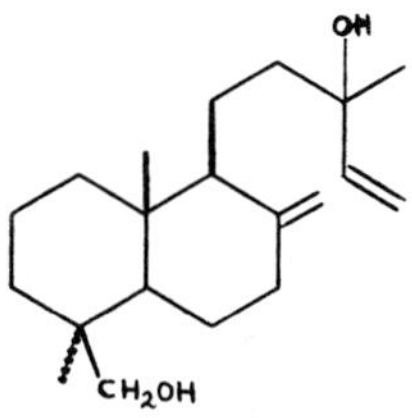

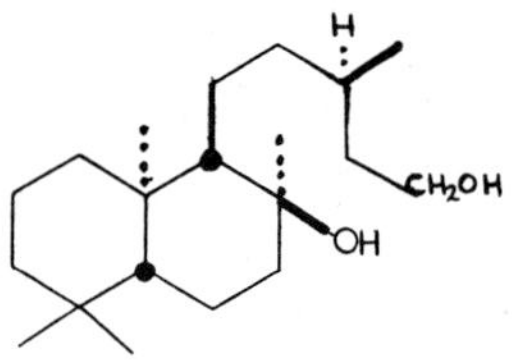

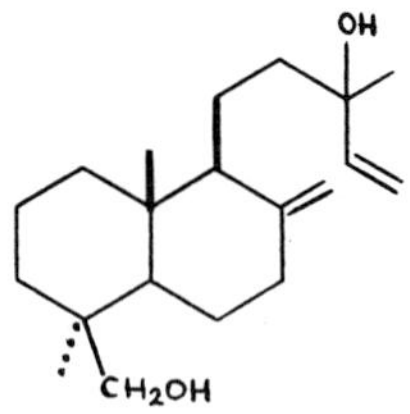

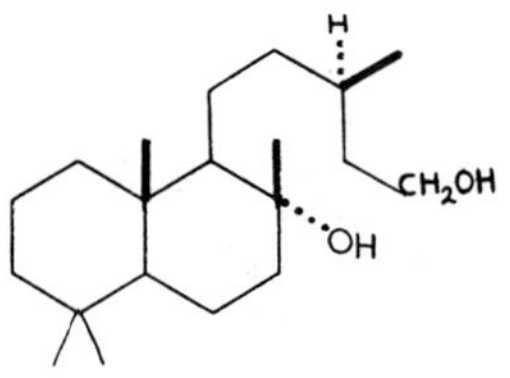

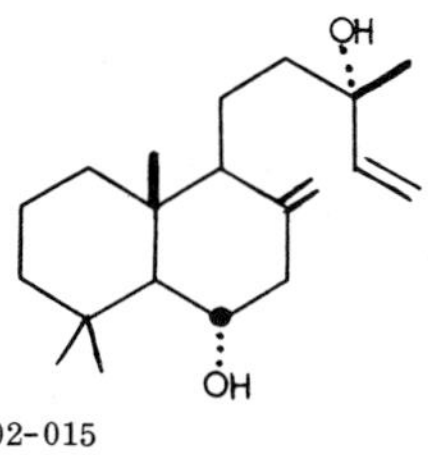

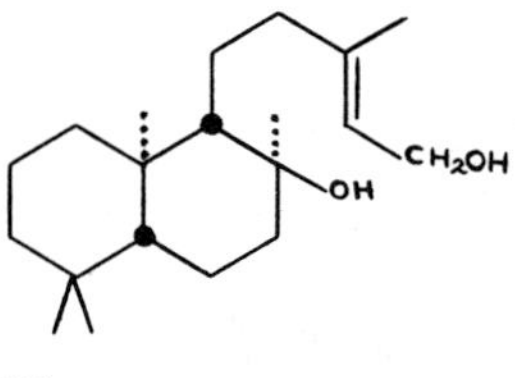

SCLAREOL $C_{20}H_{36}O_2$ 308; -3^0c; 105^0C. Chem. Ind., 1959, 1378. 4202-021	LABD-8-EN-3, 15-DIOL $C_{20}H_{36}O_2$ 308; $+29^0$; 114^0C. JCS, 1964, 3648. 4202-026
PEREGRINOL $C_{20}H_{36}O_2$ 308. Chem. Abstr., 1967 (67) 43948. 4202-022	EPERUANE-8, 15, 18-TRIOL $C_{20}H_{38}O_3$ 326; -13^0e; 121^0C. Tet., 1965, 1175. 4202-027
AGATHADIOL $C_{20}H_{34}O_2$ 306. Chem. Abstr., 1968 (68) 3015. 4202-023	LABDANE-8α, 15, 19α-TRIOL $C_{20}H_{38}O_3$ 326; 188^0C. Bull. Soc. Chim., 1963, 2299. 4202-028
CONTORTADIOL $C_{20}H_{34}O_2$ 306. Tet. Lett., 1965, 2633. 4202-024	LAGOCHILIN $C_{20}H_{36}O_5$ 356; 158^0C. Tet. Lett., 1969, 1361. 4202-029
APLYSIN-20 $C_{20}H_{35}O_2Br$ 386; -78^0m; 146^0C. Chem. Comm., 1967, 898. 4202-025	MANOYL OXIDE, (+)- $C_{20}H_{34}O$ 290; $+20^0e$; 29^0C. Chem. Ind., 1961, 1574. 4202-030

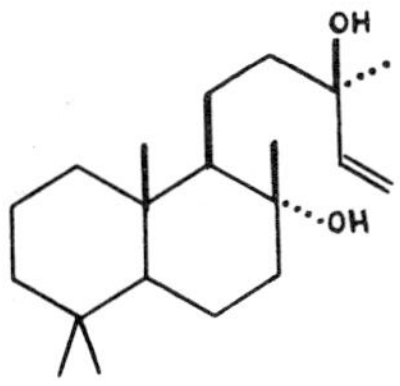

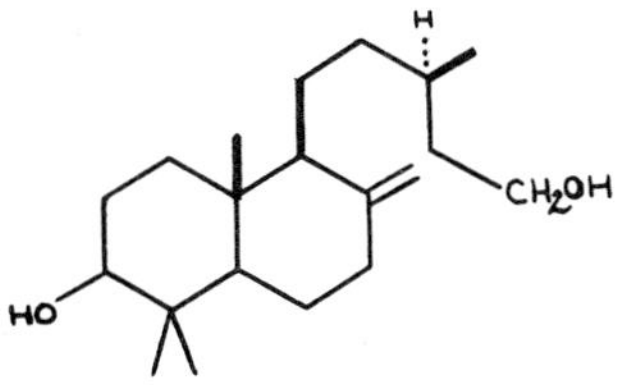

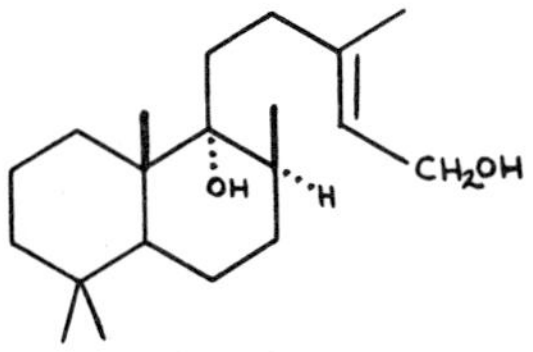

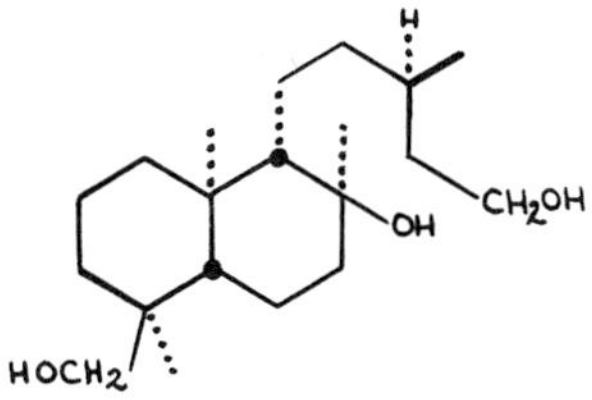

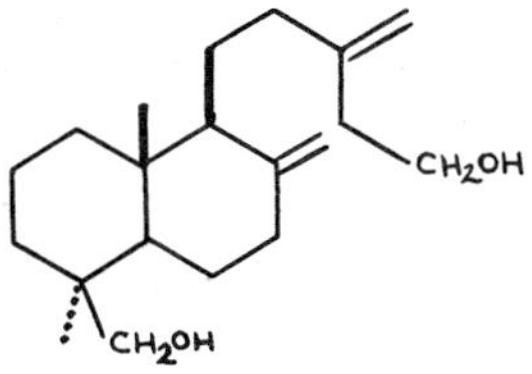

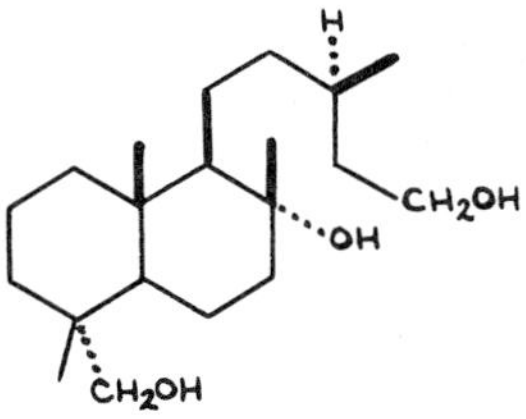

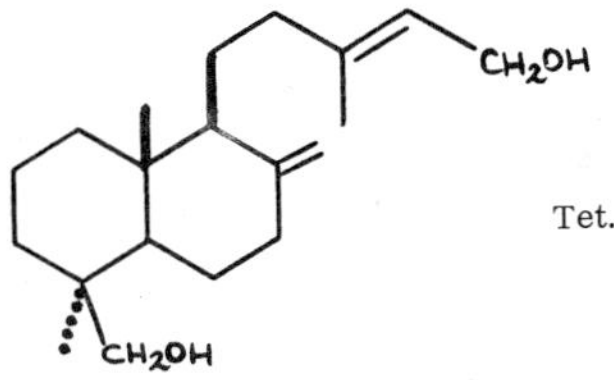

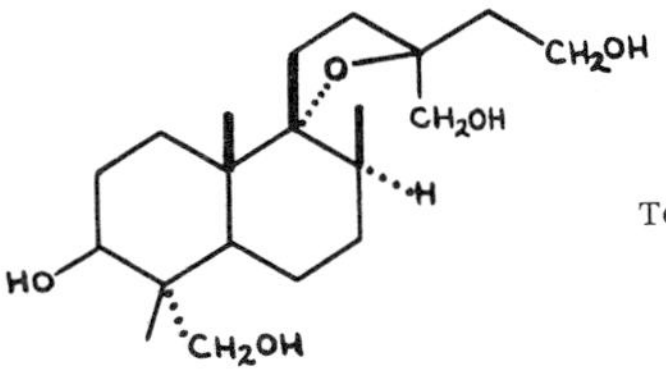

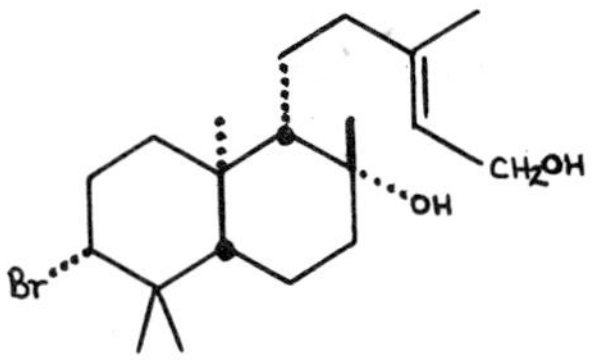

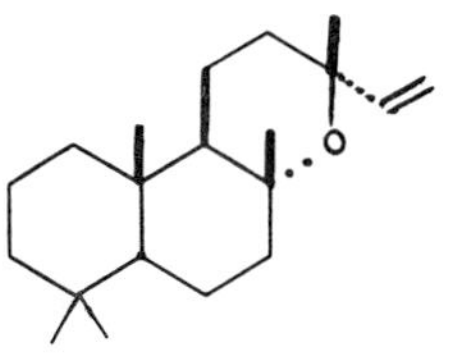

Name	Formula	Data	Reference	No.
MANOYL OXIDE, 8-epi-	$C_{20}H_{34}O$	290 -9^0c 44^0C	Tet. Lett., 1970, 1131.	4202-031
MANOYL OXIDE, 2α-OH-	$C_{20}H_{34}O_2$	306 57^0C	JCS, 1965, 3846.	4202-036
MANOYL OXIDE, (+)-13-epi-	$C_{20}H_{34}O$	290 $+38^0c$ 97^0C	Chem. Abstr., 1967, 65656.	4202-032
MANOYL OXIDE, 12α-OH-13-epi-	$C_{20}H_{34}O_2$	306 $+44^0c$ 141^0C	Tet., 1962 (18) 169.	4202-037
MANOYL OXIDE, ent-13-epi-	$C_{20}H_{34}O$	290 -37^0e 99^0C	JCS, 1963, 1937.	4202-033
MANOYL OXIDE, 2-keto-	$C_{20}H_{32}O_2$	304 $+40^0e$ 77^0C	Chem. Ind., 1960, 1300.	4202-038
MANOUL OXIDE, 8, 13-di-epi-	$C_{20}H_{34}O$	290 $+23^0c$ 82^0C	Tet. Lett., 1970, 1131.	4202-034
MANOYL OXIDE, 18-OH-2-keto-	$C_{20}H_{32}O_3$	320 114^0C	Tet., 1965, 3591.	4202-039
MANOYL OXIDE, (+)-3-keto-	$C_{20}H_{32}O_2$	304 $+54^0c$ 99^0C	JCS, 1963, 644.	4202-035
MANOYL OXIDE, 19-OH-13-epi-enantio-	$C_{20}H_{34}O_2$	306 -47^0 186^0C	Aust. J. C., 1965, 1441.	4202-040

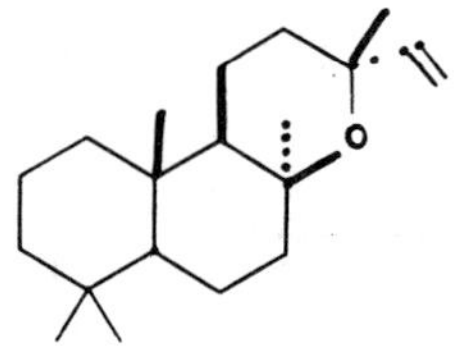

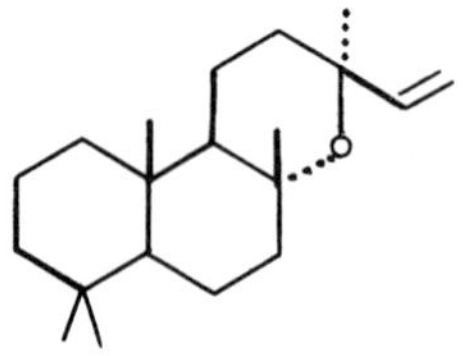

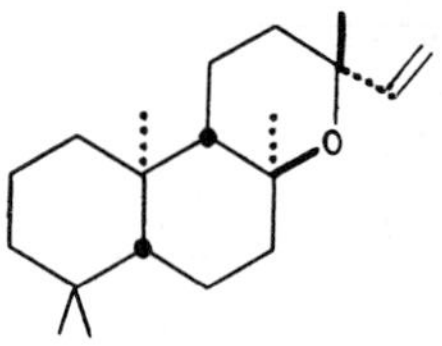

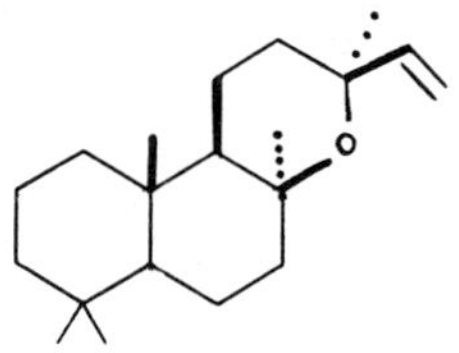

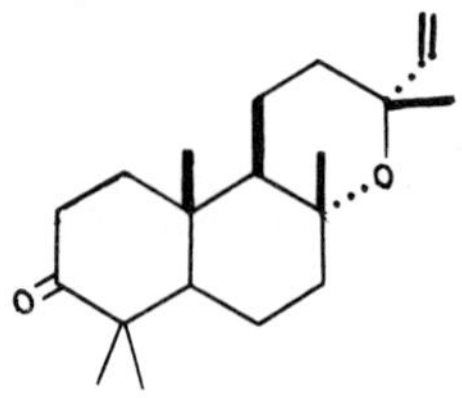

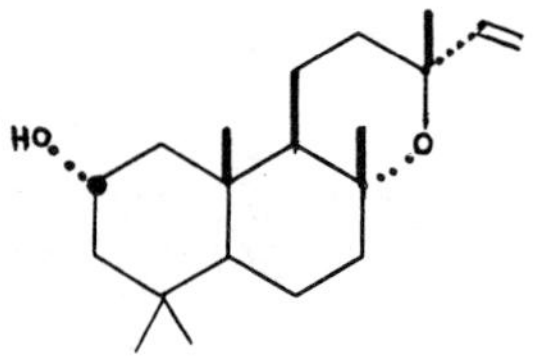

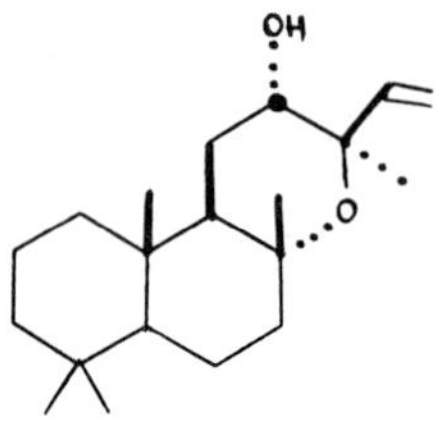

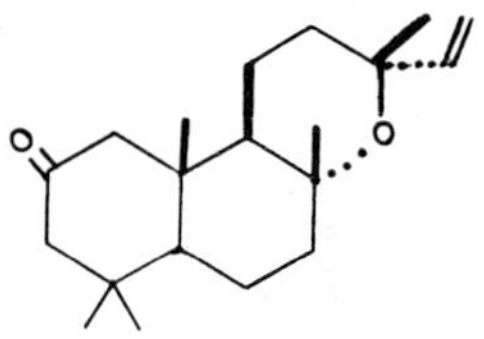

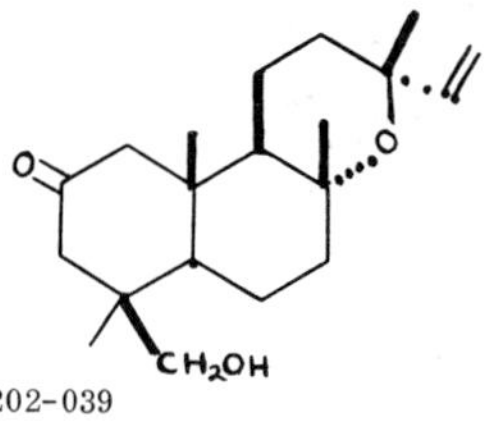

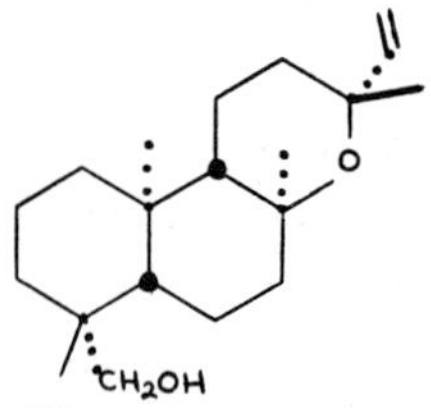

MANOYL OXIDE, 19-COOH-13-epi-enantio- $C_{20}H_{32}O_3$ 320 -44° 237°C Aust. J. C., 1965, 1441. 4202-041	TORULOSAL $C_{20}H_{32}O_2$ 304 +29°c Acta Chem. Scand., 1961, 1303. 4202-046
EPERUAN-14, 15, 18-triol, 8β, 13-epoxy- $C_{20}H_{36}O_4$ 340 -21°e 165°C Tet., 1968, 795. 4202-042	COMMUNIC ACID, cis- $C_{20}H_{30}O_2$ 302 m. e. +45°c m. e. 41°C Acta Chem. Scand., 1966 (20) 1074. 4202-047
IMBRICATOLAL $C_{20}H_{34}O_2$ 306 Tet., 1968 (24) 3417. 4202-043	COMMUNIC ACID, trans- $C_{20}H_{30}O_2$ 302 +25°m 229°C JOC, 1965 (30) 429. 4202-048
IMBRICATOLAL, acetyl- $C_{22}H_{36}O_3$ 348 +20°c Tet., 1968 (24) 3417. 4202-044	COMMUNIC ACID, iso- $C_{20}H_{30}O_2$ 302 Tet., 1970, 1935. 4202-049
CONTORTOLAL $C_{20}H_{32}O_2$ 304 Tet. Lett., 1965, 2633. 4202-045	OZIC ACID $C_{20}H_{30}O_2$ 302 -47°c 142°C JCS, 1968, 1063. 4202-050

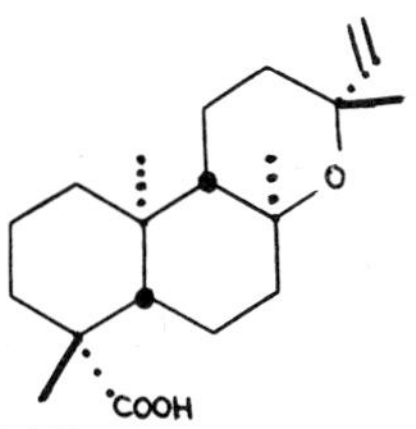

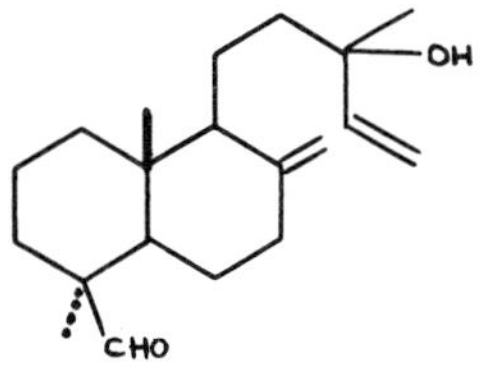

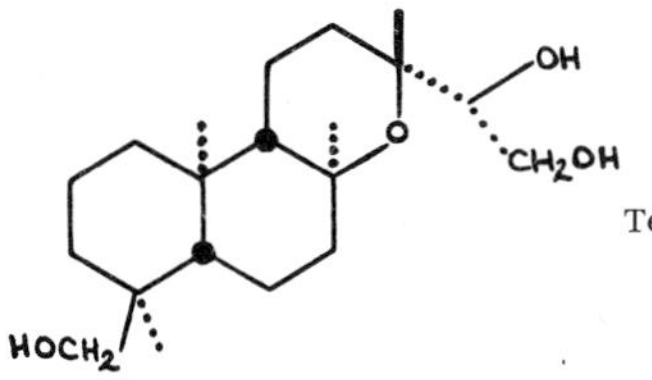

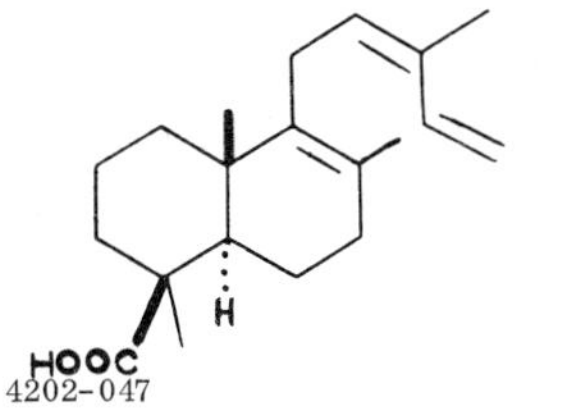

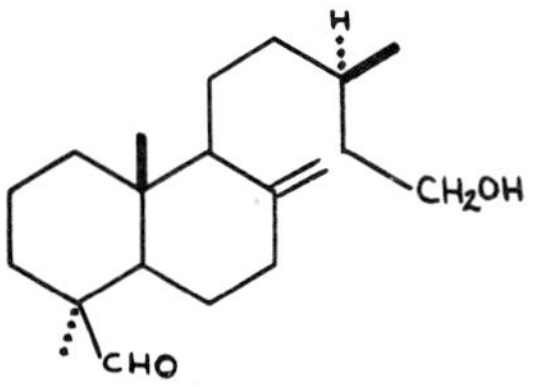

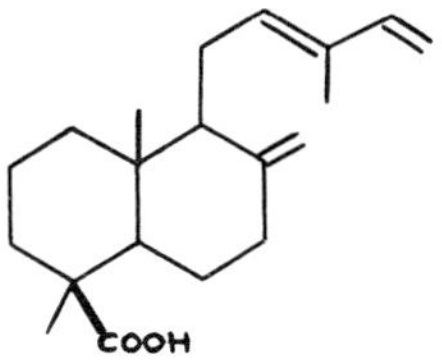

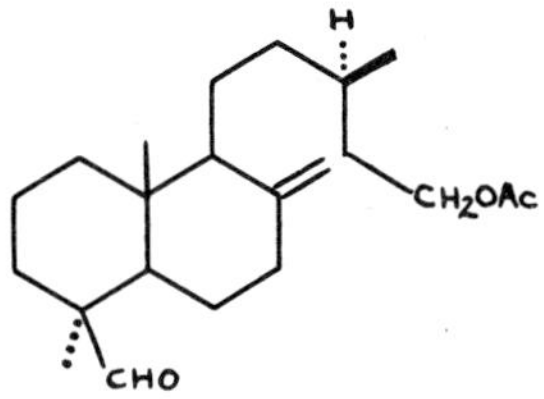

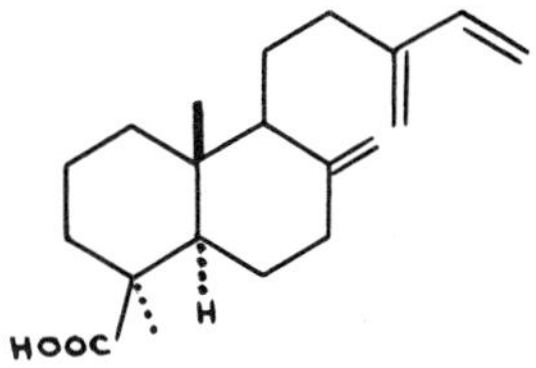

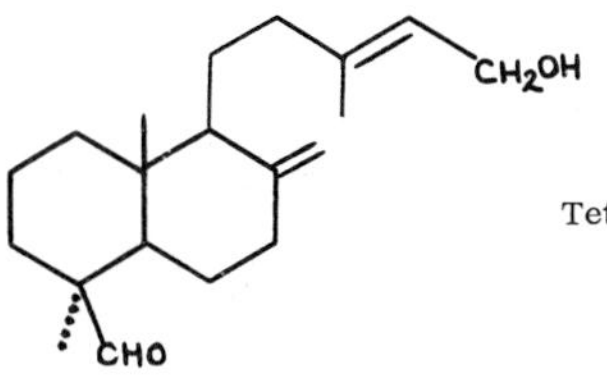

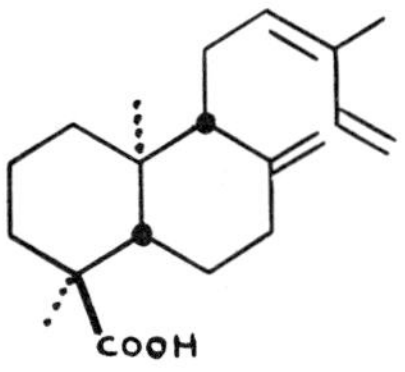

OZIC ACID, 3-OH- $C_{20}H_{30}O_3$ 318 G. Ourisson (unpublished). 4202-051	IMBRICATOLIC ACID $C_{20}H_{34}O_3$ 322 m. e. + 49°c Tet., 1968 (24) 3417. 4202-056
LABD-8(20)-ENE-19-OIC ACID, 12-OH- $C_{20}H_{34}O_3$ 322 Tet. Lett., 1968, 3141. 4202-052	IMBRICATOLIC ACID, acetyl- $C_{22}H_{36}O_4$ 364 Tet., 1968, 1314. 4202-057
ZANZIBARIC ACID $C_{20}H_{30}O_3$ 318 -17°c 204°C Bull. Soc. Chim. Fr., 1965, 2903. 4202-053	EPERU-8(20)-EN-18-OIC ACID, 15,16-diOH- $C_{20}H_{34}O_4$ 338 -38°e 152°C Tet., 1965, 1175. 4202-058
CUPRESSIC ACID $C_{20}H_{32}O_3$ 320 Gazz., 1964 (94) 1108. 4202-054	SCIADOPIC ACID, Me ester $C_{21}H_{34}O_4$ 350 108°C Tet., 1964, 1427. 4202-059
CUPRESSIC ACID, iso- $C_{20}H_{32}O_3$ 320 Gazz., 1964 (94) 1108. 4202-055	EPERU-8(20)-EN-18-OIC ACID, 15-OH- $C_{20}H_{34}O_3$ 322 -51°e 136°C Tet., 1965, 3219. 4202-060

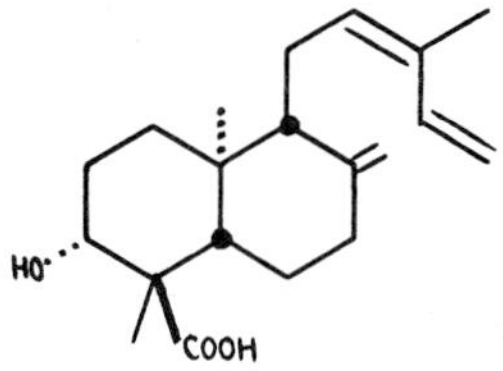

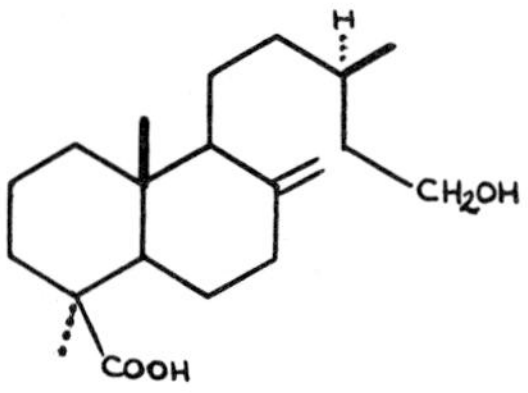

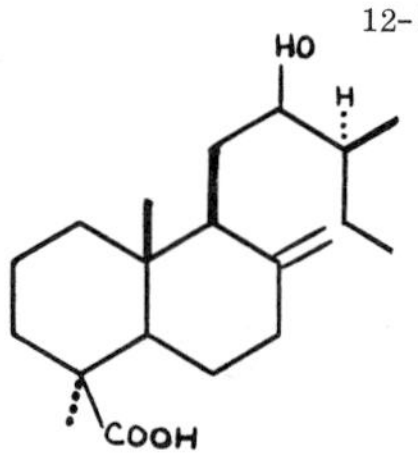

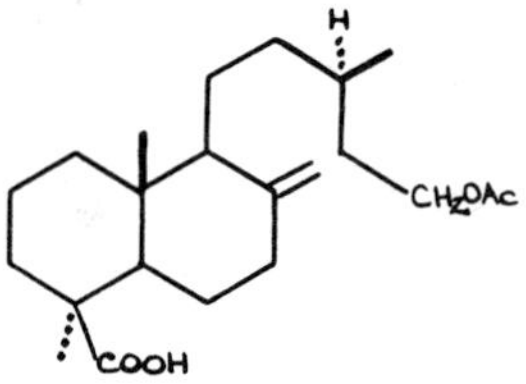

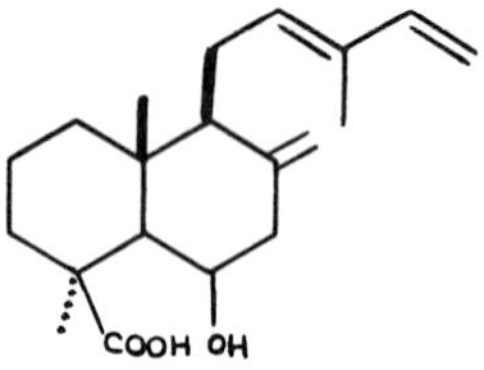

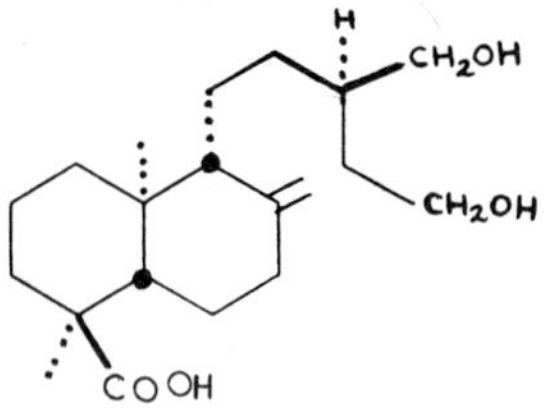

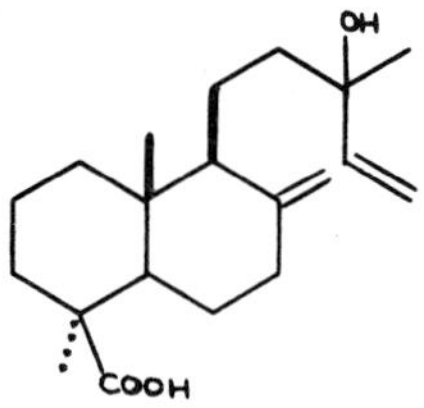

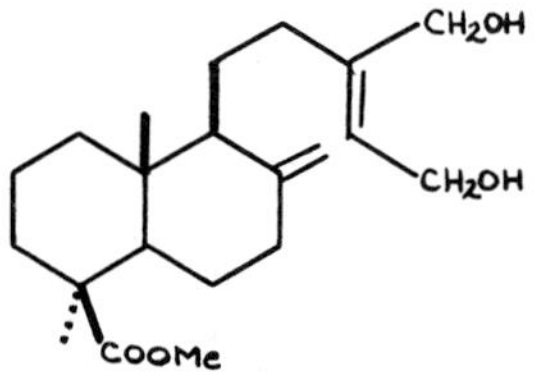

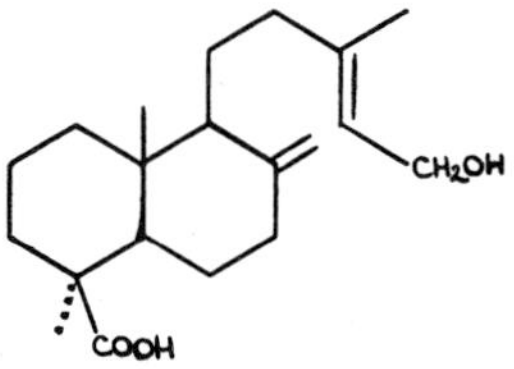

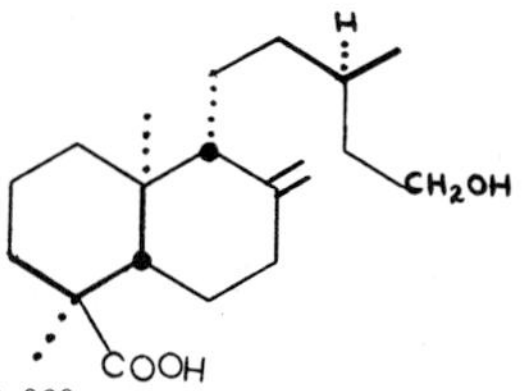

EPERUA-7, 13-DIEN-15-OIC ACID $C_{20}H_{32}O_2$ 304 m. e. -26^0 m. e. Oil JCS, 1968, 1067. 4202-061	LABD-8(20)-EN-15-OIC ACID, enantio- $C_{20}H_{34}O_2$ 306 m. e. -22^0 Tet., Supp. 8, Pt. I, 1967, 203. 4202-066
COPALIC ACID, (-)- $C_{20}H_{32}O_2$ 304 -7^0 JCS, 1968 1067. 4202-062	EPERUIC ACID $C_{20}H_{34}O_2$ 306 Oil JCS, 1967, 931. 4202-067
LABDA-8(20), 13-DIEN-OIC ACID $C_{20}H_{34}O_2$ 304 $+26^0$ JCS, 1968, 1067. 4202-063	LABD-13-EN-8β-OL-15-OIC ACID, enantio- $C_{20}H_{34}O_3$ 322 -37^0c Tet., Supp. 8, Pt. I, 1967, 203. 4202-068
EPERU-7-ENOIC ACID $C_{20}H_{34}O_2$ 306 m. e. -8^0 m. e. Oil JCS, 1968, 1067. 4202-064	LABDAN-8β-OL-15-OIC ACID, enantio- $C_{20}H_{36}O_3$ 324 m. e. -5^0 m. e. 83^0C Tet., Supp. 8, Pt. I, 1967, 203. 4202-069
CATIVIC ACID $C_{20}H_{34}O_2$ 306 -6^0e 81^0C JACS, 1954, 5001. 4202-065	LABDANOLIC ACID $C_{20}H_{36}O_3$ 324 -7^0c JCS, 1967, 931. 4202-070

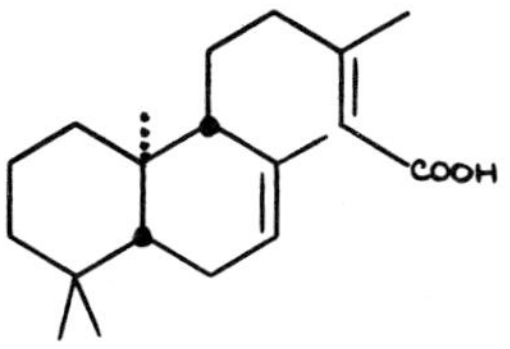

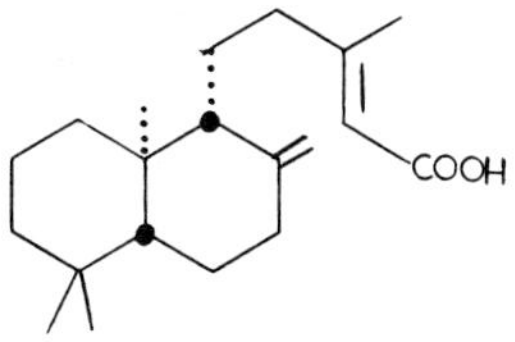

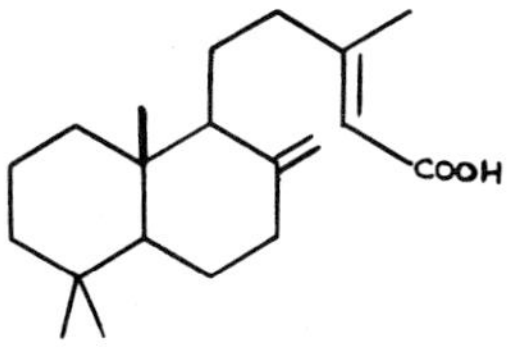

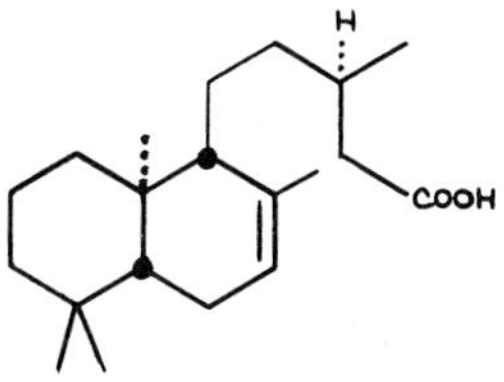

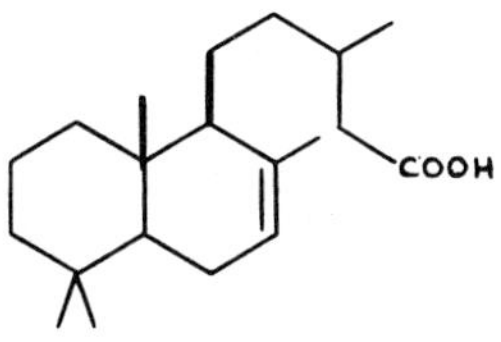

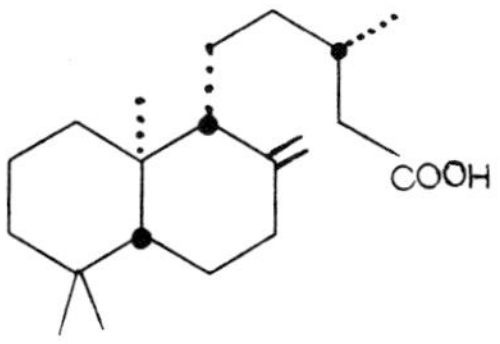

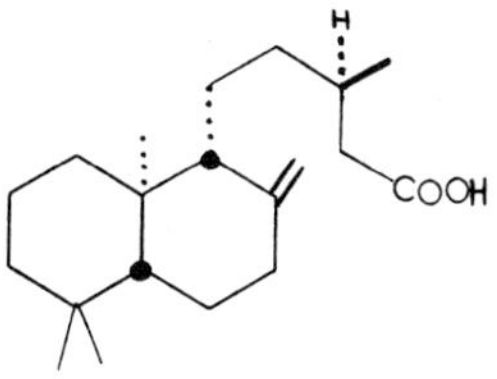

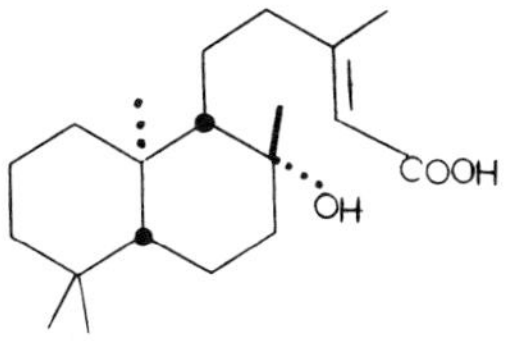

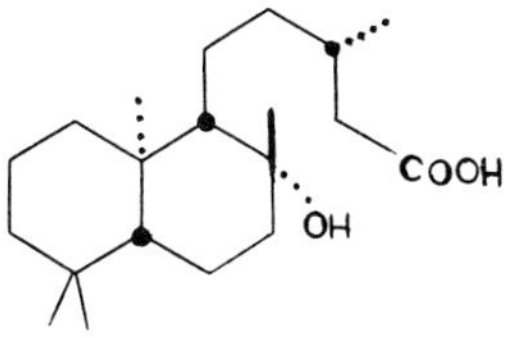

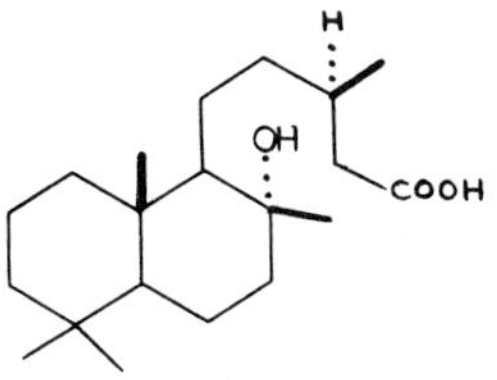

LABD-8(20)-EN-15-OIC ACID, 3-beta-hydroxy- $C_{20}H_{34}O_3$ 322 m. e. + 52^{0}c JCS, 1964, 3648. 4202-071	AGATHALIC ACID $C_{20}H_{30}O_3$ 318 + 25^0 Aust. J. C., 1968, 1923. 4202-076
LABD-8(20)-EN-18-OL-15-OIC ACID, enantio- $C_{20}H_{34}O_3$ 322 m. e. -37^0 Tet., Supp. 8, Pt. I, 1967, 203. 4202-072	AGATHIC ACID $C_{20}H_{30}O_4$ 334 + 55^{0}e 204^{0}C Helv., 1948, 2143. 4202-077
LABD-8(20)-EN-15-OIC ACID, enantio-18-OMe- $C_{21}H_{36}O_2$ 336 m. e. -26^0 Tet., Supp. 8, Pt. I, 1967, 203. 4202-073	EPERU-8(20)-EN-15, 18-DIOIC ACID $C_{20}H_{32}O_4$ 336 -36^{0}e 152^{0}C Tet., 1965, 3219. 4202-078
AGATHOLIC ACID $C_{20}H_{32}O_3$ 320 + 42^{0}e 185^{0}C Acta Chem. Scand., 1961, 1303. 4202-074	OLIVERIC ACID $C_{20}H_{32}O_4$ 336 Tet., 1970, 3461. 4202-079
LABD-8(20), 13-DIEN-15-OIC ACID, enantio-18-OAc- $C_{22}H_{34}O_4$ 362 m. e.-48^0 Tet., Supp. 8, Pt. I, 1967, 203. 4202-075	PINIFOLIC ACID $C_{20}H_{32}O_4$ 336 + 26^{0}e 195^{0}C Acta Chem. Scand., 1962, 607. 4202-080

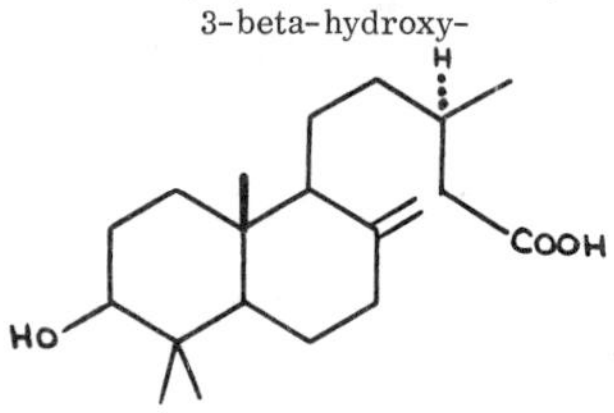

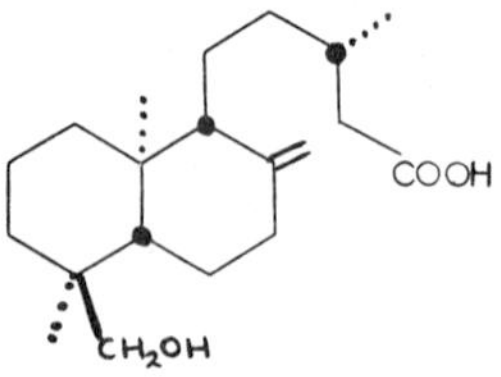

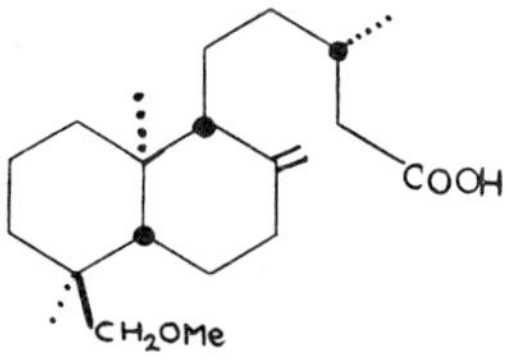

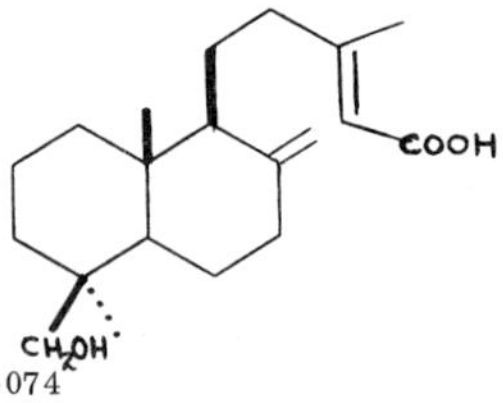

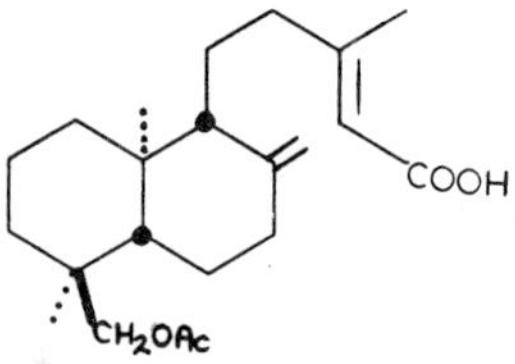

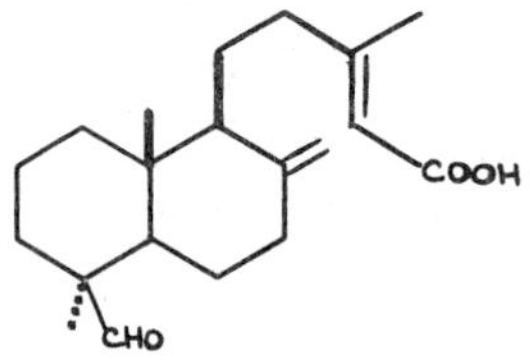

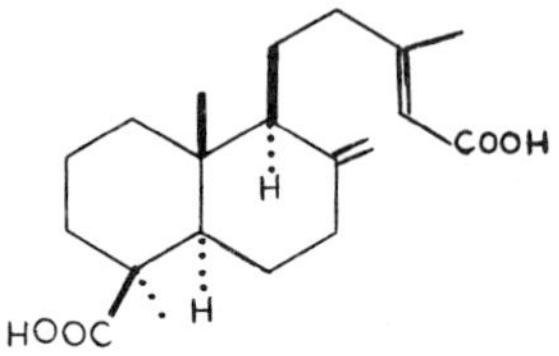

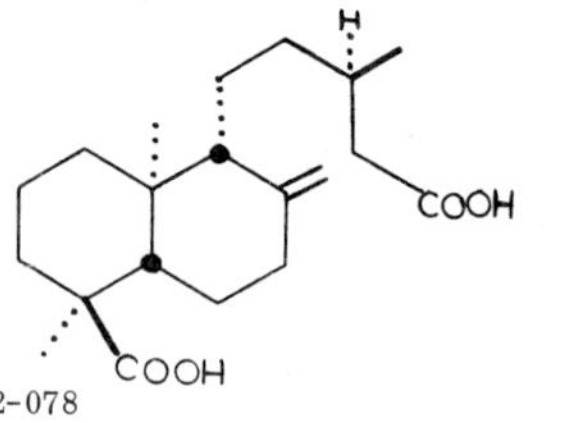

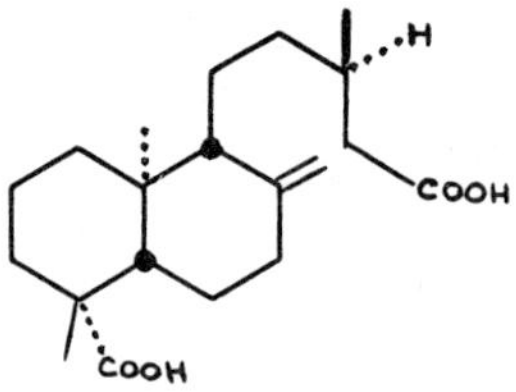

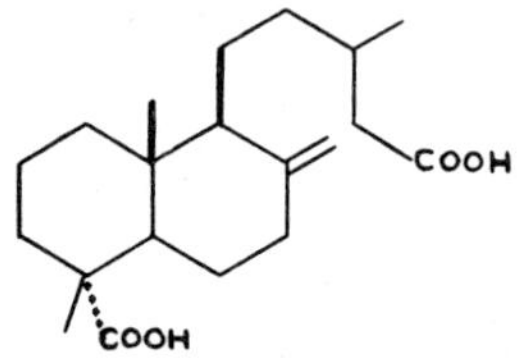

GRINDELIC ACID	$C_{20}H_{32}O_3$	320 -102⁰c 101⁰C	Chem. Abstr., 1967 (67) 54276.
4202-081			
LEONOTIS COMPOUND X	$C20H_{28}O_5$	348 233⁰C	JCS, 1968, 263.
4202-086			
GRINDELIC ACID, epoxy-	$C_{20}H_{32}O_4$	336 -82⁰c 60⁰C	Chem. Abstr., 1967 (67) 54276.
4202-082			
LEONOTIS COMPOUND Y	$C_{20}H_{28}O_3$	316 115⁰C	JCS, 1968, 262.
4202-087			
GRINDELIC ACID, 6-Oxo-	$C_{20}H_{30}O_4$	334 -95⁰c 74⁰C	Chem. Abstr., 1967 (67) 54276.
4202-083			
ANDROGRAPHOLIDE	$C_{20}H_{30}O_5$	350 -127⁰ 288⁰C	Tet., 1965, 2617.
4202-088			
CATIVIC ACID, 6-Oxo-	$C_{20}H_{32}O_3$	320 +13⁰c 115⁰C	JCS, 1960, 1324.
4202-084			
ANDROGRAPHOLIDE, neo-	$C_{26}H_{40}O_8$	-48⁰p 168⁰C	Tet. Lett., 1968, 4803.
4202-089			
SOLIDAGENONE	$C_{20}H_{28}O_3$	316 132⁰C	Acta Chem. Scand., 1967, 2289.
4202-085			
PSIADIOL	$C_{20}H_{30}O_3$	318 -58⁰ 139⁰C	Tet. Lett., 1967, 2637.
4202-090			

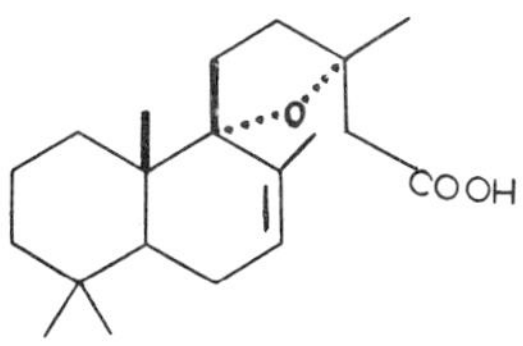

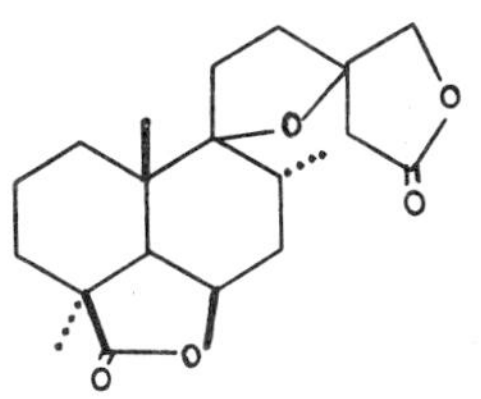

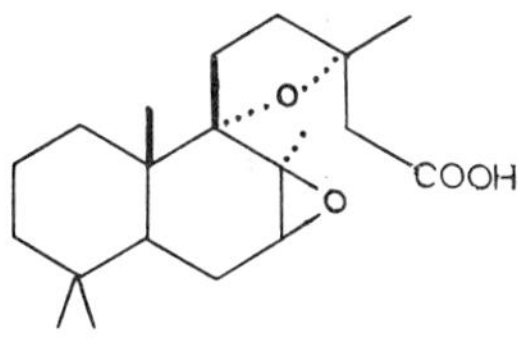

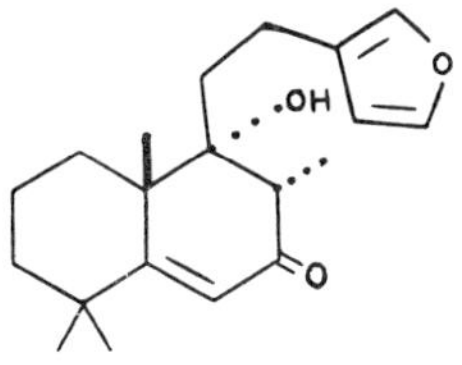

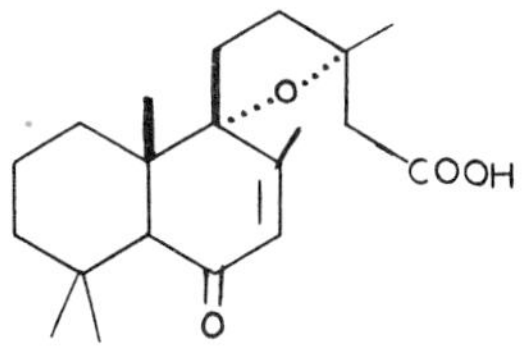

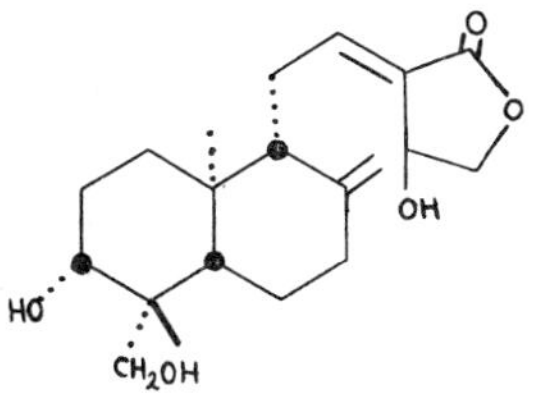

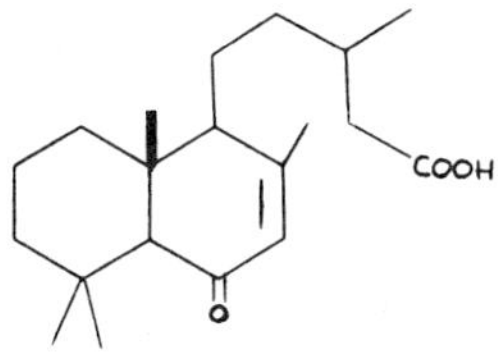

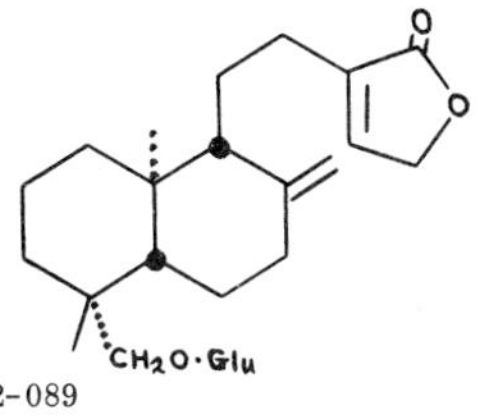

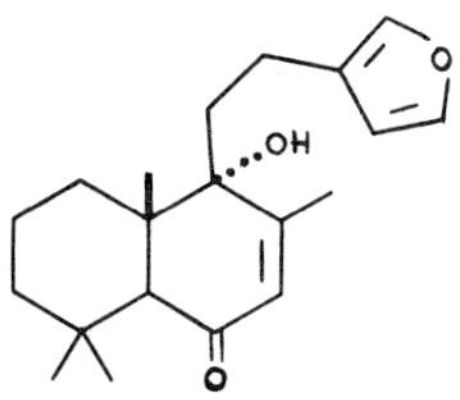

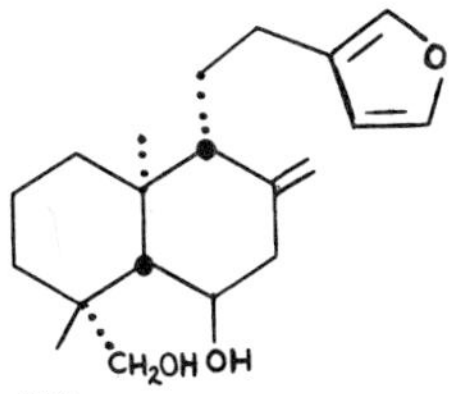

EPERU-8(20)-EN-15, 18-DIOIC ACID - BUTENOLIDE

$C_{20}H_{28}O_4$ 332 -61^0e 190^0C

Tet., 1965, 3219.

4202-091

POLYALTHIC ACID

$C_{20}H_{28}O_3$ 316 -46^0e 102^0C

Helv., 1961, 1040.

4202-092

DANIELLIC ACID, (-)-

$C_{20}H_{28}O_3$ 316 -58^0e 130^0C

JCS, 1968, 1063.

4202-093

LAMBERTIANIC ACID

$C_{20}H_{28}O_3$ 316 $+55^0$ 128^0C

Tet., 1966, 679.

4202-094

SCIADINONATE, diMe-

$C_{22}H_{28}O_6$ 388 -45^0c 122^0C

Tet. Lett., 1962, 215.

4202-095

SCIADINONE

$C_{20}H_{24}O_4$ 328 -60^0c 207^0C

Chem. Pharm. Bull. (Tokyo), 1963, 271.

4202-096

SCIADINE

$C_{20}H_{24}O_4$ 328 $+10^0$c 160^0C

Tet., 1963, 643.

4202-097

LEONOTIN

$C_{20}H_{28}O_5$ 348 $+63^0$e 175^0C

Chem. Comm., 1969, 1315.

4202-098

MARRUBIN, pre-

$C_{20}H_{28}O_4$ 332 -41^0e Oil

JCS, 1969, 2014.

4202-099

MARRUBIIN

$C_{20}H_{28}O_4$ 332 $+36^0$c 160^0C

Tet., 1967, 3909.

4202-100

PEREGRININE $C_{20}H_{26}O_5$ 346 43°m 173°C Tet. Lett., 1968, 3150. 4202-101	PEREGRININ, tetrahydro- $C_{20}H_{30}O_5$ 350 Khim. Prir. Soedin, 1970 (6) 207. 4202-106
LEVATENOLIDE, α- $C_{20}H_{30}O_3$ 318 +60°c 210°C Tet., 1961, 246. 4202-102	MANOOL, 7-OH- $C_{20}H_{34}O_2$ 306 132°C Aust. J. C., 1969, 1691. 4202-107
LEVATENOLIDE, β- $C_{20}H_{30}O_3$ 318 -60°c 209°C Tet., 1961, 246. 4202-103	MANOYL OXIDE, 2,3-diOH- $C_{20}H_{34}O_3$ 322 148°C Aust. J. C., 1969, 1691. 4202-108
LEVANTANOLIDE, α_2- $C_{20}H_{32}O_3$ 320 -44°c 165°C Tet., 1963, 107. 4202-104	MANOYL OXIDE, 3-oxo-13-epi-ent- $C_{20}H_{32}O_2$ 304 -53°c 91°C Acta Chem. Scand., 1970, 1860. 4202-109
LABD-8(20)-EN-15,18-DOIC ACID enantio- $C_{20}H_{32}O_4$ 336 m. e. -26° Tet., 1967 Suppl. 8, Pt. 1, 203. 4202-105	GRINDELIC ACID, oxy- $C_{20}H_{32}O_4$ 336 m. e.-106°c m. e. 116°C Acta Chem. Scand., 1962, 1675. 4202-110

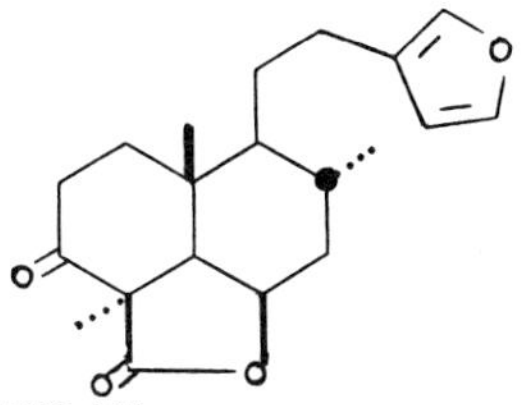

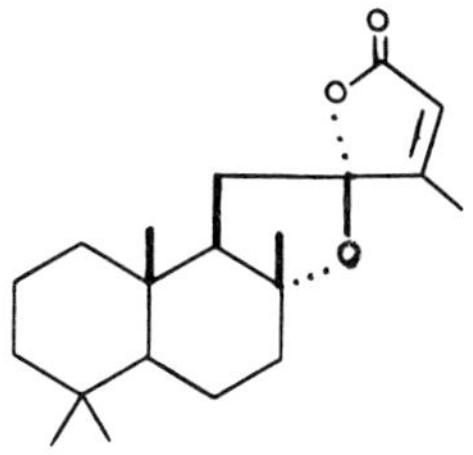

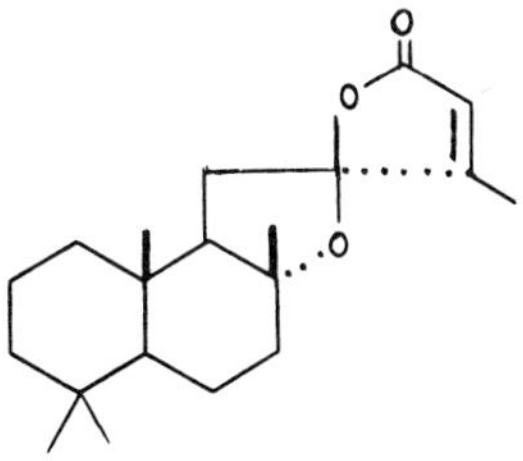

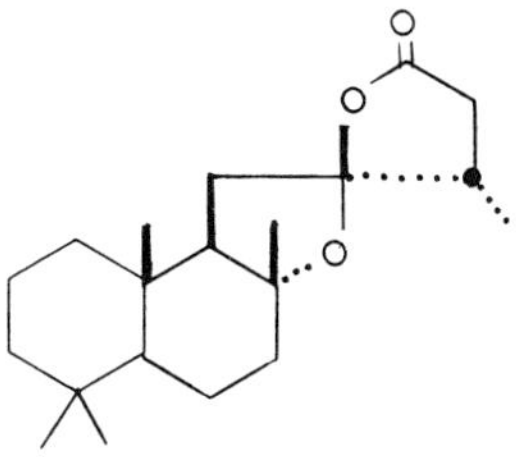

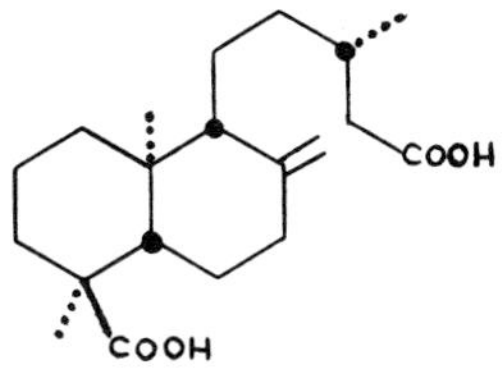

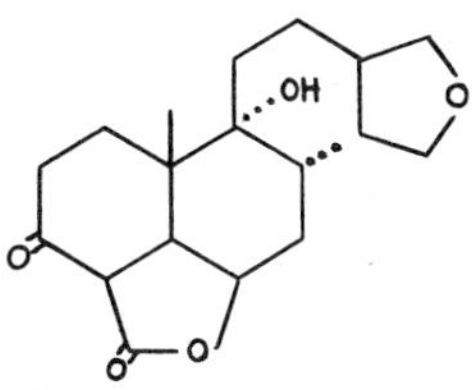

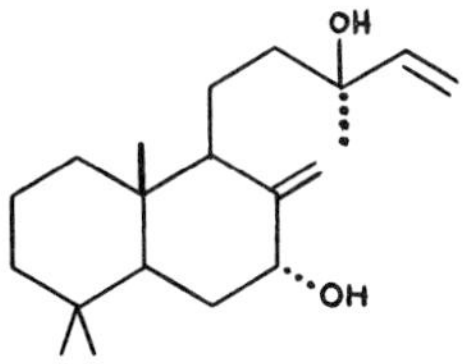

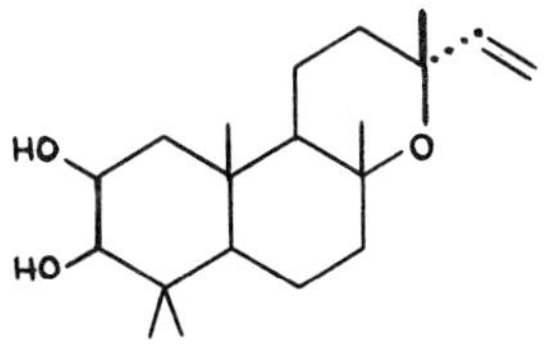

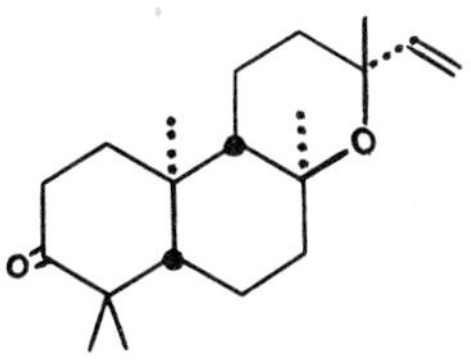

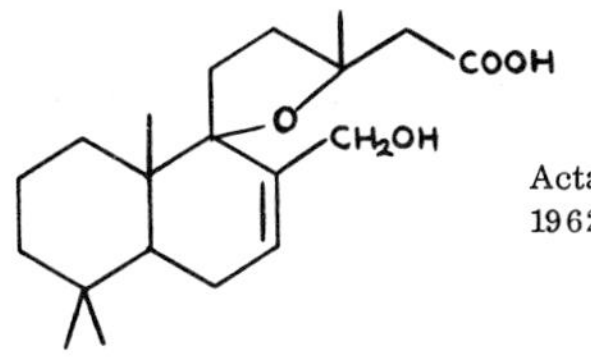

LABDA-8(20), 13-DIEN-15-OIC ACID ME ESTER, 11-OAc- $C_{23}H_{36}O_4$ 376

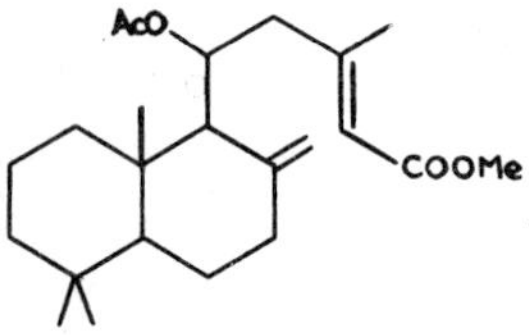

Ann. Chim. (Rome), 1970 (60) 233.

4202-111

FEROLIC ACID, Copai- $C_{20}H_{32}O_3$ 320

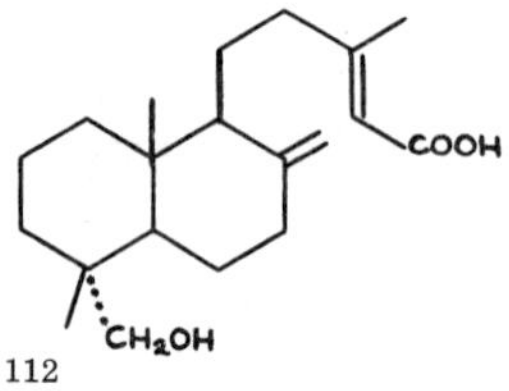

Ann. Chim. (Rome), 1970 (60) 233.

4202-112

NEPETAEFURANOL $C_{22}H_{30}O_8$ 422 $+17^{0}$m 254^{0}C

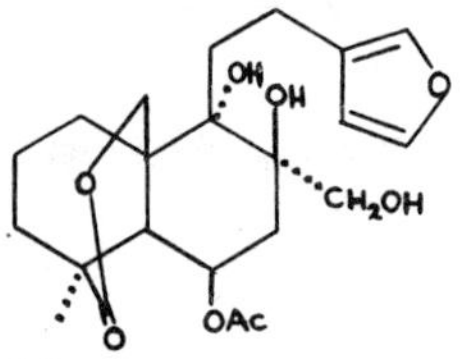

JACS., 1970, 5527.

4202-113

NEPETAEFURAN $C_{22}H_{28}O_7$ 404 $+32^{0}$m 241^{0}C

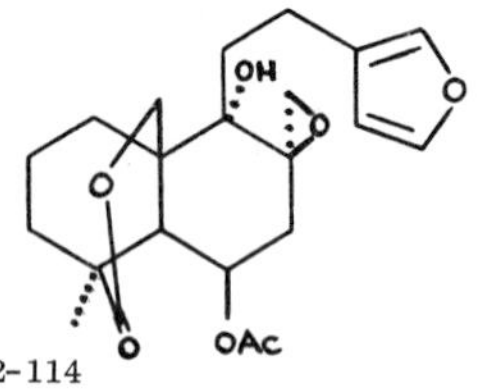

JACS., 1970, 5527.

4202-114

NEPETAEFOLIN $C_{22}H_{28}O_7$ 404 -15^{0}c 260^{0}C

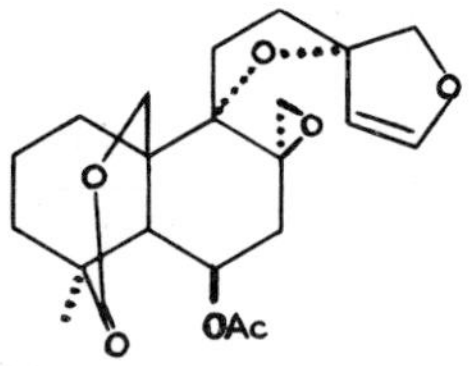

JACS., 1970, 5527.

4202-115

MARRUBENOL $C_{20}H_{32}O_4$ 336 $+20^{0}$m 138^{0}C

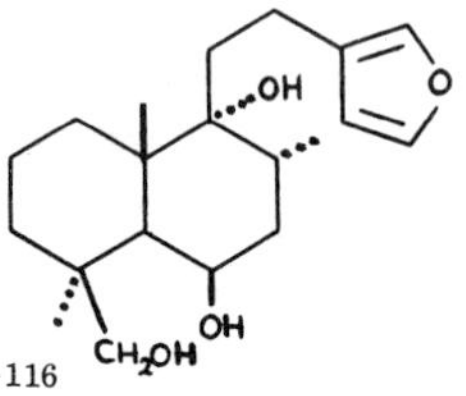

JCS., 1968, 807.

4202-116

LABDAN-15-OIC ACID, 7, 8-diOH-ent- $C_{20}H_{36}O_4$ 340 $+9^{0}$c 70^{0}C

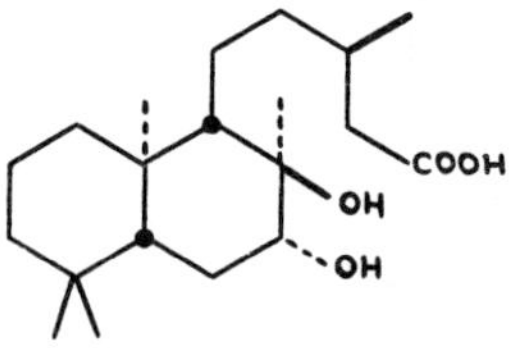

Aust. J. C., 1966, 2133.

4202-117

LABDAN-15-OIC ACID, 6, 8-diOH-ent- $C_{20}H_{36}O_4$ 340 -14^{0}e 55^{0}C

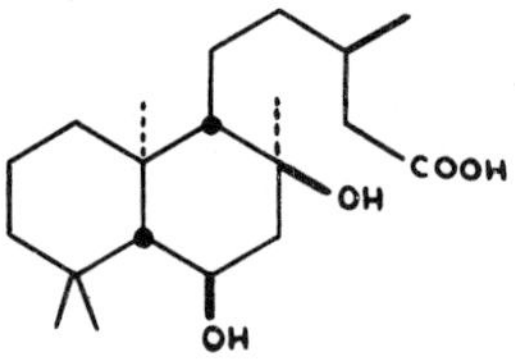

Aust. J. C., 1966, 2133.

4202-118

PIMARINOL $C_{30}H_{32}O$ 288 +94⁰ 88⁰C

Acta Chem. Scand. 1963 (17) 1826

4203-001

SANDARACOPIMARADIEN-19-OL $C_{20}H_{32}O$ 288 108⁰C

Aust. J. Chem. 1969 1265.

4203-002

(-)-PIMARA-8(14), 15-DIEN-19-OL $C_{20}H_{32}O$ 288 -96⁰c 110⁰C

Tet. Lett. 1969 1683

4203-003

SANDARACOPIMARADIEN-3-OL $C_{20}H_{32}O$ 288 -20⁰c 128⁰C

Aust. J. Chem. 1967, 969

4203-004

PIMARADIEN-3-OL, Iso $C_{20}H_{32}O$ 288 -36⁰ 147⁰C

JACS 1950 (72) 375

4203-005

SANDARACOPIMAR-15-ENE, 8β-OH, $C_{20}H_{34}O$ 290 -7⁰c 40⁰C

JCS 1967 (4) 300

4203-006

PIMARINOL, Iso- $C_{20}H_{32}O$ 288 -25⁰ 85⁰C

Aust. J. Chem. 1967, 969
Acta Chem. Scand. 1963 (17) 1826

4203-007

VIRESCENOL B $C_{20}H_{32}O_2$ 304 -25⁰c 147⁰C

Bull. Soc. Chim. Fr. 1970 1912
Chem. Comm. 1968 1404

4203-008

SANDARACOPIMARADIEN-3β, 19-DIOL $C_{20}H_{32}O_2$ 304 171⁰C

JCS 1965 3846

4203-009

SANDARACOPIMARADIENE-3, 18-DIOL $C_{20}H_{32}O_2$ 304 -19⁰c 153⁰C

JCS 1963 644

4203-010

OBLONGIFOLIOL	$C_{20}H_{32}O_2$	304 +27°c 148°C	Tet. Lett. 1968 4685 4203-011
PIMAR-8(14)-ENE, 6, 15, 16, 18-tetra-OH-enantio-	$C_{20}H_{34}O_4$	338 -22°d 193°C	Tet. Lett. 1969 4803 4203-016
VIRESCENOL A	$C_{20}H_{32}O_3$	320 -44°c 149°C	Chem. Abstr. 1968 (68) 29807 Chem. Comm. 1968 1404 4203-012
ARAUCAROLONE	$C_{20}H_{30}O_4$	334 -42°c 160°C	Acta Chem. Scand. 1965 1875 4203-017
SANDARACOPIMARA-8(14), 16-DIEN-2α, 18, 19-TRIOL	$C_{20}H_{32}O_3$	320 225°C	Tet. Lett. 1966 3173 4203-013
ARAUCARONE	$C_{20}H_{30}O_3$	318 -51°c 115°C	Acta Chem. Scand. 1965 1875 4203-018
SANDARACOPIMARA-8(14), 16-DIEN-3β, 18, 19-TRIOL	$C_{20}H_{32}O_3$	320 164°C	Tet. Lett. 1966 3173 4203-014
ARAUCARENOLONE	$C_{20}H_{28}O_4$	332 -58°c 144°C	Acta Chem. Scand. 1965 1875 4203-019
SANDARACOPIMARA-8(14), 16-DIEN-2α, 3β, 18, 19-TETROL	$C_{20}H_{32}O_4$	336 212°C	Tet. Lett. 1966 3173 4203-015
ARAUCAROL	$C_{20}H_{32}O_3$	320 -24°c 135°C	Acta Chem. Scand. 1965 1879 4203-020

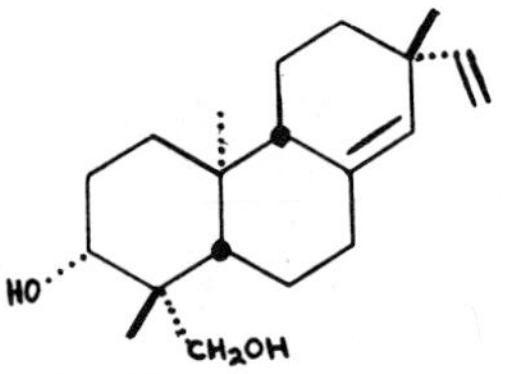

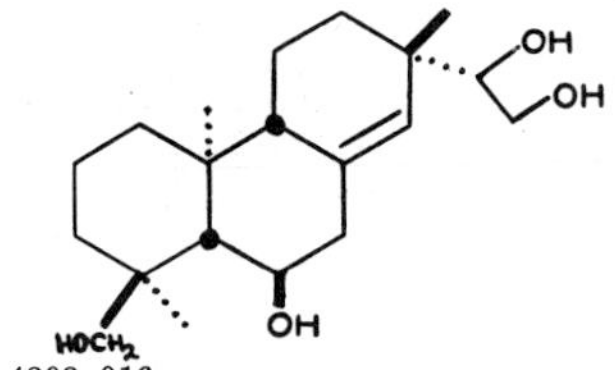

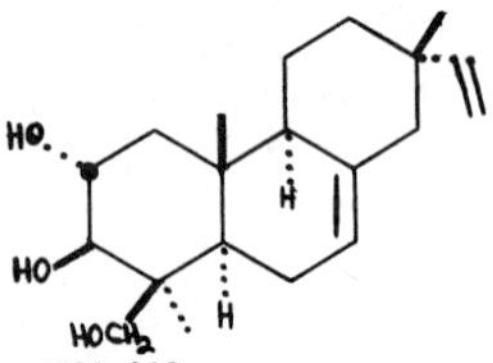

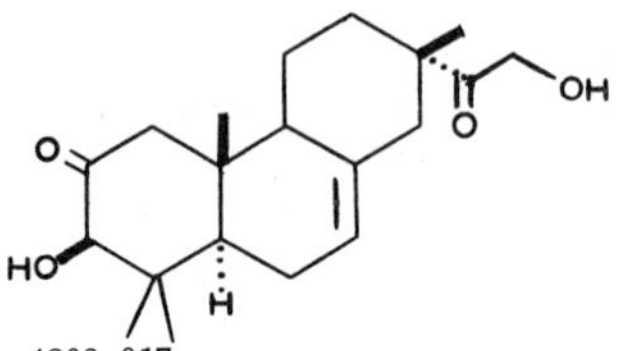

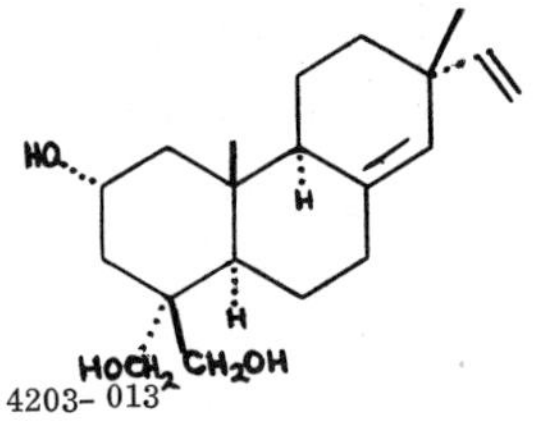

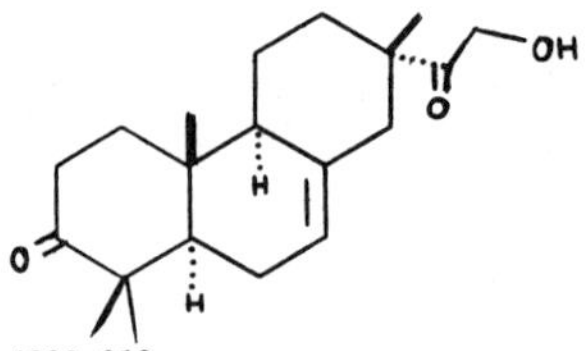

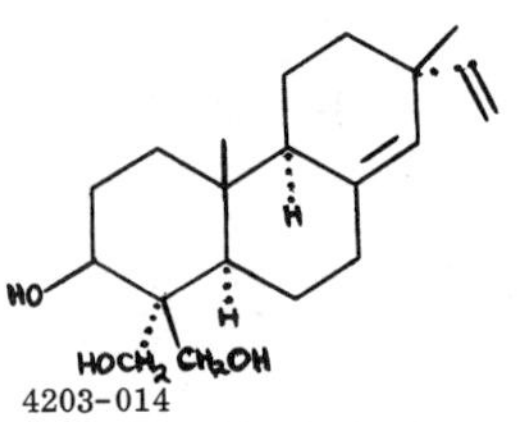

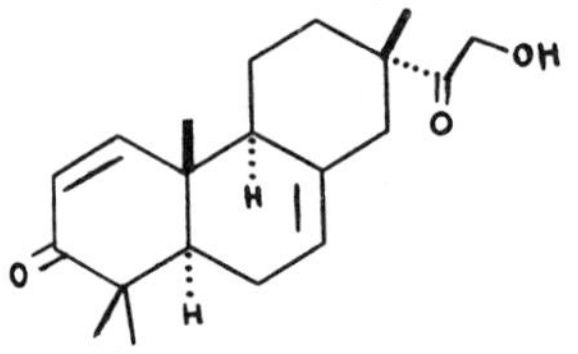

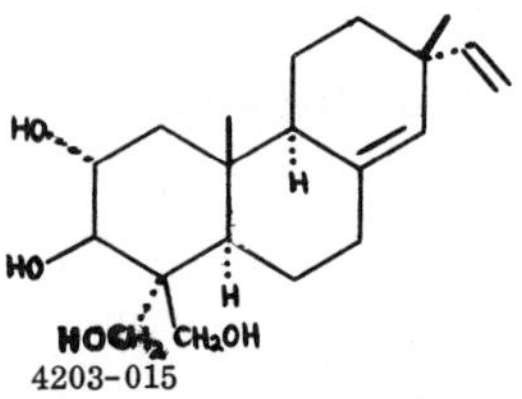

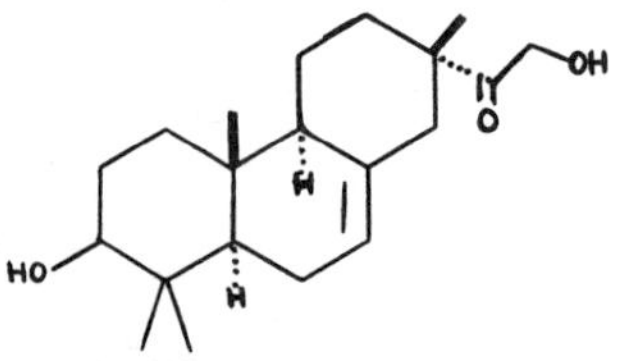

SANDARACOPIMARADIENE-3-ONE — $C_{20}H_{30}O$ — 286, -56^{0}c, 60^{0}C

JCS 1963 644

4203-021

CRYPTOPINONE — $C_{20}H_{30}O$ — 286, 51^{0}C

JACS 1956 4087

4203-022

PIMARA-8(14), 15-DIEN-18-AL — $C_{20}H_{30}O$ — 286, Oil

Aust. J. C. 1967 969

4203-023

SANDARACOPIMARIC ACID — $C_{20}H_{30}O_2$ — 30?, -20^{0}, 169^{0}C

JOC 1963 6

4203-024

PIMARIC ACID — $C_{20}H_{30}O_2$ — 302, $+60^{0}$e, 212^{0}C

Chem. Ind. 1958 629; 1961 1307; 1963 254; Tet Lett. 1960 (25) 37

4203-025

SANDARACOPIMARIC ACID, 12α-OH- — $C_{20}H_{30}O_3$ — 318, -11^{0}e, 269^{0}C

Can. J. C. 1961 2543

4203-026

PIMARA-8(9), 15-DIENOIC ACID, iso- — $C_{20}H_{30}O_2$ — 302

Tet. Lett. 1968 2053

4203-027

SANDARACOPIMARIC ACID, 6α-OH- — $C_{20}H_{30}O_3$ — 318, 266^{0}C

4203-028

PIMARIC ACID, iso- — $C_{20}H_{30}O_2$ — 302, 10^{0}, 163^{0}C

JOC 1962 (27) 1930; 1963 (28) 23.

4203-029

PIMARA-8(14), 15-DIEN-19-OIC ACID, 7α-OH-(-)- — $C_{20}H_{30}O_3$ — 318, -70^{0}p, 293^{0}C

Tet. Lett. 1969 1683

4203-030

PIMARA-8(14), 15-DIEN-10-OIC ACID, 7β-OH-(-)- $C_{20}H_{30}O_3$ 318 -63°p 218°C Tet. Lett. 1969 1683 4203-031	PIMARA-7, 15-DIEN-18-AL, Iso- $C_{20}H_{30}O$ 286 -15° 36°C Aust. J. C., 1967 (20) 969. 4203-036
PIMARA-8(14), 15-DIEN-19-OIC ACID, 7-oxo-(-)- $C_{20}H_{28}O_3$ 316 -54°p 243°C Tet. Lett. 1969, 1683 4203-032	PIMARA-8(14) 15-DIEN-19-OIC ACID $C_{20}H_{30}O_2$ 302 -121°c 164°C Tet. Lett., 1967, 5241. 4203-037
SANDARACOPIMARIC ACID, OAc- $C_{22}H_{32}O_4$ 360 158°C -50° Chem. Ind. 1964 2059 4203-033	DARUTIGENOL $C_{20}H_{34}O_3$ 322 -11°e 160°C Bull. Soc. Chim. Fr., 1963, 99. 4203-038
SANDARACOPIMARA-8(14), 15-DIENE $C_{20}H_{32}$ 272 -12°c 40°C JCS 1963 644 4203-034	OBLONGIFOLIC ACID $C_{20}H_{30}O_2$ 302 152°C Tet., 1970, 5275. 4203-039
PIMARA-7, 15-DIENE, iso- $C_{20}H_{32}$ 272 Oil Aust. J. Chem. 1967 969 4203-035	PIMARA-8(14), 15-DIEN-3-OL, ent-iso- $C_{20}H_{32}O$ 288 +12°c 126°C JCS., 1970, 2536. 4203-040

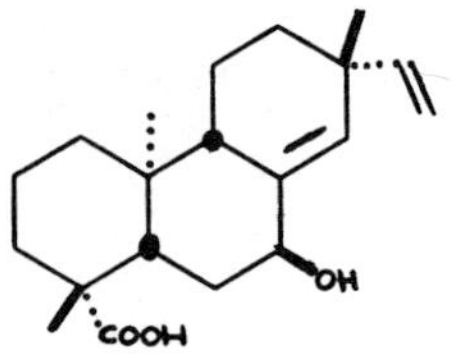

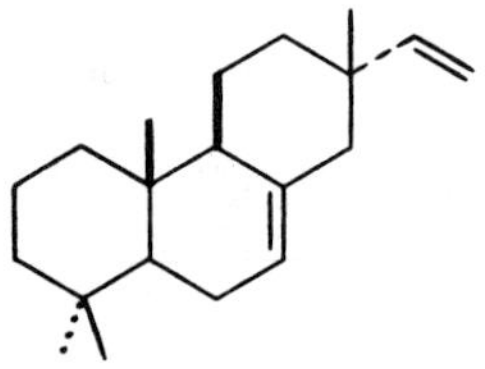

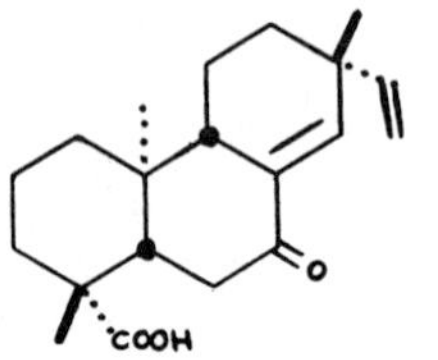

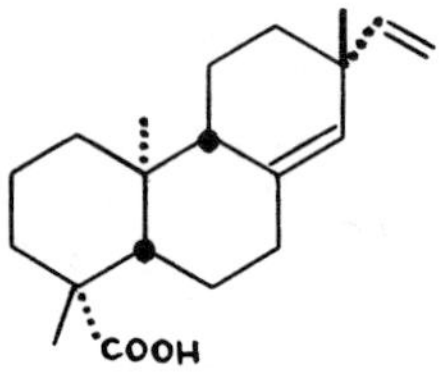

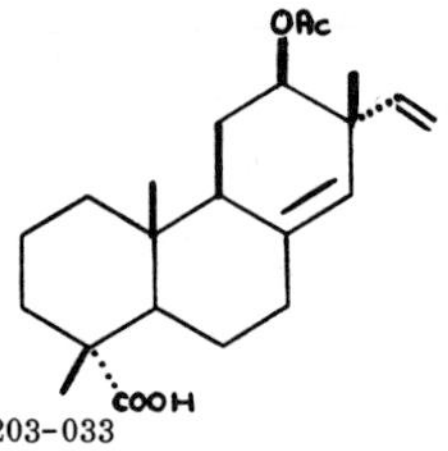

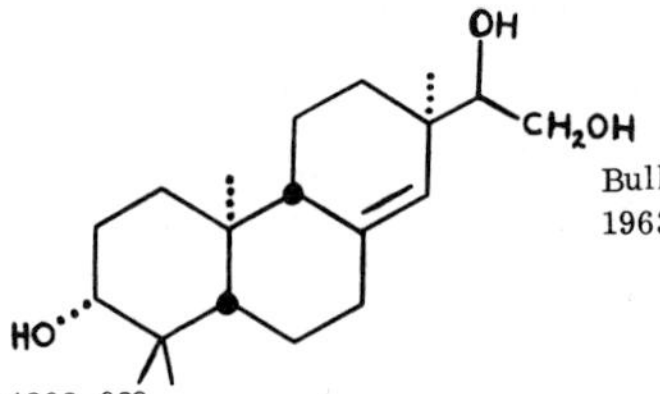

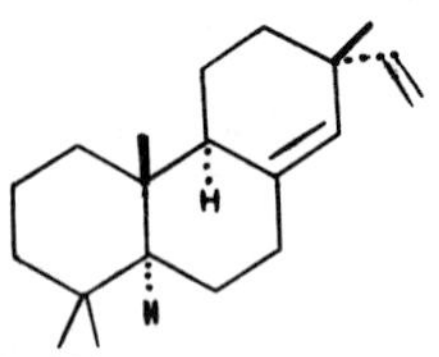

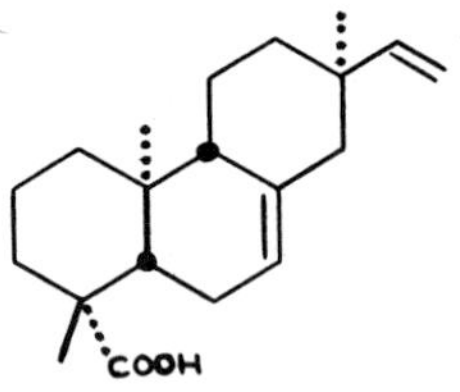

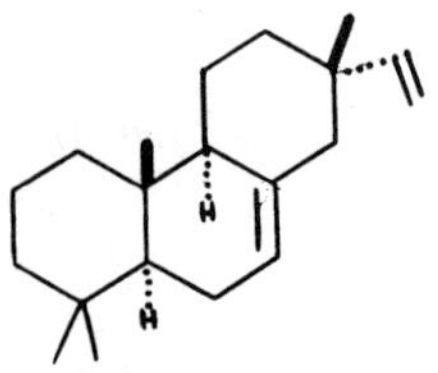

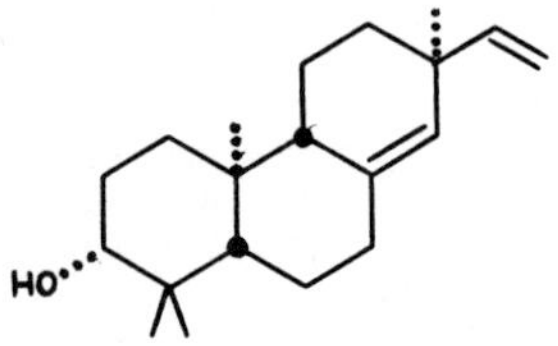

PIMARA-8(14), 15-DIEN-3, 12-DIOL, ent-iso- $C_{20}H_{32}O_2$ 304 +28°c 162°C JCS., 1970, 2536. 4203-041	PIMARA-8(14), 15-DIEN-12-ONE, ent-3-OH-iso- $C_{20}H_{30}O_2$ 302 + 290°c 158°C JCS., 1970, 2536. 4203-042

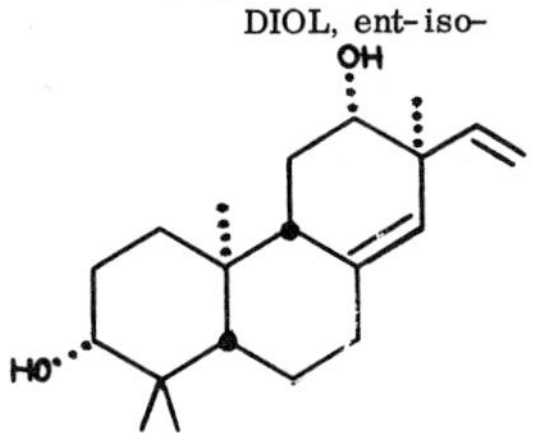

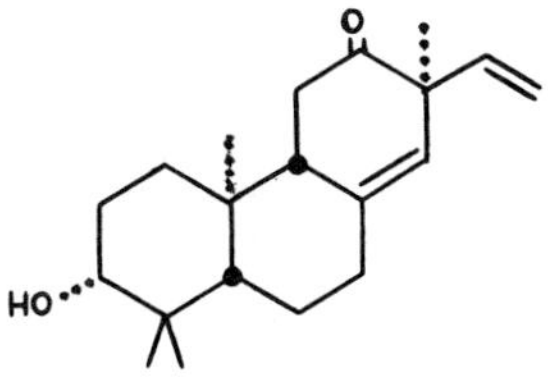

ABIETATRIENE $C_{20}H_{30}$ 270 +0.5°m 42°C Chem. Pharm. Bull. 1970 402 4204-001	HINOKIOL $C_{20}H_{30}O_2$ 302 +74° 241°C JACS 1940 1287 HO OH 4204-006
ABIETOL $C_{20}H_{36}$ 292 Chem. Abstr. 1967 (67) 32822 CH_2OH 4204-002	HINOKIONE $C_{20}H_{28}O_3$ 300 +112°e 192°C JACS 1940 1287 OH 4204-007
ABIETOL, dehydro- $C_{20}H_{30}O$ 286 CH_2OH 4204-003	SUGIOL $C_{20}H_{28}O_2$ 300 +26°e 297°C Acta Chem. Scand. 1954 1728 OH 4204-008
FERRUGINOL $C_{20}H_{30}O$ 286 + 41°e 175°C Acta Chem. Scand. 1956 1511 OH 4204-004	SUGIOL, 5-dehydro- $C_{20}H_{26}O_2$ 298 +13° 285°C JACS 1968 (90) 5923 OH 4204-009
FERRUGINOL, 6,7-dehydro- $C_{20}H_{28}O$ 284 -60°e Acta Chem. Scand. 1957, 932 OH 4204-005	CRYPTOJAPONOL $C_{21}H_{30}O_3$ 330 205°C JOC 1970 2422 HO OMe 4204-010

CALLICARPONE $C_{20}H_{28}O_4$ 332 188°c 112°C Chem. Abstr. 1967 (67) 54275 4204-011	ABIETALDEHYDE $C_{20}H_{34}O$ 290 Chem. Abstr. 1967 (67) 32822 4204-016
FERUGINOL ME ETHER, 1,3-dioxo- $C_{21}H_{28}O_3$ 328 +220° 177° Grazz., 1966 (96) 206 4204-012	ABIETINAL, dehydro- $C_{20}H_{28}O$ 284 Acta Chem. Scand. 1966 (20) 2829 4204-017
FUERSTIONE $C_{20}H_{26}O_3$ 314 Helv. 1966 1151 4204-013	ROYLEANONE $C_{20}H_{28}O_3$ 316 +134°c 182°C Can. J. C. 1962 1540 4204-018
XANTHOPEROL $C_{20}H_{26}O_3$ 314 +133°e 260°C Acta Chem. Scand. 1960 385 4204-014	ROYLEANONE, 7-OAc- $C_{22}H_{30}O_5$ 374 -14°c 213°C Can. J. C. 1962 1540 4204-019
XANTHOPEROL, 5-Epi- $C_{20}H_{26}O_3$ 314 208°C Acta Chem. Scand. 1960 385 4204-015	ROYLEANONE, 6-dehydro- $C_{20}H_{26}O_3$ 314 -620°c 167°C Can. J. C. 1962 1540 4204-020

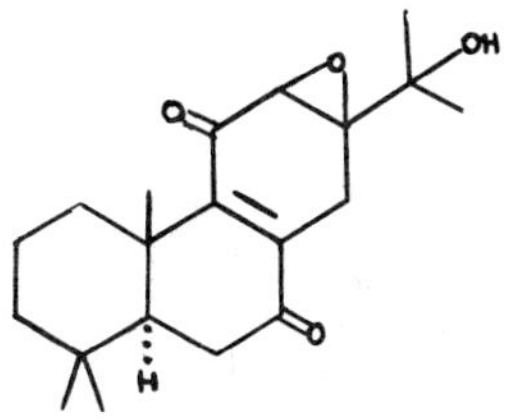

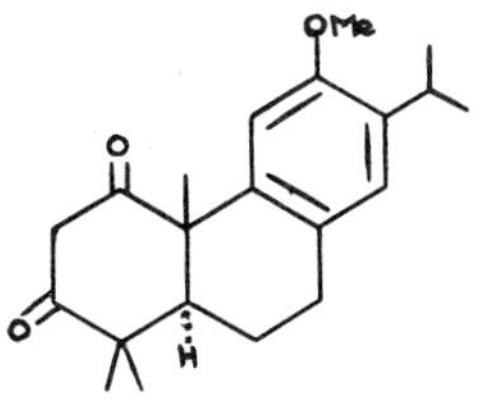

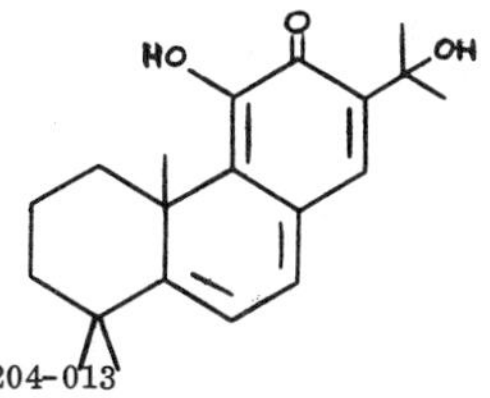

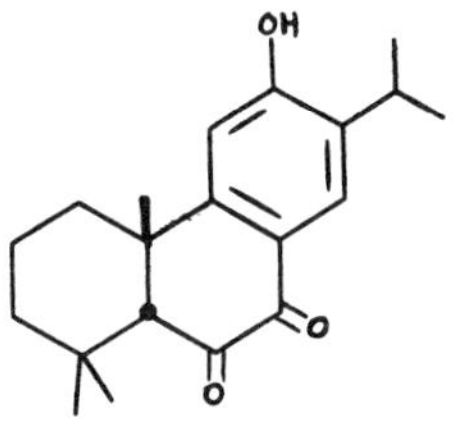

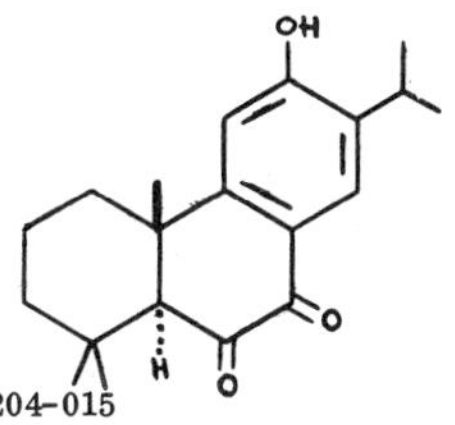

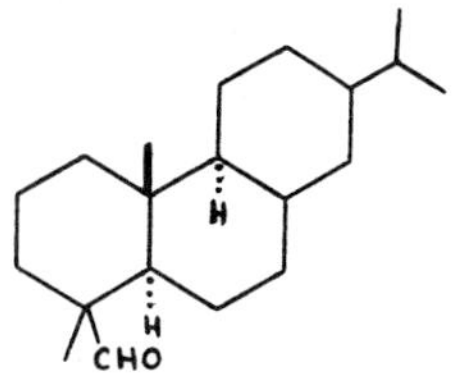

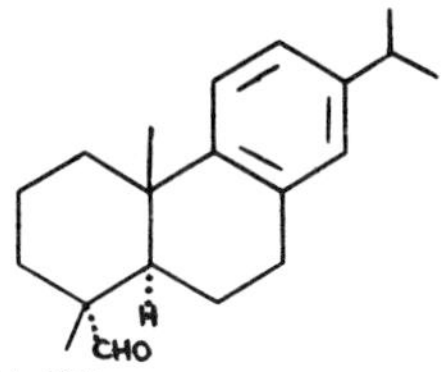

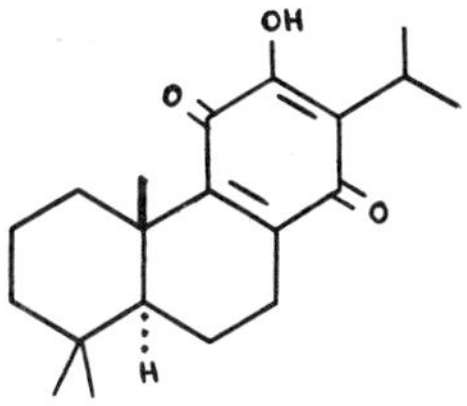

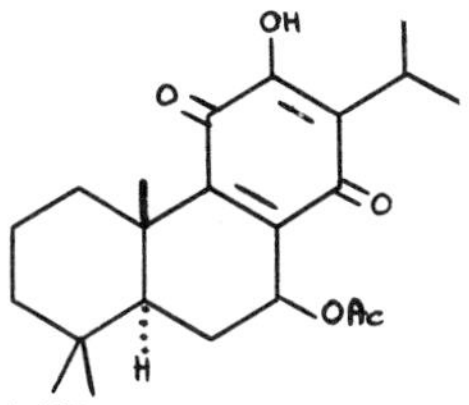

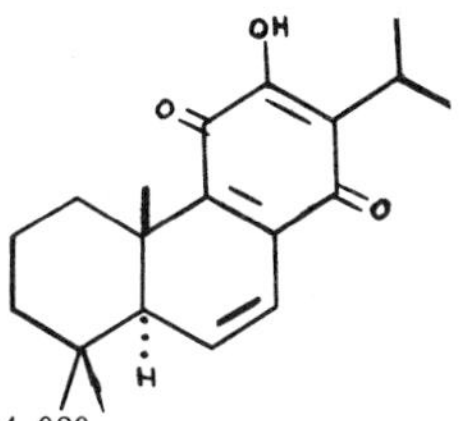

TAXOQUINONE $C_{20}H_{28}O_4$ 332 +340° 213°C JACS 1968 5923 4204-021	CALLITRISIC ACID $C_{20}H_{28}O_2$ 300 +136°e 80°C Tet. Lett. 1968 295 4204-026
TAXODONE $C_{20}H_{28}O_3$ 316 +50° 165°C JACS 1968 5923 4204-022	ABIETIC ACID, Neo- $C_{20}H_{30}O_2$ 302 +159° 168°C JACS 1948 334, 339 4204-027
TAXODIONE $C_{20}H_{26}O_3$ 314 +56° 116°C JACS 1968 5923 4204-023	(-)-SAPIETIC ACID $C_{20}H_{30}O_2$ 302 -276°e 152°C JACS 1962 4660 4204-028
ABIETIC ACID $C_{20}H_{30}O_2$ 302 109° 172°C Chem. Abstr. 1955 (49) 12385 4204-024	PALUSTRIC ACID $C_{20}H_{30}O_2$ 302 JACS 1961 2563 4204-029
DEHYDROABIETIC ACID $C_{20}H_{28}O_2$ 300 +62°e 173°C JACS 1956 250 4204-025	PALUSTRIC ACID METHYL ESTER dihydro- $C_{21}H_{34}O_2$ 318 +39° Oil Bull. Soc. Chim. Fr. 1963 2299 4204-030

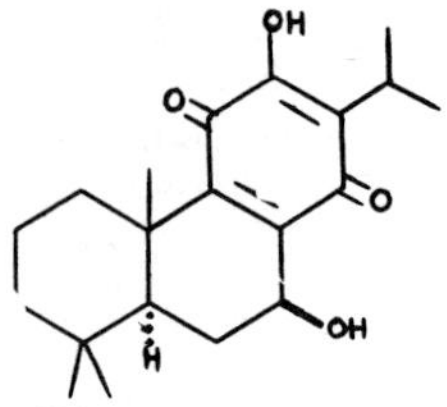

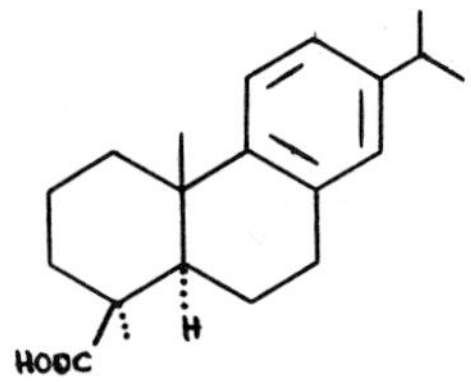

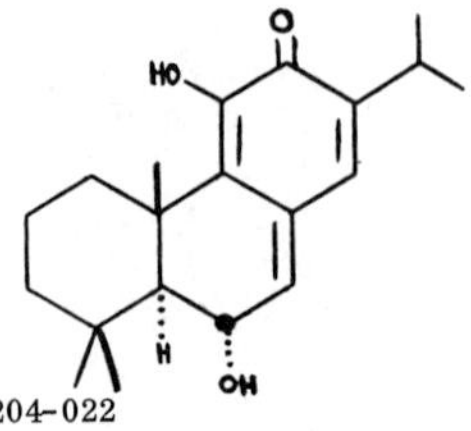

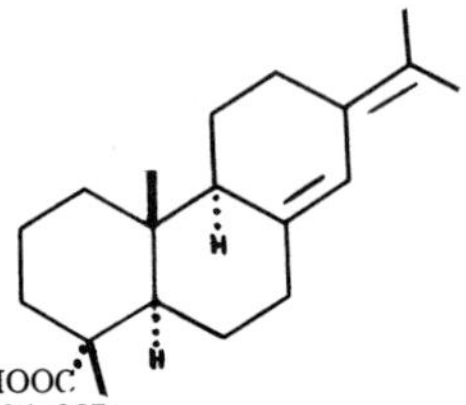

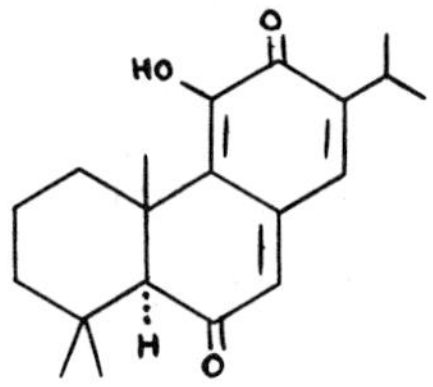

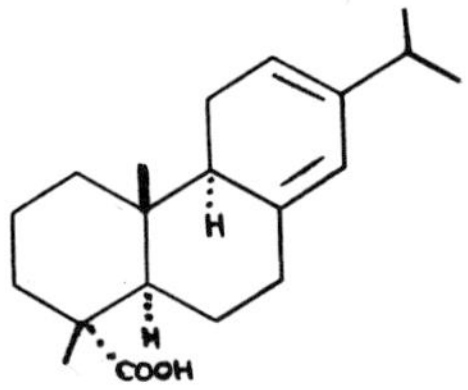

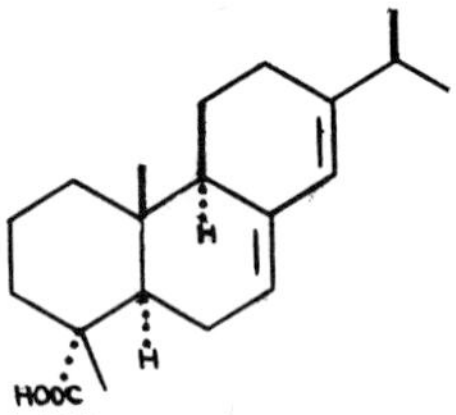

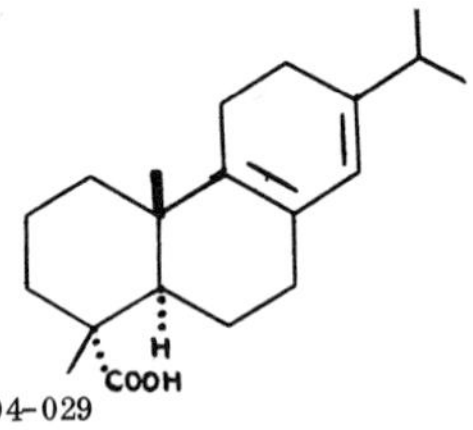

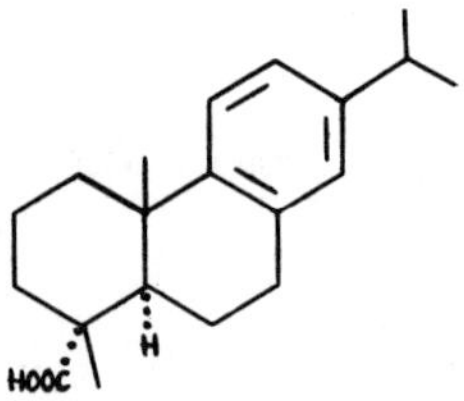

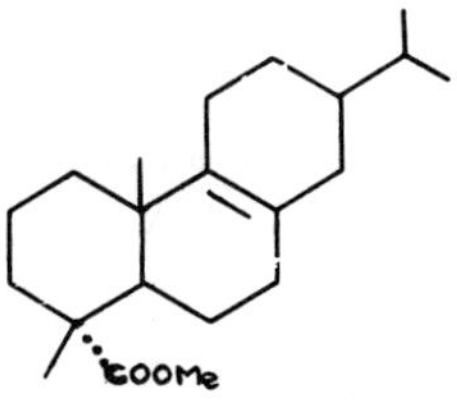

SALVIN — $C_{20}H_{28}O_4$ — 332 — $+144^0$c — 159^0C

Tet. Lett. 1965 3647

4204-031

CARNOSOL — $C_{20}H_{26}O_4$ — 330 — -66^0e — 225^0C

Tet. Lett. 1965 3647

4204-032

MAYTENONE — $C_{40}H_{60}O_4$ — 604 — $+115^0$ — 197^0C

JCS 1961 4420

4204-033

ABIETIC ACID, Dihydro- — $C_{20}H_{32}O_2$ — 304 — $+108^0$e — 175^0C

Chem. Abstr. 1955 12385

4204-034

ABIETA-7, 13-DIEN-3-OL — $C_{20}H_{32}O$ — 288

Acta Chem. Scand., 1970, 1860.

4204-035

ABIETA-7, 13-DIEN-3-ONE — $C_{20}H_{30}O$ — 286

Acta Chem. Scand., 1970, 1860.

4204-036

ABIETA-7, 13-DIEN-2-OL ACETATE — $C_{22}H_{34}O_2$ — 330

Acta Chem. Scand., 1970, 1860.

4204-037

ABIETIC ACID, 15-OH-dehydro- — $C_{20}H_{28}O_3$ — 316

Aust. J. C., 1970, 1457.

4204-038

ABIETIC ACID, 15-OH- — $C_{20}H_{30}O_3$ — 318

Aust. J. C., 1970, 1457.

4204-039

TEIDEADIOL — $C_{20}H_{30}O_2$ — 302

Chem. Abstr., 1970 (73) 88048.

4204-040

ABIETIC ACID, 4-epi- $C_{20}H_{30}O_2$ 302 m. e. -61°c m. e. 39°C [structure: COOH] Bull. Soc. Chim. Fr., 1969, 3264. 4204-041	PALUSTRIC ACID, 4-epi- $C_{20}H_{30}O_2$ 302 m. e. 57°C [structure: COOH] Bull. Soc. Chim. Fr., 1969, 3264. 4204-042

Compound	Formula	Data	Reference	No.
TOTAROL	$C_{20}H_{30}O$	286 +41°e 132°C	Nature 1952 1018	4205-001
TOTAROL, 7β-OH-	$C_{20}H_{30}O_2$	302 201°C	Tet. 1962 (18) 465	4205-006
TOTAROL, 18-OH-	$C_{20}H_{30}O_2$	302 +42°e 230°C	Tet. 1962 (18) 465	4205-002
TOTAROL Me ETHER, 1, 3, -dioxo-	$C_{21}H_{28}O_3$	328 +225° 191°C	Gazz. 1966 (96) 206	4205-007
TOTAROL, 18-Oxo-	$C_{20}H_{28}O_2$	300 +76°m 186°C	JCS 1963 1553	4205-003
PODOTOTARIN	$C_{40}H_{58}O_2$	570	Chem. Ind. 1962 1757	4205-008
TOTAROLOIC ACID, 18-	$C_{20}H_{28}O_3$	316 245°C	JCS 1963 6050; 1553	4205-004
MACROPHYLLIC ACID	$C_{40}H_{54}O_6$	630 +79° 237°C	Tet. 1963 1109	4205-009
TOTAROLONE	$C_{20}H_{28}O_2$	300 +102° 188°C	Acta Chem. Scand. 1962 (16) 1305	4205-005

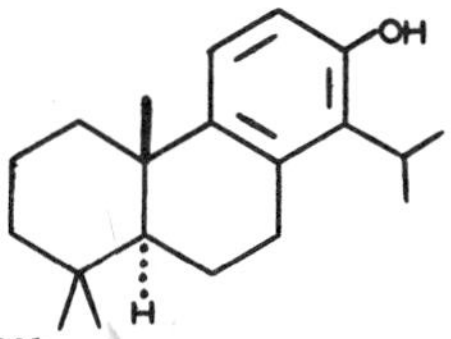

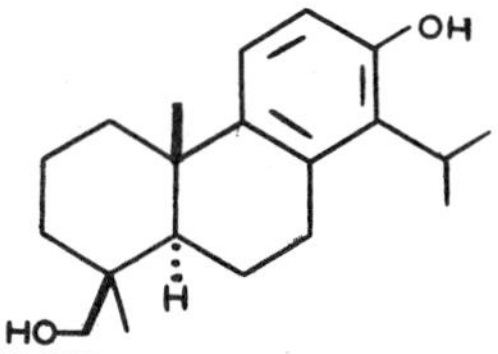

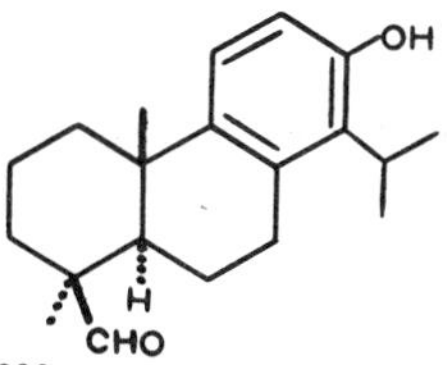

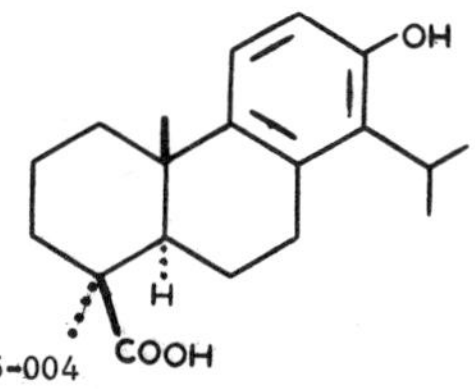

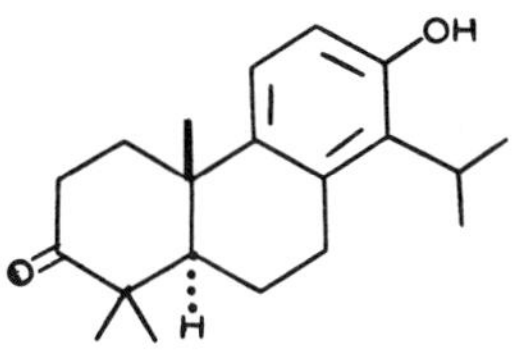

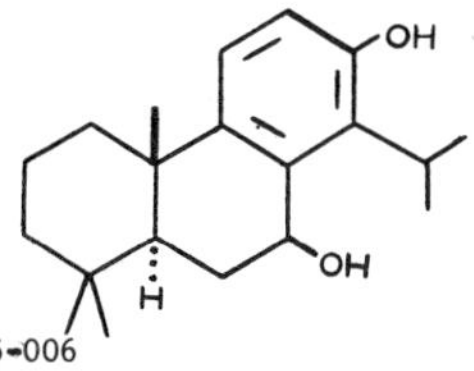

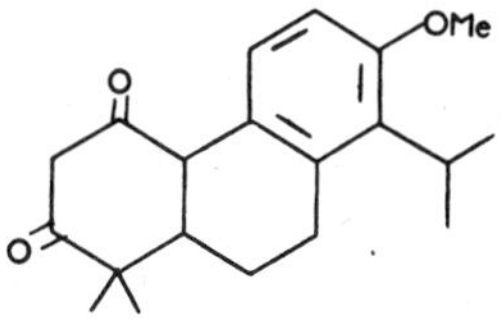

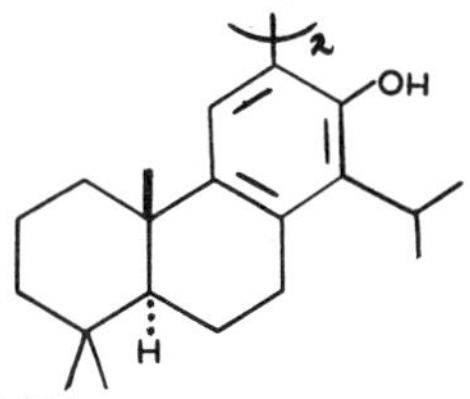

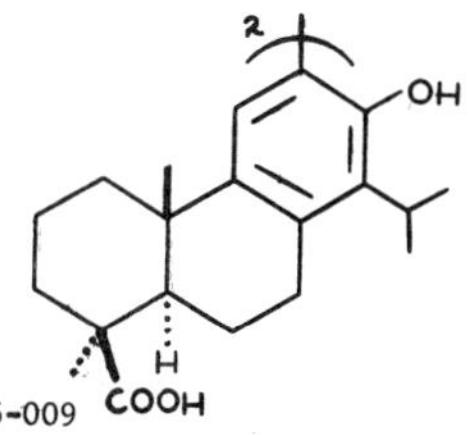

KAURENE, (+)- $C_{20}H_{32}$ 272 +101°c 49°C

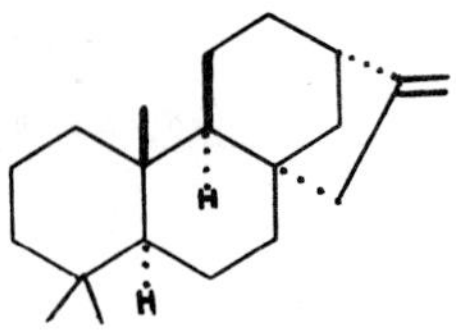

JCS 1950, 599

4206-001

KAURAN-17, 19-DIOL $C_{20}H_{34}O_2$ 306 -65° 178°C

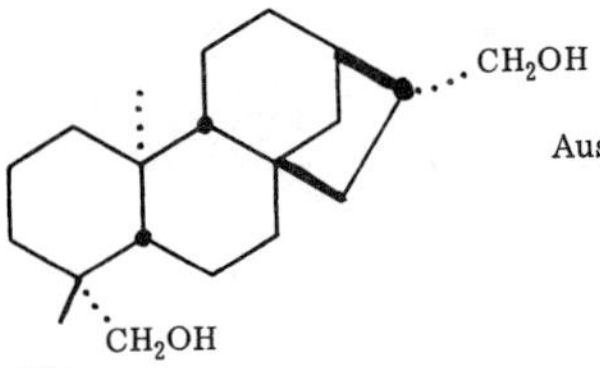

Aust. J. C. 1965 1441

4206-006

KAURENE, (-)- $C_{20}H_{32}$ 272 -80°e 50°C

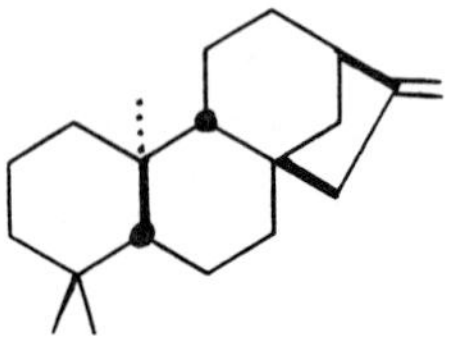

JOC 1962 3741

4206-002

(-) KAUREN-17, 19-DIOL, iso- $C_{20}H_{32}O_2$ 304 -37°e 194°C

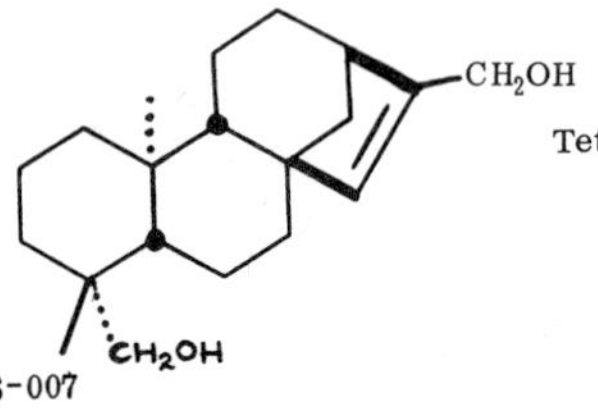

Tet. Lett. 1967 4891

4206-007

KAURENE, (+)-iso- $C_{20}H_{32}$ 272

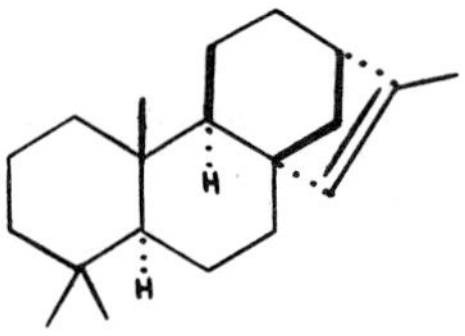

Tet. Lett. 1959 (8) 8

4206-003

(-)-KAUR-16-EN-3α, 19-diol $C_{20}H_{32}O_2$ 304 -66° 184°C

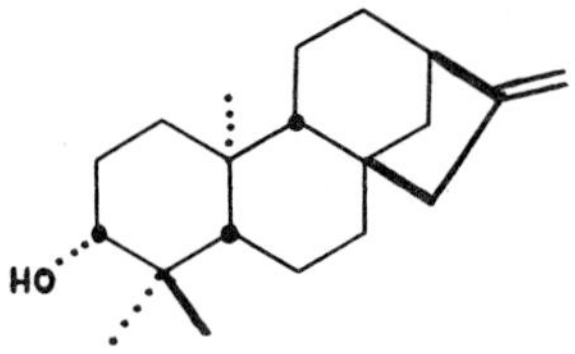

Tet. 1964 1983

4206-008

KAURANOL, (+)- $C_{20}H_{34}O$ 290 +48°c 212°C

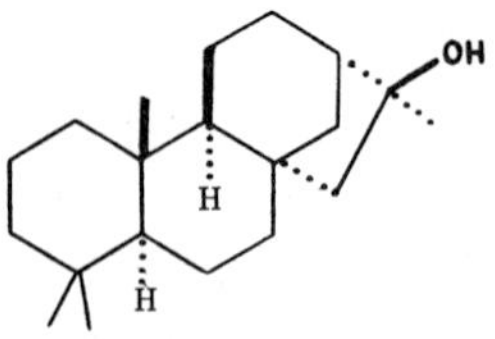

J. Appl. Chem. 1960 (10) 340

4206-004

(-)-KAUR-16-ENE-3α-OH-19-SUCCINATE $C_{24}H_{36}O_5$ 404 -62°c 156°C

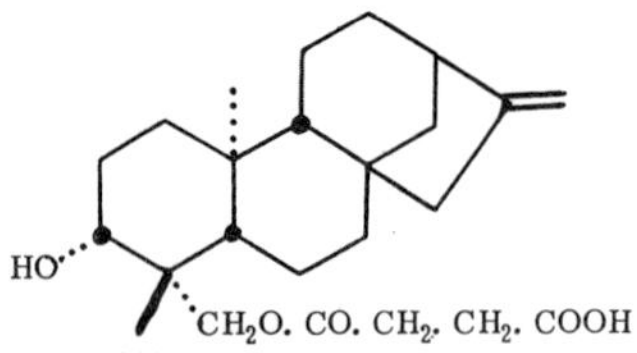

Aust. J. C. 1968 2349

4206-009

KAURANOL, (-)- $C_{20}H_{34}O$ 290 -45°c 215°C

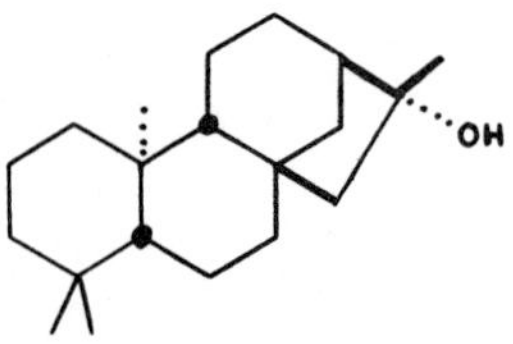

Bull. Soc. Chim. Fr. 1965 2882

4206-005

CANDICANDIOL $C_{20}H_{32}O_2$ 304 -87.5°m 219°C

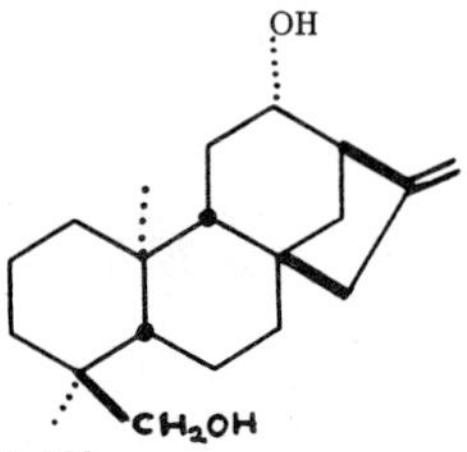

Tet. Lett. 1969 599

4206-010

SIDERIDIOL $C_{20}H_{32}O_2$ 304 +19° Tet. 1968 4073 OH 4206-011	KAURAN-16, 17, 19-TRIOL $C_{20}H_{34}O_3$ 322 -38° 226°C OH Aust. J. C. 1964 915 CH_2OH CH_2OH 4206-016
SIDEROL $C_{22}H_{34}O_3$ 346 152°C Tet. 1968 4073 OAc CH_2OH 4206-012	KAURAN-3, 17, 19-TRIOL $C_{20}H_{34}O_3$ 322 -39° 250°C CH_2OH Aust. J. C. 1964 578 HO CH_2OH 4206-017
KAURANE-3α, 16, 19-trio, 16β-(-)- $C_{20}H_{34}O_3$ 322 -47°e 235°C OH Aust. J. C. 1968 1311 HO $HOCH_2$ 4206-013	SIDEROXAL $C_{20}H_{32}O_3$ 320 246°C Chem. Abstr. 1969 (70) 29104 O OH CH_2OH 4206-018
CORYMBOL $C_{20}H_{34}O_3$ 322 -3.8°p 282°C CH_2OH OH Tet. 1966 1937 H OH 4206-014	KAURANE-3α, 16, 17, 19-TETROL, 16β-(-)- $C_{20}H_{34}O_4$ 338 -46°e 236°C OH Aust. J. C. 1968 1311 HO $HOCH_2$ 4206-019
KAURANE-3α, 16, 17-triol, 16β-(-)- $C_{20}H_{34}O_3$ 322 -47°e 235°C CH_2OH OH Aust. J. C. 1968 1311 HO 4206-015	TRICHOKAURIN $C_{24}H_{34}O_7$ 434 -93° 185°C OH Chem. Comm. 1967 (4) 148; 466 O OH OAc OAC 4206-020

ABBEOKUTONE $C_{20}H_{32}O_3$ 320 -73°c 191°C

JCS 1967 1360

4206-021

KAUR-15-EN-19-AL $C_{20}H_{30}O$ 286 -31°c 114°C

4206-022

KAURAN-19-AL, (-)-16α-OH- $C_{20}H_{32}O_2$ 304 -68° 171°C

Chem. Ind. 1968 1770

4206-023

ORIDONIN $C_{20}H_{28}O_6$ 364 -46° 249°C

Chem. Comm. 1967 252

4206-024

ENMENOL $C_{20}H_{30}O_6$ 366 -29°e 241°C

Chem. Pharm. Bull. 1970 871

4206-025

ENMELOL $C_{20}H_{30}O_5$ 350 -48°e 264°C

Chem. Pharm. Bull. 1970 884

4206-026

SODOPONIN $C_{22}H_{32}O_7$ 408 +46°p 229°C

Tet. Lett. 1970 421

4206-027

ENMEDOL $C_{22}H_{30}O_6$ 390 -45°e 298°C

Chem. Pharm. Bull. 1970 871

4206-028

KAURENIC ACID $C_{20}H_{30}O_2$ 302 -61°c 120°C

Bull. Soc. Chim. 1963 1974

4206-029

(-)-KAUR-16-EN-19-OIC ACID $C_{20}H_{30}O_2$ 302 -110°c 180°C

Tet. Lett. 1967 5241

4206-030

KAURENIC ACID, Iso- $C_{20}H_{30}O_2$ 302 $\pm 0^0$ 123^0C

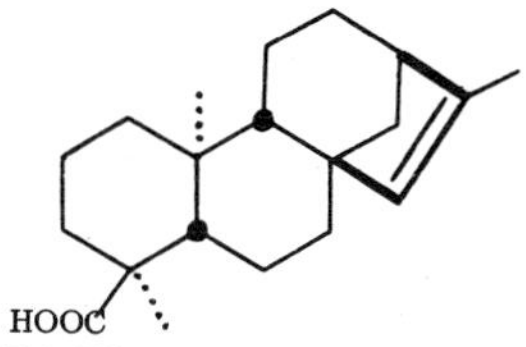

Bull. Soc. Chim. Fr. 1965 2882

4206-031

STEVIOL $C_{20}H_{30}O_3$ 318 -94^0e 214^0C

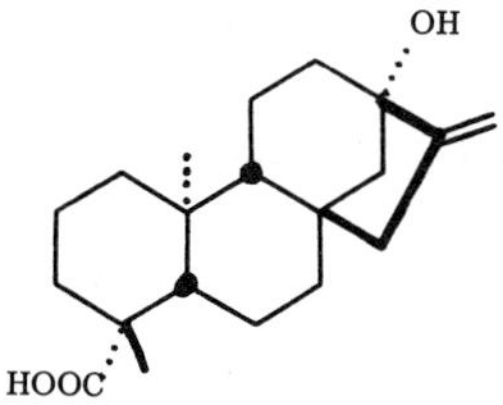

JOC 1955 884

4206-036

KAURENIC ACID, 19-ISO- $C_{20}H_{30}O_2$ 302 -53^0c 181^0C

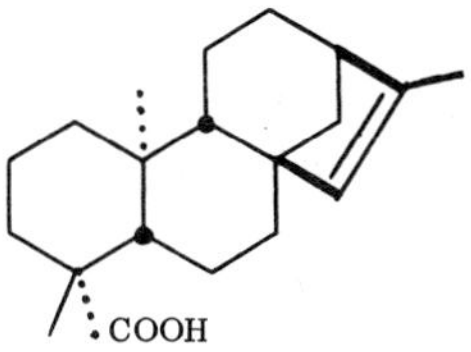

3rd Nat. Prod. Sym. Jamaica 1970

4206-032

STEVIOSIDE $C_{28}H_{60}O_{18}$ 804

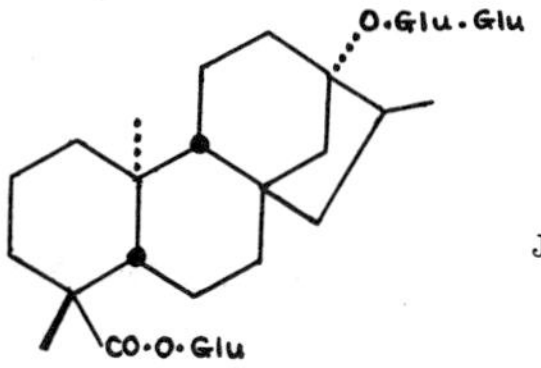

JACS 1956 (78) 4709

4206-037

(-)-KAUR-16-EN-19-OIC ACID, -$\Delta^{9(11)}$- $C_{20}H_{28}O_2$ 300 $+32^0c$ 156^0C

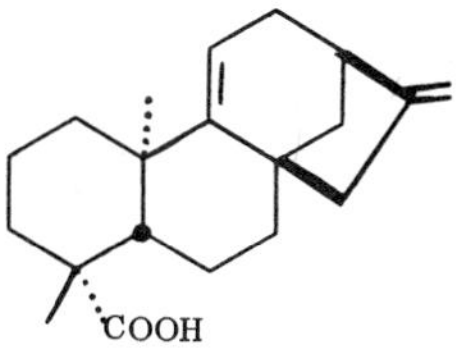

Tet. Lett. 1968 5661

4206-033

KAURENIC ACID, acetoxy- $C_{22}H_{32}O_4$ 360 -70^0 135^0C

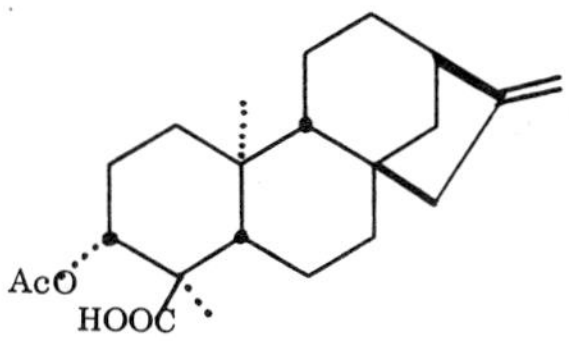

Bull. Soc. Chim. Fr. 1965 2882

4206-038

KAURANOIC ACID, 17-OH-19- $C_{20}H_{32}O_3$ 320 -96^0 212^0C

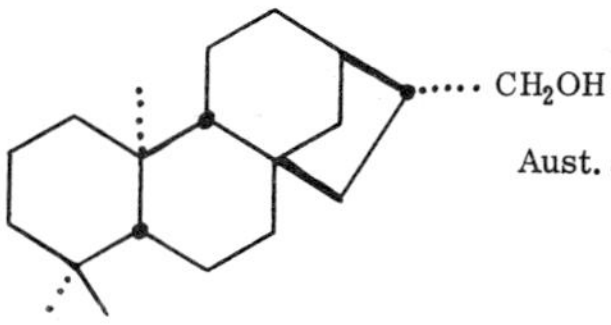

Aust. J. C. 1965 1441

4206-034

GRANDIFLORIC ACID $C_{20}H_{30}O_3$ 318 -114^0c 227^0C

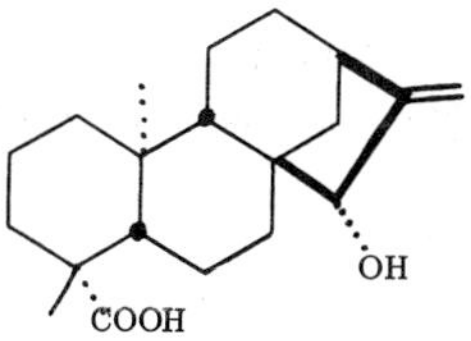

Chem. Abstr. 1969 (70) 29099

4206-039

KAURANOIC ACID, (-)-16α-OH $C_{20}H_{32}O_3$ 320 -92^0 282^0C

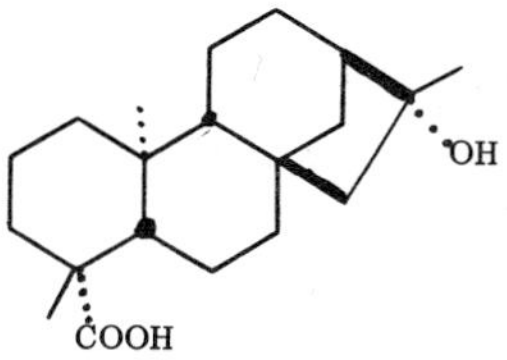

Chem. Ind. 1968 1770

4206-035

KAUR-16-EN-19-OIC ACID, 15β-OH $C_{20}H_{30}O_3$ 318 -107^0e 209^0C

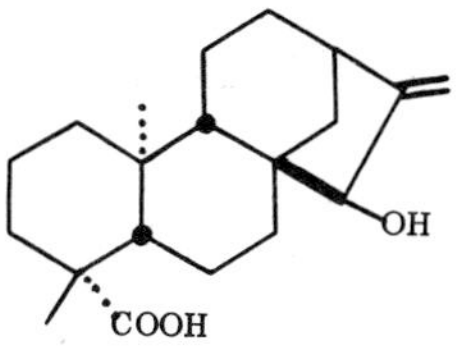

Aust. J. C. 1966 861

4206-040

XYLOPIC ACID $C_{22}H_{32}O_4$ 360 -144° 260°C JCS 1968 311 COOH OAc 4206-041	KAURANOIC ACID, 19-OH-17- $C_{20}H_{32}O_3$ 320 -66° 207°C COOH Aust. J. C. 1964 915 CH_2OH 4206-046
(-)-KAUR-9(11)-16-DIEN-19-OIC ACID, 15-OAc- $C_{22}H_{32}O_4$ 360 -85° 174°C Tet. Lett. 1968 5661 OAc COOH 4206-042	(-)-KAURAN-17-OIC ACID 1α, 19-diOH-16α- $C_{20}H_{32}O_4$ 336 -56°e 260°C OH COOH Aust. J. Chem. 1965 2005 CH_2OH 4206-047
(-)-KAUR-9(11), 16-DIEN-19-OIC ACID, 15-OH- $C_{20}H_{30}O_3$ 318 217°C Tet. Lett. 1968 5661 OH COOH 4206-043	KAURANOIC ACID, 18-OH-3-oxo-, $C_{20}H_{30}O_4$ 334 -115°c 194°C COOH Aust. J. C. 1964 578 O CH_2OH 4206-048
(-)-KAURAN-19-OIC ACID, 16α, 17-diOH-16β- $C_{20}H_{32}O_4$ 336 -88° 259°C CH_2OH OH Aust. J. C. 1965 1441; 2005 HOOC 4206-044	KAURANOIC ACID, 18-OAc-3-oxo- $C_{22}H_{32}O_5$ 376 -47°c 188°C COOH Aust. J. C. 1964 578 O CH_2OH 4206-049
(-)-KAURAN-19-OIC ACID, 12β, 17-diOH-16α- $C_{20}H_{32}O_4$ 336 -138°e 320°C OH CH_2OH Aust. J. C. 1968 2085 COOH 4206-045	KAURAN-17, 19-DIOIC ACID $C_{20}H_{30}O_4$ 334 -108° 272°C COOH Aust. J. C. 1964 915 COOH 4206-050

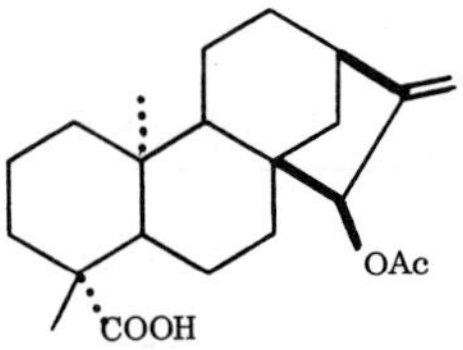

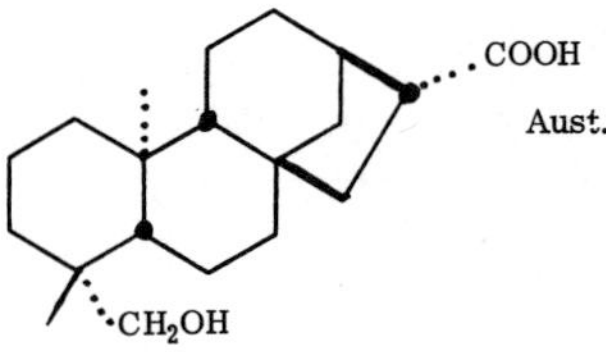

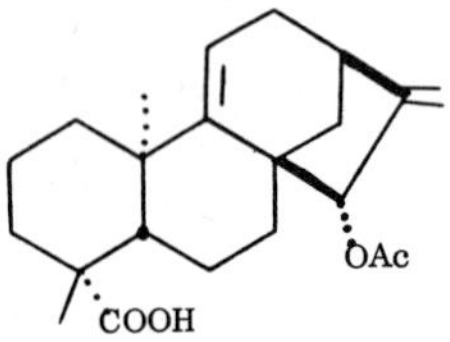

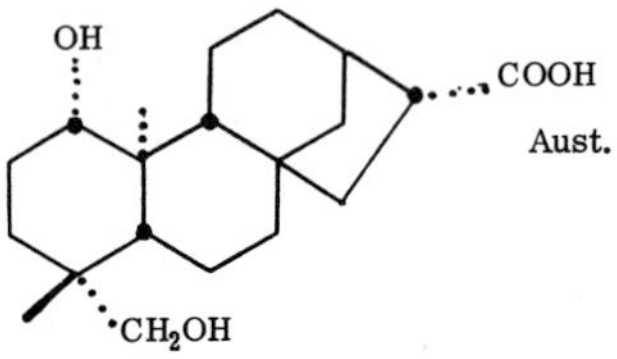

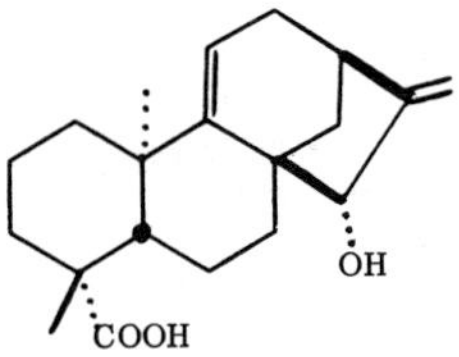

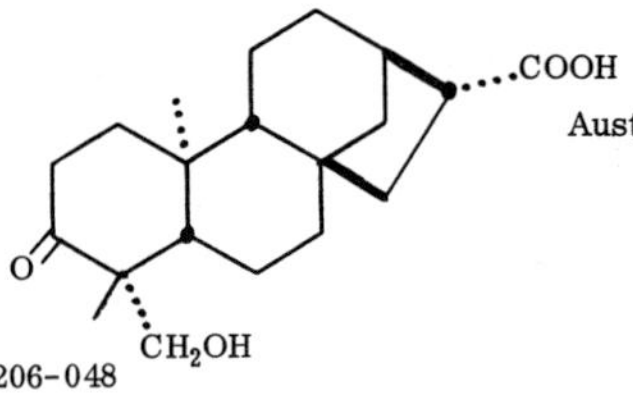

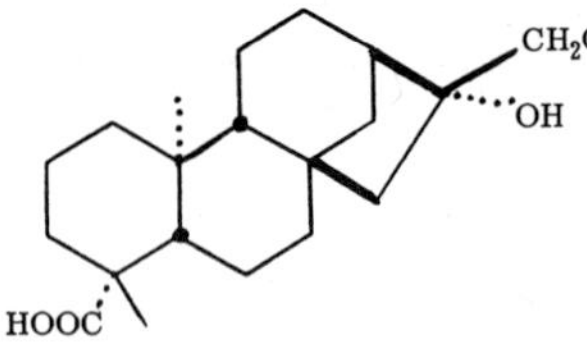

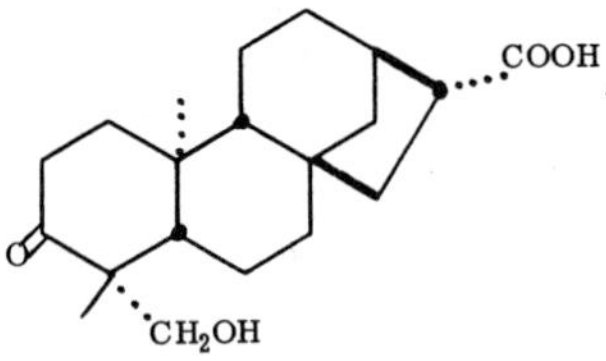

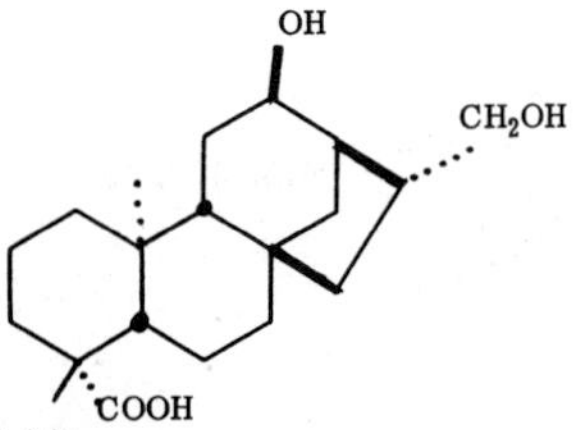

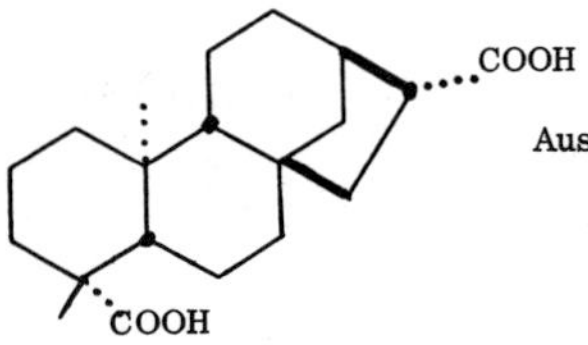

KAURENOLIDE, 7-OH- $C_{20}H_{28}O_3$ 316 -25^{0}e 188^{0}C JCS 1963 2944 4206-051	PHYLLOCLADENE (+) $C_{20}H_{32}$ 272 +16^{0}c 98^{0}C JCS, 1955, 2624. 4206-055
KAURENOLIDE, 7, 18-diOH- $C_{20}H_{28}O_4$ 332 -37^{0}e 212^{0}C JCS 1963 3783 4206-052	PHYLLOCLADENE, (+)-iso- $C_{20}H_{32}$ 272 +16^0 98^{0}C Chem. Abstr., 1964 (61) 12328. 4206-056
KAURENOLIDE, 7, 16, 18-triOH- $C_{20}H_{30}O_5$ 350 -40^{0}e 252^{0}C JCS 1963 3783 4206-053	PHYLLOCLADANOL, (+)- $C_{20}H_{34}O$ 290 +15^{0}m 185^{0}C JCS, 1963, 5374. 4206-057
KAURENOLIDE, 7, 13-diOH- $C_{20}H_{28}O_4$ 332 +2^{0}p 260^{0}C Tet., 1970, 5215. 4206-054	

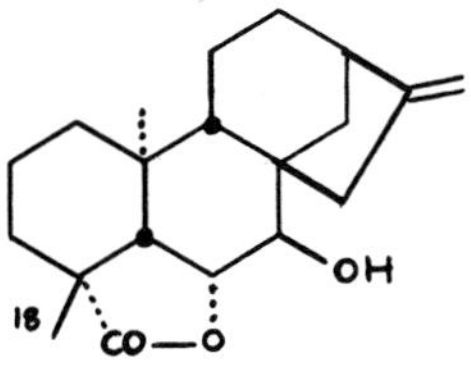

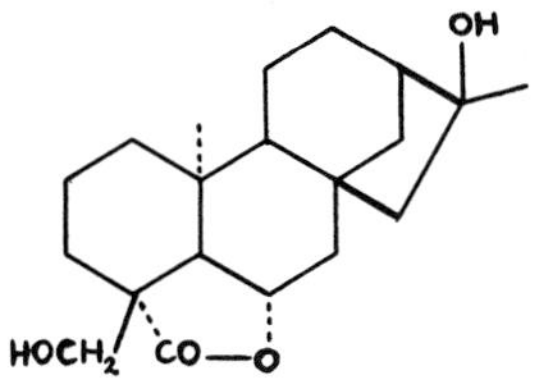

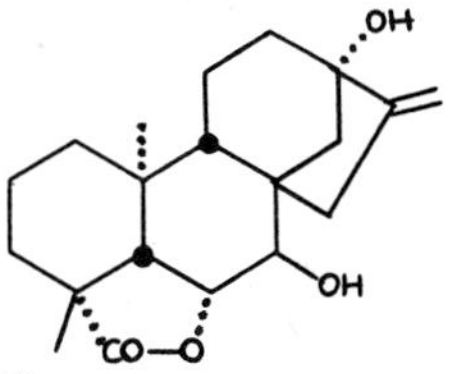

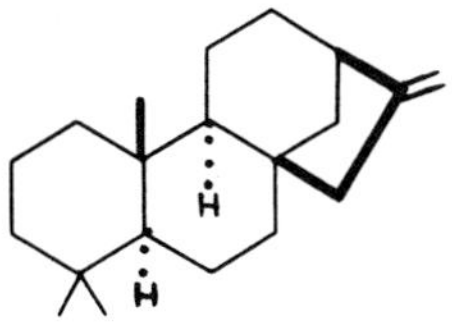

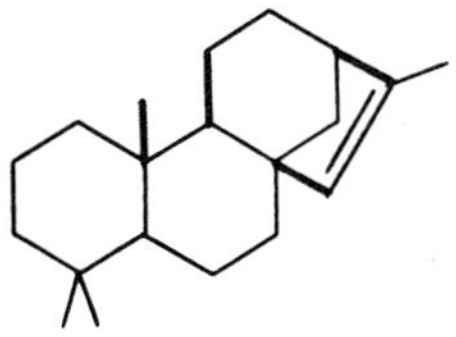

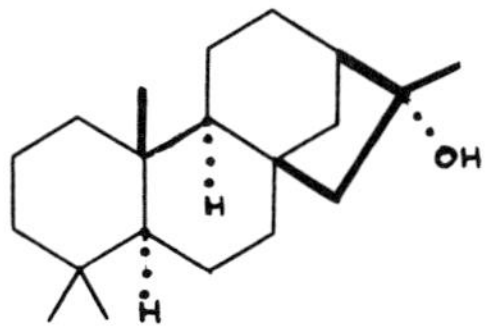

COUMINGINIC ACID $C_{25}H_{38}O_6$ 434 -81°e 200°C

R = H

CO·OR

HO, O

Helv., 1944, 1449.

4207-001

COUMINGIDINE $C_{25}H_{39}NO_5$ 433 160°C

R = NHMe

as 005

4207-006

COUMINGINE $C_{29}H_{47}O_6N$ 505 -70°e 142°C

R = NMe$_2$

Helv., 1944, 1449.

as 001

4207-002

CASSAMIC ACID $C_{21}H_{30}O_5$ 362 -62°e 218°C

R = H

COOR

COOMe

Experimentia, 1960 (16) 404.

4207-007

CASSAIC ACID $C_{20}H_{30}O_4$ 334 -126°e 203°C

R = H

CO·OR

HO, H, O

Tet. Lett., 1959, 7.

4207-003

CASSAMINE $C_{25}H_{39}O_5N$ 433 -56°e 86°C

R = NMe$_2$

Experimentia, 1960 (16) 404.

4207-008

CASSAINE $C_{24}H_{39}O_4N$ 405 -103°e 142°C

R = NMe$_2$

as 003

JACS, 1966, 5865.

4207-004

ERYTHROPHLAMIC ACID $C_{21}H_{30}O_6$ 378 -63°e 219°C

R = H

COOR

HO, COOMe, O

Experimentia, 1960 (16) 404.

4207-009

COUMINGIDINIC ACID $C_{22}H_{32}O_5$ 376

R = H

CO·OR

AcO, O

4207-005

ERYTHROPHLAMINE $C_{25}H_{39}O_6N$ 449 -62°e 150°C

R = NMe$_2$

Experimentia, 1960 (16) 404.

4207-010

Compound	Formula	Data	Reference
CASSAIDIC ACID (R = H) 4207-011	$C_{20}H_{32}O_4$	336 -100°e 276°C	Helv., 1959 (42) 1127.
CAESALPIN, δ- 4207-016	$C_{20}H_{30}O_6$	366 +40° 252°C	Gazz., 1966 (96) 698. Chem. Comm., 1970, 1244.
CASSAIDINE (R = CH2CH2NMe2) 4207-012	$C_{24}H_{41}O_4N$	407 -98°e 139°C	JACS, 1966, 5865.
CAESALPIN, ε- 4207-017	$C_{24}H_{34}O_7$	434 +2° 193°C	Tet, Lett., 1967, 5027.
ERYTHROPHLEGUINE (R = CH2CH2NMe2) 4207-013	$C_{25}H_{39}O_6N$	449 -38°m 78°C	Chem. Abstr., 1968 (68) 59771.
VOUACAPENIC ACID 4207-018	$C_{20}H_{28}O_3$	316 +108°t 230°C	JCS., 1958, 3428.
CAESALPINE, α- 4207-014	$C_{24}H_{32}O_8$	448 + 35°e 187°C	Gazz., 1966 (96) 698. Chem. Comm., 1970, 1244.
VOUACAPENIC ACID METHYL ESTER 4207-019	$C_{21}H_{30}O_3$	330 +101°c 104°C	JCS., 1955, 1117.
CAESALPINE, β- 4207-015	$C_{20}H_{28}O_6$	364 +68° 243°C	Gazz., 1966 (96) 698. Chem. Comm., 1970, 1244.
VOUACAPENYL ACETATE 4207-020	$C_{22}H_{32}O_3$	344 +63°t 115°C	JCS., 1955, 1117.

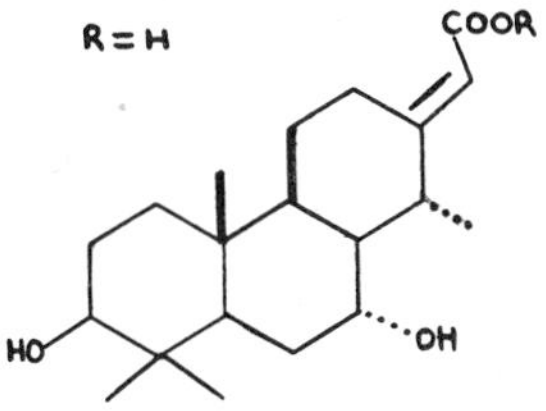

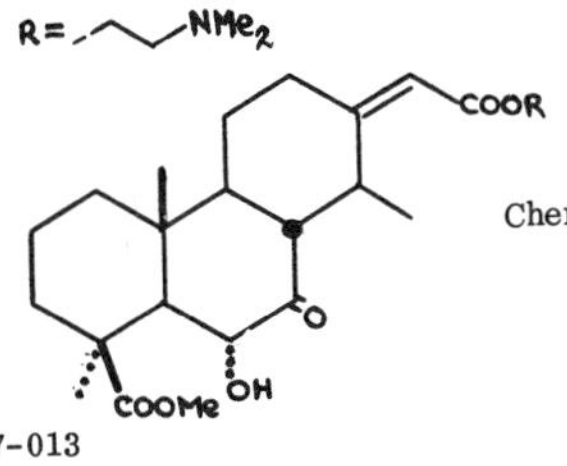

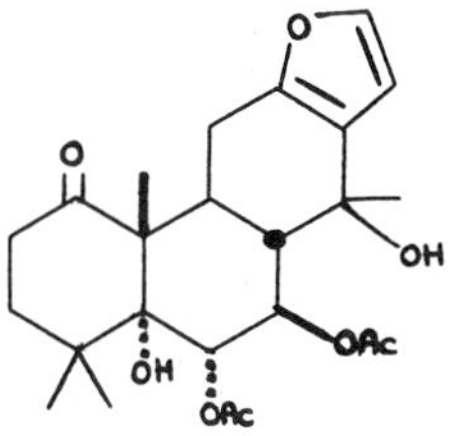

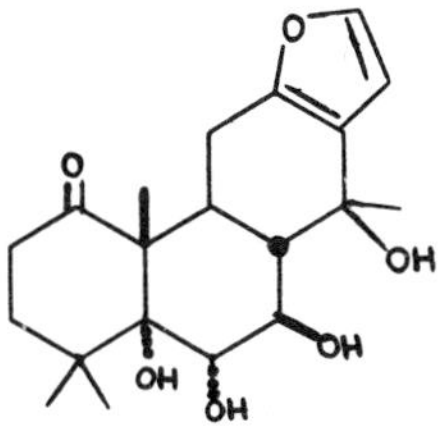

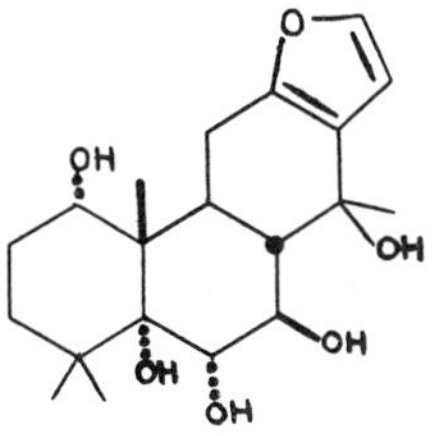

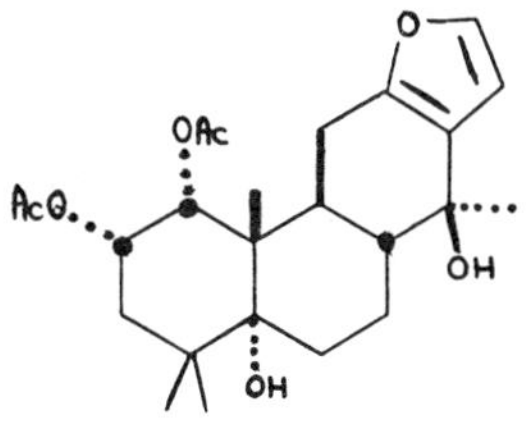

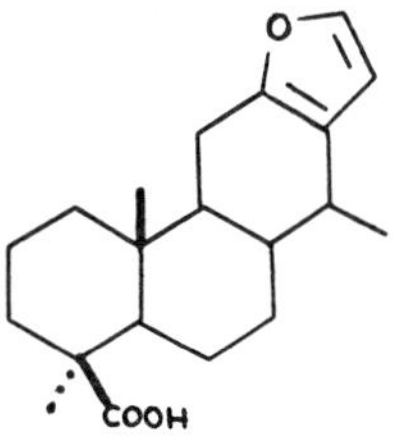

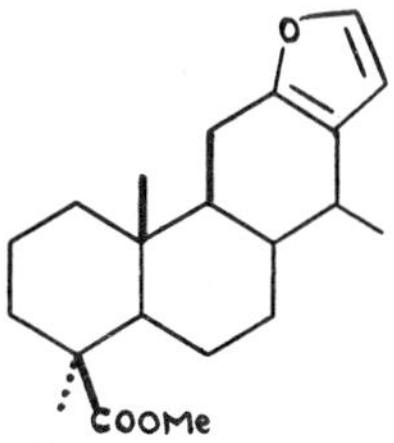

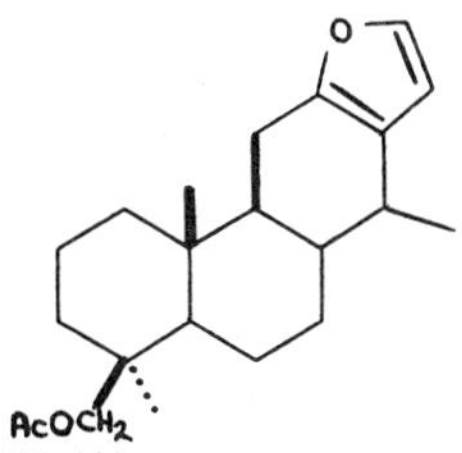

VINHATICOIC ACID ME ESTER — $C_{21}H_{30}O_3$ — 330 — $+70^0c$ — 108^0C

COOMe

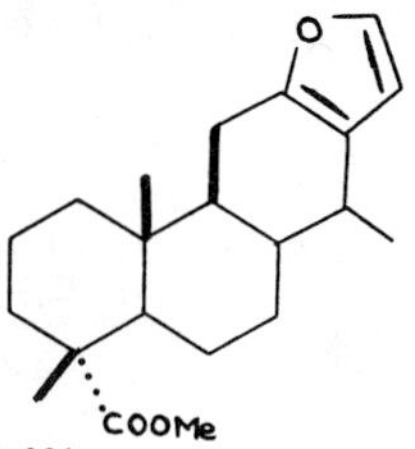

JCS., 1955, 1117.

4207-021

ERYTHROPHLEADIENOLIC ACID — $C_{21}H_{28}O_5$ — 360

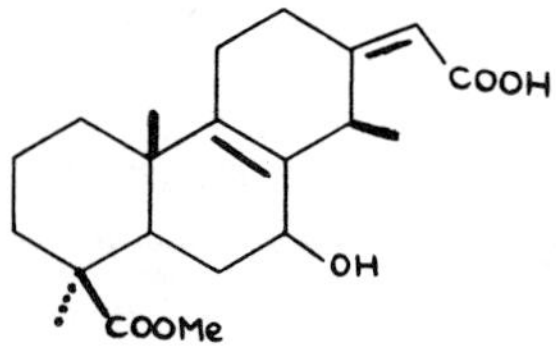

COOH

OH

COOMe

Chem. Abstr., 1962 (57) 15162.

4207-023

PICROSALVIN — $C_{20}H_{26}O_4$ — 330 — 224^0C

OH

HO

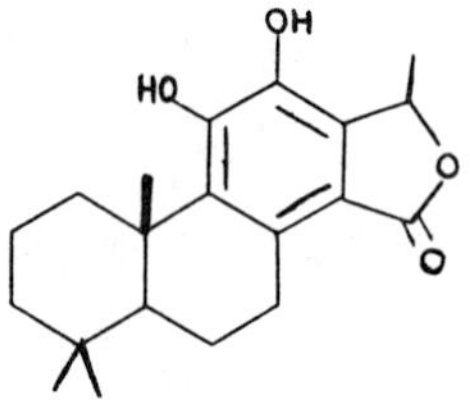

Ber., 1962, 3034.

4207-022

IVORINE — $C_{28}H_{43}NO_5$ — 473 — -43^0c — 158^0C

NHMe

Sen·O

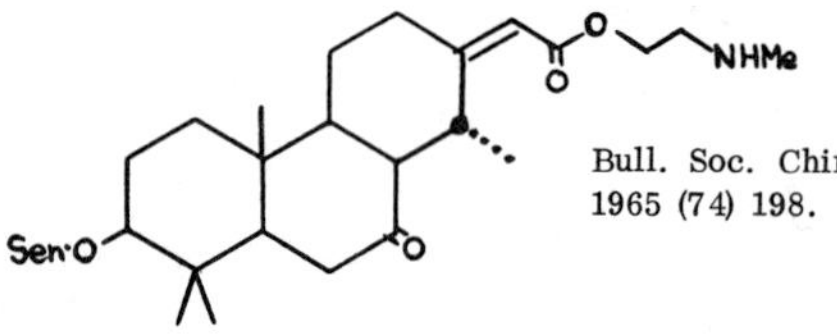

Bull. Soc. Chim. Belges, 1965 (74) 198.

4207-024

RIMUENE $C_{20}H_{32}$ 272 +53° 56°C JCS, 1966, 274. 4208-001	ROSENONOLACTONE, 6β-OH- $C_{20}H_{28}O_4$ 332 -162°c 180°C JCS. 1968, 2122. 4208-005
ROSENOLIC ACID, iso- $C_{20}H_{30}O_3$ 318 0° 193°C Tet. Lett., 1964, 849. 4208-002	ROSEIN III $C_{20}H_{24}O_4$ 328 -124°c 221°C Chem. Comm., 1970, 719. 4208-006
ROSENONOLACTONE $C_{20}H_{28}O_3$ 316 -116°c 208°C Tet., 1959 (7) 241. 4208-003	ROSOLOLACTONE $C_{20}H_{30}O_3$ 318 + 6°c 186°C JCS, 1966. 896. 4208-007
ROSENOLOLACTONE. 7-deoxo- $C_{20}H_{30}O_2$ 302 +57°c 115°C JACS, 1959, 5520. 4208-004	ERYTHROXYTRIOL P Tet. Lett., 1966 (19) 2109. 4208-008

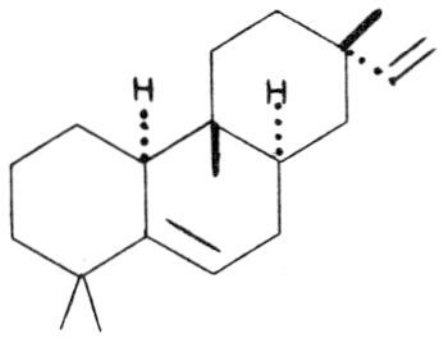

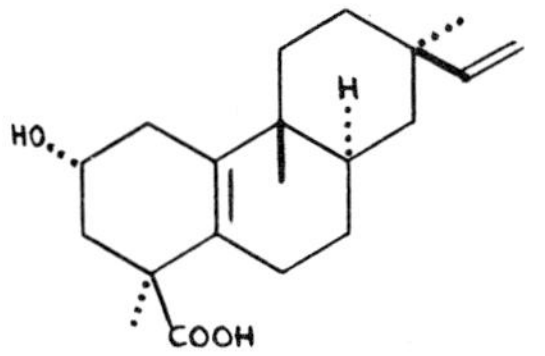

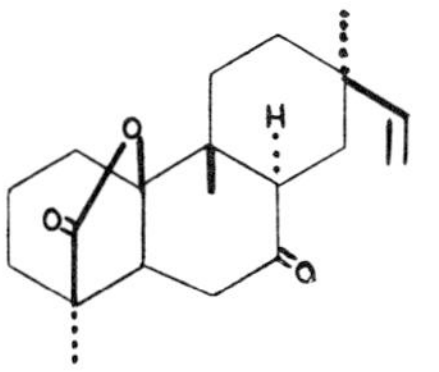

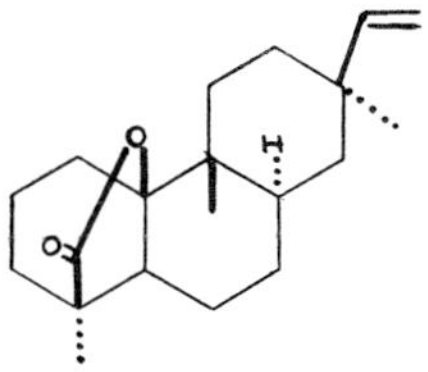

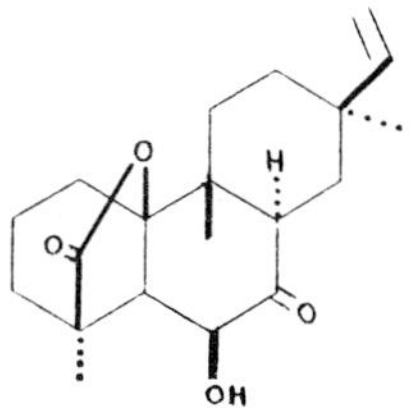

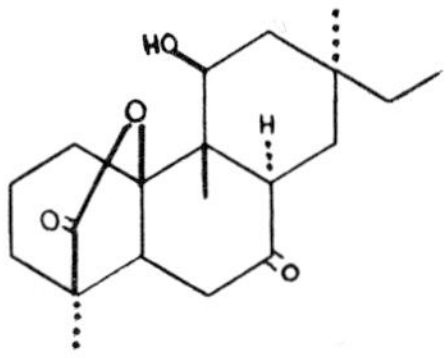

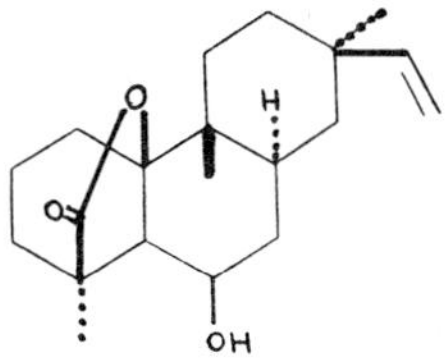

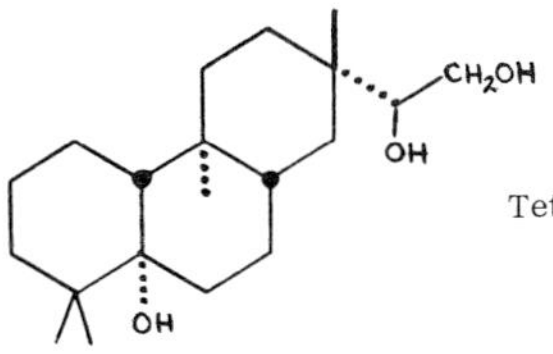

No.	Name	Formula	MW	Rotation	M.p.	Reference
4209-001	STACHENE	$C_{20}H_{32}$	272	$+33^0$c	29^0C	Chem. Ind., 1964, 500.
4209-002	HIBAENE, (-)-	$C_{20}H_{32}$	272	-50^0c	30^0C	Tet. Lett., 1964, 2223.
4209-003	HIBAENE EPOXIDE, (+)-	$C_{20}H_{32}O$	288	$+20^0$c	75^0C	JCS., 1968, 2529.
4209-004	ERYTHROXYLOL A	$C_{20}H_{32}O$	288	$+39^0$c	120^0C	JCS., 1968, 2349.
4209-005	ERYTHROXYLOL B	$C_{20}H_{32}O$	288	$+67^0$c	122^0C	JCS., 1968, 2349.
4209-006	STACHEN-3, 17-DIOL	$C_{20}H_{32}O_2$	304	$+36^0$e		Tet. Lett., 1967, 4892.
4209-007	STACHEN-3α, 19-DIOL	$C_{20}H_{32}O_2$	304	$+39^0$e	150^0C	Tet. Lett., 1967, 4891.
4209-008	ERYTHROXYDIOL A	$C_{20}H_{32}O_2$	304	$+60^0$	180^0C	JCS., 1968, 2349.
4209-009	ERYTHROXYLOL "A" epoxide	$C_{20}H_{32}O_2$	304	$+18^0$c	116^0C	JCS., 1968, 2529.
4209-010	ERYTHROXYLOL "A" acetate epoxide	$C_{22}H_{34}O_3$	346	$+14^0$c	144^0C	JCS., 1968, 2529.

BEYEROL $C_{20}H_{32}O_3$ 320 +61°p 242°C CH₂OH, HO, CH₂OH 4209-011	STACHENE DIOSPHENOL $C_{20}H_{28}O_2$ 300 +49°c 132°C HO, O JCS 1962 4046. 4209-015
BEYEROL, 17-cinnamoyl- $C_{29}H_{38}O_4$ 450 +66°c 145°C CH₂O·Cinn, HO, CH₂OH Tet. Lett., 1963, 1793. 4209-012	STACHENE KETOL $C_{20}H_{30}O_2$ 302 +30°c 129°C O, HO JCS. 1962, 4046. 4209-016
STACHENONE $C_{20}H_{30}O$ 286 +22°c 36°C O Tet. Lett., 1967, 793. 4209-013	BEYER-15-EN-17-OIC ACID, 19-OAc-3-keto- $C_{22}H_{30}O_5$ 374 m. e. +83° m. e. 136°C COOH, O, CH₂OAc Aust. J. Chem., 1968, 444. 4209-017
BEYER-15-EN-3-ONE, 6β-OAc-17-OH- $C_{22}H_{32}O_4$ 360 -104° 166°C CH₂OH, O, OAc Aust. J. Chem., 1968, 439. 4209-014	

GIBBERELLIC ACID $C_{19}H_{22}O_6$ 346
$+92^0$c
235^0C

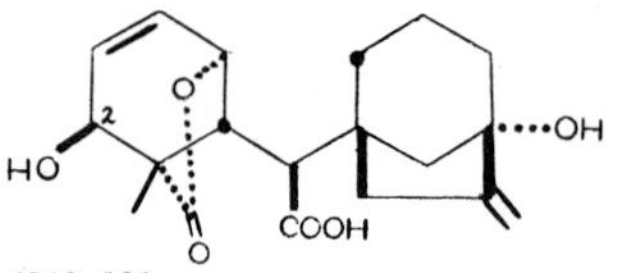

Tet. Lett., 1966, 6003.

4210-001

GIBBERELLIC ACID, 2-O-glucosyl- $C_{23}H_{32}O_{11}$ 508

Tet. Lett., 1969, 2081.

4210-002

GIBBERELLIC ACID, 2-O-Ac $C_{21}H_{24}O_7$ 388
$+150^0$c
226^0C

Phytochemistry, 1966, 5, 1221.

4210-003

GIBBERELLENIC ACID $C_{19}H_{22}O_6$ 346
185^0C

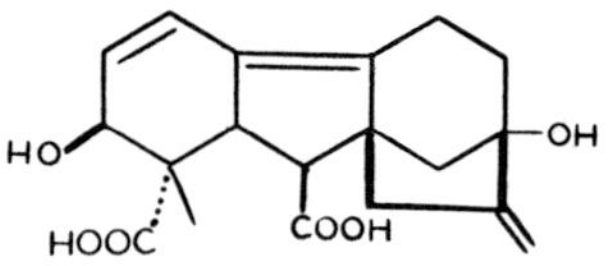

Proc. Chem. Soc., 1962, 185.

4210-004

GIBBERELLIN A_1 $C_{19}H_{24}O_6$ 348
$+36^0$
256^0C

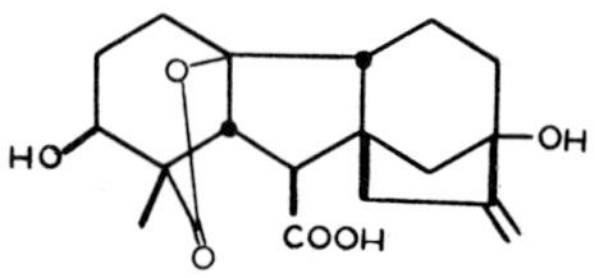

Tet., 1960, (11), 60.

4210-005

GIBBERELLIN A_2 $C_{19}H_{26}O_6$ 350
$+12^0$c
236^0C

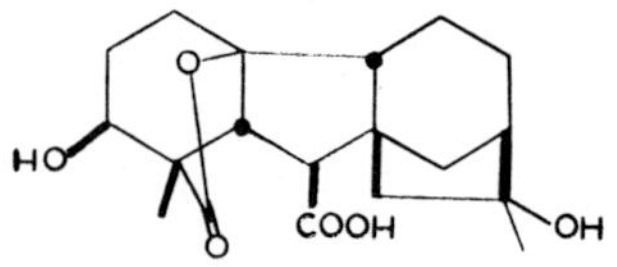

Bull. Agr. Chem. Soc. Jap., 1959, (23), 509.

4210-006

GIBBERELLIN A_4 $C_{19}H_{24}O_5$ 332
-3^0
255^0C

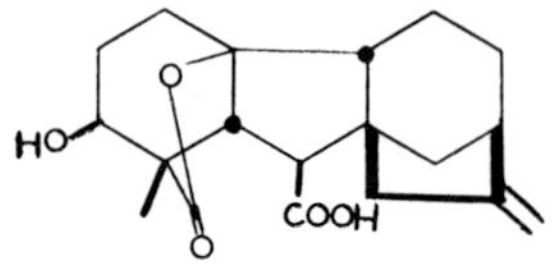

Bull. Agr. Chem. Soc. Jap., 1959, (23), 509.

4210-007

GIBBERELLIN A_5 $C_{19}H_{22}O_5$ 330
-77^0
260^0C

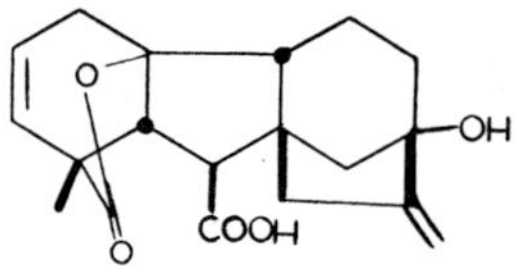

Tet., 1960, (11), 60.

4210-008

GIBBERELLIN A_6 $C_{19}H_{22}O_6$ 346
-28^0e
224^0C

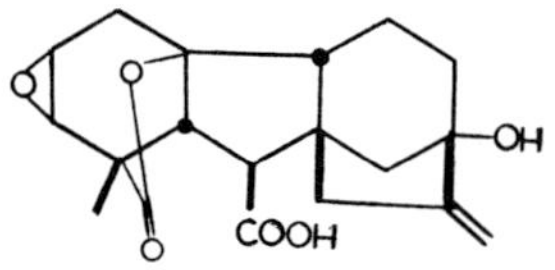

Tet., 1962, (18), 349.

4210-009

GIBBERELLIN A_7 $C_{19}H_{22}O_5$ 330
$+20^0$
202^0C

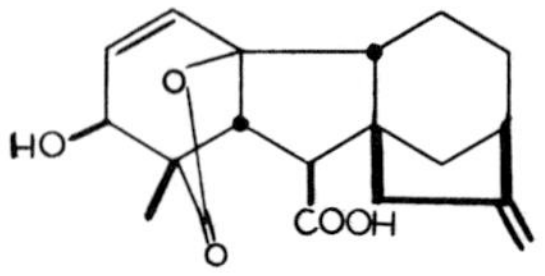

Tet., 1962, 451.

4210-010

GIBBERELLIN A_7, deoxy-iso- $C_{19}H_{22}O_4$ 314 +11° 243°C Tet., 1967, 4095 4210-011	GIBBERELLIN A_{10} $C_{19}H_{26}O_5$ 334 +3° 245°C Tet., 1966, 701. 4210-016
deleted 4210-012	GIBBERELLIN A_{11} $C_{19}H_{22}O_5$ 330 +11° 181°C Tet., 1967, 4095. 4210-017
GIBBERELLIN A_8 $C_{19}H_{24}O_7$ 364 +30°e 212°C Tet., 1962, (18) 249. 4210-013	GIBBERELLIN A_{12} $C_{20}H_{28}O_4$ 332 248°C JCS, 1965, 1570. 4210-018
GIBBERELLIN A_8, 3-O-Glucosyl $C_{25}H_{34}O_{12}$ 526 +7°m Tet. Lett., 1969, 2081. 4210-014	GIBBERELLIN A_{13} $C_{20}H_{26}O_7$ 378 -48°e 196°C JCS, 1965, 3143. 4210-019
GIBBERELLIN A_9 $C_{19}H_{24}O_7$ 364 -22°e 210°C Tet., 1962, 451. 4210-015	GIBBERELLIN A_{14} $C_{20}H_{28}O_5$ 348 -73°e 232°C JCS, 1966, 501. 4210-020

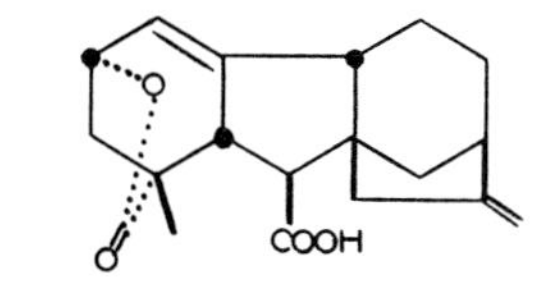

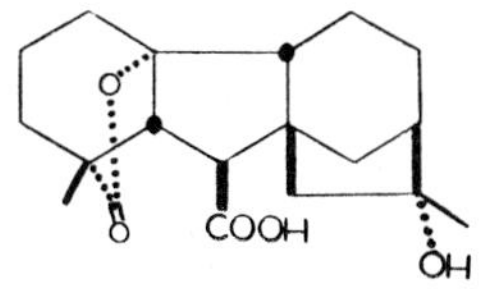

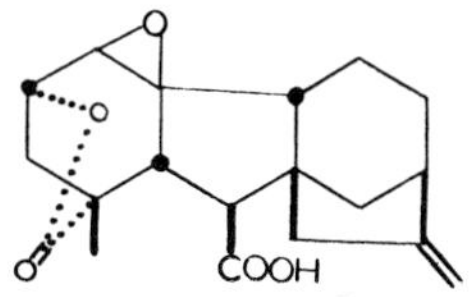

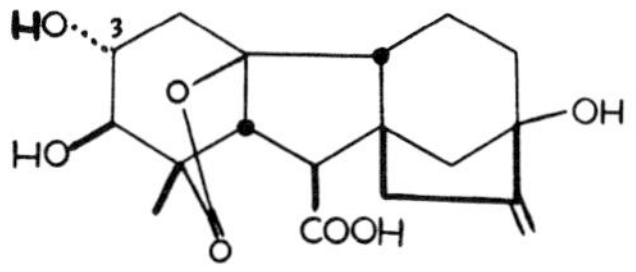

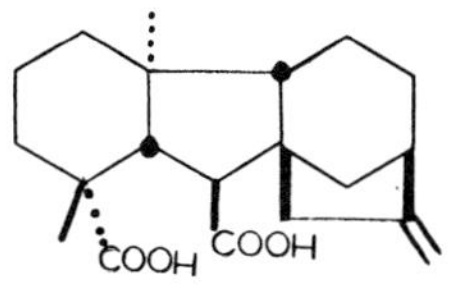

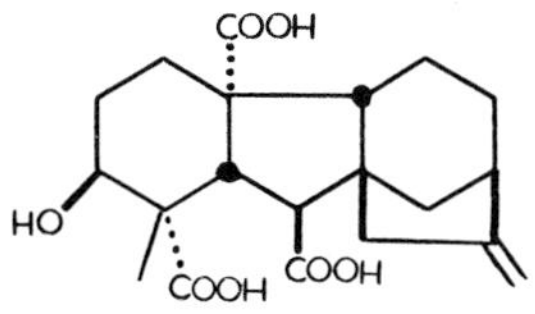

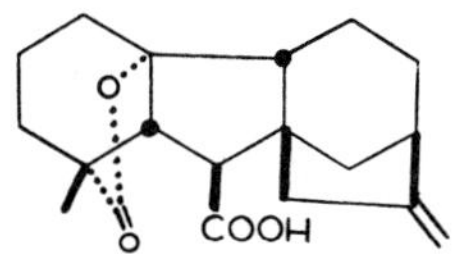

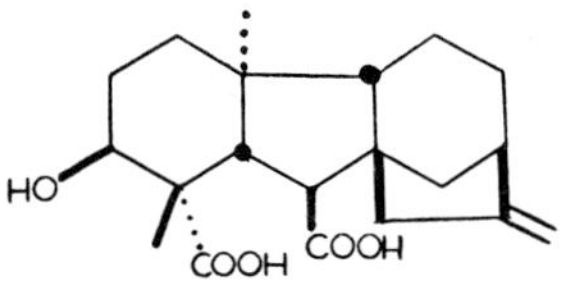

GIBBERELLIN A_{15} — $C_{20}H_{26}O_4$ — 330 — $+5^0$ — 275^0C

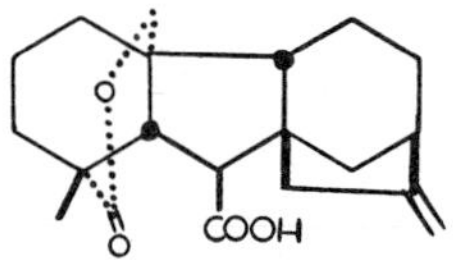

Tet., 1967(23), 733.

4210-021

GIBBERELLIN A_{16} — $C_{19}H_{24}O_6$ — 348

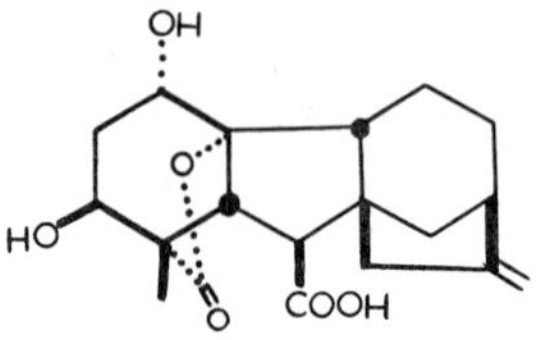

Tet., 1968, 1337.

4210-022

GIBBERELLIN A_{17} — $C_{20}H_{26}O_7$ — 378 — 145^0C

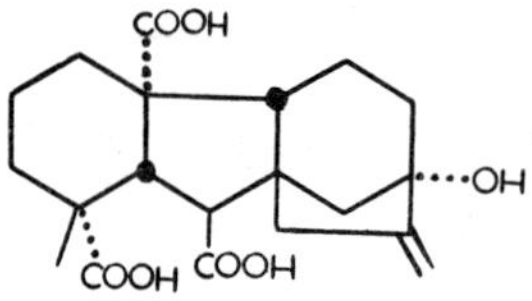

Tet. Lett., 1967, 4173.

4210-023

GIBBERELLIN A_{18} — $C_{20}H_{28}O_6$ — 364 — 241^0C

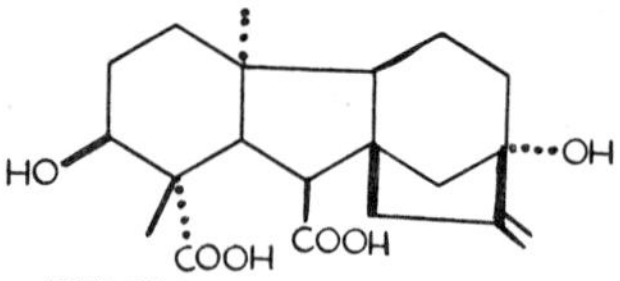

Chem. Abstr., 1969, (70) 37938.

4210-024

GIBBERELLIN A_{19} — $C_{20}H_{26}O_6$ — 362 — 236^0C

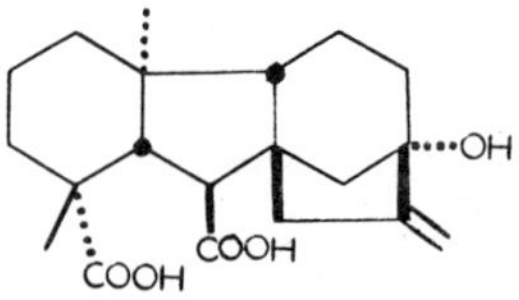

Tet. Lett., 1966, (22)2465.

4210-025

GIBBERELLIN A_{20} — $C_{19}H_{24}O_5$ — 332

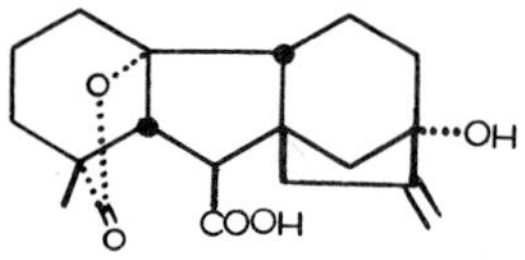

Tet. Lett., 1968, 1537.

4210-026

GIBBERELLIN A_{21} — $C_{19}H_{22}O_7$ — 362 — 245^0C

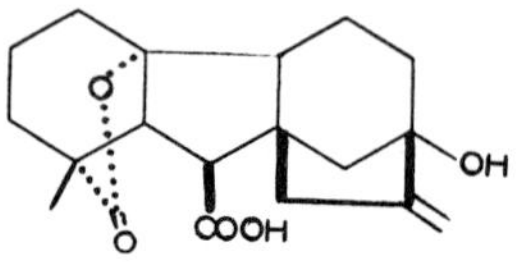

Tet. Lett., 1967, 4861.

4210-027

GIBBERELLIN A_{22} — $C_{19}H_{22}O_6$ — 346 — 214^0C

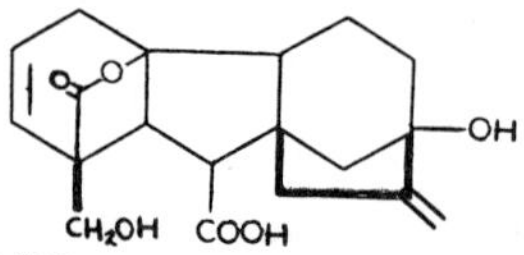

Tet. Lett., 1967, 4861.

4210-028

GIBBERELLIN A_{23} — $C_{20}H_{26}O_7$ — 378 — 187^0C

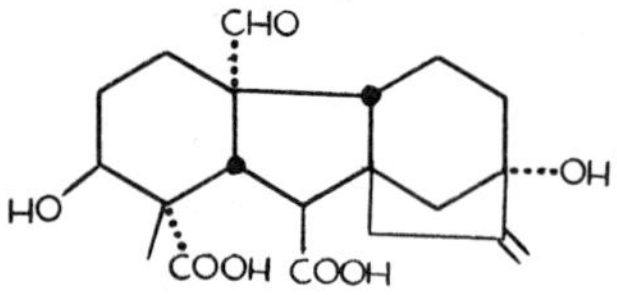

Tet. Lett., 1968, 1143.

4210-029

GIBBERELLIN A_{24} — $C_{20}H_{26}O_5$ — 346 — 201^0C

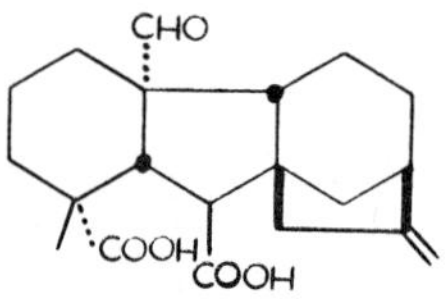

Tet. Lett., 1968, 3137.

4210-030

GIBBERELLIN A_{26} $C_{19}H_{22}O_7$ 362

256^{0}C

Tet. Lett., 1969. 2078.

4210-031

deleted

4210-035

GIBBERELLIN A_{26}, 3-O-Glucosyl- $C_{25}H_{32}O_{12}$ 524

Tet. Lett., 1969, 2081.

4210-032

GIBBERELLIN A_{29} $C_{19}H_{24}O_6$ 348

Tet. Lett., 1970, 1489.

4210-036

GIBBERELLIN A_{27} $C_{20}H_{26}O_6$ 362

164^{0}C

Tet. Lett., 1969, 2077

4210-033

GIBBERELLIN A_{29}, 3-O-Glucosyl- $C_{25}H_{34}O_{11}$ 510

Tet. Lett., 1970, 1489.

4210-037

GIBBERELLIN A_{27}, 3-O-Glucosyl- $C_{26}H_{36}O_{11}$ 524

Tet. Lett., 1969, 2081.

4210-034

TRACHYLOBANOL $C_{20}H_{32}O$ 288
-42⁰
130⁰C

Bull. Soc. Chim. Fr., 1965, 2882.

4211-001

TRACHYLOBANIC ACID, 3-OAc- $C_{22}H_{32}O_4$ 360
-58⁰c
196⁰C

Bull. Soc. Chim. Fr., 1965, 2882.

4211-004

TRACHYLOBANIC ACID $C_{20}H_{30}O_2$ 302
m. e. -41⁰
m. e. 111⁰C

Bull. Soc. Chim. Fr., 1965, 2882.

4211-002

TRACHYLOBAN-19-OIC ACID $C_{20}H_{30}O_2$ 302
m. e. -71⁰
m. e. 99⁰C

Tet., 1970, 5029.

4211-005

TRACHYLOBANIC ACID, 3-OH- $C_{20}H_{30}O_3$ 318
-72⁰
268⁰C

Bull. Soc. Chim. Fr., 1963, 1974.

4211-003

Name	Formula	Data	Reference	No.
ATISIRENE	$C_{20}H_{32}$	272 -40°c 58°C	Tet. Lett., 1965, 2729.	4212-001
SPIRADINE F	$C_{24}H_{33}NO_4$	399 115°C	Tet. Lett., 1968, 5565.	4212-006
ATISIRENE, iso-	$C_{20}H_{32}$	272 -74°c 84°C	Tet. Lett., 1965, 2729.	4212-002
SPIRADINE G	$C_{22}H_{31}NO_3$	357 -137°m 169°C	Tet. Lett., 1968, 5565.	4212-007
ATISINE	$C_{22}H_{33}NO_2$	343 60°C	JACS., 1962 (84) 2990.	4212-003
deleted				4212-008
AJACONINE	$C_{22}H_{33}NO_3$	359 -119°e 172°C	Tet., 1961 (14) 54.	4212-004
DENUDATIN	$C_{22}H_{33}NO_2$	343 +0.15°e 248°C	Chem. Comm., 1970. 359.	4212-009
ATIDINE	$C_{22}H_{33}NO_3$	359	JACS. 1965 799.	4212-005
HETIDINE	$C_{21}H_{27}NO_4$	357 221°C	Chem. Comm., 1970, 393.	4212-010

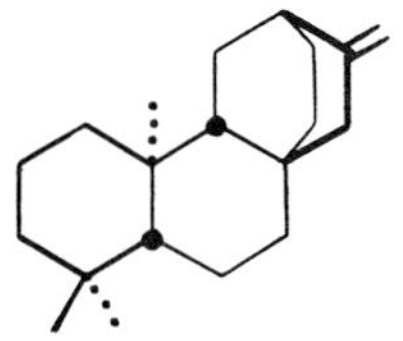

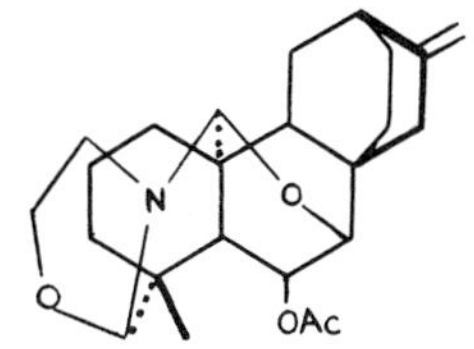

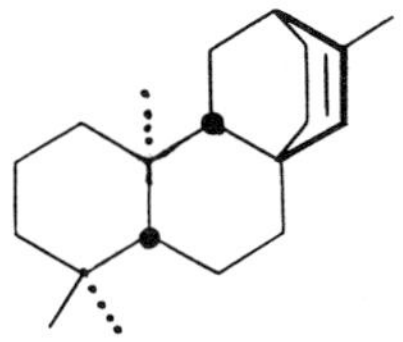

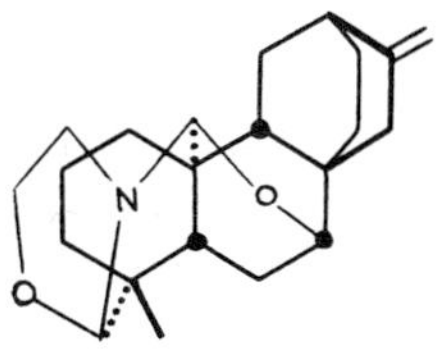

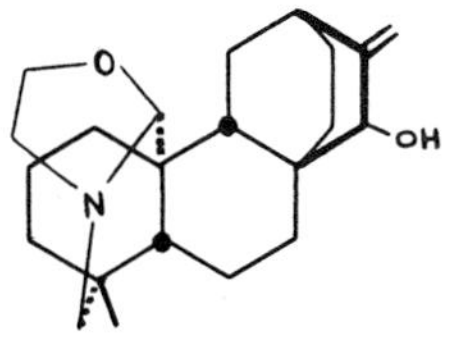

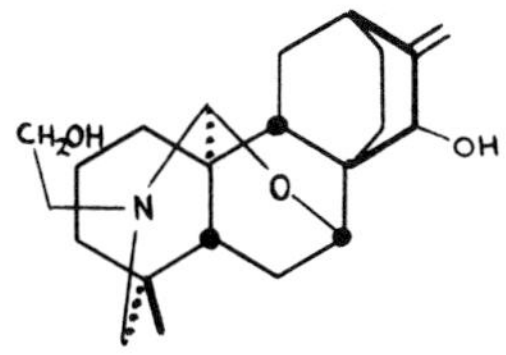

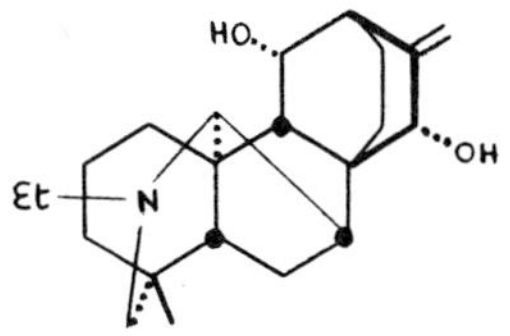

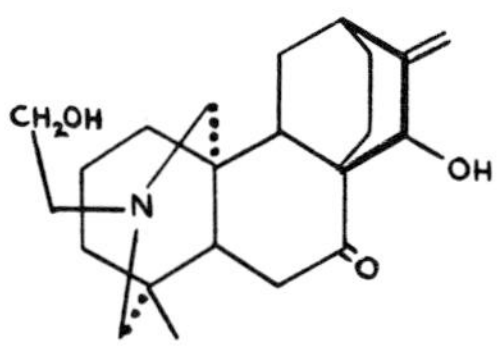

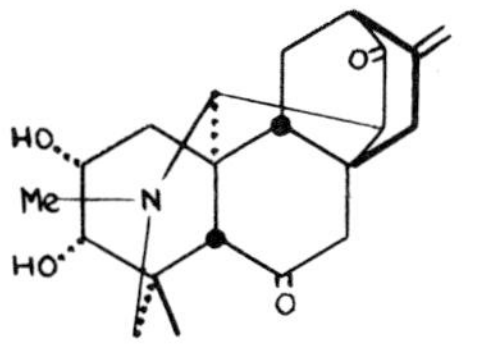

MIYACONITINE $C_{23}H_{29}NO_6$ 415

Tet. Lett., 1970, 2327.

4212-011

SPIRADINE A $C_{20}H_{25}NO_2$ 311

281°C

Tet. Lett., 1968, 1369.

4212-016

MIYACONITINONE $C_{23}H_{27}NO_6$ 413

Tet. Lett., 1970, 2323.

4212-012

SPIRADINE B $C_{20}H_{27}NO_2$ 313

260°C

Tet. Lett., 1968, 1369.

4212-017

HETISINE $C_{20}H_{27}NO_3$ 329

Can. J. C., 1968 2635.

4212-013

SPIRADINE C $C_{22}H_{29}NO_3$ 355

249°C

Tet. Lett., 1968, 1369.

4212-018

HETISINONE $C_{20}H_{25}NO_3$ 327

Can. J. C., 1968, 2635.

4212-014

SPIRADINE D $C_{22}H_{29}NO_2$ 339

135°C

Tet. Lett., 1968, 2989.

4212-019

KOBUSINE $C_{20}H_{27}NO_2$ 313

Chem. Comm., 1970, 98.

4212-015

KOBUSINE, pseudo- $C_{20}H_{27}NO_3$ 329

271°C

Chem. Pharm. Bull., 1962, 879.

4212-020

IGNAVINE $C_{27}H_{31}NO_5$ 449 Tet. Lett., 1970, 4825. 4212-021	HYPOGNAVINE $C_{27}H_{31}NO_5$ 449 Tet. Lett., 1971, 795. 4212-023
HYPOGNAVINE, iso- $C_{27}H_{31}NO_4$ 433 135^{0}C J. Pharm. Soc. Jap., 1956 (76) 550. 4212-022	

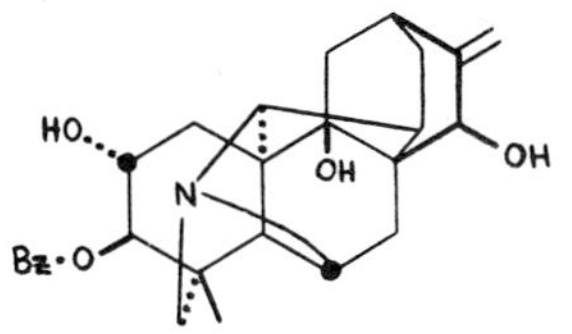

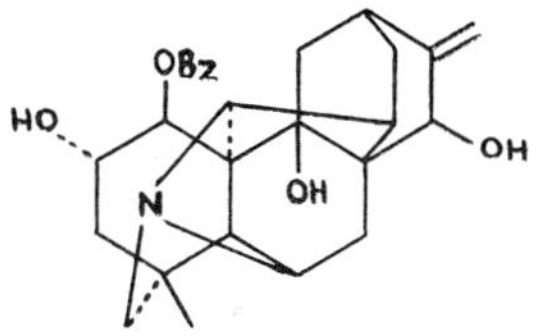

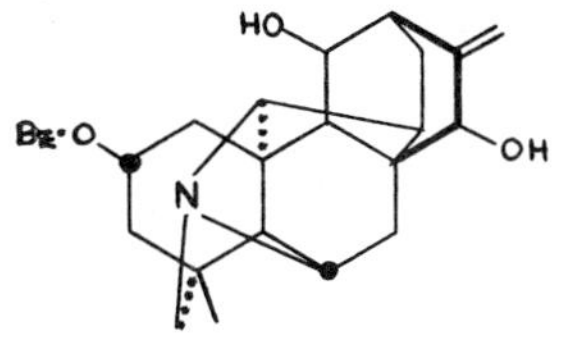

TALATIZIDINE — $C_{23}H_{37}NO_5$ — 407 — -20^0m — 219^0C

JACS., 1967. 4146.

4213-001

CHASMANINE — $C_{25}H_{41}NO_6$ — 451

R = H

Tet., Suppl. 8, Pt. 1, 101.

4213-006

TALATIZIDINE, iso- — $C_{23}H_{37}NO_5$ — 407 — 116^0C

JACS., 1967 4146.

4213-002

CHASMANINE, HOMO- — $C_{26}H_{43}NO_6$ — 465 — $+19^0$e — 106^0C

R = Me

Can. J. C., 1965, 1093.

4213-007

TALATIZAMINE — $C_{24}H_{39}NO_5$ — 421 — 146^0C

Chem. Abstr., 1968 (69) 77562.

4213-003

NEOPELLINE — $C_{33}H_{45}NO_8$ — 583

Arch. Pharm., 1924 (262) 553.

4213-008

CONDELPHINE — $C_{25}H_{39}NO_6$ — 449 — $+21^0$ — 157^0C

JACS., 1967, 4146.

4213-004

DELPHONINE — $C_{24}H_{39}NO_7$ — 453 — $+38^0$ e — 78^0C

$R_1 = R_2 = H$

JOC., 1957, 1428.

4213-009

NEOLINE — $C_{24}H_{39}NO_6$ — 437

Tet., Suppl. 8, Pt. 1, 101.

4213-005

DELPHININE — $C_{33}H_{45}NO_9$ — 599

R_1 = Benzoyl

R_2 = Ac

Can. J. C., 1969, 2734.

4213-010

BIKHACONITINE $C_{36}H_{51}NO_{11}$ 673 164°C R = Veratroyl Can. J. C., 1963 (41) 3055. 4213-011	INDACONITINE $C_{34}H_{47}NO_{10}$ 629 R = H; R' = Bz Tet. Lett., 1962, 923. 4213-016
CHASMANTHININE $C_{36}H_{49}NO_9$ 639 +10°e 160°C Can. J. C., 1964 (42) 154. 4213-012	PSEUDACONITINE $C_{37}H_{53}NO_{12}$ 703 R = Me; R' = Veratroyl Can. J. Chem., 1963, 1485. 4213-017
CHASMACONITINE $C_{34}H_{47}NO_9$ 613 +10°e 181°C Can. J. C., 1964 (42) 154. 4213-013	MESACONITINE $C_{33}H_{45}NO_{11}$ 631 R = OH; R' = Me Ann., 1964 (680) 88. 4213-018
ACONITINE $C_{34}H_{47}NO_{11}$ 645 Can. J. C., 1969 (47) 2734. 4213-014	HYPACONITINE $C_{33}H_{45}NO_{10}$ 615 +22°c 198°C R = H; R' = Me Ann., 1964 (680) 88. 4213-019
JESACONITINE $C_{35}H_{49}NO_{12}$ 675 130°C JOC., 1968, 2497. 4213-015	ACONITINE, deoxy- $C_{34}H_{47}NO_{10}$ 629 +16°e 178°C R = H; R' = Et Ann., 1964 (680) 88. 4213-020

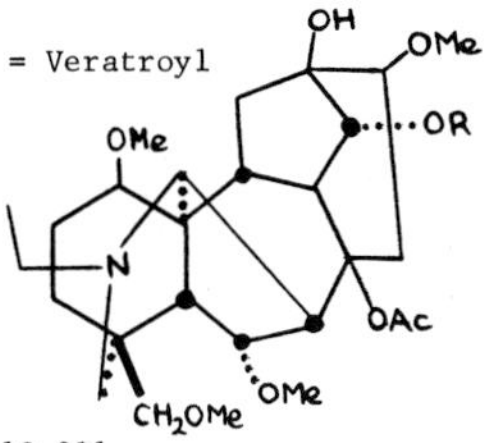

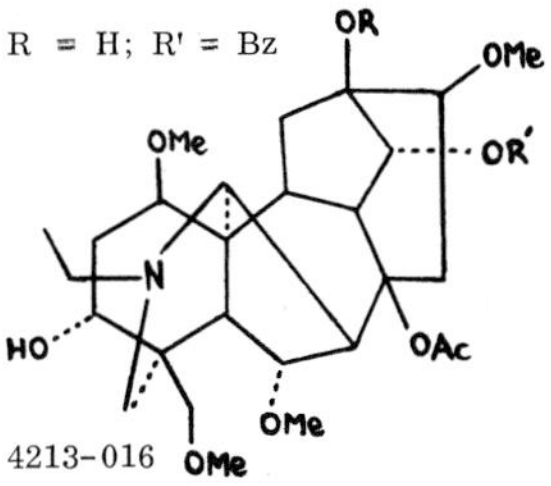

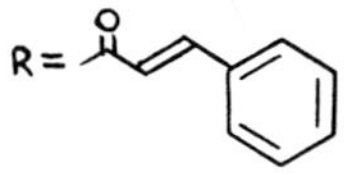

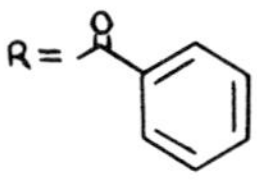

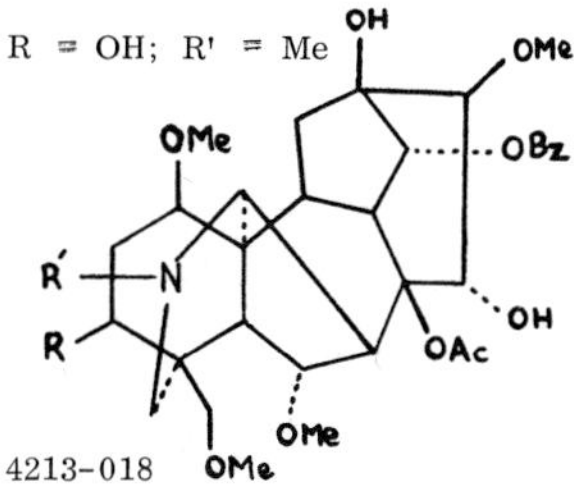

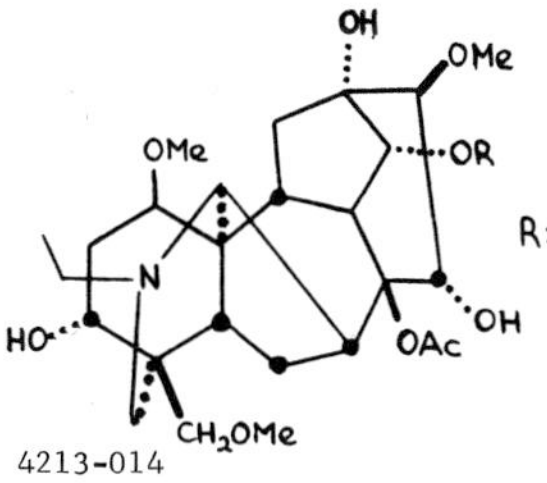

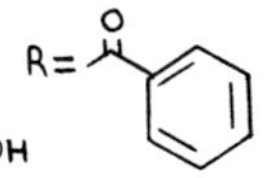

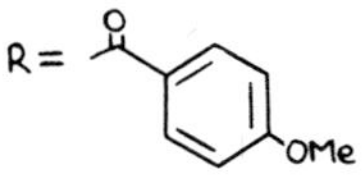

DELCOSINE — $C_{24}H_{39}NO_7$ — 453 — +57°c — 204°C

$R_1 = R_2 = H$

Can. J. C., 1960, 2433.

4213-021

DELCOSINE, 14-dehydro- — $C_{24}H_{37}NO_7$ — 451

14-oxo ; R_2 = H

Can. J. C., 1960, 2433.

4213-022

DELCOSINE, Ac; — $C_{26}H_{41}NO_8$ — 495 — 192°C

R_1 = Ac ; R_2 = H

Can. J. C., 1954, 780.

4213-023

DELSOLINE — $C_{25}H_{41}NO_7$ — 467 — 214°C

R_1 = Me

R_2 = H

Tet., 1958 (4) 157.

4213-024

BROWNIINE — $C_{25}H_{41}NO_7$ — 467

R_1 = H

R_2 = Me

Can. J. C., 1966, 583.

4213-025

BROWNIINE, 14-dehydro- — $C_{25}H_{39}NO_7$ — 465 — +19°e — 163°C

14-oxo ; R_2 = Me

Can. J. C., 1966, 1.

4213-026

LYCOCTONINE — $C_{25}H_{41}NO_7$ — 467 — +50°e — 153°C

R = H

Chem. Ind., 1959, 633.

4213-027

DELPHATINE — $C_{26}H_{43}NO_7$ — 481

R = Me

Chem. Abstr., 1970 (73) 77443.

4213-028

LYCOCTONINE, Anthaniloyl- — $C_{31}H_{46}N_2O_8$ — 574 — + 51°c — 135°C

R =

JACS, 1952 (74) 1411.

4213-029

AJACINE — $C_{34}H_{48}N_2O_9$ — 628 — +50°e — 154°C

R =

Chem. Abstr., 1964 (61) 8627.

4213-030

AVADHARIDINE — $C_{36}H_{51}N_3O_{10}$ — 685 — +45°e — 89°C

R =

Chem. Abstr., 1960, (54) 11064.

4213-031

DELPHELINE — $C_{25}H_{39}NO_6$ — 449 — -26°c — 227°C

R = R' = H

JACS., 1958, 497.

4213-036

DELSEMINE — $C_{37}H_{53}N_3O_{10}$ — 699 — +43°e — 125°C

R =

Planta Med., 1965 (13) 200.

4213-032

DELTAMINE — $C_{25}H_{39}NO_7$ — 465

R = OH ; R' = H

JACS., 1959, 4110.

4213-037

LYCACONITINE — $C_{30}H_{48}N_2O_{10}$ — 596 — +43°e — amorph.

R =

Chem. Abstr., 1960 (54) 11064.

4213-033

DELTALINE — $C_{27}H_{41}NO_8$ — 507 — -28°m — 194°C

R = OH ; R' = Ac

JACS., 1959, 4110.

4213-038

LYCACONITINE, Me– — $C_{37}H_{50}N_2O_{10}$ — 682 — +49°e — 130°C

R =

Chem. Abstr., 1966 (64) 1007.

4213-034

ELATINE — $C_{38}H_{50}N_2O_{10}$ — 694 — +3°c — 224°C

Chem. Abstr., 1960 (54) 11064.

R =

4213-039

LAPPACONINE — $C_{23}H_{37}NO_6$ — 423

Tet. Lett., 1969, 2189.

4213-035

SONGORINE $C_{22}H_{31}NO_3$ 357 -140° 212°C Chem. Pharm. Bull., 1965. 1270. 4214-001	VEATCHINE $C_{23}H_{33}O_2N$ 355 120°C JACS., 1964 (86) 929. 4214-006
LUCICULINE $C_{22}H_{33}NO_3$ 359 -11°e 116°C R = H Chem. Pharm. Bull., 1965, 1270. 4214-002	CUAUCHICHICINE $C_{22}H_{33}NO_2$ 343 -71°m 154°C Cf. Veatchine JACS., 1955, 6633. 4214-007
LUCIDUSCULINE $C_{24}H_{35}NO_4$ 401 170°C R = Ac Chem. Pharm. Bull., 1965, 1270. 4214-003	SONGORAMINE $C_{22}H_{29}NO_3$ 355 212°C Khim. Prir. Soedin, 1970 (6) 101. 4214-008
NAPELLINE, iso- $C_{22}H_{33}NO_3$ 359 121°C Cf. Luciculine Chem. Ind., 1957, 173. 4214-004	GARRYFOLINE $C_{22}H_{33}NO_2$ 343 -60° 132°C C_{15} epimer of Veatchine (006) JACS., 1962, 2990. 4214-009
GARRYINE $C_{22}H_{33}O_2N$ 343 -84°e Oil JACS., 1954 (76) 6068. JACS., 1964, 290. 4214-005	

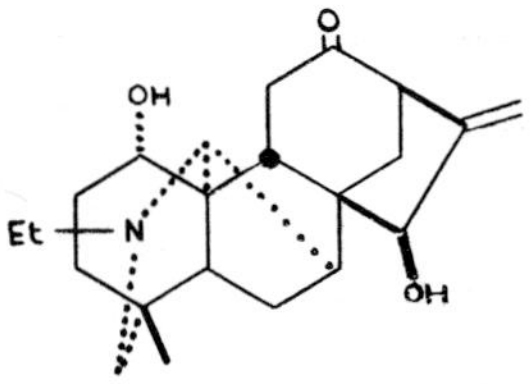

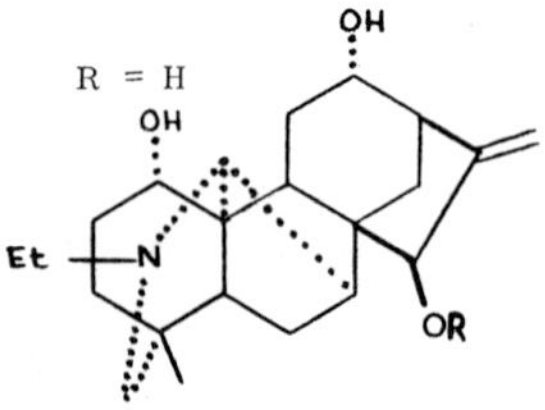

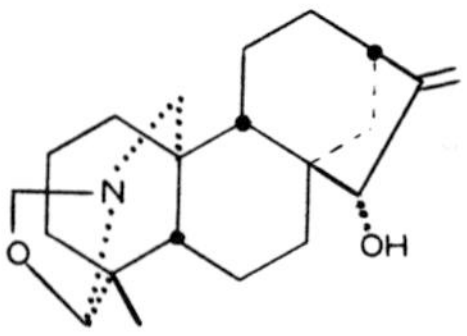

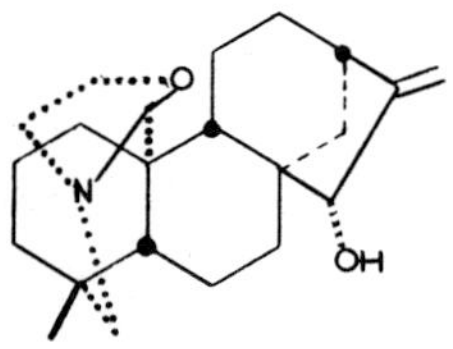

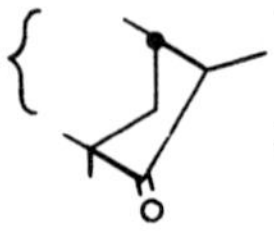

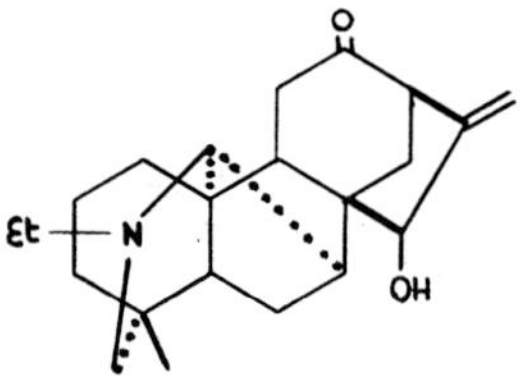

TAXA-4(16). 11-DIENE-5. 9. 10, 13-TETRAOL — $C_{20}H_{32}O_4$ — 336, +134°, 198°C OR2, OR3, R1O, OR4 Chem. Comm., 1966 (24) 923. R1 = R2 = R3 = R4 = H 4215-001	TAXA-4(20), 11-DIENE-2. 5, 7. 9. 10-PENTAOL. 5, 7. 9. 10-tetra-O-Ac-2-i-butyroyl- — $C_{32}H_{46}O_{10}$ — 590, +56°, 155°C R = Ac Chem. Comm., 1969. 1282. 4215-006
TAXA-4(20), 11-DIENE-5, 9. 10, 13-TETRAOL. 9. 10-di-O-Ac- — $C_{24}H_{36}O_6$ — 420, +146°, 235°C R1 = R4 = H R2 = R3 = Ac Chem. Comm., 1969. 1282. 4215-002	TAXA-4(20). 11-DIENE-5. 7. 9. 10, 13-PENTAOL. penta-O-Ac- — $C_{30}H_{42}O_{10}$ — 562, +92°, 207°C OAc, OAc, OAc, AcO, OAc Chem. Comm., 1969. 1282. 4215-007
TAXUSIN — $C_{28}H_{40}O_8$ — 504, 132°C R1 = R2 = R3 = R4 = Ac Chem. Pharm. Bull., 1968. 546. 4215-003	TAXA-4(20), 11-DIENE-2, 5, 9. 10, 13-PENTAOL. penta-O-Ac- — $C_{30}H_{42}O_{10}$ — 562, +46°, 165°C OAc, OAc, AcO, OAc, OAc Chem. Comm., 1969. 1282. 4215-008
TAXA-4(20), 11-DIENE, 2, 5, 7, 10-TETRAOL, 5, 7, 10-tri-O-Ac-2-(α-Me-butyroyl)- — $C_{31}H_{46}O_8$ — 546, +45°, 115°C OAc, OAc, OAc Chem. Comm., 1969. 1282. 4215-004	TAXA-4(20), 11-DIENE-2, 5, 7. 9, 10, 13-HEXAOL, hexa-O-Ac- — $C_{32}H_{44}O_{12}$ — 620, +31°, 197°C OAc, OAc, OAc, AcO, OAc, OAc Chem. Comm., 1969. 1282. 4215-009
TAXA-4(20), 11-DIENE-2, 5,7, 9, 10-PENTAOL, 7, 9, 10-tri-0-Ac-2-i-butyroyl- — $C_{30}H_{44}O_9$ — 548, +63°, 227°C OAc, OAc, OAc, OR, O-i-But Chem. Comm., 1969, 1282. R = H 4215-005	BACCATIN — $C_{33}H_{42}O_{13}$ — 646 OAc, O, OAc, AcO, OH, OH, OH, OAc, $CH_2O·CO·Ph$ Chem. Comm., 1970. 216. 4215-010

BACCATIN-I $C_{32}H_{44}O_{13}$ 636 + 86° 298°C

OAc OAc OAc AcO OR1 R2 OAc

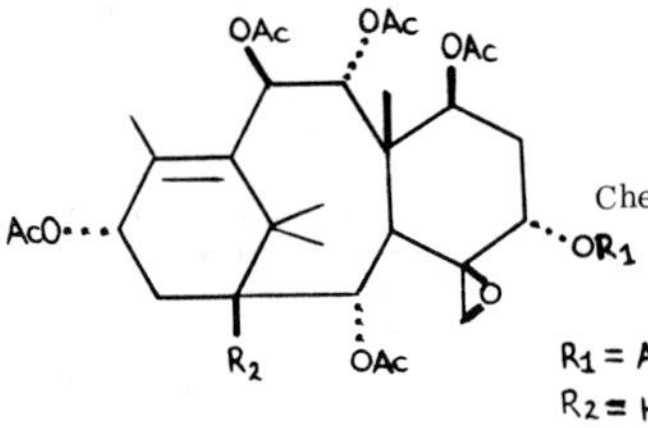

R_1 = Ac
R_2 = H

Chem. Comm., 1970, 1381.

4215-011

TAXININE $C_{35}H_{42}O_9$ 606 + 137°c 265°C

R_1 = Cinnamoyl
R_2 = Ac

Chem. Comm., 1966, 77.

4215-016

BACCATIN-I, 5-de-Ac- $C_{30}H_{42}O_{12}$ 594 257°C

R_1 = R_2 = H

Chem. Comm., 1970, 1381.

4215-012

TAXININE A $C_{26}H_{36}O_8$ 476 + 106°c 254°C

R_1 = H
R_2 = Ac

Chem. Comm., 1967, 1201.

4215-017

BACCATIN-I, 1-OH- $C_{32}H_{44}O_{14}$ 652 + 102° 273°C

R_1 = Ac
R_2 = OH

Chem. Comm., 1970, 1381.

4215-013

TAXININE H $C_{28}H_{38}O_9$ 518 + 96°c 166°C

R_1 = R_2 = Ac

Chem. Comm., 1967, 1201.

4215-018

BACCATIN-V $C_{31}H_{38}O_{11}$ 588 -87° 255°C

OAc O OH HO OH OBz OAc

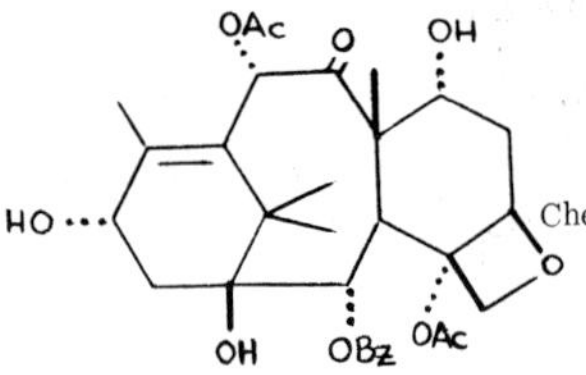

Chem. Comm., 1970, 1382.

4215-014

TAXICIN-I, O-Cinnamoyl- $C_{29}H_{36}O_7$ 496 + 285°c 233°C

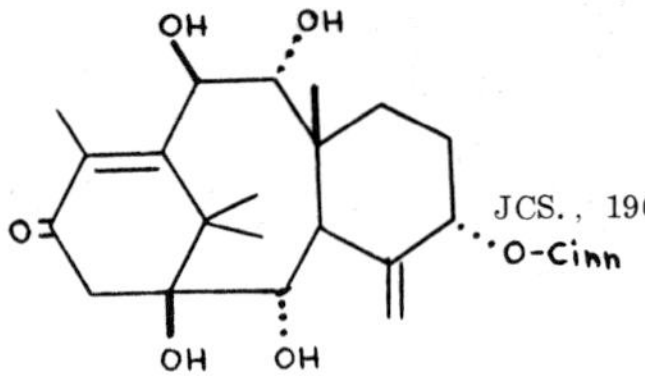

OH OH O O-Cinn OH OH

JCS., 1967, 452.

4215-019

TAXICIN-II, O-Cinnamoyl- $C_{29}H_{36}O_6$ 480

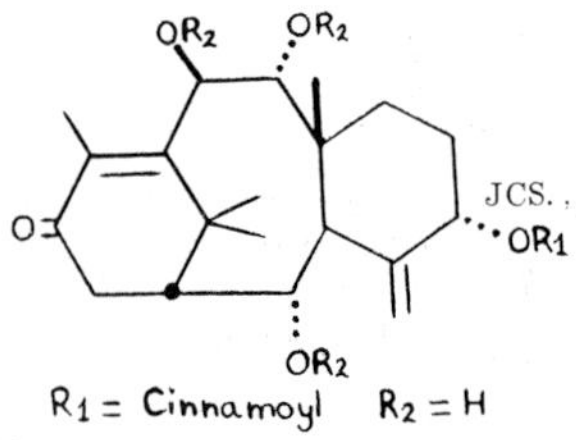

OR2 OR2 O OR1 OR2

JCS., 1967, 452.

R_1 = Cinnamoyl R_2 = H

4215-015

INCENSOLE	$C_{20}H_{34}O_2$	306 -77.5⁰ Oil	Tet., 1967. 1977.
4216-001			
CEMBROL	$C_{20}H_{34}O$	290 +56⁰c Oil	Chem. Abstr., 1966 (64) 5145.
4216-006			
INCENSOLE OXIDE (*epoxide)	$C_{20}H_{34}O_3$	322 -48⁰ 165⁰C	Tet., 1968, 6519.
4216-002			
THUNBERGOL	$C_{20}H_{34}O$	290 +74⁰c Oil	Acta Chem. Scand., 1968 (22) 943.
4216-007			
OVATODIOLIDE	$C_{20}H_{24}O_4$	328 149⁰C	Tet., 1965, 2117.
4216-003			
DUVATRIEN-1,5-DIOL, α-3,8,13-	$C_{20}H_{34}O_2$	306	JOC., 1963 (28) 1165.
4216-008			
CEMBRENE	$C_{20}H_{32}$	272 +238⁰c 59⁰C	Acta Cryst., 1969 (25B) 261.
4216-004			
DUVATRIEN-1,5-DIOL, β-3,8,13- (stereoisomer)	$C_{20}H_{34}O_2$	306	JOC., 1963 (28) 1165.
4216-009			
CEMBRENE, iso-	$C_{20}H_{32}$	272 +60⁰c Oil	Chem. Abstr., 1968 (69) 44055.
4216-005			
DUVATRIEN-1-OL 5,8-OXIDE, 3,9,13-alpha-	$C_{20}H_{32}O_2$	304 +86⁰c 96⁰C	JOC., 1964, 16.
4216-010			

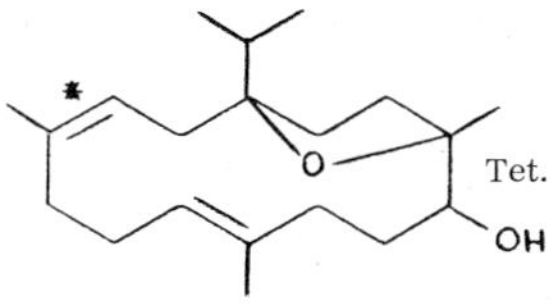

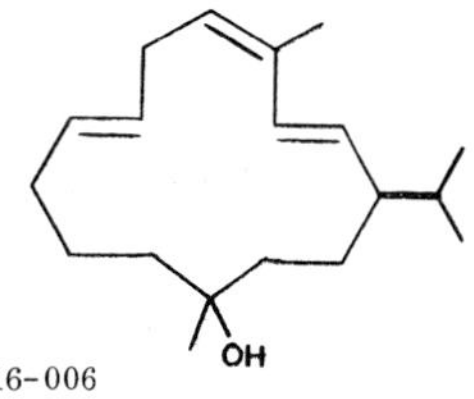

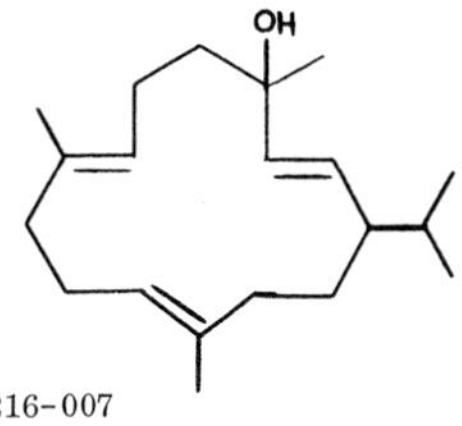

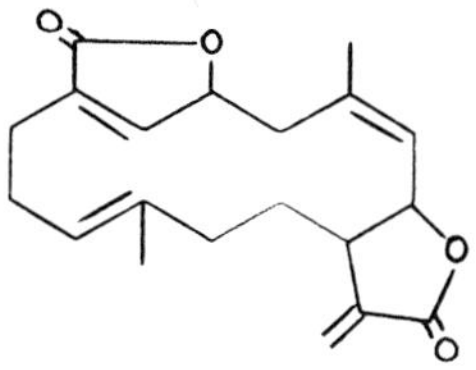

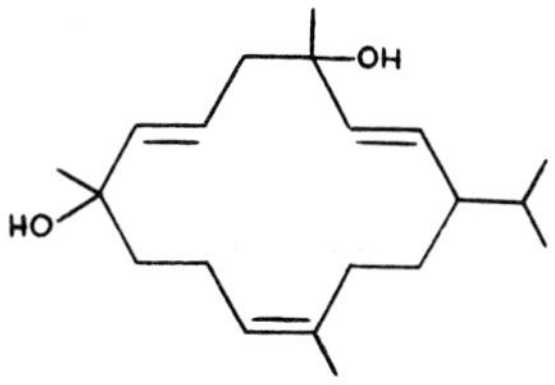

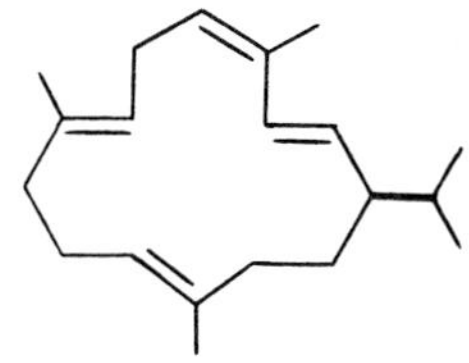

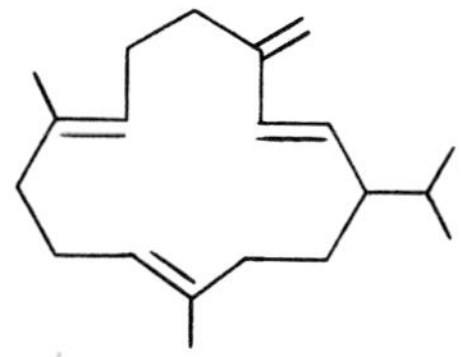

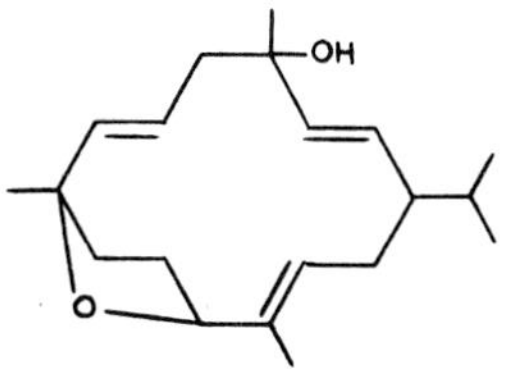

DUVATRIEN-1-OL 5,8-OXIDE, 3,9(17),13-beta- $C_{20}H_{32}O_2$ 304 +73°c 109°C 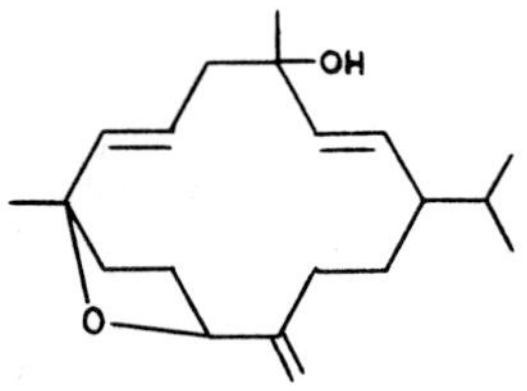 JOC., 1964, 16. 4216-011	CRASSIN ACETATE $C_{22}H_{34}O_5$ 378 +70° 145°C 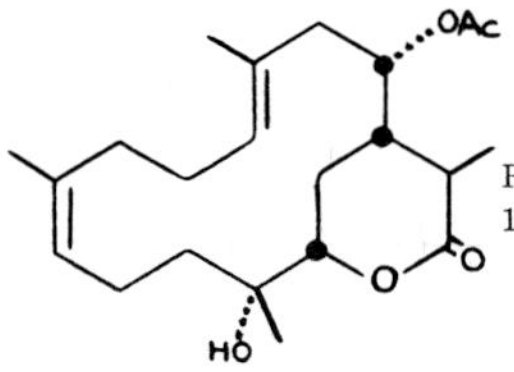 Rec. Trav. Chim., 1969 (88) 1413. 4216-014
DUVATRIEN-1,3-DIOL, α-4,8,13- $C_{20}H_{34}O_2$ 306 +282°c 66°C 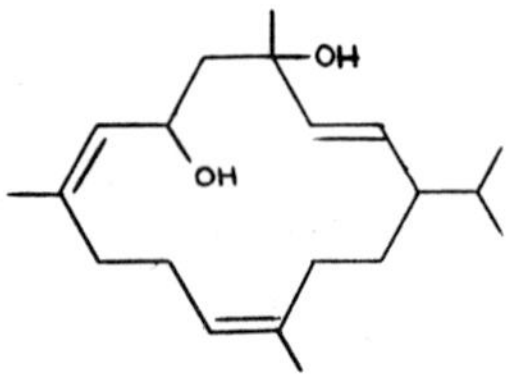 JOC., 1962 (27) 3989. 4216-012	EUNICIN $C_{20}H_{30}O_4$ 334 -89° 155°C 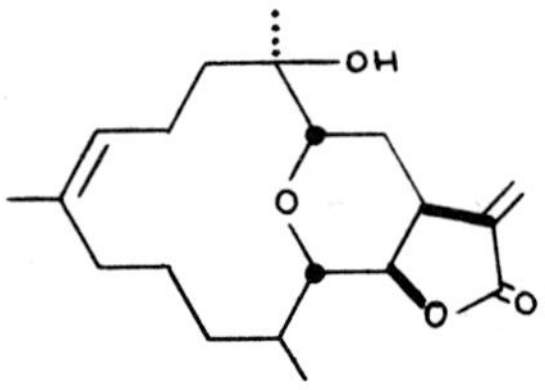 Chem. Comm., 1968, 384. 4216-015
DUVATRIEN-1,3-DIOL, β-4,8,13- 306 +162°c 127°C stereoisomer JOC., 1962, 3989. 4216-013	

PLATHYTERPOL — $C_{20}H_{34}O$ — 290, -28^0, Oil — Chem. Comm. 1969 683. — 4217-001	CISTODIOL — $C_{20}H_{36}O_2$ — 308, $+48^0$, $86\text{-}88^0C$ — R = CH_2OH — Tet. Lett. 1970 1401. — 4217-006
KOLAVELOOL — $C_{20}H_{34}O$ — 290 — Tet. Lett. 1968 2685. — 4217-002	CISTODIOC ACID — $C_{20}H_{32}O_4$ — 336, $+64^0e$, 256^0C — R = COOH — Tet. Lett. 1970 1401. — 4217-007
KOLAVENOL — $C_{20}H_{34}O$ — 290 — R = CH_2OH — Tet. Lett. 1964 3751. — 4217-003	AGBANINDIOL A — $C_{20}H_{34}O_2$ — 306, -92^0, $85\text{-}7^0C$ — JCS. 1969 2153. — 4217-008
KOLAVENIC ACID — $C_{20}H_{32}O_2$ — 304, -65^0c, Oil — R = COOH — Tet. Lett. 1968 2681. — 4217-004	SOLIDAGONIC ACID — $C_{22}H_{34}O_4$ — 362, -98^0, 144^0C — Tet. Lett. 1968 4325. — 4217-009
KOLAVIC ACID — $C_{20}H_{30}O_4$ — 334 — Tet. Lett. 1964 3751. — 4217-005	AGBANINOL — $C_{20}H_{30}O_2$ — 302, -4.9^0, 70^0C — JCS. 1969 2153. — 4217-010

AGBANINDIOL B	$C_{20}H_{30}O_3$	318 +26°	SOLIDAGO GLYCOL	$C_{20}H_{30}O_3$	318 -51°e Oil
HO, HO	JCS. 1969 2153.		R = CH_2OH	Acta Chem. Scand. 1968 351.	
4217-011			4217-016		
SOLIDAGO ALDEHYDE	$C_{20}H_{28}O_2$	300 -161°e Oil	SOLIDAGO DIALDEHYDE	$C_{20}H_{26}O_3$	314 -46°e 104°C
R = CHO; R	Acta Chem. Scand. 1968 351		CHO, CHO	Acta Chem. Scand. 1968 351	
4217-012			4217-017		
SOLIDAGOIC ACID A	$C_{20}H_{28}O_3$	316 -58°e 170°C	SOLIDAGO LACTOL	$C_{20}H_{28}O_3$	316 -30°e 20°C
R = COOH	Acta Chem. Scand. 1968 351.		*, OH	Acta Chem. Scand. 1968 351.	
4217-013			4217-018		
SOLIDAGOIC ACID B	$C_{25}H_{36}O_5$	416 -28°e 135°C	SOLIDAGO EPOXYLACTOL	$C_{20}H_{28}O_4$	332 -47°e Oil
COOH, CH_2O·Ang	Acta Chem. Scand. 1968 351.		*epoxide	Acta Chem. Scand. 1968 351	
4217-014			4217-019		
SOLIDAGO ALCOHOL	$C_{20}H_{30}O_2$	302 -45°e Oil	HARDWICKIIC ACID, (-)-	$C_{20}H_{28}O_3$	316 -100° 144°C
R = CH_3; R, CH_2OH	Acta Chem. Scand. 1968 351.			Tet. Lett. 1968 2685.	
4217-015			4217-020		

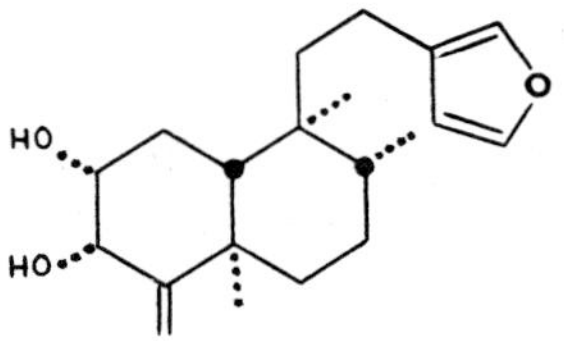

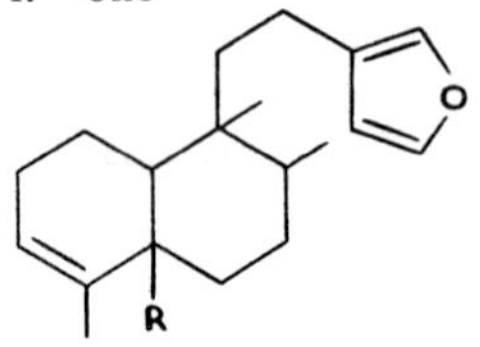

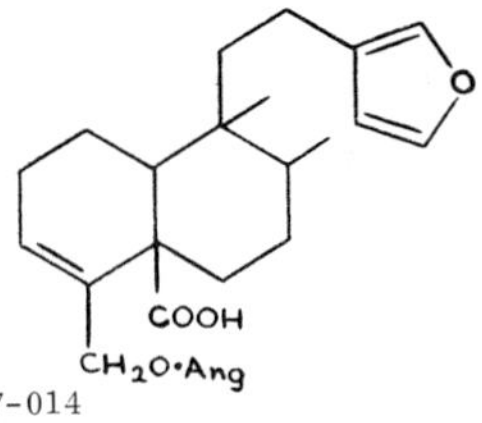

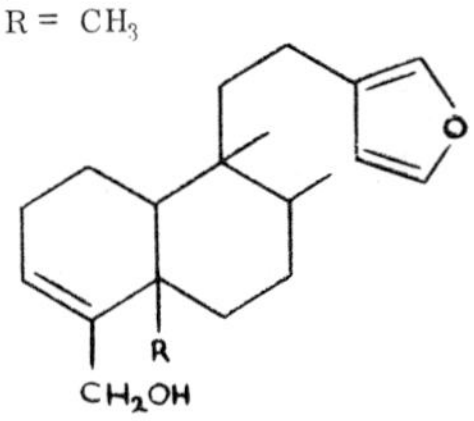

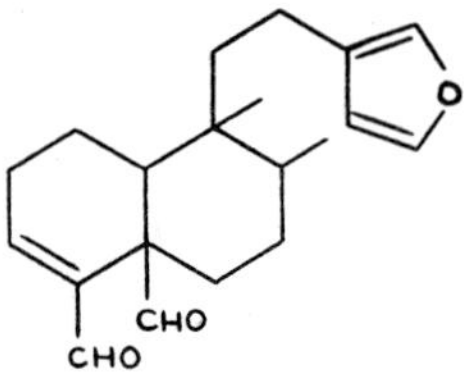

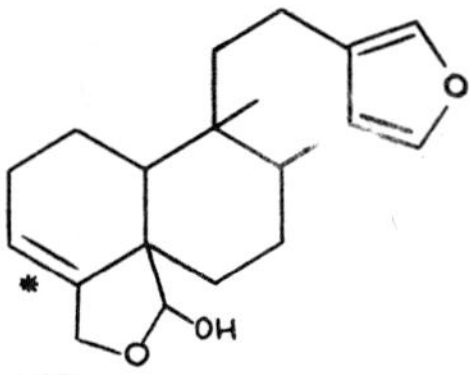

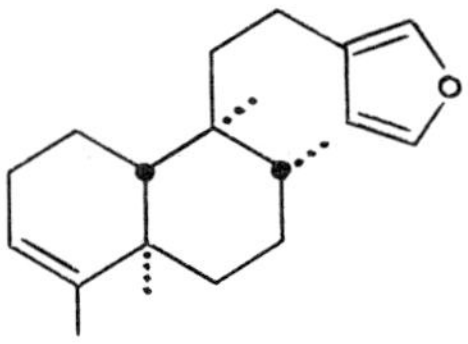

HARDWICKIIC ACID, (+)- $C_{20}H_{28}O_3$ 316 +25^0 106^{0}C

Tet. Lett. 1965 1983.

4217-021

HAUTRIWAIC ACID $C_{20}H_{26}O_3$ 314 -156^0 120^{0}C

Tet. Lett. 1967 4777.

4217-022

OLEARIN $C_{20}H_{26}O_5$ 346 -120^{0}c 213^{0}C

Chem. Comm. 1967 (1) 9.

4217-023

DODONAEA DITERPENE $C_{22}H_{30}O_6$ 390 -109^0 161^{0}C

Tet. Lett. 1967 4777.

4217-024

CASCARILLIN $C_{22}H_{32}O_7$ 408

JCS. 1966 1482.

4217-025

TINOPHYLLONE $C_{21}H_{24}O_6$ 372 -98^{0}c 175^{0}C

Chem. Inc. 1965 1074.

4217-026

COLUMBIN $C_{20}H_{22}O_6$ 358 +52^0

JCS., 1956. 2090.

4217-027

CHASMANTHIN $C_{20}H_{22}O_7$ 374 225^{0}C

*epoxide

JCS. 1966 1483.

4217-028

JATEORIN $C_{20}H_{22}O_7$ 374

JCS. 1966 1483.

4217-029

JATEORIN, Iso- $C_{20}H_{22}O_7$ 374 +30^{0}p 166^{0}C

C_{13} epimer

JCS. 1962 4809

4217-030

PALMARIN $C_{20}H_{22}O_7$ 374 $+17^0$p 246^0C JCS. 1966 1483. H O OH 4217-031	ELONGATOLIDE-B $C_{22}H_{32}O_4$ 360 Oil R = AC Acta Chem. Scand. 1969 1068. 4217-036
FIBLEUCIN $C_{20}H_{20}O_6$ 356 -28^0p $169-72^0$C R = H Chem. Comm. 1969 653. H * O OH R 4217-032	ELONGATOLIDE-C $C_{25}H_{36}O_4$ 400 Oil R = Angeloyl Acta Chem. Scand. 1969 1068. 4217-037
FIBRAURIN $C_{20}H_{20}O_7$ 372 -28^0p 288^0C *epoxide R = H Tet. 1966 2649. 4217-033	ELONGATOLIDE-E $C_{22}H_{32}O_5$ 376 Oil R = AC Acta Chem. Scand. 1969 1068. OR 4217-038
FIBRAURIN, 6-OH- $C_{20}H_{20}O_8$ 388 *epoxide R = α-OH Tet. 1966 2649. 4217-034	ELONGATOLIDE-E $C_{25}H_{36}O_5$ 416 Oil R = Angeloyl Acta Chem. Scand. 1969 1068. 4217-039
ELONGATOLIDE-A $C_{20}H_{30}O_3$ 318 Oil R = H Acta Chem. Scand. 1969 1068. OR 4217-035	PICROPOLIN $C_{22}H_{26}O_8$ 418 R = H Ber. 1967 1998. OR CH_2OAc 4217-040

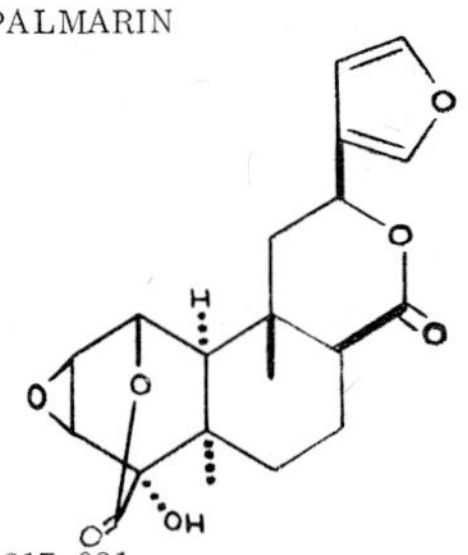

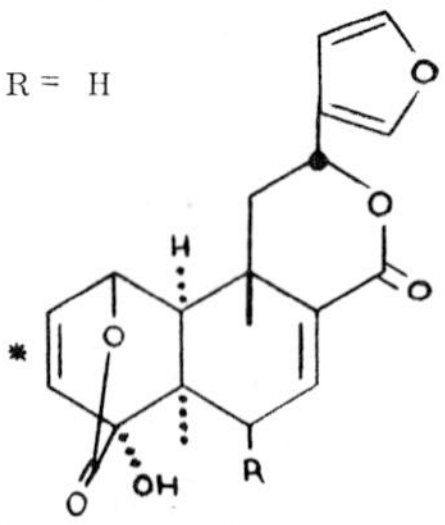

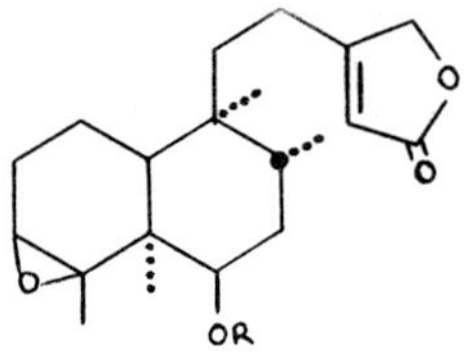

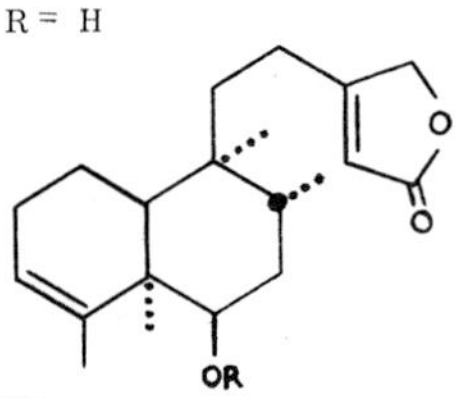

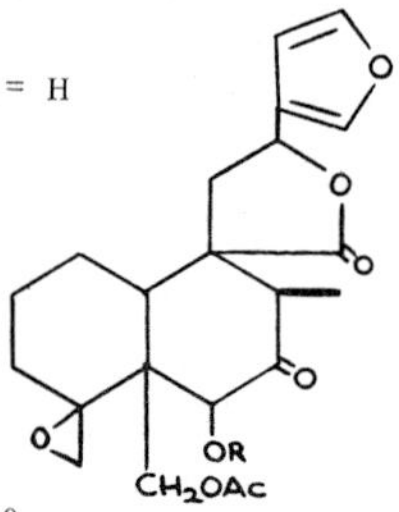

PICROPOLIN', Acetyl- $C_{24}H_{28}O_9$ 460

R = Ac

Ber. 1967 1998.

4217-041

CASCARILLIN A $C_{29}H_{28}O_5$ 348
-61^0
187-203^{0}C

Chem. Comm. 1965 218.

4217-043

PICROPOLIN-hemi-acetal. Ac- $C_{22}H_{26}O_8$ 418

Ber., 1967, 1998.

4217-042

CLERODIN $C_{24}H_{3497}$ 434
-47^{0}c
164^{0}C

JCS. 1966 1482.

4217-044

ASEBOTOXIN-I	$C_{23}H_{38}O_7$	426 198°C

Chem. Pharm. Bull., 1970, 1071.

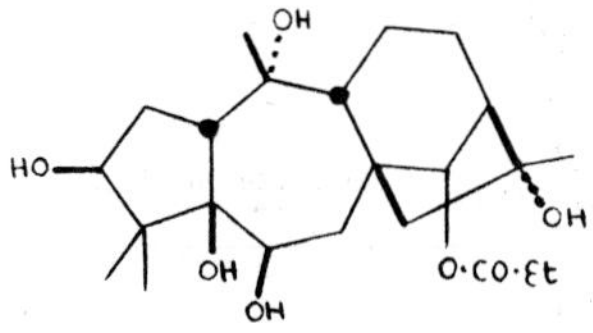

4218-001

ASEBOTOXIN-II	$C_{23}H_{36}O_6$	408 143°C

Chem. Pharm. Bull., 1970, 1071.

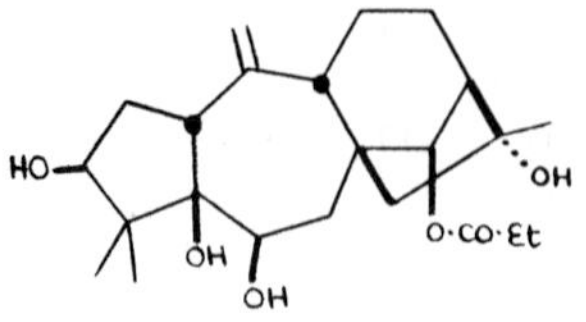

4218-002

ASEBOTOXIN-III	$C_{23}H_{36}O_8$	440 259°C

Chem. Pharm. Bull., 1970, 1071.

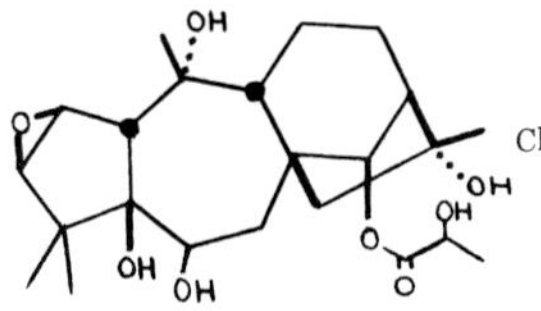

4218-003

GRAYANOTOXIN-I	$C_{22}H_{36}O_7$	412 -9°e 272°C

Chem. Pharm. Bull., 1970, 1071.
Tet. Lett., 1970, 3943.

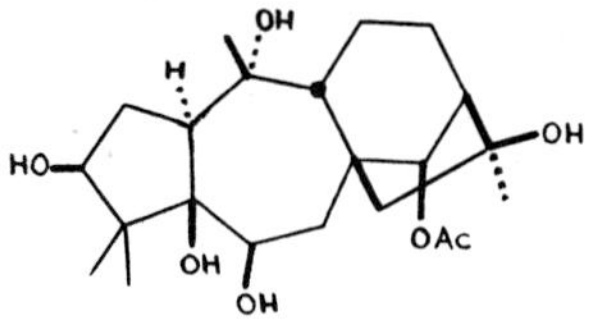

4218-004

GRAYANOTOXIN-II	$C_{20}H_{32}O_5$	352 -42° 198°C

Chem. Pharm. Bull., 1970, 1071.

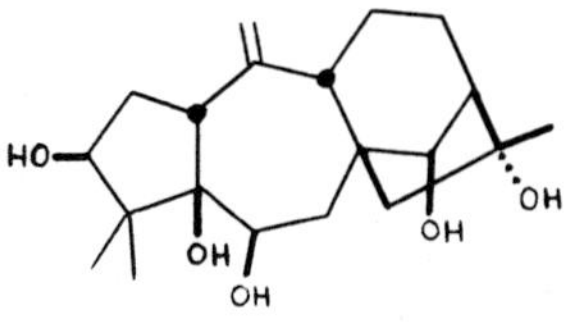

4218-005

GRAYANOTOXIN-III	$C_{20}H_{34}O_6$	370 -17°e 225°C

Chem. Pharm. Bull., 1970, 1071.

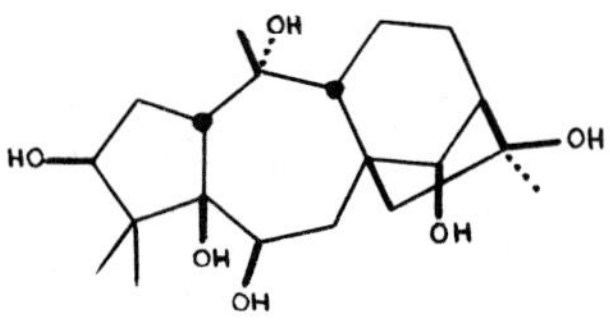

4218-006

RHODOJAPONIN-I	$C_{24}H_{36}O_8$	452 249°C

Chem. Pharm. Bull., 1970, 1071.

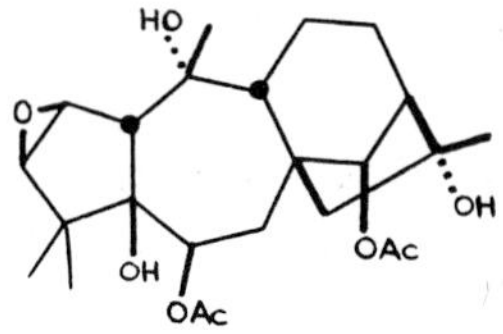

4218-007

RHODOJAPONIN-II	$C_{22}H_{34}O_7$	410

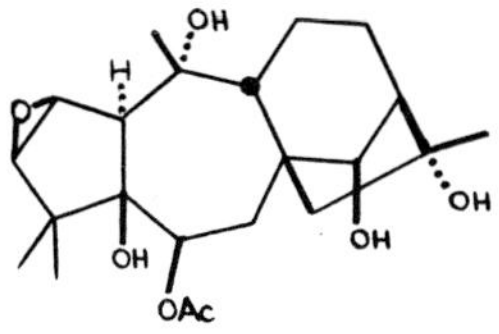

Chem. Pharm. Bull., 1970, 1071.

4218-008

RHODOJAPONIN-III	$C_{20}H_{32}O_6$	368 286°C

Chem. Pharm. Bull., 1970, 1071.

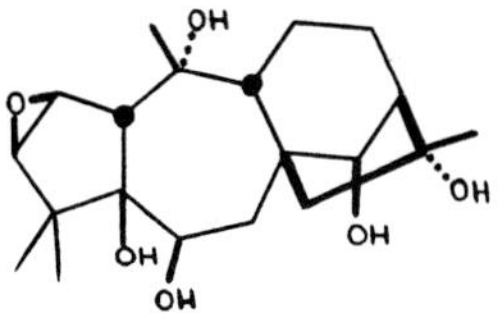

4218-009

LYONIATOXIN	$C_{22}H_{34}O_7$	410 253°C

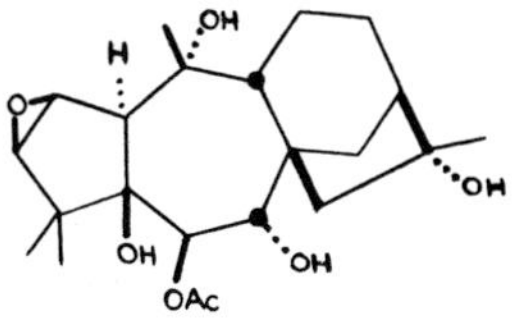

Chem. Pharm. Bull., 1970, 852; 2586.

4218-010

GRAYANOTOXIN-IV $C_{22}H_{34}O_6$ 394 -19^{0}m 175^{0}C HO OH OH OH OAc Tet., 1970, 4765. 4218-011	GRAYANOTOXIN-V $C_{20}H_{32}O_6$ 368 -61^{0}m 231^{0}C OH O OH OH OH OH Tet., 1970, 4765. 4218-012

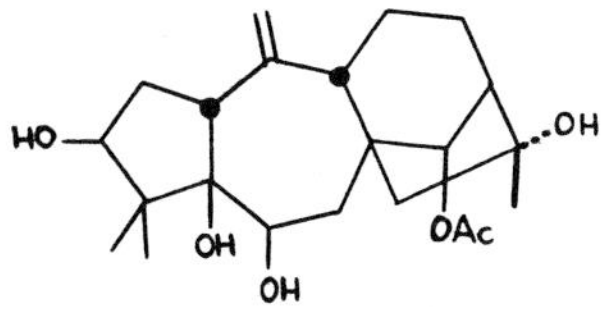

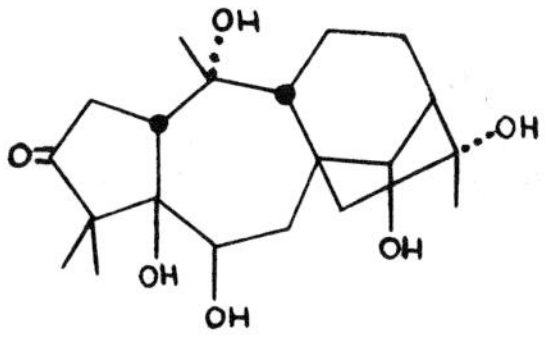

FUJENAL $C_{20}H_{26}O_4$ 330, -74°a, 170°C R = CHO CHO JCS., 1963, 2937. 4219-001	ENMEIN $C_{20}H_{26}O_6$ 362, -131°p, 298°C R = H RO, OH Tet., 1966 (22) 3423. 4219-006
FUJENOIC ACID $C_{20}H_{26}O_5$ 346, -59°e, 206°C R = COOH JCS., 1963, 2937. 4219-002	ENMEIN-3-ACETATE $C_{22}H_{28}O_7$ 404, -112°, 269°C R = Ac Tet. Lett., 1966, 3153. 4219-007
GIBBERELLA DICARBOXYLIC ACID $C_{20}H_{28}O_5$ 348, 170°C CHO, COOH, COOH JCS., 1963, 2937. 4219-003	ENMEIN, 16,17-dihydro- $C_{20}H_{28}O_6$ 364, 257°C R = H * dihydro- Chem. Comm., 1966, 297. 4219-008
EMEMODIN $C_{20}H_{26}O_6$ 362, -131°e, 238°C HO Chem. Pharm. Bull., 1970, 884. 4219-004	ISODOCARPIN $C_{20}H_{26}O_5$ 346, -172°, 272°C R = H R, OH Chem. Pharm. Bull., 1968, 1573. 4219-009
ISODOTRICIN $C_{21}H_{30}O_7$ 394, -114°, 245°C CH_2OMe, HO, OH Tet. Lett., 1966, 3153. 4219-005	NODOSIN $C_{20}H_{26}O_6$ 362, -203°, 280°C R = β- OH Chem. Pharm. Bull., 1968, 509. 4219-010

NODOSIN, epi- $C_{20}H_{26}O_6$ 362 -173^0p 246^0C

R = α - OH

Chem. Comm., 1968, 763.

4219-011

ISODONAL $C_{22}H_{28}O_7$ 404 $+92^0$p 246^0C

R = α - OH

Tet. Lett., 1967, 3781.

4219-014

NODOSINOL $C_{20}H_{28}O_6$ 364 -86^0p 246^0C

Tet. Lett., 1970, 421.

4219-012

TRICHODONIN $C_{22}H_{28}O_7$ 404 $+32^0$ 237^0C

R = β- OH

Chem. Comm., 1968, 763.

4219-015

ISODOACETAL $C_{22}H_{28}O_6$ 388 -134^0c 300^0C

13th Symp. Chem. Nat. Prod. 1969 (Japan).

4219-013

PHOROBOL

$C_{20}H_{28}O_6$ 364 $+102^0$w 250^0C

$R_1 = H$
$R_2 = H$

JCS., 1968. 1347.

4220-001

COCARCINOGEN A1

$C_{36}H_{56}O_8$ 616 $+49^0$ Resin

R_1 = Acetyl
R_2 = Tetradecanoyl

Chem. Ind., 1965. 943.

4220-002

COCARCINOGEN A3

$C_{38}H_{60}O_8$ 644 Oil

R_1 = Acetyl
R_2 = Hexadecanoyl

Experimentia, 1966 (22) 482.

4220-003

COCARCINOGEN B1

$C_{37}H_{58}O_8$ 630 $+54^0$c Oil

R_1 = Dodecanoyl
R_2 = Me-Butyroyl

Angew. Chem. Int. Ed., 1964 (3) 747.

4220-004

COCARCINOGEN B2

$C_{35}H_{54}O_8$ 602 $+50^0$c Oil

R_1 = Decanoyl
R_2 = Me-Butyroyl

Angew. Chem. Int. Ed., 1964 (3) 747.

4220-005

PHORBOL-13-isobutyrate-20-Acetate, 12-desoxy-

$C_{26}H_{36}O_7$ 460

R_1 = i-Butyroyl
R_2 = Ac

Tet. Lett., 1969. 3510.

4220-006

PHORBOL-13-α-Me-Butyrate-20-Acetate, 12-desoxy-

$C_{27}H_{38}O_7$ 474

R_1 = α-Me-Butyroyl
R_2 = Acetyl

Tet. Lett., 1969. 3510.

4220-007

PHORBOL-13-tiglate-20-acetate, 12-desoxy-

$C_{27}H_{36}O_7$ 472

R_1 = Tigloyl
R_2 = Acetyl

Tet. Lett.. 1969. 3510.

4220-008

PHORBOL-13-isobutyrate, 12-desoxy-

$C_{24}H_{34}O_6$ 418

R_1 = Isobutyroyl
R_2 = H

Tet. Lett., 1969, 3510.

4220-009

PHORBOL-13-α-Me-butyrate, 12-desoxy-

$C_{25}H_{36}O_6$ 432

R_1 = α-Me-Butyroyl
R_2 = H

Tet. Lett., 1969 3510.

4220-010

PHORBOL-13-tiglate, 12-desoxy- $C_{25}H_{34}O_6$ 430

R_1 = Tigloyl
R_2 = H

Tet. Lett., 1969 3510.

4220-011

EUPHORBIA COMPOUND C' $C_{31}H_{42}O_9$ 558

R = Ac

Tet. Lett., 1970, 567.

4220-013

EUPHORBIA COMPOUND C $C_{29}H_{40}O_8$ 516

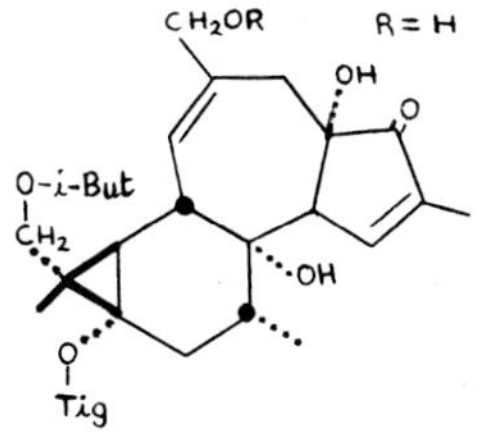

CH2OR R = H

OH

O

O-i-But

CH2

OH

O

Tig

Tet. Lett., 1970, 567.

4220-012

BIFLORIN	$C_{20}H_{22}O_3$	310 97°C	Helv., 1963 (46) 409.
4291-001			
VERTICILLOL	$C_{20}H_{34}O$	290 +168°c 105°C	Chem. Pharm. Bull., 1964, 1510.
4291-006			
DEVADARENE	$C_{20}H_{32}$	272 +20°c Oil	Tet. Lett., 1965, 2729.
4291-002			
TAXININE K (R = H)	$C_{26}H_{36}O_8$	476 168°C	Chem. Comm., 1967, 1201.
4291-007			
ERYTHROXYDIOL X (R = H)	$C_{20}H_{34}O_2$	306 +16°c 125°C	JCS., 1966, 268.
4291-003			
TAXININE L (R = Ac)	$C_{28}H_{38}O_9$	518 160°C	Chem. Comm., 1967, 1201.
4291-008			
ERYTHROXYTRIOL Q (R = OH)	$C_{20}H_{34}O_3$	322	Tet. Lett., 1967 (19) 2109.
4291-004			
EUNICELLIN	$C_{28}H_{42}O_9$	522 -36° 187°C	Tet. Lett., 1968, 2879.
4291-009			
DEVADAROOL, OH-	$C_{20}H_{34}O_3$	322 +3°e 182°C	Tet. Lett. 1964 3767.
4291-005			
RYANODINE	$C_{25}H_{35}NO_9$	493 +26°e 220°C	Coll. Czech. Chem. Comm., 1968, 2656.
4291-010			

EREMOLACTONE — $C_{20}H_{26}O_2$ — 298, -251^0, 139^0C

Tet. Lett., 1966, 4749.

4291-011

LATHYROL, 6, 20-epoxy- — $C_{20}H_{30}O_5$ — 350, 200^0C

$R_1 = R_2 = R_3 = H$

Tet. Lett., 1970, 3071.

4291-016

FUSICOCCIN — $C_{36}H_{56}O_{12}$ — 680

$R_1 = R_2 = Ac$

Chem. Comm., 1968, 1198.

4291-012

LATHYROL-diacetate-phenylacetate, expoxy- — $C_{32}H_{40}O_8$ — 552, 200^0C

$R_1 = R_3 = H$;
$R_2 = CO \cdot CH_2 \cdot Ph$

Tet. Lett., 1970, 2241.

4291-017

FUSICOCCIN, mono-des-Ac- — $C_{34}H_{54}O_{11}$ — 638, $+18^0e$

$R_1 = Ac$
$R_2 = H$

Experimentia, 1970 (26) 349.

4291-013

BERTYADIONOL — $C_{20}H_{26}O_3$ — 314, 160^0C

Tet. Lett., 1970, 4599.

4291-018

FUSICOCCIN, di-des-Ac- — $C_{32}H_{52}O_{10}$ — 596, $+9^0e$

$R_1 = R_2 = H$

Experimentia, 1970 (26) 350.

4291-014

JATROPHONE — $C_{20}H_{24}O_3$ — 312, $+292^0e$, 152^0C

JACS., 1970, 4476.

4291-019

ARTEMISENE — $C_{20}H_{30}$ — 270, -19^0

Chem. Abstr., 1953 (47) 9944.

4291-015

INGENOL-3-HEXADECANOATE — $C_{36}H_{58}O_6$ — 586

R = Hexadecanoyl

Tet. Lett., 1970, 4075.

4291-020

CASBENE $C_{20}H_{32}$ 272

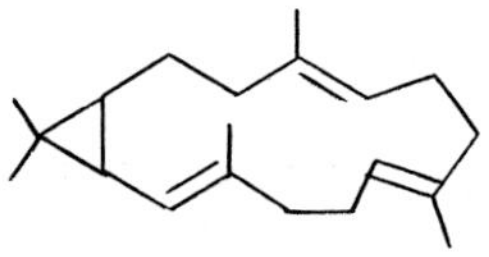

Biochemistry, 1970, 70 & 80.

4291-021

CHETTAPHANIN I — $C_{21}H_{26}O_6$ — 374, -73^0a, 158^0C Tet. Lett., 1970, 1095. 4292-001	THELEPOGINE — $C_{20}H_{31}NO$ — 301, 185^0C Tet. Lett., 1960, 18. 4292-006
ERYTHROXYDIOL Y — $C_{20}H_{34}O_2$ — 306, $+87^0$c, 145^0C JCS., 1966, 268. 4292-002	PORTULAL — $C_{20}H_{32}O_4$ — 336, -50^0e, 113^0 C Tet. Lett., 1969, 359. 4292-007
ERYTHROXYDIOL Z — $C_{20}H_{34}O_2$ — 306, -35^0c, 137^0C JCS., 1966, 268. 4292-003	PLEUROMUTILIN — $C_{22}H_{34}O_5$ — 378, $+20^0$e, 170^0C Chem. Ind., 1963, 374. 4292-008
DOLABRADIENE — $C_{20}H_{32}$ — 272, -70^0 Tet. Lett., 1964, 1755. 4292-004	SEMPERVIROL — $C_{20}H_{30}O$ — 286, ac. $+51^0$c, ac. 92^0C Tet. Lett. 1967 (8) 673. 4292-009
CLEISTANTHOL — $C_{20}H_{28}O_3$ — 316, -40^0e, 194^0C Chem. Comm., 1969, 1074. 4292-005	DACRYDIUM LACTONE — $C_{21}H_{32}O_4$ — 348, 208^0C Tet., 1970, 1619. 4292-010

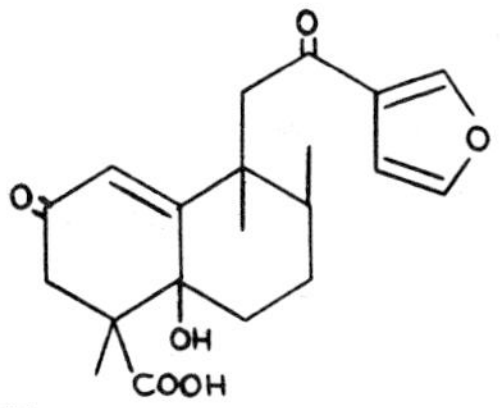

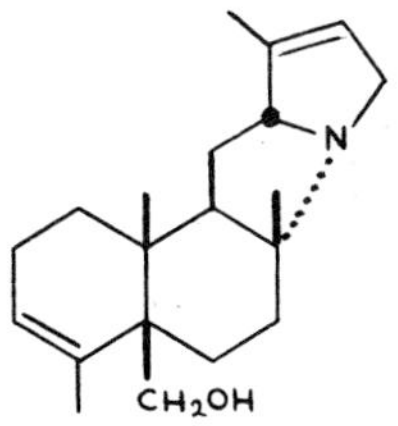

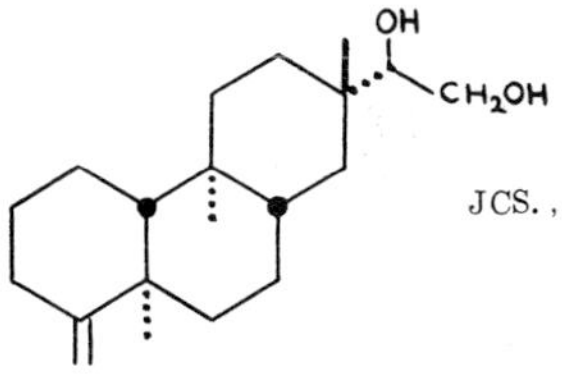

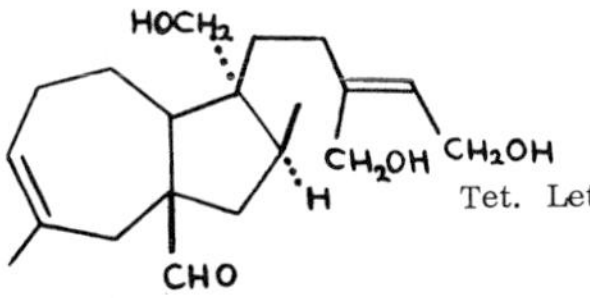

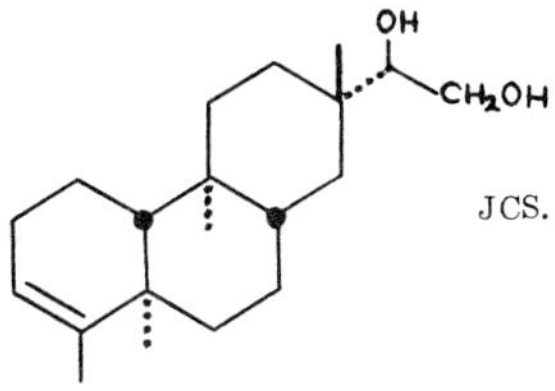

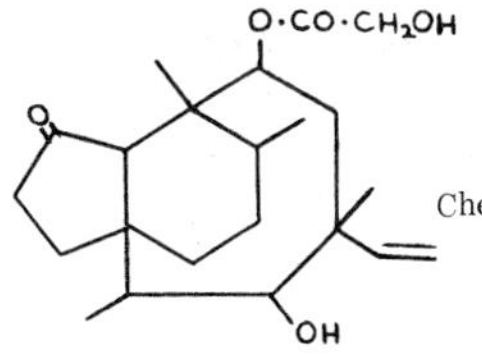

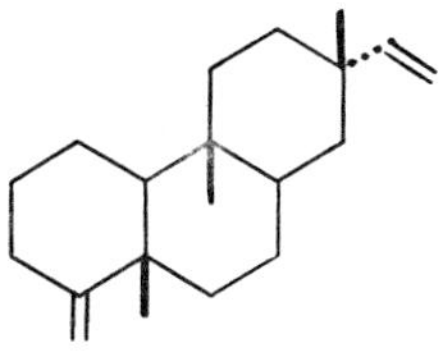

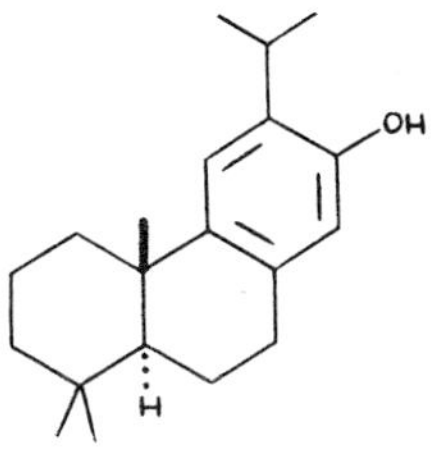

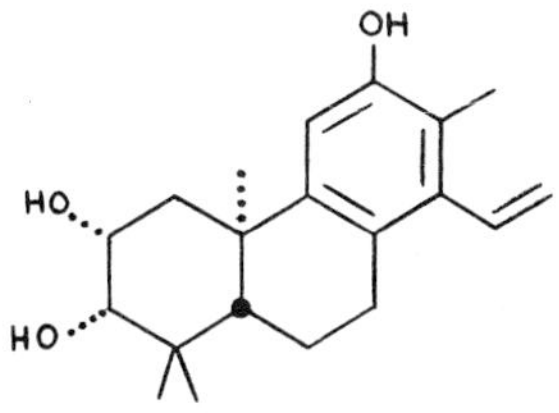

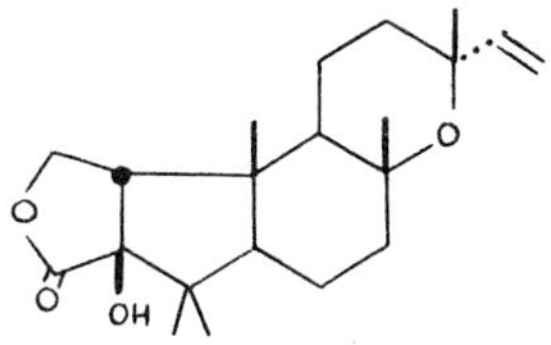

COLENSEN-2-OL, 2-alpha-carboxy- $C_{20}H_{32}O_4$ 336

m. e. 67^{0}C

Aust. J. C., 1969, 1265.

4292-011

KAHWEOL $C_{20}H_{26}O_3$ 314

Tet., 1964, 1339.

4292-014

COLENSENONE $C_{19}H_{30}O_2$ 290
+39^0
100^{0}C

JCS., 1962, 3740.

4292-012

DELNUDINE $C_{20}H_{25}NO_3$ 327

236^{0}C

Tet. Lett., 1969, 5335.

4292-015

CAFESTOL $C_{20}H_{28}O_3$ 316
-101^{0}c
158^{0}C

Tet., 1964, 1339.

4292-013

BEYERA-4(18), 15-DIENE-3-OIC ACID, 17-OH-3, 4-seco- $C_{20}H_{30}O_3$ 318

Aust. J. C., 1968, 442

4293-001

PIMARA-4(18)-7-DIEN-3-OIC ACID, 15, 16-diOH-3, 4-seco-ent- $C_{20}H_{32}O_4$ 336 -40^{0}c 148^{0}C

AJC., 1968, 2529

4293-002

MANOYL OXIDE, seco-epi- $C_{20}H_{32}O_5$ 352 192^{0}C

JCS., 1965, 3846

4293-003

$C_{20}H_{26}O_3$ 314 +36^0 182^{0}C

Chem. Abstr., 1970 (73) 25689b.

4293-004

NAGILACTONE A $C_{19}H_{24}O_6$ 348 + 89^0 305^{0}C

R = H

Tet. Lett., 1968, 2071

4293-005

NAGILACTONE B $C_{19}H_{24}O_7$ 364 +92^0 258-61^{0}C

R = OH

Tet. Lett., 1968, 2071

4293-006

NAGILACTONE C $C_{19}H_{22}O_7$ 362 +111^0 290^{0}C

Tet. Lett., 1969, 2951

4293-007

PODOLACTONE A $C_{19}H_{22}O_8$ 378 291-3^{0}C

R = H

Chem. Comm., 1970, 170.

4293-008

PODOLACTONE B $C_{19}H_{22}O_9$ 394 272-5^{0}C

R = OH

Chem. Comm., 1970, 170

4293-009

NAGILACTONE D $C_{18}H_{20}O_6$ 332 +90^0 265-6^{0}C

Tet. Lett. 1969 2951

4293-010

Compound	Formula	Data	Reference	No.
INUMAKILACTONE A	$C_{18}H_{20}O_8$	364 0^0 251-3^{0}C	Tet. Lett. 1968 2065	4293-011
MILTIRONE	$C_{19}H_{22}O_2$	282 OI 100^{0}C	Chem. Comm. 1970 299	4293-016
NIMBIOL	$C_{18}H_{24}O_2$	272 +32^{0}c 250^{0}C	Tet. 1960 (10) 45	4293-012
TANSHINONE IIA	$C_{19}H_{18}O_3$	294 OI 199^{0}C	JCS. 1968 48	4293-017
PODOCARPIC ACID	$C_{17}H_{22}O_3$	274 +144^{0}e 194^{0}C	Chem. Ind. 1956 113	4293-013
TANSHINONE II B	$C_{19}H_{18}O_4$	310 -48^{0}a 200-4^{0}C	JCS. 1968 48	4293-018
18-nor-iso-PIMARADIEN-4α-OL	$C_{19}H_{30}O$	274	Aust. J. C. 1967 969	4293-014
Crypto-TANSHINONE	$C_{19}H_{20}O_3$	296 184^{0}C	JCS. 1968 48	4293-019
PIMARA-4(19), 7, 15-TRIENE, 18-NOR-ISO-	$C_{19}H_{28}$	256 +53c 63^{0}C	Chem. Comm. 1969 1407	4293-015
iso-TANSHINONE-II	$C_{19}H_{18}O_3$	294 OI 208^{0}C	Tet. Lett. 1969 301	4293-020

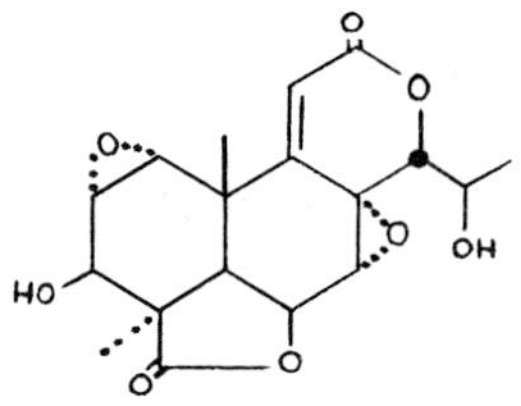

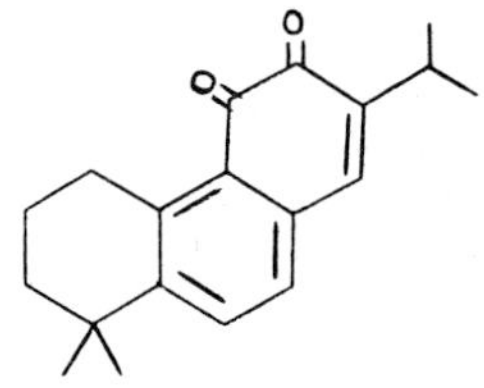

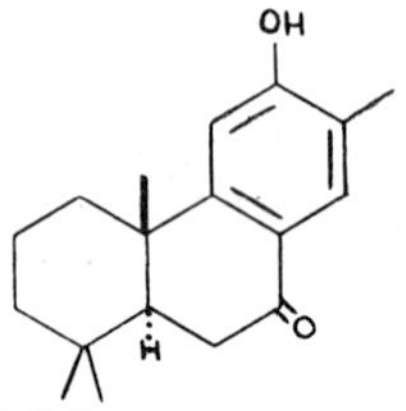

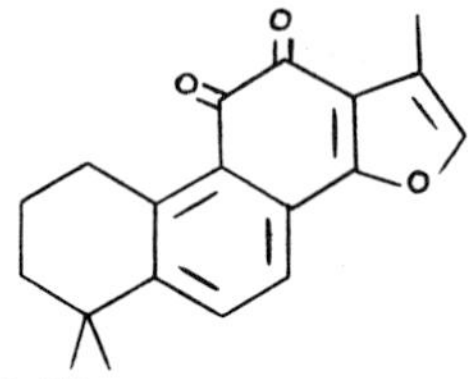

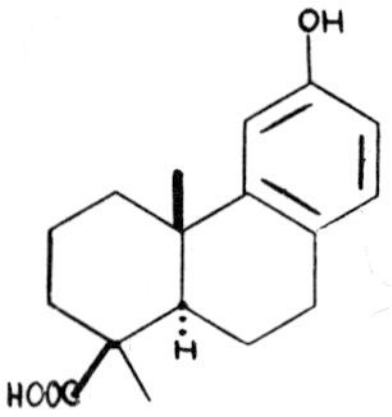

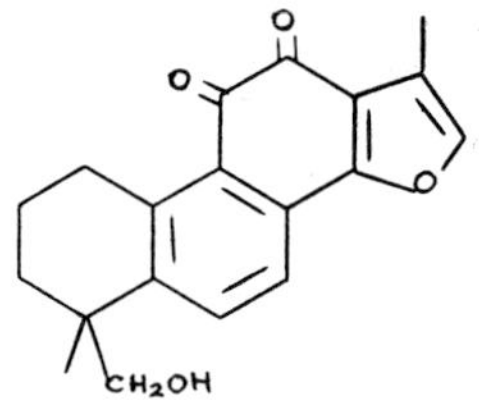

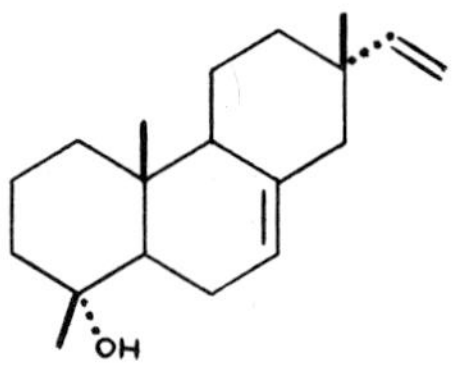

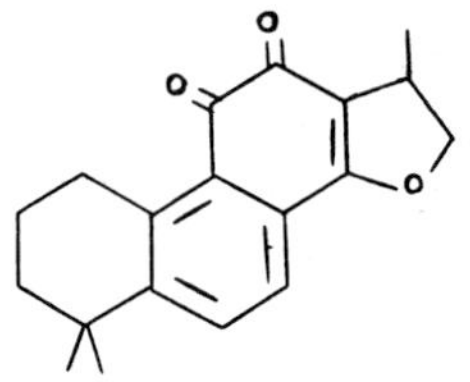

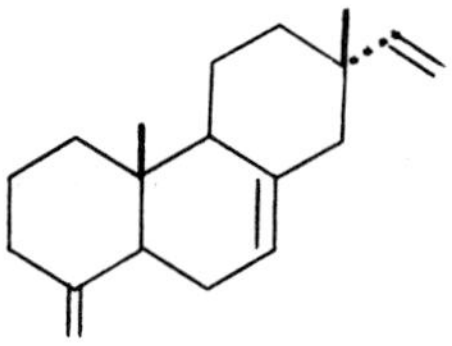

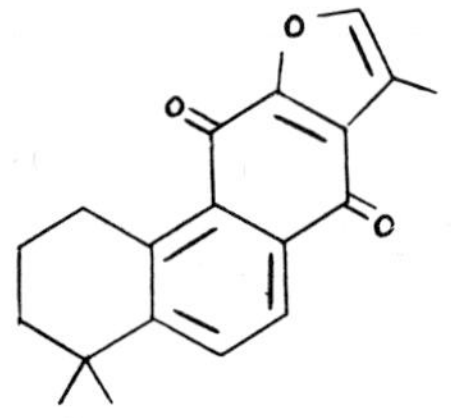

Name	Formula	Data	Reference	No.
iso-crypto-TANSHINONE	$C_{19}H_{20}O_3$	296 $+56^0$d 121^0C	Tet. Lett. 1969 301	4293-021
FICHTELITE	$C_{19}H_{34}$	262 $+19^0$c 46.5^0C	Helv. 1935 (18) 611	4293-026
OH-TANSHINONE	$C_{19}H_{18}O_4$	310 0^0 187^0C	Tet. Lett. 1968 3230	4293-022
deleted				4293-027
Methyl TANSHINONATE	$C_{20}H_{18}O_5$	338 -139^0 176^0C	Tet. Lett. 1968 3230	4293-023
4β-OH-18-nor-HIBAENE	$C_{19}H_{30}O$	274 $+49^0$c 115^0C	JCS. 1968 2529	4293-028
iso-TANSHINONE-I	$C_{18}H_{12}O_3$	276 OI 219^0C	Tet. Lett. 1969 301	4293-024
4α-OH-18-nor-HIBAENE	$C_{19}H_{30}O$	274 $+25^0$c Oil	JCS. 1968 2529	4293-029
TANSHINONE I	$C_{18}H_{12}O_3$	OI 233^0C	JCS. 1968 48	4293-025
COLEON B	$C_{19}H_{20}O_6$	344 $+130^0$e 259^0C	Helv. 1969 (52) 1685	4293-030

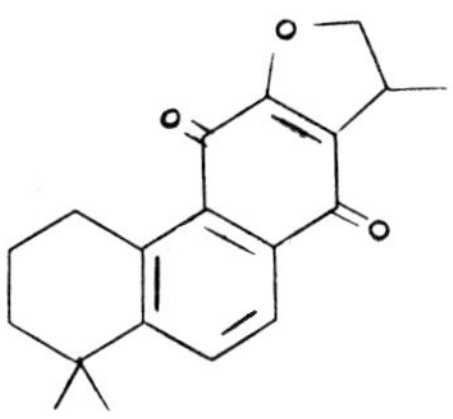

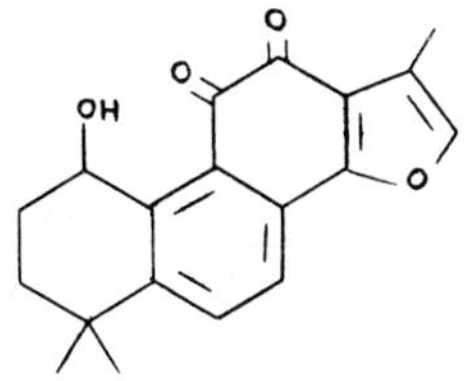

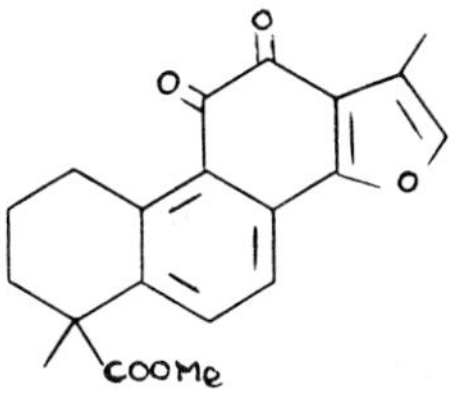

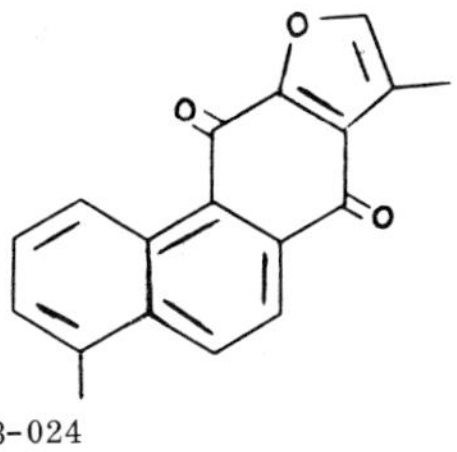

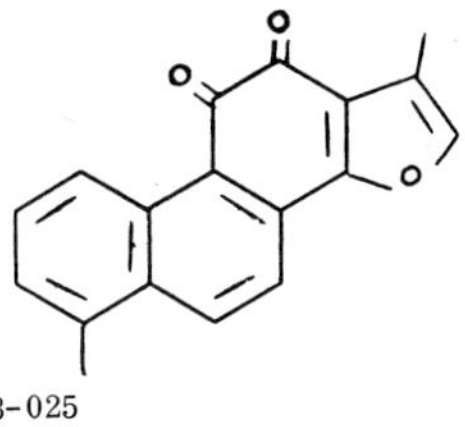

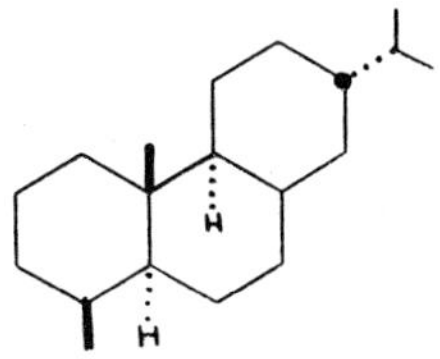

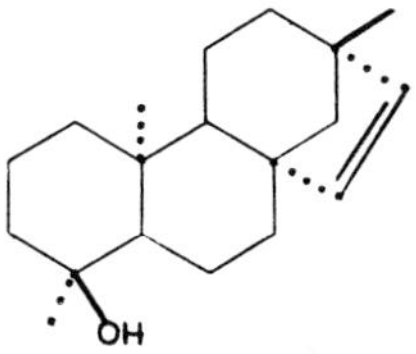

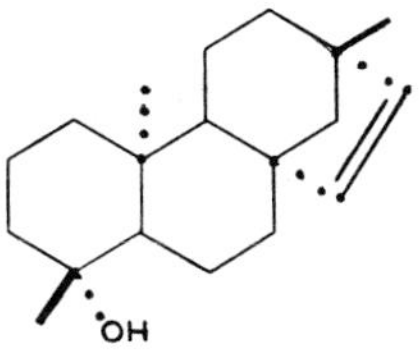

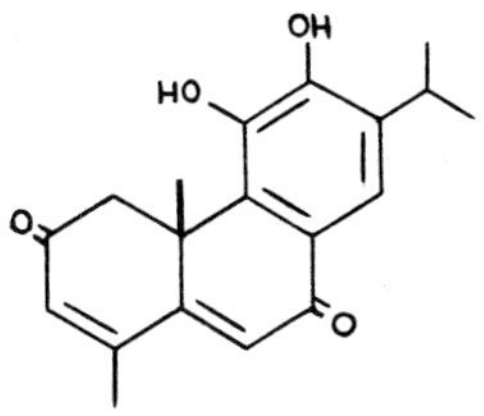

$C_{20}H_{32}O$ 288 Tet. Lett. 1967 (5) 459. 4293-031	GINKOLIDE C $C_{20}H_{24}O_{11}$ 440 -19°d R = OH Tet. Lett. 1967 (4) 299-326. 4293-036
$C_{20}H_{34}O_5$ 354 JOC. 1966 1797. 4293-032	LABD-11-EN-13-ONE, 14, 15-bis-nor-8-OH- $C_{18}H_{30}O_2$ 278 0°c 126°C Chem. Abstr. 1967 (67) 64577 4293-037
GINKOLIDE A $C_{20}H_{24}O_9$ 408 -39°d R = H Tet. Lett. 1967 (4) 299-326 4293-033	DAPHNETOXIN $C_{27}H_{30}O_8$ 482 +63° 195°C JACS. 1970 1070. R = H 4293-038
GINKOLIDE B $C_{20}H_{24}O_{10}$ 424 -63°d R = OH Tet. Lett. 1967 (4) 299-326 4293-034	MEZEREIN $C_{38}H_{38}O_{10}$ 654 +118°c 269°C R = O. CO. CH = CH. CH = CH. Ph Tet. Lett. 1970 4261 4293-039
GINKOLIDE M $C_{20}H_{24}O_{10}$ 424 -39°d R = H Tet. Lett. 1964 (4) 299-326 4293-035	$C_{22}H_{28}O_5$ 372 -259°c 171-4°C Tet. Lett. 1969 2399. 4293-040

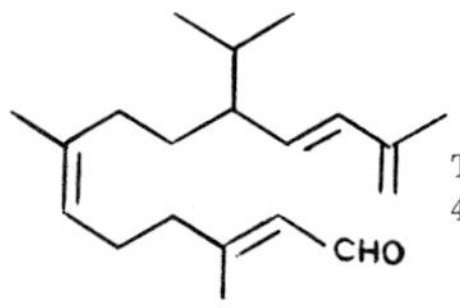

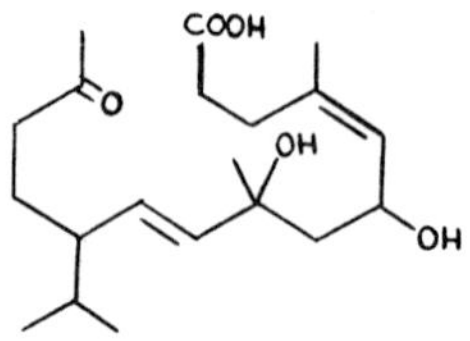

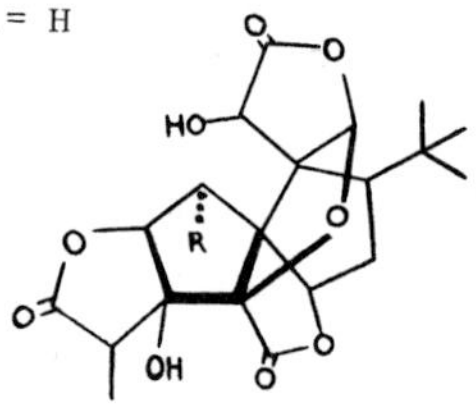

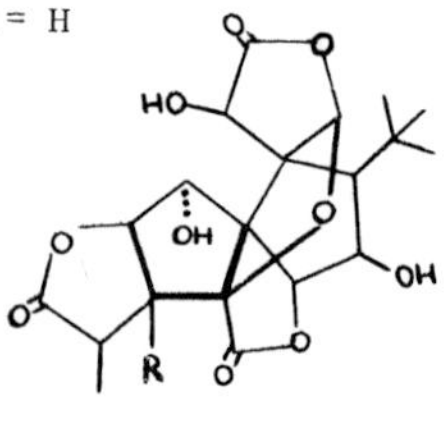

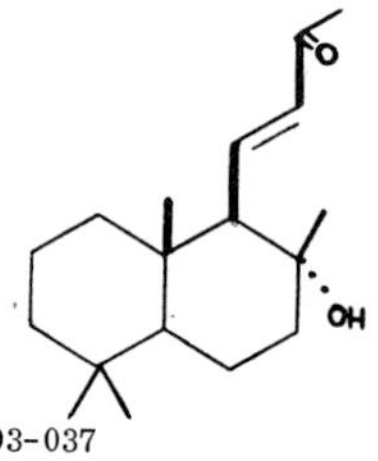

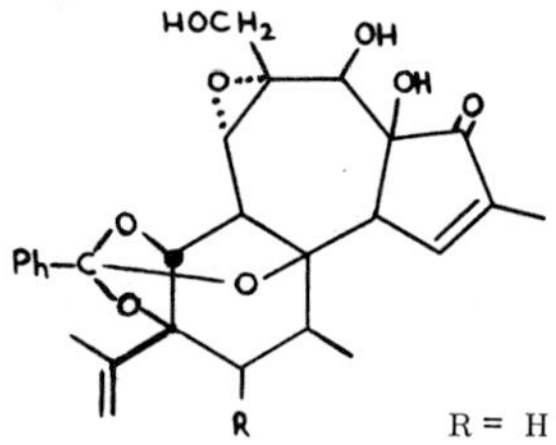

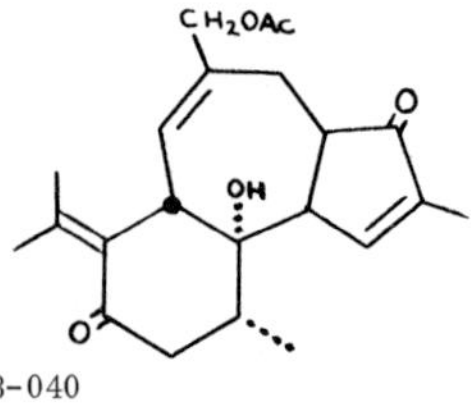

HETEROPHYLLINE R = H	$C_{21}H_{31}NO_4$	361 +10.5°m 222°C	Tet. Lett. 1967 (6) 557.
4293-041			
ATRACYLIGENIN R = H	$C_{19}H_{28}O_4$	320 -146° 189°C	Aust. J. C. 1968 459
4293-045			
HETEROPHYLLISINE R = CH_3	$C_{22}H_{33}NO_4$	375 +15.5°m 178°C	Tet. Lett. 1967 (6) 557.
4293-042			
ATRACTYLOSIDE R = (i-Val-O, OSO_3K, OSO_3K, CH_2OH sugar)	$C_{30}H_{44}O_{16}S_2K_2$	802 -53°w 174°C	Gazz., 1967 (97) 935.
4293-046			
HETEROPHYLLIDINE R = H	$C_{21}H_{31}NO_5$	377 +42°m 270°C	Tet. Lett. 1967 (6) 557.
4293-043			
MANOYL OXIDE, nor-seco-epi-	$C_{19}H_{30}O_3$	306 174°C	JCS., 1965, 3846.
4293-047			
HETERATISINE R = CH_3	$C_{22}H_{33}NO_5$	391 +40°m 268°C	Tet. Lett. 1965 215.
4293-044			
PODOCARPUS LACTOL	$C_{20}H_{28}O_4$	332 169°C	JOC., 1970, 2422.
4293-048			

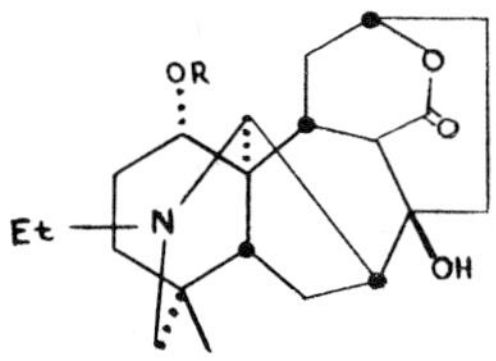

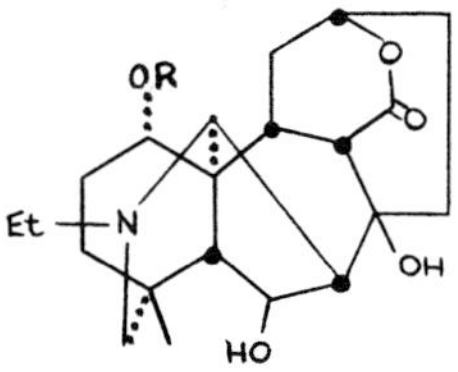

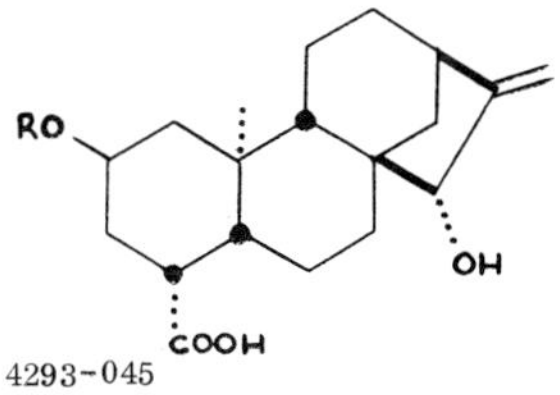

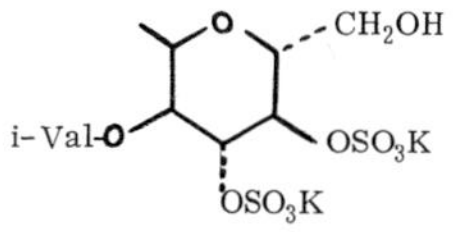

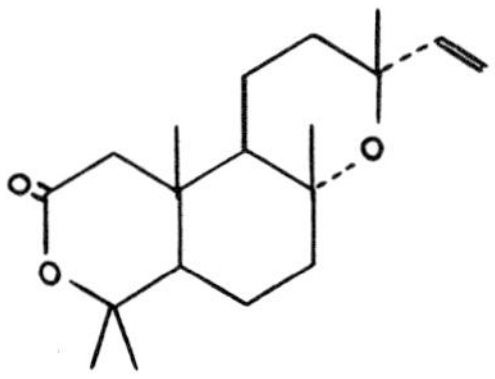

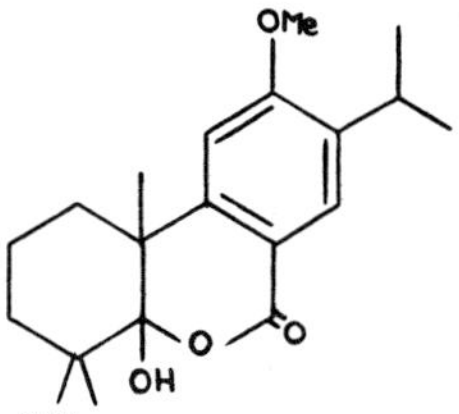

CAMPHORENE, α-	$C_{20}H_{32}$	272 OI Oil	Chem. Abstr. 1956 5854. 4299-001
DIOSBULBIN A	$C_{19}H_{20}O_6$	344	Ber. 1968 3096. 4299-005
CORYMBOSITIN (R_1 = OH; R_2 = H)	$C_{23}H_{40}O_6$	412 221⁰C	Tet. 1967 1557. 4299-002
DIOSBULBIN B (R = CH_3)	$C_{20}H_{24}O_7$	376	Ber. 1968 3096. 4299-006
TURBICORYTIN (R_1 = H; R_2 = OH)	$C_{23}H_{40}O_6$	412	Tet. 1966 1937. 4299-003
DIOSBULBIN C (R = H)	$C_{19}H_{22}O_7$	362 +65⁰ 250⁰C	Ber. 1968 3096. 4299-007
CROTONIN	$C_{19}H_{24}O_4$	316 -1.4⁰ 147⁰	Chem. Comm. 1967 191. 4299-004

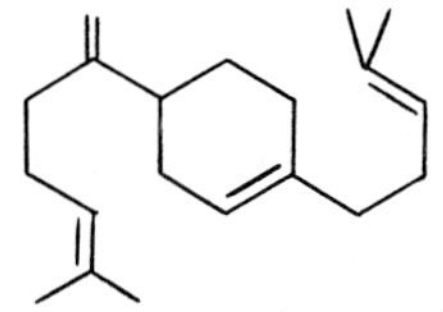

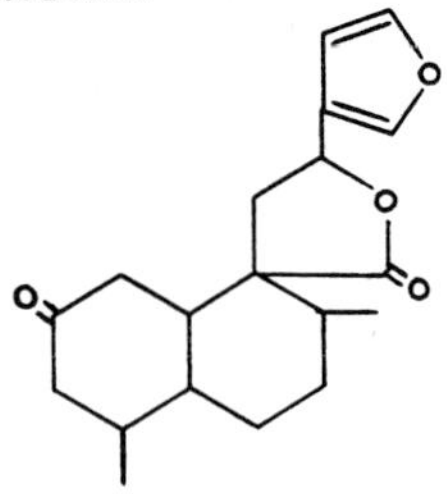

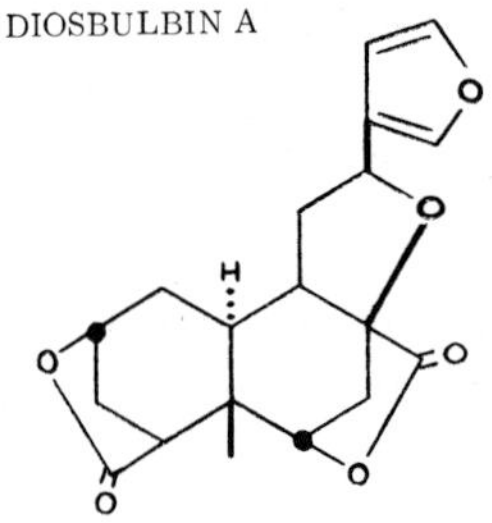

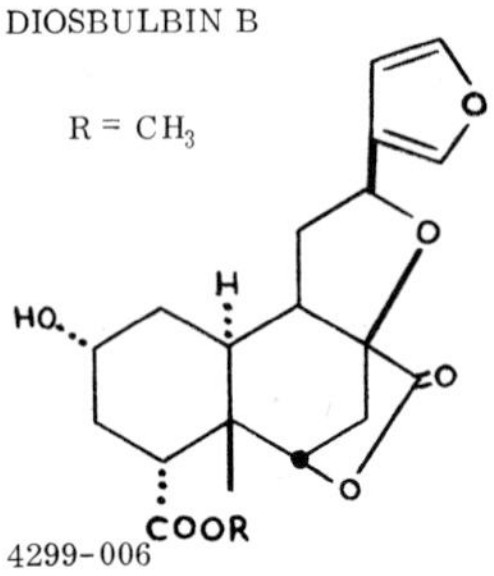

Structure labels: OH, OH, CH_2OH, HO, CH_2OH, CH_2R_1, R_2; H, O, HO, COOR.

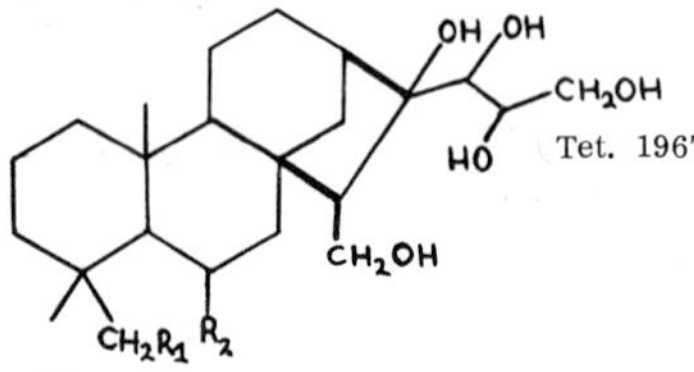

43-- THE SESTERTERPENES

The sesterterpenes are comparatively recent additions to the terpene family and as yet only a few have been discovered. Apart from the acyclic compounds two cyclic skeletons have been identified—that of the ophiobolins and that of gascardic acid, the latter at present being the sole representative of a potential class of sesterterpenes. Particularly elegant tracer studies have elucidated the biosynthesis of both cyclic skeletons.

Sesterterpenes

4301 Linear

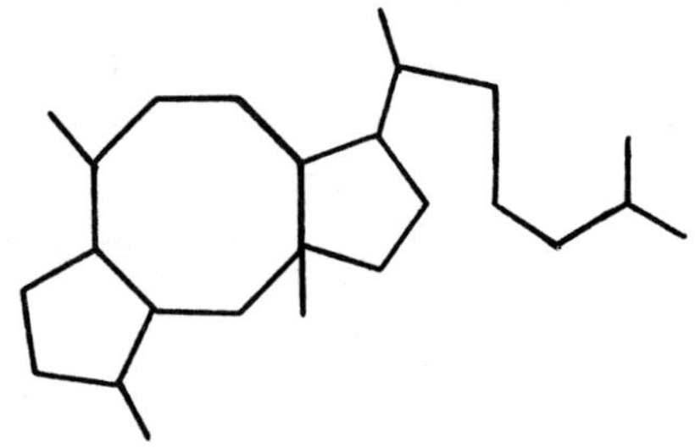

4302 Ophiobolane

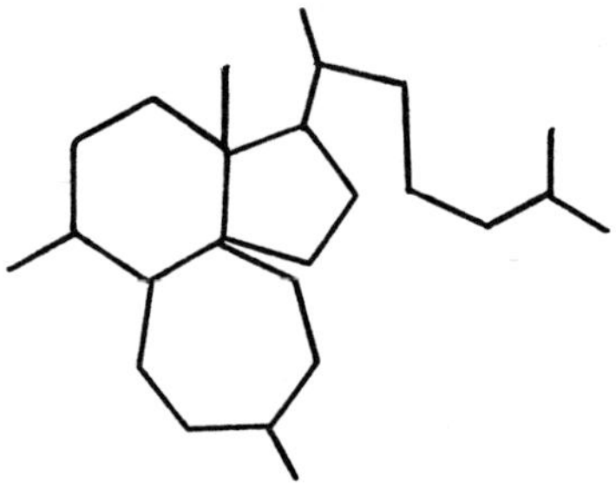

4303 Gascardic acid

GERANYL FARNESOL $C_{25}H_{42}O$ 358
OI

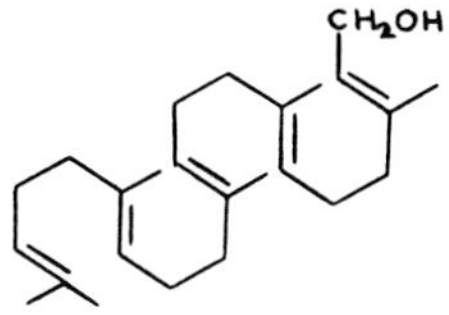

Chem. Comm., 1969, 214.

4301-001

GERANYL FARNESOL, 10, 11, 14, 15, 18, 19-hexahydro- $C_{25}H_{48}O$ 364
Oil

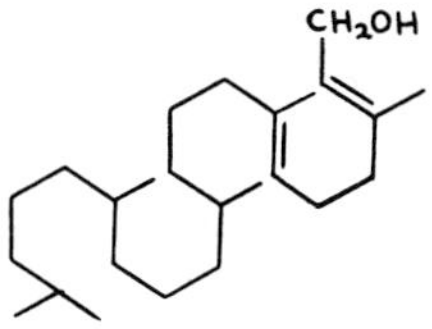

Tet. Lett., 1969, 4880.

4301-003

GERANYL-NEROLIDOL $C_{25}H_{42}O$ 358

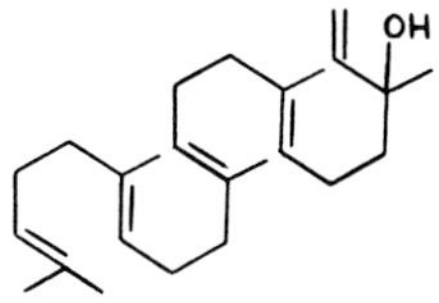

Tet. Lett., 1968, 4457.

4301-002

OPHIOBOLIN A $C_{25}H_{36}O_4$ 400 +270° 182°C OHC, O, OH Tet. Lett., 1966, 3035. 4302-001	CEROPLASTOL I $C_{25}H_{40}O$ 356 JACS., 1968, 1092. 4302-006
OPHIOBOLIN B $C_{25}H_{38}O_4$ 402 +300°c 175°C R = OH OHC, R, O, OH Tet. Lett., 1966, 2211. 4302-002	CEROPLASTOL II $C_{25}H_{40}O$ 356 Tet. Lett., 1969, 1317. 4302-007
OPHIOBOLIN C $C_{25}H_{38}O_3$ 386 +363°c 121°C R = H Tet. Lett., 1966, 2211. 4302-003	CEROPLASTERIC ACID $C_{25}H_{38}O_2$ 370 +87° HOOC * JACS., 1968, 1092. 4302-008
OPHIOBOLIN D $C_{25}H_{36}O_4$ 400 +76°c 139°C HO, HOOC, O Tet. Lett., 1967, 4111. 4302-004	ALBOLIC ACID $C_{25}H_{38}O_2$ 370 +139°c *Geometric Isomer Tet. Lett., 1969, 2929. 4302-009
OPHIOBOLIN F $C_{25}H_{42}O$ 358 +23° 81°C HO Chem. Comm., 1969, 1319. 4302-005	

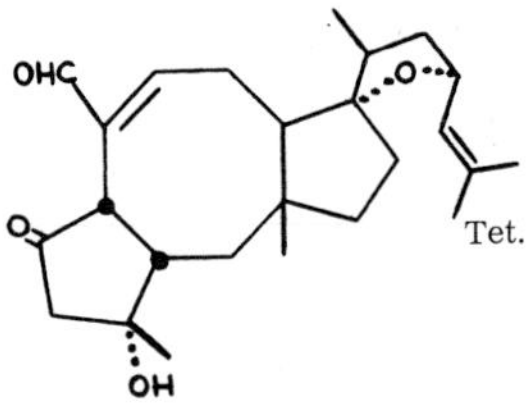

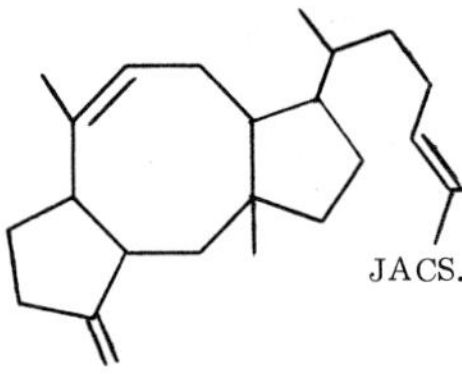

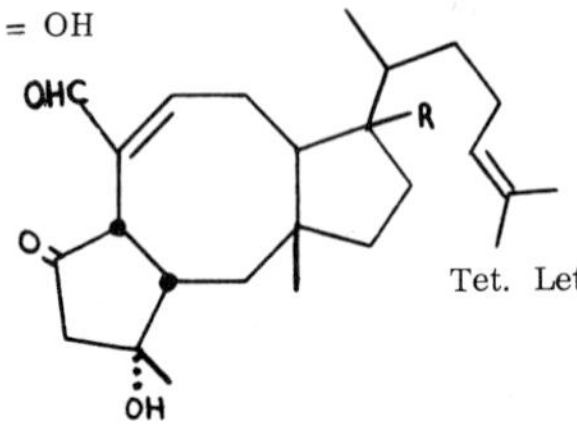

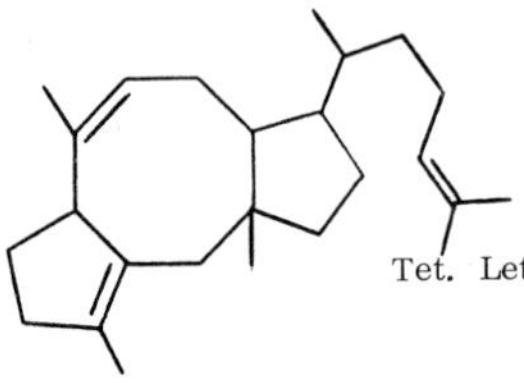

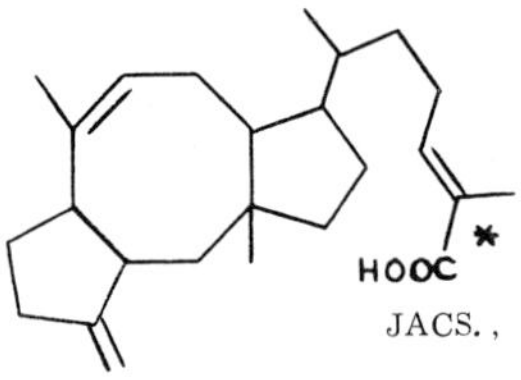

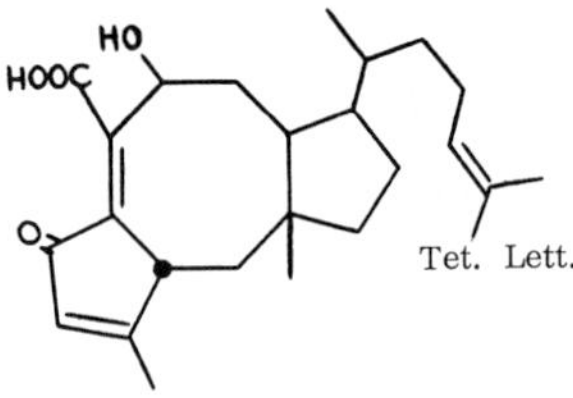

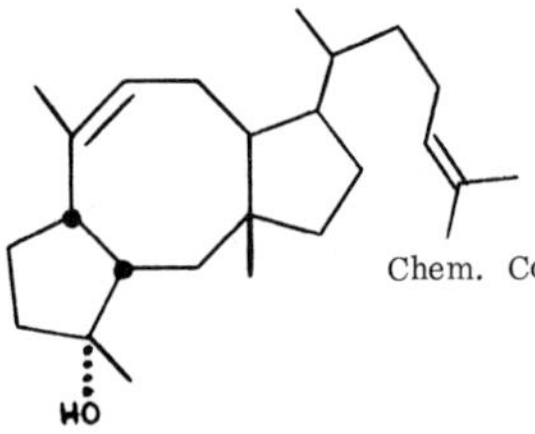

GASCARDIC ACID

$C_{25}H_{38}O_2$

370

+145°c

124°C

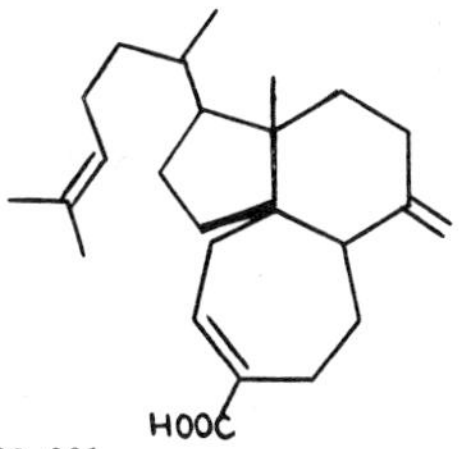

HOOC

Chem. Soc. (G. B.) Mtg., 1965, Nottingham.

4303-001

44-- THE TRITERPENES

Rapidly approaching the size of the sesquiterpene section, the 750 naturally occurring triterpenes listed here can be suitably placed into 29 main classes (see Main Skeleton Key) and 5 feature classes. As with the diterpenes these feature classes are biogenetic in character: 4491 unusually cyclized compounds; 4492 rearranged skeletons; 4493 degraded skeletons; 4494 ring- A contracted skeletons; and 4499 the miscellaneous section.

Proposed biogenetic interrelationships of the triterpene skeletons are presented on the chart following.

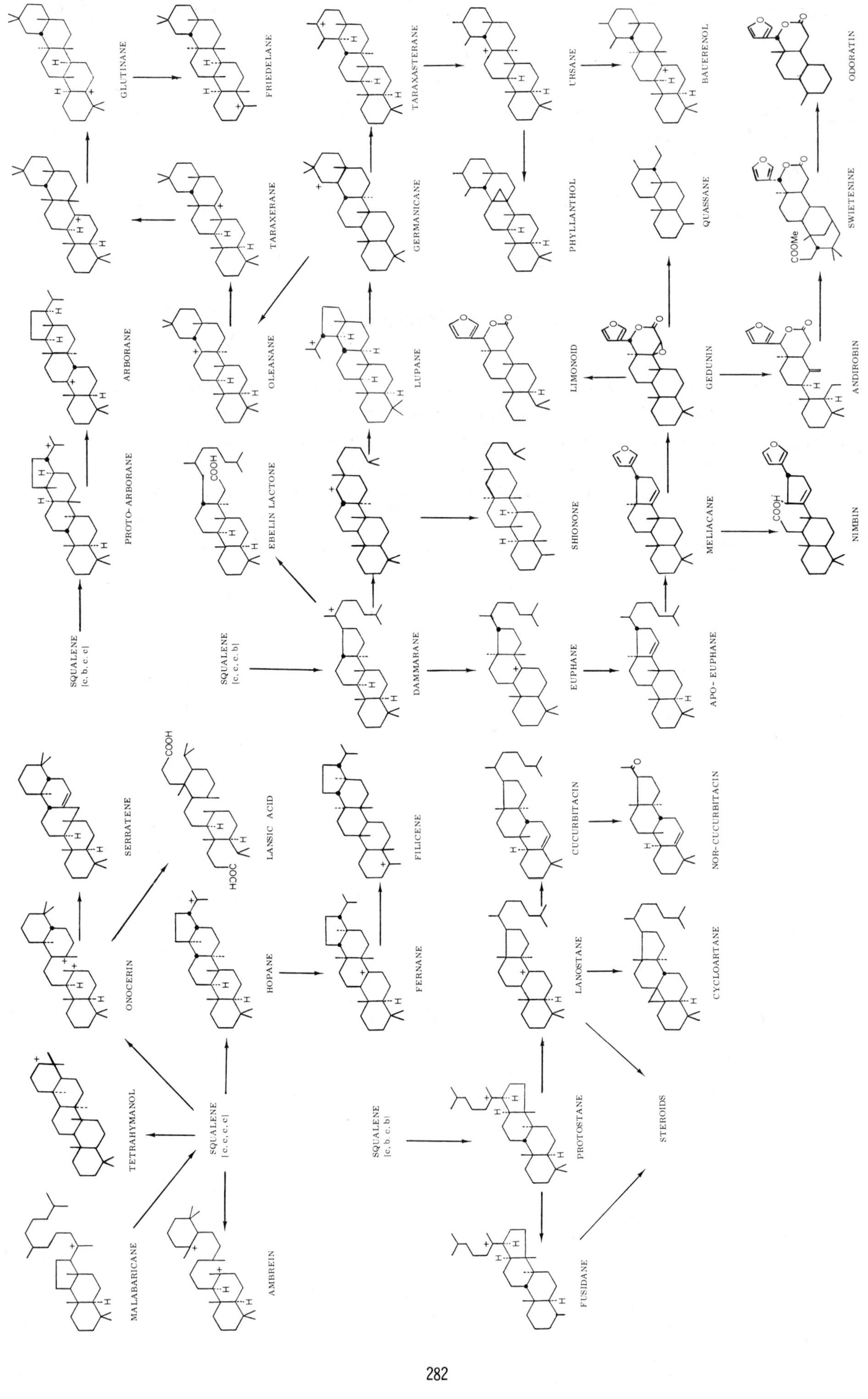

MALABARICANE

TETRAHYMANOL

ONOCERIN

SERRATENE

SQUALENE
[c, b, c, c]

PROTO-ARBORANE

ARBORANE

GLUTINANE

AMBREIN

SQUALENE
[c, c, c, c]

HOPANE

LANSIC ACID

COOH

HOOC

SQUALENE
[c, c, c, b]

EBELIN LACTONE

COOH

OLEANANE

TARAXERANE

FRIEDELANE

SQUALENE
[c, b, c, b]

FERNANE

FILICENE

DAMMARANE

LUPANE

GERMANICANE

TARAXASTERANE

FUSIDANE

PROTOSTANE

LANOSTANE

CUCURBITACIN

EUPHANE

SHIONONE

LIMONOID

PHYLLANTHOL

URSANE

STEROIDS

CYCLOARTANE

NOR-CUCURBITACIN

APO-EUPHANE

MELIACANE

GEDUNIN

QUASSANE

BAUERENOL

NIMBIN

COOH

ANDIROBIN

SWIETENINE

COOMe

ODORATIN

44__ Triterpenes - Main Skeleton Key

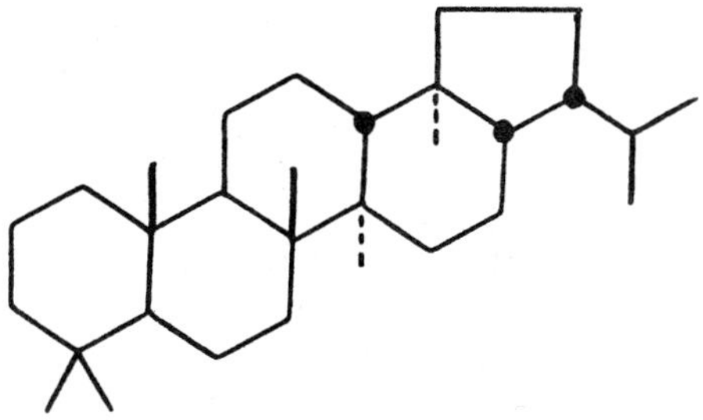

01 Hopane	05 Euphane
02 Fernane	06 Lupane
03 Friedelane	07 Germanicane
04 Dammarane	08 Oleanane

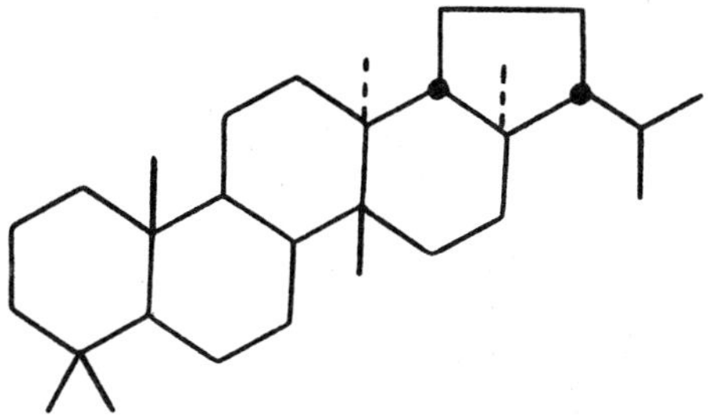

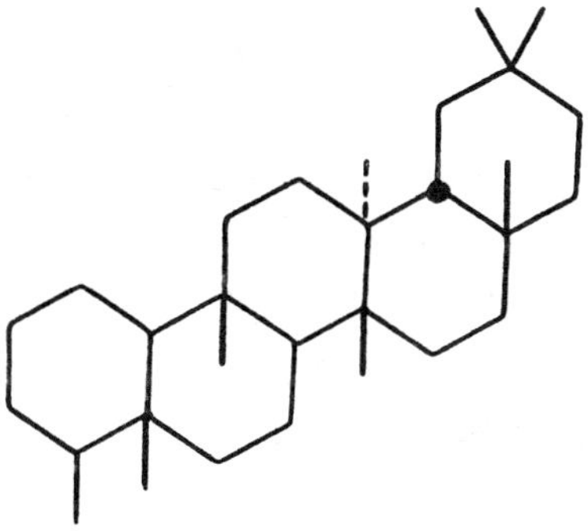

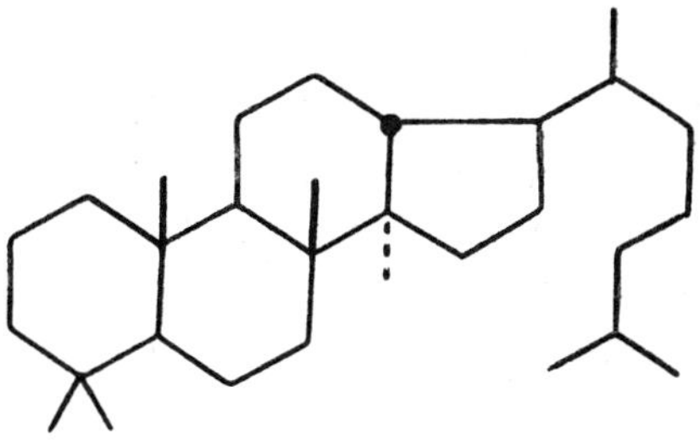

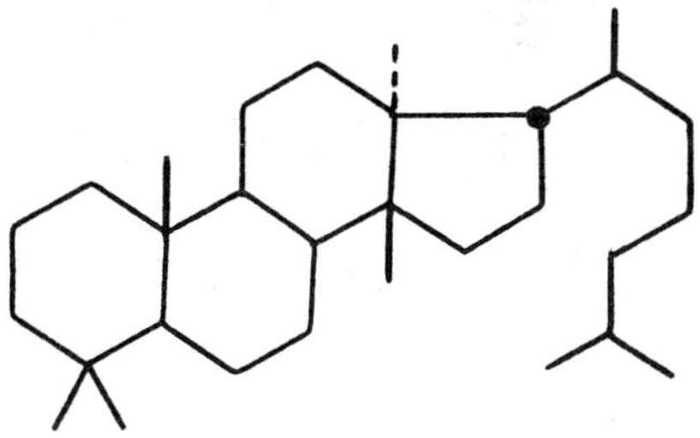

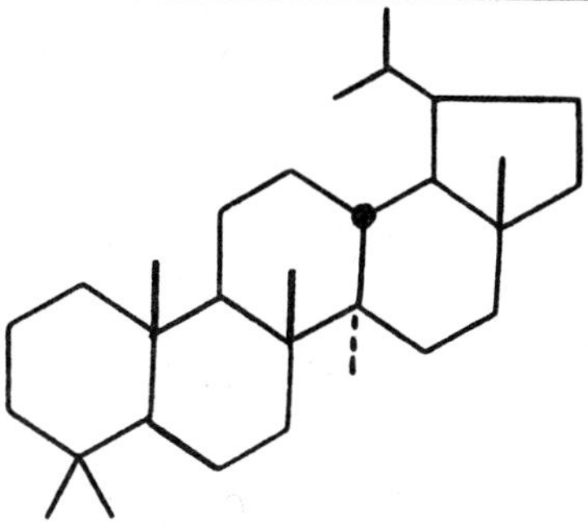

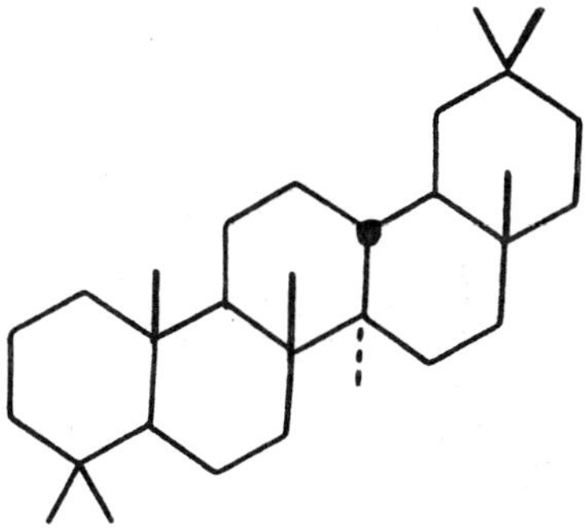

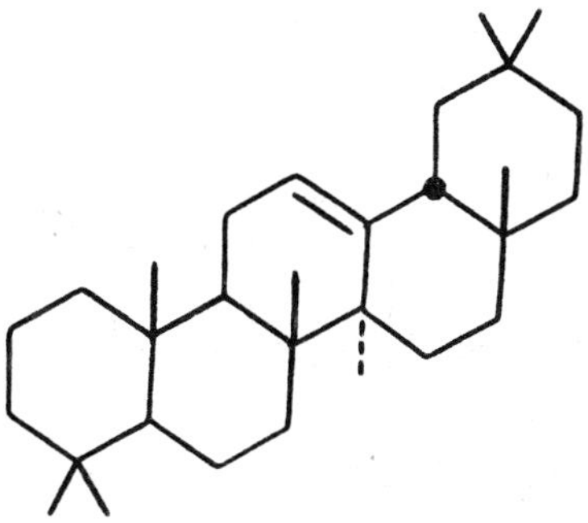

44__ Triterpenes - Main Skeleton Key

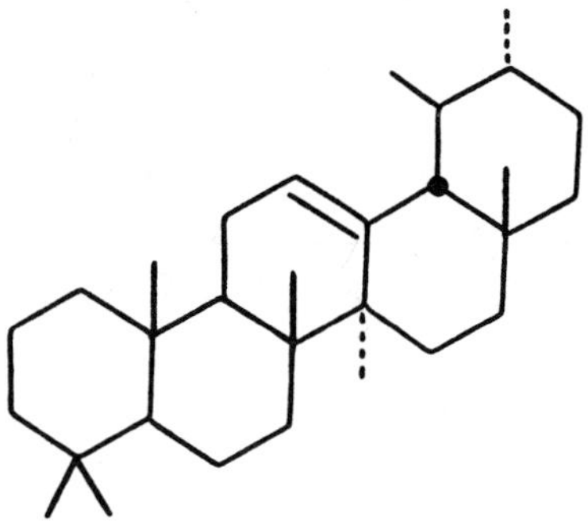

09 Ursane	13 Taraxasterane
10 Lanostane	14 Glutinane
11 Eburicane	15 Serratane
12 Cucurbitacin	16 Taraxerane

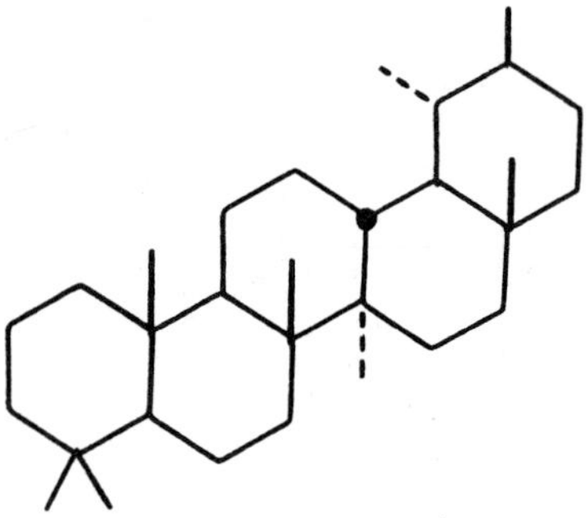

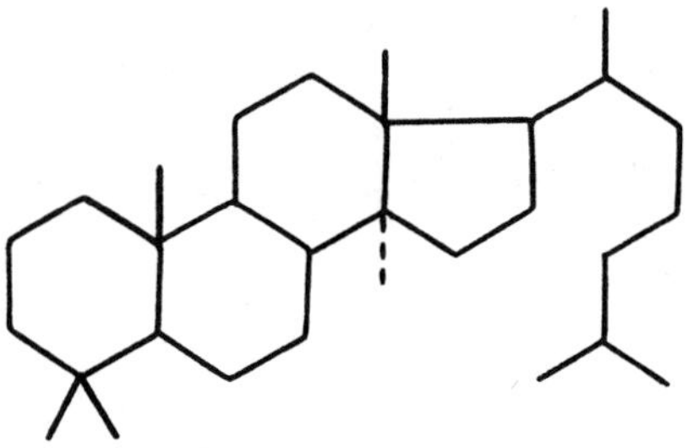

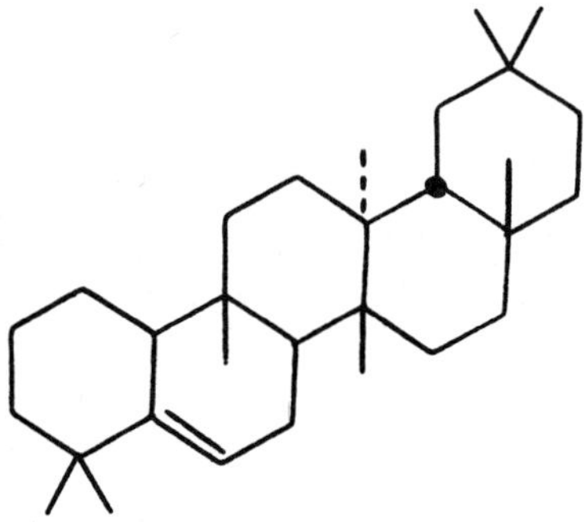

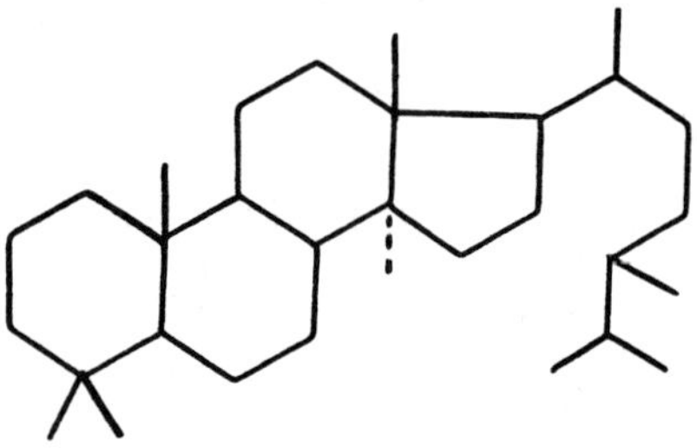

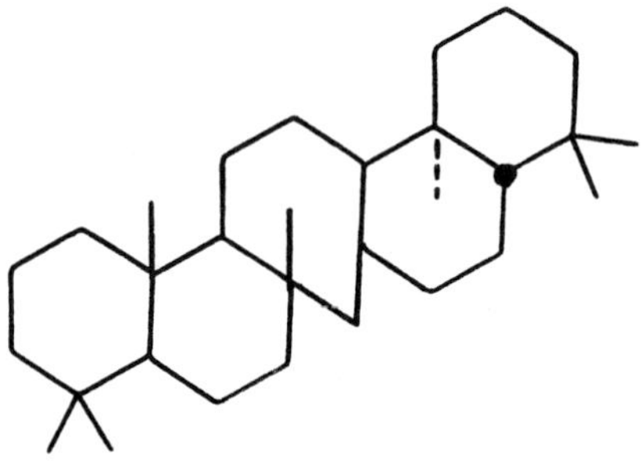

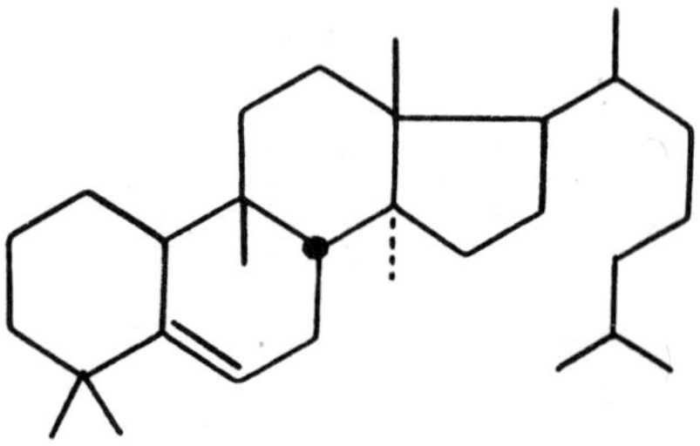

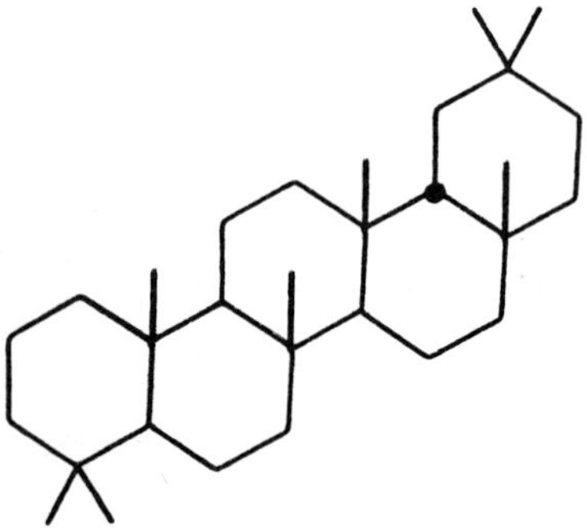

44__ Triterpenes - Main Skeleton Key

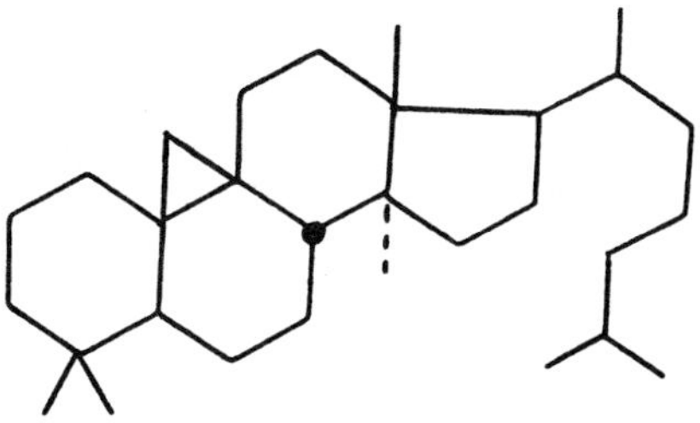

17 Cycloartane	21 Protostane
18 Buxane (3-amino)	22 Arborene
19 Buxane (3,20-diamino)	23 Limonin group
20 Malabaricane	24 Swietenine group

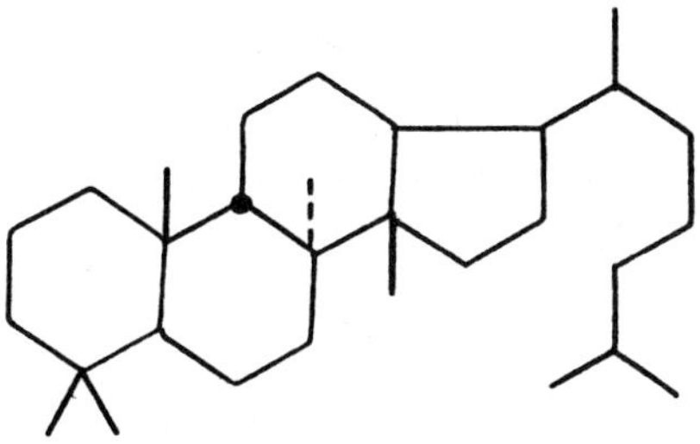

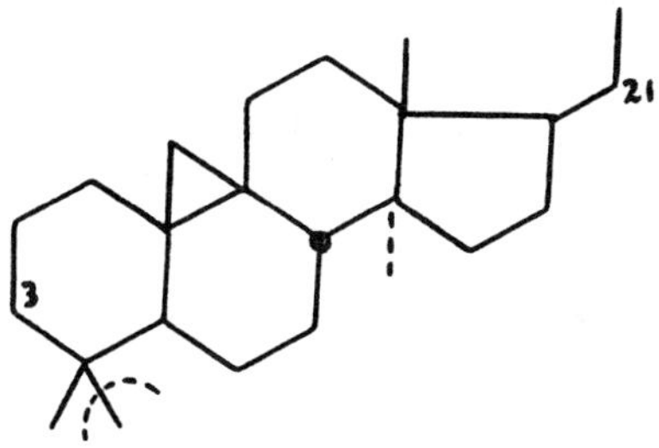

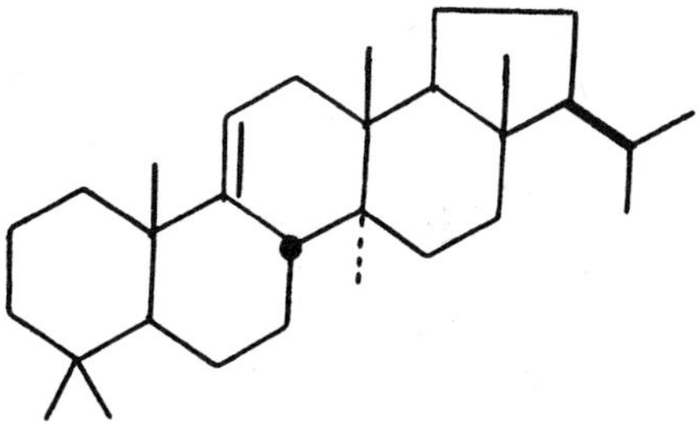

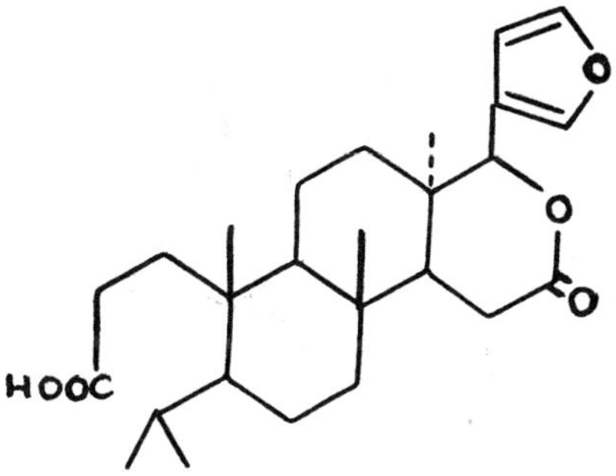

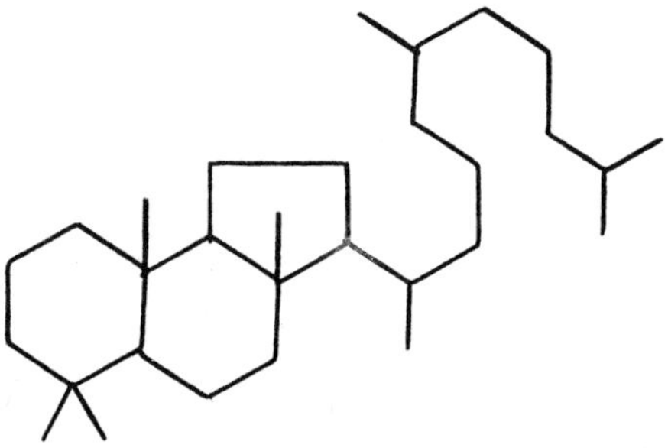

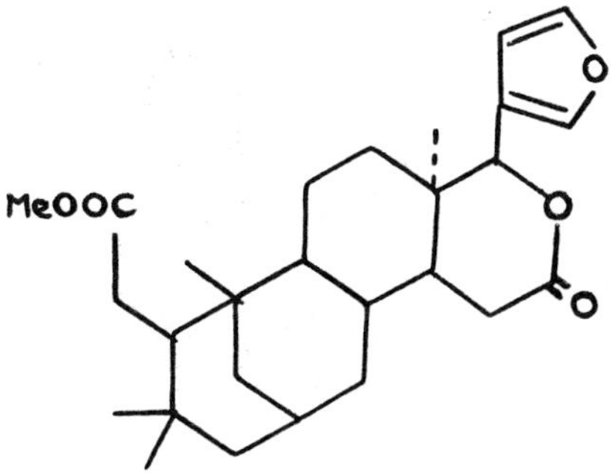

44__ Triterpenes - Main Skeleton Key

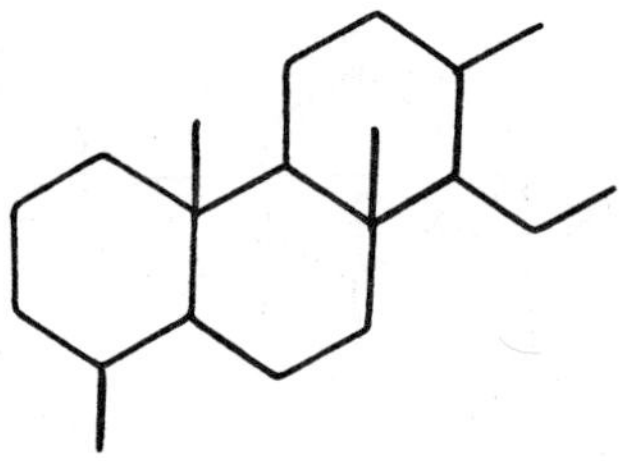

25 Quassin group COOMe	29 Fusidane
26 Meliacane	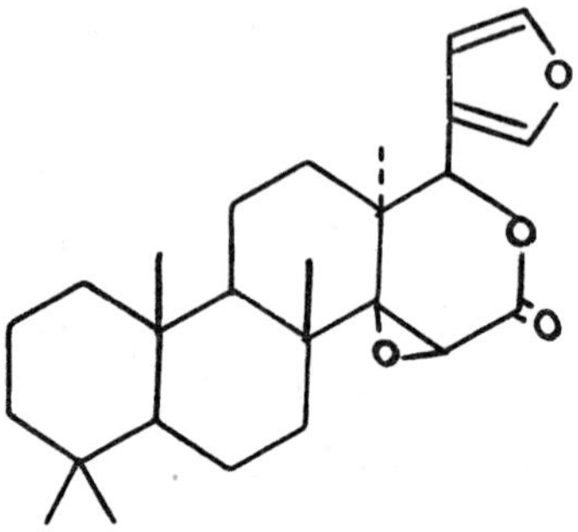
27 Gedunin group	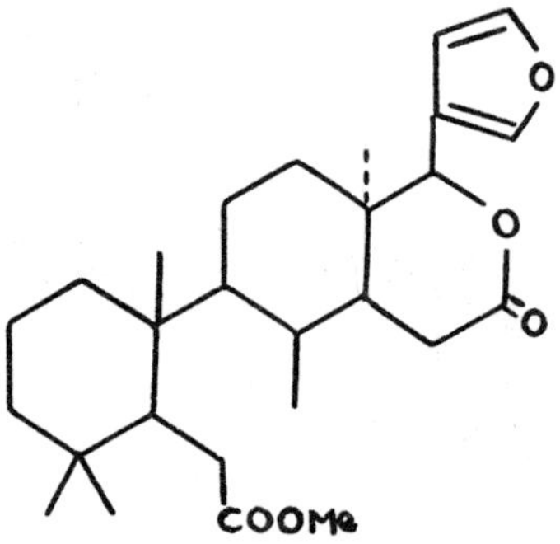
28 Andirobin group	

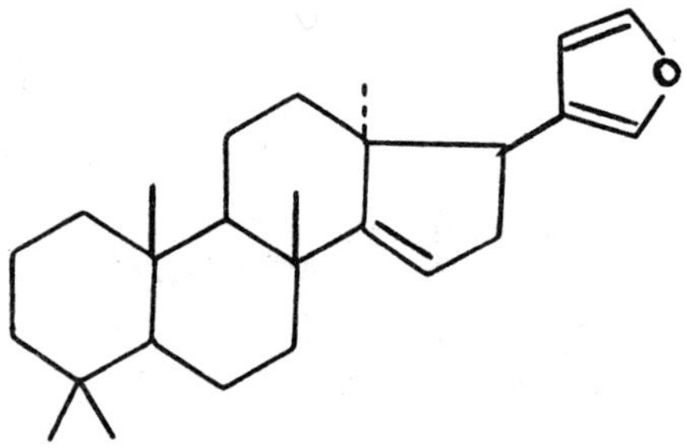

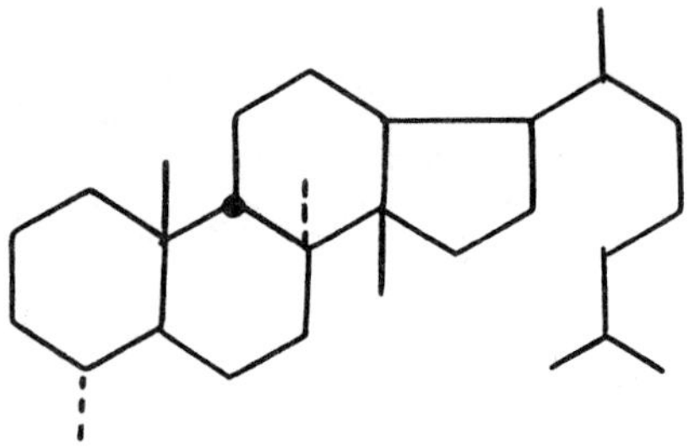

4491	Triterpenes—Less Common Skeletons-Unusually Cyclized

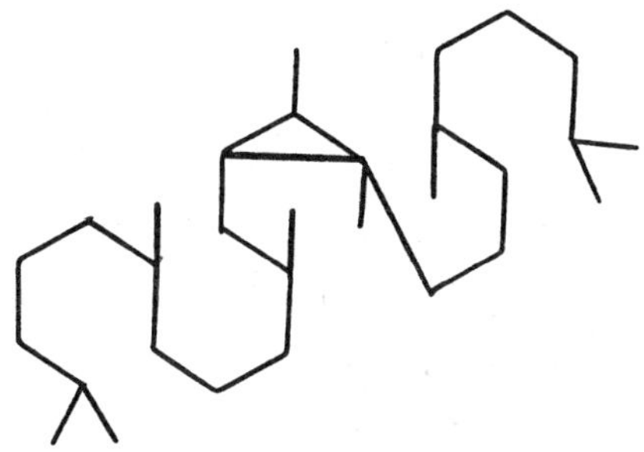

001

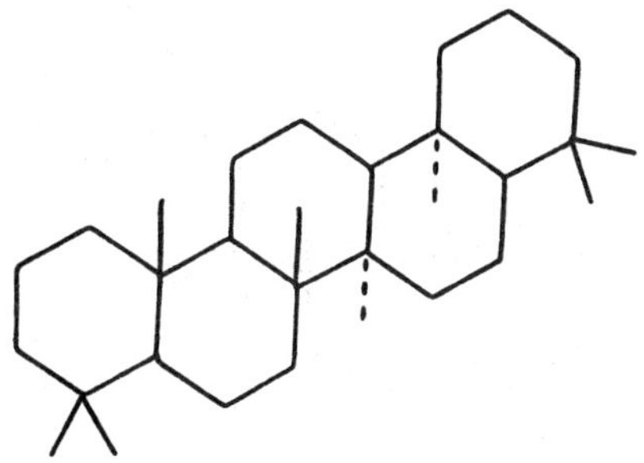

007

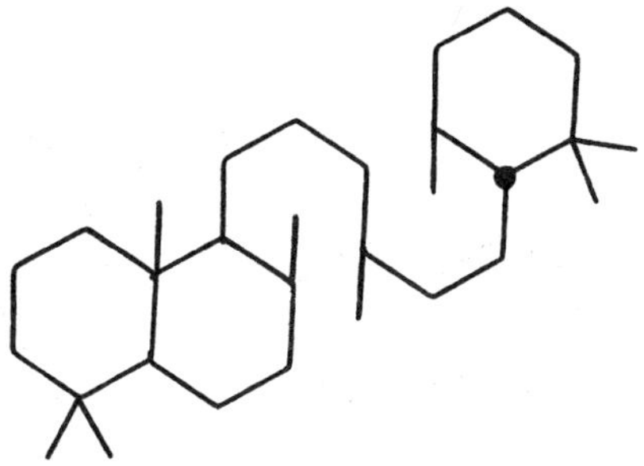

002 003 004

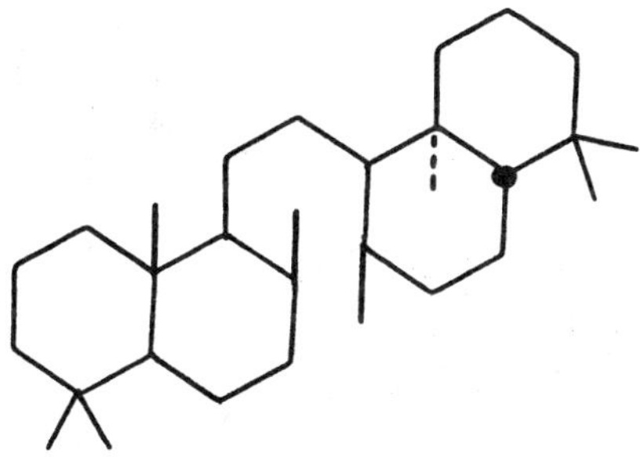

005

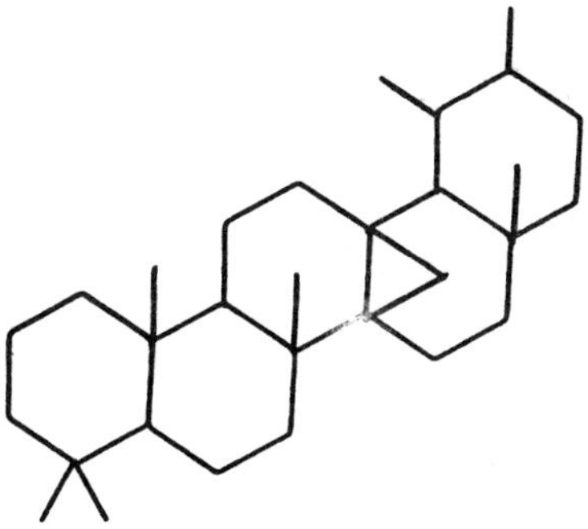

006

4492	Triterpenes – Less Common Skeletons-Rearranged

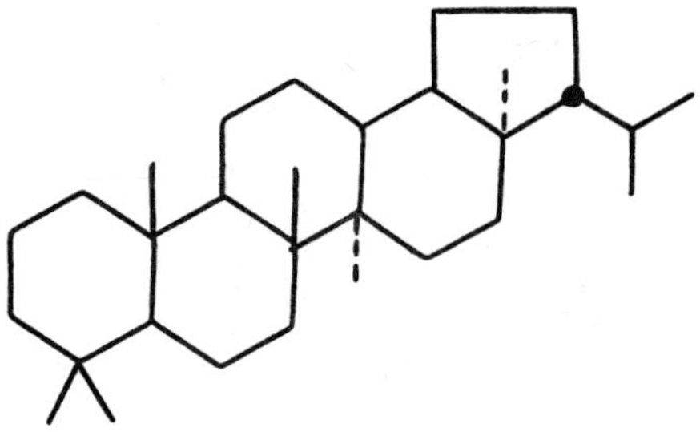

001 002 003 004

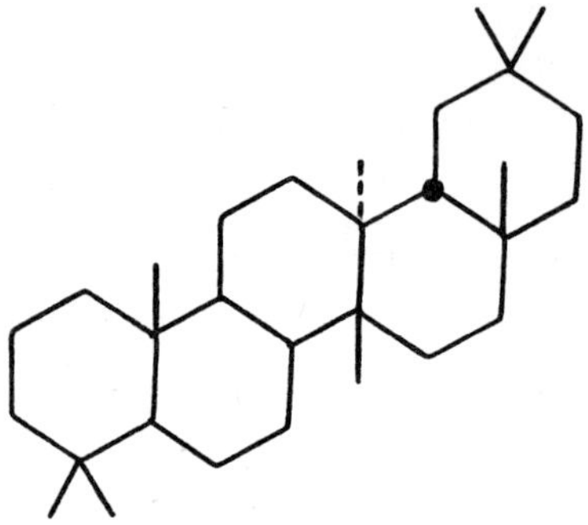

012 013 014 015

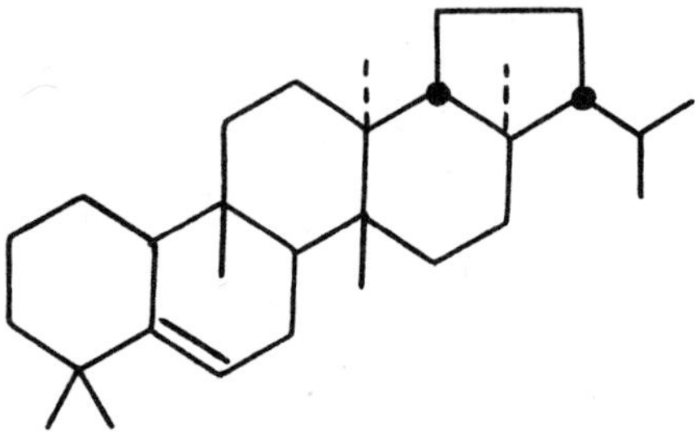

005 006 007

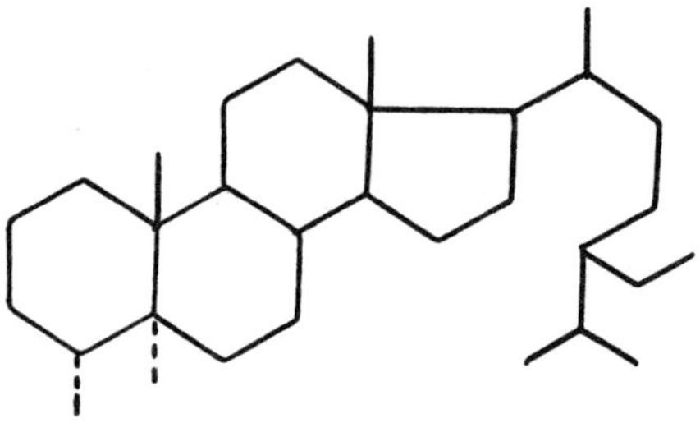

016

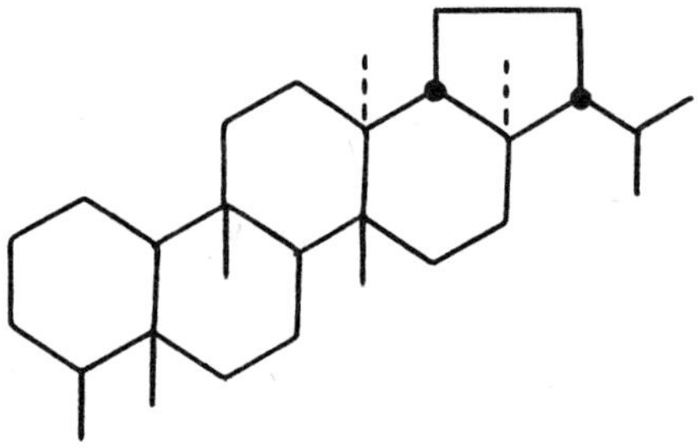

008 009 010

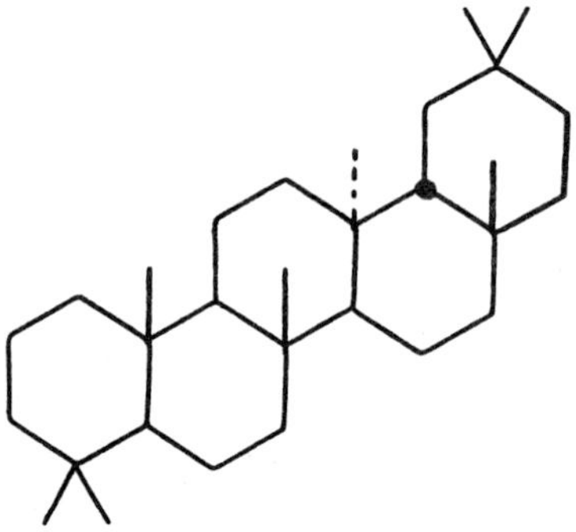

018 019

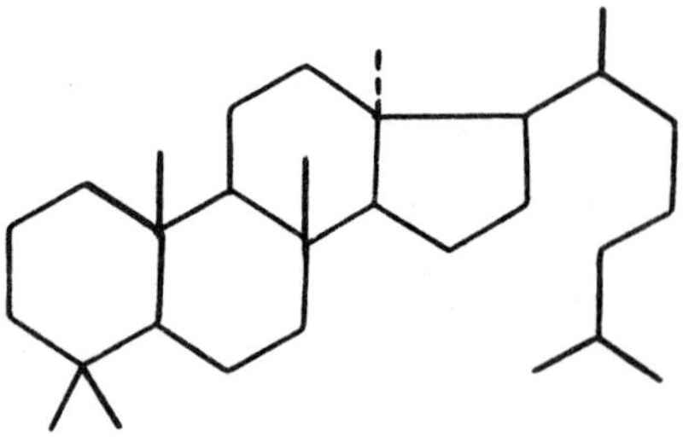

011

4493	Triterpenes-Less Common Skeletons-Degraded

Nor-Lanostane 001 002 003 004 005 006 013 061	Nor-Hopane 020 021 022
Nor-Eburicane 007 008 009	Nor-Oleanane (023) 024 025 026 062 071
Nor-Cycloartane 015	Nor-Friedelane 027 028
Nor-Lupane 016 017 018 019	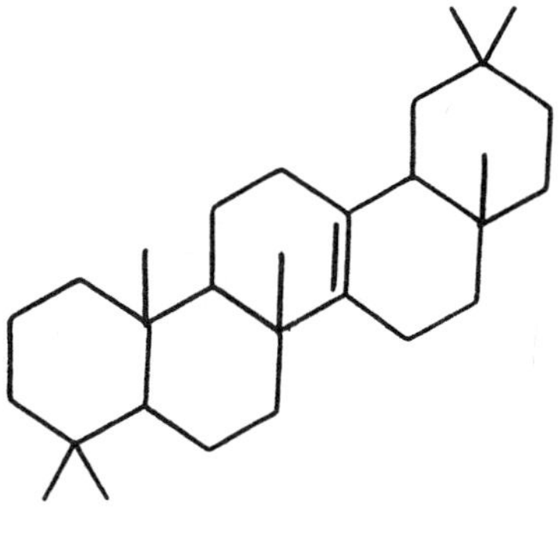 023

4493	Triterpenes-Less Common Skeletons-Degraded

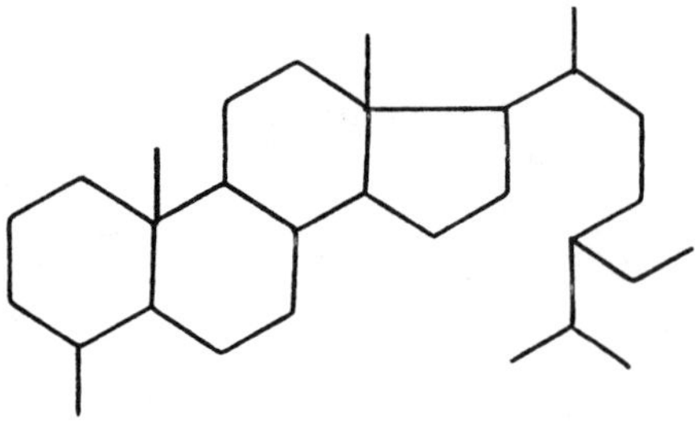

010 011 012	032
014	031
029	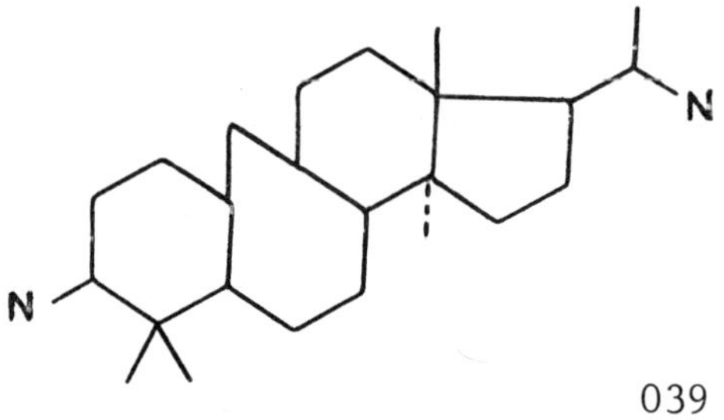 N, N; 039; 033 034 035 036 037 038
030	050

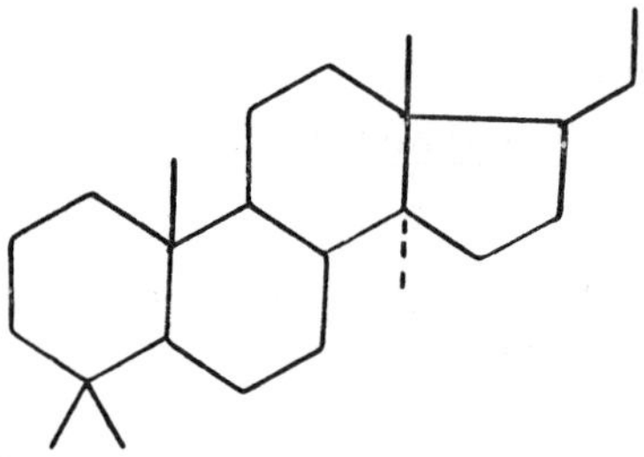

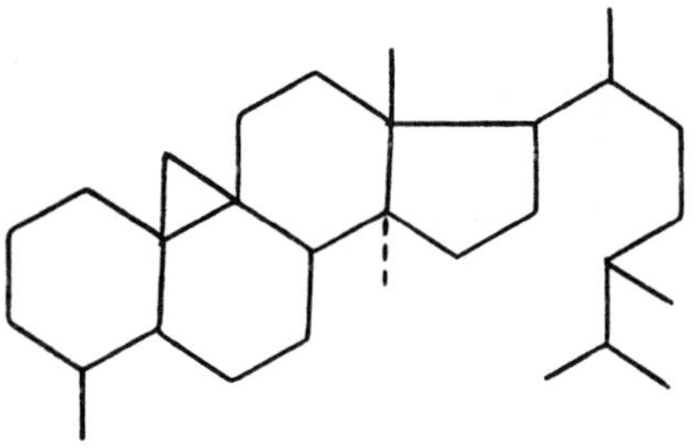

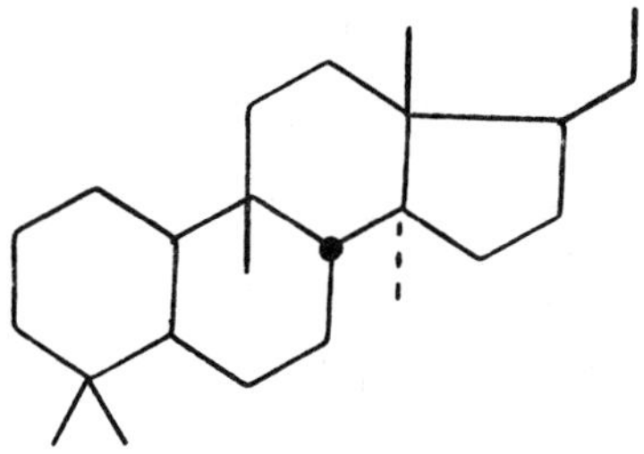

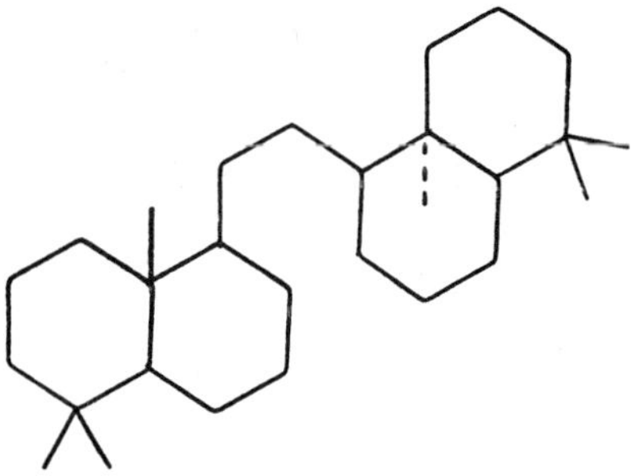

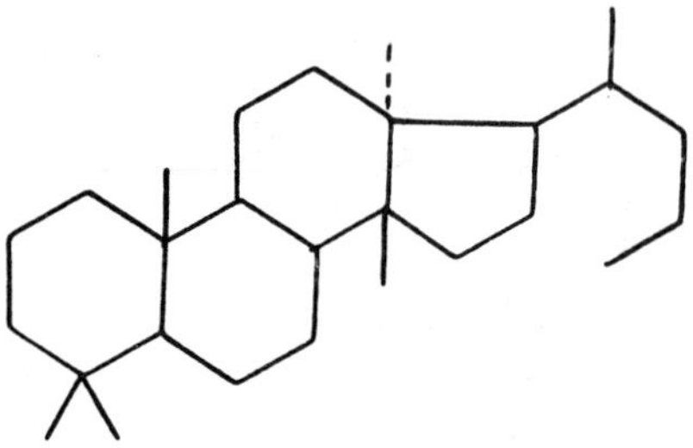

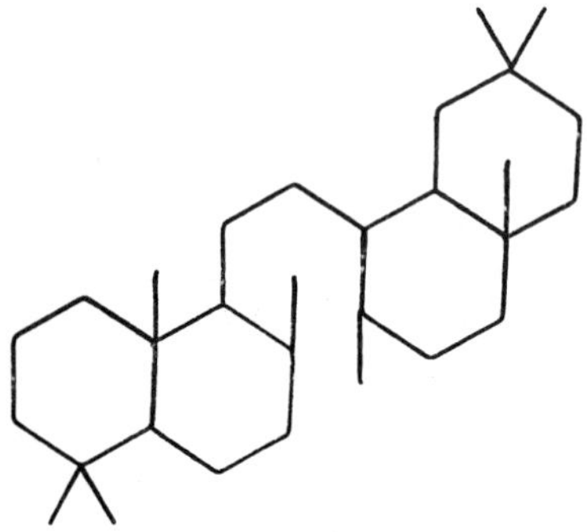

4493	Triterpenes– Less Common Skeletons-Degraded

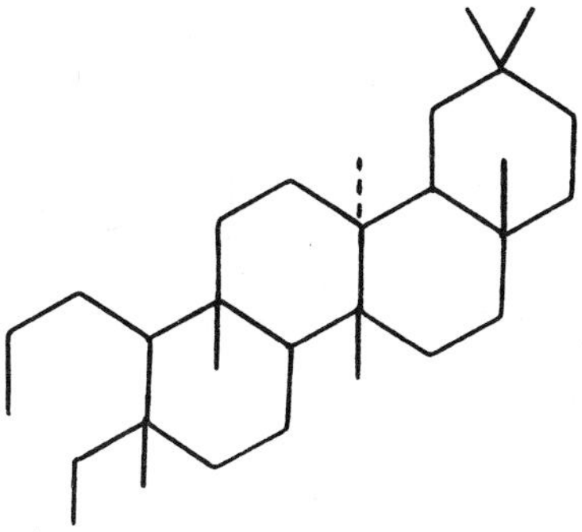

040 041 042	047
043	046
044	045
048 049	051 052

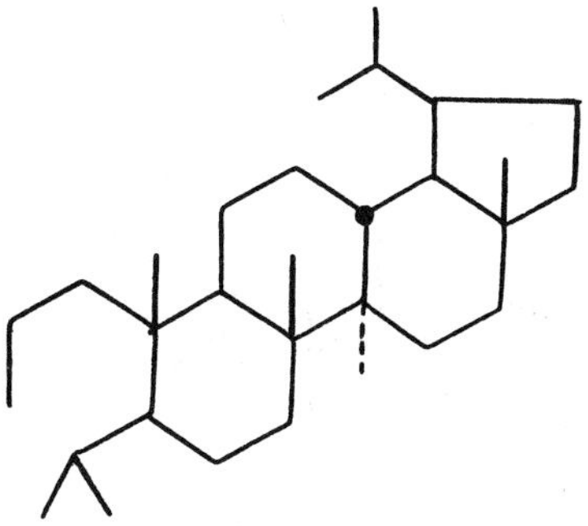

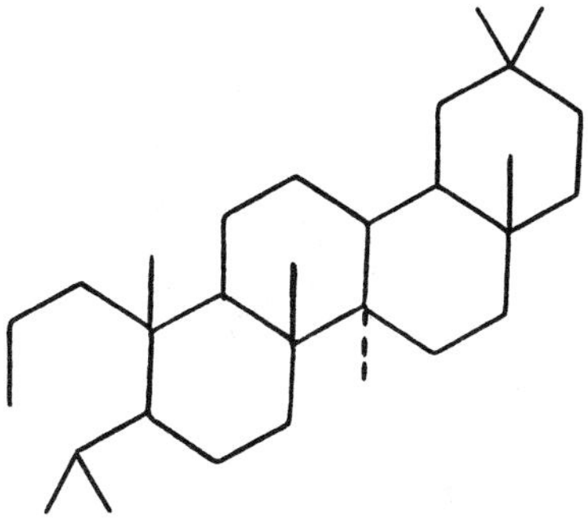

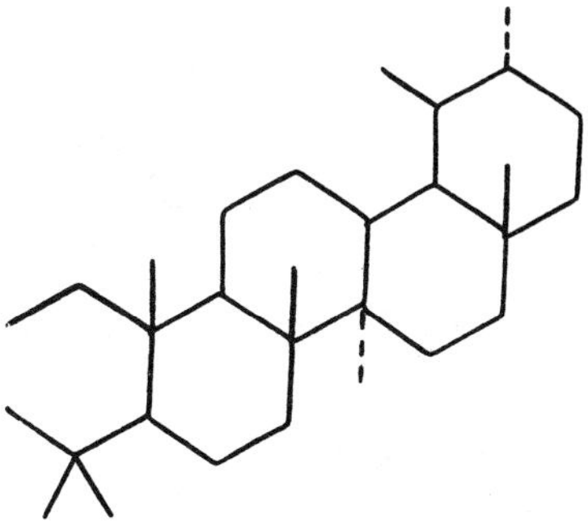

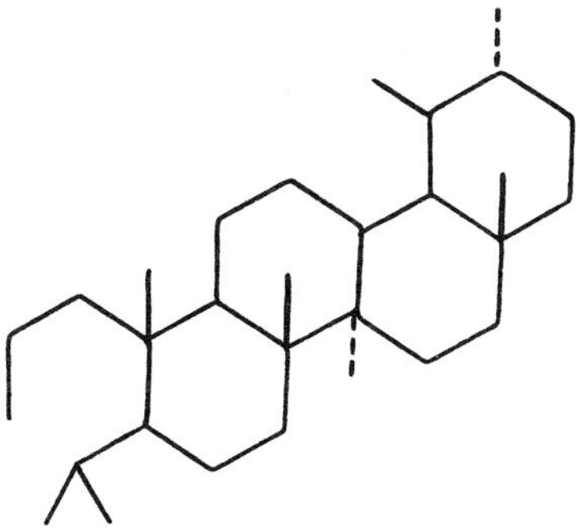

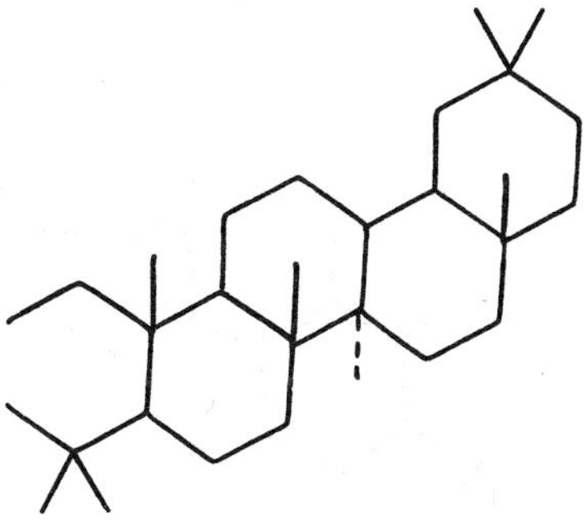

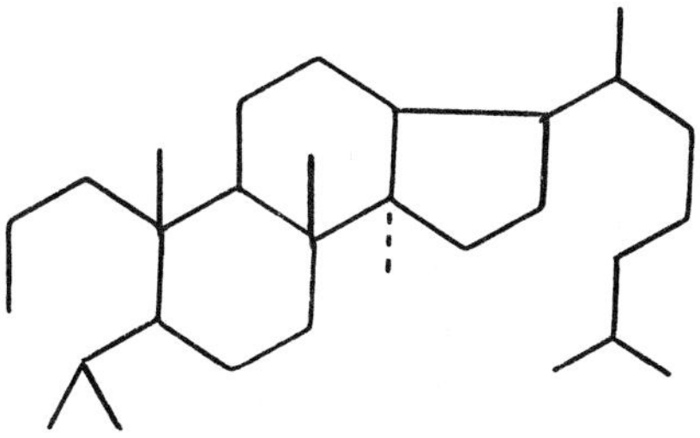

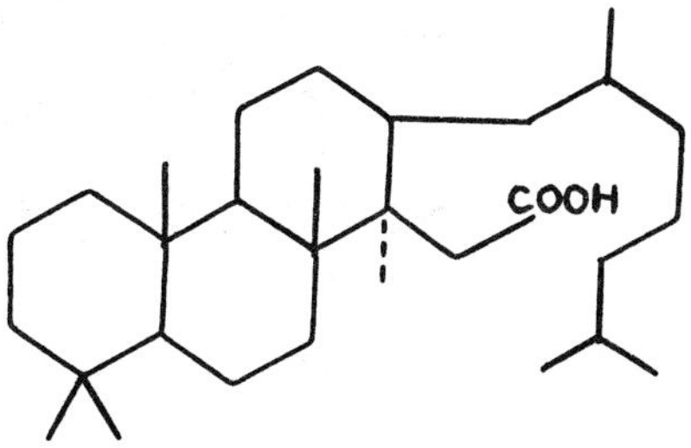

COOH

4493	Triterpenes — Less Common Skeletons-Degraded

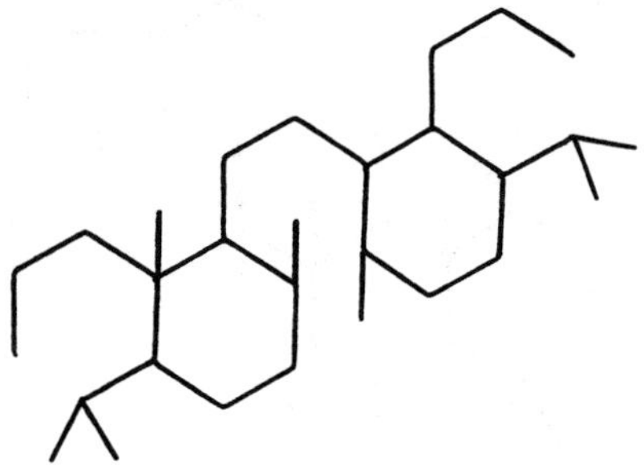

055

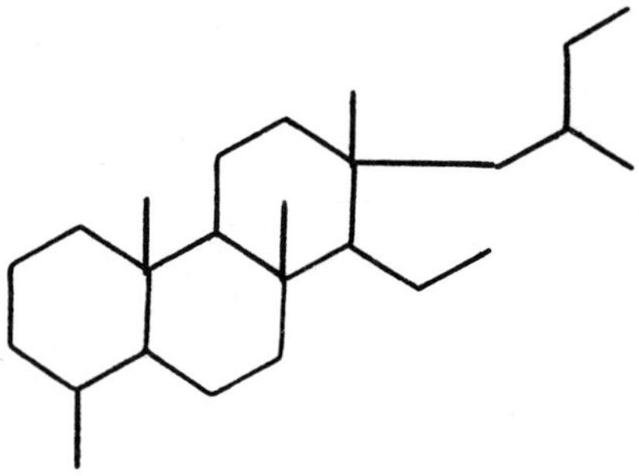

063 064

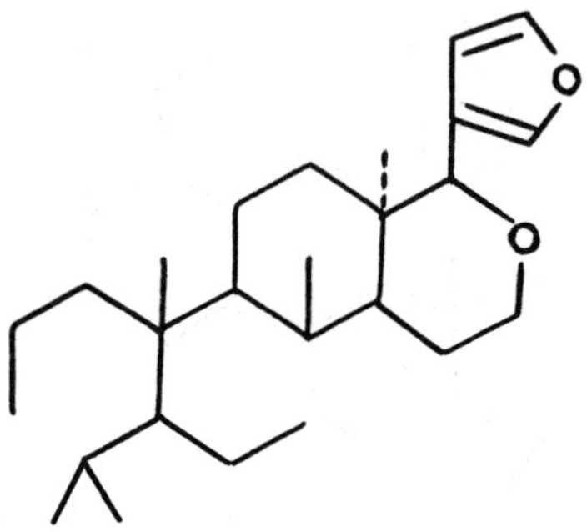

053

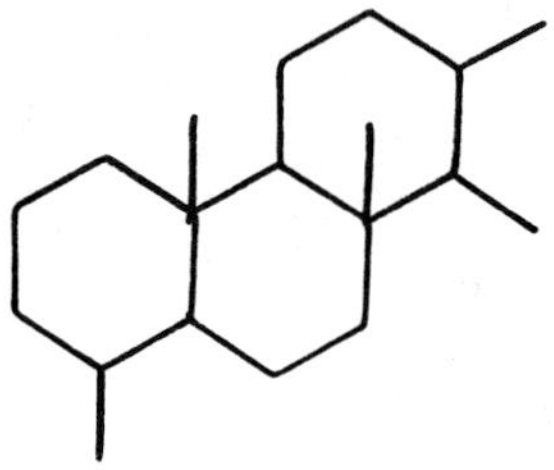

065 066 067 068 069 070

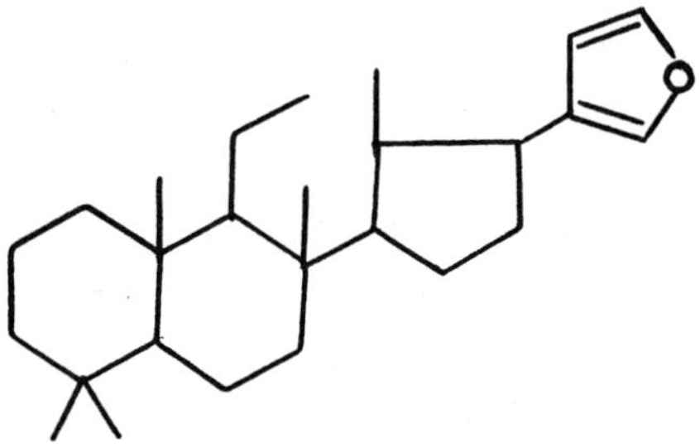

057 058 059 060

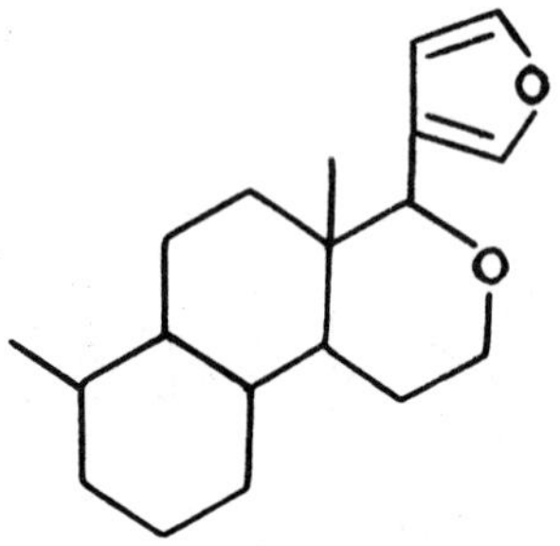

054

4494	Triterpenes – Less Common Skeletons-Ring A contracted

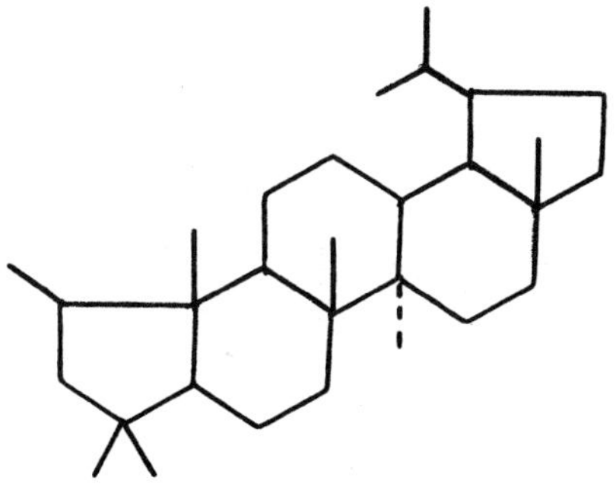

005	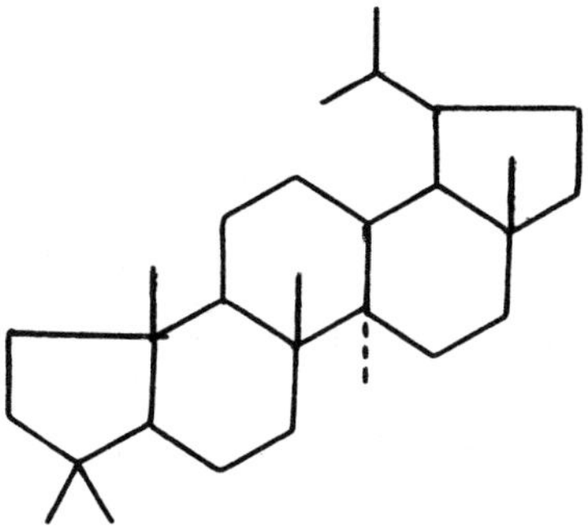
001 002	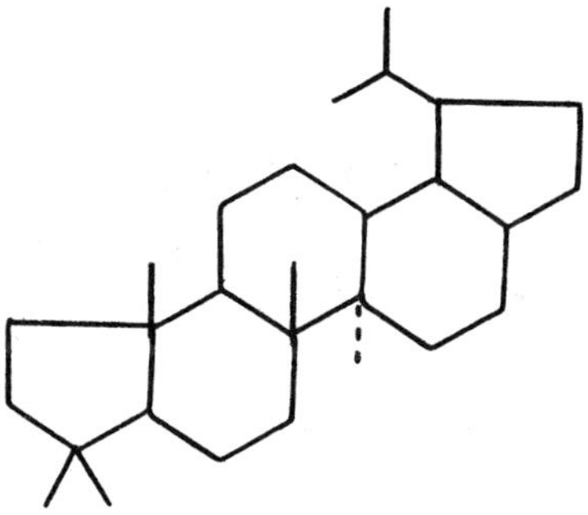
003 004	

4499	Triterpenes - Less Common Skeletons-Misc.

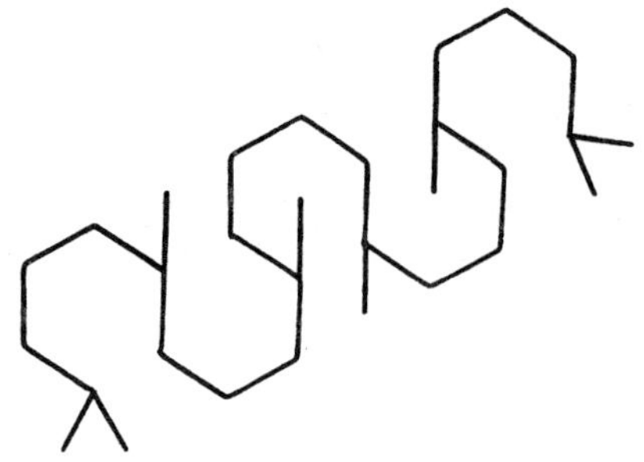

001 002

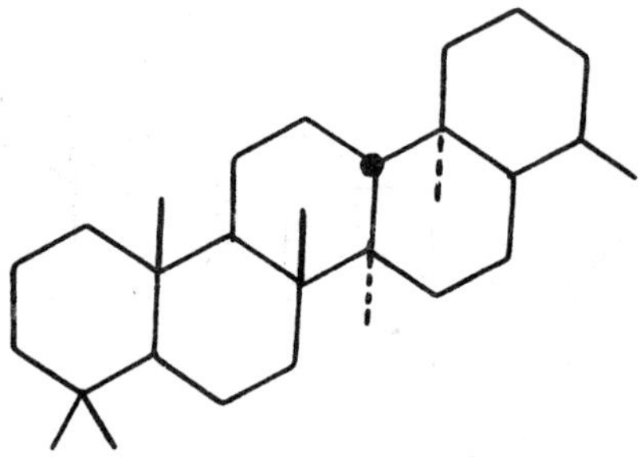

005

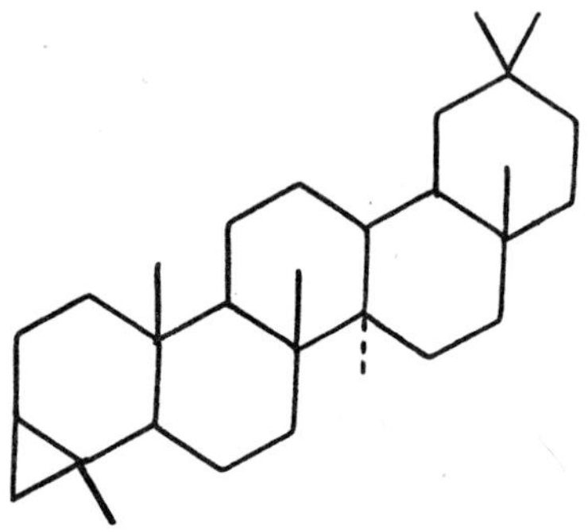

003

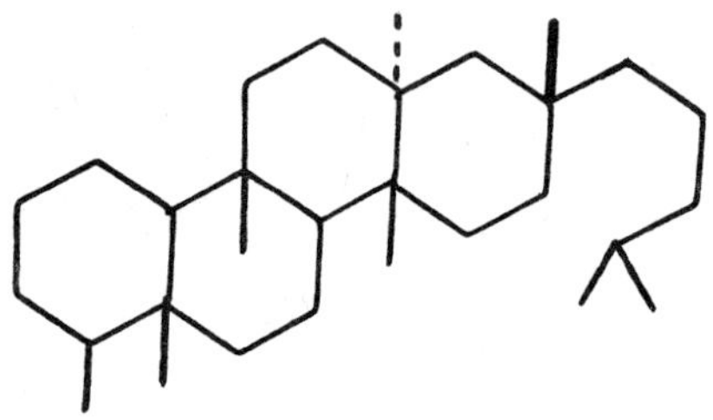

006

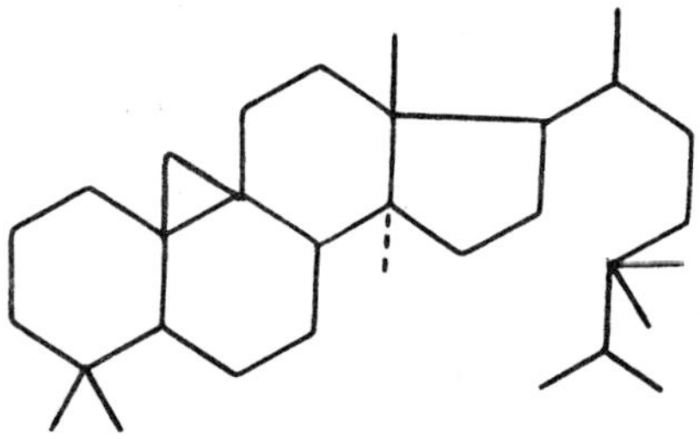

004

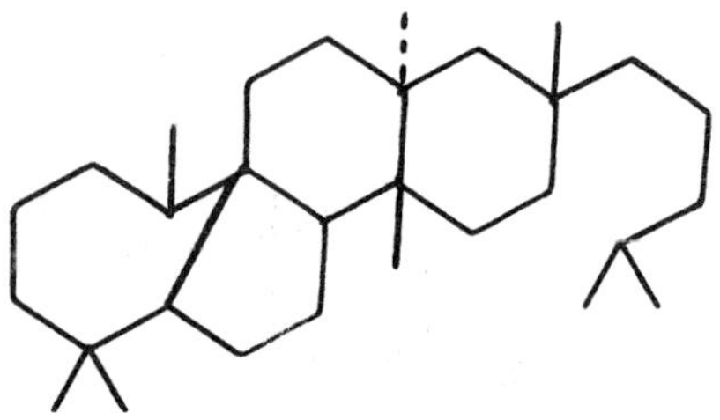

008

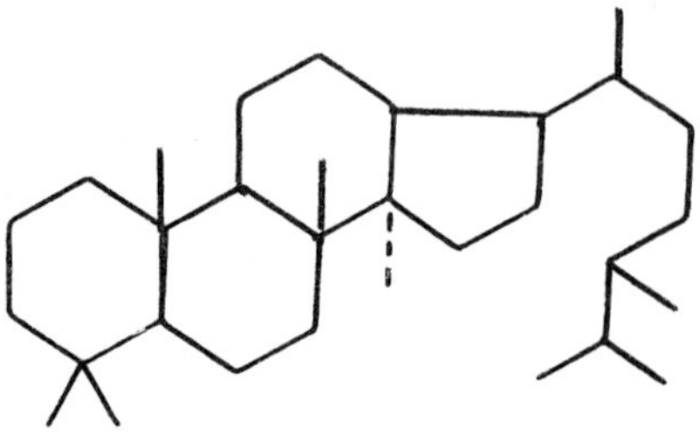

007

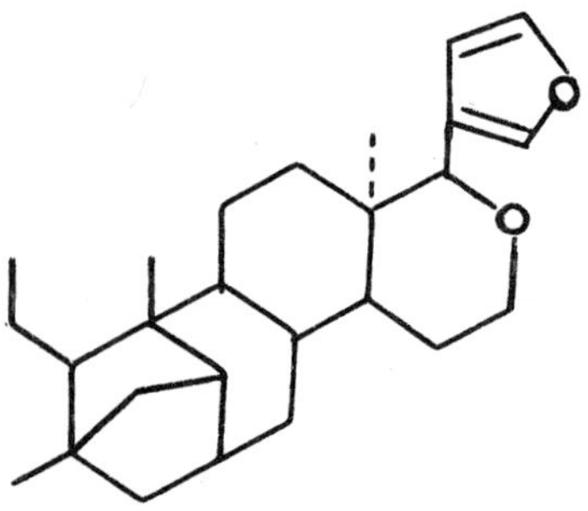

009

DIPLOPTENE $C_{30}H_{50}$ 410

Tet. Lett., 1967 (1) 23.

4401-001

MORETENOL $C_{30}H_{50}O$ 426 + 29°c 236°C

HO

Chem. Comm., 1967, 1217.

4401-006

DIPLOPTEROL $C_{30}H_{52}O$ 428

OH

Chem. Comm., 1968, 1372.

4401-002

MORETENOL, 3-epi- $C_{30}H_{50}O$ 426 -2°c 223°C

HO

Chem. Comm., 1967, 1217.

4401-007

NERIFOLIOL $C_{30}H_{52}O$ 428 + 35° 243°C

CH_2OH

Chem. Comm., 1968, 1372.

4401-003

HOPANE, 3β, 22-diOH- $C_{30}H_{52}O_2$ 444 +37°c 285° C

OH

HO

Aust. J. Chem., 1968 (21) 2583.

4401-008

HOP-17(21)-EN-3β-OL $C_{30}H_{50}O$ 426 + 45°c 229°C

HO

Aust. J. Chem., 1964, 697.

4401-004

ZEORIN $C_{30}H_{52}O_2$ 444 +54°c 225° C A/C

OH

OH

Chem. Pharm. Bull., 1969, 279.
Tet. Lett., 1971, 1161.

4401-009

HOP-17(21)-EN-3β-OL ACETATE $C_{32}H_{52}O_2$ 468 + 55°c 260°C

AcO

Aust. J. Chem., 1964, 697.

4401-005

HOPANE, 7-OAc-22-OH- $C_{32}H_{54}O_3$ 486 +26° 248°C

OH.

OAc

JCS., 1966, 1557.

4401-010

DUSTANIN $C_{30}H_{52}O_2$ 444 +20°c 264°C OH OH Tet. Lett., 1967, 23. 4401-011	LEUCOTYLIN, 6α, 16β-OAc- $C_{34}H_{56}O_5$ 544 +109° 232°C OH OAc OAc Chem. Pharm. Bull., 1966, 804. 4401-016
LEUCOTYLIN, 6-deoxy- $C_{30}H_{52}O_2$ 444 +68° 268°C OH OH Chem. Pharm. Bull., 1966, 804. 4401-012	MOLLUGOGENOL A $C_{30}H_{52}O_4$ 476 +58°p 252°C OH OH HO OH Tet., 1969, 3301. 4401-017
LEUCOTYLIN, 6-deoxy-16β-OAc- $C_{32}H_{54}O_3$ 486 +52° 228°C OH OAc Chem. Pharm. Bull., 1966, 804. 4401-013	MORETENONE $C_{30}H_{48}O$ 424 +61°c 204°C O Aust. J. Chem., 1968, 1675. 4401-018
LEUCOTYLIN $C_{30}H_{52}O_3$ 460 +49° 333°C OH OH OH Chem. Pharm. Bull., 1969, 279. 4401-014	HOPANONE, OH- $C_{30}H_{50}O_2$ 442 250°C OH O Chem. Comm., 1968, 1372. 4401-019
LEUCOTYLIN, 6-OAc- $C_{32}H_{54}O_4$ 502 +36° 225°C OH OH OAc Chem. Pharm. Bull., 1966, 804. 4401-015	HOPANE, 17(21)epoxy- $C_{30}H_{50}O$ 426 +47°c 270°C O Tet. Lett., 1966, 979. 4401-020

LEUCOTYLIC ACID $C_{30}H_{50}O_4$ 474 +330°c 260°C Tet. Lett., 1966, 607. 4401-021	DOLICHORRHIZIN $C_{32}H_{54}O_3$ 486 200°C Phytochem., 1970, 2037. 4401-024
LEUCOTYLIC ACID, 16-OAc- $C_{33}H_{54}O_5$ 530 m.e. +95°c m.e. 176°C Chem. Pharm. Bull., 1966, 804. 4401-022	PHLEBIC ACID A $C_{32}H_{52}O_4$ 500 Phytochem., 1969, 2345. 4401-025
PYXINIC ACID $C_{30}H_{50}O_4$ 474 +62°e 254°C Tet. Lett., 1966, 613. 4401-023	PHLEBIC ACID B $C_{30}H_{50}O_3$ 458 +53° 259°C Phytochem., 1970, 2037. 4401-026

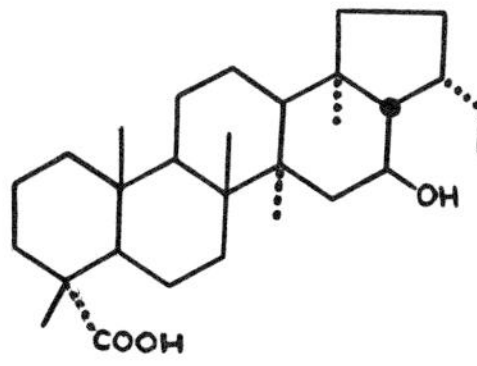

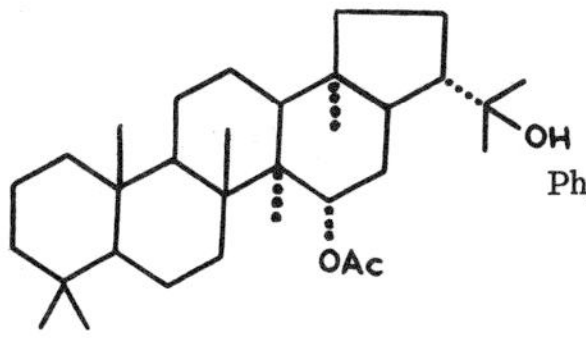

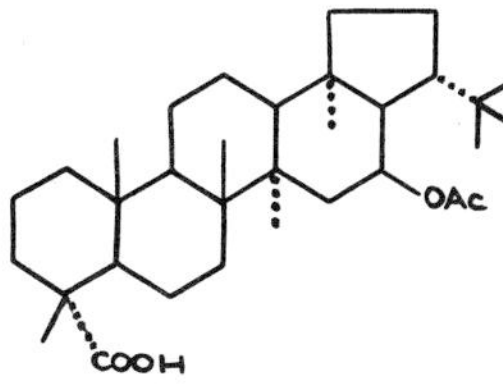

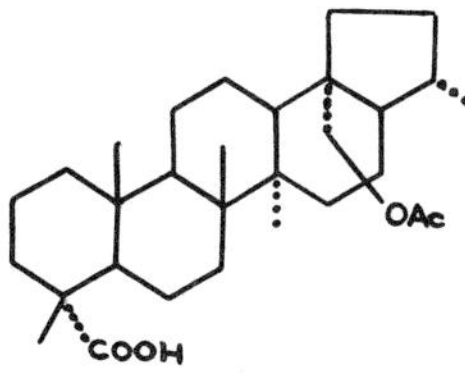

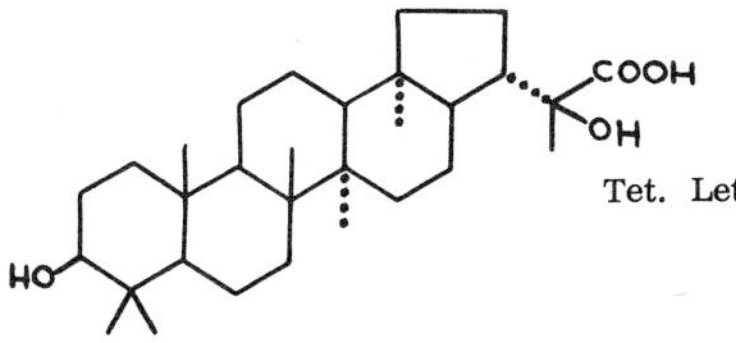

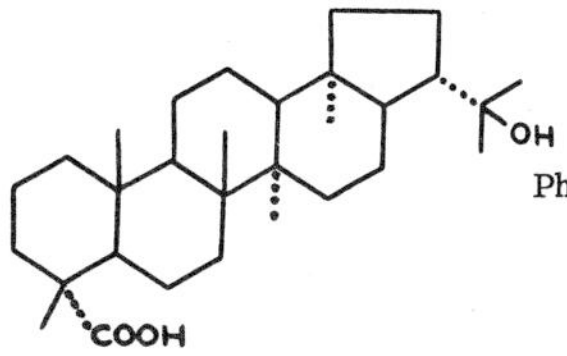

FERN-7-ENE $C_{30}H_{50}$

410
-27°c
212°C

Tet. Lett., 1966 6069.

4402-001

FERN-8-ENE $C_{30}H_{50}$

410
+18°c
192°C

Tet. Lett., 1966 6069.

4402-002

FERN-9-ENE $C_{30}H_{50}$

410

172°C

Tet. Lett., 1966, 6089.

4402-003

FERNA-7,9(11)-DIENE $C_{30}H_{48}$

408
-157°c
199°C

Chem. Comm., 1968, 1105.

4402-004

MOTIOL $C_{30}H_{50}O$

426
-44°
217°C

Tet. Lett., 1965, 2017.

4402-005

FERNENOL $C_{30}H_{50}O$

426
-24°c
194°C

Chem. Pharm. Bull., 1966 (14) 97.

4402-006

ARUNDOIN $C_{31}H_{52}O$

440
-5°c
242°C

Chem. Comm., 1969, 601.

4402-007

MOTIDIOL $C_{30}H_{50}O_2$

442
-22°
233°C

Tet. Lett., 1965, 2017.

4402-008

FERNENEDIOL $C_{30}H_{50}O_2$

442
-34°
204°

Tet. Lett., 1966, 3461.

4402-009

FERN-9(11)-EN-3-ONE $C_{30}H_{48}O$

424

Chem. Comm., 1969, 601.

4402-010

ARUNDOIN, 12-oxo- $C_{31}H_{50}O_2$ 454 -5^{0}c 291^{0}C MeO Chem. Comm., 1969, 601. 4402-011	DAVALLIC ACID $C_{30}H_{48}O_2$ 440 283^{0}C COOH Tet. Lett., 1963, 1451. 4402-012

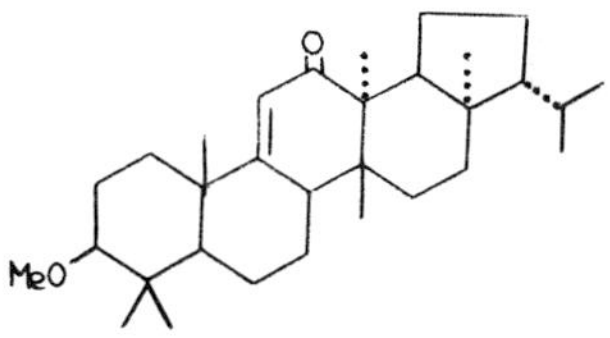

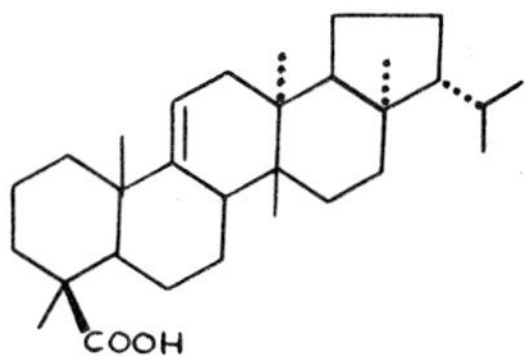

FRIEDELINOL $C_{30}H_{52}O$ 428 +18^{0}c 304^{0}C Aust. J. Chem., 1968, 1675. HO 4403-001	CERIN $C_{30}H_{50}O_2$ 442 -44^{0}c 249^{0}C JCS., 1968, 2342. HO O 4403-006
FRIEDELAN-3α-YL-ACETATE $C_{32}H_{54}O_2$ 470 -12^0 300^{0}C Aust. J. Chem., 1968, 1675. AcO 4403-002	ROXBURGHOLONE $C_{30}H_{50}O_2$ 442 -8^0 315^{0}C Tet., 1968, 6259. HO O 4403-007
FRIEDELINOL, epi- $C_{30}H_{52}O$ 428 +20^{0}c 280^{0}C Chem. Abstr., 1967 (67) 73702. HO 4403-003	PUTRANJIVADIONE $C_{30}H_{48}O_2$ 440 -37^0 287^{0}C Tet., 1968, 1205. O O 4403-008
FRIEDELIN $C_{30}H_{50}O$ 426 -21^{0}c 268^{0}C Acta Chem. Scand., 1954 (8) 71. O 4403-004	CANOPHYLLOL $C_{30}H_{50}O_2$ 442 -21^{0}p 281^{0}C CH_2OH Tet., 1967, 1901. O 4403-009
PACHYSANDIOL A $C_{30}H_{52}O_2$ 444 +14^0 279^{0}C JCS., 1968 2342. HO HO 4403-005	CANOPHYLLAL $C_{30}H_{48}O_2$ 440 -16^{0}p 264^{0}C CHO Tet., 1967, 1901. O 4403-010

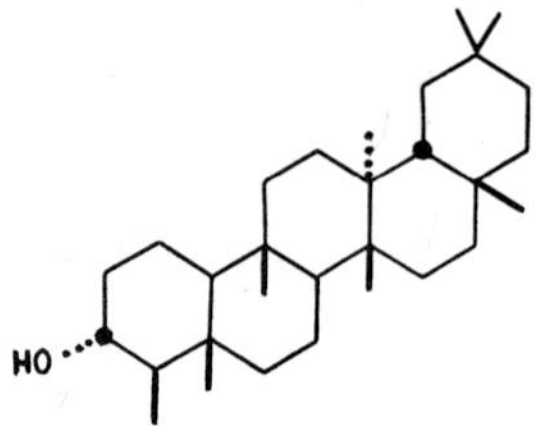

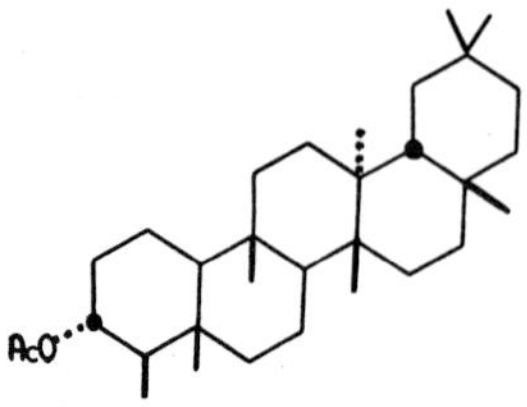

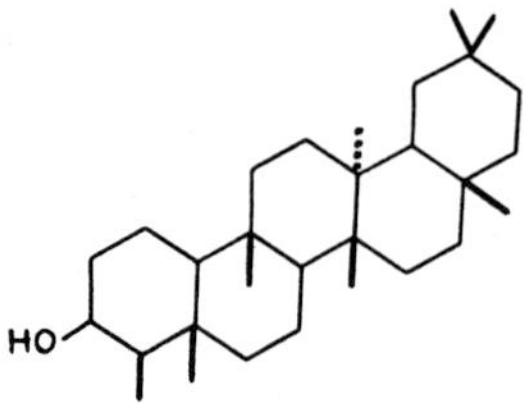

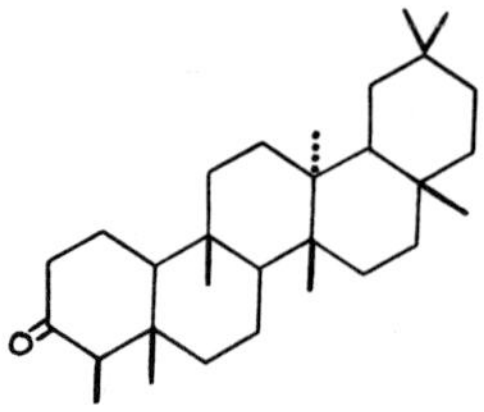

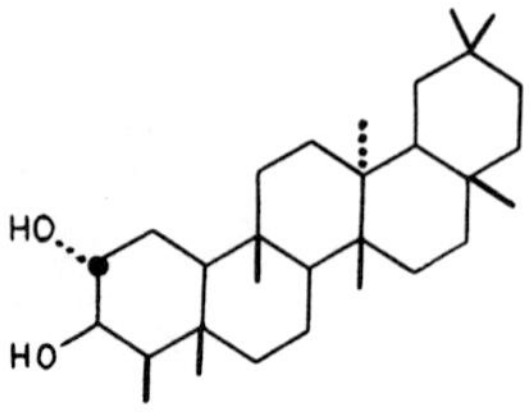

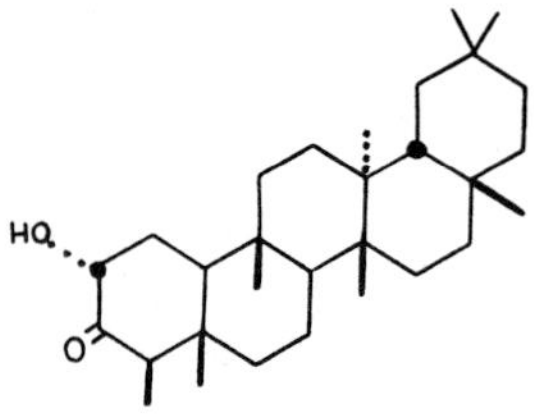

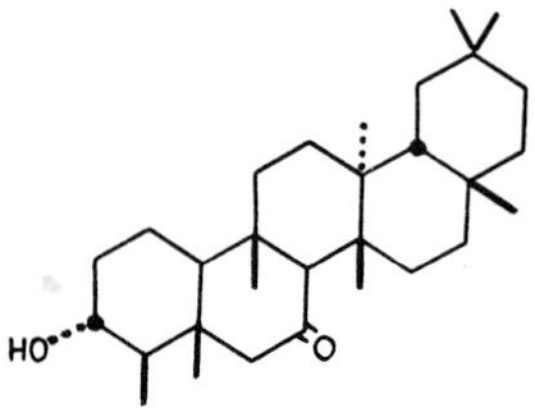

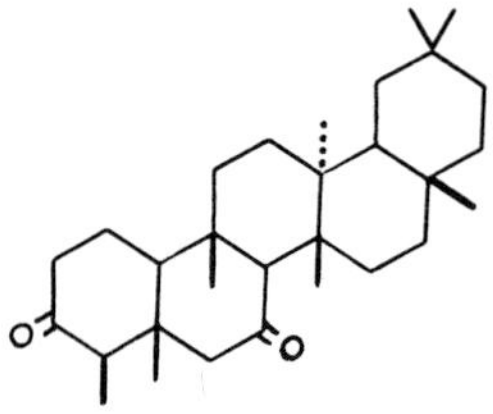

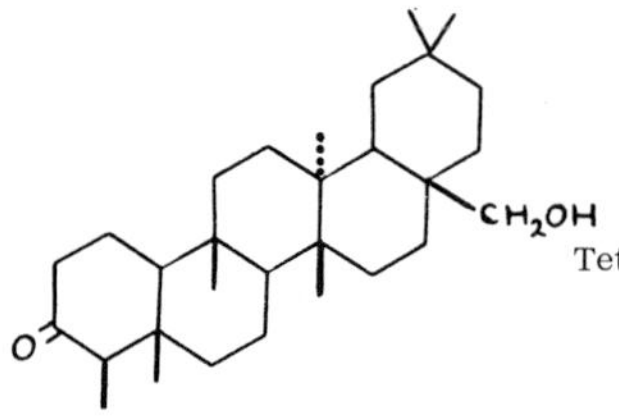

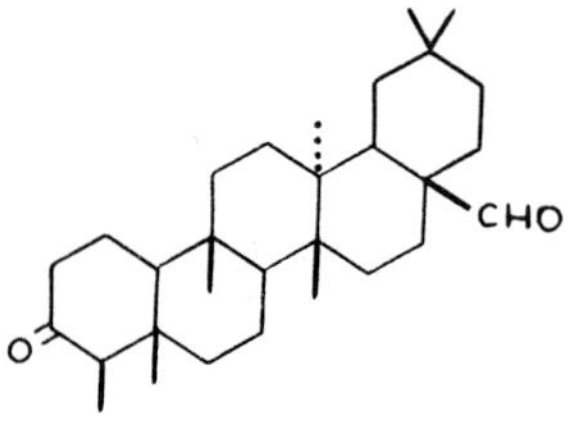

CANOPHYLLIC ACID $C_{30}H_{50}O_3$ 458 21°p 312 HO COOH Tet., 1967, 1901. 4403-011	FRIEDELANE-3, 21-DIONE, 25-OH- $C_{30}H_{48}O_3$ 456 +109° 310°C R = oxo Aust. J. C., 1970, 1651. 4403-016
FRIEDELAN-3-ONE, 21α-OH- $C_{30}H_{50}O_2$ 442 -32°c 268°C R = α - OH R O Aust. J. C., 1970, 1651. 4403-012	FRIEDELAN-25-AL $C_{30}H_{50}O$ 426 -34°c 290°C OHC Tet. Lett., 1963, 173. 4403-017
FRIEDELAN-3-ONE, 21β-OH- $C_{30}H_{50}O_2$ 442 -15°c 275°C R = β - OH Aust. J. C., 1970, 1651. 4403-013	FRIEDELANE-3, 25-DIONE $C_{30}H_{48}O_2$ 440 -62°c 309°C R = H R OHC O Tet. Lett., 1963, 173. 4403-018
FRIEDELANE-3, 21-DIONE $C_{30}H_{48}O_2$ 440 +115°c 250°C R = oxo Aust. J. C., 1970, 1651. 4403-014	FRIEDELANE-3, 25-DIONE, 21-OH- $C_{30}H_{48}O_3$ 456 -72°c 292°C R = α - OH Aust. J. C., 1970, 1651. 4403-019
FRIEDELAN-3-ONE, 25-OH- $C_{30}H_{50}O_2$ 442 -20°c 308°C R = H R HOCH2 O Tet. Lett., 1963, 173. 4403-015	FRIEDELANE-3, 21, 25-TRIONE $C_{30}H_{46}O_3$ 454 +72°c 303°C R = oxo Aust. J. C., 1970, 1651. 4403-020

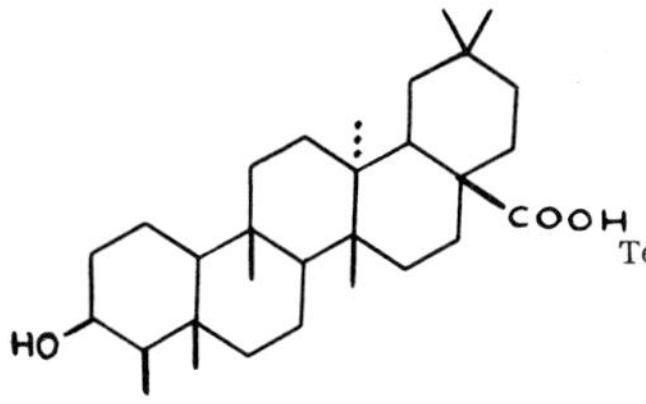

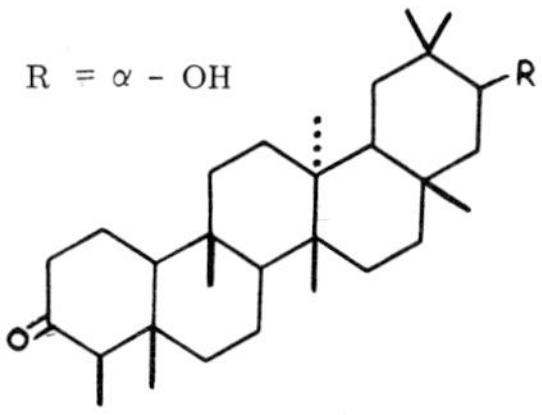

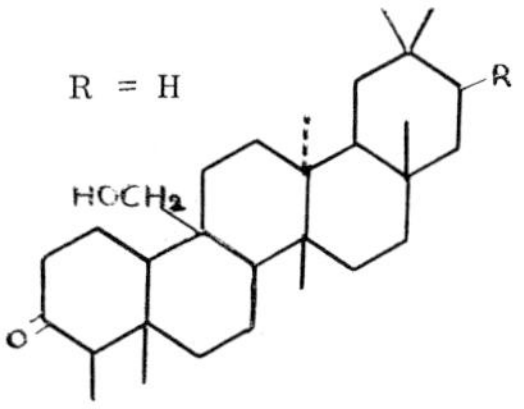

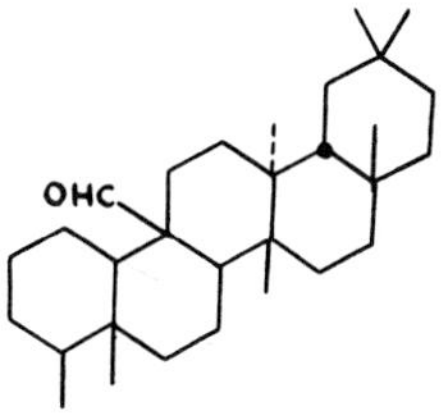

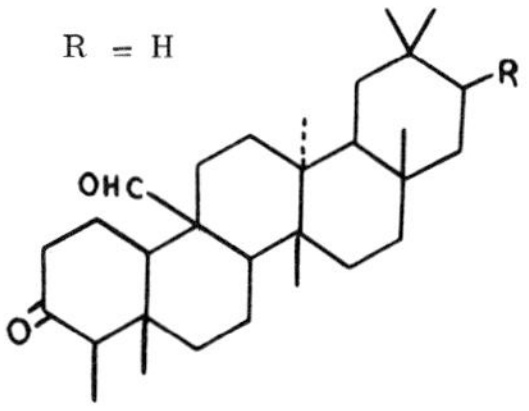

DAMMADIENOL $C_{30}H_{50}O$ 426 +47°c 137°C JCS 1956 2196. 4404-001	PANAXADIOL, proto- $C_{30}H_{52}O_3$ 460 +21°c 237°C OH OH HO Chem. Pharm. Bull., 1966. 1150. 4404-006
DAMMARENEDIOL I $C_{30}H_{52}O_2$ 444 +27°c 143°C OH HO Tet. Lett., 1968, 4235. 4404-002	PANAXADIOL, 20-epi-proto- $C_{30}H_{52}O_3$ 460 200°C *epimer Tet. Lett., 1967, 3579. 4404-007
DAMMARENEDIOL II $C_{30}H_{52}O_2$ 444 +33°c 132°C * epimer JCS., 1956, 2196. 4404-003	deleted 4404-008
AGLAIOL $C_{30}H_{50}O_2$ 442 +53° 114°C HO O Tet., 1965, 917. 4404-004	BETULAFOLIENETRIOL $C_{30}H_{52}O_3$ 460 +12°c 197°C OH HO Ann., 1961 (644) 146. 4404-009
CARNAUBADIOL $C_{31}H_{54}O_2$ 458 +34°c 168°C OH HO Aust. J. C., 1965, 1411. 4404-005	BETULAFOLIENETETRAOL $C_{30}H_{52}O_4$ 476 +12°c 169°C OH OH OH HO Ann., 1961 (644) 146. 4404-010

PANAXATRIOL, proto- $C_{30}H_{52}O_4$ 476

R = H

OH OR HO OR

Tet. Lett., 1965, 207.

4404-011

DAMMAR-24-EN-12, 20-DIOL-3-ONE $C_{30}H_{50}O_3$ 458 197°C

OH OH O

Tet. Lett., 1968, 4239.

4404-016

GINSENOSIDE $C_{42}H_{72}O_{14}$ 800 +32°p 195°C

R = glucosyl

Tet. Lett., 1968, 5449.

4404-012

DRYOBALANONE $C_{30}H_{50}O_3$ 458 +6°c 81°C

CH₂OH OH O

Tet. Lett., 1967, 2807.

4404-017

DAMMADIENONE $C_{30}H_{48}O$ 424 +90°c 80°C

O

Tet., 1958 (3) 279.

4404-013

BACOGENIN $C_{30}H_{50}O_4$ 474 -70° 242°C

R = H

OH HOCH₂ O RO

Ind. J. C., 1965, 24.

4404-018

DIPTEROCARPOL $C_{30}H_{50}O_2$ 442 +66° 136°C

OH O

Tet. Lett., 1966, 4797.

4404-014

BACOSIDE A $C_{41}H_{68}O_{13}$ 768 -42°e 250°C

R = arabinosido-glucosyl

Ind. J. C., 1965, 24.

4404-019

ALNUSFOLIENEDIOLONE $C_{30}H_{50}O_3$ 458 +52° 199°C

OH OH O

Ann., 1961 (644) 162.

4404-015

DRYOBALANONOLIC ACID $C_{30}H_{48}O_4$ 472 m.e.+76°c m.e.148°C

COOH OH O

Tet. Lett., 1969, 5077.

4404-020

PANAXADIOL $C_{30}H_{52}O_3$ 460 +4⁰ 250⁰C R = H (Artifact ?) Tet. Lett., 1963, 795. 4404-021	KAPUROL $C_{30}H_{52}O_3$ 460 207⁰C (Cf. Ocotillol ?) Chem. Abstr., 1968 (69) 10582. 4404-026
PANAXATRIOL $C_{30}H_{52}O_4$ 476 +14⁰ 238⁰C R = OH (Artifact ?) Tet. Lett., 1966, 141. 4404-022	KAPURONE $C_{30}H_{50}O_3$ 458 184⁰C (Cf. Ocotillone ?) Chem. Abstr., 1968 (69) 10582. 4404-027
OCOTILLOL $C_{30}H_{52}O_3$ 460 +28⁰ 200⁰C Can. J. C., 1965, 3311. 4404-023	PYXINOL, diacetyl- $C_{34}H_{56}O_6$ 560 +10⁰c 247⁰C Tet. Lett., 1969, 4245. 4404-028
OCOTILLONE, 20(R)- $C_{30}H_{50}O_3$ 458 Chem. Abstr., 1968 (69) 10582. 4404-024	PYXINOL, 3-epi- $C_{30}H_{52}O_4$ 476 +3⁰c 240⁰C Tet. Lett., 1968, 4239. 4404-029
OCOTILLONE, 20(S)- $C_{30}H_{50}O_3$ 458 Chem. Abstr., 1968 (69) 10582. 4404-025	

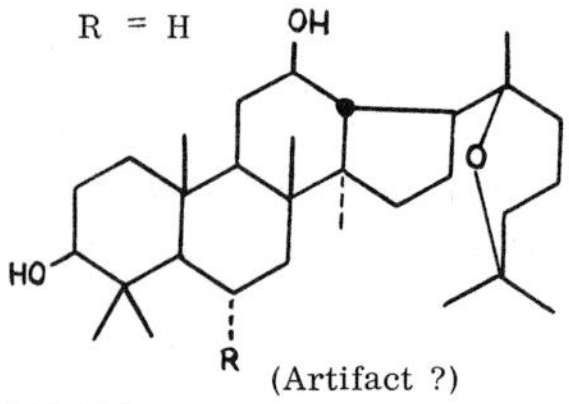

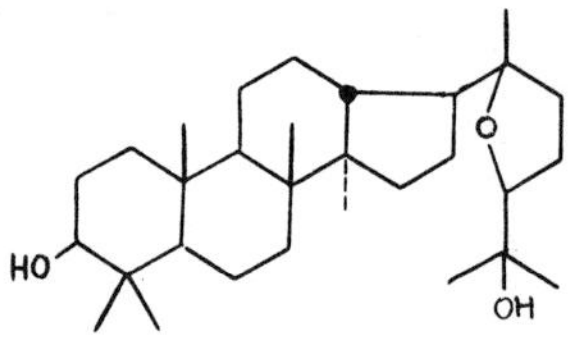

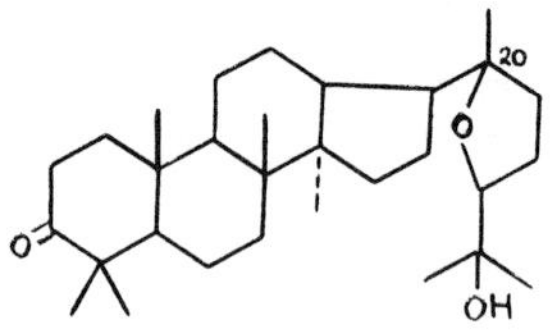

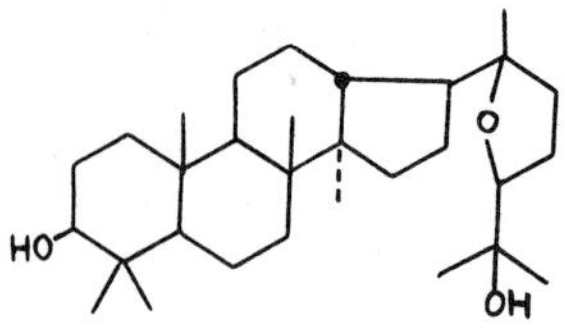

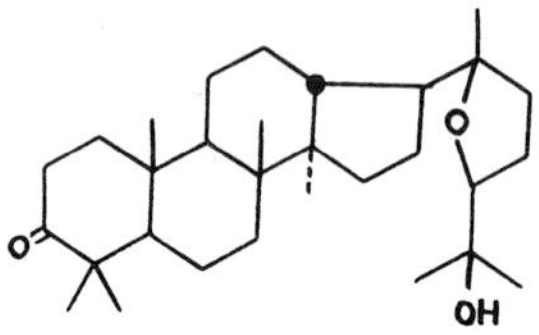

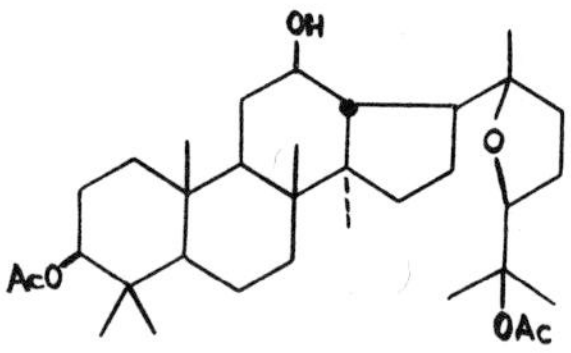

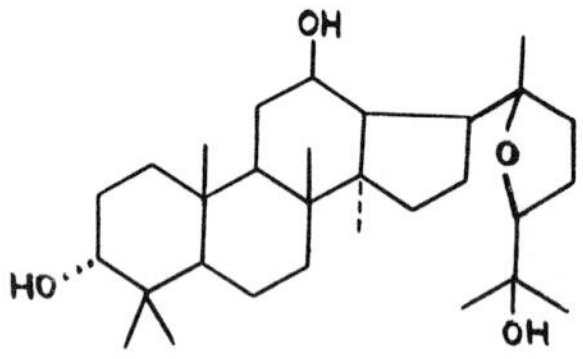

EUPHOL	$C_{30}H_{50}O$	426 +32°c 116°C
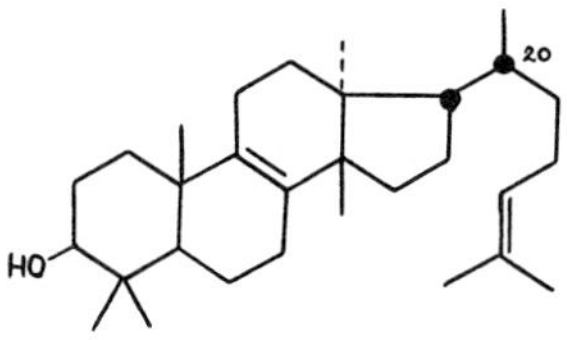	JCS., 1958, 179.	
4405-001		

EUPHORBOL HEXACOSANOATE	$C_{57}H_{102}O_2$	818 111°C
R = $\cdot CO \cdot (CH_2)_{24} \cdot CH_3$	Khim. Prir. Soedin, 1970 (6) 136.	
4405-006		

TIRUCALLOL	$C_{30}H_{50}O$	426 +5°c 134°C
20-epimer	Helv., 1955, 222.	
4405-002		

SCHINOL	$C_{30}H_{48}O_3$	456 -40°c 145°C
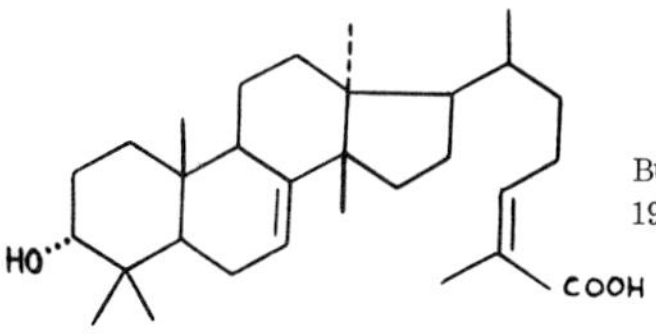	Bull. Soc. Chim. Fr., 1963, 911.	
4405-007		

NERIFOLIOL	$C_{30}H_{50}O$	426 +20° 130°C
	Chem. Abstr., 1967 (66) 11062.	
4405-003		

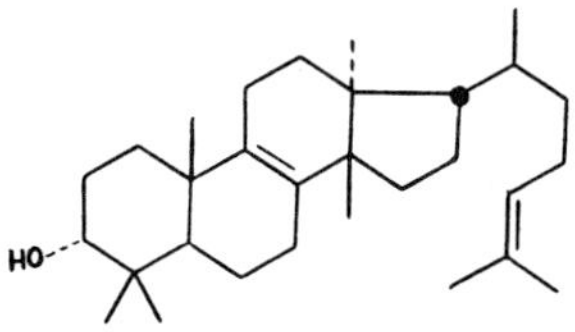

ELEMOLIC ACID, α-	$C_{30}H_{48}O_3$	456 -21°c 222°C
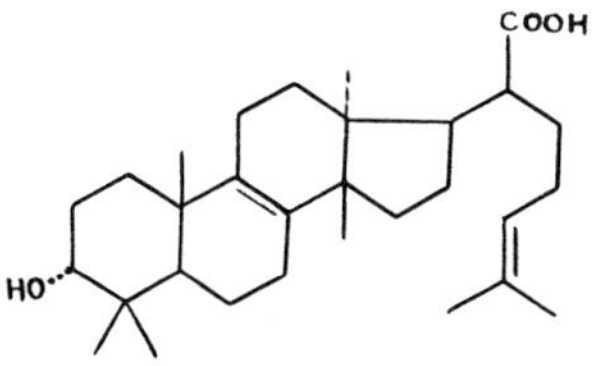	JCS., 1963, 2762.	
4405-008		

BUTYROSPERMOL	$C_{30}H_{50}O$	426 -12°c 112°C
	JCS., 1956, 3272.	
4405-004		

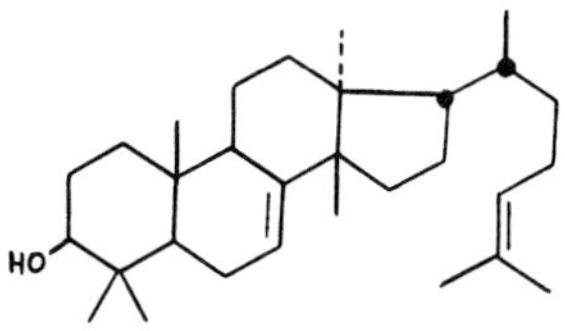

ELEMOLIC ACID, α-iso-	$C_{30}H_{48}O_3$	456 m.e.-34° m.e. 148°C
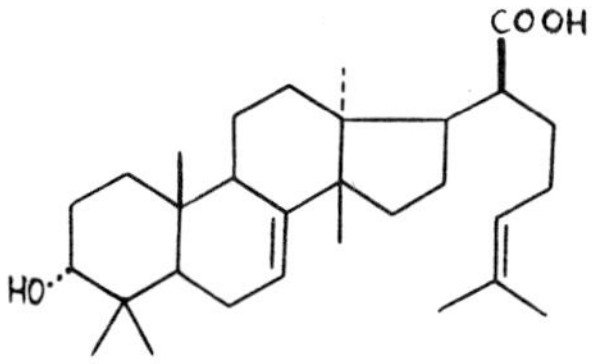	JCS., 1970, 740.	
4405-009		

EUPHORBOL	$C_{31}H_{52}O$	440 0°c 126°C
R = H	Helv., 1955, 1517.	
4405-005		

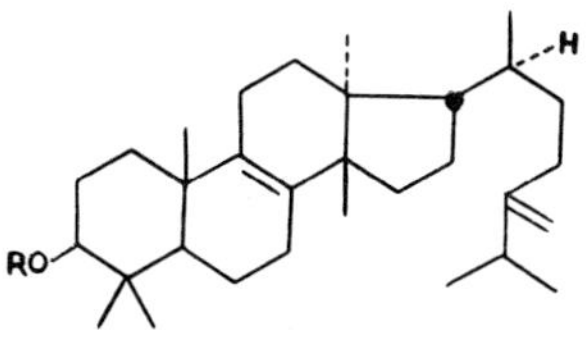

ELEMOLIC ACID, β-	$C_{30}H_{48}O_3$	456 m.e.-2°c m.e. 118°C
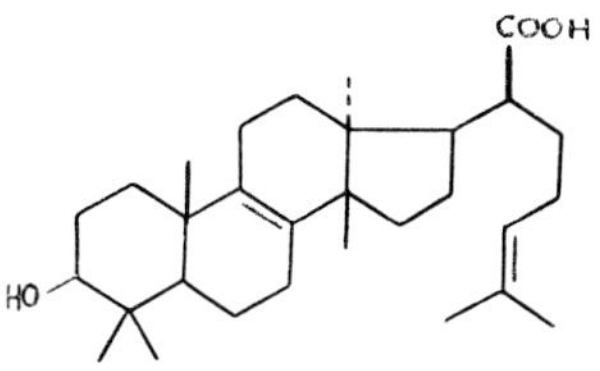	JCS., 1970, 740.	
4405-010		

ELEMONIC ACID, β- — $C_{30}H_{46}O_3$ — 454 — $+37^{0}$ — $211^{0}C$

JCS., 1963, 2762.

4405-011

BOURJOTINOLONE B — $C_{30}H_{48}O_3$ — 456 — -89^{0} — $202^{0}C$

Aust. J. C., 1966, 455.

4405-012

$C_{30}H_{52}O_5$ — 492

JCS., 1969, 2435.

4405-013

MASTICADIENONIC ACID — $C_{30}H_{46}O_3$ — 454 — $-76^{0}c$ — $178^{0}C$

JCS., 1956, 4150.

4405-014

MASTICADIENONIC ACID, iso- — $C_{30}H_{46}O_3$ — 454 — $+34^{0}c$ — $166^{0}C$

JCS., 1956 4158.

4405-015

KULINONE — $C_{30}H_{48}O_2$ — 440 — $-20^{0}c$ — $138^{0}C$

Chem. Comm., 1968, 1156.

4405-016

BOURJOTINOLONE A — $C_{30}H_{48}O_4$ — 472 — -34^{0} — $176^{0}C$

Aust. J. C., 1966, 455.

4405-017

SAPELIN A — $C_{30}H_{50}O_4$ — 474 — $-5^{0}c$ — $216^{0}C$

JCS., 1970, 311.

4405-018

SAPELIN B — $C_{30}H_{50}O_4$ — 474 — $-18^{0}c$ — $175^{0}C$

JCS., 1970, 311.

4405-019

FLINDISSOL — $C_{30}H_{48}O_3$ — 456 — $-46^{0}c$ — $198^{0}C$

JCS., 1963, 2762.

4405-020

MELIANOL — $C_{30}H_{48}O_4$ — 472, -38^0, 194^0C

R = H

JCS., 1967, 1347.

4405-021

TURRAEANTHIN — $C_{32}H_{50}O_5$ — 514, $+3^0$, 220^0C

R = Ac

JCS., 1967, 820.

4405-022

APHANAMIXIN — $C_{32}H_{50}O_5$ — 514, -45^0, 233^0C

C_{21} - epimer

R = Ac

Tet. Lett., 1967, 1741.

4405-023

MELIANONE — $C_{30}H_{48}O_4$ — 470, -62^0, 224^0C

JCS., 1967, 1347.

4405-024

MELIANTRIOL — $C_{30}H_{50}O_5$ — 490, -23^0, 177^0C

Chem. Comm., 1967, 910.

4405-025

MELIANODIOL — $C_{30}H_{48}O_5$ — 488, -60^0c, 231^0C

Chem. Comm., 1967, 910.

4405-026

KULOLACTONE — $C_{30}H_{46}O_3$ — 454, -42^0, amorph.

R = α - OH

Tet. Lett., 1969, 891.

4405-027

KULACTONE — $C_{30}H_{44}O_3$ — 452, -58^0, 164^0C

R = O

Tet. Lett., 1969, 891.

4405-028

ODORATOL — $C_{30}H_{50}O_4$ — 474, -45^0, 237^0C

JCS., 1968, 2485.

4405-029

ODORATOL, iso- — $C_{30}H_{50}O_4$ — 474, -48^0, 247^0C

24-epimer

JCS., 1968, 2485.

4405-030

ODORATONE $C_{30}H_{48}O_4$ 472 -90^0 230^0C JCS., 1968, 2485. 4405-031	

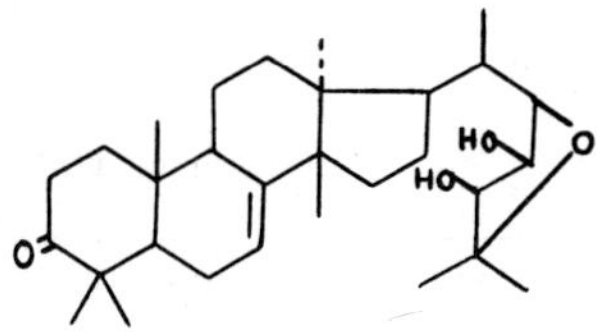

LUPEOL $C_{30}H_{50}O$ 426 +27⁰ 215⁰C HO Acta Chem. Scand., 1966 (20) 1720. 4406-001	BETULIN $C_{30}H_{50}O_2$ 442 19⁰p 255⁰C CH_2OH HO Chem. Abstr., 1968 (68) 29900. 4406-006
LUPEOL METHYL ETHER $C_{31}H_{52}O$ 440 +36⁰c 250⁰C MeO Chem. Comm., 1969, 601. 4406-002	MONOGYNOL A $C_{30}H_{52}O_2$ 444 +24⁰ 233⁰C HO HO Acta Chem. Scand., 1966 (20) 1720. 4406-007
JASMINOL $C_{30}H_{50}O$ 426 +42⁰c 209⁰C CH_2OH Experimentia 1970 (26) 10. 4406-003	LUPAN-3β, 20, 28-TRIOL $C_{30}H_{52}O_3$ 460 HO CH_2OH HO Acta Chem. Scand., 1966 (20) 1720. 4406-008
GLOCHIDIOL $C_{30}H_{50}O_2$ 442 OH HO Tet., 1966, 1513. 4406-004	GLOCHIDONE $C_{30}H_{46}O$ 422 +73⁰ 164⁰C O Tet., 1966, 1513. 4406-009
LUP-20-ENE-3, 16-DIOL $C_{30}H_{50}O_2$ 442 218⁰C OH HO Aust. J. Chem., 1968 (21) 2529. 4406-005	GLOCHIDONOL $C_{30}H_{48}O_2$ 440 228⁰C OH O JCS., 1969, 1710. 4406-010

CLERODOLONE — $C_{30}H_{48}O_3$ — 456 — $+8^0$c — 283^0C

Tet., 1966, 2377.

4406-011

CLERODONE — $C_{30}H_{50}O$ — 426

Tet., 1966, 2377.

4406-012

BETULINIC ACID — $C_{30}H_{48}O_3$ — 456 — $+12^0$ — 320^0C

Acta Chem. Scand., 1966 (20) 1720.

4406-013

ALPHITOLIC ACID — $C_{30}H_{48}O_4$ — 472

JCS., 1968, 1047.

4406-014

MELALEUCIC ACID — $C_{30}H_{46}O_5$ — 486 — $+19^0$p — 364^0C

Tet., 1965, 1529.

4406-015

STELLATOGENIN — $C_{30}H_{46}O_3$ — 472 — $+36^0$c — 318^0C

JOC., 1967, 3150.

4406-016

THURBEROGENIN — $C_{30}H_{48}O_4$ — 454 — $+11^0$c — 284^0C

JOC., 1967, 3150.

4406-017

THURBERIN — $C_{30}H_{50}O_2$ — 442 — $+12^0$m — 207^0C

JOC., 1969, 1367.

4406-018

LUPEOL, epi- — $C_{30}H_{50}O$ — 426 — $+15^0$c — 202^0C

Tet., 1966, 1513.

4406-019

LUPENONE — $C_{30}H_{48}O$ — 424 — $+67^0$c — 170^0C

J. Ind. C.S., 1966, 63.

4406-020

BETULONIC ACID $C_{30}H_{46}O_3$ 454 253°C COOH O JCS., 1963. 3269. 4406-021	LUP-20(29)-ENE-1, 3-DIOL $C_{30}H_{50}O_2$ 442 247°C OH HO Phytochem., 1970, 1099. 4406-023
SORBICORTOL II $C_{30}H_{50}O_3$ 458 +25°c 260°C CH_2OH HO CH_2OH JCS., 1960, 4303. 4406-022	

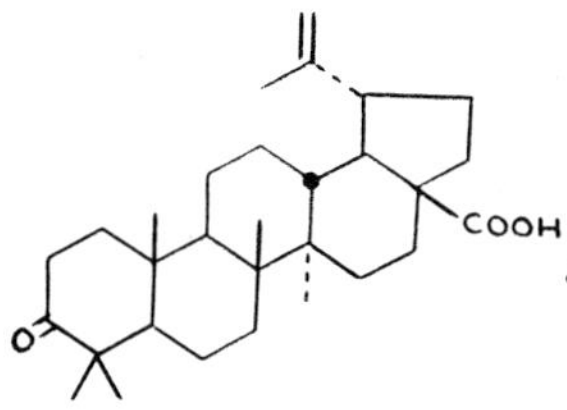

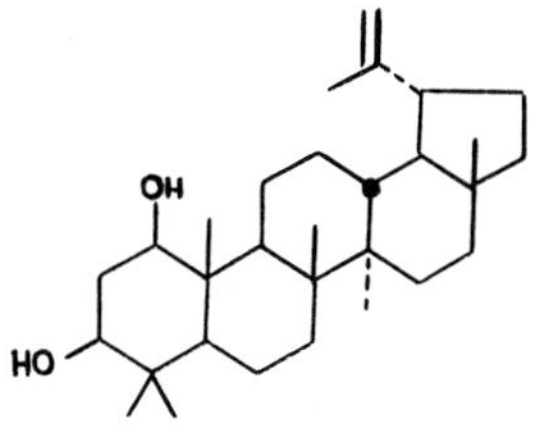

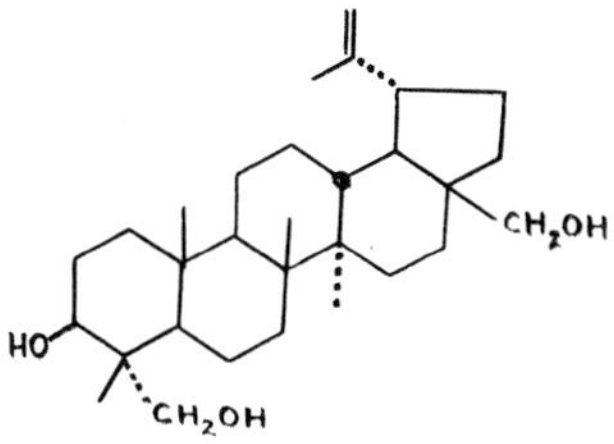

GERMANICOL — $C_{30}H_{50}O$ — 426, +6°c, 178°C

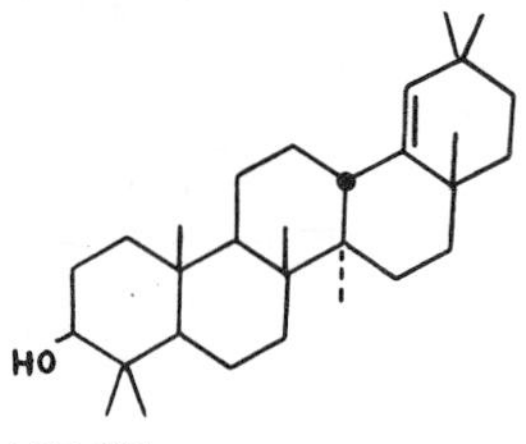

JCS., 1944, 283.

4407-001

MILIACIN — $C_{31}H_{52}O$ — 440, +27°b, 285°C

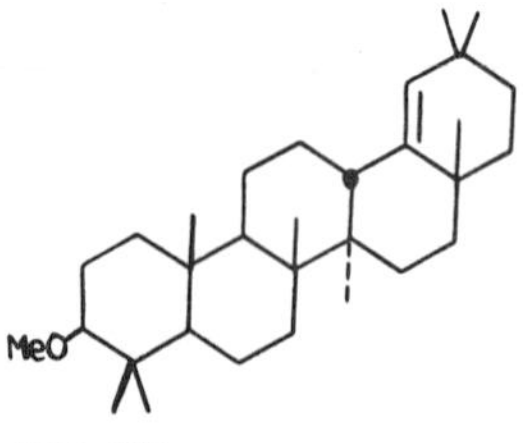

Chem. Abstr., 1968 (69) 59446.

4407-002

GERMANIDIOL — $C_{30}H_5O_2$ — 442, +37°, 275°C

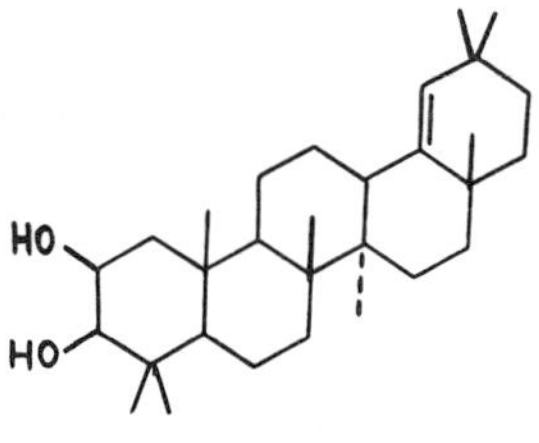

Tet. Lett., 1965, 2017.

4407-003

GERMANIDIOL, epi- — $C_{30}H_{50}O_2$ — 442, -1°, 221°C

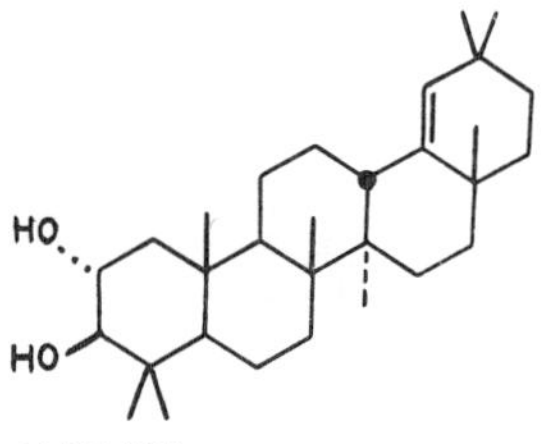

Tet. Lett., 1966, 3461.

4407-004

MOROLIC ACID — $C_{30}H_{48}O_3$ — 456, +31°c, 273°C

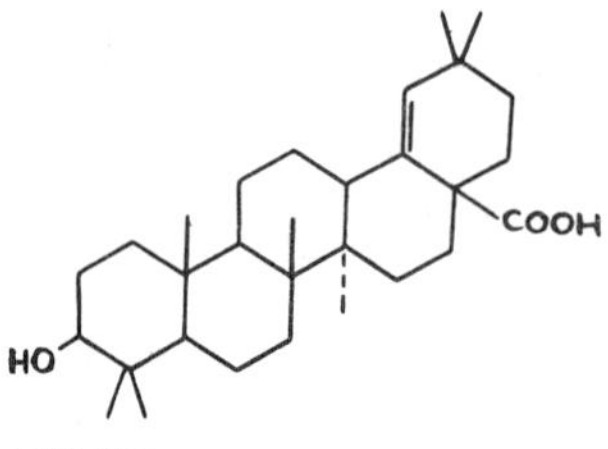

JCS., 1951, 278.

4407-005

MOROLIC ACID, acetyl- — $C_{32}H_{50}O_4$ — 498, +48°, 315°C

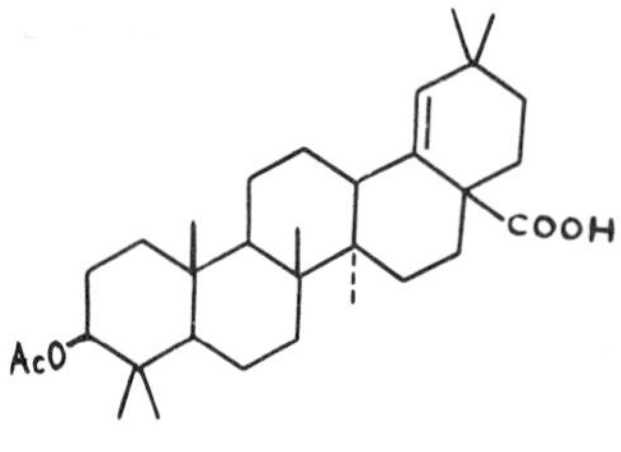

Bull. Soc. Chim. Biol., 1951, 1672.

4407-006

AMYRIN, β- — $C_{30}H_{50}O$ — 426, +88°c, 199°C

Chem. Abstr., 1951 (45) 3912.

4408-001

SAWAMILLETIN, iso- — $C_{31}H_{52}O$ — 440, 248°C

Phytochemistry, 1967, 727.

4408-002

MANILADIOL — $C_{30}H_{50}O_2$ — 442, +68°, 221°C

JACS., 1956, 2312.

4408-003

ERYTHRODIOL — $C_{30}H_{50}O_2$ — 442, +80°c, 235°C

R= H

Tet. Lett., 1968, 2643.

4408-004

ERYTHRODIOL-3-0-PALMITATE — $C_{46}H_{80}O_3$ — 680

R = Palmitoyl

Phytochemistry, 1968, 637.

4408-005

AEGICERADIOL — $C_{30}H_{48}O_2$ — 440, +74°c, 186°C

Tet., 1962, 461.

4408-006

SOYASAPOGENOL C — $C_{32}H_{48}O_2$ — 440, +71°c, 240°C

Tet., 1958 (4) 111.

4408-007

PRIMULA-GENIN A — $C_{30}H_{50}O_3$ — 458, +52°c, 254°C

Helv., 1949, 1911.

4408-008

LONGISPINOGENIN — $C_{30}H_{50}O_3$ — 458, +53°, 248°C

JACS., 1953, 5940.

4408-009

ERYTHRODIOL, 22α-OH- — $C_{30}H_{50}O_3$ — 458, 280°C

Chem. Ind., 1968, 745.

4408-010

SOYASAPOGENOL B — $C_{30}H_{50}O_3$ — 458 — +90°c — 260°C

Tet., 1958 (4) 111.

HO HOCH₂ OH

4408-011

BARRINGTOGENOL D — $C_{30}H_{48}O_4$ — 472 — +57°d — 308°C

Tet. Lett., 1967, 1675.

4408-016

SAIKOGENIN E — $C_{30}H_{48}O_3$ — 456 — +112° — 289°C

Tet. Lett., 1966, 4725.

4408-012

SOYASAPOGENOL A — $C_{30}H_{50}O_4$ — 474 — +103°c — 312°C

Tet., 1958 (4) 111.

4408-017

AEGICERIN — $C_{30}H_{48}O_3$ — 456 — -23° — 255°C

Tet., 1964, 973.

4408-013

BARRINGTOGENOL — $C_{30}H_{50}O_4$ — 474 — +18°p — 290°C

JCS., 1956, 4369.

4408-018

GLYCYRRHETOL — $C_{30}H_{48}O_3$ — 456 — +87° — 253°C

Chem. Abstr., 1968 (68) 49810.

4408-014

JEGOSAPOGENOL B — $C_{30}H_{48}O_4$ — 472 — 304°C

Chem. Abstr., 1968 (69) 27555.

4408-019

BARRINGTOGENOL C, 16-desoxy- — $C_{30}H_{50}O_4$ — 574 — +50°e — 289°C

Tet. Lett., 1967, 2577.

4408-015

CAMELLIAGENIN A — $C_{30}H_{50}O_4$ — 474 — +33°e — 282°C

R = H

Chem. Pharm. Bull., 1969, 474.

4408-020

CAMELLIAGENIN - A-16-CINNAMATE — $C_{39}H_{56}O_5$ — 604 — amorph.

R = Cinnamoyl

Chem. Pharm. Bull., 1968, 1846.

4408-021

SAIKOGENIN A — $C_{30}H_{48}O_4$ — 472 — -43^0e — 290^0C

Tet., 1967, 3333.

4408-022

PRIVEROGENIN-B-ACETATE — $C_{32}H_{52}O_5$ — 516 — -23^0 — 252^0C

Ann., 1966 (696) 160.

4408-023

SAIKOGENIN F — $C_{30}H_{48}O_4$ — 472 — $+96^0$c — 266^0C

Tet. Lett., 1966, 5045.

4408-024

SAIKOGENIN G — $C_{30}H_{48}O_4$ — 472 — $+45^0$ — 245^0C

Chem. Pharm. Bull., 1968, 643.

4408-025

BARRIGENOL, A_1- — $C_{30}H_{50}O_5$ — 490 — $+4^0$ — 301^0C

Tet. Lett., 1967, 2289.

4408-026

BARRINGTOGENOL C — $C_{30}H_{50}O_5$ — 490 — $+30^0$p — 327^0C

R= H

Tet. Lett., 1967, 1675.

4408-027

JEGOSAPOGENIN — $C_{35}H_{56}O_6$ — 572 — $+9^0$p — 264^0C

R = Tigloyl

Tet. Lett., 1967, 2353.

4408-028

CAMELLIAGENIN C — $C_{30}H_{50}O_5$ — 490 — $+37^0$e — 262^0C

Chem. Pharm. Bull., 1969, 474.

4408-029

GYMNESTROGENIN — $C_{30}H_{50}O_5$ — 490

Helv., 1968, 1235.

4408-030

BARRINGTOGENOL B — $C_{35}H_{56}O_6$ — 572, $+3^0$, 246^0C

$R_1 = R_2 = H$; R_3 = Tigloyl

Tet., 1968, 1113.

4408-031

BARRIGENOL, R1- — $C_{30}H_{50}O_6$ — 506

Tet. Lett., 1967, 2289.

4408-036

BARRINGTOGENOL E — $C_{44}H_{58}O_7$ — 698, -20^0c, 220^0C

$R_1 = R_2$ = Benzoyl

R_3 = H

Chem. Abstr., 1968 (68) 49885.

4408-032

TANGINOL — $C_{30}H_{50}O_6$ — 506, $+9^0$, 284^0C

Tet., 1967, 3837.

4408-037

AESCIGENIN, iso- — $C_{30}H_{48}O_5$ — 488, $+23^0e$, 331^0C

Tet. Lett., 1967 (17) 1675.

4408-033

THEASAPOGENOL A — $C_{30}H_{50}O_6$ — 506, $+14^0p$, 302^0C

Tet. Lett., 1967 (7) 637.

4408-038

AESCIGENIN — $C_{30}H_{48}O_5$ — 488

Tet. Lett., 1967 (17) 1675.

4408-034

GYMNEMAGENIN — $C_{30}H_{50}O_6$ — 506

Chem. Comm., 1968, 1681.

4408-039

BARRIGENOL, 7β-OH-A1- — $C_{30}H_{50}O_6$ — 506, $+37^0$, 309^0C

Tet. Lett., 1961, 100.

4408-035

AESCIGENIN, proto- — $C_{30}H_{50}O_6$ — 506

$R_1 = R_2 = H$

Tet. Lett., 1967 (17) 1675.

4408-040

AESCIGENIN-21-tiglate-22-acetate, proto- $C_{37}H_{58}O_8$ 630

R_1 = Tigloyl

R_2 = Ac

Chem. Abstr., 1968 (69) 59442.

4408-041

deleted

4408-046

OLEANOLIC ALDEHYDE $C_{30}H_{48}O_2$ 440 $+72^0$ 170^0C

R = H

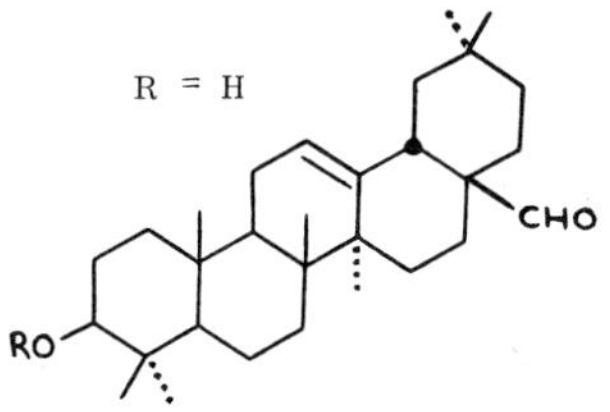

Chem. Abstr., 1967 (67) 116988.

4408-042

CAMELLIAGENIN B $C_{30}H_{48}O_5$ 488 $+48^0$ 203^0C

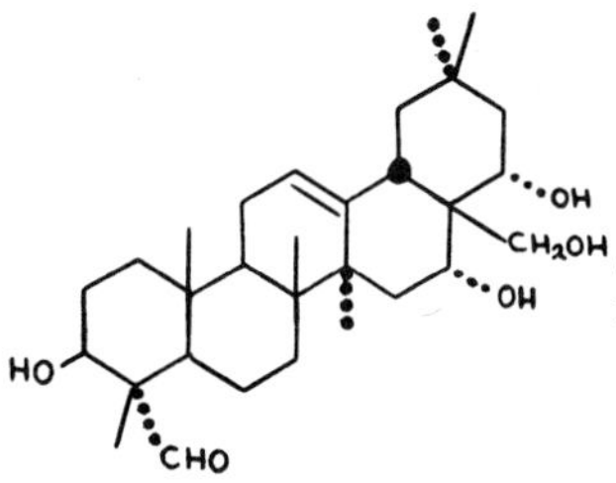

Chem. Pharm. Bull., 1969, 474.

4408-047

OLEANOLIC ALDEHYDE, O-Ac- $C_{32}H_{50}O_3$ 482 226^0C

R = Ac

Tet., 1965, 2699.

4408-043

THEASAPOGENOL E $C_{30}H_{48}O_6$ 504 $+34^0p$ 238^0C

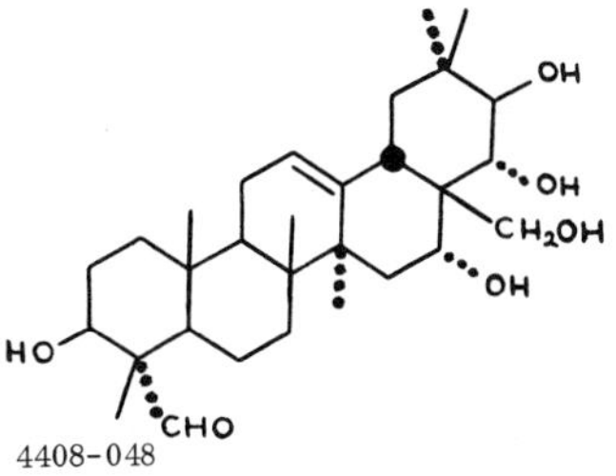

Tet. Lett., 1967, 1127.

4408-048

GUMMOSOGENIN $C_{30}H_{48}O_3$ 456 $+28^0c$ 252^0C

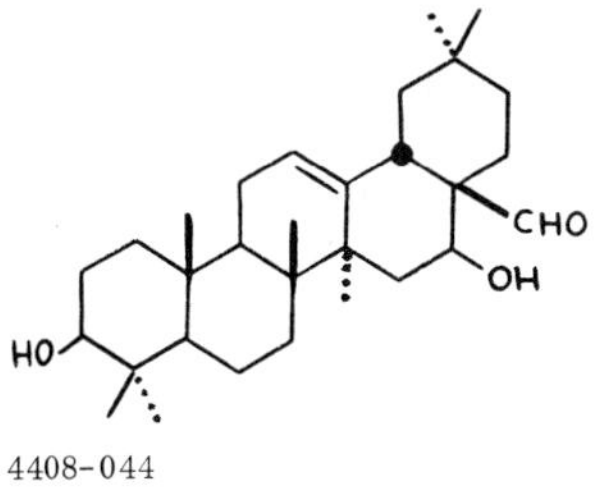

JACS., 1954, 4089.

4408-044

CAMELLIAGENIN D $C_{30}H_{48}O_6$ 504 $+39^0$ 254^0C

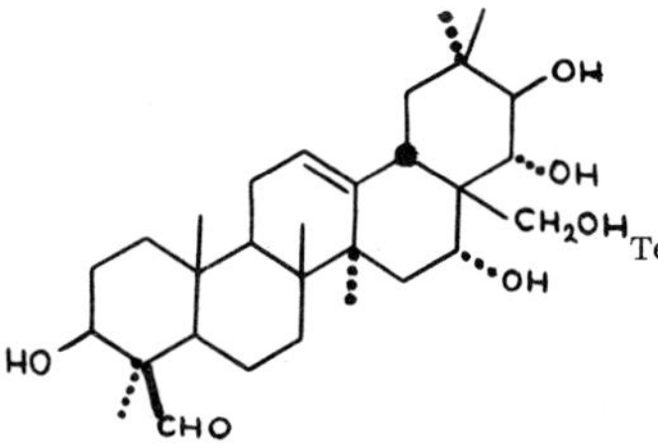

Tet. Lett., 1967 (12) 1127.

4408-049

deleted

4408-045

CYCLAMIRETIN A $C_{30}H_{48}O_4$ 472

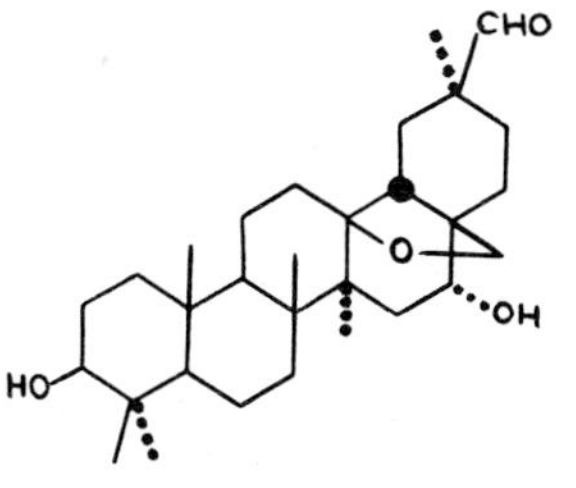

Ann., 1966, 165.

4408-050

CYCLAMIRETIN D — $C_{30}H_{48}O_4$ — 472

Ann., 1966, 165.

4408-051

CYCLAMIGENIN B — $C_{30}H_{46}O_4$ — 470, -10^0, 295^0C

Tet., 1968, 1377.

4408-052

CYCLAMIGENIN A1 — $C_{32}H_{52}O_4$ — 500, $+42^0c$, 216^0C

Tet., 1968, 5649.

4408-053

CYCLAMIGENIN C — $C_{32}H_{52}O_4$ — 500, -114^0c, 218^0C

C - 30 - Epimer

Tet., 1968, 5649.

4408-054

CYCLAMIGENIN D — $C_{32}H_{52}O_5$ — 516, -5^0c, 326^0C

Tet., 1968, 5649.

4408-055

BOSWELLINIC ACID, α- — $C_{30}H_{48}O_3$ — 456, $+115^0c$, 289^0C

Helv., 1951, 2321.

4408-056

COMMIC ACID C — $C_{30}H_{48}O_4$ — 472, 344^0C

Tet., 1961 (15) 212.

4408-057

JACQUINIC ACID — $C_{30}H_{48}O_5$ — 488, $+49^0$, 311^0C

Tet., 1965, 1735.

4408-058

LIQUORIC ACID — $C_{31}H_{46}O_5$ — 498, m. c. $+155^0c$, m. e. 262^0C

Tet., 1965, 2109.

4408-059

MERISTOTROPIC ACID — $C_{30}H_{44}O_4$ — 468, m. e. 278^0C

Chem. Abstr., 1968 (69) 87231.

4408-060

GLYCYRRHETIC ACID, 11-deoxo-	$C_{30}H_{48}O_3$	456	GLYCYRRHETIC ACID, 18α-OH-	$C_{30}H_{46}O_5$	486
COOH; HO	Chem. Abstr., 1966, 15435.		HO; HO	Chem. Abstr., 1968 (68) 13206.	
4408-061			4408-066		
GLYCYRRHETIC ACID, 24-OH-11-deoxo-	$C_{30}H_{48}O_4$	472	GLYCYRRHETIC ACID, 24-OH-	$C_{30}H_{46}O_5$	486
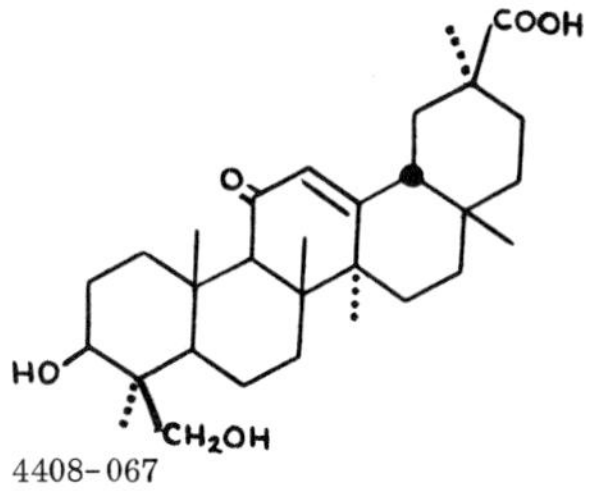 COOH; HO; CH₂OH	Chem. Abstr., 1968 (68) 49811.		COOH; HO; CH₂OH	Chem. Abstr., 1968 (68) 49811.	
4408-062			4408-067		
GLYCYRRHETIC ACID	$C_{30}H_{46}O_4$	470 +162°c 298°C	GLABROLIDE, deoxy-	$C_{30}H_{46}O_3$	454 +60°e 276°C
R = H; COOH; 20; RO	Chem. Abstr., 1969 (70) 29115.		HO	Chem. Abstr., 1966, 15435.	
4408-063			4408-068		
GLYCYRRHINIC ACID	$C_{42}H_{62}O_{16}$	822 +46°e	GLABROLIDE	$C_{30}H_{44}O_4$	468 +76°e
R = Glucurono-Glucuronosyl-	JCS., 1950, 1986.		HO	Chem. Abstr., 1966 (65) 15435.	
4408-064			4408-069		
LIQUIRITIC ACID	$C_{30}H_{46}O_4$	470 +85°e 300°C	GLABROLIDE, iso-	$C_{30}H_{44}O_4$	468 +46°e 320°C
R = H, C - 20 - epimer	Chem. Abstr., 1966, 15435.		HO	Chem. Abstr., 1966, 15436.	
4408-065			4408-070		

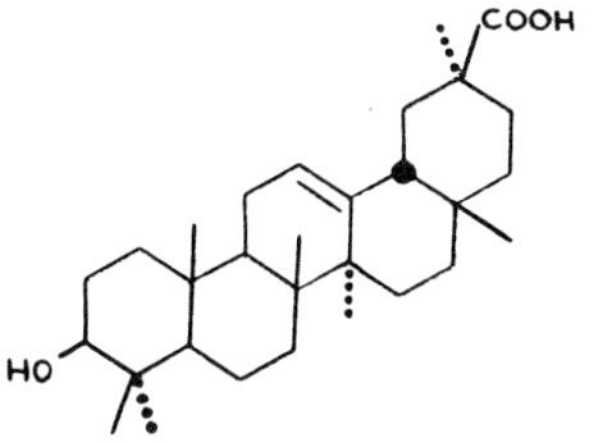

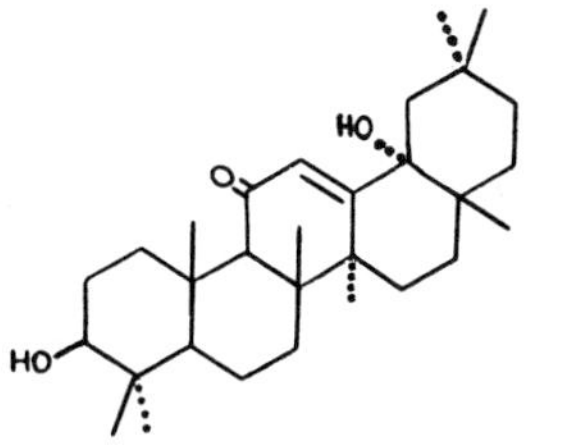

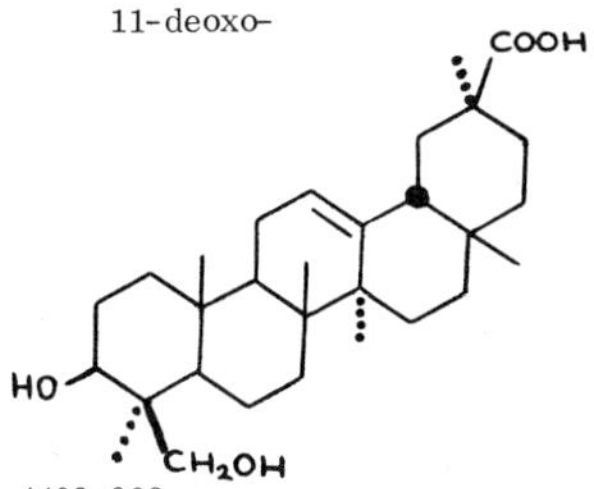

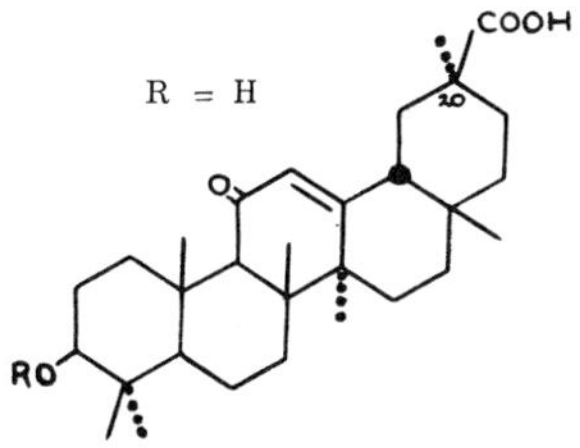

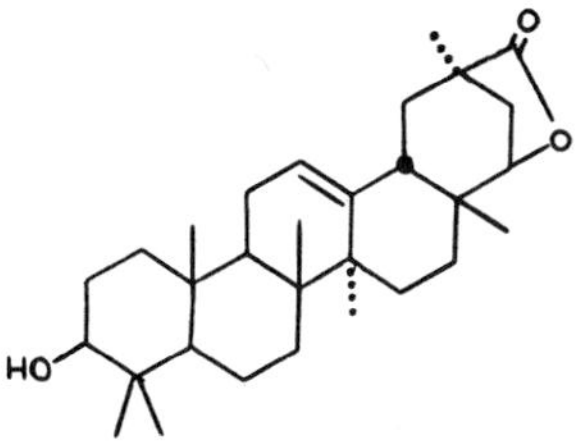

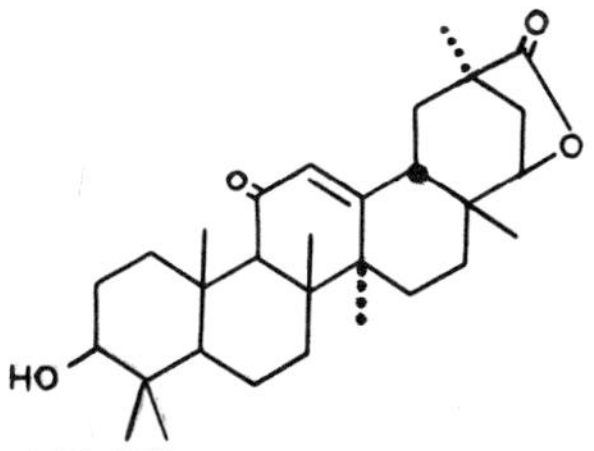

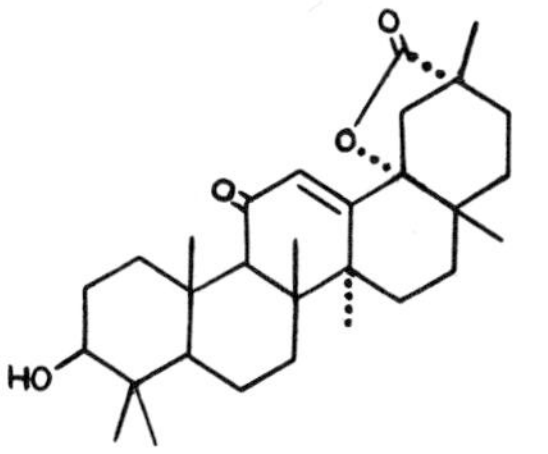

GLABROLIDE, 21α-OH-iso- $C_{30}H_{44}O_5$ 484 -11°p 304°C Chem. Abstr., 1968 (68) 49810. 4408-071	SAPINDIC ACID $C_{30}H_{46}O_3$ 454 Experimentia, 1968 (24) 1091. 4408-076
OLEANOLIC ACID $C_{30}H_{48}O_3$ 456 R = H JCS., 1968, 1647. 4408-072	MOMORDIC ACID $C_{30}H_{46}O_4$ 470 275°C Tet. Lett., 1966, 5137. 4408-077
OLEANOLIC ACID, Ac- $C_{32}H_{50}O_4$ 498 258°C R = Ac Chem. Abstr., 1956, 3492. 4408-073	CRATAEGOLIC ACID $C_{30}H_{48}O_4$ 472 JCS., 1968, 1047. 4408-078
OLEANOLIC ACID, 3-epi- $C_{30}H_{48}O_3$ 456 +68° c 298°C R = H 3 - epimer Tet., 1963, 479. 4408-074	SUMARESINOL $C_{30}H_{48}O_4$ 472 +54°c 298°C Helv., 1955, 1304. 4408-079
OLEANONIC ACID $C_{30}H_{46}O_3$ 454 JCS., 1968, 1047. 4408-075	SIARESINOL $C_{30}H_{48}O_4$ 472 +39°e 280°C Helv., 1943. 1218. 4408-080

ECHINOCYSTIC ACID — $C_{30}H_{48}O_4$ — 472; $+28^0$e; 309^0C

Helv., 1956, 413.

4408-081

COCHALIC ACID — $C_{30}H_{48}O_4$ — 472; $+58^0$d; 304^0C

JACS., 1955, 3579.

4408-082

ALBIGENIC ACID — $C_{30}H_{48}O_4$ — 472; -13^0e; 247^0C

Tet., 1959 (7) 19.

4408-083

LANTADEN A — $C_{35}H_{52}O_5$ — 552; $+84^0$c; 298^0C

R = Tigloyl

JCS., 1956, 4160.

4408-084

LANTADEN B — $C_{35}H_{52}O_5$ — 552; $+85^0$m; 294^0C

R = Senecoyl

JCS., 1954, 3689.

4408-085

MACHAERINIC ACID — $C_{30}H_{48}O_4$ — 472

JACS., 1955, 1825.

4408-086

MACHAERIC ACID — $C_{30}H_{44}O_4$ — 468

Festsch. Arth. Stoll. 1957 330.

4408-087

QUERETAROIC ACID — $C_{30}H_{48}O_4$ — 472; 321^0C

JACS., 1956, 3783.

4408-088

MESEMBRYANTHEMOIDIGENIN — $C_{30}H_{48}O_4$ — 472; $+71^0$; 306^0C

Tet. Lett., 1965, 4161.

4408-089

HEDERAGENIN — $C_{30}H_{48}O_4$ — 472; $+67^0$; 320^0C

JCS., 1968, 1047.

4408-090

deleted

4408-091

ACEROCIN $C_{41}H_{62}O_7$ 666

206°C

OR, OAc, COOH, HO

$R = CO\ CH\overset{c}{=}CH\ CH\overset{t}{=}CH\ CH_2CHMe_2$

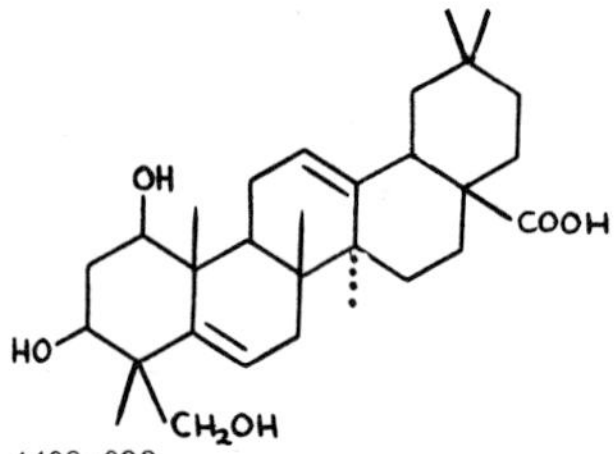

Chem. Comm., 1970, 969.

4408-096

deleted

4408-092

ACEROTIN $C_{41}H_{62}O_7$ 666

242°C

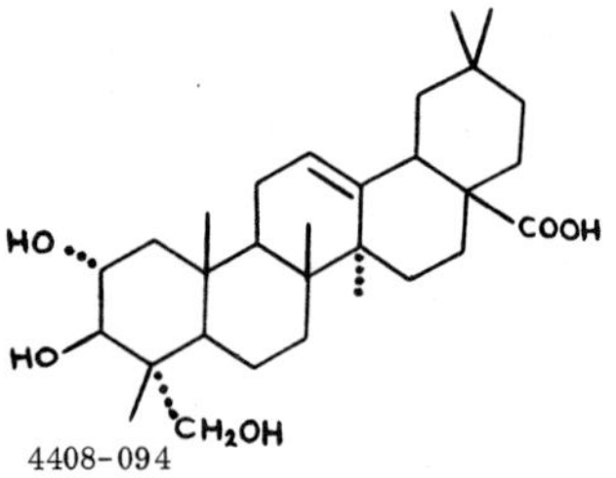

$R = \cdot CO(CH\overset{t}{=}CH)_2 \cdot CH_2CHMe_2$

Chem. Comm., 1970, 969.

4408-097

BASSIAIC ACID $C_{30}H_{46}O_5$ 486

+82°p

316°C

OH, COOH, HO, CH_2OH

JCS., 1955, 1338.

4408-093

ACACIC ACID $C_{30}H_{48}O_5$ 488

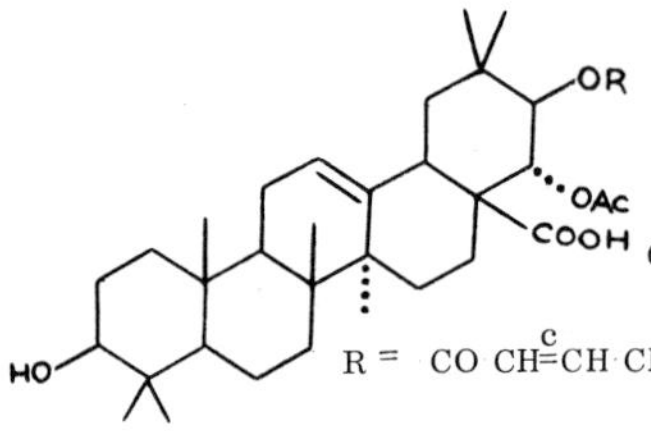

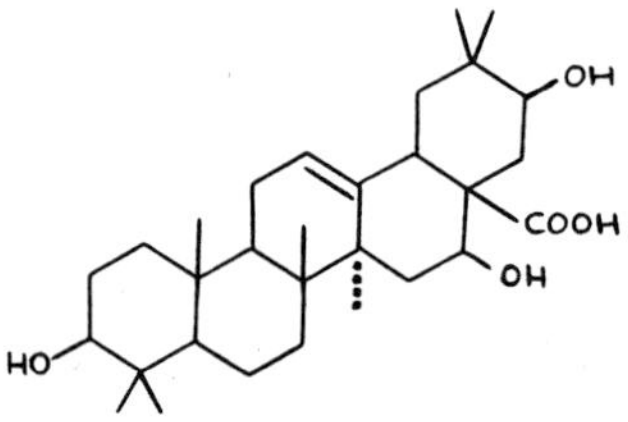

OH, COOH, OH, HO

Tet. Lett., 1965, 1187.

4408-098

ARJUNOLIC ACID $C_{30}H_{48}O_5$ 488

+64°e

339°C

HO, COOH, HO, CH_2OH

JCS., 1968, 1047.

4408-094

DUMORTIERIGENIN $C_{30}H_{46}O_4$ 470

-19°c

294°C

OH, O, O, HO

JACS., 1956, 5685.

4408-099

ENTAGENIC ACID $C_{30}H_{48}O_5$ 488

+35°e

313°C

OH, OH, COOH, HO

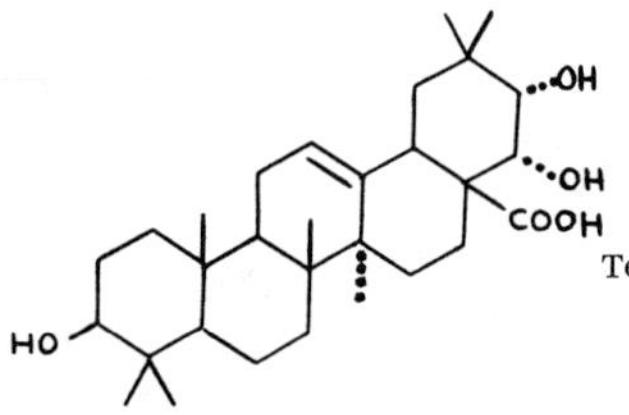

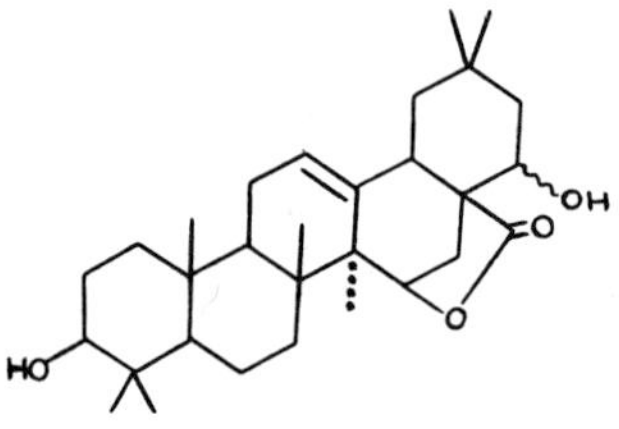

Tet., 1967 (23) 1499.

4408-095

PROCERAGENIN A $C_{30}H_{46}O_4$ 470

-13°c

295°C

O, OH, O, HO

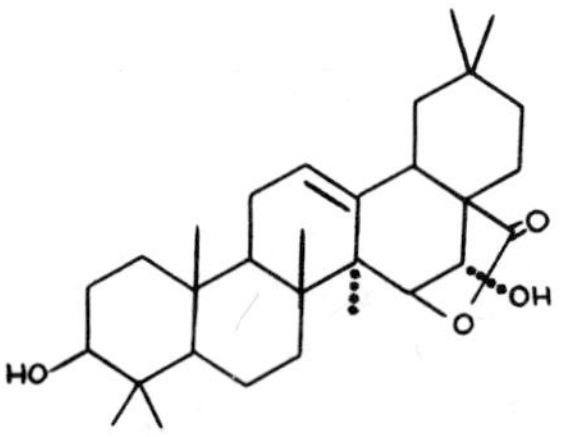

Tet. Lett., 1966, 5743.

4408-100

No.	Name	Formula	MW	Rotation	m.p.	Reference
4408-101	PLATYCOGENIC ACID C	$C_{30}H_{48}O_6$	504	+78°e	285°C	Chem. Comm., 1969, 1313.
4408-102	TERMINOLIC ACID	$C_{30}H_{48}O_6$	504	+42°e	347°C	Tet. Lett., 1960 (27) 12.
4408-103	POLYGALACIC ACID	$C_{30}H_{48}O_6$	504	+ 47°p	298°C	Chem. Comm., 1968, 1005.
4408-104	CACCIGENIN	$C_{30}H_{48}O_6$	504			Tet. Lett., 1966, 1947.
4408-105	PLATICODIGENIN	$C_{30}H_{48}O_7$	520	+35°p	241°C	Tet. Lett., 1968, 5577.
4408-106	GYPSOGENIN	$C_{30}H_{46}O_4$	470	+91°e	275°C	JACS., 1956, 2312.
4408-107	QUILLAIC ACID	$C_{30}H_{46}O_5$	486	+ 56°p	293°C	JCS., 1941, 552.
4408-108	MEDICAGENIC ACID	$C_{30}H_{46}O_6$	502	+106°	352°C	JOC., 1963 (28) 240.
4408-109	BARRINGTOGENIC ACID	$C_{30}H_{46}O_6$	502	+ 72°m	333°C	JCS., 1956, 4369.
4408-110	PRESENEGENIN	$C_{30}H_{46}O_7$	518			Chem. Ind., 1965, 2098.

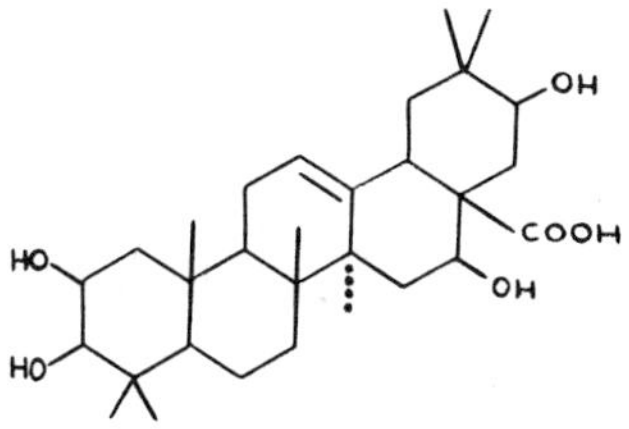

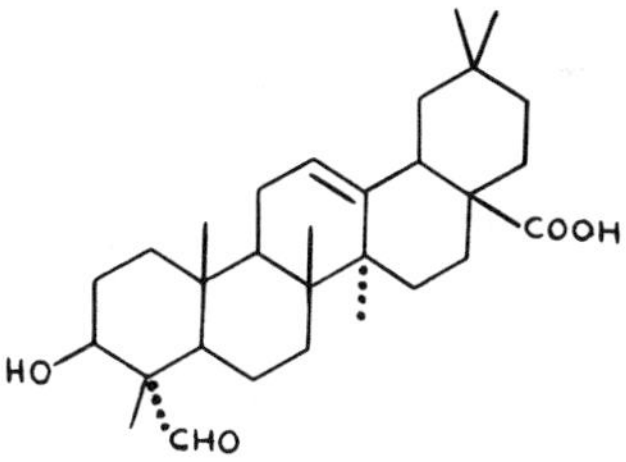

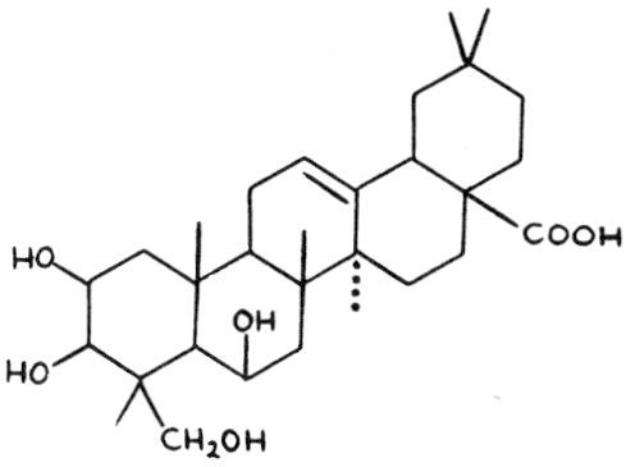

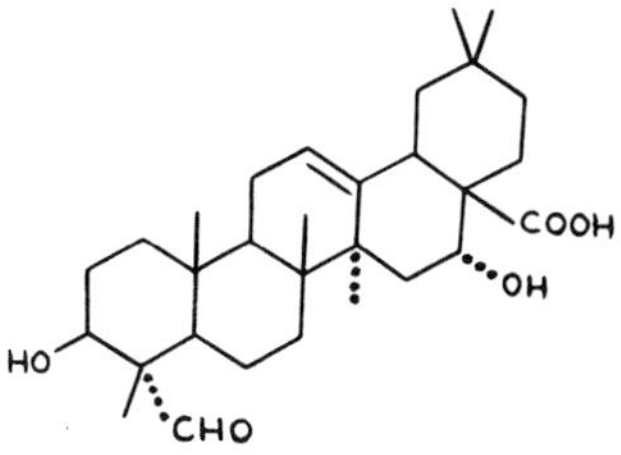

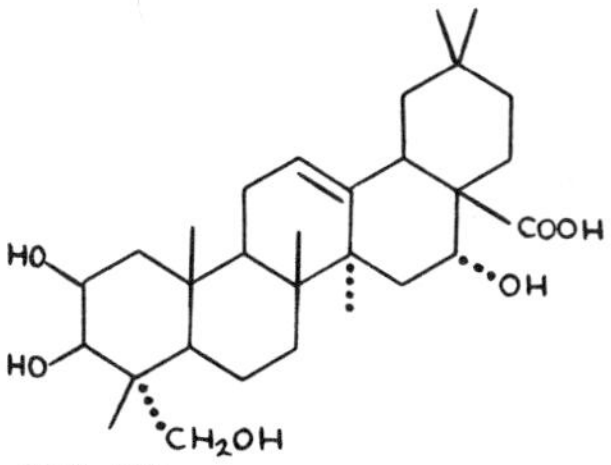

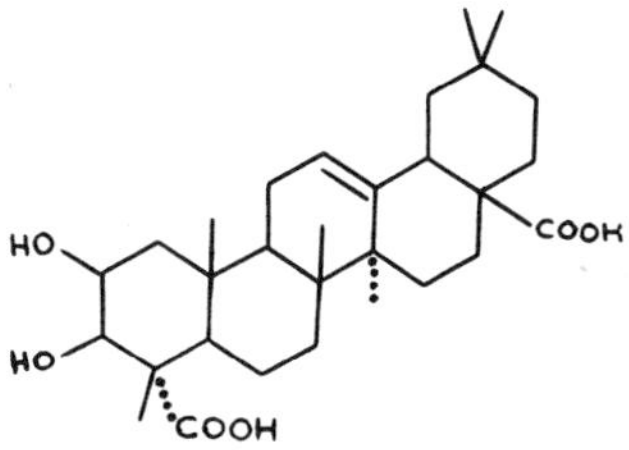

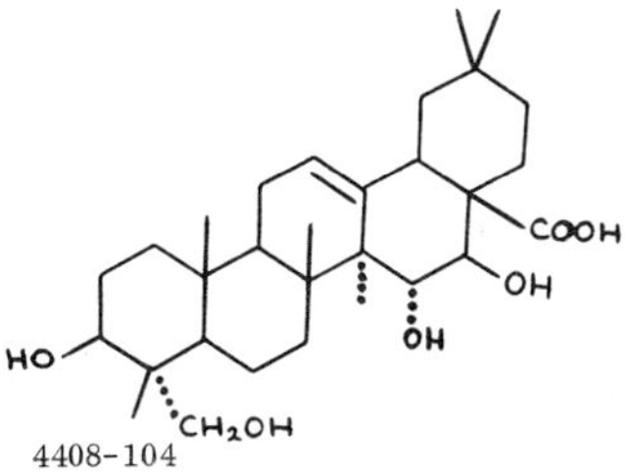

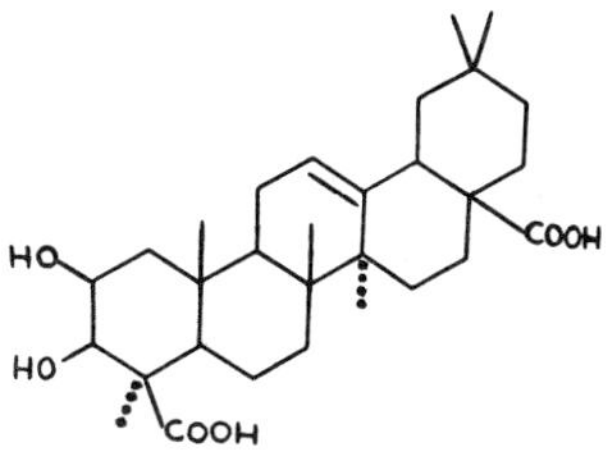

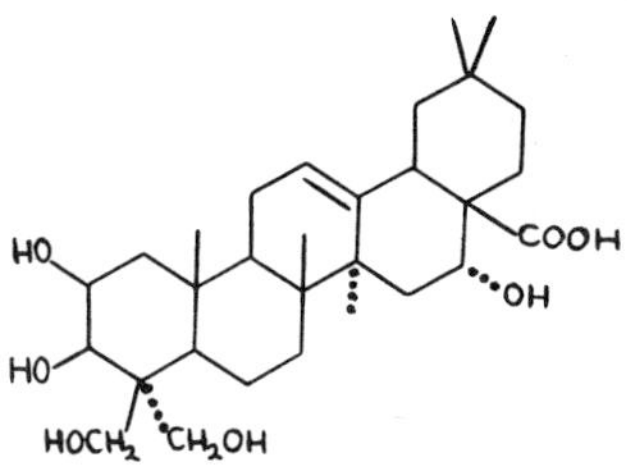

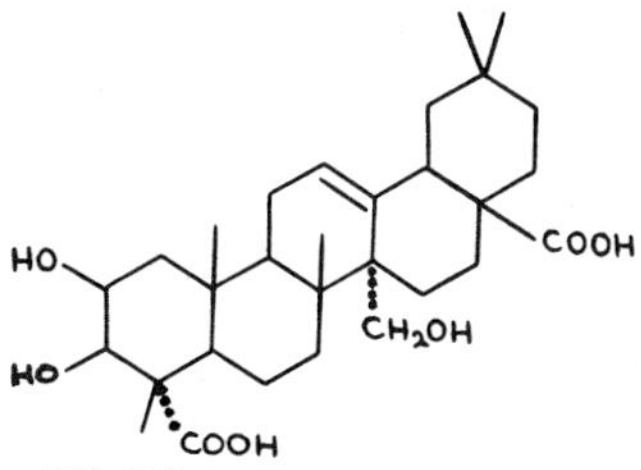

PLATYCOGENIC ACID A — $C_{30}H_{46}O_8$ — 534, +74°e, 246°C

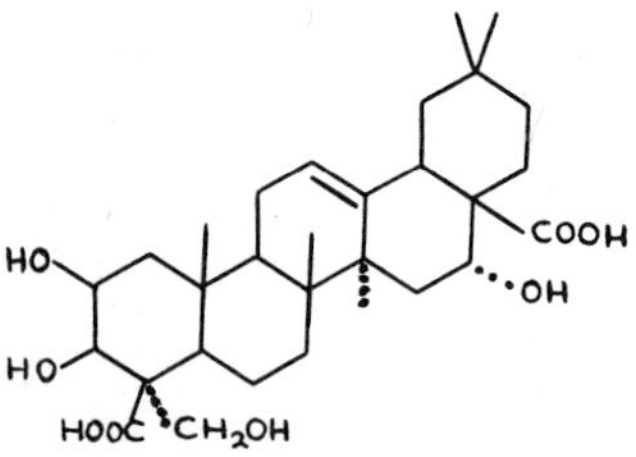

Chem. Comm., 1969, 1313.

4408-111

AMYRIN, 11-OH-β- — $C_{30}H_{50}O_2$ — 442, 230°C

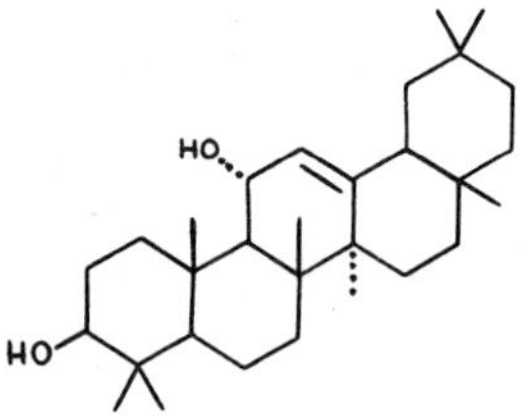

JCS., 1967, 490.

4408-116

PLATYCOGENIC ACID B — $C_{30}H_{46}O_8$ — 534, +101°e, 276°C

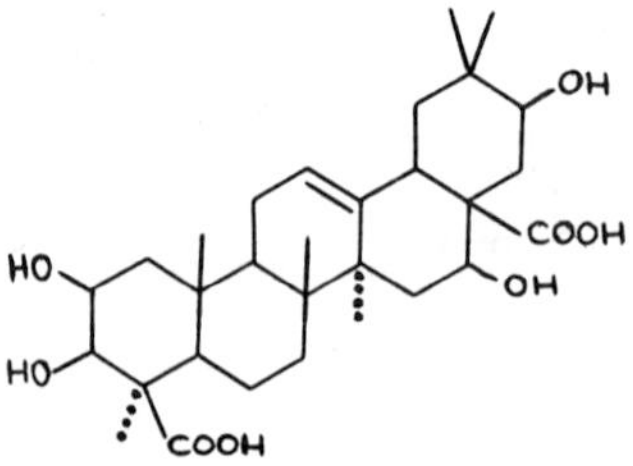

Chem. Comm., 1969, 1313.

4408-112

OLEANOLIC ACID, 3-METHYL ETHER — $C_{31}H_{50}O_3$ — 470, 249°C

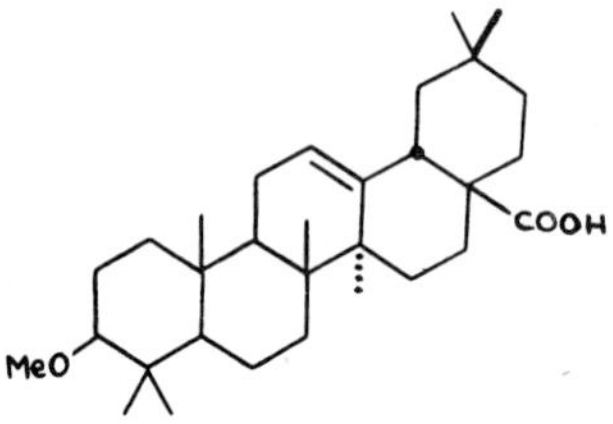

Khim. Prir. Soedin, 1970 (6) 272.

4408-117

CINCHOLIC ACID — $C_{30}H_{46}O_5$ — 486, +118°p, 266°C

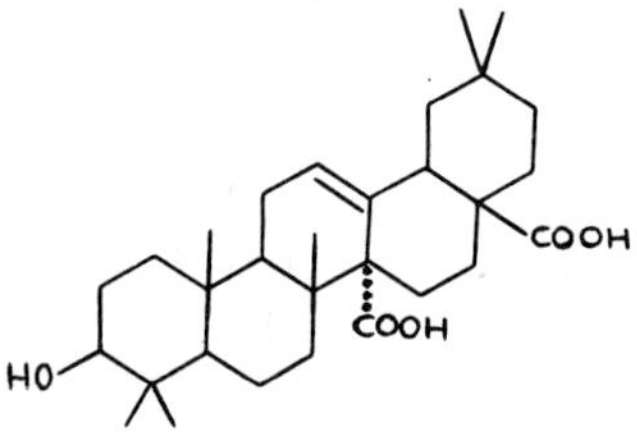

Ann., 1963 (667) 151.

4408-113

GYMNOSPOROL — $C_{30}H_{46}O_3$ — 454

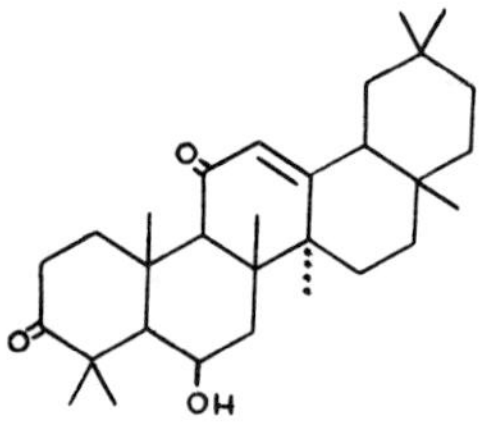

Ind. J. C., 1970, 395.

4408-118

SERRATAGENIC ACID — $C_{30}H_{46}O_5$ — 486, +37°p, 310°C

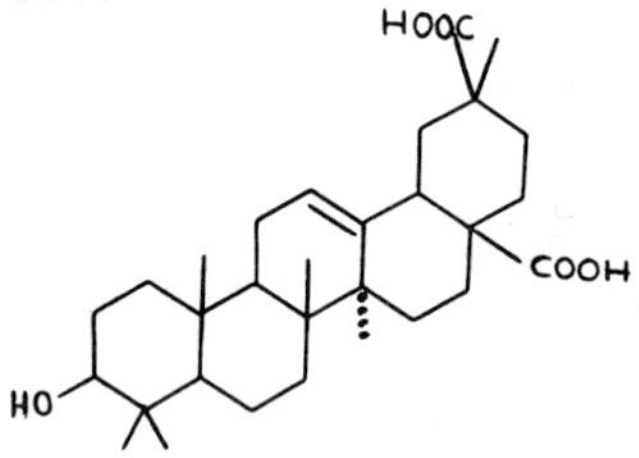

Tet., 1969, 3701.

4408-114

ARMILLARIGENIN — $C_{30}H_{48}O_4$ — 472, +11°e, 300°C

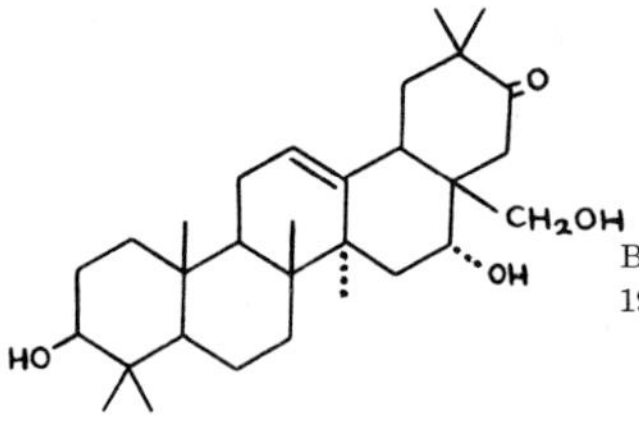

Bull. Soc. Chim. Fr., 1969, 226.

4408-119

SERJANIC ACID — $C_{31}H_{48}O_5$ — 500

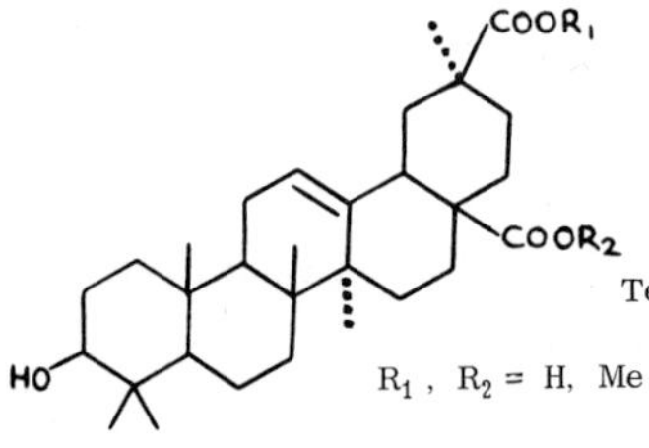

Tet. Lett., 1967, 2129.

R_1, R_2 = H, Me (?)

4408-115

SOYASAPOGENOL-E — $C_{30}H_{48}O_3$ — 456

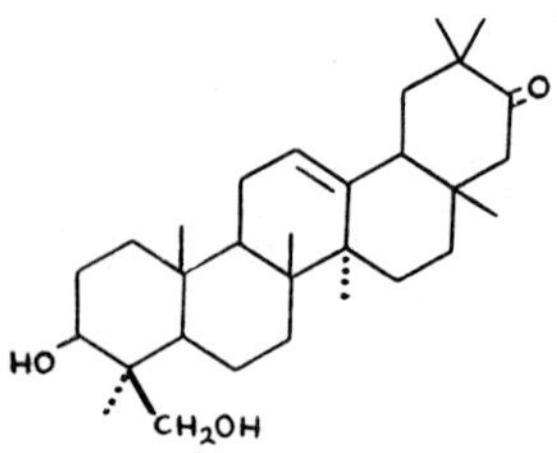

JCS., 1964, 5885.

4408-120

PHYTOLACCAGENIN $C_{31}H_{48}O_7$ 532
318°C

COOMe
COOH
HO
HO
CH_2OH

JACS., 1964, 957.

4408-121

SAIKOSAPONIN C $C_{48}H_{76}O_{17}$ 924
+4°e
202°C

Saikogenin E (012)
3-0-glucosido-rhamnosido-glucosyl-

Tet. Lett., 1968, 303.

4408-125

MACHAERINIC ACID LACTONE, 2-OH- $C_{30}H_{46}O_4$ 470
+5°
266°C

HO
HO
O
O

JOC., 1963 (28) 2390.

4408-122

SAIKOSAPONIN D $C_{42}H_{68}O_{13}$ 780
+37°e
212°C

Saikogenin G (025)
3-0-fucosido-glucosyl-

Tet. Lett., 1968, 303.

4408-126

MUSENNIN $C_{51}H_{82}O_{21}$ 1030

Echinocystic Acid (081)
3-0-(arabinosyl)$_3$-glucoside

Ann., 1969 (724) 183.

4408-123

OLEAN-11, 13(18)-DIENE-3, 23, 28-TRIOL $C_{30}H_{48}O_3$ 456
+65°p
297°C

CH_2OH
HO
CH_2OH

JCS., 1963, 1401.

4408-127

SAIKOSAPONIN A $C_{42}H_{68}O_{13}$ 780
+46°e
225°C

Saikogenin F (024)
3-0-fucosido-glucosyl-

Tet. Lett., 1968, 303.

4408-124

AMYRIN, α- R = H JCS. 1956 456. 4409-001	$C_{30}H_{50}O$	426 +83°c 186°C	
URSOLIC ACID ME ESTER, O-Ac- R_1 = Ac R_2 = Me Chem. Ind. 1966 1920. 4409-006	$C_{33}H_{52}O_4$	512	
AMYRIN METHYL ETHER, α- R = Me Phytochemistry 1967 727. 4409-002	$C_{31}H_{52}O$	440 221-3°C	
EMPETROSIDE A R_1 = glucosyl R_2 = H Chem. Abstr. 1970 (73) 4146. 4409-007	$C_{36}H_{58}O_8$	618 +20° 209-11°	
UVAOL Helv. 1954 2145. 4409-003	$C_{30}H_{50}O_2$	442 +69°c 232°C	
URSOLIC ACID LACTONE R = H Planta. Med. 1970 (18) 326. 4409-008	$C_{30}H_{46}O_3$	454	
AMYRENONE, α- Chem. Abstr. 1968 (68) 29900. 4409-004	$C_{30}H_{48}O$	424 114°c 125°C	
URSOLIC ACID LACTONE, Ac- R = Ac Aust. J. C. 1964 477. 4409-009	$C_{32}H_{48}O_4$	496 +39°c 262°C	
URSOLIC ACID R_1 = R_2 = H Helv. 1954 2145. 4409-005	$C_{30}H_{48}O_3$	456 +72° 292°C	
MICROMERIC ACID Chem. Abstr. 1968 (69) 19326. 4409-010	$C_{30}H_{46}O_3$	454	

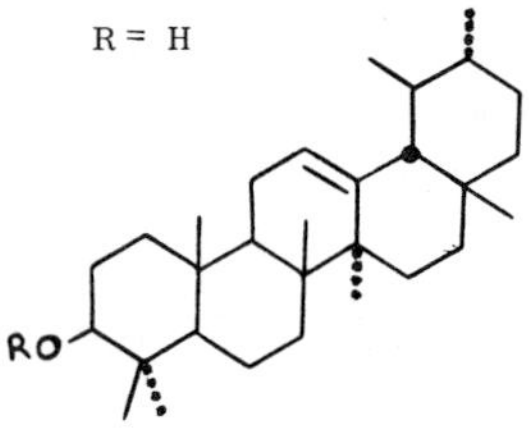

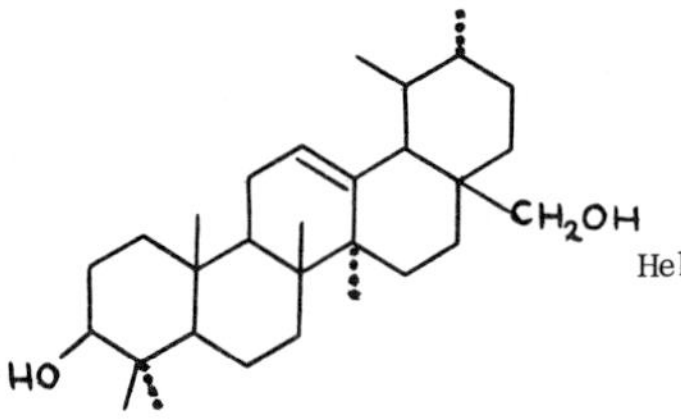

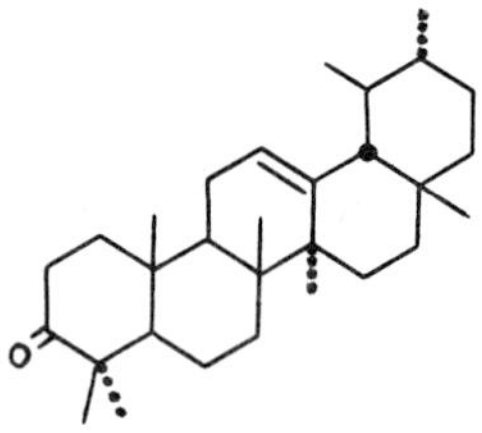

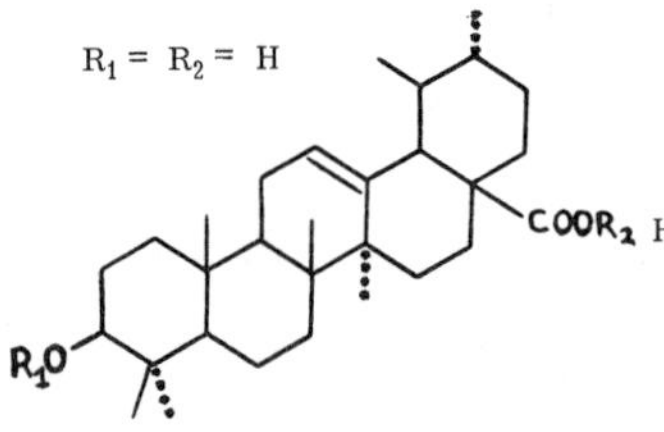

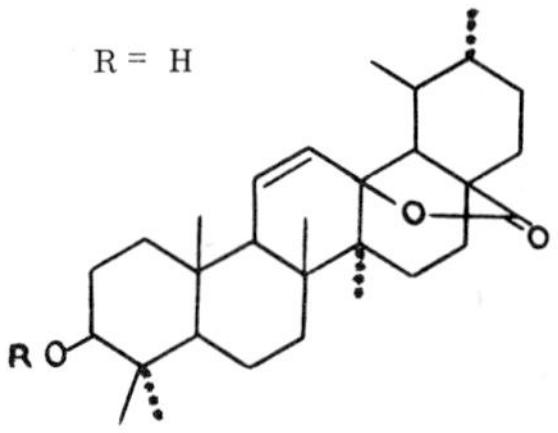

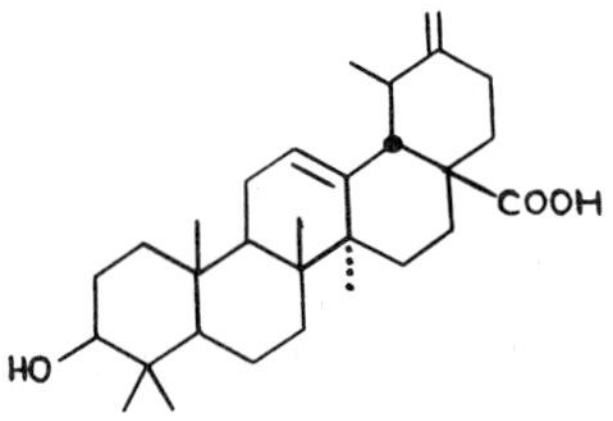

VANGUEROLIC ACID $C_{30}H_{46}O_3$ 454 +308^{0}c 272-3^{0}C JCS. 1962 5163. 4409-011	POMONIC ACID $C_{30}H_{46}O_4$ 470 Chem. Abstr. 1967 76185. 4409-016
URSOLIC ACID, 2α-OH- $C_{30}H_{48}O_4$ 472 +55^0 213^{0}C JCS. 1967 510. 4409-012	ASIATIC ACID $C_{30}H_{48}O_5$ 488 +51^{0}e 300-5^{0}C JCS. 1968 1047. 4409-017
BENTHAMIC ACID $C_{30}H_{48}O_4$ 472 m.e. + 54^0 m.e. 201^{0}C Tet. Lett., 1967, 4649. 4409-013	DRYOBALANOLIDE $C_{30}H_{46}O_5$ 486 Tet. Lett. 1968 4363. 4409-018
URSOLIC ACID, 20β-OH- $C_{30}H_{48}O_4$ 472 Chem. Ind. 1966 1720. 4409-014	ROTUNDIC ACID $C_{30}H_{48}O_5$ 488 Tet. Lett. 1968 4639. 4409-019
URSOLIC ACID, 2, 19-diOH- $C_{30}H_{48}O_5$ 488 An. Quim. 1970 (66) 293. 4409-015	URSONIC ACID, 2, 20-diOH- $C_{30}H_{46}O_5$ 486 JCS. 1967 851. 4409-020

THANKUNIC ACID, Iso- $C_{30}H_{48}O_6$ 504 +24^0 288-900^{0}C

COOH Chem. Abstr. 1969 (70) 47795.

HO OH OH HOCH$_2$

4409-021

COMMIC ACID E $C_{30}H_{48}O_5$ 488 +104^{0}p 329^{0}C

OH HO HO COOH

Tet. 1961 (15) 266.

4409-026

MADECASSIC ACID $C_{30}H_{48}O_6$ 504 +31^{0}m 265-8^{0}C

COOH Chem. Abstr. 1967 (67) 32828.

HO OH HO CH$_2$OH

4409-022

BRYONOLIC ACID $C_{30}H_{48}O_3$ 456

HOOC

Chem. Abstr. 1967 32785.

HO

4409-027

ASIATIC ACID, 11-oxo- $C_{30}H_{46}O_6$ 502

O COOH Tet. Lett. 1968 4363.

HO HO CH$_2$OH

4409-023

QUINOVAIC ACID $C_{30}H_{46}O_5$ 486 +88^0 300^{0}C

COOH JCS. 1953 3111.

COOH HO

4409-028

COMMIC ACID B $C_{30}H_{48}O_3$ 456 +39^{0}c 239-41^{0}C

Tet. Lett. 1964 3177.

HO COOH

4409-024

PHILLYRIGENIN $C_{30}H_{48}O_4$ 472

O CO

Tet. 1965 2585.

CH$_2$OH HO

4409-029

COMMIC ACID D $C_{30}H_{46}O_4$ 472

Tet. 1961 (15) 212.

HO HO COOH

4409-025

TORMENTIC ACID $C_{30}H_{48}O_5$ 488 -20^{0}p 273^{0}C

OH COOH Bull. Soc. Chim. Fr. 1966 3458.

HO HO

4409-030

MADASIATIC ACID $C_{30}H_{48}O_5$ 488 249^{0}C Bull. Soc. Chim. Fr. 1969 3592. 4409-031	BOSWELLIC ACID, β- $C_{30}H_{48}O_3$ 456 + 107^{0}c 239^{0}C JCS., 1956, 2904. 4409-034
IFFLAIONIC ACID $C_{30}H_{46}O_3N_2$ 454 88^{0}c 259-60^{0}C Aust. J. C. 1963 683. 4409-032	PYRETHROL $C_{30}H_{50}O$ 426 +92^{0}c 220^{0}C Chem. Abstr., 1967 (66) 65658. 4409-035
TOMENTOSOLIC ACID $C_{30}H_{46}O_3$ 454 + 18^{0}c 285^{0}C JCS., 1962, 5163. 4409-033	

LANOSTEROL $C_{30}H_{50}O$ 426 +58°c 140°C JCS., 1953, 2159. 4410-001	LANOSTEROL, 23-OH- $C_{30}H_{50}O_2$ 442 +66° 159°C Tet., 1969, 1449. 4410-006
LANOSTEROL, dihydro- $C_{30}H_{52}O$ 428 +61°c 145°C JCS., 1950, 1017. 4410-002	LANOST-8-EN-7-ONE, 3-OH- $C_{30}H_{50}O_2$ 442 +30°c 136°C JCS., 1956, 1740. 4410-007
PARKEOL $C_{30}H_{50}O$ 426 +77° 160°C Chem. Ind., 1956, 1458. 4410-003	LANOST-8-EN-7, 11-DIONE, 3-OH- $C_{30}H_{48}O_3$ 456 +98°c 113°C JCS., 1956, 1740. 4410-008
AGNOSTEROL $C_{30}H_{48}O$ 424 +69° 169°C Chem. Ind., 1956, 550. 4410-004	LANOSTANE-7, 11-DIONE, 3-OH- $C_{30}H_{50}O_3$ 458 +58° 190°C Tet. Lett., 1968, 6237. 4410-009
AGNOSTEROL, dihydro- $C_{30}H_{50}O$ 426 +68° 159°C JCS., 1950, 1017. 4410-005	LANOSTA-8, 24-DIEN-7-ONE, 3-OH- $C_{30}H_{48}O_2$ 440 Tet. Lett., 1968, 6237. 4410-010

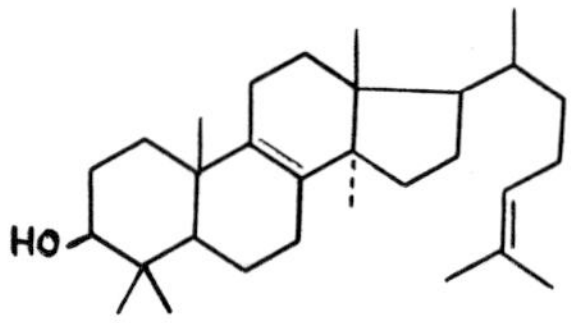

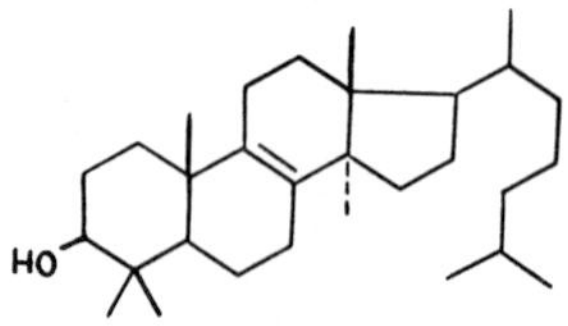

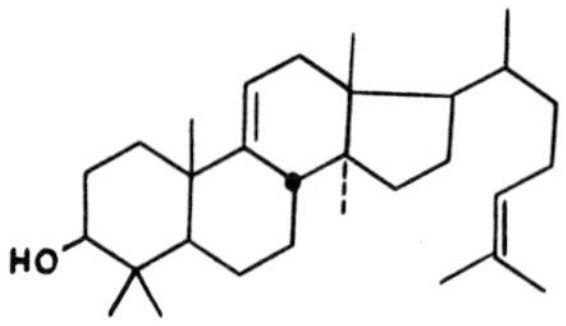

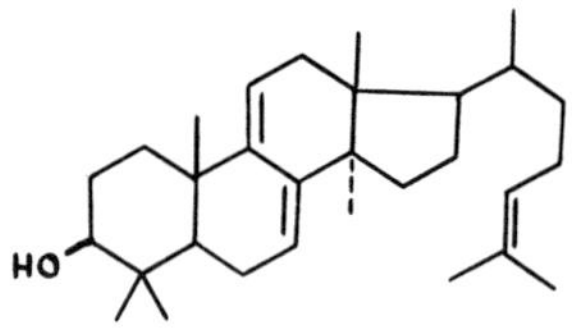

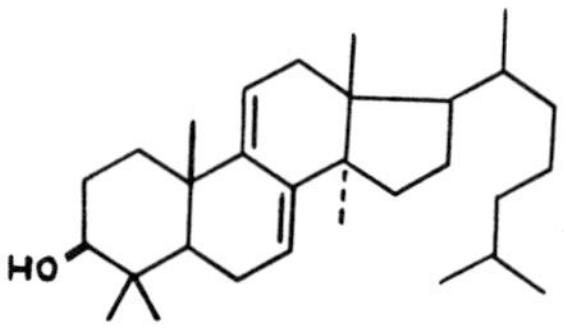

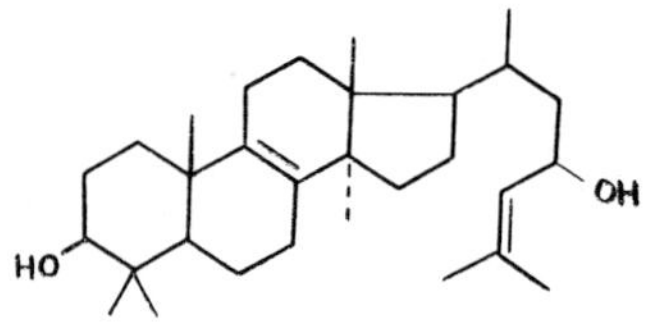

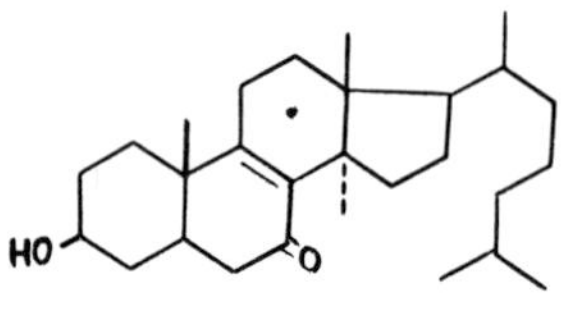

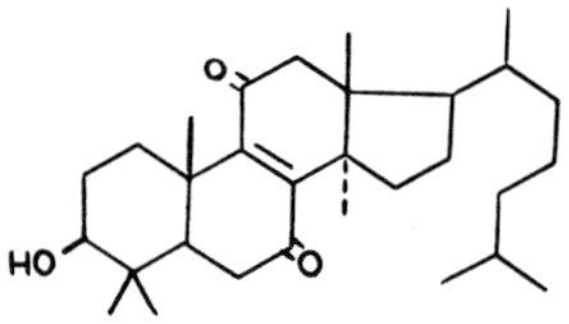

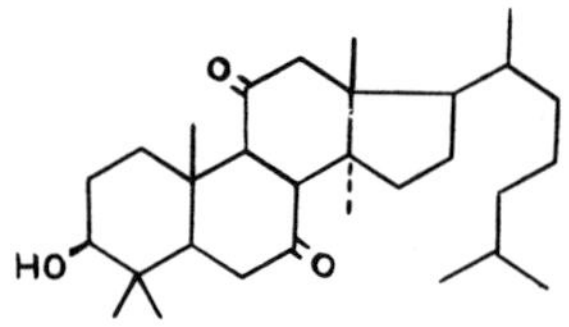

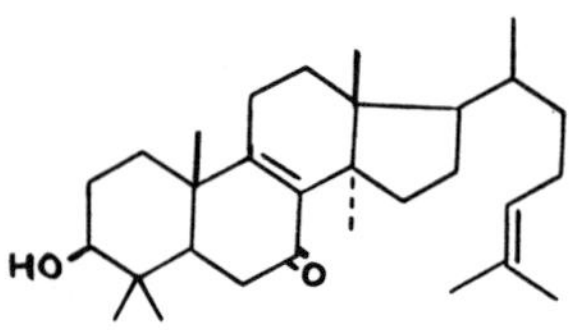

LANOSTA-7, 9(11), 24-TRIENE-3β, 21-DIOL - $C_{30}H_{48}O_2$ 440 +72° 195°C JCS., 1959, 2031. 4410-011	TRAMETENOLIC ACID B Me Ester $C_{31}H_{50}O_3$ 470 +52° 134°C R = CH_3 JCS., 1959, 2036. 4410-016
LANOSTA-7, 9(11)-24-TRIEN-3--ONE, 21-OH- $C_{30}H_{46}O_2$ 438 +56°c 120°C JCS., 1959, 2031. 4410-012	TRAMETENOLIC ACID, 15α-OH- $C_{30}H_{48}O_4$ 472 +45°p 255°C JCS., 1967, 1776. 4410-017
LANOSTA-8, 24-DIEN-3-ONE, 21-OH- $C_{30}H_{48}O_2$ 440 +69° 116°C JCS., 1959, 2031. 4410-013	LANOSTA-8, 24-DIEN-21-OIC ACID ME ESTER, 3-α-OAc- $C_{33}H_{52}O_4$ 512 +7° 136°C Gazz., 1959, 818. 4410-018
LANOSTA-7, 9(11), 24-TRIEN-21--OIC ACID, 3-β-OH- $C_{30}H_{46}O_3$ 454 +105°c 258°C Chem. Pharm. Bull., 1970, 779. 4410-014	PINICOLIC ACID A $C_{30}H_{46}O_3$ 454 +68° 198°C Gazz., 1959, 818. 4410-019
TRAMETENOLIC ACID B $C_{30}H_{48}O_3$ 456 +43°p 258°C R = H JCS., 1959, 2036. 4410-015	TYROMYCIC ACID $C_{30}H_{44}O_3$ 452 +45°c 174°C Bull. Soc. Chim. Fr., 1967, 1844. 4410-020

ABIESLACTONE $C_{31}H_{48}O_3$ 468 -187° 253°C MeO Chem. Comm., 1965, 525. 4410-021	GRISEOGENIN $C_{30}H_{46}O_5$ 486 -22°c 286°C OH OH HO Tet., 1967 (23) 761. 4410-026
SEYCHELLOGENIN $C_{30}H_{46}O_3$ 454 HO JACS., 1969, 4918. 4410-022	HOLOTHURINOGENIN, 17-desoxy-22, 25-oxido- $C_{30}H_{44}O_4$ 468 -9° 286°C R = H R HO Tet., 1966, 1857. 4410-027
KOELLIKERIGENIN $C_{30}H_{46}O_4$ 470 OR HO R = H JACS., 1969, 4918. 4410-023	HOLOTHURINOGENIN, 22, 25-oxido- $C_{30}H_{44}O_5$ 484 -21° 316°C R = OH Tet., 1966, 1857. 4410-028
TERNAYGENIN $C_{31}H_{48}O_4$ 484 R = CH_3 JACS., 1969, 4918. 4410-024	STICHOPOGENIN A_2 $C_{30}H_{44}O_4$ 468 -48°c 239°C OH HO Tet. Lett., 1969, 1151. 4410-029
PRASLINOGENIN $C_{31}H_{48}O_5$ 500 291°C OH OMe HO Tet. Lett., 1970, 467. 4410-025	STICHOPOGENIN A_4 $C_{30}H_{46}O_5$ 486 239°C OH OH HO Tet. Lett., 1969, 1151. 4410-030

ECHINODOL $C_{32}H_{50}O_4$ 498 +48^{0}c 237^{0}C

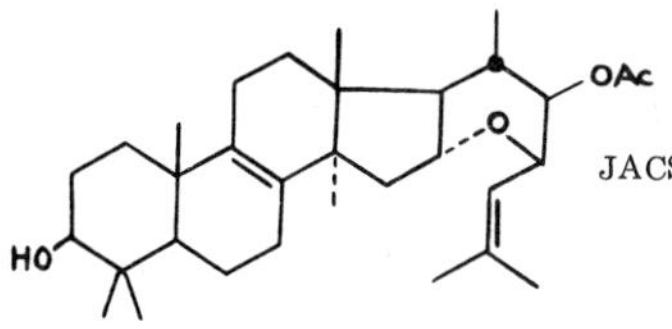

JACS., 1966, 3882.

4410-031

INOTODIOL $C_{30}H_{50}O_2$ +56^{0}c 191^{0}C

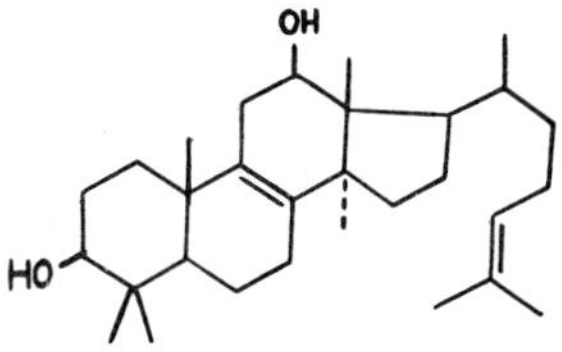

Chem. Abstr., 1969 (71) 50270.

4410-033

ACERINOL $C_{30}H_{46}O_5$ 486

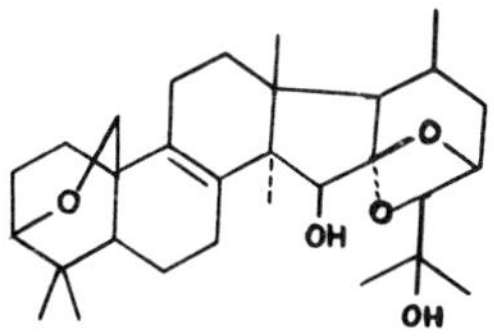

Chem. Abstr., 1968 (69) 36278.

4410-032

OBTUSIFOLDIENOL $C_{31}H_{52}O$ 440 +72°c 140°C HO Chem. Abstr. 1959 (53) 20123. 4411-001	TUMULOSIC ACID $C_{31}H_{50}O_4$ 486 +10°p 306-10°C R = H OH RO Chem. Pharm. Bull. 1958 606. 4411-006
EBURICOIC ACID $C_{31}H_{50}O_3$ 470 +50° 292°C R = H COOH RO JCS. 1954 3713. 4411-002	PACHYMIC ACID $C_{33}H_{52}O_5$ 528 +18°c 299°C R = Ac Chem. Pharm. Bull. 1958 608. 4411-007
EBURICOIC ACID, acetyl- $C_{33}H_{52}O_4$ 512 +42° 259°C R = Ac JCS. 1954 3713. 4411-003	TUMULOSIC ACID, dehydro- $C_{31}H_{48}O_4$ 484 COOH OH HO JCS. 1954 4713. 4411-008
EBURICOIC ACID, dehydro- $C_{31}H_{48}O_3$ 468 +40°c 287°C COOH HO JCS. 1954 3713. 4411-004	SULPHURENIC ACID $C_{31}H_{50}O_4$ 486 +42°p 253°C COOH HO OH Tet. 1964 2297. 4411-009
EBURICONIC ACID, dehydro- $C_{31}H_{46}O_3$ 466 +28°c 241°C COOH O Tet. Lett. 1967 4823. 4411-005	POLYPORENIC ACID A $C_{31}H_{50}O_4$ 486 +64°p 200°C HO COOH JCS. 1953 457. 4411-010

POLYPORENIC ACID C $C_{31}H_{46}O_4$ 482 +6°p 275°C COOH, OH JCS 1953 2548. 4411-011	QUERCINIC ACID, Carboacetyl $C_{34}H_{52}O_7$ 572 -31°p 211°C R = CO·CH_2·COOH RO, O, COOH Tet. Lett. 1967 1461. 4411-013
POLYPORENIC ACID C, 6α-OH- $C_{31}H_{46}O_5$ 498 COOH, O, OH Chem. Comm. 1969 1150. 4411-012	

BRYOGENIN — $C_{30}H_{48}O_3$ — 456 — $+180^0$ — 157^0C

(Artifact ?)

Planta Med., 1966, 86.

4412-001

BRYODULCOSIGENIN — $C_{30}H_{50}O_4$ — 474 — $+190^0c$ — 182^0C

R = OH

Ann. Chim., 1965 (55) 164.

4412-002

BRYOSIGENIN — $C_{30}H_{48}O_4$ — 472 — $+144^0$ — 190^0C

R = oxo

Natwiss., 1965, 661.

4412-003

CUCURBITACIN A — $C_{32}H_{46}O_9$ — 574 — $+97^0e$ — 208^0C

R = CH_2OH

JCS., 1963, 3828.

4412-004

CUCURBITACIN B — $C_{32}H_{46}O_8$ — 558 — $+88^0y$ — 185^0

R = CH_3

JOC., 1963, 1790.

4412-005

CUCURBITACIN B, 2-epi- — $C_{32}H_{46}O_8$ — 558 — $+41^0c$ — 230^0C

R = CH_3

2 - epimer

JCS., 1962, 3259.

4412-006

CUCURBITACIN C — $C_{32}H_{48}O_8$ — 560 — $+95^0e$ — 208^0C

R = CH_2OH

JCS., 1963, 3828.

4412-007

CUCURBITACIN D — $C_{30}H_{44}O_7$ — 516 — 46^0c — 151^0C

R = oxo

JOC., 1963, 1790.

4412-008

CUCURBITACIN D, 22-de-oxo- — $C_{30}H_{46}O_6$ — 502 — $+103^0c$ — amorph

R = H

JCS., 1967, 964.

4412-009

CUCURBITACIN D, 22-de-oxo-iso- — $C_{30}H_{46}O_6$ — 502 — 76^0c — amorph

JCS., 1967, 964.

4412-010

CUCURBITACIN E $C_{32}H_{44}O_8$ 556
-62^{0}c
234^{0}C

JOCS., 1963, 1790.

4412-011

CUCURBITACIN L $C_{30}H_{44}O_7$ 516
-49^{0}e
140^{0}C

* dihydro-

JCS., 1964, 529.

4412-016

CUCURBITACIN F $C_{30}H_{46}O_7$ 518
+38^{0}e
245^{0}C

JCS., 1963, 4275.

4412-012

CUCURBITACIN J $C_{30}H_{44}O_8$ 532
-36^0
202^{0}C

JCS., 1964, 529.

4412-017

CUCURBITACIN G $C_{30}H_{46}O_8$ 534
+84^{0}c
149^{0}C

Planta Med., 1966, 86.

4412-013

CUCURBITACIN K $C_{30}H_{44}O_8$ 532
-74^0
195^{0}C

24-epimer

JCS., 1964, 529.

4412-018

CUCURBITACIN H $C_{30}H_{46}O_8$ 534
+57^{0}c

24-epimer

Planta Med., 1966, 86.

4412-014

CUCURBITACIN O $C_{30}H_{46}O_7$ 518

R = H

JOC., 1970, 2892.

4412-019

CUCURBITACIN I $C_{30}H_{42}O_7$ 514
-52^{0}e
149^{0}C

JCS., 1963, 3828.

4412-015

CUCURBITACIN P $C_{30}H_{48}O_7$ 520

*dihydro

R = H

JOC., 1970, 2891.

4412-020

CUCURBITACIN Q $C_{32}H_{48}O_8$ 560 126°C R = Ac JOC., 1970, 2891. 4412-021	GRATIOSIDE $C_{36}H_{58}O_9$ 634 +81°e 289°C R = glucosyl Ann., 1964 (674) 185/196. 4412-023
GRATIOGENIN $C_{30}H_{48}O_4$ 472 + 175°c 203°C R = H RO OH Ann., 1964 (674) 185/196. 4412-022	GRATIOGENIN, 16-OH- $C_{30}H_{48}O_5$ 488 +133°c 250°C HO OH OH Ann., 1964 (674) 185/196. 4412-024

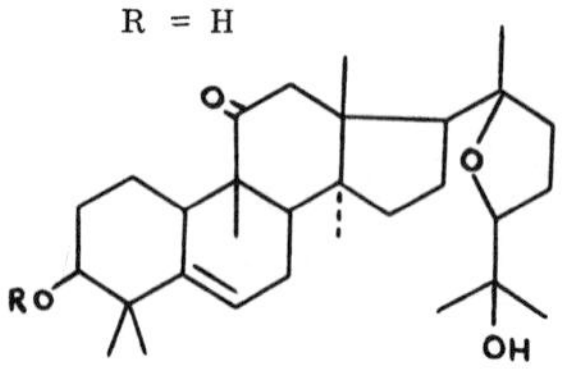

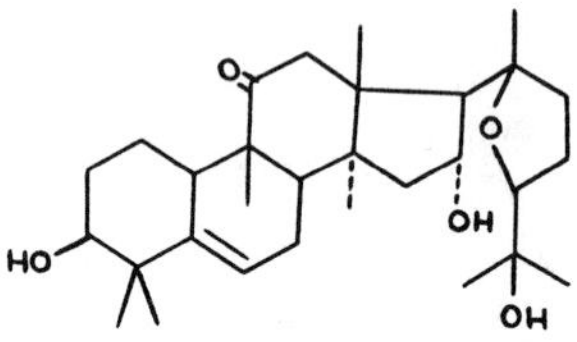

TARAXASTEROL	$C_{30}H_{50}O$	426 +96°c 221°C
JCS. 1954 1905.		
4413-001		

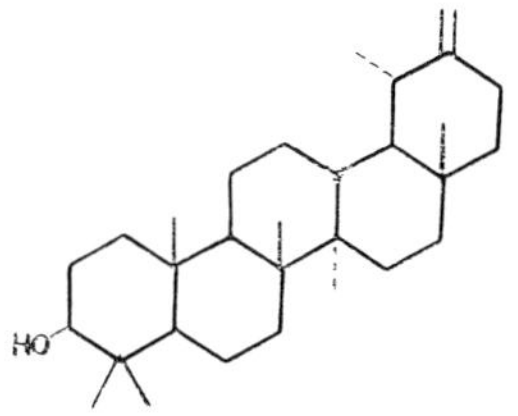

FARADIOL	$C_{30}H_{50}O_2$	442 +45°c 236°C
JOC. 1962 3204.		
4413-005		

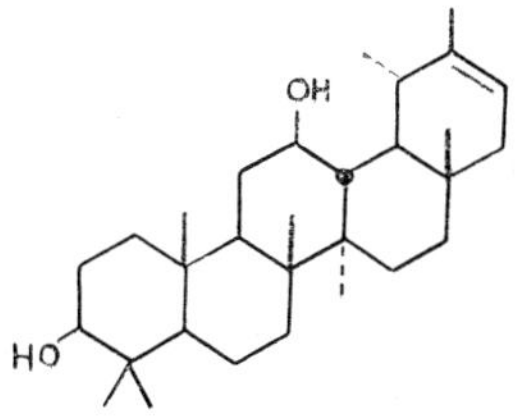

TARAXASTEROL, pseudo-	$C_{30}H_{50}O$	426 +50°c 218°C
JCS. 1954 1905.		
4413-002		

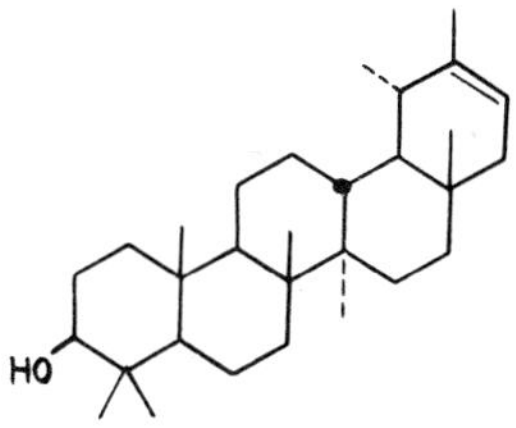

ARNIDIOL	$C_{30}H_{50}O_2$	442 +83°c 257°C
JOC. 1962 3204.		
4413-006		

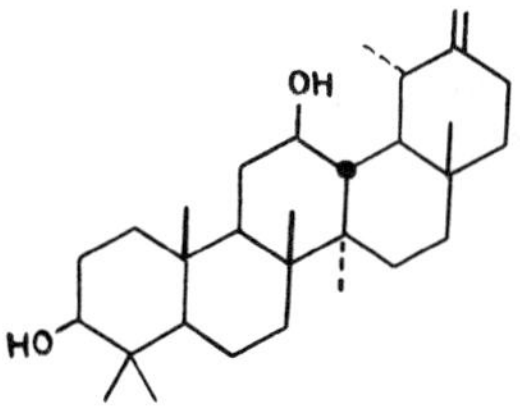

TARAXASTAN-3, 20-DIOL, pseudo-	$C_{30}H_{52}O_2$	444 -11°c 271°C
JCS. 1941 181.		
4413-003		

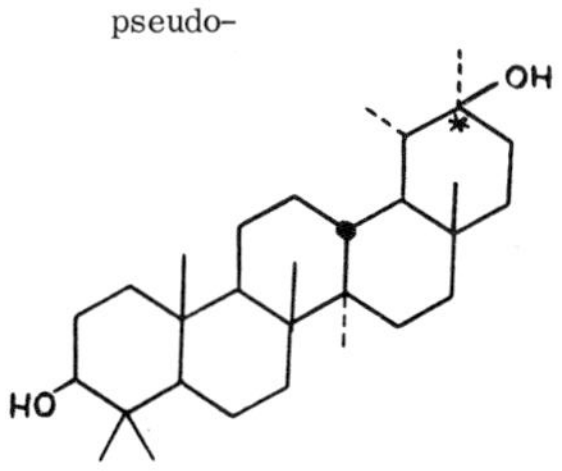

TARAXASTANONOL, epi- pseudo-	$C_{30}H_{50}O_2$	442 +25°c 258°C
Tet. 1965 3197.		
4413-007		

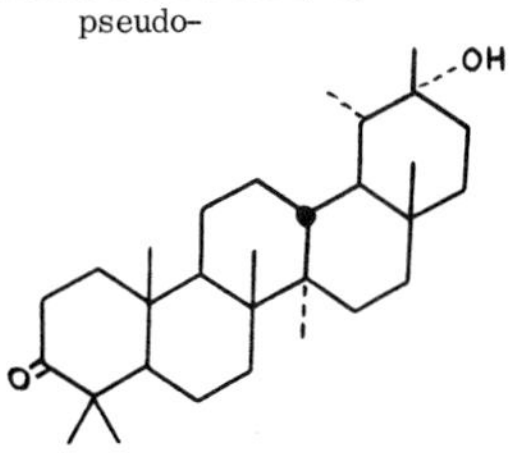

TARASTAN-3, 20-DIOL, epi- pseudo-	$C_{30}H_{52}O_2$	444 0° 262°C
*epimer		
Tet. 1966 2861.		
4413-004		

PHYLLIRIGENIN	$C_{30}H_{48}O_4$	472 +23°c 341°C
Tet. 1965 2585.		
4413-008		

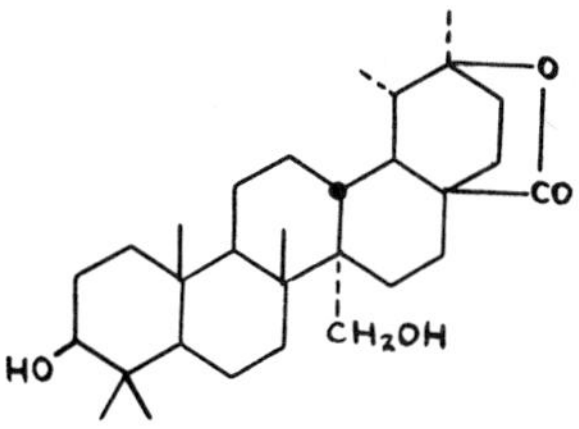

GLUTINOL $C_{30}H_{50}O$ 426 +55°c 210-12°C JCS. 1967 490. HO 4414-001	ALNUSENONE $C_{30}H_{48}O$ 424 +31°c 247°C Tet. 1958 (2) 246. O 4414-003
GLUTINOL, 3-epi- $C_{30}H_{50}O$ 426 +61°c 204°C Phytochem. 1966 1341. HO 4414-002	DENDROPANOXIDE $C_{30}H_{50}O$ 426 +68°c 207°C Chem. Pharm. Bull. 1960 1145. O 4414-004

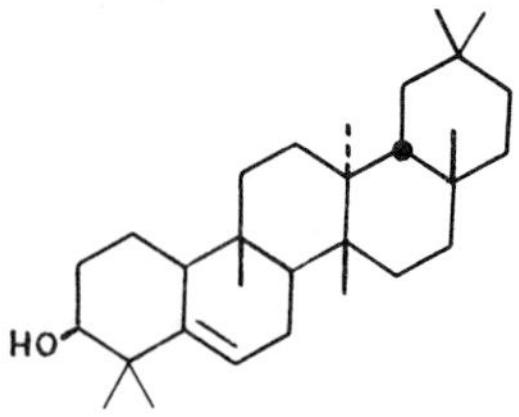

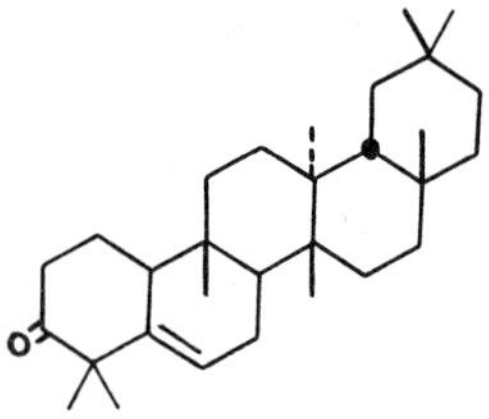

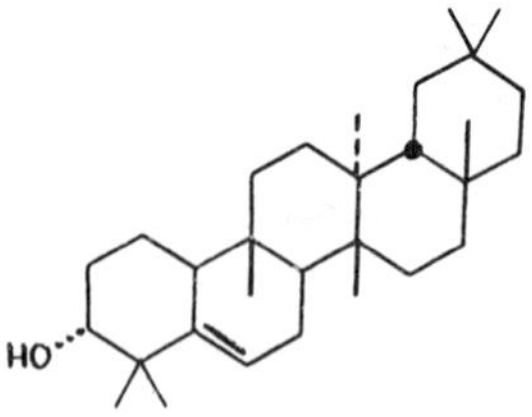

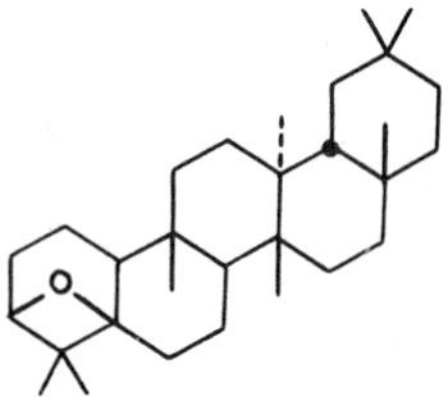

SERRATENE $C_{30}H_{50}$ 410
-13^{0}c
238^{0}C

Chem. Comm. 1967 (1) 50.

4415-001

SERRATENE, iso- $C_{30}H_{50}$ 410

Chem. Comm. 1967 (1) 50.

4415-002

SERRATENEDIOL $C_{30}H_{50}O_2$ 442
-22^{0}c
283^{0}C

R_1O OR_2

R_1 = R_2 = H

Chem. Pharm. Bull. 1967 1153.

4415-003

SERRATENEDIOL-3-ME ETHER $C_{31}H_{52}O_2$ 456

R_1 = Me; R_2 = H

Tet. Lett. 1965 2745.

4415-004

PHLEGMANOL A $C_{39}H_{58}O_5$ 606

R_1 =

O OH OH

R_2 = H

Chem. Comm. 1970 1118.

4415-005

SERRATENEDIOL DI-ME ETHER $C_{32}H_{54}O_2$ 470

R_1 = R_2 = CH_3

Tet. Lett. 1965 2745.

4415-006

SERRATENEDIOL, 21-epi- $C_{39}H_{50}O_2$ 442
-32^{0}c
301^{0}C

RO OH

R = H

Tet. Lett. 1968 761.

4415-007

SERRATENEDIOL-3-ME-
-ETHER, 21-epi- $C_{31}H_{52}O_2$ 456
+3^0
308^{0}C

R = CH_3

Tet. Lett. 1968 761.

4415-008

SERRATENEDIOL, 3, 21-di-epi- $C_{39}H_{50}O_2$ 442

R_1O OR_2

R_1 = R_2 = H

Tet. Lett. 1965 2745.

4415-009

SERRATENEDIOL 3-ME ETHER,
3, 21-di-epi- $C_{31}H_{52}O_2$ 456
-55^0
377^{0}C

R_1 = CH_3; R_2 = H

Tet. Lett. 1968 761.

4415-010

SERRATENEDIOL DI-ME ETHER, 3, 21-di-epi- $C_{32}H_{54}O_2$ 470 254-6^{0}C $R_1 = R_2 = CH_3$ Can. J. C. 1970 1021. 4415-011	SERRATRIOL $C_{30}H_{50}O_3$ 458 336^{0}C OH 21 HO CH$_2$OH Tet. Lett. 1966 5933. 4415-016
SERRATENEDIOL 3-ME ETHER, 3, 21-di-epi-iso- $C_{31}H_{52}O_2$ 456 OH MeO Tet. 1969 3737. 4415-012	SERRATRIOL, 21-epi- $C_{30}H_{50}O_3$ 458 330-3^{0}C Chem. Comm. 1969 1040. 4415-017
SERRATENE, 3β-OH-21-oxo- $C_{30}H_{48}O_2$ 440 R = H O RO 3 Tet. Lett. 1965 2745. 4415-013	LYCOCLAVANOL $C_{30}H_{50}O_3$ 458 309^{0}C OH HO CH$_2$OH Tet. Lett. 1966 5933. 4415-018
SERRATENE, 3β-OMe-21-oxo- $C_{31}H_{50}O_2$ 454 R = CH$_3$ Tet. 1969 3732. 4415-014	SERRATENEDIOL, 16-oxo- $C_{30}H_{48}O_3$ 456 294-7^{0}C OH 21 O HO 3 Chem. Comm. 1969 1042. 4415-019
SERRATENE, 3α-OMe-21-oxo- $C_{31}H_{50}O_2$ 454 -71^{0}c 240-2^{0}C R = CH$_3$ 3-epimer Can. J. C. 1970 1021. 4415-015	SERRATENEDIOL, 16-oxo-epi- $C_{30}H_{48}O_3$ 456 300-4^{0}C 21-epimer Chem. Comm 1969 1042. 4415-020

SERRATENEDIOL, 16-oxo-di-epi- $C_{30}H_{48}O_3$ 456

318-23°C

3, 21-di-epimer

Chem. Comm. 1969 1042.

4415-021

SERRATRIOL, 16-oxo- $C_{30}H_{48}O_4$ 472

294-8°C

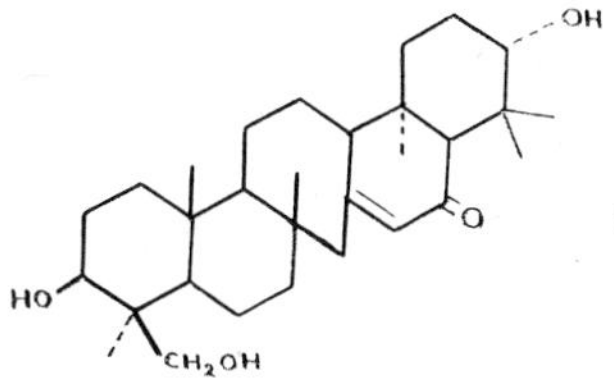

Chem. Comm 1970 261.

4415-024

TOHOGENOL $C_{30}H_{52}O_3$ 460

R = CH_3

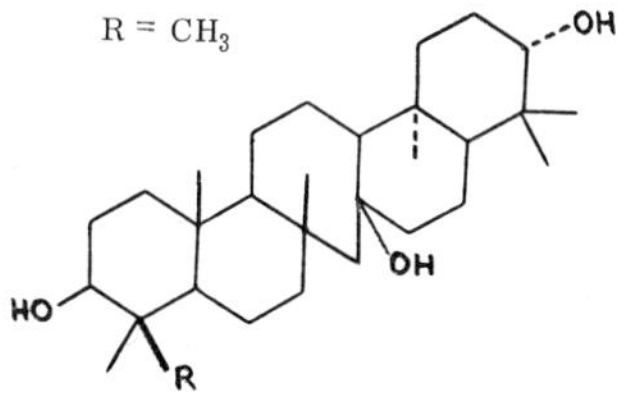

Aust. J. C. 1967 387.

4415-022

LYCOCLAVANOL, 16-oxo- $C_{30}H_{48}O_4$ 472

328-33°C

R = H

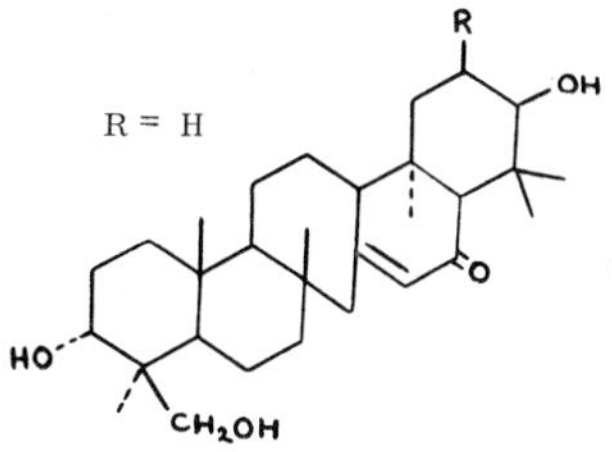

Chem. Comm 1970 261.

4415-025

TOHOGENINOL $C_{30}H_{52}O_4$ 476

312°C

R = CH_2OH

Chem. Pharm. Bull. 1965 (13) 750.

4415-023

LYCOCLAVANINE $C_{30}H_{48}O_5$ 488

344-6°C

R = OH

Chem Comm. 1970 260.

4415-026

TARAXERENE $C_{30}H_{50}$ 410 $+1^0c$ 237^0C Acta Chem. Scand. 1954 (8) 71. 4416-001	TARAXERONE $C_{30}H_{48}O$ 424 $+12^0c$ 240^0C Chem. Pharm. Bull. 1965 87. 4416-005
TARAXEROL $C_{30}H_{50}O$ 426 $+3^0c$ 280^0C R = H RO 3 JCS. 1955 2131. 4416-002	RETIGERADIOL $C_{30}H_{52}O_2$ 444 $+28^0c$ $272\text{-}4^0C$ HO HO Chem. Abstr. 1969 (70) 4315. 4416-006
SAWAMILLETIN $C_{31}H_{52}O$ 440 $+8^0$ 278^0C R = CH_3 JCS. 1959 (80) 1487. 4416-003	MYRICADIOL $C_{30}H_{50}O_2$ 442 $+10^0c$ 274^0C HO R R = CH_2OH Chem. Abstr. 1960 (54) 8889. 4416-007
TARAXEROL, 3-epi- $C_{30}H_{50}O$ 426 -0.5^0 258^0C R = H Aust. J. C. 1968 2137. 4416-004	MYRICOLAL $C_{30}H_{48}O_2$ 440 288^0C R = CHO Chem. Abstr. 1960 (54) 15431. 4416-008

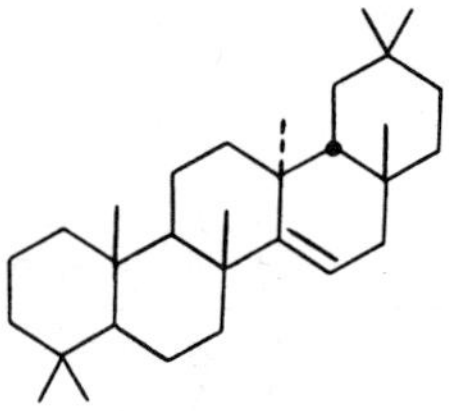

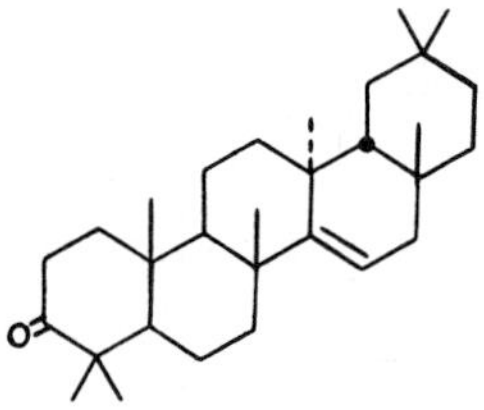

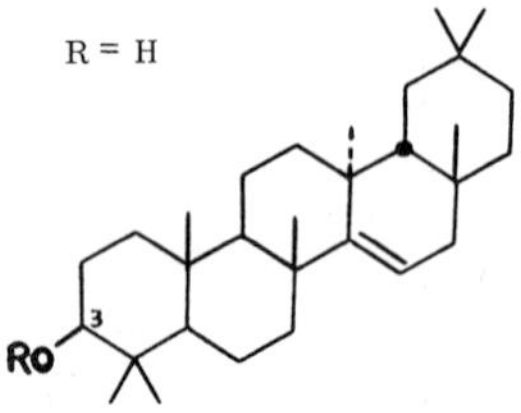

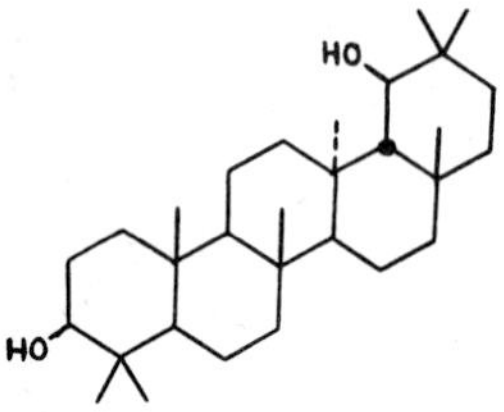

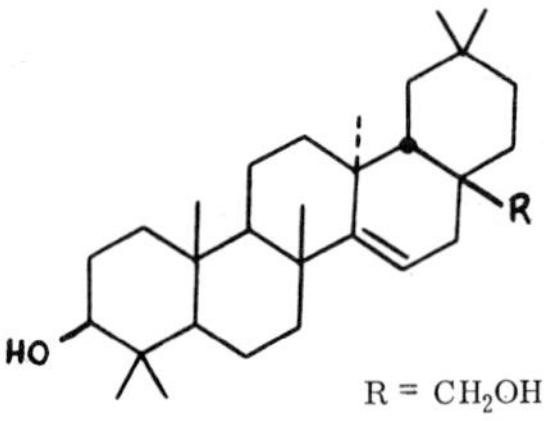

CYCLOARTANOL $C_{30}H_{52}O$ 428 +45⁰c 100⁰C HO Chem. Pharm. Bull., 1960, 108. 4417-001	CYCLOART-23-EN-3, 25-DIOL $C_{30}H_{50}O_2$ 442 +38⁰c 202⁰C R_1 = R_2 = H R_1O OR_2 JCS., 1962, 4034. 4417-006
CYCLOARTENOL $C_{30}H_{50}O$ 426 +48⁰ 85⁰C HO JCS., 1953, 3673. 4417-002	CYCLOART-23-ENE, 3-OAc-25-OMe- $C_{33}H_{54}O_3$ 498 +48⁰ 154⁰C R_1 = Ac R_2 = Me JCS., 1962, 4034. 4417-007
CYCLOARTANOL, 25-OH- $C_{30}H_{52}O_2$ 444 +49⁰ 182⁰C R = H HO OR Chem. Abstr., 1970 (73) 45627. 4417-003	CYCLOARTENONE $C_{30}H_{48}O$ 424 +24⁰c 109⁰C O JCS., 1951, 1444. 4417-008
CYCLOARTANOL, 25-OEt- $C_{32}H_{56}O_2$ 472 +48⁰ 109⁰C R = Ethyl Chem. Abstr., 1970 (73) 45627. 4417-004	CYCLOART-25-EN-24-ONE, 3-OAc- $C_{32}H_{50}O_3$ 482 +58⁰c 136⁰C AcO O JCS., 1962, 4034. 4417-009
CYCLOART-25-EN-3, 24-DIOL $C_{30}H_{50}O_2$ 442 +48⁰c 186⁰C HO OH JCS., 1962, 4034. 4417-005	CIMIGENOL $C_{30}H_{48}O_5$ 488 +38⁰c 216⁰C R = H HO OH O O OR Chem. Abstr., 1968 (69) 36348. 4417-010

CIMIGENOL 25-ME ETHER $C_{31}H_{50}O_5$ 502

R = CH_3

Chem. Abstr., 1968 (69) 59440.

4417-011

ACTEIN $C_{30}H_{46}O_6$ 502

R_1 = H; R_2 = OH

Chem. Abstr., 1968 (68) 39918.

4417-012

ACTEOL, 27-desoxy-12-O-Ac- $C_{32}H_{48}O_6$ 528 -87°c 290°C

R_1 = Ac; R_2 = H

Chem. Abstr., 1968 (69) 67559.

4417-013

MANGIFEROLIC ACID $C_{30}H_{48}O_3$ 456 +49°c 182°C

R = β- OH

Tet. Lett., 1965, 2377.

4417-014

MANGIFERONIC ACID $C_{30}H_{46}O_3$ 454 +23° 188°C

R = oxo-

Ann. Chim., 1967 (57) 508.

4417-015

CYCLOARTANOL, 24-methylene- $C_{31}H_{52}O$ 440 +43°c 122°C

Chem. Pharm. Bull., 1960, 108.

4417-016

CYCLOBRANOL $C_{31}H_{52}O$ 440 +38°c 157°C

Chem. Abstr., 1970 (73) 4058.

4417-017

CYCLOLAUDENOL $C_{31}H_{52}O$ 440 +46°c 125°C

JCS., 1955, 1607.

4417-018

CYCLOSADOL ACETATE $C_{33}H_{54}O_2$ 482 +50°c 121°C

Bull. Soc. Chim. Fr., 1969, 2037.

4417-019

CYCLOARTANOL, 24-ME-24-OMe- $C_{32}H_{56}O_2$ 472 +43° 151°C

Chem. Abstr., 1970 (73) 45627.

4417-020

CYCLOARTAN-3, 27-DIOL, 24-methylene- $C_{31}H_{52}O_3$ 472 +54° 156°C HO, CH_2OH Chem. Abstr., 1967 (67) 116988. 4417-021	CYCLOARTANONE, 24-methylene- $C_{31}H_{50}O$ 438 O Chem. Comm., 1969, 1166. 4417-023
CYCLOART-24-EN-27-AL, 3-OH- $C_{31}H_{50}O_2$ 454 +41° 166°C HO, CHO Chem. Abstr., 1967 (67) 116988. 4417-022	AMBONIC ACID $C_{31}H_{48}O_3$ 452 +9° 150°C O, COOH Chem. Abstr., 1968 (68) 13210. 4417-024

BUXTAUINE — $C_{24}H_{37}NO_2$ — 371; $+154^0$c; 178^0C

R = H

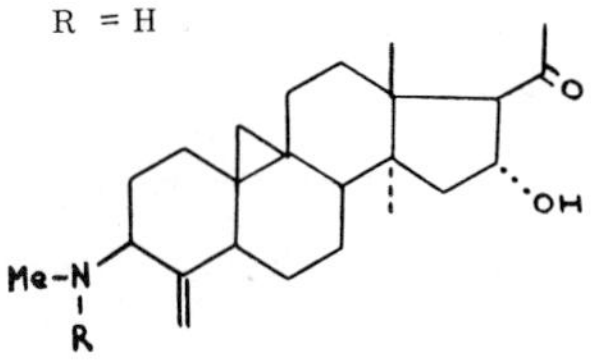

Tet. Lett., 1965, 358.
JOC., 1965, 3931.
JCS., 1965, 6688.

4418-001

BUXPIINE — $C_{25}H_{39}NO_2$ — 385; $+158^0$c; 173^0C

R = Me

JOC., 1965. 6688.
Tet. Lett., 1965, 3579.

4418-002

CYCLOMICROBUXEINE — $C_{25}H_{37}NO$ — 367

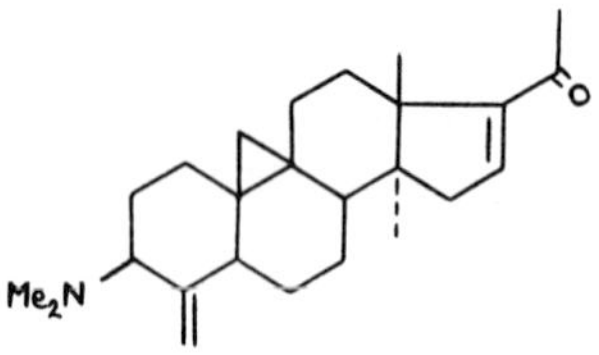

JCS., 1966, 1805.

4418-003

CYCLOBUXOMICREINE — $C_{25}H_{39}NO$ — 369; $+37^0$c; 196^0C

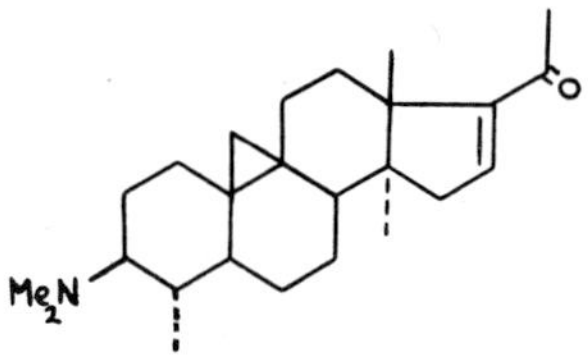

JCS., 1966, 1412.

4418-004

CYCLOSUFFROBUXININE — $C_{24}H_{35}NO$ — 353

R = H

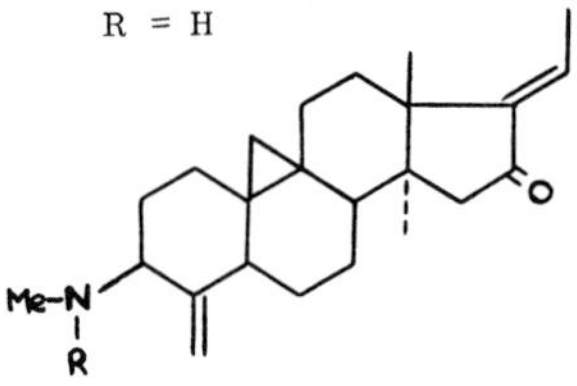

JCS., 1966, 1805.

4418-005

CYCLOSUFFROBUXINE — $C_{25}H_{37}NO$ — 367; 197^0C

R = Me

JCS., 1966, 1805.

4418-006

CYCLOBUXOSUFFRINE — $C_{25}H_{39}NO$ — 369; -62^0c; 203^0C

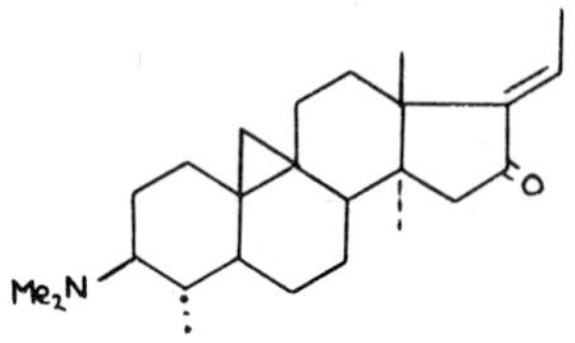

JCS., 1966, 1412.

4418-007

BUXENONE — $C_{25}H_{39}NO$ — 369; -48^0c; 174^0C

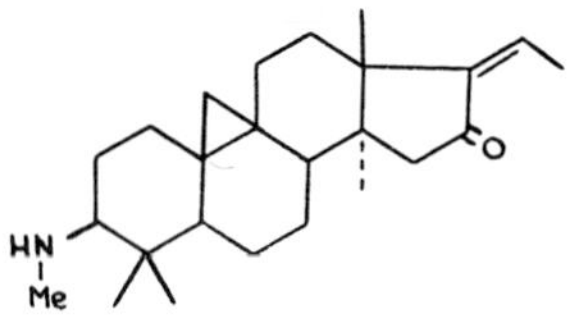

Pharmazie 1966 (21) 643.

4418-008

CYCLOBUXOPHYLLININE — $C_{25}H_{39}NO$ — 369; -51^0c; 182^0C

R_1 = H; R_2 = Me

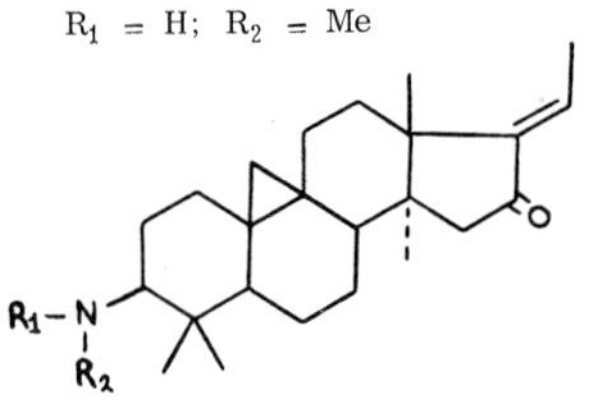

JCS., 1966, 1412.

4418-009

CYCLOBUXOPHYLLINE — $C_{26}H_{41}NO$ — 383; -72^0c; 195^0C

R_1 = R_2 = Me

JCS., 1966, 1412.

4418-010

BUXANINE $C_{32}H_{45}NO_2$ 473 -38°c 167°C Pharmazie 1966 769. 4418-011	BUXANDONINE $C_{26}H_{43}NO$ 385 +24°c 158°C R = H Pharmazie, 1968, 666. 4418-015
BUXENE $C_{27}H_{41}NO_3$ 427 203°C R_1 = H; R_2 = COOEt Tet. Lett., 1969, 4423. 4418-012	CYCLOMIKURANINE $C_{26}H_{43}NO_2$ 401 -3°c 210°C R = α - OH JCS., 1966, 1412. 4418-016
BUXENE. N-Me- $C_{28}H_{43}NO_3$ 441 -104°c 181°C R_1 = Me; R_2 = COOEt Tet. Lett., 1969, 4423. 4418-013	CYCLOBUXOVIRIDINE $C_{26}H_{41}NO$ 383 +16°c 182°C JCS., 1966, 1412. 4418-017
CYCLOROLFEINE $C_{25}H_{41}NO_2$ 387 +119°c 253°C Bull. Soc. Chim. Fr., 1966, 1216. 4418-014	CYCLOROLFOXAZINE $C_{26}H_{41}NO_3$ 415 +106°c 239°C Bull. Soc. Chim. Fr., 1966, 1216. 4418-018

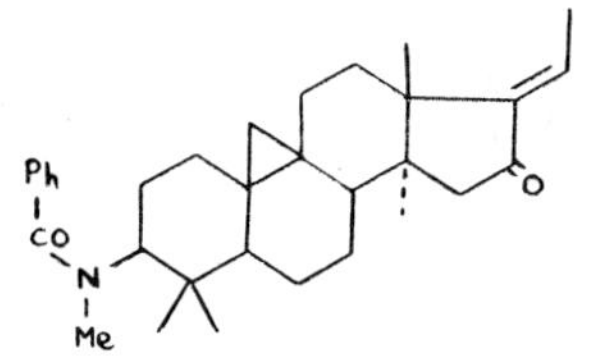

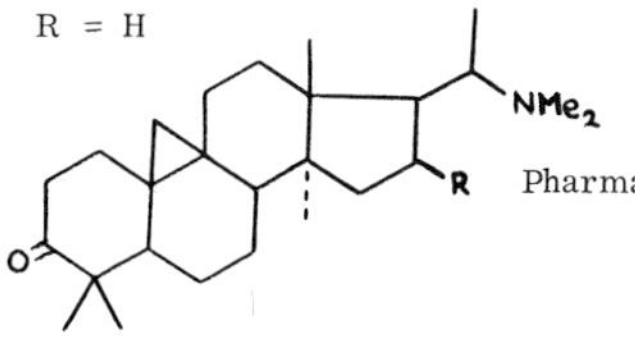

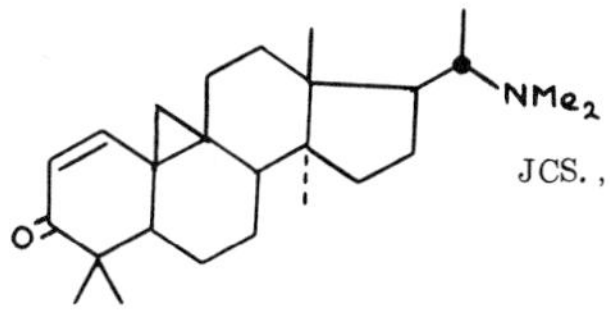

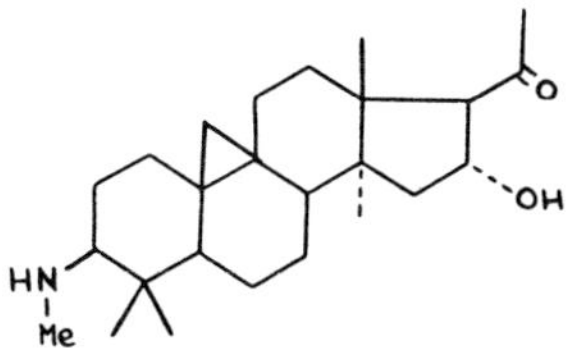

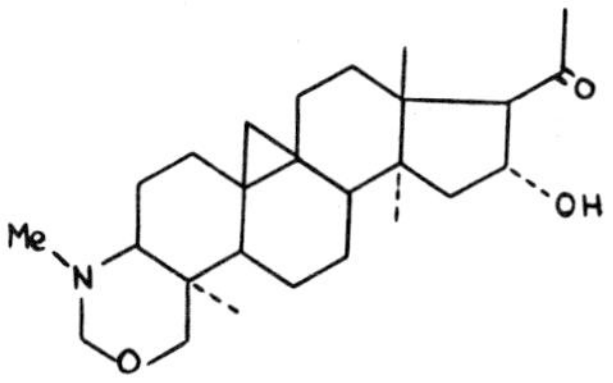

CYCLOPROTOBUXINE-D $C_{26}H_{46}N_2$ 386 +112°c 141°C

R = H

NMe
R

HN
Me

Tet. 1966 3321.

4419-001

CYCLOPROTOBUXOLINE-C, 3-N-Benzoyl- $C_{34}H_{52}N_2O_2$ 520 +43° 253°C

R = Me

HO
Ph
N
Me
NMe
R

Tet. 1967 4563.

4419-006

CYCLOPROTOBUXINE-D, 21-N-Ac- $C_{28}H_{48}N_2O$ 428 +53° 222°C

R = Ac

Tet. 1967 4563.

4419-002

CYCLOPROTOBUXOLINE-D, 3-N-Benzoyl- $C_{33}H_{50}N_2O_2$ 506 +42° 237°C

R = H

Tet. 1967 4563.

4419-007

CYCLOPROTOBUXINE-C $C_{27}H_{48}N_2$ 400 +78°c 212°C

R = Me

JCS. 1966 1805.

4419-003

CYCLOVIROBUXINE-D $C_{26}H_{46}N_2O$ 402 +63°c 220°C

R = H

N R Me
OH
MeN
R

Bull. Soc. Chim. Fr. 1966 758.

4419-008

CYCLOPROTOBUXINE-A $C_{28}H_{50}N_2$ 414 +31°c 206°C

NMe₂
Me₂N

JCS. 1966 1805.

4419-004

CYCLOVIROBUXINE A $C_{28}H_{50}N_2O$ 430 +44°c 242°C

R = Me

JCS. 1966 1412.

4419-009

BUXERIDINE $C_{34}H_{52}N_2$ 488 +14°c 208-11°C

NMe₂
Ph
N
Me

Tet. Lett. 1967 4247.

4419-005

CYCLOVIROBUXEINE- B $C_{27}H_{46}N_2O$ 414 -75°c 203°C

R = H

NHMe
OR
Me
N
Me

JOC. 1966 608.

4419-010

CYCLOVIROBUXEINE-B, O-Tigloyl- $C_{32}H_{52}N_2O_2$ 496 -150° 178-183°C

R = Tigloyl

Tet. 1967 4567.

4419-011

BALEABUXINE $C_{30}H_{50}N_2O_2$ 470 +115°c 258°C

R = isobutyroyl

Bull. Soc. Chim. Fr. 1966 3478.

4419-016

CYCLOMALAYANINE B $C_{36}H_{52}N_2O_3$ 560 -61°c 170°C

R = p-coumaroyl-

Bull. Soc. Chim. Fr. 1966 758.

4419-012

BUXATINE $C_{33}H_{48}N_2O_2$ 504 +112°c 214°C

R = benzoyl

Tet. Lett. 1967 4247.

4419-017

CYCLOVIROBUXEINE A $C_{28}H_{48}N_2O$ 428 -87°c 220°C

Bull. Soc. Chim. Fr. 1966 758.

4419-013

BUXARINE $C_{33}H_{48}N_2O_3$ 520 +98°c 211°C

Chem. Abstr. 1967 65668.

4419-018

BUXANDRINE $C_{35}H_{50}N_2O_4$ 562 289°C

Chem. Abstr. 1968 (68) 69171.

4419-014

BUXEPIDINE $C_{33}H_{50}N_2O_2$ 506 -20°c 278°C

Pharmazie 1968 666.

4419-019

CYCLOKOREANINE-B $C_{27}H_{46}N_2O$ 414 +109°c 235°C

JCS. 1966 1805.

4419-015

CYCLOMICROPHYLLINE-F. dihydro- $C_{26}H_{46}N_2O_2$ 418 +5°c 260°C

R = H

JCS. 1965 4512.

4419-020

CYCLOMICROPHYLLINE-C, dihydro- $C_{27}H_{48}N_2O_2$ 432 +74°c 256°C

R = Me

JCS. 1965 4512.
Bull. Soc. Chim. Fr. 1966 1216.

4419-021

CYCLOMICROPHYLLINE-A $C_{28}H_{48}N_2O_2$ 444 -92°c 233°C

R = H

NMe2, OR, Me2N, CH2OH

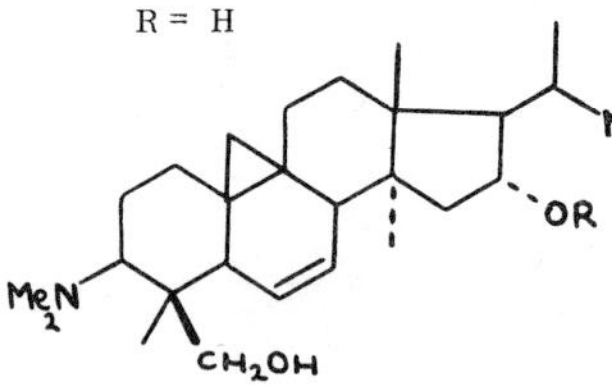

JCS. 1965 4512.

4419-026

CYCLOMICROPHYLLINE-F, N-Benzoyl-dihydro- $C_{33}H_{50}N_2O_3$ 522 +19° 293°C

R = Benzoyl

Tet. 1967 4563.

4419-022

CYCLOMICROPHYLLIDINE-A $C_{35}H_{52}N_2O_3$ 548 -160°c amorph

R = Benzoyl

JCS. 1965 4512.

4419-027

CYCLOMICROPHYLLINE-A, dihydro- $C_{28}H_{50}N_2O_2$ 446 +37°c 272°C

NMe2, OH, Me2N, CH2OH

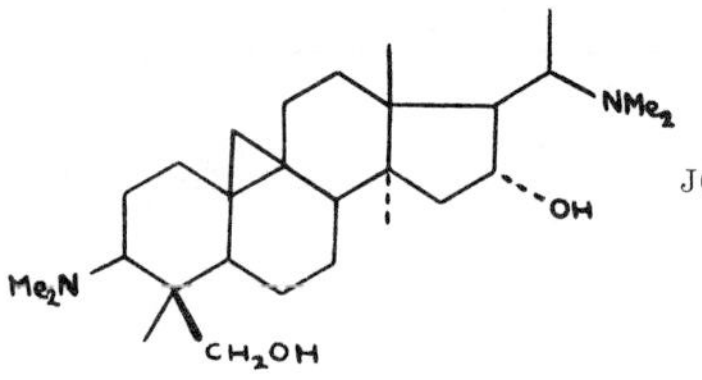

JCS. 1965 4512.

4419-023

CYCLOMICROSINE $C_{34}H_{50}N_2O_3$ 534 -33°c 283°C

Ph, CO, N, Me, NMe2, OH, CH2OH

Tet. Lett. 1967 4247.

4419-028

CYCLOMICROPHYLLINE B $C_{27}H_{46}N_2O_2$ 430 -66°c 252°C

NHMe, OH, Me2N, CH2OH

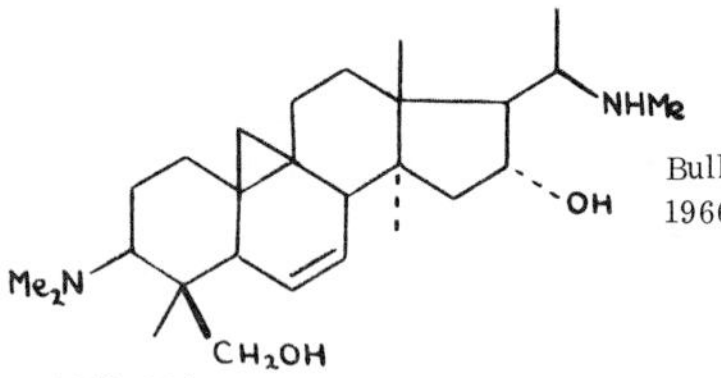

Bull. Soc. Chim. Fr. 1966 758.

4419-024

BUXIDINE $C_{33}H_{48}N_2O_3$ 520 +67.5°c 154-7°C

Ph, CO, NH, NMe2, OH, CH2OH

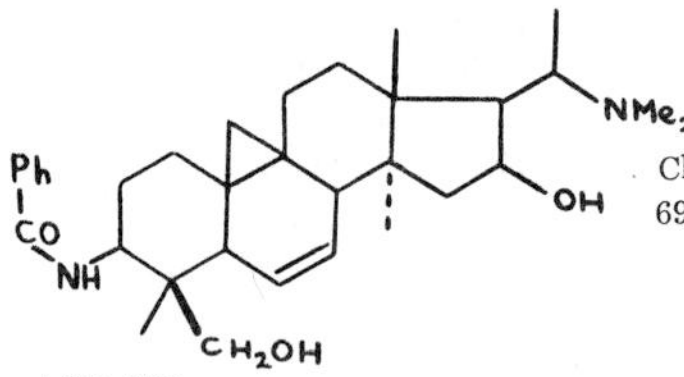

Chem. Abstr. 1968 (68) 69171.

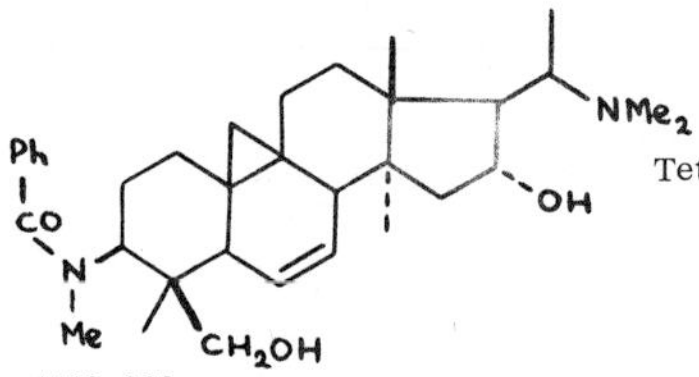

4419-029

CYCLOMICROPHYLLINE-C $C_{27}H_{46}N_2O_2$ 430 -40°c 237°C

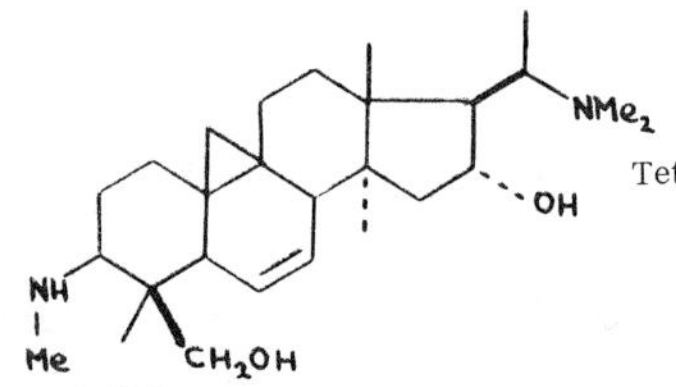

NMe2, OH, NH, Me, CH2OH

Tet. Lett. 1967 4247.

4419-025

BUXAZIDINE B $C_{27}H_{46}N_2O_2$ 430 -31°c 235°C

NHMe, O, Me2N, CH2OH

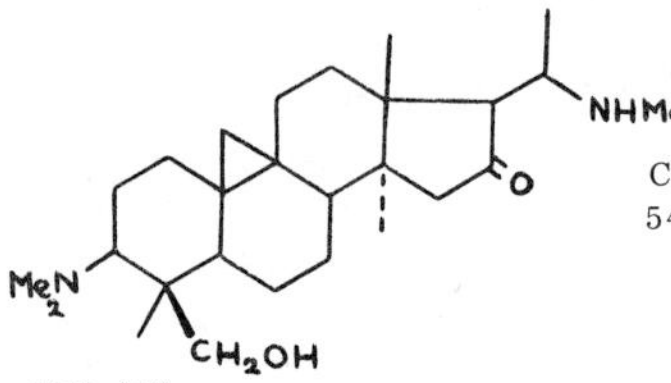

Chem. Abstr. 1967 (67) 54344.

4419-030

CYCLOXOBUXOLINE-F, N-Benzoyl- — $C_{33}H_{48}N_2O_3$ — 520 — +76° — 255°C

R = H

Ph, CO, NH, O, CH_2OR, NMe_2

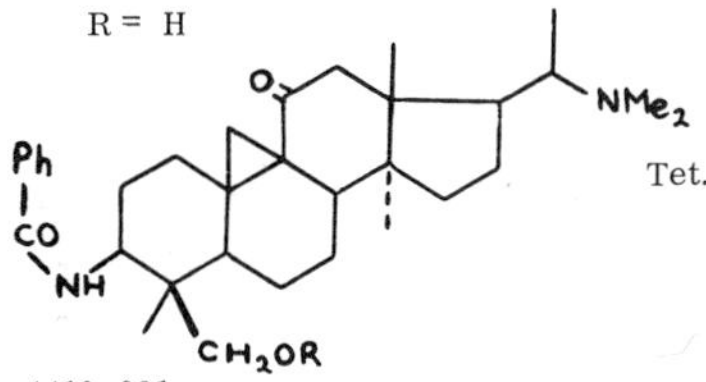

Tet. 1967 4563.

4419-031

CYCLOBUXOXAZINE-C — $C_{27}H_{46}N_2O_2$ — 430 — +29°c — 246°C

R = H

R-N, O, NMe_2, OH

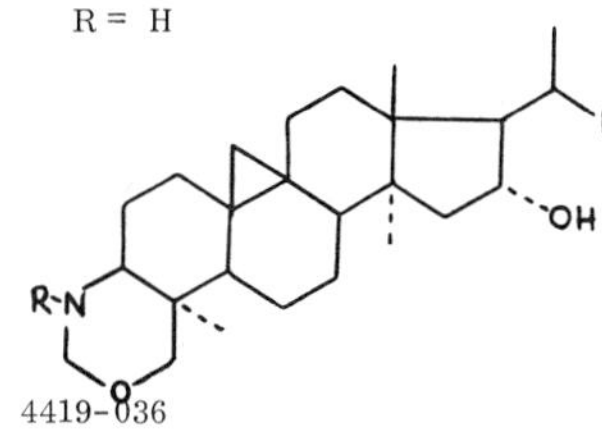

JCS. 1965 4537.

4419-036

CYCLOXOBUXOLINE-F, N-Benzoyl-O-Ac- — $C_{35}H_{50}N_2O_4$ — 562 — +114° — 217°C

R = Ac

Tet. 1967 4563.

4419-032

CYCLOBUXOXAZINE-A — $C_{28}H_{48}N_2O_2$ — 444 — +45°c — 197°C

R = Me

Bull. Soc. Chim. Fr. 1966 1216.

4419-037

CYCLOXOBUXIDINE-H, N-i-butyroyl- — $C_{29}H_{48}N_2O_4$ — 488 — +76°c — 285°C

R_1 = isobutyroyl; R_2 = H

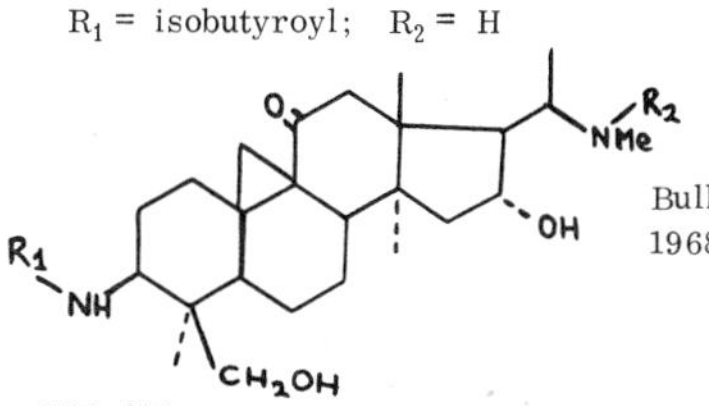

R_1, NH, O, CH_2OH, R_2, NMe, OH

Bull. Soc. Chim. Fr. 1968 763.

4419-033

CYCLOMETHOXAZINE-B — $C_{28}H_{48}N_2O_2$ — 444 — +54°c — 195°C

Me-N, O, NHMe, OH

Bull. Soc. Chim. Fr. 1966 1216.

4419-038

BALEABUXIDINE — $C_{30}H_{50}N_2O_4$ — 502 — +150°c — 257°C

R_1 = isobutyroyl

R_2 = Me

Tet. 1966 3321.

4419-034

BUXOXAZINE — $C_{27}H_{46}N_2O_3$ — 446 — +116°c — 292°C

HN, O, O, NMe_2, OH

Bull. Soc. Chim. Fr. 1968 763.

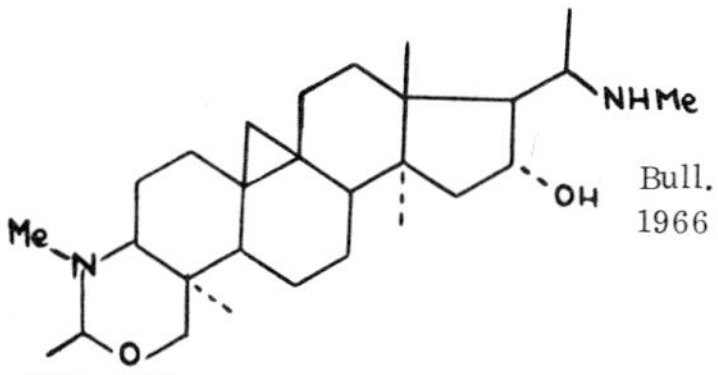

4419-039

CYCLOXOBUXIDINE-F, N-Benzoyl- — $C_{33}H_{48}N_2O_4$ — 536 — +52°c — 227°C

R_1 = benzoyl

R_2 = Me

Bull. Soc. Chim. Fr. 1968 763.

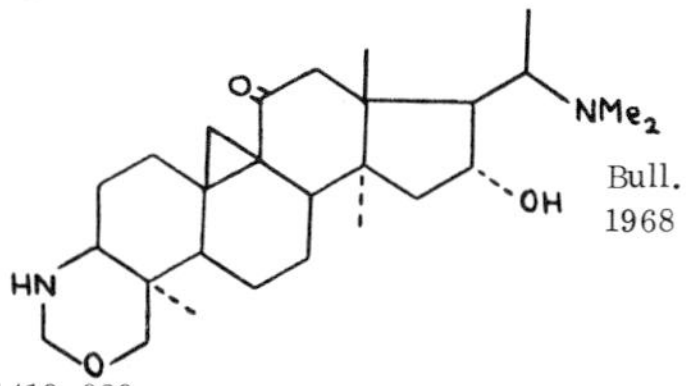

4419-035

BUXOCYCLAMINE A — $C_{27}H_{48}N_2$ — 400 — +87°c — 188°C

Me_2N, NMe_2

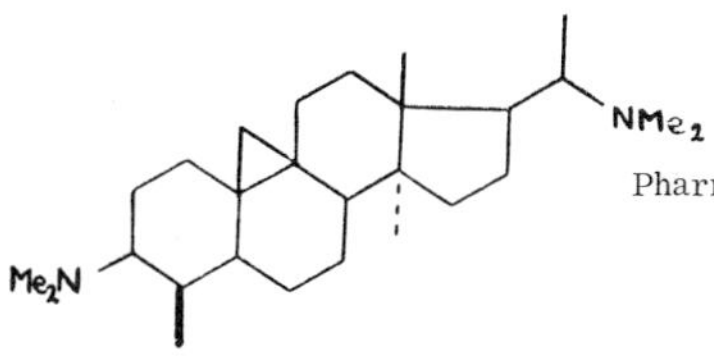

Pharmazie 1968 37.

4419-040

CYCLOBUXAMINE $C_{24}H_{42}N_2O$ 374 +30°c 210°C

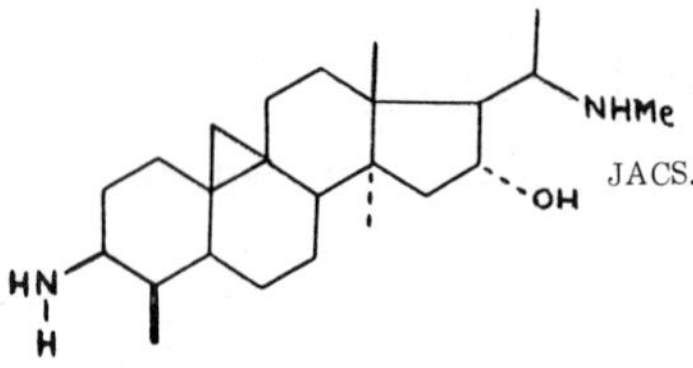

JACS. 1964 4430.

4419-041

CYCLOBUXINE-D $C_{25}H_{42}N_2O$ 386 +98°c 246°C

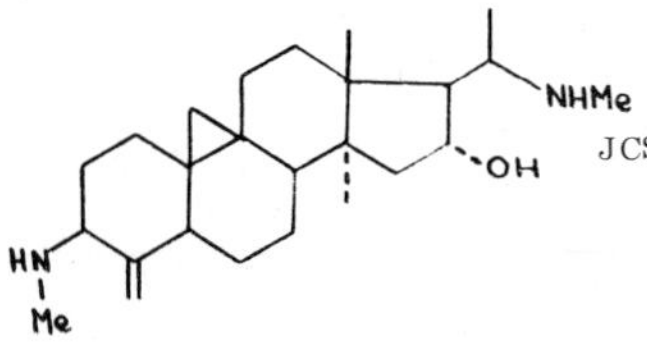

JCS. 1966 1805.

4419-042

MALABARICANEDIOL $C_{30}H_{52}O_3$ 460 +23°c HO OH Tet. Lett. 1967 4841. 4420-001	MALABARICOL, epoxy- $C_{30}H_{50}O_4$ 474 +25°c 144°C OH O O Tet. Lett. 1967 4841. 4420-003
(+)-MALABARICOL $C_{30}H_{50}O_3$ 458 +36°c 69°C O O OH Tet. Lett. 1968 2215. 4420-002	

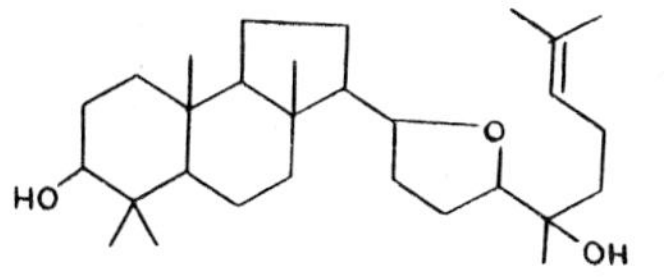

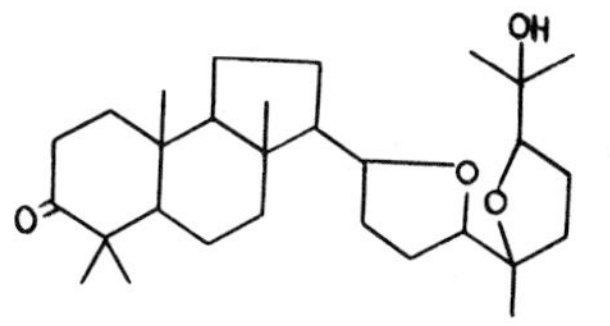

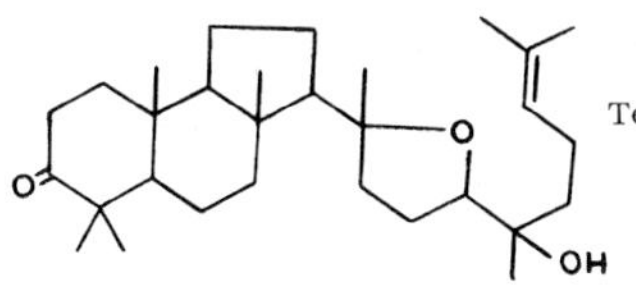

PROTOSTA-17(20), 24-DIENE, 4-OH- $C_{30}H_{50}O$ 426 +10°c 104°C HO Tet. Lett. 1969 1023. 4421-001	ALISOL A, mono-Ac- $C_{32}H_{52}O_6$ 532 +86° 195°C R = Ac Tet. Lett. 1968 105. 4421-006
PROTOSTA-13(17), 24-DIENE, 3-OH- $C_{30}H_{50}O$ 426 HO Tet. Lett. 1969 1023. 4421-002	ALISOL B $C_{30}H_{48}O_4$ 472 +130° 167°C HO, OH, O Tet. Lett. 1968 105. 4421-007
PROTOSTA-17(20), 24-DIEN-3, 31-DIOL $C_{30}H_{50}O_2$ 442 +19° 143°C HO, CH_2OH Tet. Lett. 1968 4769. 4421-003	ALISOL B, mono-Ac- $C_{32}H_{50}O_5$ 514 +121° 162°C O, OAc Tet. Lett. 1968 105. 4421-008
ALISOL A $C_{30}H_{50}O_5$ 490 +99° R = H HO, OH, OR, OH, O Tet. Lett. 1968 105. 4421-004	ALISOL C ACETATE $C_{32}H_{48}O_6$ 528 +101°c 233°C HO, O, OAc Chem. Pharm. Bull. 1970 1369. 4421-009
ALISOL A, epi $C_{30}H_{50}O_5$ 490 +81° *epimer (?) Tet. Lett. 1968 103. R = H 4421-005	

ARBORINOL $C_{30}H_{50}O$ 426 +34°c 274°C R = H RO 3 Tet. 1967 131. 4422-001	CYLINDRIN $C_{31}H_{52}O$ 440 R = CH_3 3-epimer Tet. 1967 131. 4422-004
ARBORINOL, Me ether $C_{31}H_{52}O$ 440 +11°c 284°C R = CH_3 Chem. Comm. 1969 601. 4422-002	ARBORINONE $C_{30}H_{48}O$ 424 +29° 216°C O Phytochemistry 1967 727. 4422-005
ISO-ARBORINOL $C_{30}H_{50}O$ 426 +66° 330°C R = H 3-epimer Tet. 1967 131. 4422-003	

LIMONIN, desoxy- $C_{26}H_{30}O_7$ 454 -39^0 336^0C JOC. 1965 749. 4423-001	OBACUNONE $C_{26}H_{30}O_7$ 454 -50^0c 230^0C R = H JCS. 1961 255. 4423-006
LIMONIN $C_{26}H_{30}O_8$ 470 -126^0a 300^0C *epoxide JCS. 1961 255. 4423-002	ZAPOTERIN $C_{26}H_{30}O_8$ 470 270^0C R = OH JOC. 1968 3577. 4423-007
EVODOL $C_{26}H_{28}O_9$ 484 -199^0a 282^0C Chem. Pharm. Bull. 1963 535. 4423-003	OBACUNOIC ACID $C_{26}H_{32}O_8$ 472 Chem. Abstr. 1965 (62) 12157. 4423-008
RUTAEVIN $C_{26}H_{30}O_9$ 486 $285\text{-}91^0C$ JOC. 1967 3442. 4423-004	VEPRISONE $C_{27}H_{34}O_8$ 486 -18^0c 180^0C Tet. 1964 2985. 4423-009
OBACUNOL, 7α- $C_{26}H_{32}O_7$ 456 $242\text{-}5^0C$ JOC. 1968 3577. 4423-005	NOMILIN $C_{28}H_{34}O_9$ 514 -96^0a 178^0C R = Ac JCS. 1961 255. 4423-010

NOMILIN, des-Ac- $C_{26}H_{32}O_8$ 472 -112⁰ 265⁰C R = H JOC. 1965 749. 4423-011	ICHANGIN $C_{26}H_{32}O_9$ 488 210⁰C JOC. 1966 2279. 4423-013
SPATHELIN $C_{29}H_{34}O_{10}$ 542 -7⁰ 180⁰C MeOOC OAc Tet. Lett. 1968 5299. 4423-012	

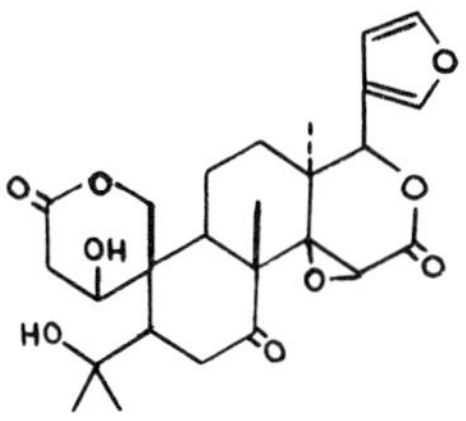

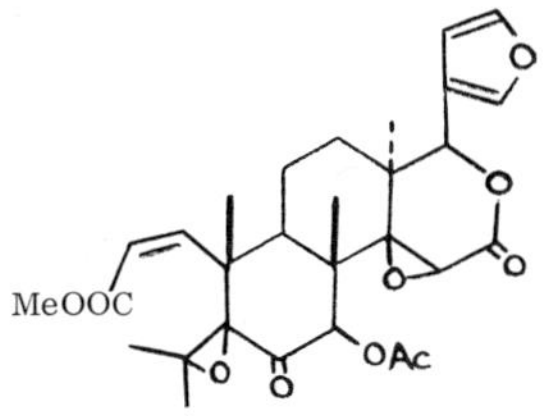

SWIETENINE. 6-deoxy-des-tigloyl- — $C_{27}H_{34}O_7$ — 470 — 260-5°C

R = H

JCS. 1968 1974.

4424-001

SWIETENINE — $C_{29}H_{36}O_8$ — 512 — -150°c — 223-6°C

R = Ac

JCS. 1968 1974.

4424-002

XYLOCARPIN — $C_{29}H_{36}O_9$ — 528 — -88°c — 210-12°C

JCS. 1970 211.

4424-003

SWIETENINE. 12-OAc-6--deoxy-des-tigloyl-Ac- — $C_{31}H_{38}O_{10}$ — 570 — -131°c — 251°C

JCS. 1968 1974.

4424-004

SWIETENINE — $C_{32}H_{42}O_9$ — 570 — -168° — 260°C

Tet. Lett. 1964 2593.

4424-005

MEXICANOLIDE — $C_{27}H_{32}O_7$ — 468 — -100°c — 226°C

Tet. 1968 1489

4424-006

MEXICANOLIDE. 3β-dihydro- — $C_{27}H_{34}O_7$ — 470

R = H

Chem. Comm 1967 500.

4424-007

FISSINOLIDE — $C_{29}H_{36}O_8$ — 512 — -165° — 170°C

R = Ac

Tet. Lett. 1970 2798.

4424-008

KHAYASIN — $C_{31}H_{40}O_8$ — 540

R = isobutyroyl

Chem. Comm. 1966 27.

4424-009

KHAYASIN T — $C_{32}H_{40}O_8$ — 552 — -146°c — 146°C

R = tigloyl

JCS. 1968 1974.

4424-010

Name	Formula	Data	Reference	No.
KHAYASIN B (R = benzoyl)	$C_{34}H_{38}O_8$	574, -156^{0}c, 198^{0}C	JCS. 1968 1474.	4424-011
KHAYASIN dihydro-	$C_{31}H_{42}O_8$	542, -116^0	JCS. 1970 205.	4424-012
SWIETENOLIDE	$C_{27}H_{34}O_8$	486, -136^{0}c, 222^{0}C	Tet. 1968 1503: 1507.	4424-013
CARAPIN	$C_{27}H_{32}O_7$	468, +64^0, 178^{0}C	Chem. Comm. 1966 (16) 576.	4424-014
CARAPIN ACETATE, 3-β--dihydro-	$C_{29}H_{34}O_8$	510, 235-7^{0}C	JCS. 1964 2440.	4424-015
CARAPIN, 6-OH-	$C_{27}H_{32}O_8$	484	Tet. Lett. 1967 3449.	4424-016
ANGUSTINOLIDE	$C_{29}H_{36}O_8$	512, -185^{0}c, 168-71^{0}C	Tet. 1970 222.	4424-017
ANGUSTIDIENOLIDE	$C_{29}H_{34}O_8$	510, +133^{0}c, 248-54^{0}C	Tet. 1970 222.	4424-018
ANGUSTIDIENOLIDE, 2-hydroxy-	$C_{29}H_{34}O_9$	526, +101^{0}c, 210-20^{0}C	Tet. 1970 222.	4424-019

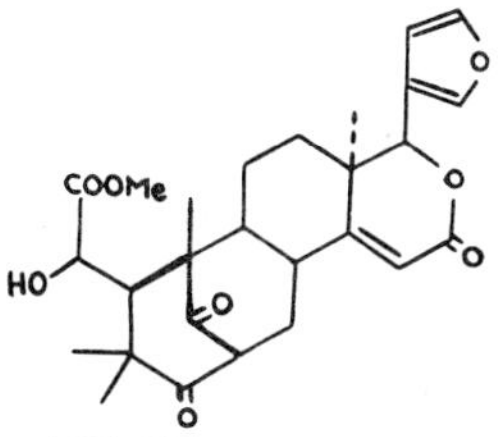

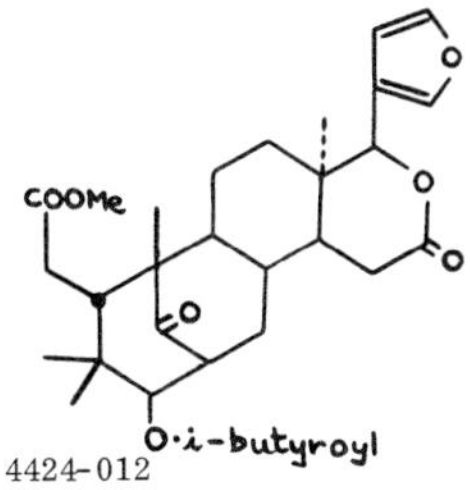

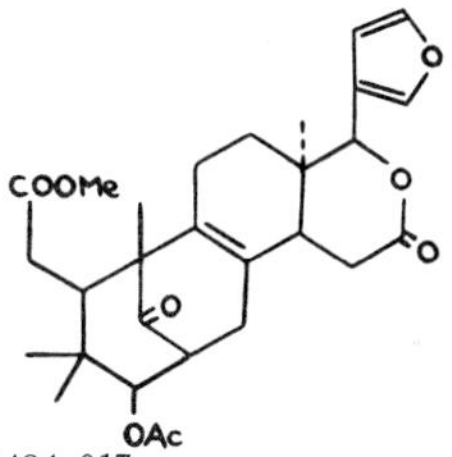

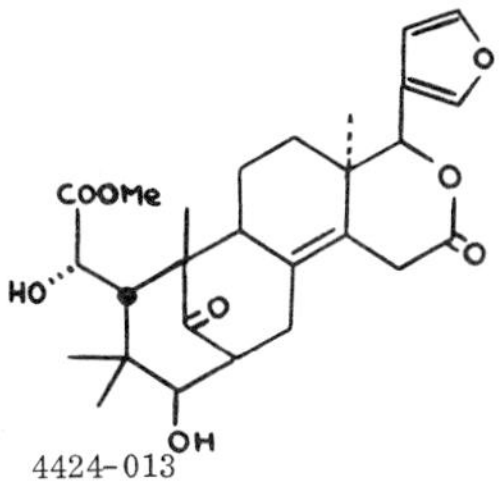

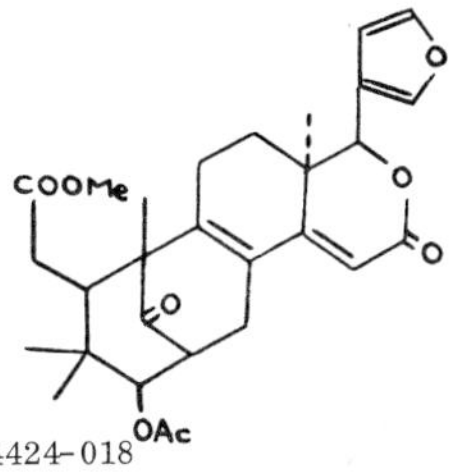

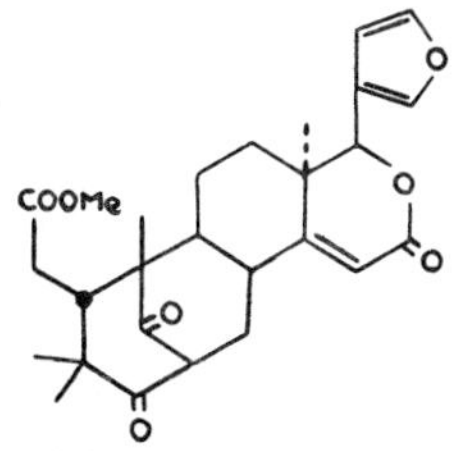

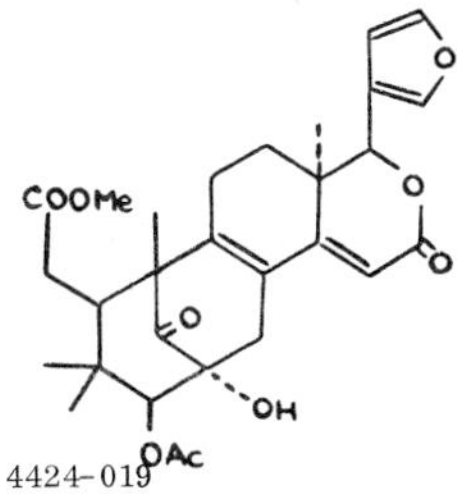

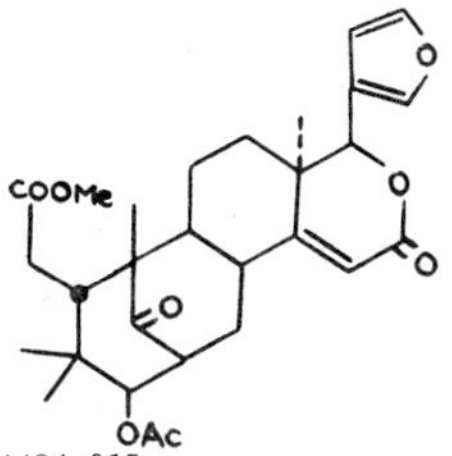

QUASSIN | $C_{22}H_{28}O_6$ | 388, $+35^0$m, 222^0C

R = Me

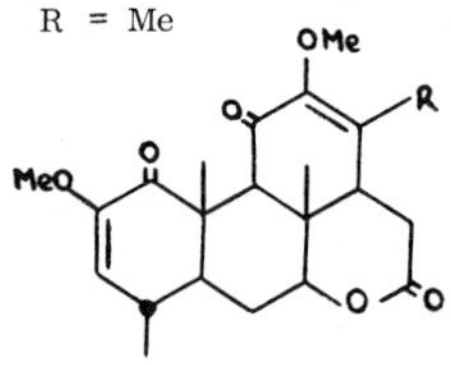

Tet. Lett., 1960 (20) 25.

4425-001

QUASSIN, 17-OH- | $C_{22}H_{28}O_7$ | 404, $+24^0$c, 232^0C

R = CH_2OH

Chem. Abstr., 1967 (66) 75879.

4425-002

QUASSIN, neo- | $C_{22}H_{30}O_6$ | 390, $+41^0$m, 231^0C

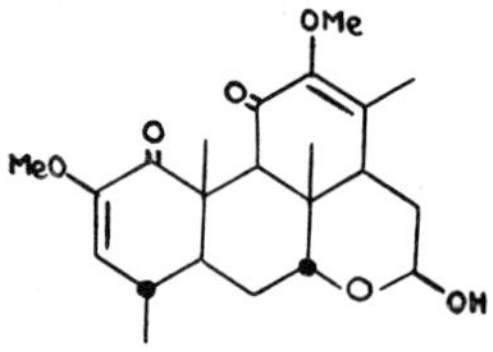

Tet. Lett., 1960 (20) 25.

4425-003

NIGAKILACTONE A | $C_{21}H_{30}O_6$ | 378, $+35^0$e, 238^0C

R_1 = R_2 = H

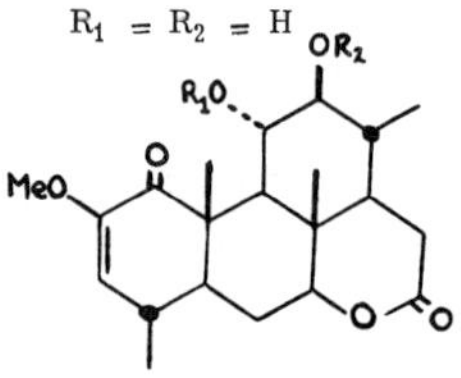

Tet. Lett., 1969, 3013.

4425-004

NIGAKILACTONE B | $C_{22}H_{32}O_6$ | 392, $+17^0$e, 278^0C

R_1 = H; R_2 = Me

Tet. Lett., 1969, 3013.

4425-005

NAGAKILACTONE C | $C_{24}H_{34}O_7$ | 434, $+9^0$e, 253^0C

R_1 = Ac; R_2 = Me

Tet. Lett., 1969. 3013.

4425-006

NIGAKILACTONE E | $C_{24}H_{34}O_8$ | 450, $+36^0$e, 280^0C

R = Ac

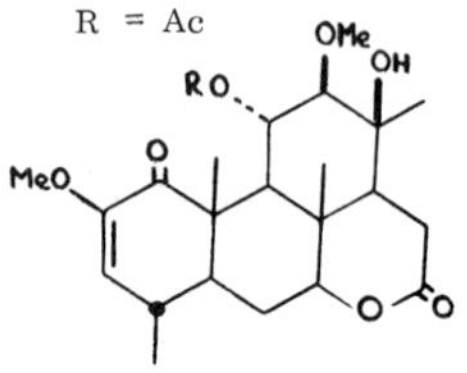

Bull. Chem. Soc. Jap., 1970 (43) 969.

4425-007

NIGAKILACTONE F | $C_{22}H_{32}O_7$ | 408, $+46^0$e, 265^0C

R = H

Bull. Chem. Soc. Jap., 1970 (43) 969.

4425-008

NIGAKILACTONE H | $C_{22}H_{32}O_8$ | 424, $+67^0$e, 275^0C

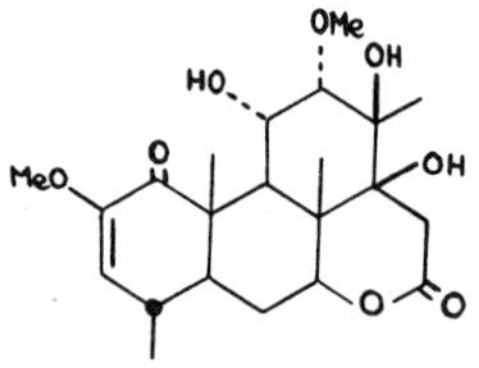

Bull. Chem. Soc. Jap., 1970, 3021.

4425-009

NIGAKIHEMIACETAL A | $C_{22}H_{34}O_7$ | 410, $+20^0$e, 263^0C

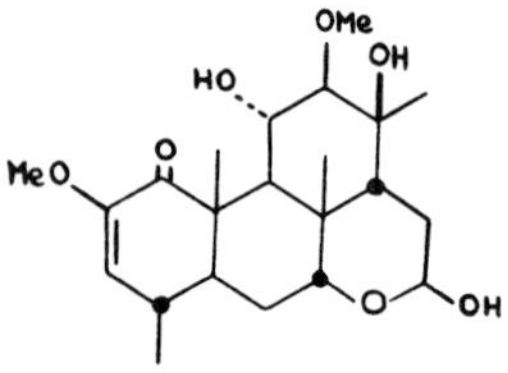

Bull. Chem. Soc. Jap. 1970, 3021.

4425-010

NIGAKIHEMIACETAL C — $C_{21}H_{32}O_6$ — 380, $+49^0$m, 265^0C — Chem. Pharm. Bull., 1970, 2590. — 4425-011	GLAUCARUBOLONE — R = H — $C_{20}H_{26}O_8$ — 394, -34^0p, 256^0C — Tet. Lett., 1968, 385. — 4425-016
KLAINEANONE — $C_{20}H_{28}O_6$ — 364, $+32^0$c, 268^0C — Bull. Soc. Chim. Fr., 1965, 2739. — 4425-012	GLAUCARUBINONE — R = — $C_{25}H_{34}O_{10}$ — 494, $+50^0$p, 230^0C — Phytochemistry, 1965, 149. — 4425-017
AILANTHONE — $C_{20}H_{24}O_7$ — 376, $+13^0$, 238^0C — Tet. Lett., 1964, 3983. — 4425-013	BRUCEINE D — $C_{20}H_{26}O_9$ — 410, -21^0p, 286^0C — Compt. Rend., 1968 (267) 1346. — 4425-018
deleted — 4425-014	BRUCEINE G — $C_{20}H_{26}O_8$ — 394, $+59^0$p, 256^0C — Experimentia, 1968, 768. — 4425-019
CHAPARRINONE — $C_{20}H_{26}O_7$ — 378, -47^0p, 240^0C — * dihydro- — Bull. Soc. Chim. Fr., 1965, 2793. — 4425-015	BRUSATOL — R = — $C_{26}H_{32}O_{11}$ — 520, $+44^0$, 277^0C — JOC., 1968, 429. — 4425-020

BRUCEINE A R = isopentyl 4425-021	$C_{26}H_{34}O_{11}$ 522 -82°p 270°C JOC., 1968, 429.	GLAUCARUBIN $C_{25}H_{36}O_{10}$ 496 +69° 255°C Chem. Comm., 1965, 204. 4425-026
BRUCEINE B R = acetyl 4425-022	$C_{23}H_{28}O_{11}$ 480 -77°p 266°C Experimentia, 1967, 424.	CHAPARRIN $C_{20}H_{28}O_7$ 380 +14° 308°C Can. J. C., 1965, 3013. 4425-027
BRUCEINE C R = 4425-023	$C_{28}H_{36}O_{12}$ 564 -34° Experimentia, 1967, 424.	PICRASIN B $C_{21}H_{28}O_6$ 376 255°C Chem. Pharm. Bull., 1970, 219. 4425-028
BRUCEINE E R = Me 4425-024	$C_{20}H_{28}O_9$ 412 +25°p 263°C Compt. Rend., 1968 (267) 1346.	AMAROLIDE $C_{20}H_{28}O_6$ 364 260°C R = H Tet. Lett., 1970, 2399. 4425-029
BRUCEINE F R = CH_2OH 4425-025	$C_{20}H_{28}O_{10}$ 428 225°C Compt. Rend., 1969 (268) 1392.	AMAROLIDE, acetyl- $C_{22}H_{30}O_7$ 406 265°C R = Ac Tet. Lett., 1965, 2273. 4425-030

TRICHILENONE ACETATE	$C_{28}H_{36}O_5$	452	Chem. Comm 1967 720.
4426-001			
HAVANENSIN, 1,3,7-triAcetate	$C_{32}H_{44}O_8$	556	Chem. Comm 1967 720.
*epoxide; R_1 = R_2 = Ac			
4426-006			
MELDENIN	$C_{28}H_{38}O_5$	454 240-4^{0}C	Tet. Lett. 1968 437.
4426-002			
AZADIRONE	$C_{28}H_{36}O_4$	436 +26^{0}c	Chem. Comm. 1967 278.
R = H			
4426-007			
HAVANENSIN-3,7-DIACETATE, deoxy-	$C_{30}H_{42}O_6$	498 -18^0 200-4^{0}C	JCS. 1970 205.
R_1 = H; R_2 = Ac			
4426-003			
AZADIRADIONE	$C_{28}H_{34}O_5$	450 -24^{0}c	Chem. Comm. 1967 278.
R = oxo			
4426-008			
HAVANENSIN, 1,7-diAcetate-	$C_{30}H_{42}O_7$	514	Chem. Comm 1967 720.
*epoxide; R_1 = Ac; R_2 = H			
4426-004			
AZADIRADIONE, epoxy-	$C_{28}H_{34}O_6$	466 -75^{0}c 199	Chem. Comm. 1967 278.
*epoxide; R = oxo			
4426-009			
HAVANENSIN, 3,7-diAcetate-	$C_{30}H_{42}O_7$	514	Chem. Comm. 1967 720.
*epoxide; R_1 = H; R_2 = Ac			
4426-005			
CEDRELONE	$C_{26}H_{30}O_5$	422 -64^{0}c 204^{0}C	JCS. 1963 2515.
4426-010			

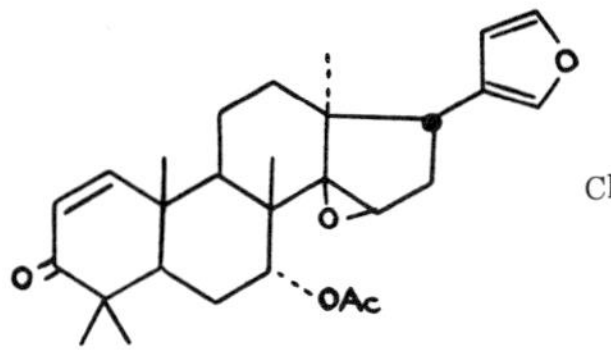

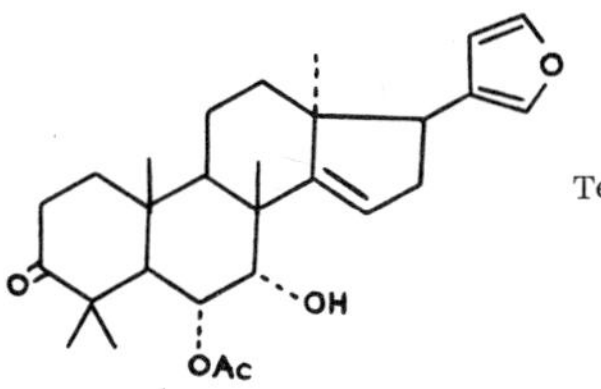

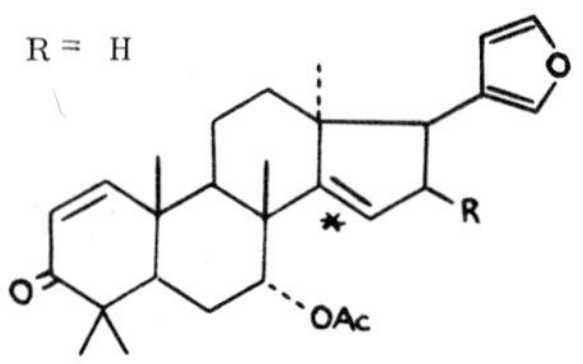

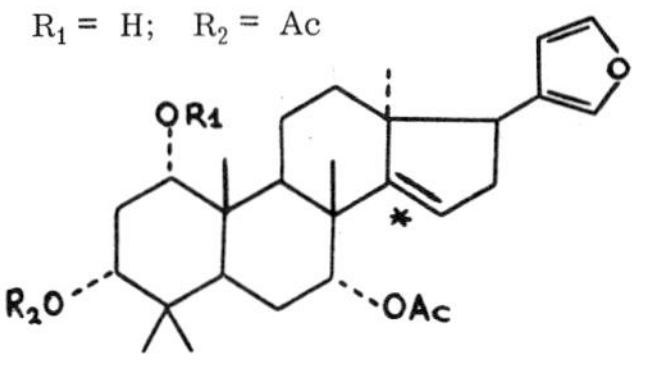

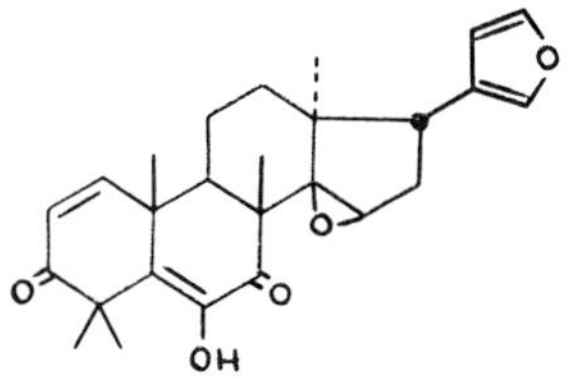

ANTHOTHECOL | $C_{28}H_{32}O_7$ | 480 -63°c 225°C

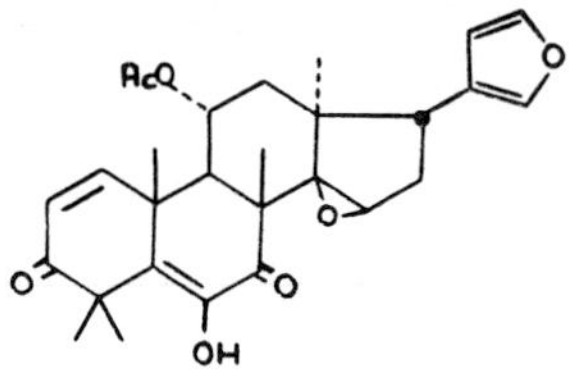

JCS. 1963 983.

4426-011

NIMBININ | $C_{28}H_{34}O_6$ | 466 +45°c 203°C

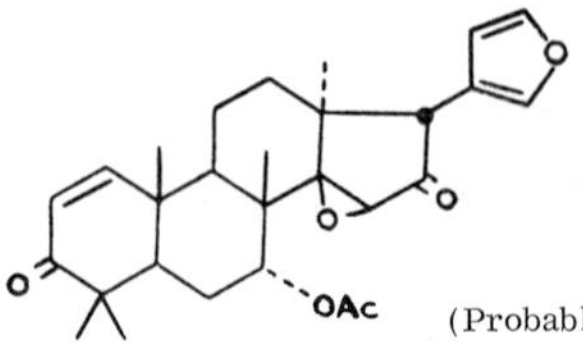

Tet. Lett. 1967 3563.

(Probable Structure)

4426-012

HEUDELOTTIN D | $C_{35}H_{46}O_{11}$ | 642

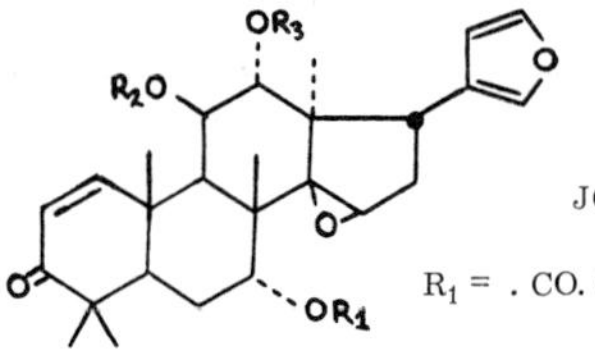

JCS. 1968 1828.

R_1 = . CO. $C_5H_{11}O_2$

R_2 = . CO. H R_3 = Ac

4426-013

HEUDELOTTIN E | $C_{38}H_{52}O_{11}$ | 684 +55°c 125°C

R_1 = . CO. $C_5H_{11}O_2$

R_2 = α- Me-butyroyl

JCS. 1968 1828.

R_3 = formyl

4426-014

HEUDELOTTIN C | $C_{39}H_{54}O_{11}$ | 698 +45°c 183°C

R_1 = . CO. $C_5H_{11}O_2$

R_2 = α- Me-butyroyl

JCS. 1968 1828.

R_3 = Acetyl

4426-015

HEUDELOTTIN F | $C_{41}H_{58}O_{11}$ | 726 +62° 161°C

R_1 = . CO. $C_{11}H_{23}O_2$

R_2 = formyl

JCS. 1968 1828.

R_3 = Acetyl

4426-016

GRANDIFOLIONE | $C_{30}H_{10}O_8$ | 528 -35°c 233-5°C

R = H

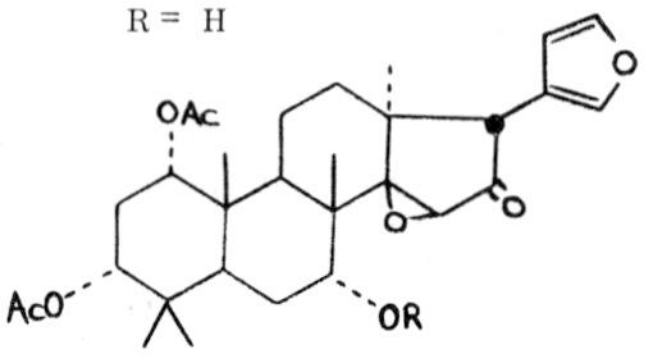

JCS. 1968 2227.

4426-017

KHAYANTHONE | $C_{32}H_{42}O_9$ | 570 -57° 216°C

R = Ac

JCS. 1970 205.

4426-018

HIRTIN, desAc- | $C_{30}H_{34}O_{10}$ | 554

R = H

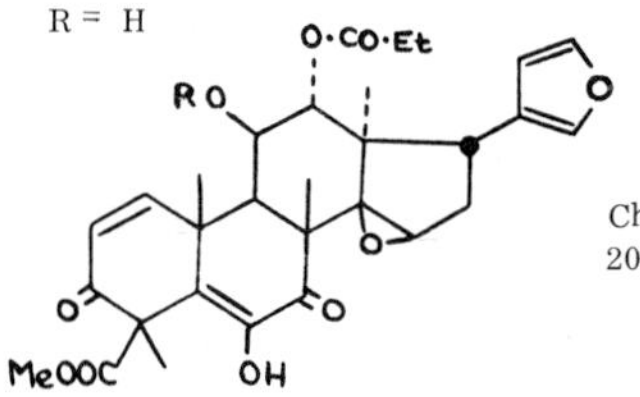

Chem. Comm. 1966 (7) 206.

4426-019

HIRTIN | $C_{32}H_{36}O_{11}$ | 596 +26°c 160°C

R = Ac

Chem. Comm. 1966 (7) 206.

4426-020

NIMBOLIN A $C_{39}H_{46}O_8$ 642
-39^0
180-3^{0}C

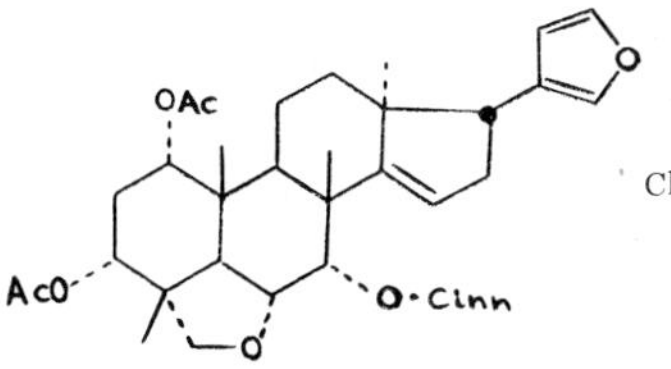

Chem. Comm. 1969 1166.

4426-021

GEDUNIN, 7-oxo-

$C_{26}H_{32}O_6$ 440

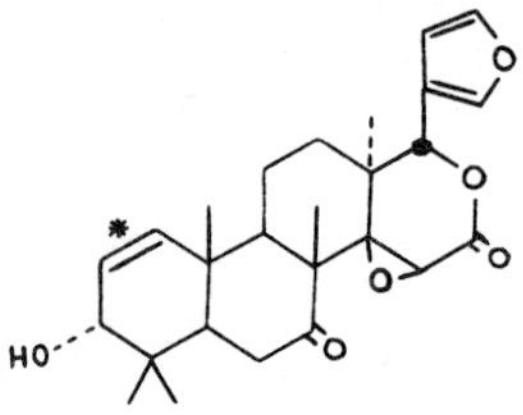

JCS. 1963 980.

4427-001

GEDUNOL, 7-oxo-dihydro-

$C_{26}H_{34}O_6$ 442
-126°c
301°C

*dihydro

JCS. 1962 5095.

4427-002

GEDUNIN, 7-desAc-

$C_{26}H_{32}O_6$ 440
+75°c
250°C

R = H

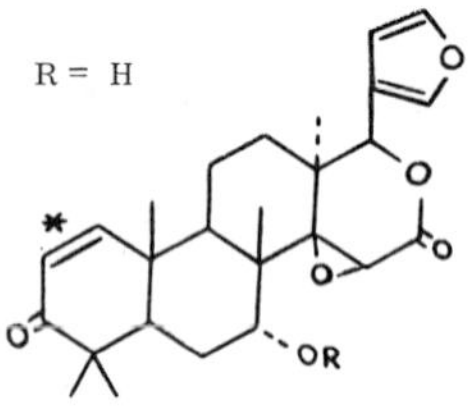

JCS. 1961 3705.

4427-003

GEDUNIN

$C_{28}H_{34}O_7$ 482
-44°
218°C

R = Ac

JCS. 1966 944.

4427-004

GEDUNIN, dihydro-

$C_{28}H_{36}O_7$ 484
+4°c
238°C

*dihydro

R = Ac

JCS. 1963 980.

4427-005

GEDUNIN, 11β-OAc-

$C_{30}H_{36}O_9$ 540
+33°
177°C

R = H

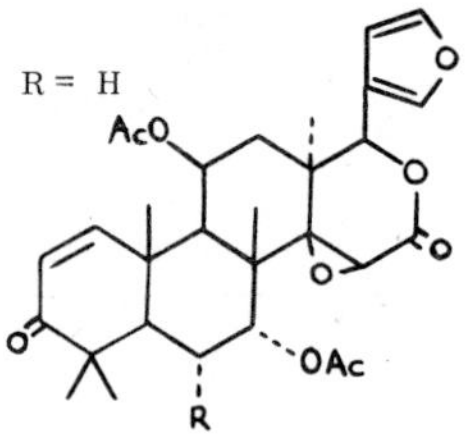

Tet. 1966 891.

4427-006

GEDUNIN, 6α, 11β-diOAc-

$C_{32}H_{38}O_{11}$ 598
+120°

R = OAc

Tet. 1966 891.

4427-007

KHIVORIN

$C_{32}H_{42}O_{10}$ 586
-42°c
256-63°C

$R_1 = R_2 =$ Ac

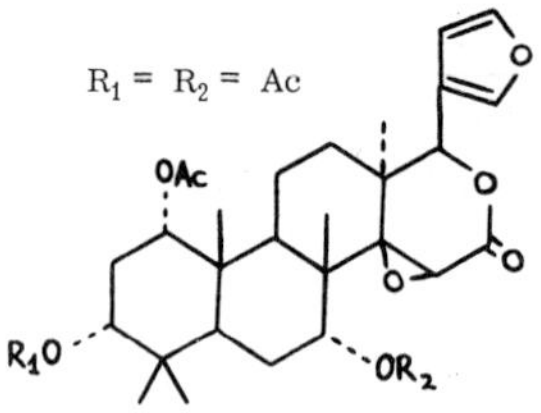

JCS. 1966 506.

4427-008

KHIVORIN, 3-deacetyl-

$C_{30}H_{40}O_9$ 544
-38°
180

$R_1 =$ H; $R_2 =$ Ac

JCS. 1967 554.

4427-009

KHIVORIN, 7-desAc-

$C_{30}H_{40}O_9$ 544
-32°c
amorph.

$R_1 =$ Ac; $R_2 =$ H

Chem. Comm. 1966 867.

4427-010

KHIVORIN, 3, 7-di-des-Ac-7-oxo- $C_{28}H_{36}O_8$ 500 R = H JCS. 1967 554. 4427-011	GEDUNIN, photo- $C_{28}H_{34}O_9$ 514 289-92^{0}C Tet. 1969 5007. 4427-014
KHIVORIN, 7-des-OAc-7-oxo- $C_{30}H_{38}O_9$ 542 -106^{0}c 229^{0}C R = AC 4427-012	ENTANDROPHRAGMIN $C_{43}H_{56}O_{17}$ 844 -4^{0}c 256^{0}C Chem. Comm. 1967 (2) 81. 4427-015
KHIVORIN, 11β-OAc- $C_{34}H_{44}O_{12}$ 644 -27^{0}c 268^{0}C Chem. Comm. 1968 1172. 4427-013	NYASIN $C_{32}H_{42}O_{10}$ 586 -42^0 302^{0}C Chem. Comm. 1967 500. 4427-016

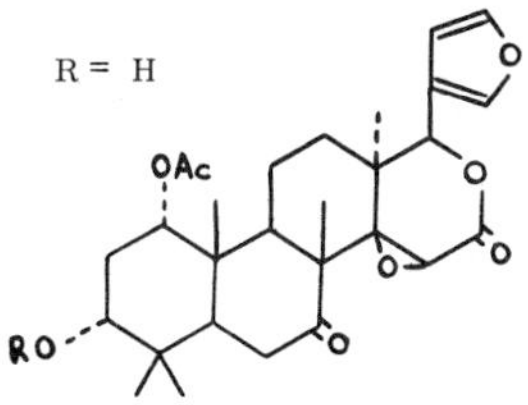

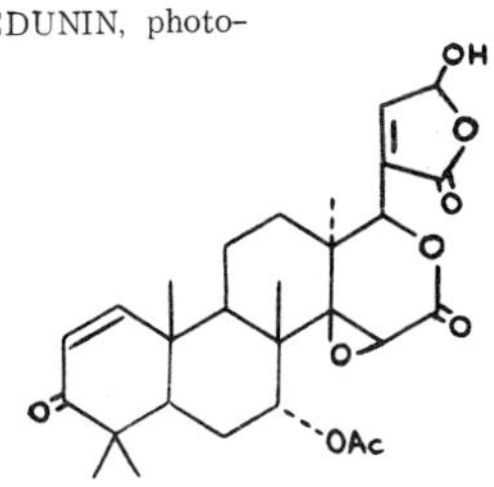

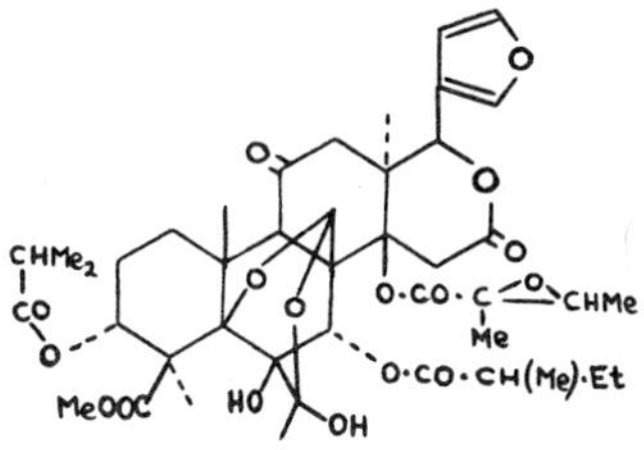

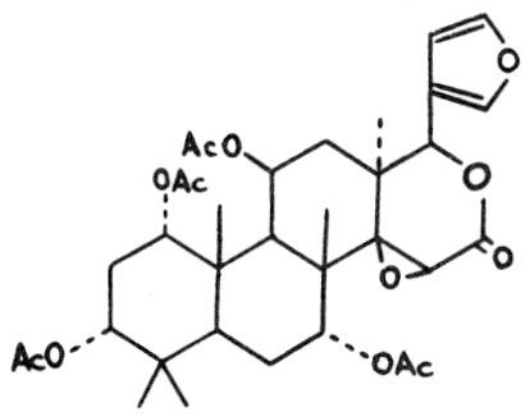

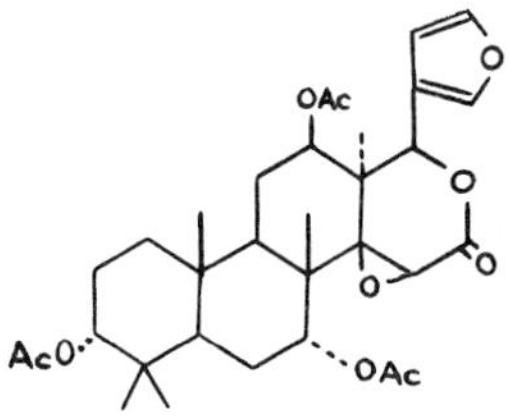

ANDIROBIN, desoxy-	$C_{27}H_{32}O_6$	452	METHYL ANGOLENSATE	$C_{27}H_{34}O_7$	470 -48° 201°C
COOMe	Chem. Ind. 1967 1365.		COOMe	JCS. 1967 171.	
4428-001			4428-004		
ANDIROBIN	$C_{27}H_{32}O_7$	468 +38°c 196°C	6-OH-ANGOLENSATE, Me-	$C_{27}H_{34}O_8$	486 -83° 252°C
* epoxide	Tet. 1970 1637.		R = H OR COOMe	JCS. 1968 1974.	
4428-002			4428-005		
APHANAMIXININ	$C_{27}H_{34}O_7$	470 -120°c 208°C	ANGOLENSATE, 6-OAc-Methyl-	$C_{29}H_{36}O_9$	528 -82° 173°C
COOMe	Tet. 1970 1859.		R = Ac	JCS. 1968 1974.	
4428-003			4428-006		

FUSIDIC ACID | $C_{31}H_{48}O_6$ | 516 | -9^0 | 193^0C

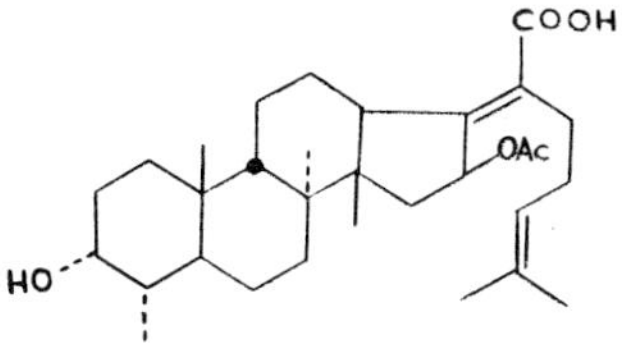

Tet. 1965 3505.

4429-001

HELVOLIC ACID, 6-des-OAc | $C_{31}H_{42}O_6$ | 510 | -52^0c | 215^0C

R = H

Chem. Pharm. Bull. 196
1966 (14) 436.

4429-005

FUSIDA-1, 17(20), 24-TRIEN-21-OIC ACID, 3-oxo-16-OAc- | $C_{31}H_{44}O_5$ | 496 | $+35^0$ | 204^0C

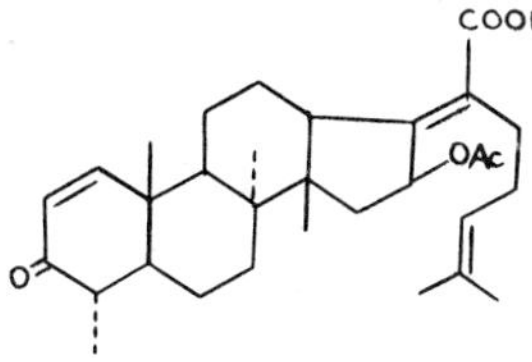

Tet. Lett. 1968 4847.

4429-002

CEPHALOSPORIN P_1 | $C_{33}H_{50}O_8$ | 574 | $+28^0$ | 147^0C

R_1 = Ac; R_2 = H

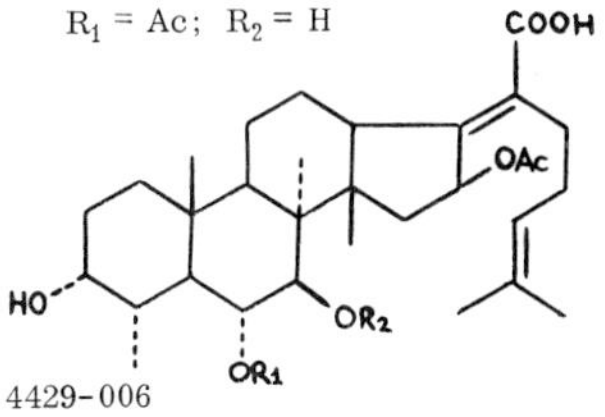

Chem. Comm. 1966 685.

4429-006

HELVOLIC ACID | $C_{33}H_{44}O_8$ | 568 | -124^0 | $211-26^0C$

R = OAc

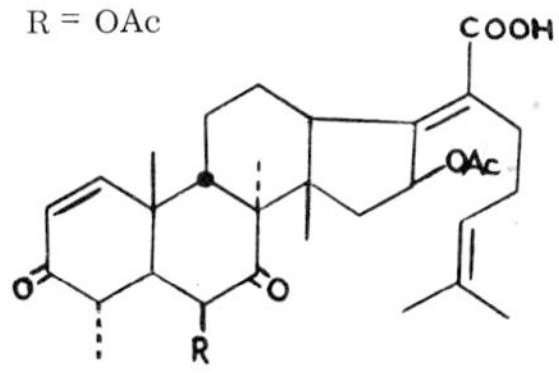

Chem. Comm. 1970 1119.

4429-003

CEPHALOSPORIN P_1, monodes-Ac- | $C_{31}H_{48}O_7$ | 532 | $+38^0m$ | $197-8^0C$

R_1 = R_2 = H

Tet. 1969 3341.

4429-007

HELVOLINIC ACID | $C_{31}H_{42}O_7$ | 526 | -89^0c | 203^0C

R = OH

Chem. Pharm. Bull. 1966 436.

4429-004

CEPHALOSPORIN P_1, iso- | $C_{33}H_{50}O_8$ | 574

R_1 = H; R_2 = Ac

Tet. 1969 3341.

4429-008

SQUALENE, pre- $C_{30}H_{50}O$ 426

Chem. Comm. 1971 218.

4491-001

AMBREIN $C_{30}H_{52}O$ 428 $+14^0$ 84^0C

Helv. 1946 912.

4491-005

ONOCERIN, α- $C_{30}H_{50}O_2$ 442 $+18^0c$ 202^0C

JCS. 1955 2639.

4491-002

PHYLLANTHOL $C_{30}H_{50}O$ 426 $+43^0c$ 234^0C

JCS. 1954 3810.

4491-006

ONOCERIN, β- $C_{30}H_{50}O_2$ 442 $+111^0c$ 234^0C

JCS. 1955 2639.

4491-003

TETRAHYMANOL $C_{30}H_{52}O$ 428 313^0C

JACS. 1968 3563.

4491-007

ONOCERADIENEDIONE $C_{30}H_{46}O_2$ 438 128^0C

Tet. Lett. 1968 3731.

4491-004

WALLICHIENENE	$C_{30}H_{48}$	408 $+42^0c$ 211^0C	Chem. Comm. 1968 1105.
4492-001			
ADIANENEDIOL $R_1 = R_2 = OH$	$C_{30}H_{50}O_2$	442 $+3^0$ 230^0C	Tet. Lett. 1965 2017.
4492-006			
WALLICHIENE	$C_{30}H_{50}$	410 ±0 196^0C	Chem. Comm. 1968 1105.
4492-002			
SIMIARENOL $R_1 = H;\ R_2 = OH$	$C_{30}H_{50}O$	426 $+51^0$ 210^0C	Chem. Comm. 1969 601.
4492-007			
HOPENE, Neo-	$C_{30}H_{50}$	410 $+42^0c$ 211^0C	Chem. Comm. 1968 1105.
4492-003			
FILICENE $R = CH_3$	$C_{30}H_{50}$	410 225^0C	Tet. Lett. 1966 6069.
4492-008			
MOTIOL, neo-	$C_{30}H_{50}O$	426 -44^0 218^0C	Tet. Lett. 1965 2017.
4492-004			
FILICENAL R = CHO	$C_{30}H_{48}O$	424 $+74^0$ 272^0C	Tet. Lett. 1966 6069.
4492-009			
ADIANENE $R_1 = R_2 = H$	$C_{30}H_{50}$	410 $+51^0$ 191^0C	Tet. Lett. 1964 3413.
4492-005			
ADIANTOXIDE $R = CH_3$ *epoxide	$C_{30}H_{50}O$	426 228^0C	Tet. 1969 2939.
4491-010			

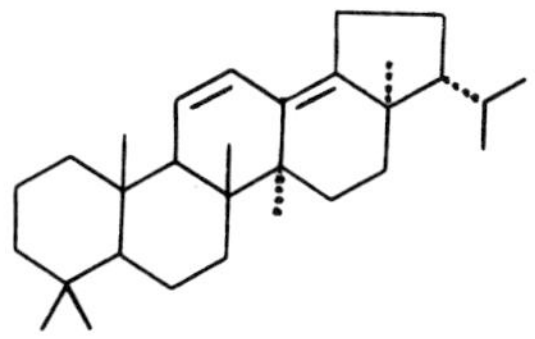

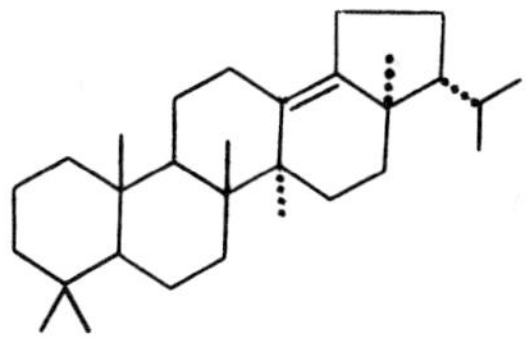

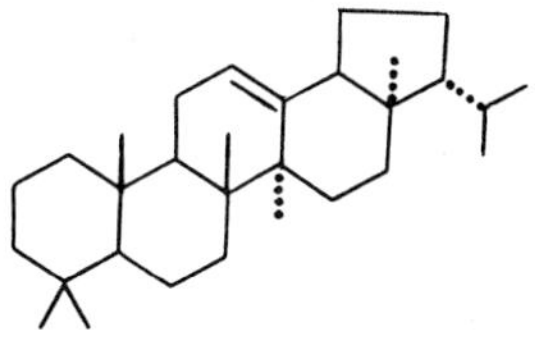

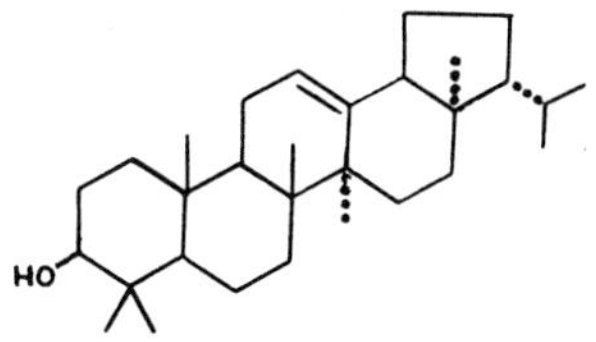

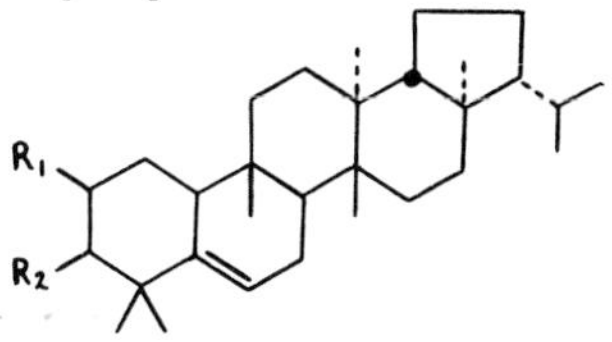

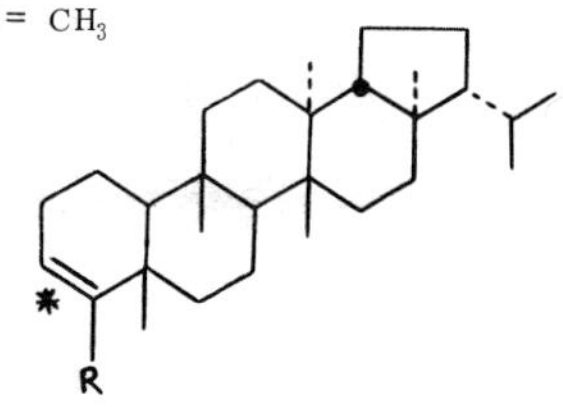

GRANDIFOLIOLENONE $C_{32}H_{48}O_6$ 528

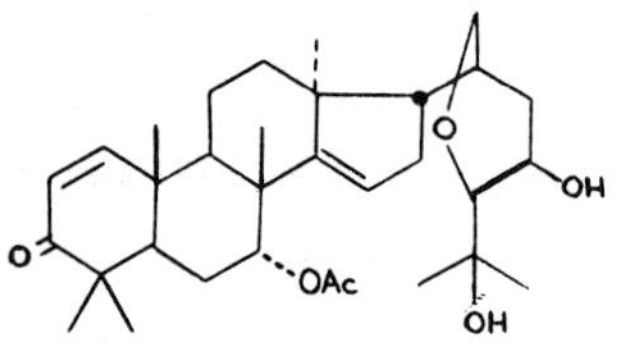

Chem. Comm. 1967 1193.

4492-011

CARPESTEROL $C_{37}H_{54}O_4$ 562 +67° 251°C

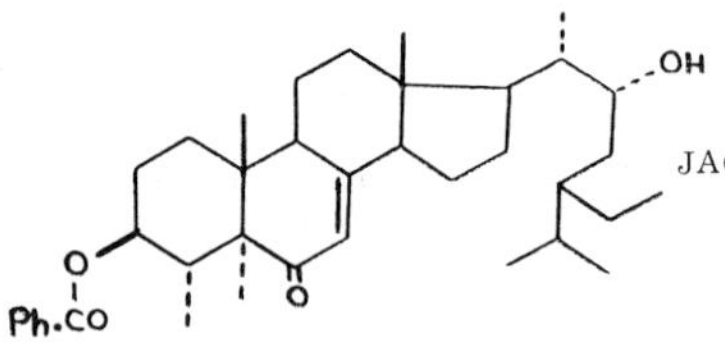

JACS. 1970 7005.

4492-016

WALSURENOL $C_{30}H_{50}O$ 426

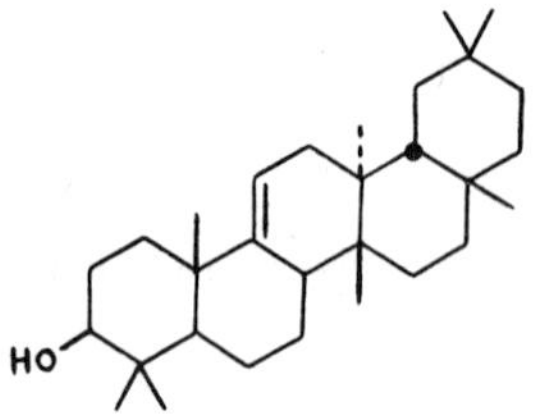

Chem. Comm. 1968 418.

4492-012

deleted

4492-017

BAUERENOL $C_{30}H_{50}O$ 426 -25°c 206°C

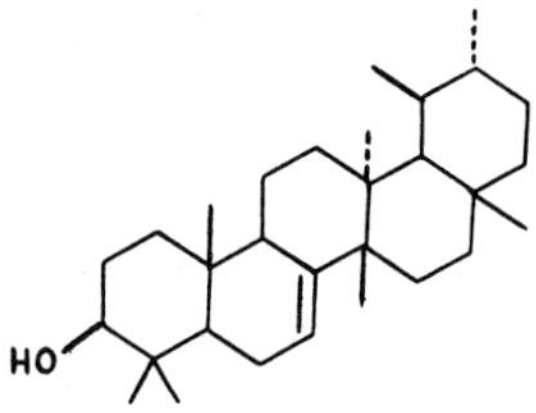

Aust. J. C. 1966 451.

4492-013

ALEURITOLIC ACID $C_{30}H_{48}O_3$ 456 301°C

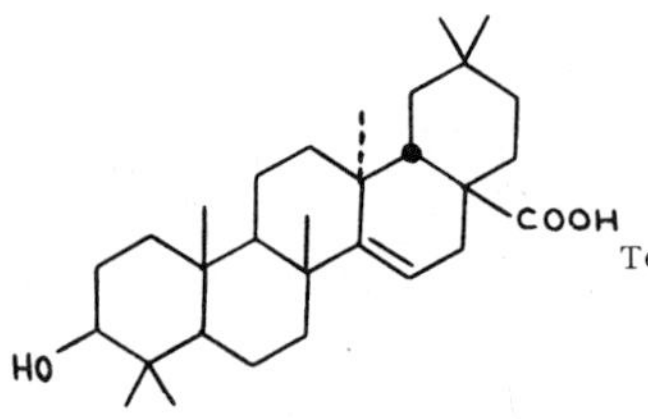

Tet. 1970 3017.

4492-018

BAUERENOL, iso- $C_{30}H_{50}O$ 426 +37°c 162°C

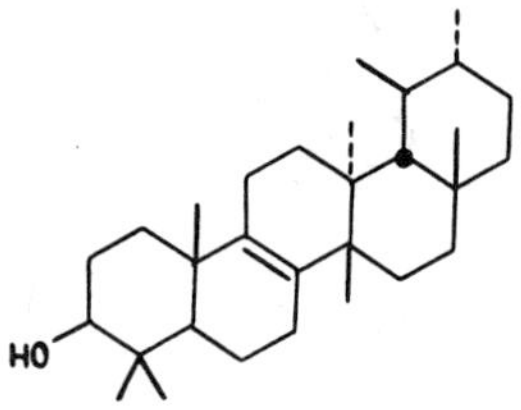

Tet. Lett. 1968 5965.

4492-014

URSENOL, iso- $C_{30}H_{50}O$ 426 +30°c 201°C

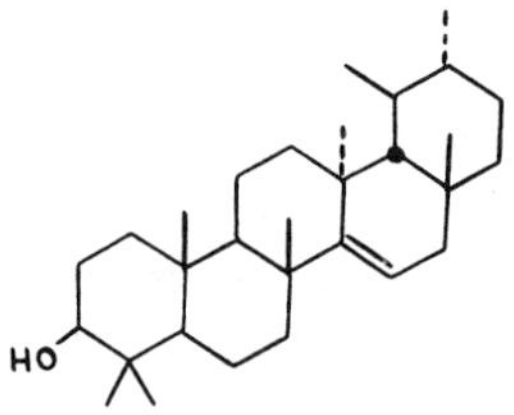

JCS. 1966 1814.

4492-019

MULTIFLORENOL $C_{30}H_{50}O$ 426 -28°c 189°C

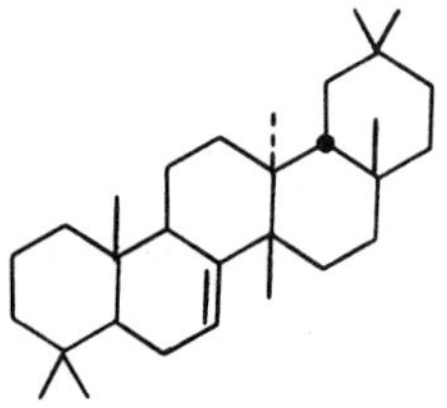

Tet. 1963 123.

4492-015

LANOSTEROL, 30-NOR- $C_{29}H_{48}O$ 412
+42°c
140°C

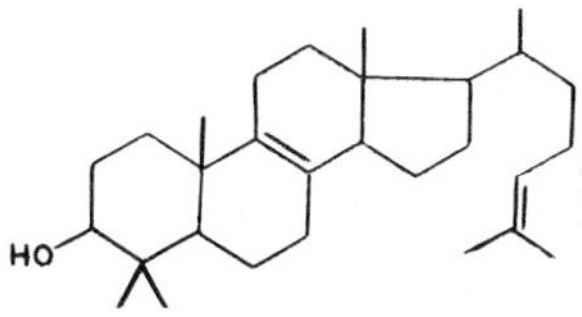

Bull. Soc. Chim. Fr., 1965, 3682.

4493-001

LOPHENOL, 24-dehydro- $C_{28}H_{46}O$ 398
139°C

Compt. Rend., 1969 (268) 2105.

4493-006

LANOSTEROL, 29-NOR- $C_{29}H_{48}O$ 412
+ 60°
110°C

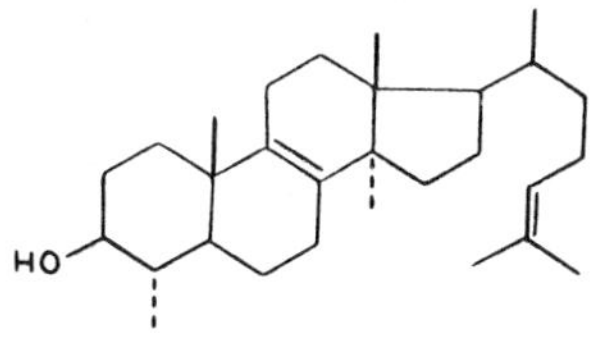

Compt. Rend., 1969 (268) 2105.

4493-002

OBTUSIFOLIOL $C_{30}H_{50}O$ 426

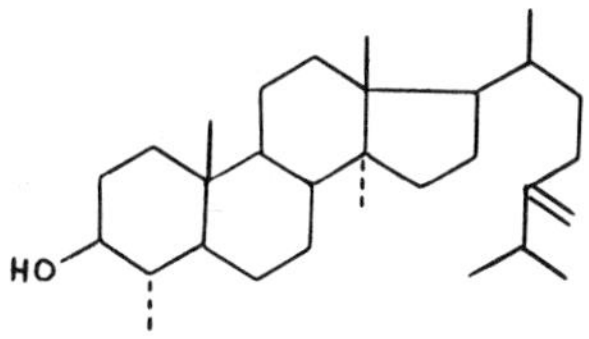

Chem. Abstr., 1967, 108792.

4493-007

LANOSTEROL, 28,30-bisnor- $C_{28}H_{46}O$ 398

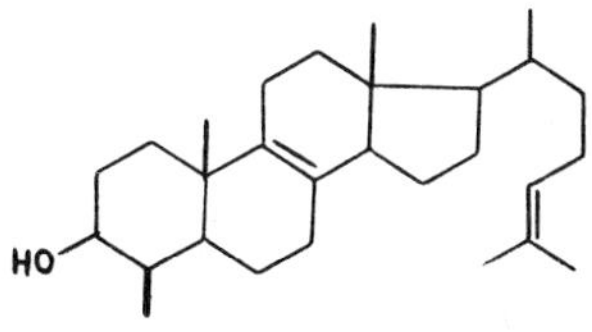

Biochemistry, 1967 (6) 869.

4493-003

LOPHENOL, 24-methylene- $C_{29}H_{48}O$ 412
+2.4°
166°C

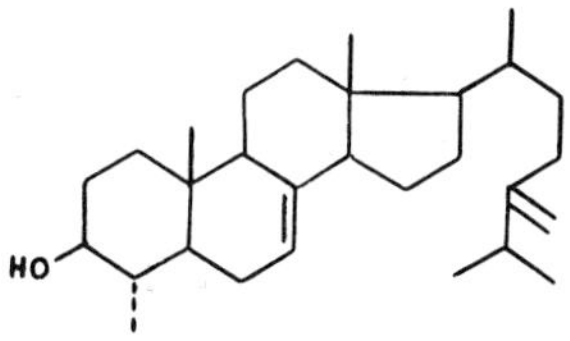

Tet., 1965, 1559.

4493-008

LANOSTEROL, 29,30-bisnor- $C_{28}H_{46}O$ 398
ac. +61°C
ac. 138°c

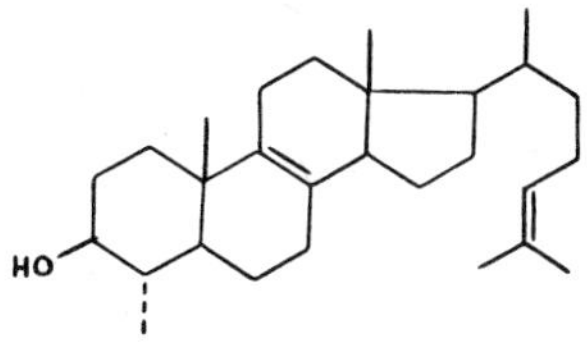

Bull. Soc. Chim. Fr., 1965, 3682.

4493-004

ZYMOSTEROL, 24,25-dihydro-
4α-Me-24-methylene- $C_{29}H_{48}O$ 412
+55°c
144°C

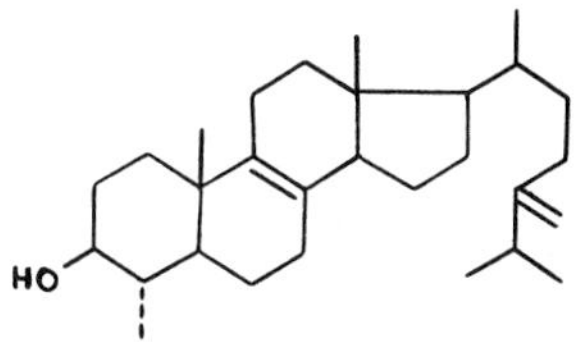

JCS., 1970, 780.

4493-009

LOPHENOL $C_{28}H_{48}O$ 400
+5°c
149°C

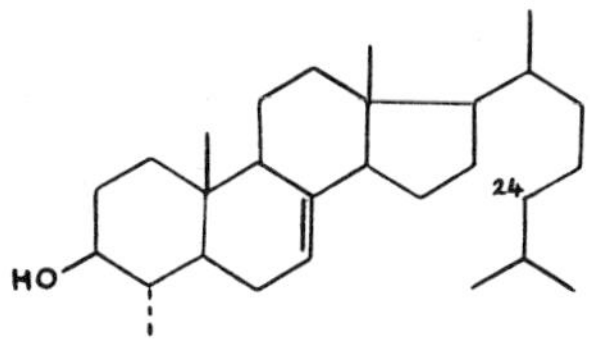

JACS., 1958, 6284.

4493-005

STIGMASTA-7, 24(28)-DIENOL,
4β-Me- $C_{30}H_{50}O$ 426
+6°
165°C

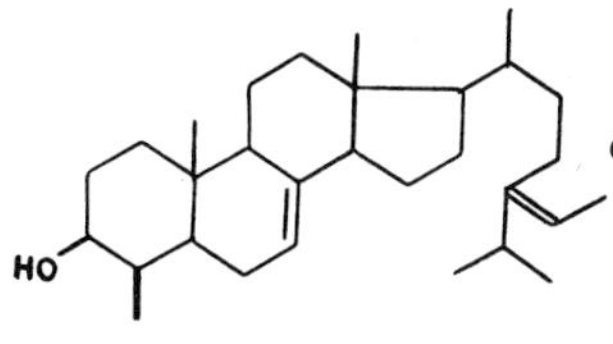

Chem. Comm., 1969, 107.

4493-010

STIGMASTA-7, 25-DIENOL, 4α-Me- $C_{30}H_{50}O$ 426 Tet. Lett., 1970, 3043. 4493-011	PUTROL $C_{29}H_{48}O_2$ 428 R = OH Chem. Abstr., 1968 (69) 93685. 4493-016
CITROSTADIENOL $C_{30}H_{50}O$ 426 +24°c 163°C Tet. Lett., 1968, 6163. 4493-012	PUTRONE $C_{29}H_{46}O_2$ 426 R = O Chem. Abstr., 1968 (69) 93685. 4493-017
MACDOUGALLIN $C_{28}H_{48}O_2$ 416 +72°c 174°C JACS., 1966, 790. 4493-013	PLATANIC ACID $C_{29}H_{46}O_4$ 458 -51°c 286°C JCS., 1963, 3269. 4493-018
CYCLOEUCALENOL $C_{30}H_{50}O$ 426 +45°c 138°C JCS., 1956, 1384. 4493-014	PLATANIC ACID, 3-dehydro- $C_{29}H_{44}O_4$ 456 m. c. + 2°c m. e. 163°C JCS., 1963, 3269. 4493-019
CYCLOARTANOL, 29-NOR- $C_{29}H_{50}O$ 414 +42°c 130°C Tet. Lett., 1967, 3567. 4493-015	ADIANTONE $C_{29}H_{48}O$ 412 +81° Chem. Comm., 1968, 1372. 4493-020

ADIANTONE, 21-OH- $C_{29}H_{48}O_2$ 428 $+50^0$p 281^0C

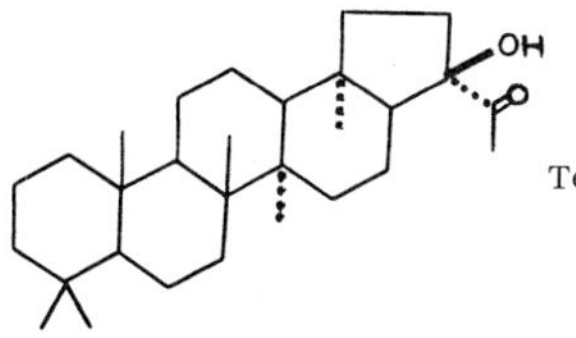

Tet. Lett., 1966, 5679.

4493-021

ALBIGENIN $C_{29}H_{46}O_2$ 426 -114^0 227^0C

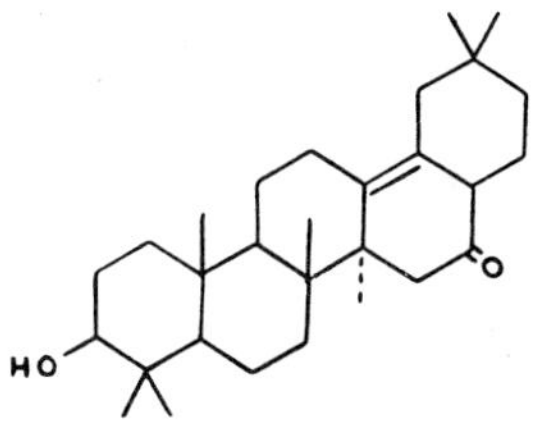

Tet., 1962 (18) 155.

4493-026

ADIPEDATOL $C_{29}H_{48}O_2$ 428 $+88^0$ 186^0C

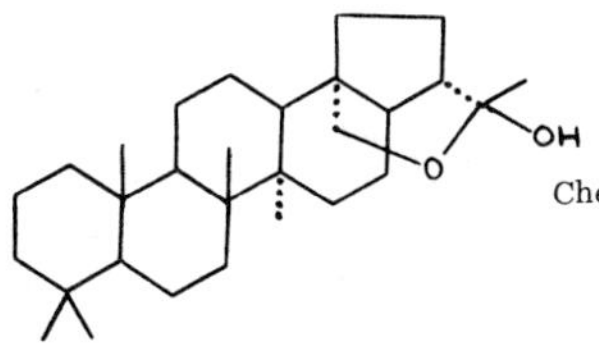

Chem. Comm., 1968, 1372.

4493-022

CELASTROL $C_{29}H_{38}O_4$ 450 205^0C

R = H

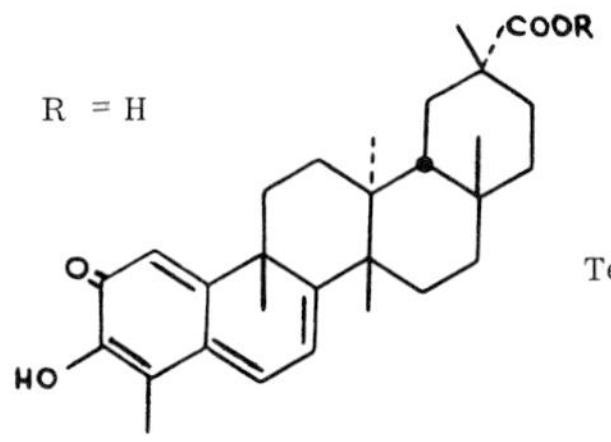

Tet. Lett., 1962, 603.

4493-027

POLYGALIC ACID $C_{29}H_{44}O_6$ 488 $+18^0$e 300^0C

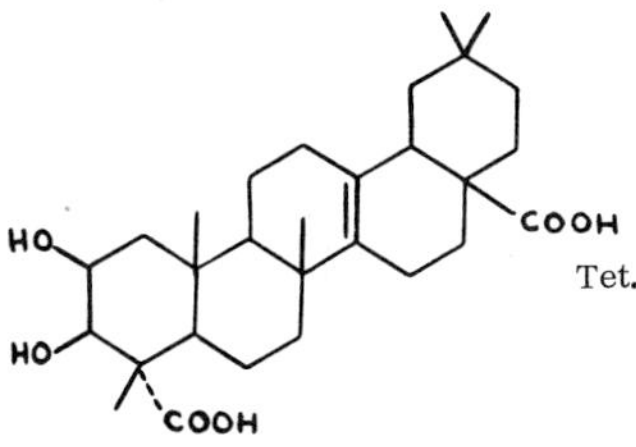

Tet. Lett., 1964, 3065.

4493-023

PRISTIMERIN $C_{30}H_{40}O_4$ 464 220^0C

R = CH_3

Tet. Lett., 1962, 603.

4493-028

AEGICERADIENOL $C_{29}H_{46}O$ 410 $+79^0$c 190^0C

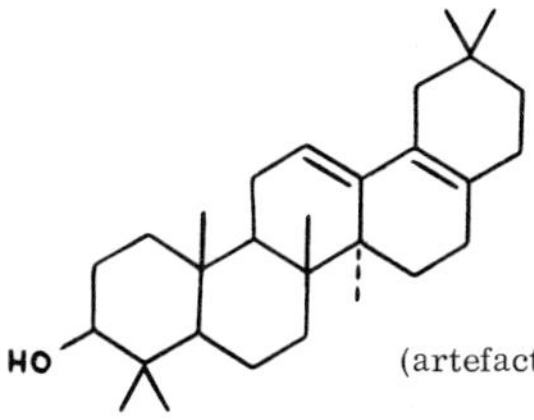

Tet. Lett., 1965, 4639.

(artefact ?)

4493-024

CLAVATOL $C_{28}H_{50}O_4$ 450 278^0C

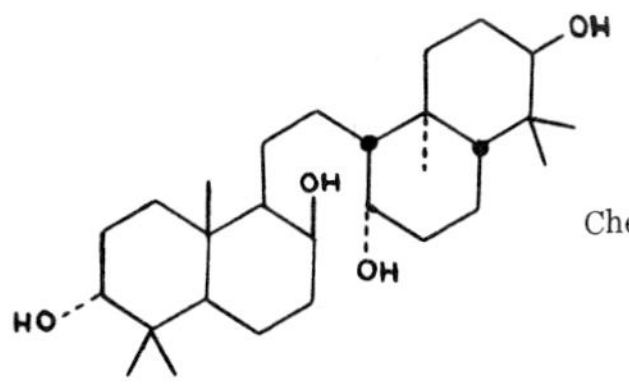

Chem. Comm., 1970, 1275.

4493-029

CYCLAMIGENIN A_2 $C_{29}H_{46}O_3$ 442 -17^0 208^0C

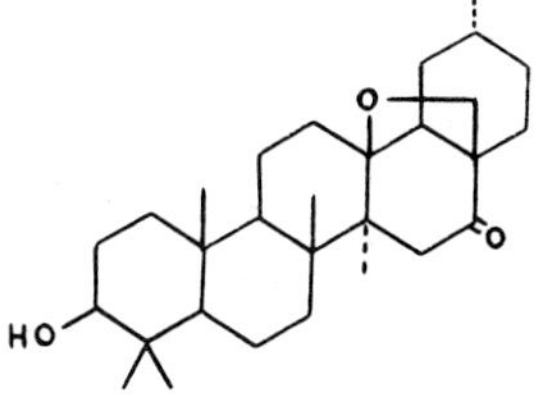

Tet., 1968, 5649.

4493-025

BOURJOTONE $C_{27}H_{42}O_2$ 398 -37^0 151^0C

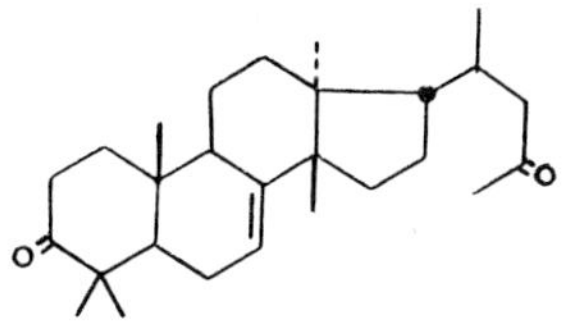

Aust. J. C., 1966, 455.

4493-030

CUCURBITACIN D, hexanor- $C_{24}H_{34}O_5$ 402

217^{0}C

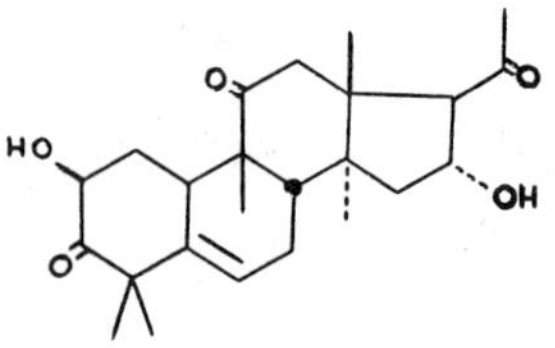

Can. J. Chem., 1970, 1787.

4493-031

LANOSTEROL, 3-epi-hexanor- $C_{24}H_{38}O_2$ 358

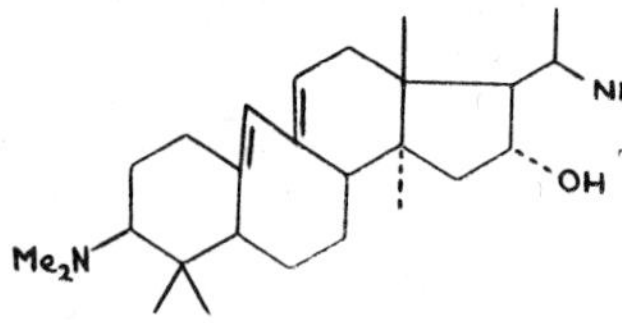

+109^{0}d

231^{0}C

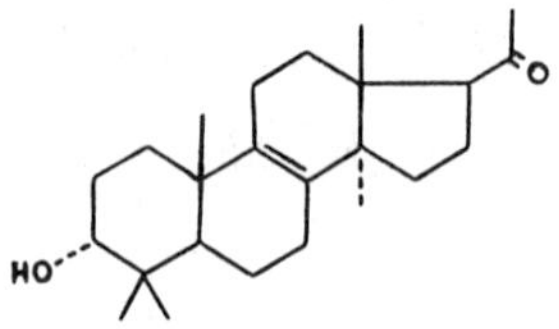

4493-032

BUXPSIINE $C_{26}H_{39}NO$ 381

+105^{0}c

177^{0}C

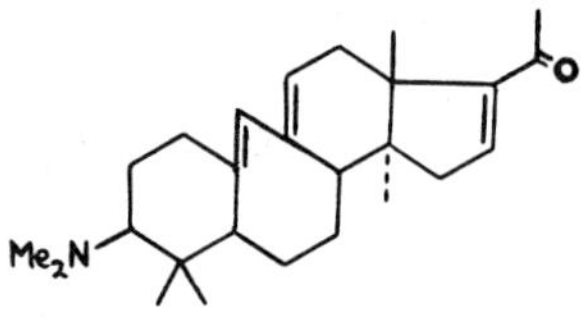

Tet. Lett., 1966, 915.

4493-033

BUXENINE G $C_{25}H_{42}N_2$ 370

+20^{0}c

R = H

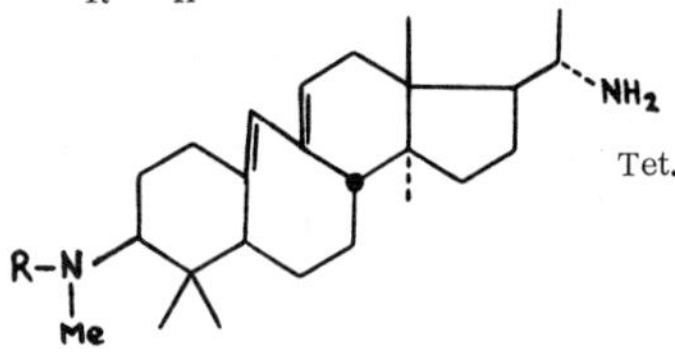

Tet. Lett., 1966, 3815.

4493-034

BUXAMINE E $C_{26}H_{44}N_2$ 384

+32^{0}c

ac. 237^{0}C

R = Me

Tet., 1966, 3321.

4493-035

BUXAMINOL E $C_{26}H_{44}N_2O$ 400

+38^{0}c

202^{0}C

Tet., 1966, 3321.

4493-036

BUXIDIENE F, N-isobutyryl- $C_{30}H_{50}N_2O_3$ 486

-67^{0}c

253^{0}C

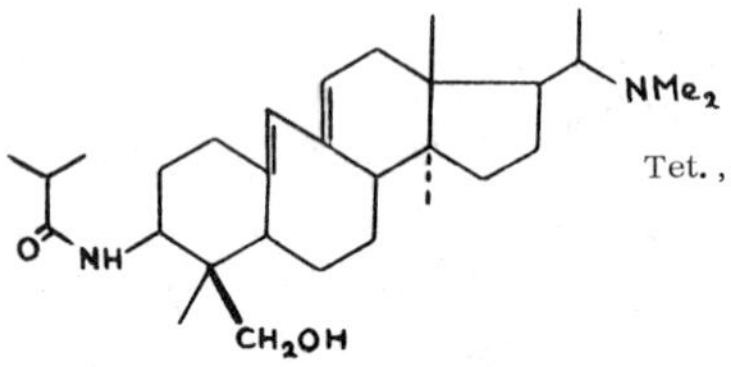

Tet., 1966, 3321.

4493-037

BUXIDIENINE F, N-Benzoyl- $C_{33}H_{48}N_2O_3$ 520

-36^0

287^{0}C

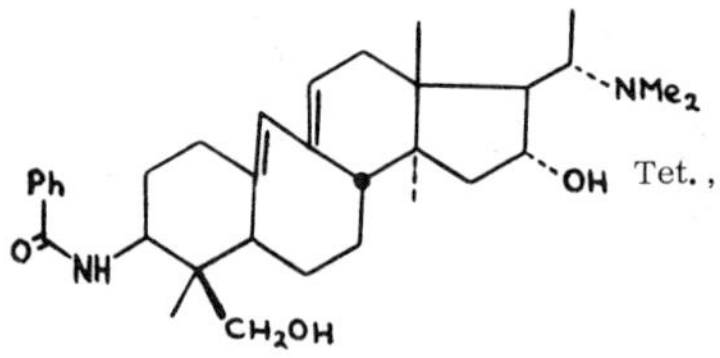

Tet., 1967, 4563.

4493-038

BUXALINE F, N-isobutyryl- $C_{30}H_{52}N_2O_4$ 504

-32^{0}c

275^{0}C

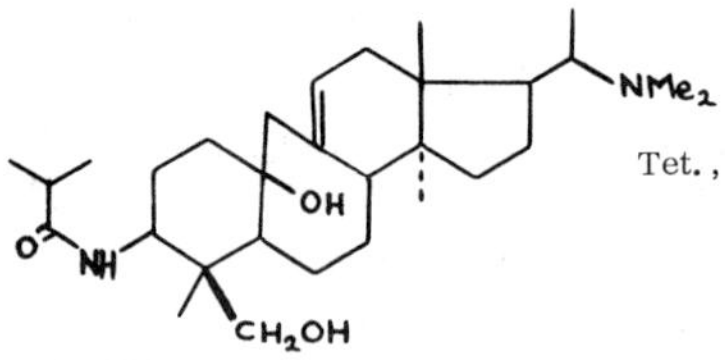

Tet., 1966, 3321.

4493-039

PUTRANJIVIC ACID $C_{30}H_{50}O_2$ 442

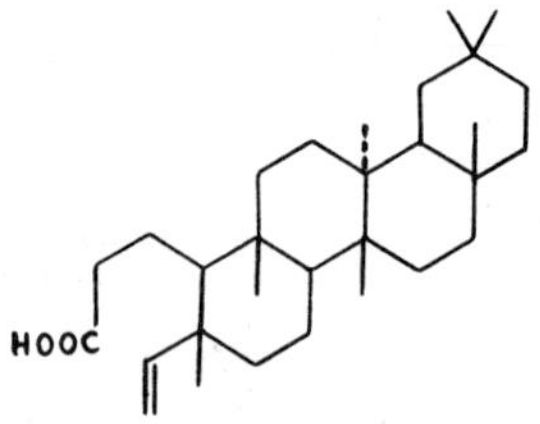

Chem. Abstr., 1968 (69) 93685.

4493-040

Name	Formula	Data	Reference	No.
PUTRIC ACID	$C_{30}H_{52}O_3$	460	Tet. Lett., 1969, 231.	4493-041
ASIATIC ACID ME ESTER, 2, 3-seco-	$C_{32}H_{48}O_6$	528; +117°; 235°C	Chem. Comm., 1970, 891.	4493-046
APETALACTONE	$C_{30}H_{50}O_3$	458; +37°c; 335°C	JCS., 1968, 1323.	4493-042
CANARIC ACID	$C_{30}H_{52}O_2$	444; +56°c; 216°C	Aust. J. C., 1965, 213.	4493-047
NYCTANTHIC ACID	$C_{30}H_{50}O_2$	442	JCS., 1960, 2016.	4493-043
DAMMARENOLIC ACID	$C_{30}H_{48}O_3$	456	JCS., 1960, 1100.	4493-048
ROBURIC ACID	$C_{30}H_{48}O_2$	440; +78°c; 182°C	Tet. Lett., 1963, 921.	4493-044
SHOREIC ACID	$C_{30}H_{48}O_4$	472	Bull. Soc. Chim. Fr., 1968, 2131.	4493-049
OLEAN-12-EN-2, 3, 28-TRIOIC - ACID, 2, 3-SECO-	$C_{30}H_{46}O_6$	502; +65°p; 296°C	Proc. Chem. Soc., 1962, 27.	4493-045
SOYA SAPOGENOL D	$C_{30}H_{50}O_3$	458; -61°c; 298°C	Helv., 1950, 1385.	4493-050

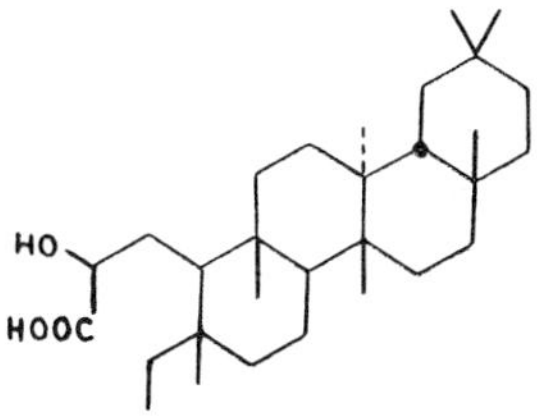

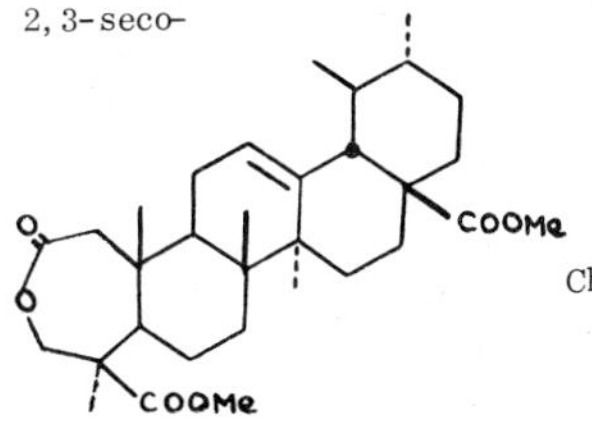

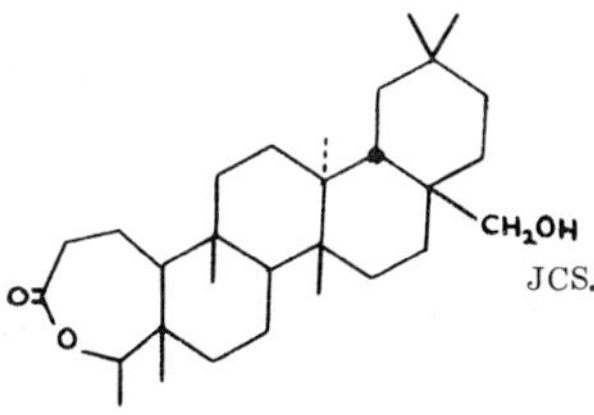

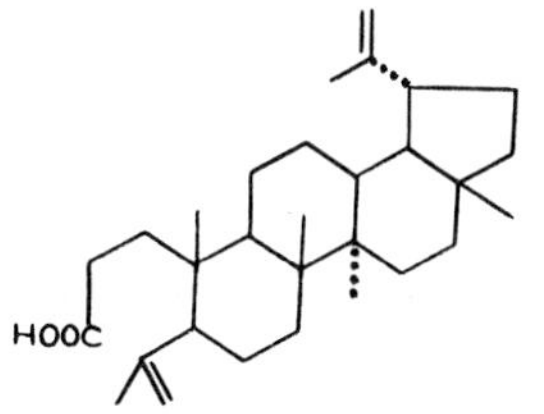

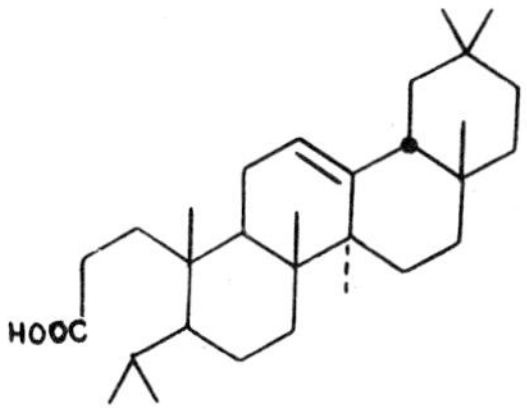

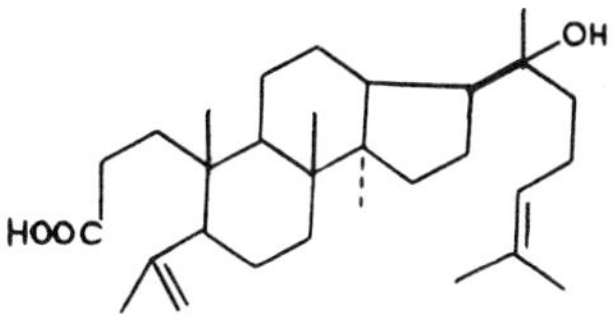

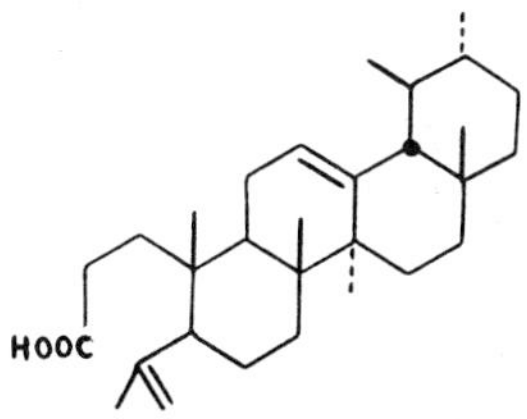

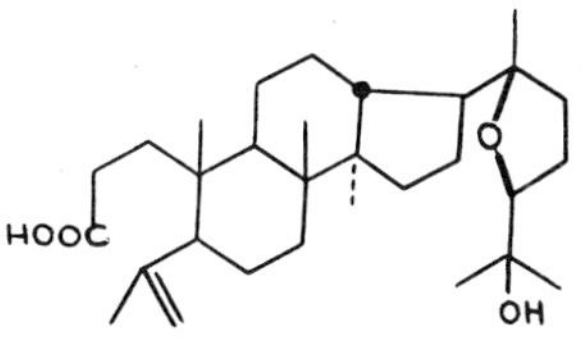

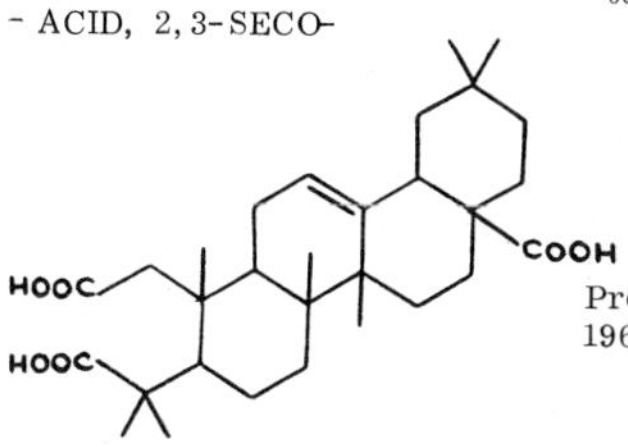

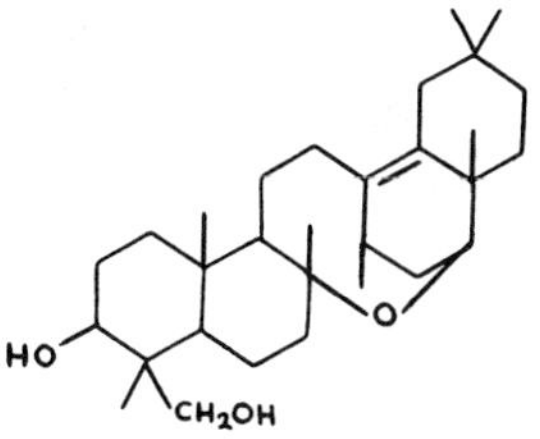

EBELIN LACTON, cis- — $C_{30}H_{46}O_3$ — 454

ac. 247^{0}C

(artefact?)

Aust. J. C., 1970, 2087.

4493-051

EBELIN LACTONE, trans- — $C_{30}H_{46}O_3$ — 454

-14^{0}c

184^{0}C

* trans

Chem. Ind., 1963, 1206.

4493-052

IVORENSIC ACID, Me ester — $C_{27}H_{34}O_8$ — 486

-98^0

280^{0}C

Chem. Comm., 1969, 889.

4493-053

ODORATIN — $C_{19}H_{22}O_4$ — 314

+155^{0}c

220^{0}C

Chem. Comm., 1966, 576.

4493-054

LANSIC ACID — $C_{30}H_{46}O_4$ — 470

-7^{0}c

183^0

Tet. Lett., 1968, 3731.

4493-055

deleted

4493-056

NIMBOLIN B — $C_{39}H_{46}O_{10}$ — 674

-93^0

244^{0}C

Chem. Comm., 1969, 1166.

4493-057

NIMBOLIDE — $C_{27}H_{30}O_7$ — 466

+206^0

246^{0}C

Chem. Comm., 1967, 808.

4493-058

NIMBIN — $C_{30}H_{36}O_9$ — 540

203^{0}C

Tet., 1968, 1517.

4493-059

SALANNIN — $C_{34}H_{44}O_9$ — 596

+167^0

169^{0}C

Tet., 1968, 1525.

4493-060

LANOSTEROL, 29,30-bisnor-dihydro- — $C_{28}H_{48}O$ — 400; $+55^0$; 107^0C

J. Biol. Chem., 1960 (235) 2253.

4493-061

ECHINOCYSTENOLONE, nor- — $C_{29}H_{46}O_2$ — 426

Chem. Abstr., 1967 (67) 116987.

4493-062

PICRASIN A — $C_{26}H_{34}O_8$ — 474; 298^0C

Chem. Pharm. Bull., 1970, 1082.

4493-063

SIMAROLIDE — $C_{27}H_{36}O_9$ — 504; $+74^0$; 270^0C

Proc. Chem. Soc., 1964, 293.

4493-064

CEDRONYLINE — $C_{19}H_{26}O_7$ — 366; $+17^0$; 267^0C

R = R' = OH

Bull. Soc. Chim. Fr., 1964, 2016.

4493-065

CEDRONINE — $C_{19}H_{24}O_7$ — 364; -13^0; 280^0C

R = oxo; R' = OH

Bull. Soc. Chim. Fr., 1964, 2016.

4493-066

SAMADERIN B — $C_{19}H_{22}O_7$ — 362; $+68^0$; 240^0C

R = R' = oxo

Bull. Soc. Chim. Fr., 1964, 2016.

4493-067

SAMADERIN C — $C_{19}H_{24}O_7$ — 364; $+59^0$; 268^0C

R = OH; R' = oxo

Bull. Soc. Chim. Fr., 1964, 2016.

4493-068

EURYCOMALACTONE — $C_{19}H_{24}O_6$ — 348; $+100^0$; 269^0C

Chem. Comm., 1969, 821.

4493-069

EURYCOMALACTONE, dihydro- — $C_{19}H_{26}O_6$ — 350; 248^0C

* dihydro

JOC., 1970, 1104.

4493-070

EUPTELEOGENIN $C_{29}H_{42}O_4$ 454
+83^0
272^{0}C

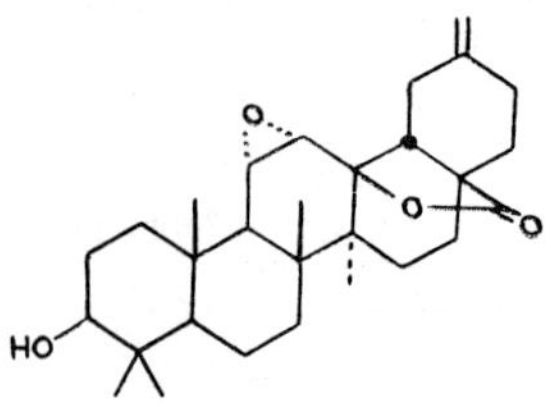

HO O O O

Tet. Lett., 1965, 3218.

4493-071

JINGULLIC ACID $C_{29}H_{40}O_4$ 452 +129^{0}c 180^{0}C Chem. Comm. 1969 579. 4494-001	EMMOLACTONE, 22-deoxy $C_{28}H_{38}O_2$ 406 +106^{0}c 227^{0}C R = H Chem. Comm. 1970 814. 4494-004
CEANOTHENIC ACID $C_{29}H_{42}O_4$ 454 +39^0 354^{0}C Can. J. C. 1962 1632. 4494-002	CEANOTHIC ACID $C_{30}H_{46}O_5$ 486 +38^{0}e 346^{0}C JOC. 1962 4070. 4494-005
EMMOLACTONE $C_{28}H_{38}O_3$ 422 +123^{0}c 256^{0}C R = OH Aust. J. C. 1967 2737. 4494-003	

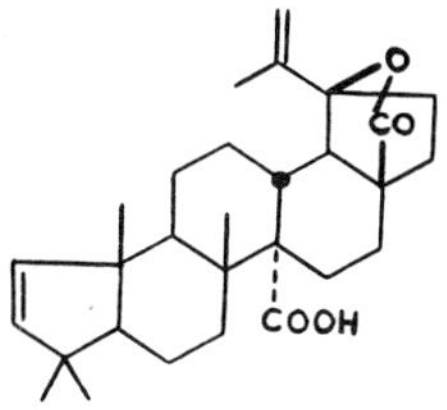

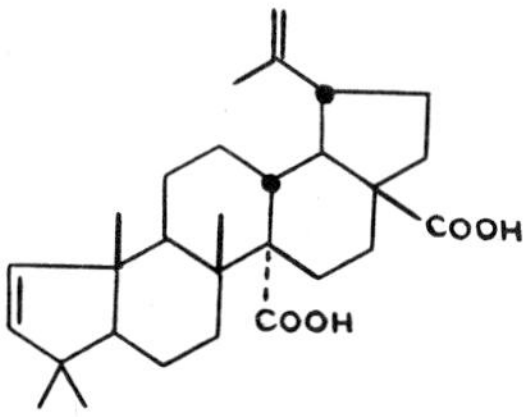

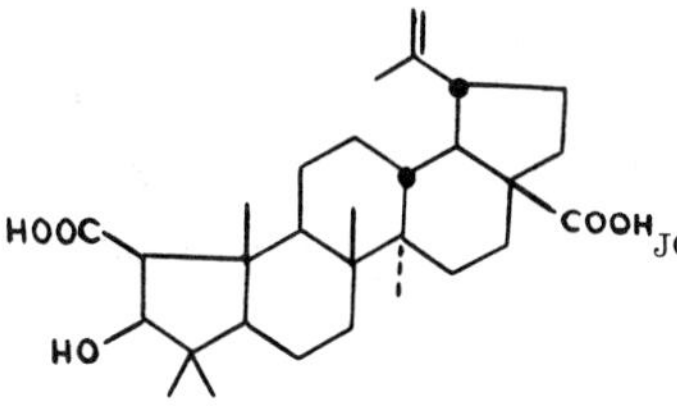

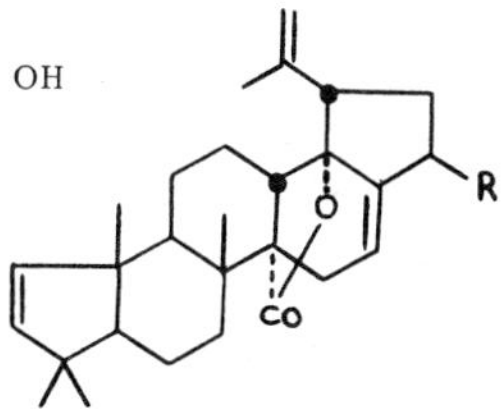

SQUALENE $C_{30}H_{50}$ 410
OI
Oil

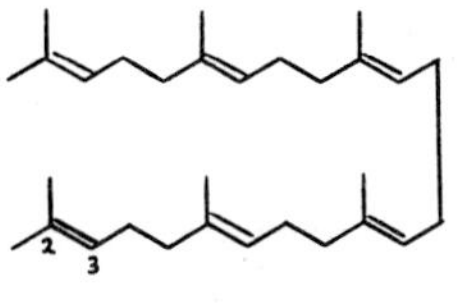

JACS. 1967 7150.

4499-001

SQUALENE, 2, 3-epoxy- $C_{30}H_{50}O$ 426

Tet. Lett. 1967 3533.

4499-002

SIARESINOLIC ACID $C_{30}H_{46}O_4$ 472
+99^{0}c
274^{0}C

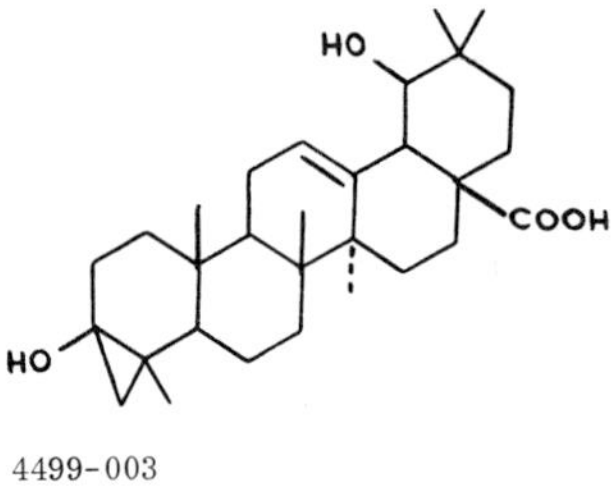

JCS. 1952 78.

4499-003

NEOLITSIN, cyclo- $C_{33}H_{56}O$ 468
+63^{0}c
172^{0}C

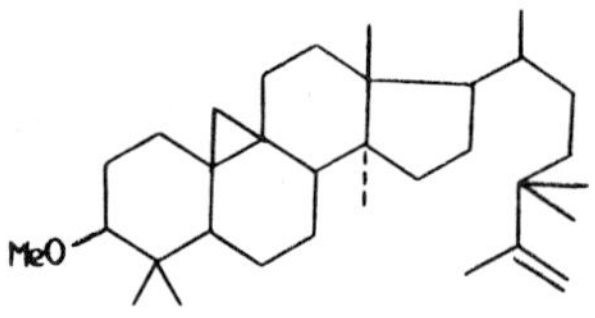

Aust. J. C. 1969 2371.

4499-004

HAKONANOL, keto- $C_{29}H_{48}O_2$ 428
+8^{0}c
296^{0}C

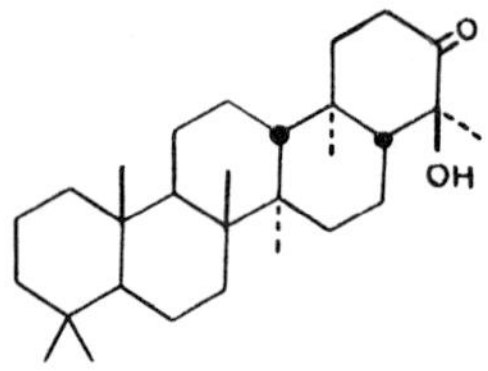

Tet. Lett. 1966 5679.

4499-005

SHIONONE $C_{30}H_{50}O$ 426
+1^{0}c
118^{0}C

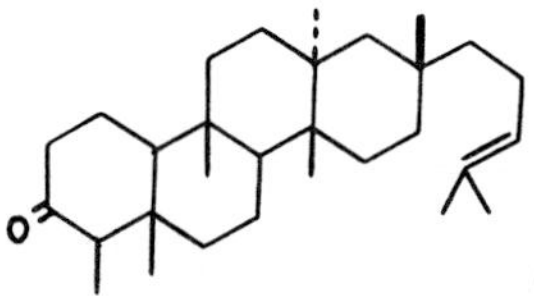

Tet. Lett. 1967 2991.

4499-006

ALNINCANONE $C_{31}H_{52}O_2$ 456

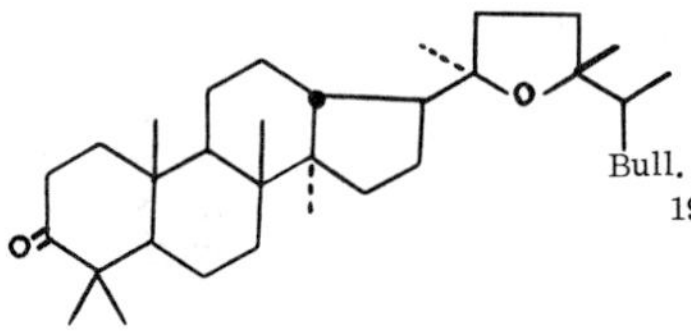

Bull. Soc. Chim. Fr. 1968 1089.

4499-007

BACCHARIS OXIDE $C_{30}H_{50}O$ 426
+42^{0}c
149^{0}C

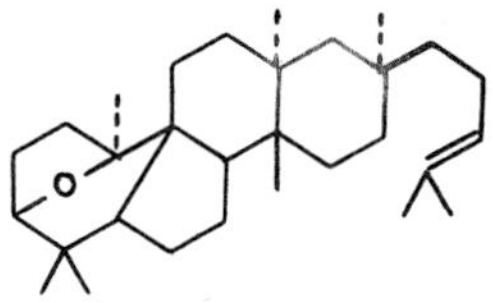

Acta Chem. Scand. 1970 2479.

4499-008

UTILIN $C_{41}H_{52}O_{17}$ 816
-356^{0}c
278^{0}C

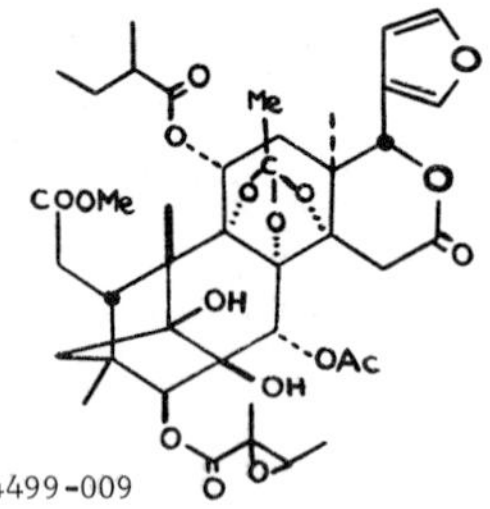

Chem. Comm., 1970, 1388.

4499-009

45-- THE STEROIDS

From a biogenetic standpoint the steroids can be considered as part of the triterpene class, but as sufficiently distinctive steroidal skeletons are evident, their separate classification seems justified. There are 19 major skeleton types (see Main Skeleton Key), 2 feature subclasses (4591 rearranged skeletons and 4592 degraded skeletons) and a miscellaneous section 4599.

45-- Steroids — Main Skeleton Key

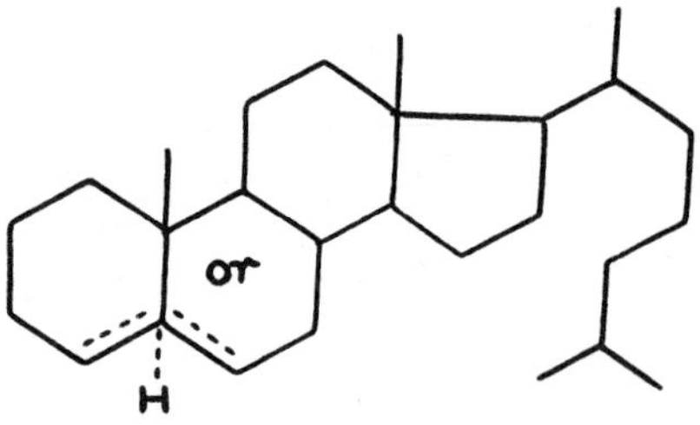

or H 01 Cholestane (Cf 09)	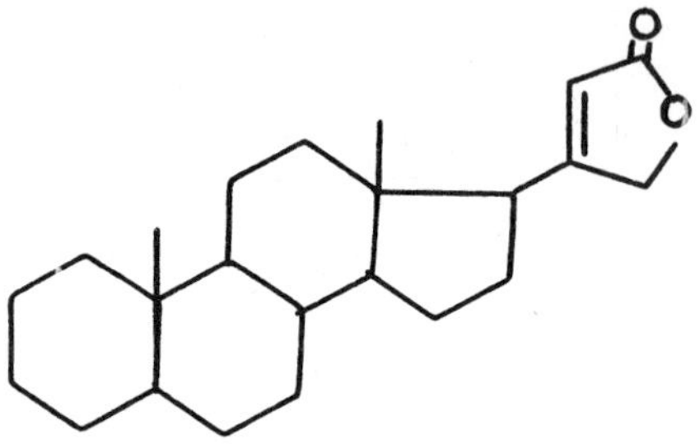 O O 05 Cardenolide
02 Ergostane	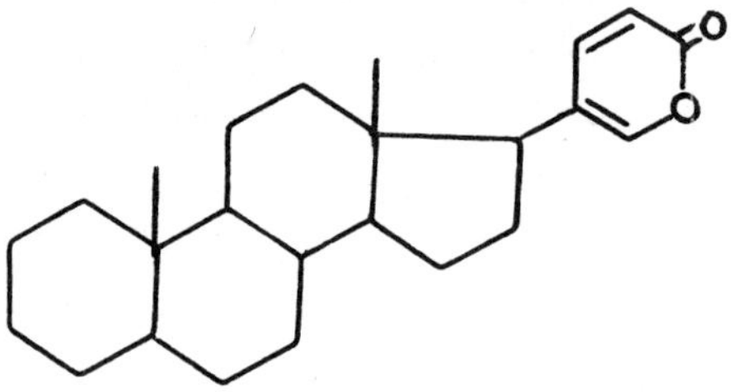 O O 06 Bufadienolides
03 Stigmastane	07 Pregnane
O O 04 Spirostane	08 Androstane

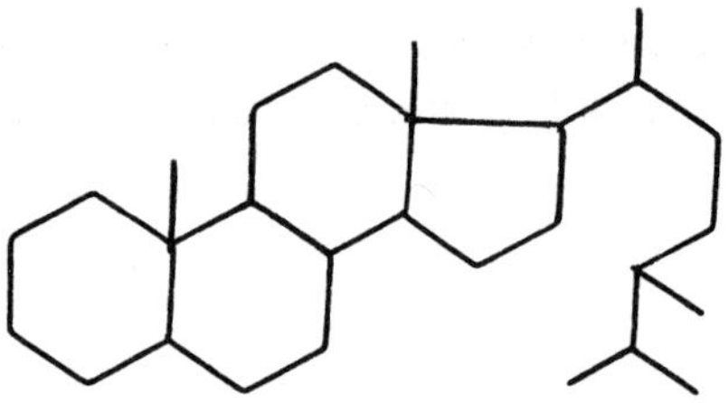

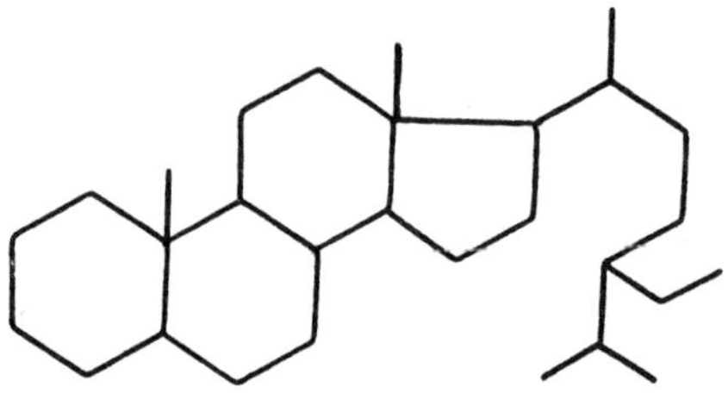

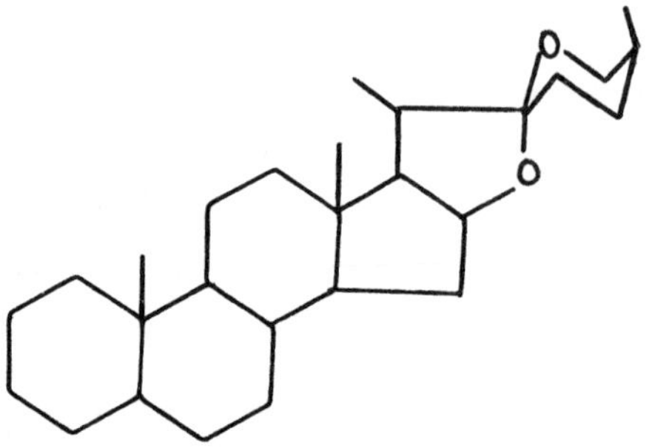

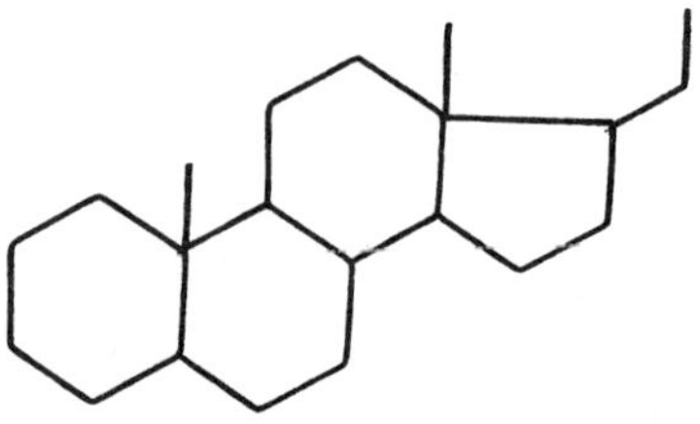

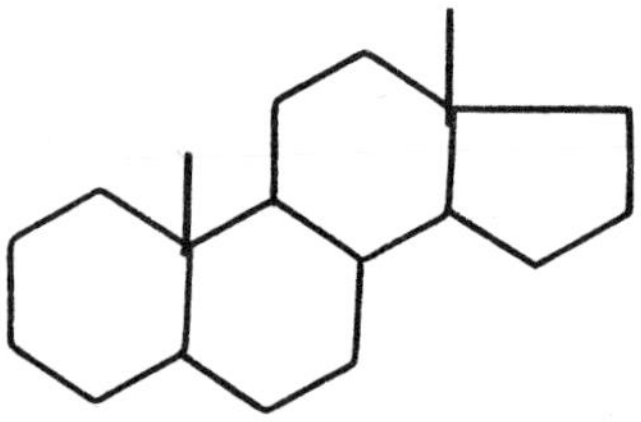

45-- Steroids — Main Skeleton Key

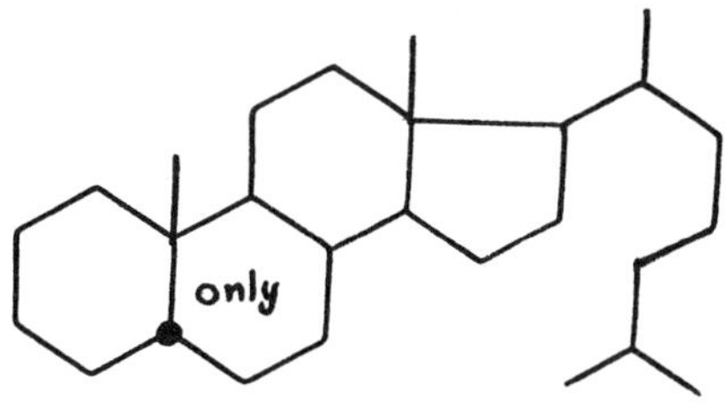

only 09 Coprostane	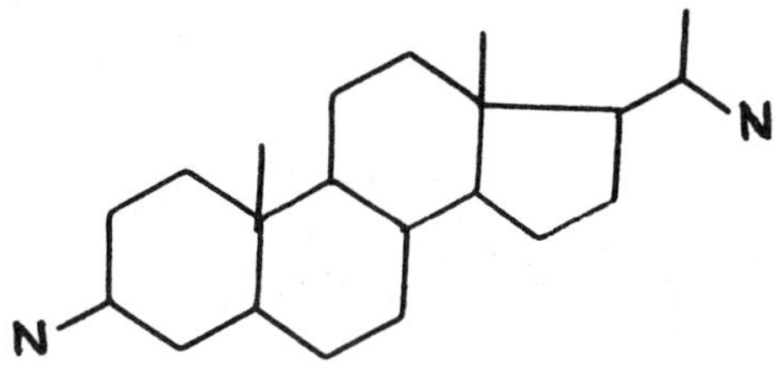 N N 13 Pachysandra Alkaloids
COOH 10 Cholane	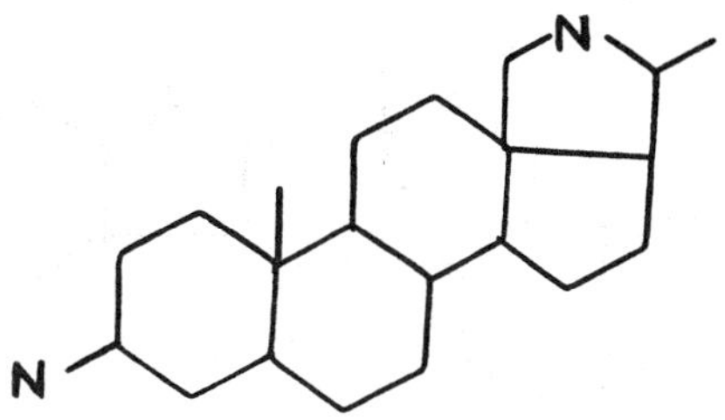 N N 14 Holarrhena Alkaloids
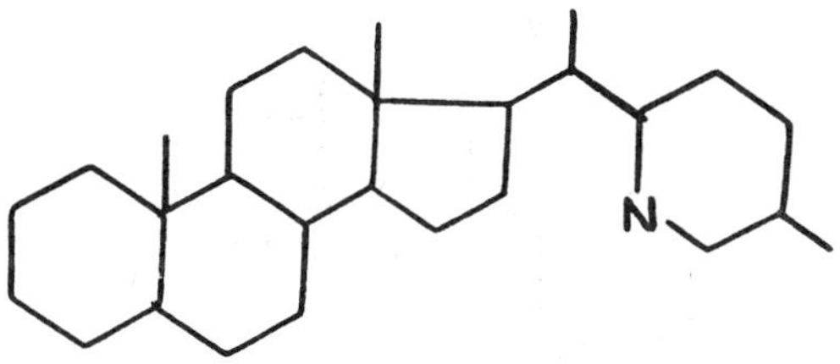 N 11 3-Amino-Pregnane	N 15 Solanum Alkaloids (I)
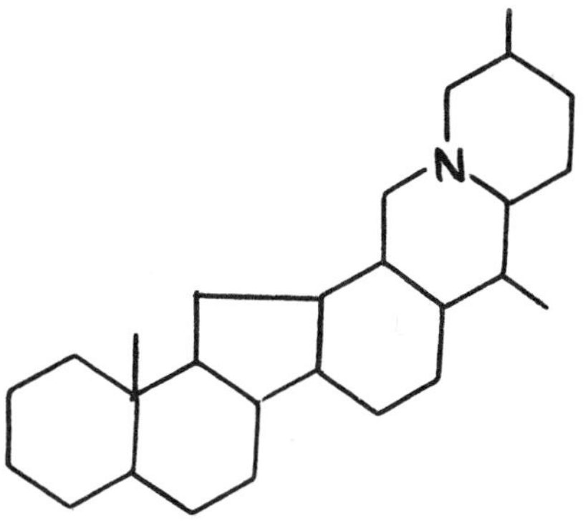 N 12 20-Amino-Pregnanes	N 16 Cevine Alkaloids (II)

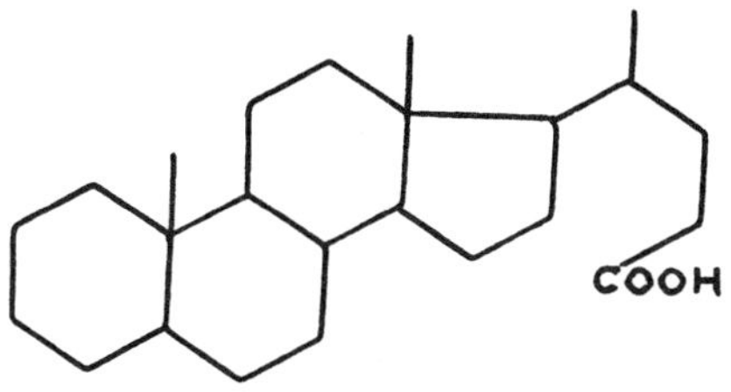

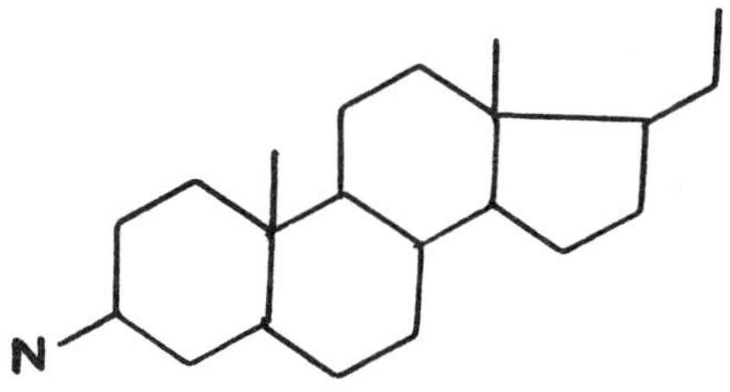

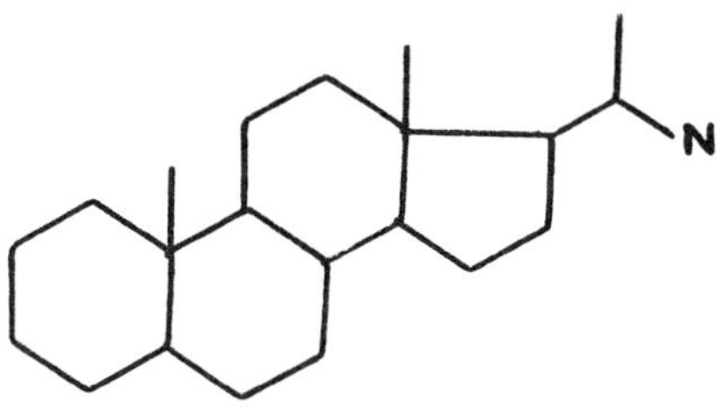

45-- Steroids — Main Skeleton Key

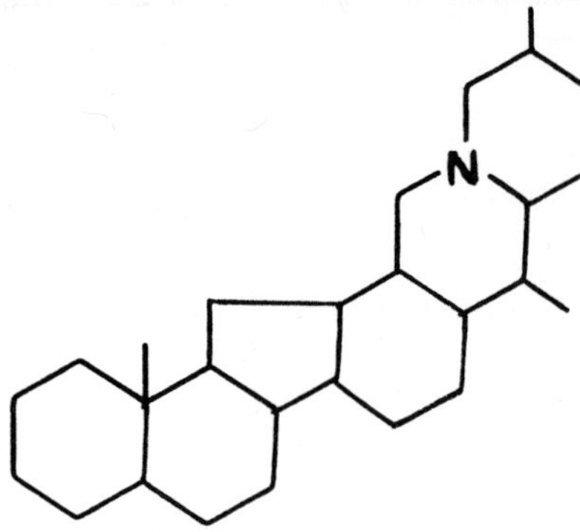

17 Cevine Alkaloids (I)

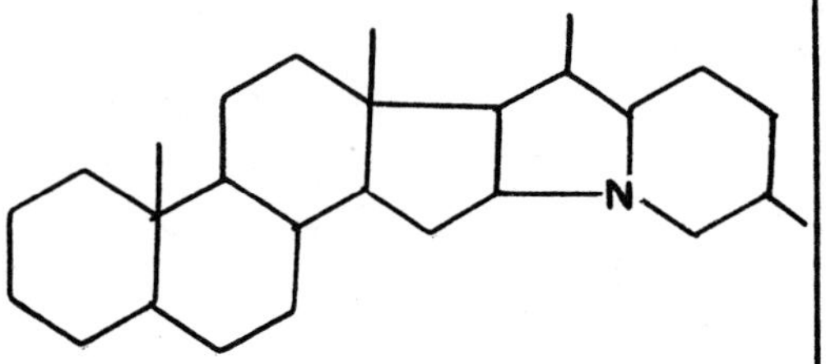

18 Solanum Alkaloids (II)

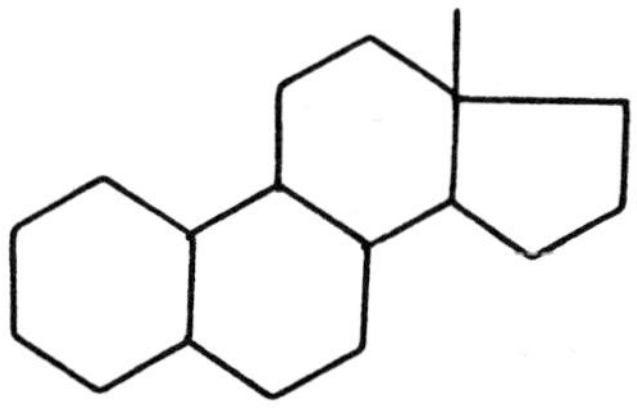

19 Oestrane

4591	Steroids — Less Common Skeletons - Rearranged

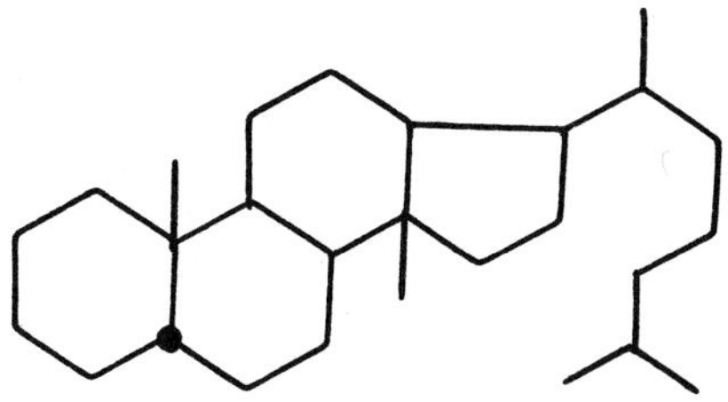

001	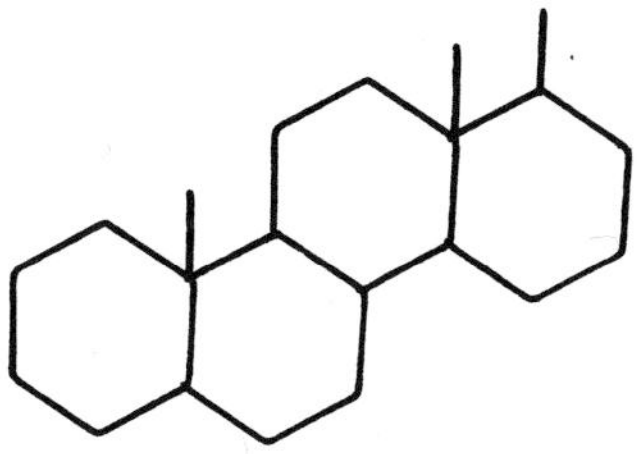
002 003	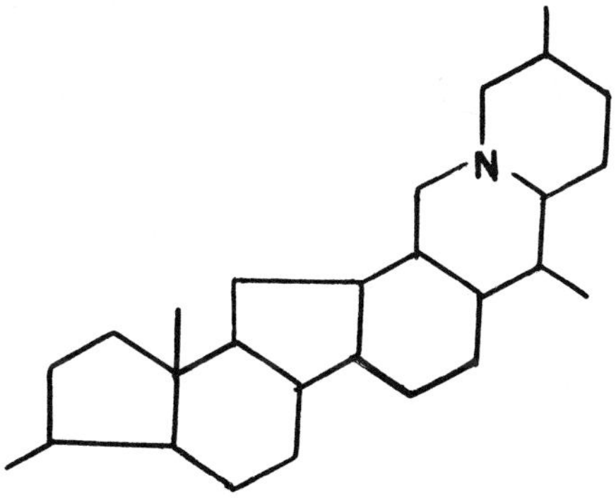
004	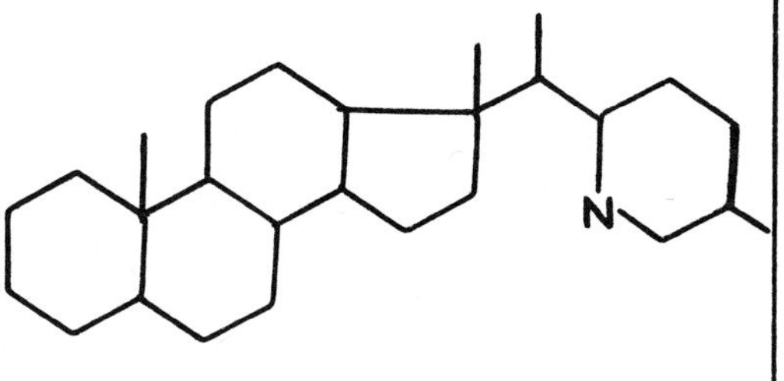
005 006 007	

4592	Steroids — Less Common Skeletons - Degraded

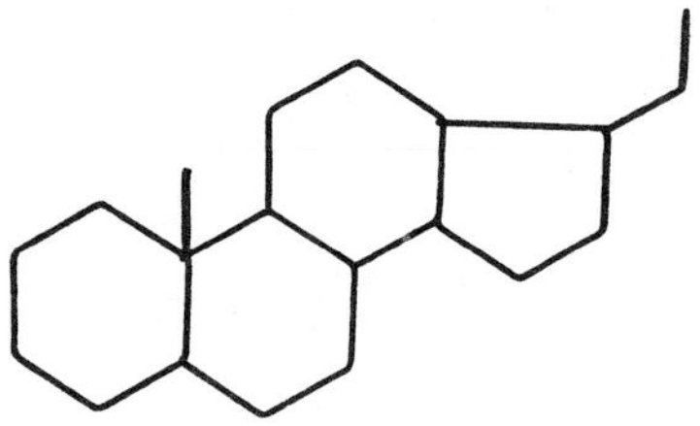

001	024 025 026
002	018 019
006 007	020 021
014	015

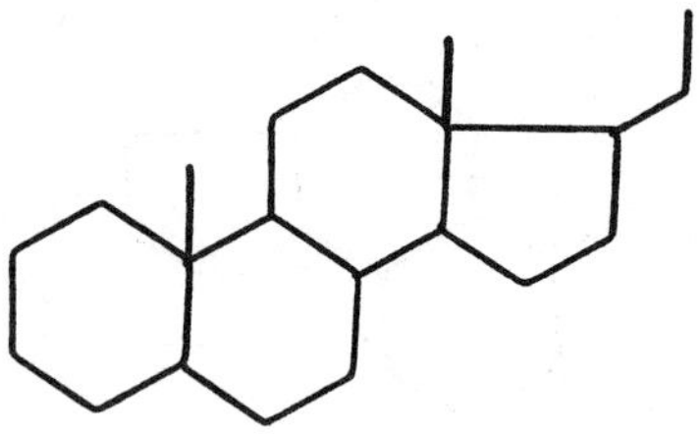

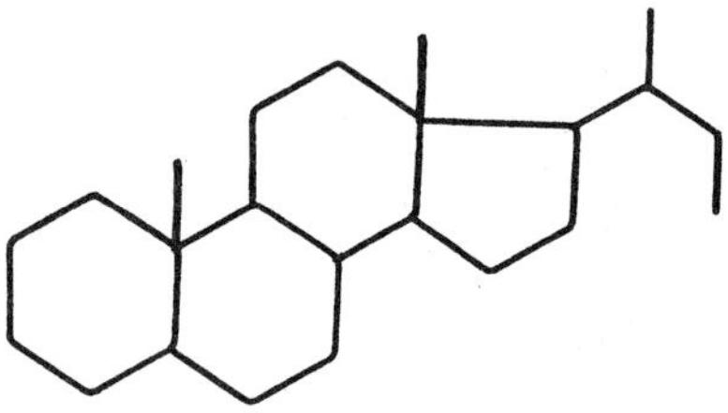

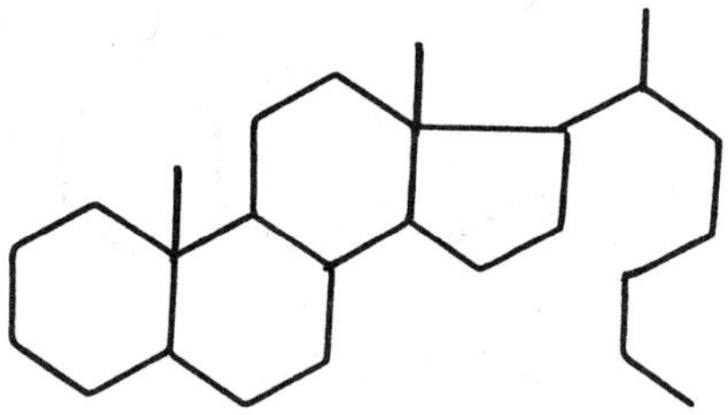

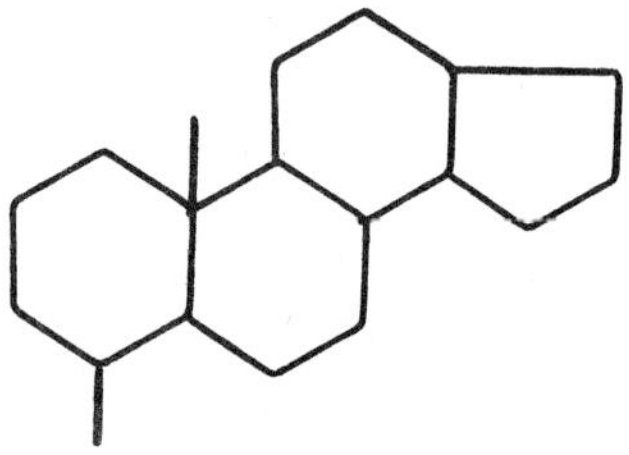

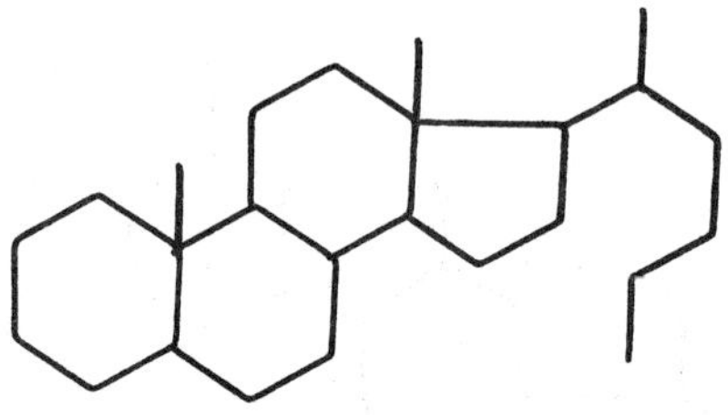

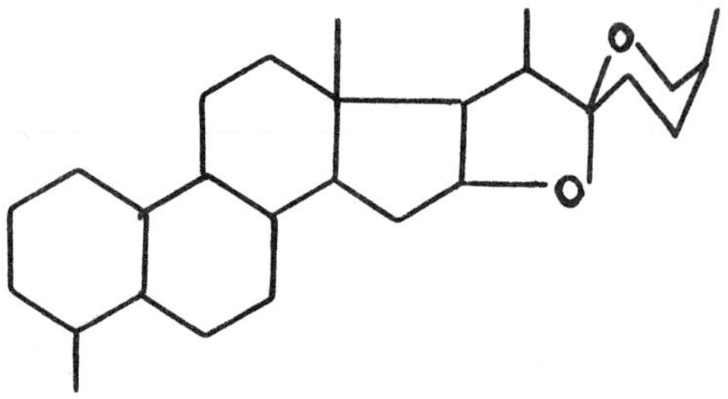

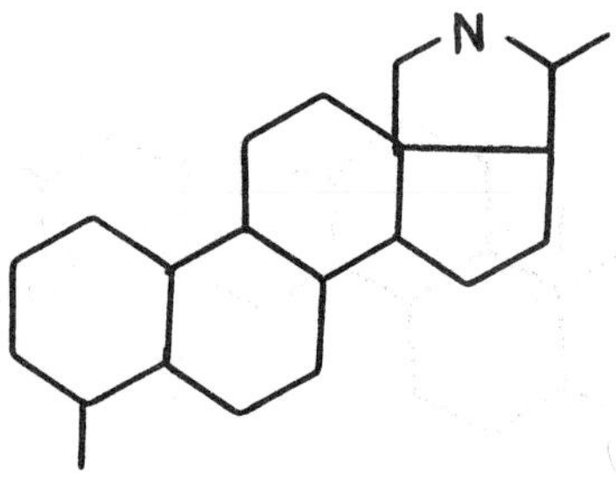

4592	Steroids – Less Common Skeletons - Degraded

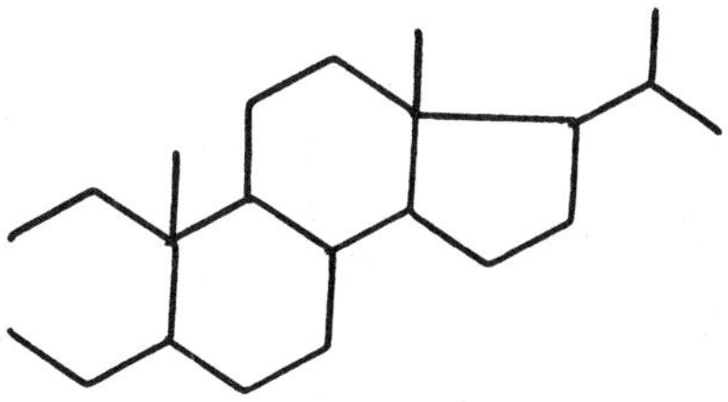

016	008 009 022
017 023	010 011 012
005	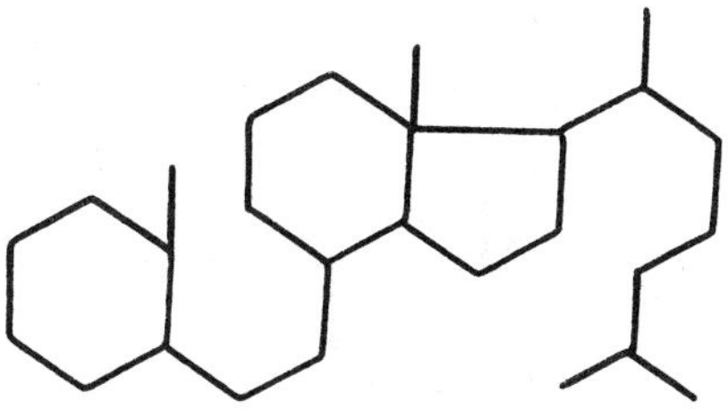
003 004	

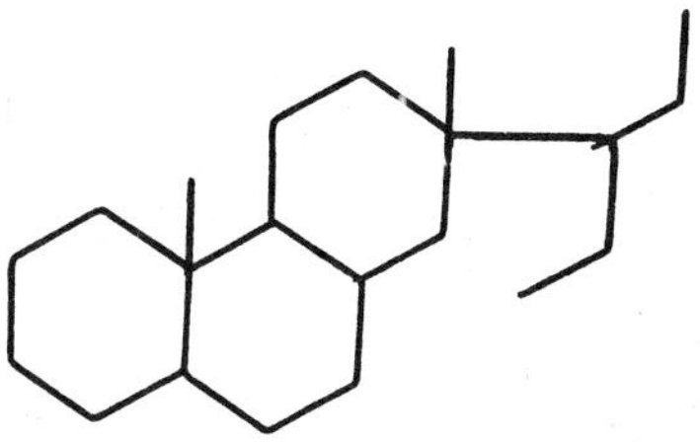

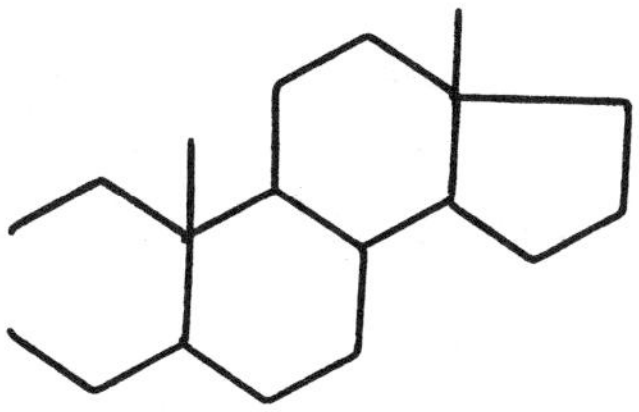

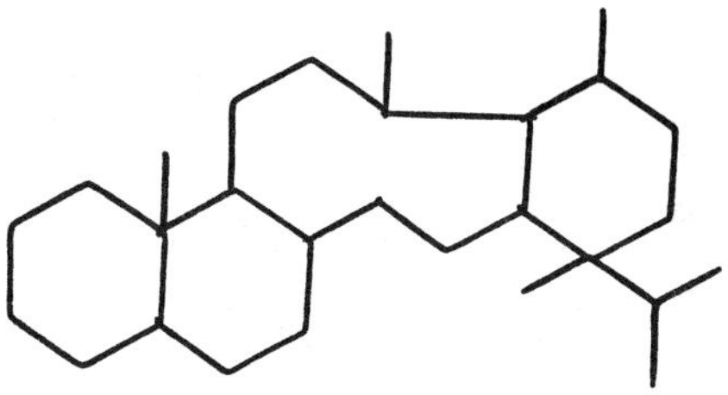

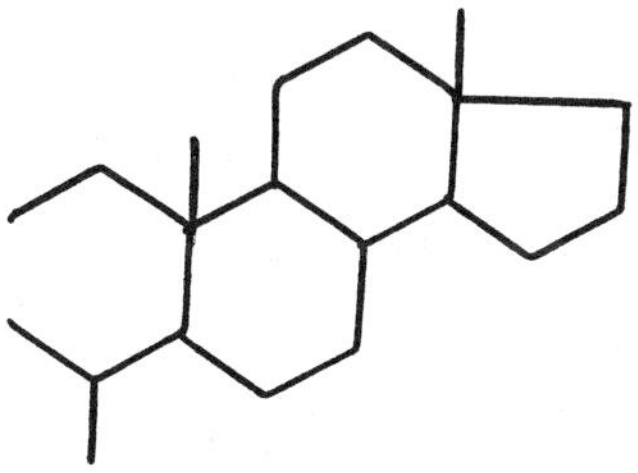

4599	Steroids - Less Common Skeletons - Misc.

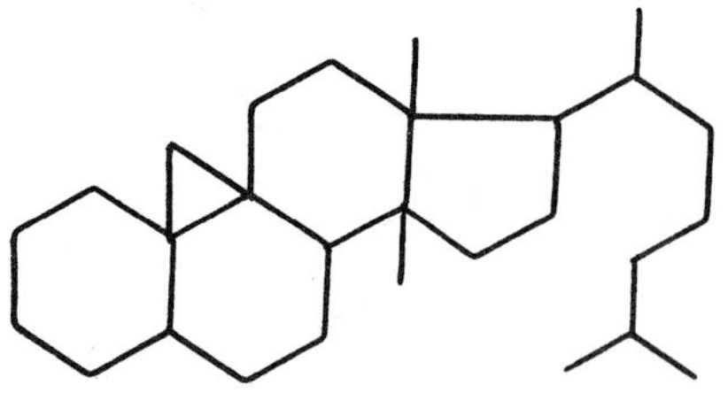

N

001

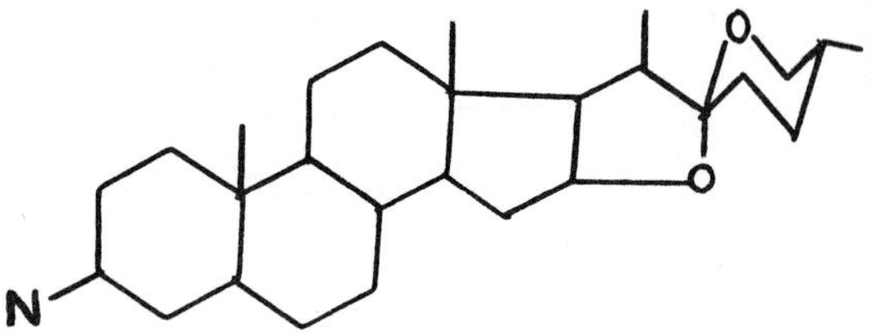

005

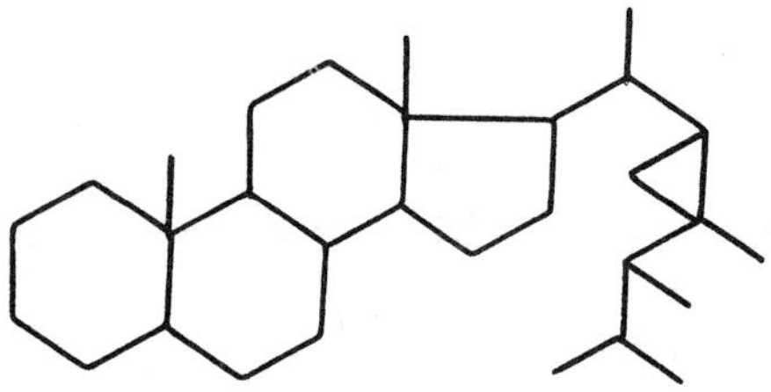

002

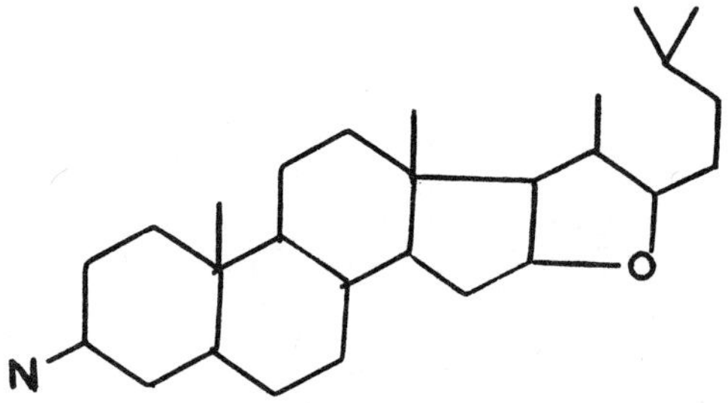

N

006

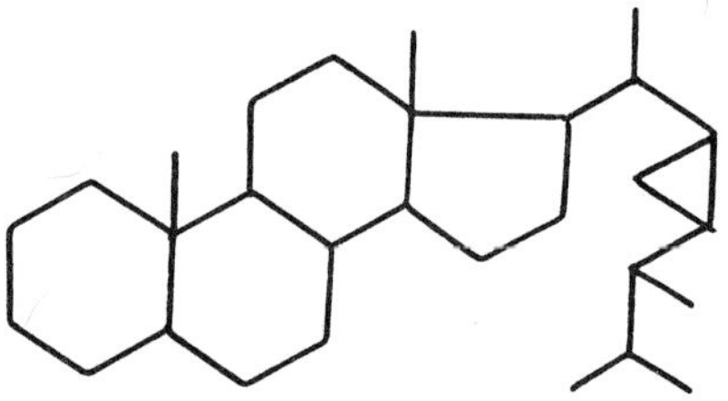

008

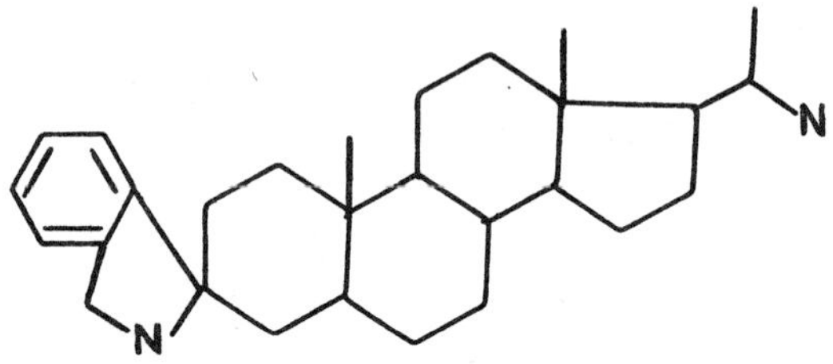

N

007

N

N

N

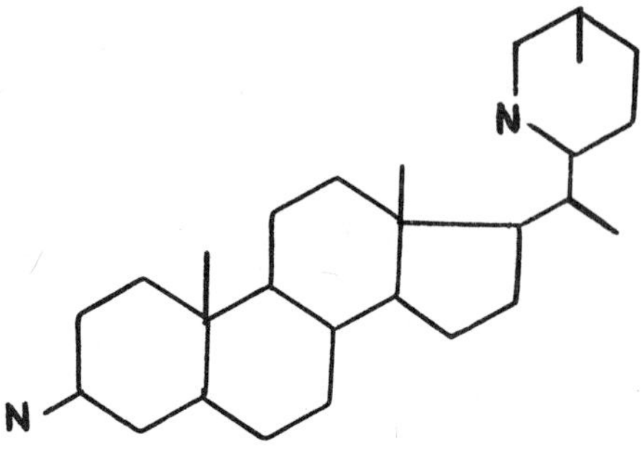

003

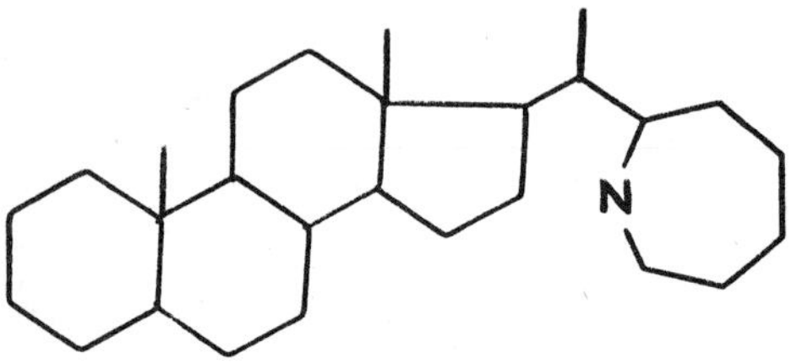

004

CHOLEST-5-ENE $C_{27}H_{46}$ 370

Phytochemistry, 1968, 657.

4501-001

CHOLESTEROL $C_{27}H_{46}O$ 386 -31^{0}r 148^{0}C

Chem. Comm., 1969, 45.

4501-002

CARCINOLIPIN $C_{44}H_{78}O_2$ 638 75^{0}C

$CO\cdot(CH_2)_{12}\cdot CH\cdot Et$ (Me)

Biochem. J., 1968 (107) 129.

4501-003

DESMOSTEROL $C_{27}H_{44}O$ 384 -41^{0} 121^{0}C

Compt. Rend., 1969 (268) 2105.

4501-004

CHOLESTEROL, 22-dehydro- $C_{27}H_{44}O$ 384 -58^{0} 135^{0}C

JACS., 1960, 1442.

4501-005

CHOLESTEROL, 7, 22-bis-dehydro- $C_{27}H_{42}O$ 382 -111^{0}c 121^{0}C

Tet. Lett., 1968, 6103.

4501-006

ZYMOSTEROL $C_{27}H_{44}O$ 384 $+49^{0}$c 110^{0}C

JCS., 1949, 214.

4501-007

CHOLEST-7-EN-3-ONE $C_{27}H_{44}O$ 384 $+25^{0}$c 147^{0}C

Nature, 1966 (210) 416.

4501-008

CHOLESTEROL, 22-OH- $C_{27}H_{46}O_2$ 402 -39^{0}c 186^{0}C

Acta Chem. Scand., 1953, 1220.

4501-009

PENIOCEROL $C_{27}H_{46}O_2$ 402 $+59^{0}$c 170^{0}C

Proc. Chem. Soc., 1961, 450.

4501-010

CHOLESTEROL, 20α, 22-diOH- $C_{27}H_{46}O_3$ 418

J. Biol. Chem., 1961 (236) 695.

4501-011

CYPRINOL $C_{27}H_{48}O_5$ 452 $+29^0$e 243^0C

Biochem. J., 1964 (90) 303.

4501-012

CHIOGRASTEROL A $C_{27}H_{46}O_5$ 450 -7^0 208^0C

Chem. Comm., 1968, 104.

4501-013

CHIOGRASTEROL B $C_{27}H_{48}O_5$ 452 $+16^0$ 241^0C

Chem. Comm., 1968, 105.

4501-014

CHIOGRASTERONE $C_{27}H_{42}O_4$ 430 -65^0 140^0C

JCS., 1970, 910.

4501-015

CHIOGRASTERONE, iso- $C_{27}H_{42}O_4$ 430 -21^0 149^0C

JCS., 1970, 910.

4501-016

NOLOGENIN $C_{27}H_{44}O_5$ 448 265^0C

JACS., 1947, 2167; 2208; 2224.

4501-017

RICOGENIN $C_{27}H_{40}O_5$ 444 226^0C

JACS., 1949, 3856.

4501-018

CRYPTOGENIN $C_{27}H_{42}O_4$ 430 188^0C

Tet. Lett., 1966, 4319.

4501-019

CHOLESTEROL, 24-oxo- $C_{27}H_{44}O$ 400 134^0C

Acta Chem. Scand., 1970, 1846.

4501-020

Name	Formula	Data	Reference	No.
ERGOSTANOL	$C_{28}H_{50}O$	402 +16^0 144	Helv., 1951, 832.	4502-001
BRASSICASTEROL	$C_{28}H_{46}O$	398 -64^{0}c 148^{0}C	JOC., 1947, 67.	4502-006
SPONGESTEROL	$C_{28}H_{48}O$	400 +10^{0}c 153^{0}C	JOC., 1945, 570.	4502-002
CRINOSTEROL	$C_{28}H_{46}O$	398 -47^0 138^{0}C	Nature, 1967 (213) 905.	4502-007
CAMPESTEROL	$C_{28}H_{48}O$	400 -33^{0}c 158^{0}C	JACS, 1954, 4192.	4502-003
CHOLESTEROL, 24-methylene-	$C_{28}H_{46}O$	398 -35^{0}c 142^{0}C	JOC., 1956, 372.	4502-008
BRASSICASTEROL, 22-dihydro-	$C_{28}H_{48}O$	400	Chem. Abstr., 1966 (64) 3637.	4502-004
EPISTEROL	$C_{28}H_{46}O$	398 -5^{0}c 150^{0}C	JCS., 1946, 512.	4502-009
FUNGISTEROL	$C_{28}H_{48}O$	400 - 1^{0}c 149^{0}C	Ann., 1941 (548) 270.	4502-005
ERGOSTEROL, 5, 6-dihydro-	$C_{28}H_{46}O$	398 -19^{0}c 176^{0}C	JCS., 1948, 1354.	4502-010

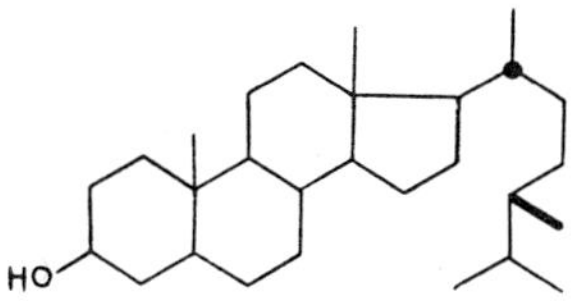

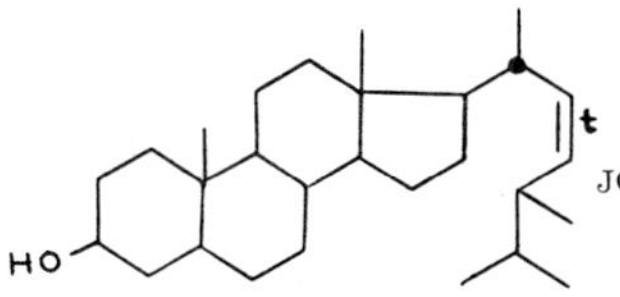

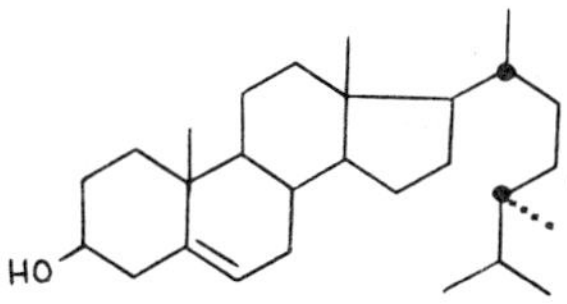

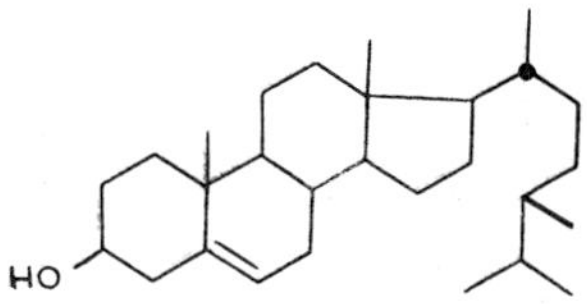

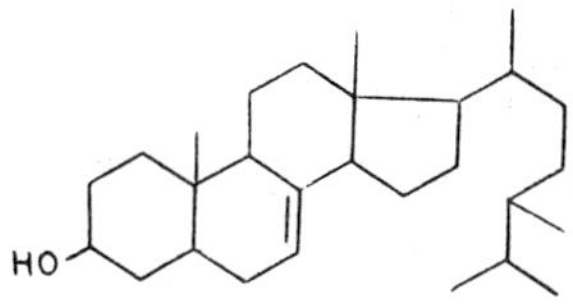

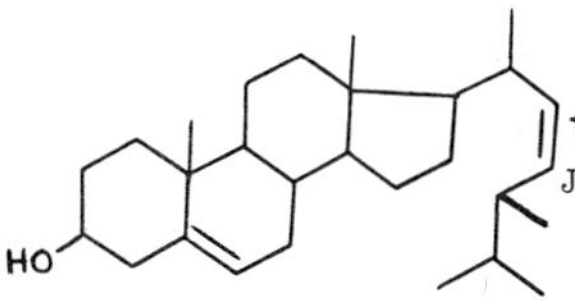

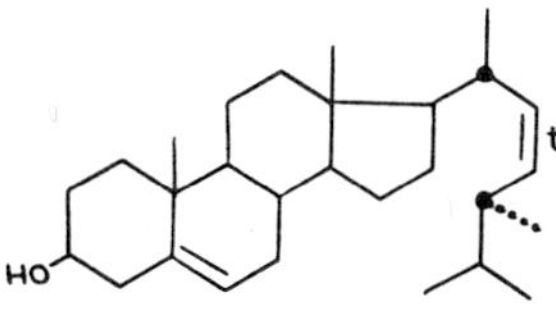

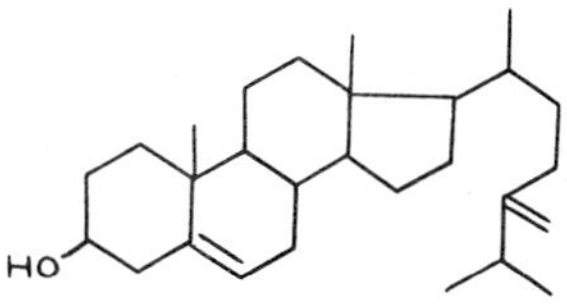

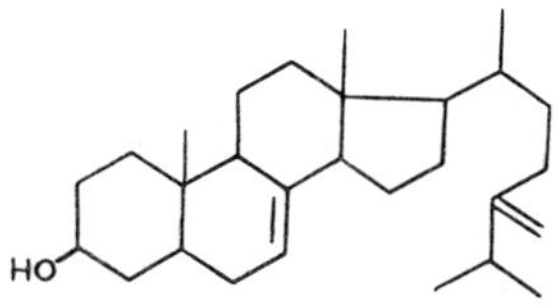

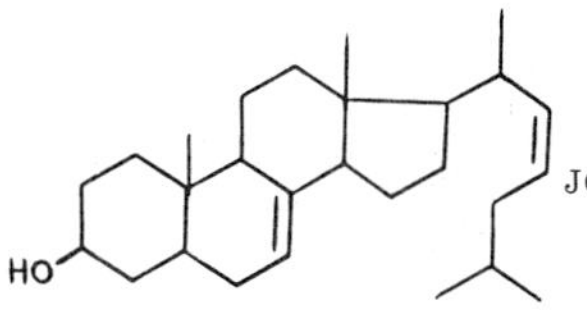

FECOSTEROL $C_{28}H_{46}O$ 398 +42°c 162°C JCS., 1949, 214. 4502-011	ERGOSTEROL, 14-dehydro- $C_{28}H_{42}O$ 394 -396°t 199°C JCS., 1951, 2728. 4502-016
ASCOSTEROL $C_{28}H_{46}O$ 398 +45°c 146 JCS., 1949, 214. 4502-012	ERGOSTEROL, 24(28)-dehydro- $C_{28}H_{42}O_2$ 410 -78°c 119°C JOC., 1954, 1734. 4502-017
ERGOSTEROL $C_{28}H_{44}O$ 396 -133°c 166°C Chem. Comm., 1966, 595. 4502-013	CEREVISTEROL $C_{28}H_{46}O_3$ 430 -83°p 258°C JCS., 1954, 1356. 4502-018
PYROCALCIFEROL $C_{28}H_{44}O$ 396 +502°e 94°C JCS., 1959, 1159. 4502-014	CAMPEST-4-EN-3-ONE $C_{28}H_{46}O$ 398 Chem. Pharm. Bull., 1969, 163. 4502-019
ERGOSTA-5, 8, 17(20)-TRIEN-3-OL- $C_{28}H_{44}O$ 396 136°C Chem. Abstr., 1969 (70) 862. 4502-015	ERGOST-7, 22-DIENE-3-ONE $C_{28}H_{44}O$ 396 +6°c 185°C Ind. J. C., 1965, 575. 4502-020

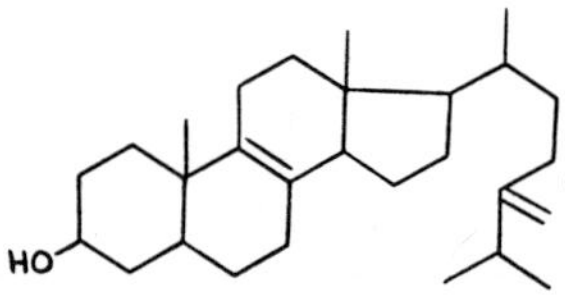

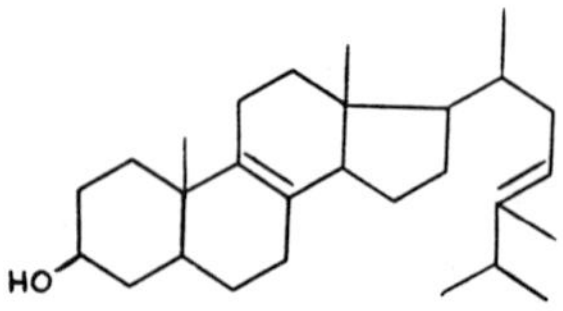

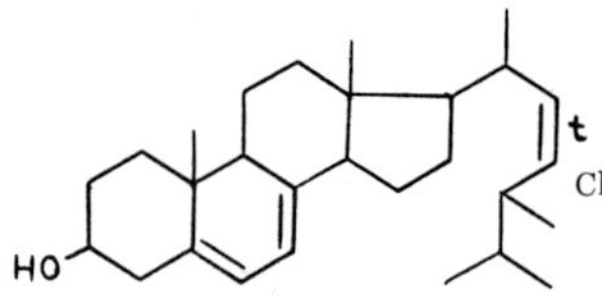

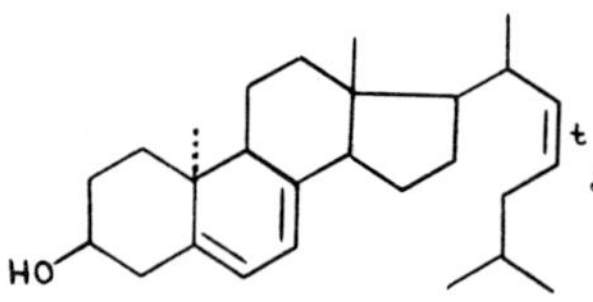

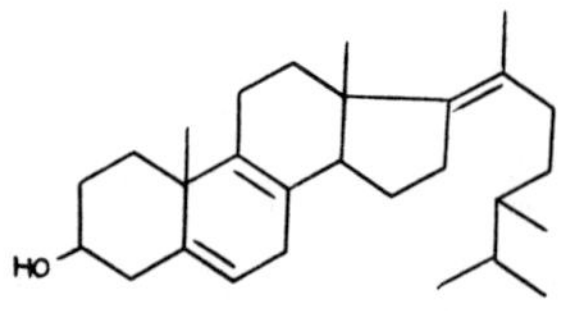

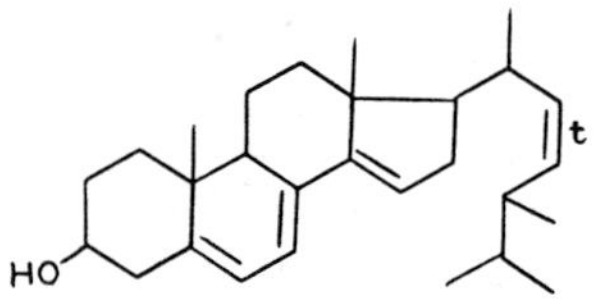

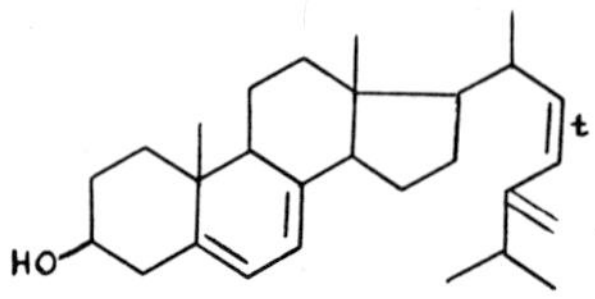

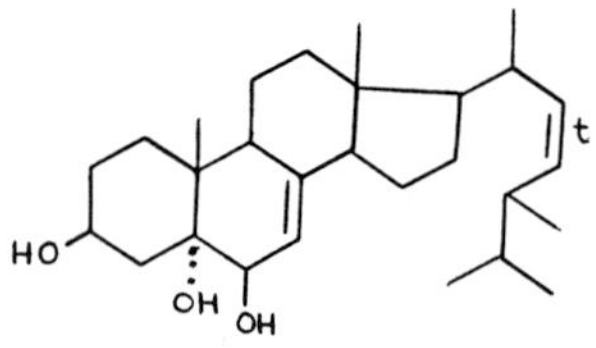

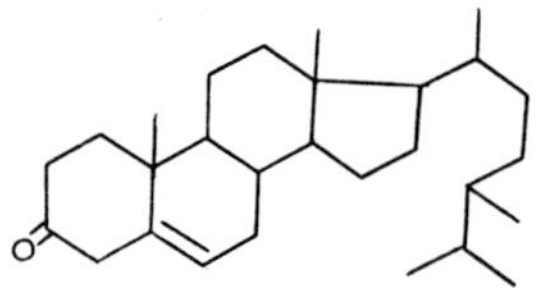

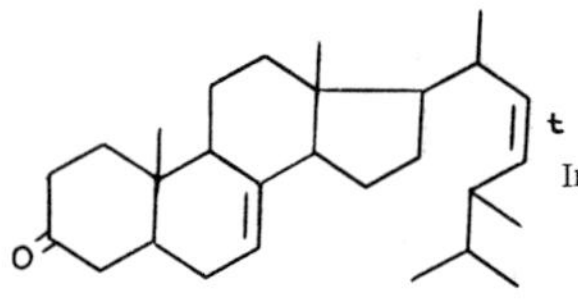

ERGOSTERONE, iso- $C_{28}H_{42}O$ 394 -26° 108°C Ind. J. C., 1965, 575. 4502-021	JABOROSALACTONE A $C_{28}H_{38}O_5$ 454 +95°c 218°C Tet., 1966, 1121. 4502-026
ERGOSTA-4, 6, 8(14), 22-TETRA-EN-3-ONE $C_{28}H_{40}O$ 392 +610°c 114°C Tet. Lett., 1968, 4763. 4502-022	JABOROSALACTONE B $C_{28}H_{38}O_5$ 454 228°C Tet., 1966, 1121. 4502-027
ERGOSTEROL PEROXIDE $C_{28}H_{44}O_3$ 428 -36° 178°C Helv., 1947 (30) 1028. 4502-023	JABOROSALACTONE C $C_{28}H_{39}O_5Cl$ 219°C Tet., 1968, 5169. 4502-028
MAKISTERONE A $C_{28}H_{46}O_7$ 494 264°C Tet. Lett., 1968, 3883. 4502-024	JABOROSALACTONE D $C_{28}H_{40}O_6$ 472 283°C Tet., 1968, 5169. 4502-029
MAKISTERONE B $C_{28}H_{46}O_7$ 494 172°C Tet. Lett., 1968, 3887. 4502-025	JABOROSALACTONE E $C_{28}H_{39}O_5Cl$ 257°C Tet., 1968, 5169. 4502-030

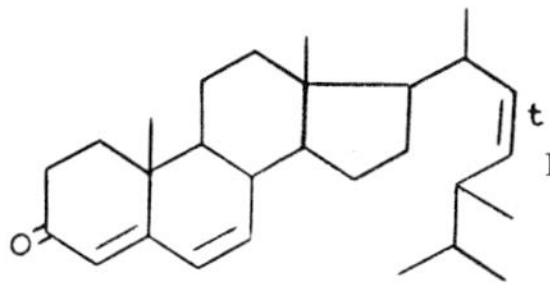

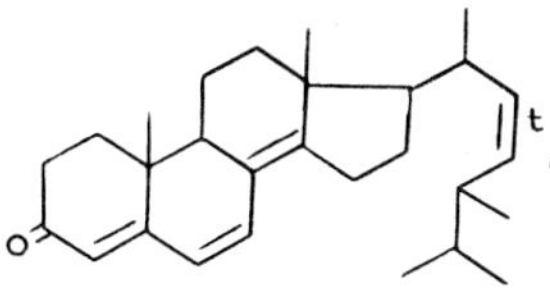

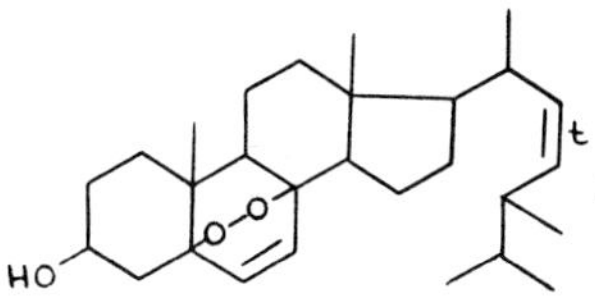

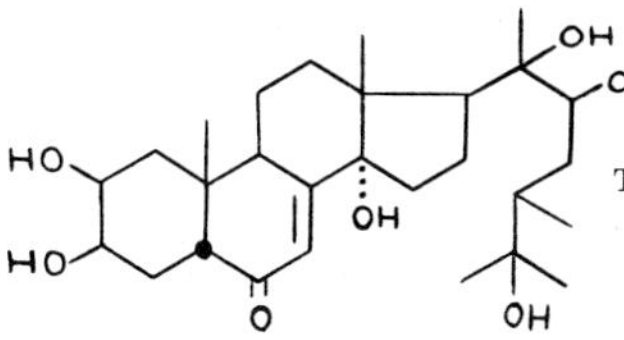

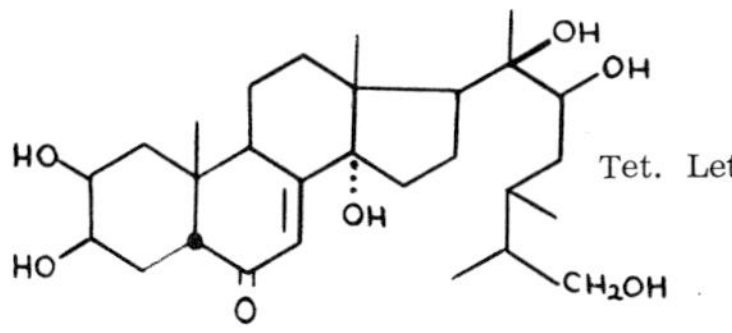

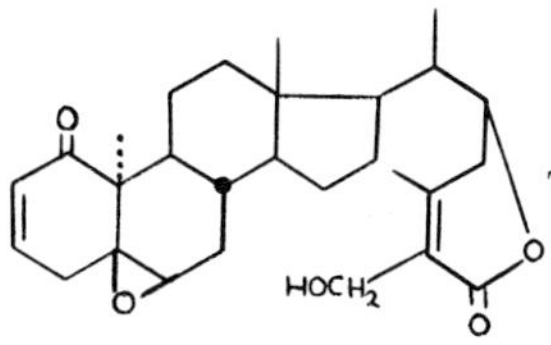

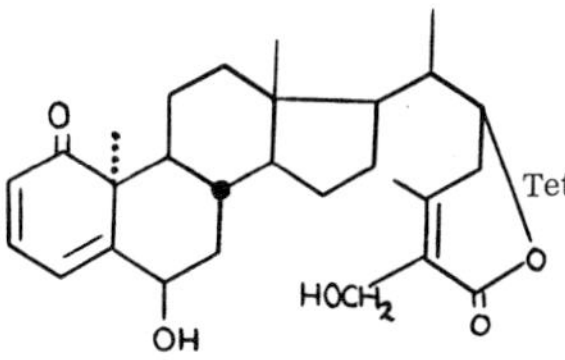

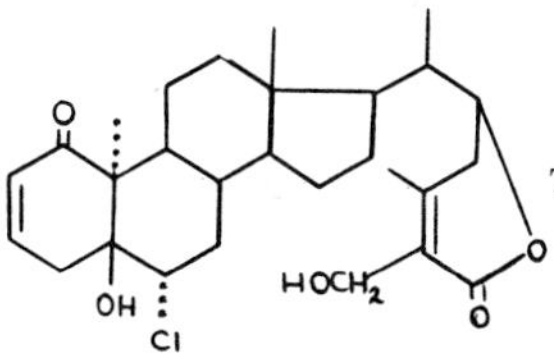

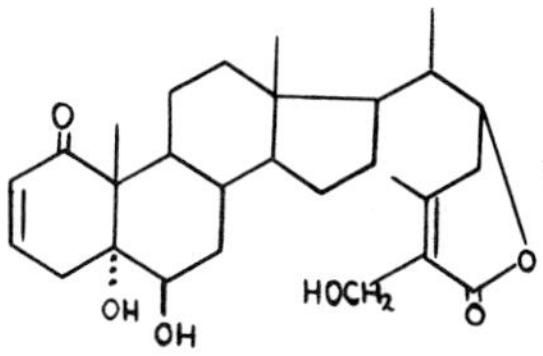

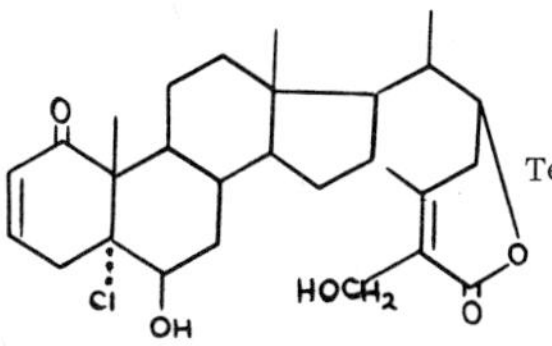

WITHAFERIN A $C_{28}H_{38}O_6$ 470 JCS., 1966, 1753. 4502-031	WITHANOLIDE D $C_{28}H_{38}O_6$ 470 Tet., 1970, 2215. 4502-035
WITHAFERIN A, 2, 3-dihydro-3-OMe- $C_{29}H_{42}O_7$ 402 JCS., 1965, 7517. 4502-032	WITHANOLIDE D, 24, 25-dihydro- $C_{28}H_{40}O_6$ 472 +14^{0}c 275^{0}C Tet. 1970, 2215. 4502-036
WITHAFERIN A, 27-deoxy- $C_{28}H_{38}O_5$ 454 +101^{0}c 268^{0}C Tet., 1970, 2215. 4502-033	WITHANOLIDE D, 20-deoxy-24, 25-dihydro- $C_{28}H_{40}O_5$ 456 +32^{0}c 252^{0}C Tet., 1970, 2215. 4502-037
WITHAFERIN A, 27-deoxy-14-OH- $C_{28}H_{38}O_6$ 470 +67^{0}c 266^{0}C JCS., 1966, 1765. 4502-034	

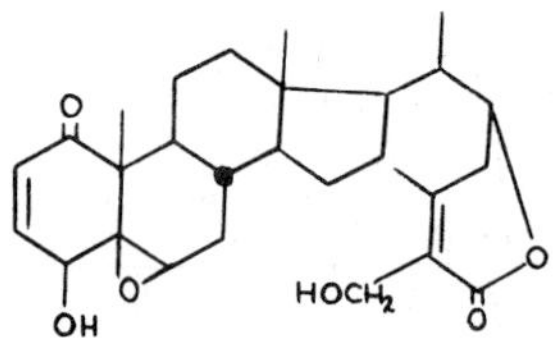

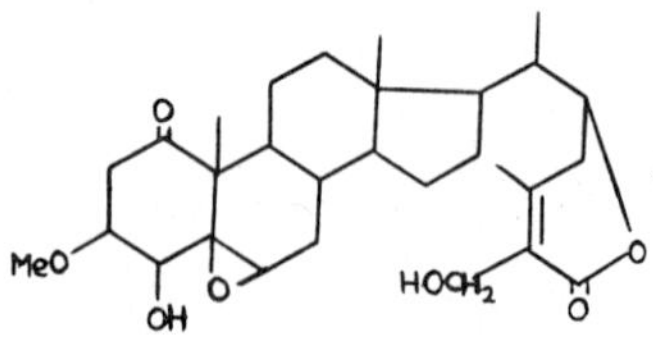

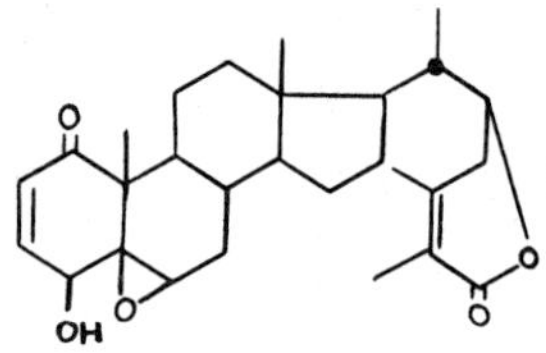

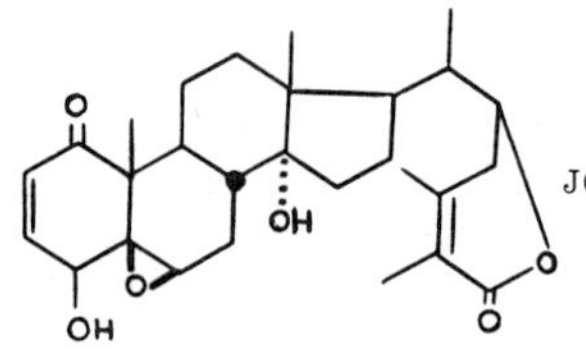

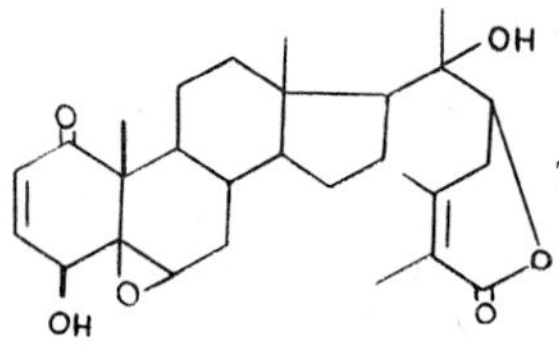

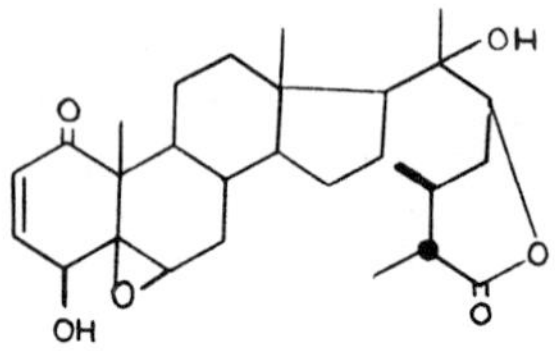

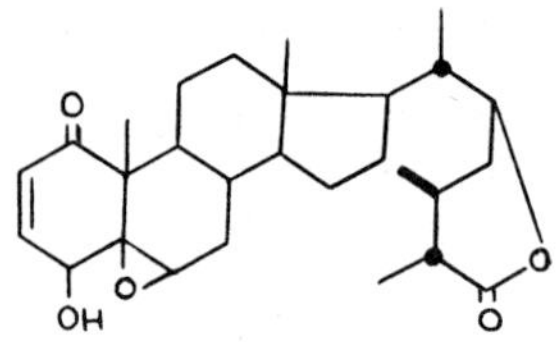

SITOSTANOL $C_{29}H_{52}O$ 416
$+28^{0}$c
140^{0}C

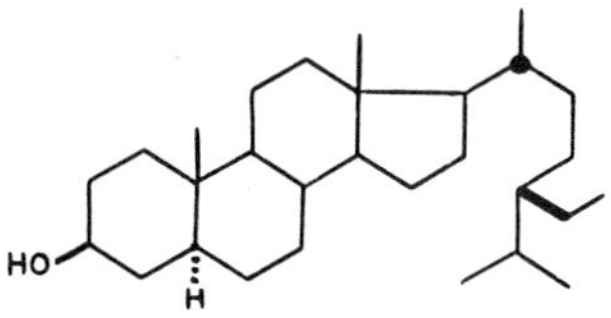

JCS. 1944 336.

4503-001

PORIFERASTANOL $C_{29}H_{52}O$ 416
$+25^{0}$c
143^{0}C

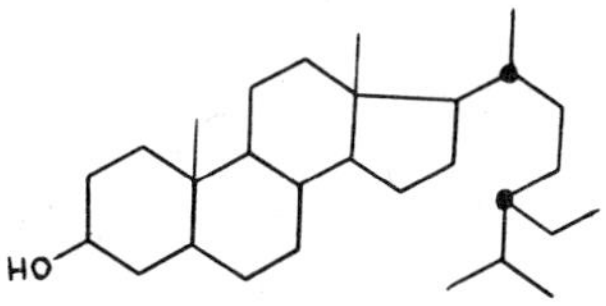

Chem. Comm. 1968 1698.

4503-002

$C_{29}H_{50}O$ 414
-36^{0}c
139^{0}C

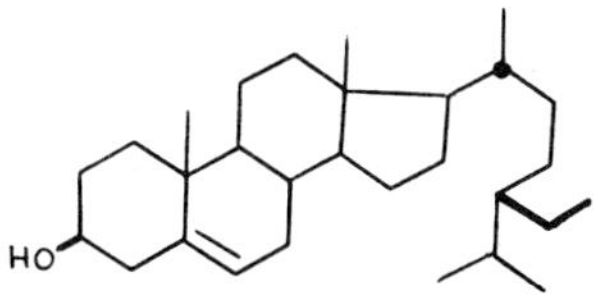

Can. J. C. 1968 2325.

4503-003

SITOSTEROL, γ- $C_{29}H_{50}O$ 414
-37^{0}c
140^{0}C

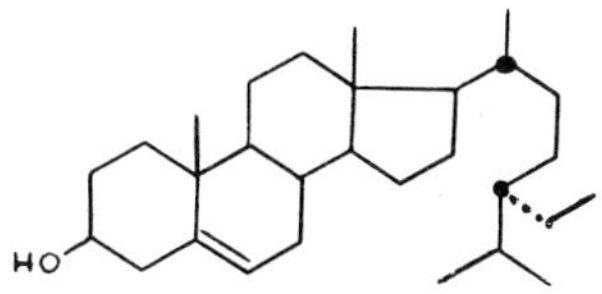

Chem. Abstr. 1966 (64) 3637.

4503-004

STIGMAST-7-EN-3-OL $C_{29}H_{50}O$ 414
$+11^{0}$c
146^{0}C

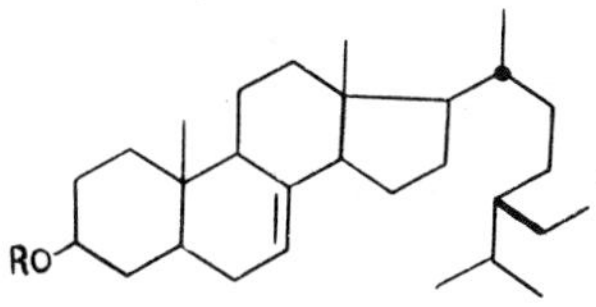

JACS. 1954 4192.

R = H

4503-005

STIGMAST-7-EN-3-OL ACETATE $C_{31}H_{52}O_2$ 456
$+12.5^{0}$c
155^{0}C

R= Ac

Chem. Pharm. Bull. 1970 213.

4503-006

STIGMAST-22-EN-3-OL $C_{29}H_{50}O$ 414
$+5^{0}$c
163^{0}C

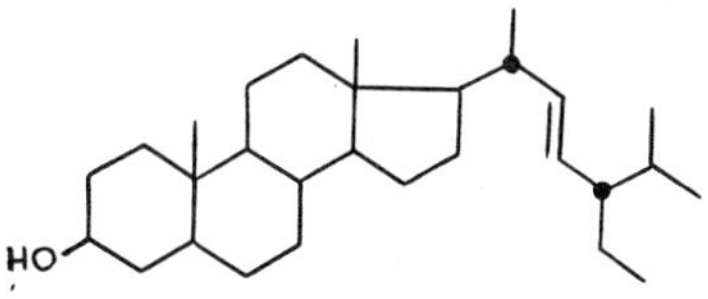

Arch. Biochem. Biophys. 1960 (91) 266.

4503-007

STIGMASTEROL $C_{29}H_{48}O$ 412
-51^{0}c
169^{0}C

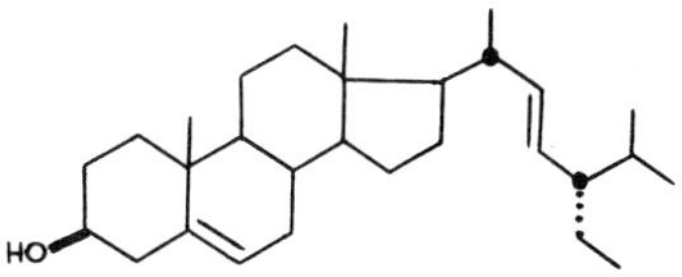

JACS. 1955 5442.

4503-008

PORIFERASTEROL $C_{29}H_{48}O$ 412
-49^{0}
156^{0}C

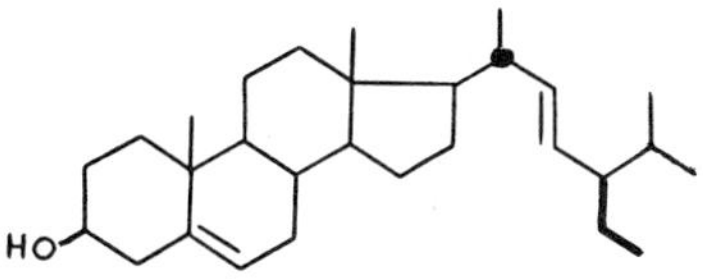

Plant Cell Physiol. 1965 (6) 211.

4503-009

FUCOSTEROL $C_{29}H_{48}O$ 412
-42^{0}
124^{0}C

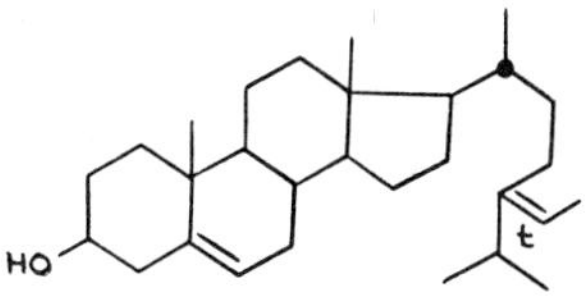

JCS. 1950 2881.

4503-010

FUCOSTEROL, iso- $C_{29}H_{48}O$ 412 -38° 137°C Tet. Lett. 1968 6163. 4503-011	SPINASTERY ACETATE, α- $C_{31}H_{50}O_2$ 454 -3.5°c 175°C R = Ac Chem. Pharm. Bull. 1970 213. 4503-016
SARGASTEROL $C_{29}H_{48}O$ 412 -47°c 133°C JACS. 1958 921. 4503-012	AVENASTEROL, 7-dehydro- $C_{29}H_{48}O$ 412 +9°c 145°C Tet. Lett. 1968 6163. 4503-017
CLEROSTEROL $C_{29}H_{48}O$ 412 -50°c 147°C Tet. 1966 2377. 4503-013	STIGMASTA-7, 24(28)-DIEN-3-OL $C_{29}H_{48}O$ 412 +13° 148-51°C Tet. Lett. 1968 2443. 4503-018
CHONDRILLASTEROL $C_{29}H_{48}O$ 412 -2°c 169°C JOC. 1950 812. 4503-014	STIGMASTA-7, 25-DIEN-3-OL $C_{29}H_{48}O$ 412 +1° 139°C Ber. 1966 (99) 3559. 4503-019
SPINASTEROL, α- $C_{29}H_{48}O$ 412 -4°c 170°C JCS 1951 5051. R = H 4503-015	STIGMASTA-8(14), 22-DIEN-3-OL $C_{29}H_{48}O$ 412 Tet. Lett. 1968 5727. 4503-020

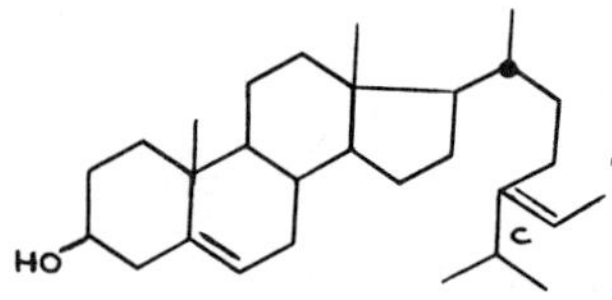

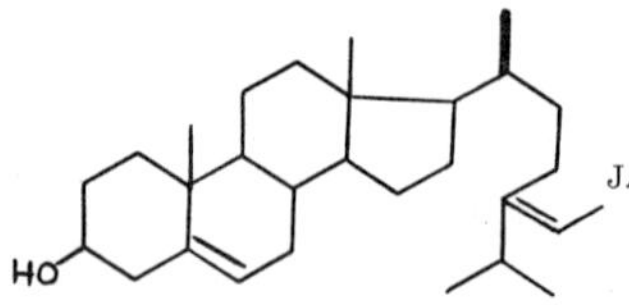

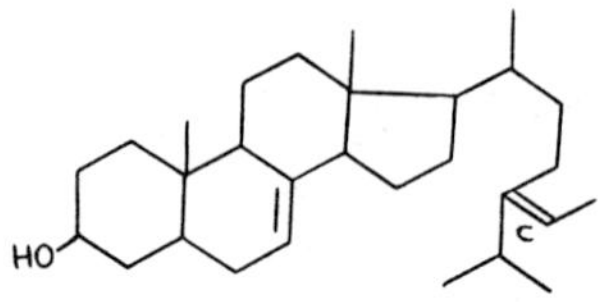

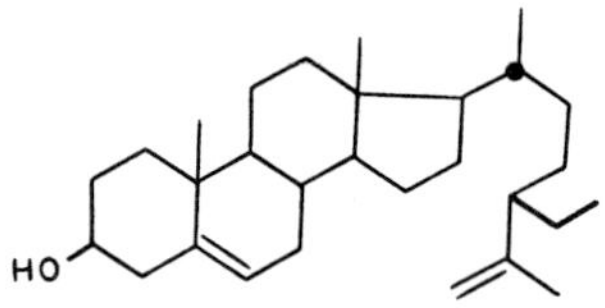

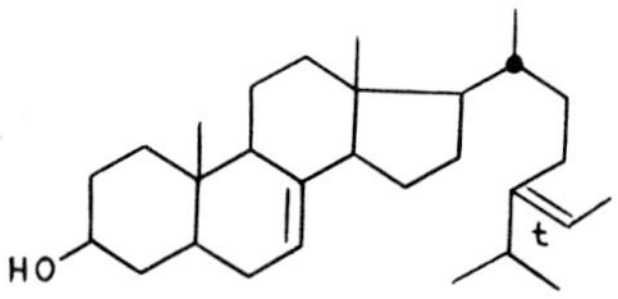

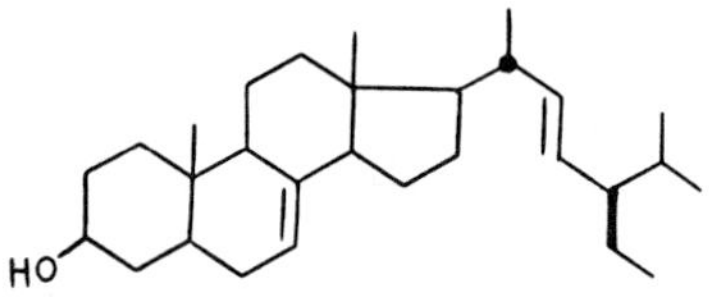

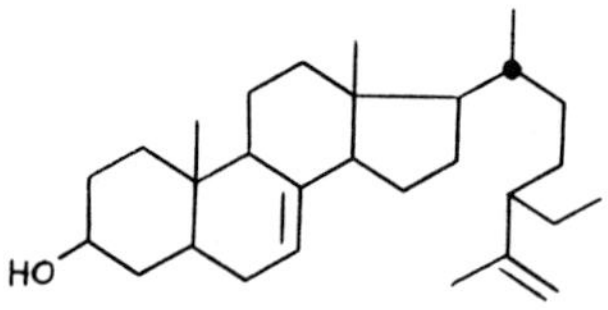

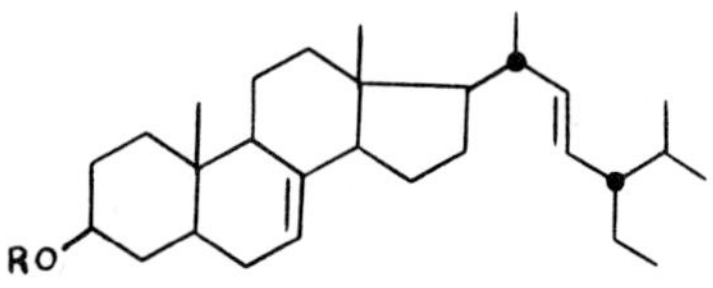

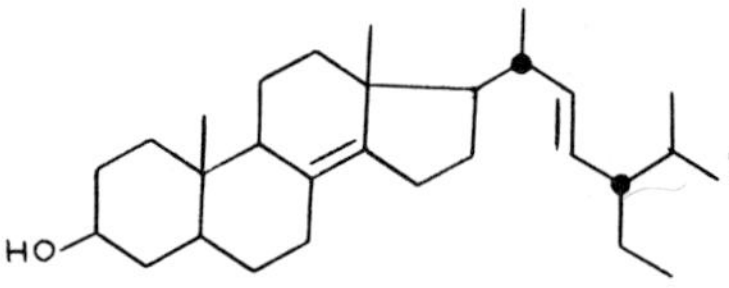

SITOSTEROL, α- $C_{29}H_{48}O$ 412 -2°c 165°C

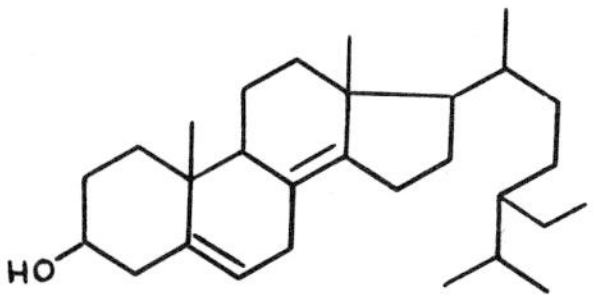

JACS. 1939 3700.

4503-021

SARINGOSTEROL $C_{29}H_{48}O_2$ 428 -31°c 160°C

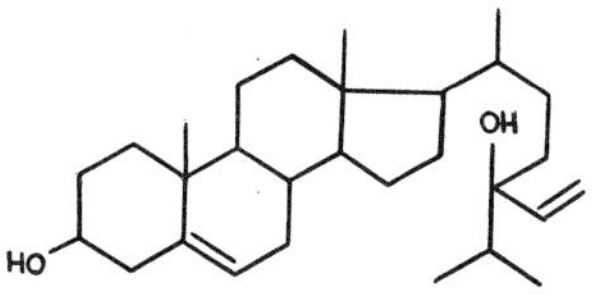

Chem. Ind. 1966 1179.

4503-026

STIGMASTA-5, 22, 25-TRIEN-3-OL $C_{29}H_{46}O$ 410 -38°c 146°C

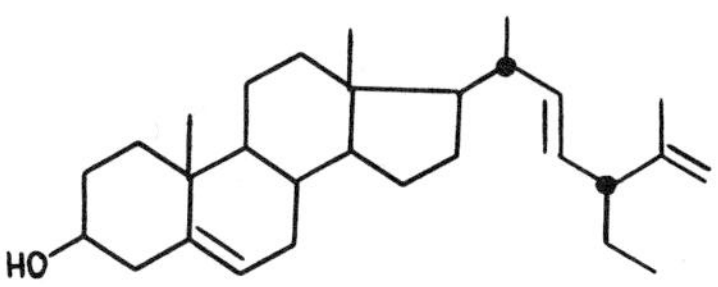

Tet. Lett. 1970 3043.

4503-022

SITOSTENONE, β- $C_{29}H_{48}O$ 412 +81. 3°c 95-96. 5°

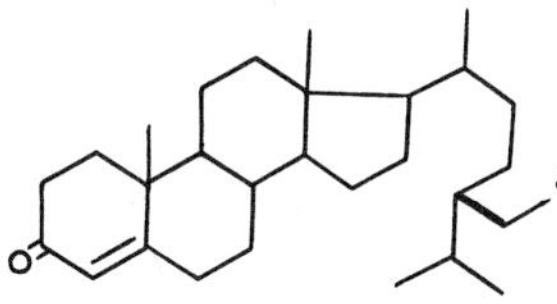

JCS. 1963 5001.

4503-027

STIGMASTA-7, 16, 25-TRIEN-3-OL $C_{29}H_{46}O$ 410 +8. 2°c 167°C

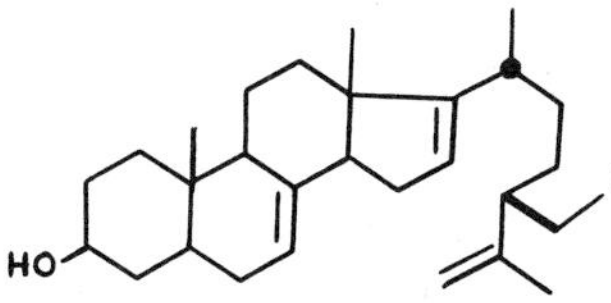

Chem. Abstr. 1968 (68) 78485.

4503-023

STIGMAST-7-EN-3-ONE $C_{29}H_{48}O$ 412 +130°c 154-8°C

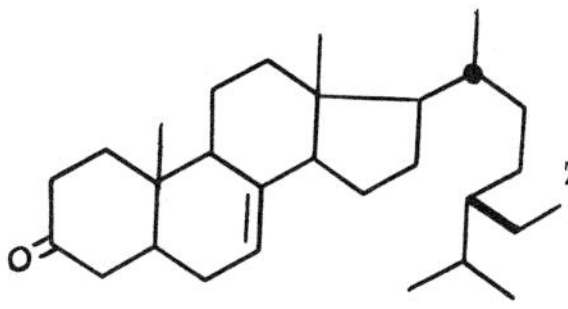

Z. Nat. 1968 42.

4503-028

STIGMASTA-7, 22, 25-TRIEN-3β-OL $C_{29}H_{46}O$ 410 +13°c 161-5°C

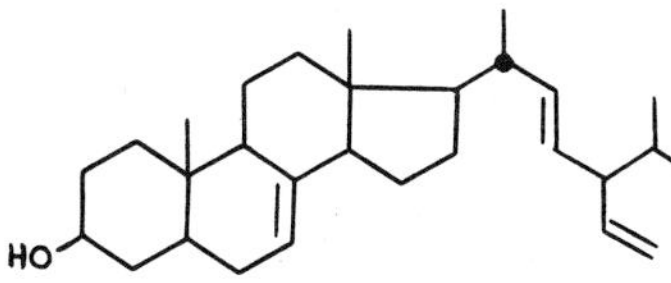

Z. Nat. 1968 42.

4503-024

TREMULONE $C_{29}H_{46}O$ 410 -288°c 111°

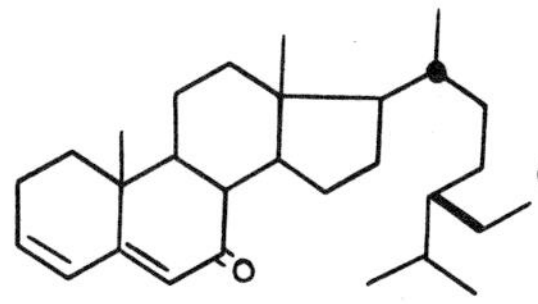

Can. J. C. 1962 (40) 2017.

4503-029

CORBISTEROL $C_{29}H_{46}O$ 410 -105°e 151°C

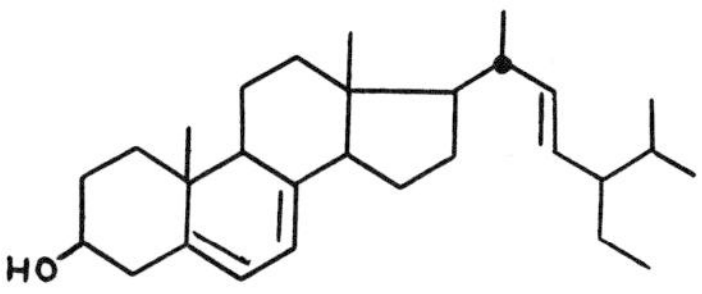

Chem. Abstr. 1959 (53) 22071.

4503-025

STIGMASTA-4, 22-DIEN-3-ONE $C_{29}H_{46}O$ 410

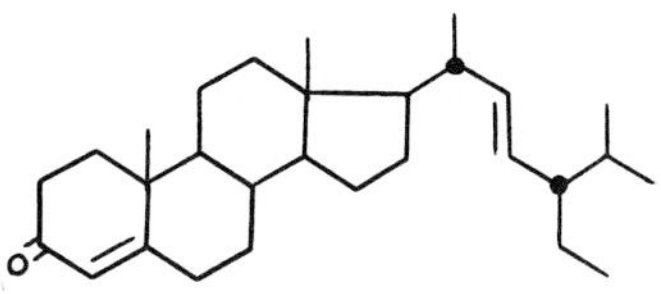

Chem. Pharm. Bull., 1969, 163.

4503-030

STIGMASTAN-3, 6-DIONE, 5α- $C_{29}H_{48}O_2$ 428

199°C

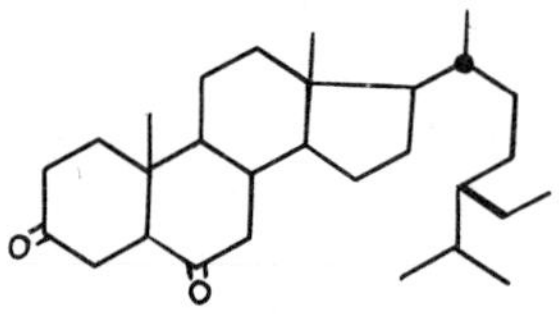

Chem. Pharm. Bull. 1969 163.

4503-031

STIGMAST-4-EN-3-ONE $C_{29}H_{48}O$ 412

+89°c

90°C

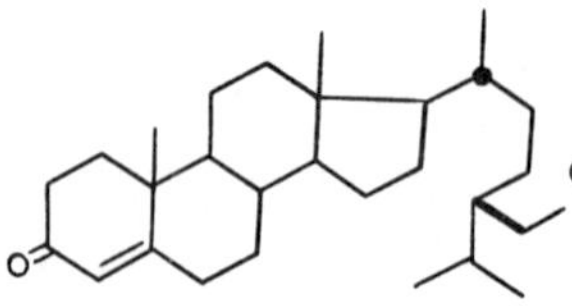

Chem. Pharm. Bull. 1969 163.

4503-032

AMARASTERONE A $C_{29}H_{48}O_7$ 508

210°C

R_1 = CH_2OH

R_2 = CH_3

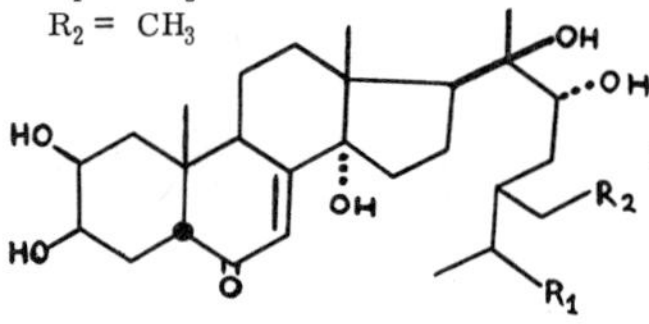

Tet. Lett. 1968 4953.

4503-033

AMARASTERONE B $C_{29}H_{48}O_7$ 508

285°C

R_1 = CH_3

R_2 = CH_2OH

Tet. Lett. 1968 4953.

4503-034

AJUGASTERONE B $C_{29}H_{46}O_7$ 506

+55°d

240°C

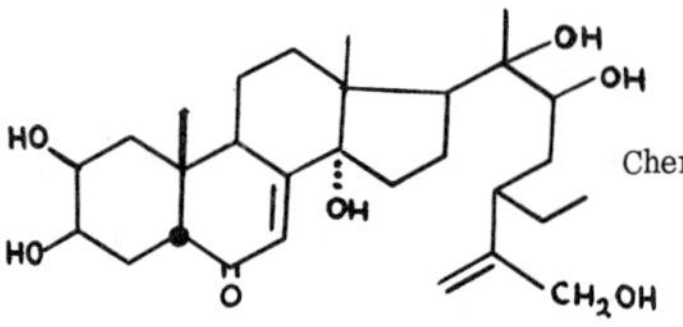

Chem. Comm. 1969 82.

4503-035

PODECDYSONE A $C_{29}H_{48}O_7$ 508

263°C

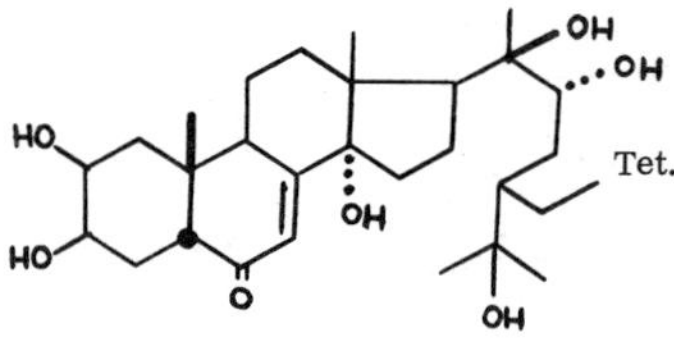

Tet. Lett. 1968, 4061.

4503-036

MAKISTERONE D $C_{29}H_{48}O_7$ 508

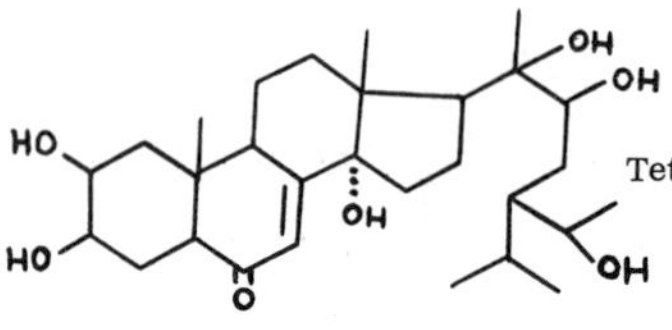

Tet. Lett. 1968 3887

4503-037

ANTHERIDIOL $C_{29}H_{42}O_5$ 470

250-55°C

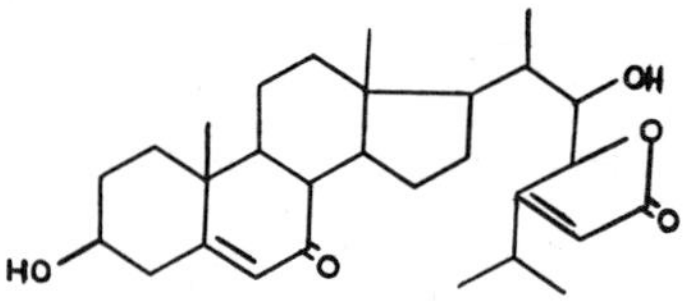

JACS. 1968 5635.

4503-038

CYASTERONE $C_{29}H_{44}O_8$ 520

+65°p

165°C

R = H

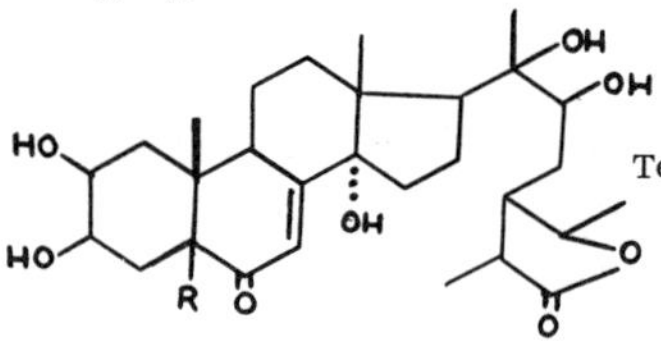

Tet. Lett. 1967 3191.

4503-039

SENGOSTERONE $C_{29}H_{44}O_9$ 536

+40°p

160°C

R = OH

Tet. 1970 887.

4503-040

CAPITASTERONE $C_{29}H_{44}O_7$ 504 235°C Tet. Lett. 1968 4929. 4503-041	CYASTERONE, pre- $C_{29}H_{44}O_8$ 520 +39°m Chem. Pharm. Bull., 1970, 1078. 4503-043
AJUGALACTONE $C_{29}H_{40}O_8$ 516 230°C JACS., 1970, 7512. 4503-042	VERNOSTEROL $C_{29}H_{46}O$ 410 Tet. Lett., 1970, 2971. 4503-044

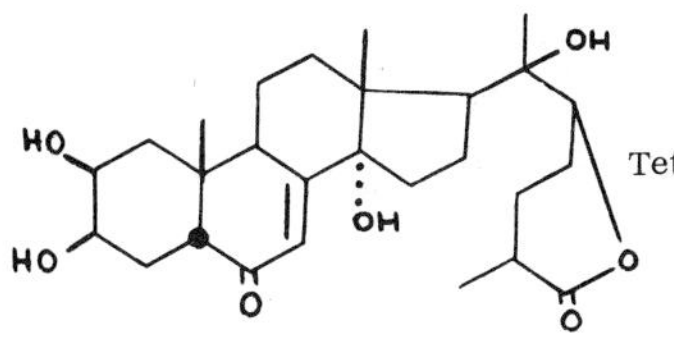

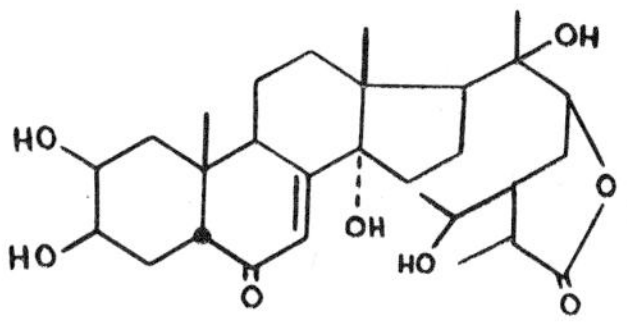

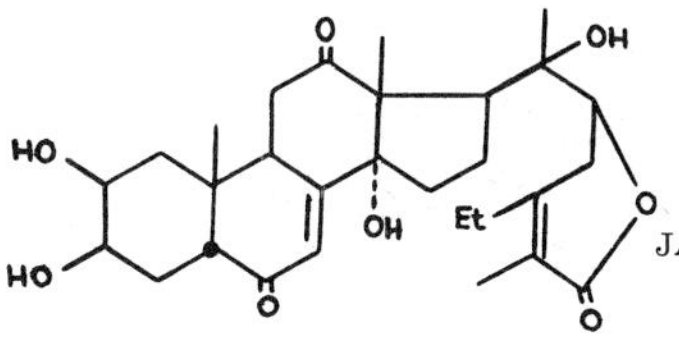

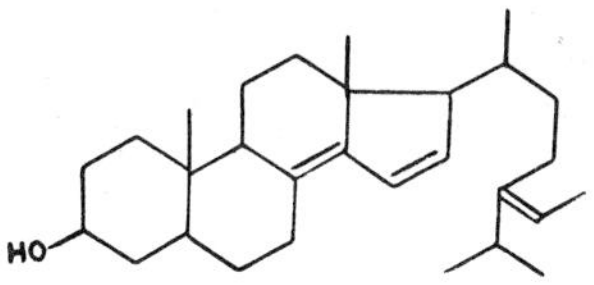

TIGOGENIN $C_{27}H_{44}O_3$ 416
-70°c
207°C

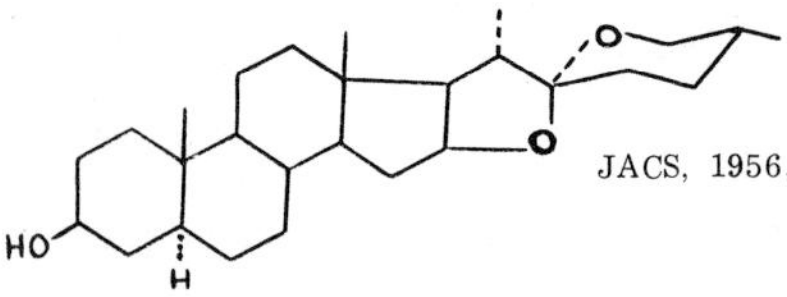

JACS, 1956, 446.

4504-001

TIGOGENIN, neo- $C_{27}H_{44}O_3$ 416
-65°c
203°C

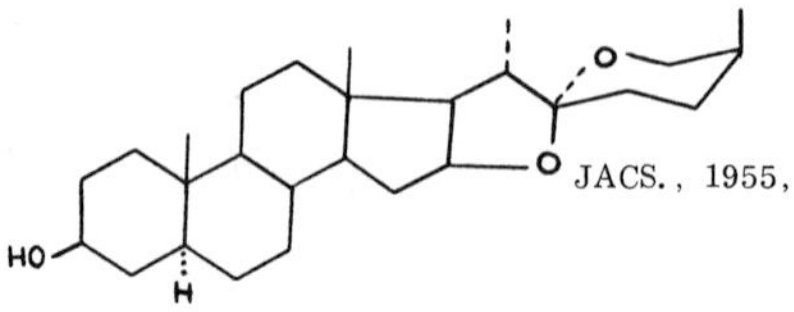

JACS., 1955, 1671.

4504-002

SMILAGENIN $C_{27}H_{44}O_3$ 416
-69°c
186°C

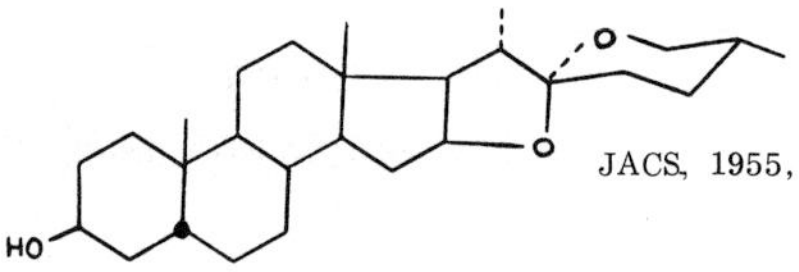

JACS, 1955, 641.

4504-003

SARSASAPOGENIN $C_{27}H_{44}O_3$ 416
-78°c
200°C

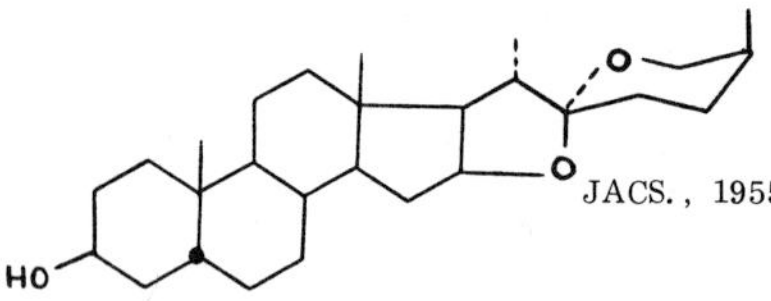

JACS., 1955, 641.

4504-004

DIOSGENIN $C_{27}H_{42}O_3$ 414
-121°c
206°C

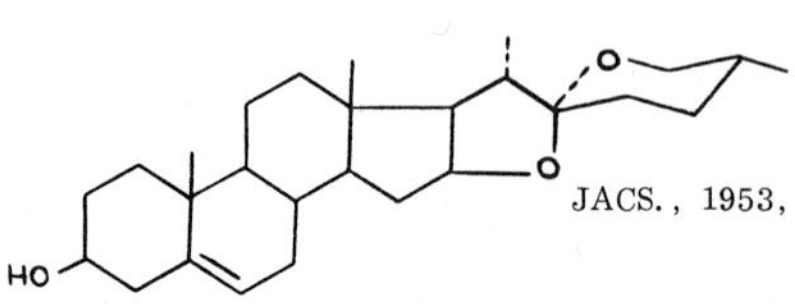

JACS., 1953, 6067.

4504-005

YAMOGENIN $C_{27}H_{42}O_3$ 414
-129°c
200°C

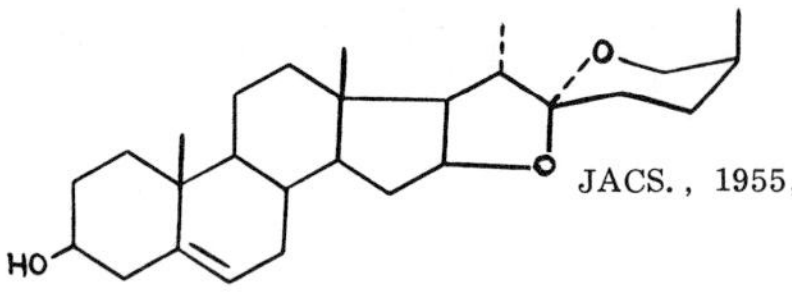

JACS., 1955, 645.

4504-006

TIGOGENIN, 25(27)-dehydro- $C_{27}H_{42}O_3$ 414

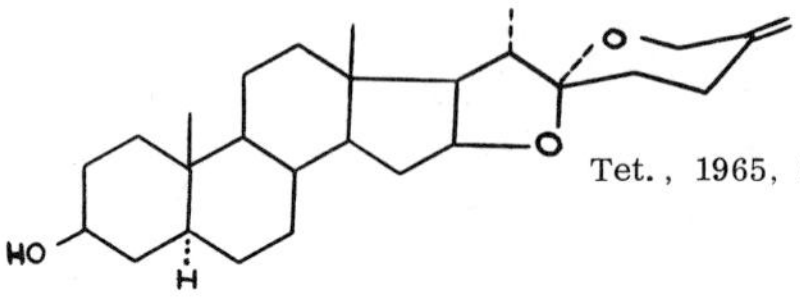

Tet., 1965, 2089.

4504-007

SCEPTRUMGENIN $C_{27}H_{40}O_3$ 412
-122°c
183°C

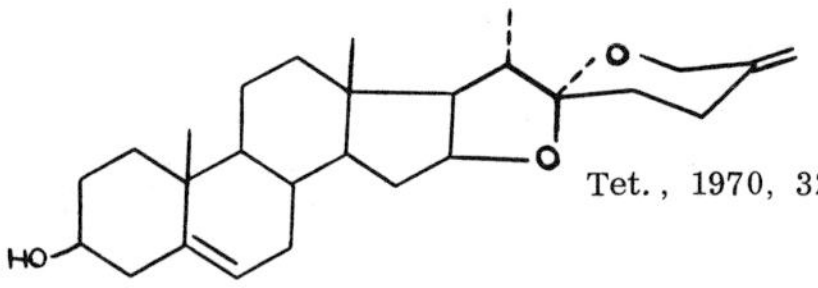

Tet., 1970, 3233.

4504-008

RHODEASAPOGENIN $C_{27}H_{44}O_4$ 432
-72°
294°C

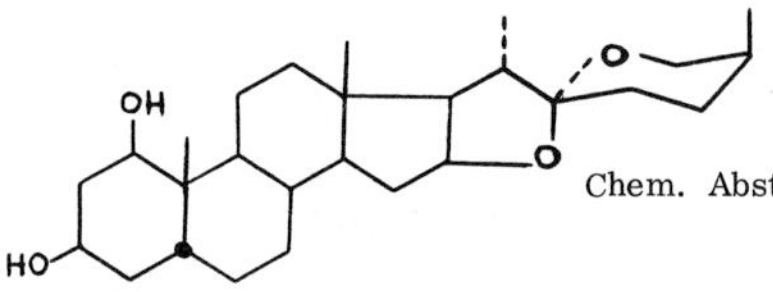

Chem. Abstr., 1959, 2282.

4504-009

RHODEASAPOGENIN, iso- $C_{27}H_{44}O_4$ 432
-76°
239°C

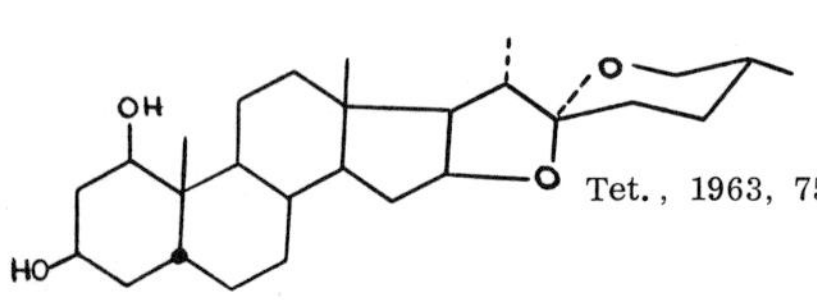

Tet., 1963, 759.

4504-010

RUSCOGENIN — $C_{27}H_{42}O_4$ — 430, -127^{0}c, 207^{0}C

Planta Med., 1964, 228.

4504-011

RUSCOGENIN, neo- — $C_{27}H_{42}O_4$ — 430, -118^{0}c, 200^{0}C

Tet. Lett., 1960 (19) 25.

4504-012

GITOGENIN — $C_{27}H_{44}O_4$ — 432, -75^{0}c, 272^{0}C

R = H

JACS., 1954, 5531.

4504-013

GITONIN, F- — $C_{50}H_{82}O_{23}$ — 1050, -66^{0}p, 253^{0}C

R = Lycotetraosyl

Tet., 1965, 299.

4504-014

GITOGENIN, neo- — $C_{27}H_{44}O_4$ — 432, -, 248^{0}C

JACS., 1947, 2375.

4504-015

SAMOGENIN — $C_{27}H_{44}O_4$ — 432, -74^{0}c, 206^{0}C

JACS., 1955, 4291.

4504-016

MARKOGENIN — $C_{27}H_{44}O_4$ — 432, -70^{0}c, 256^{0}C

JACS., 1955, 4291.

4504-017

YONOGENIN — $C_{27}H_{44}O_4$ — 432, -53^{0}c, 236^{0}C

Tet., 1965, 3633.

4504-018

CHLOROGENIN — $C_{27}H_{44}O_4$ — 432, -45^{0}c, 274^{0}C

JACS., 1947, 2399.

4504-019

ROCKOGENIN — $C_{27}H_{44}O_4$ — 432, -64^{0}a, 220^{0}C

JACS., 1952, 2693.

4504-020

DIGALOGENIN $C_{27}H_{44}O_4$ 432 -75°c 219°C

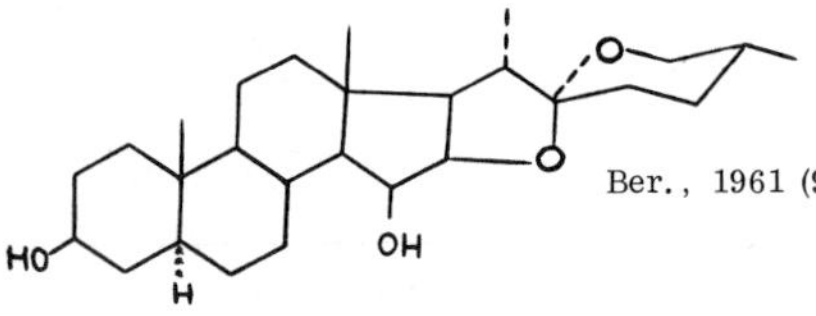

Ber., 1961 (94) 2019.

4504-021

YUCCAGENIN $C_{27}H_{42}O_4$ 430 -120°c 247°C

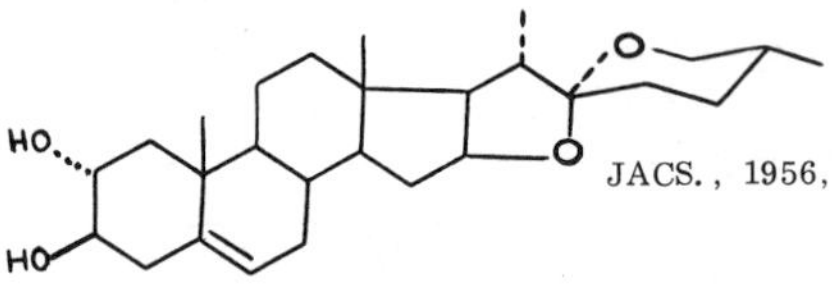

JACS., 1956, 1167.

4504-026

PENNOGENIN $C_{27}H_{42}O_4$ 430 246°C

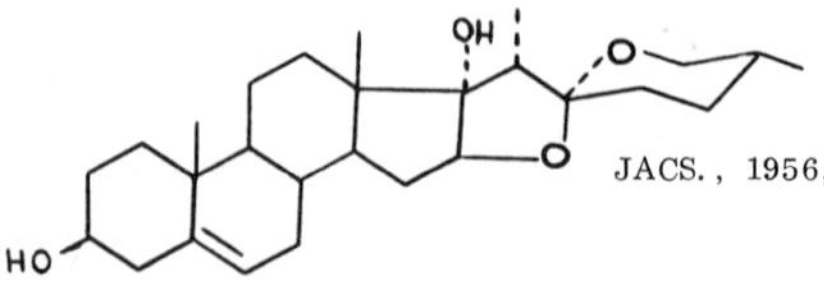

JACS., 1956, 440/5.

4504-022

LILAGENIN $C_{27}H_{42}O_4$ 430 - 245°C

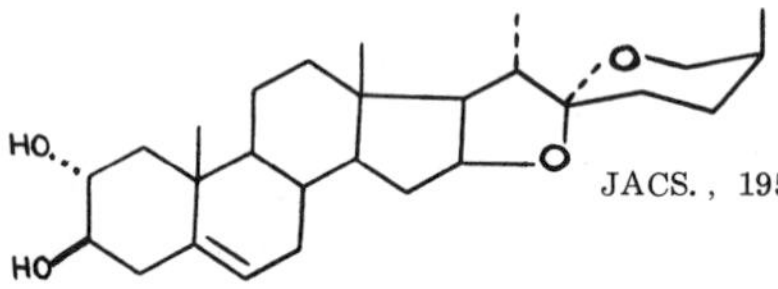

JACS., 1951, 5375.

4504-027

ISOPLEXIGENIN A $C_{27}H_{44}O_4$ 432 -61° 227°C

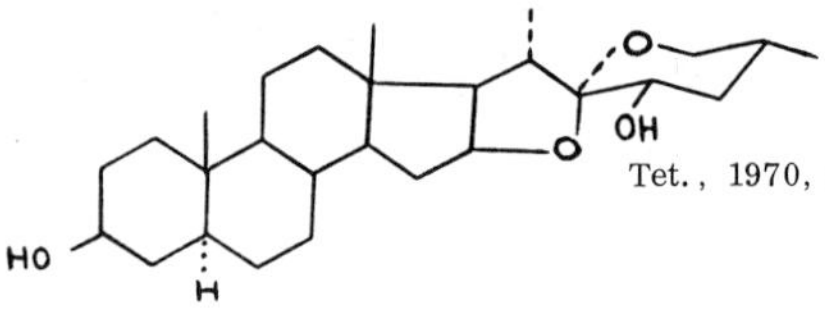

Tet., 1970, 3233.

4504-023

GITOGENIN, 25(27)-dehydro- $C_{27}H_{42}O_4$ 430 -80°c 242°C

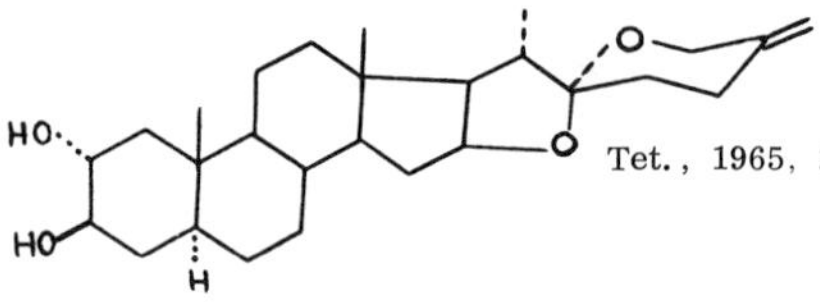

Tet., 1965, 2089.

4504-028

CHOLEGENIN, iso- $C_{27}H_{44}O_4$ 432 -67°c 256°C

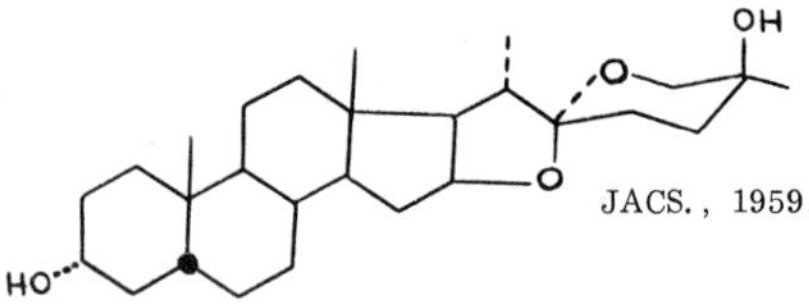

JACS., 1959, 5225.

4504-024

BETHOGENIN $C_{28}H_{44}O_4$ 444 -98° 194°C

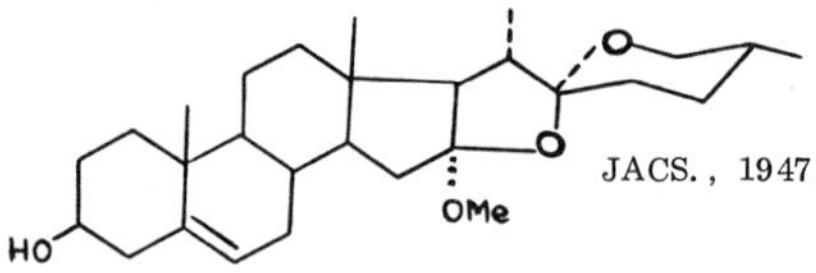

JACS., 1947, 2395.

Artifact ?

4504-029

CONVALLAMAROGENIN $C_{27}H_{42}O_4$ 430 -79° 260°C

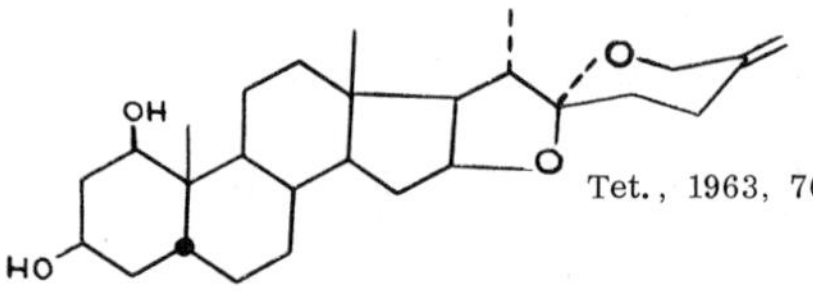

Tet., 1963, 760.

4504-025

ISOPLEXIGENIN B $C_{27}H_{42}O_4$ 430 -96°d 206°C

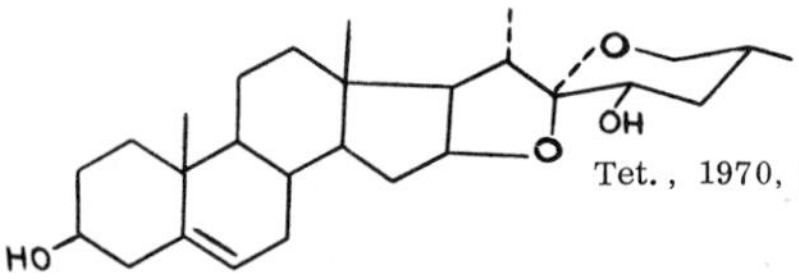

Tet., 1970, 3233.

4504-030

NARTHOGENIN, iso- $C_{27}H_{42}O_4$ 430 JCS., 1963, 4815. 4504-031	GARNEAGENIN, iso- $C_{27}H_{44}O_5$ 448 -63^0 243^0C Tet., 1963, 759. 4504-036
TOKOROGENIN $C_{27}H_{44}O_5$ 448 -39^0p 266^0C Chem. Abstr., 1959, 16206. 4504-032	DIOTIGENIN $C_{27}H_{44}O_5$ 448 -60^0 280^0C Chem. Pharm. Bull., 1968, 421. 4504-037
CONVALLAGENIN A $C_{27}H_{44}O_5$ 448 $-28^0c/m$ 268^0C Chem. Pharm. Bull., 1967, 1204. 4504-033	AGAPANTHAGENIN $C_{27}H_{44}O_5$ 448 285^0C JCS., 1956, 1167. 4504-038
REINECKIAGENIN $C_{27}H_{44}O_5$ 448 -71^0 279^0C Tet., 1963, 759. 4504-034	MAGOGENIN $C_{27}H_{44}O_5$ 448 284^0C JACS., 1947, 2399. 4504-039
REINECKIAGENIN, iso- $C_{27}H_{44}O_5$ 448 -66^0 241^0C Tet., 1963, 759. 4504-035	METAGENIN $C_{27}H_{44}O_5$ 448 -82^0 Tet. Lett., 1960 (3) 5. 4504-040

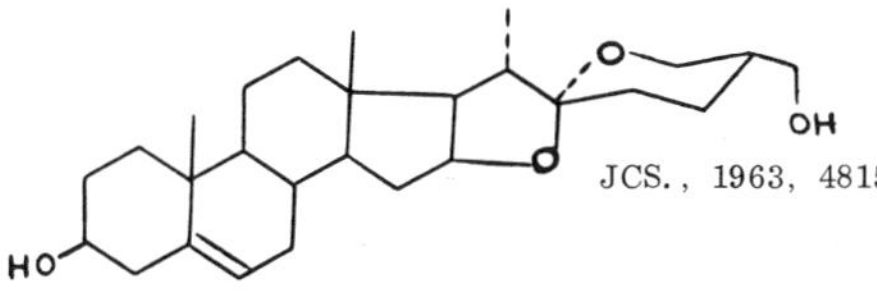

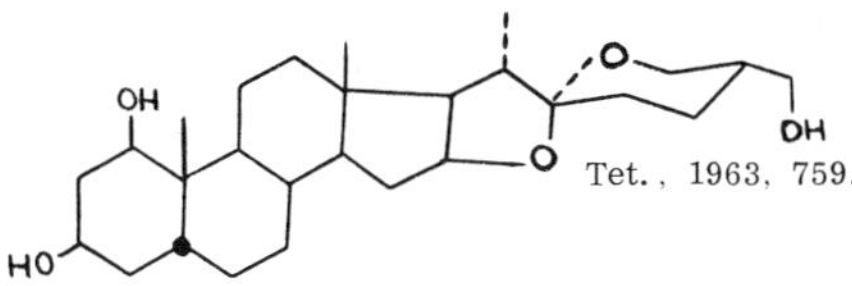

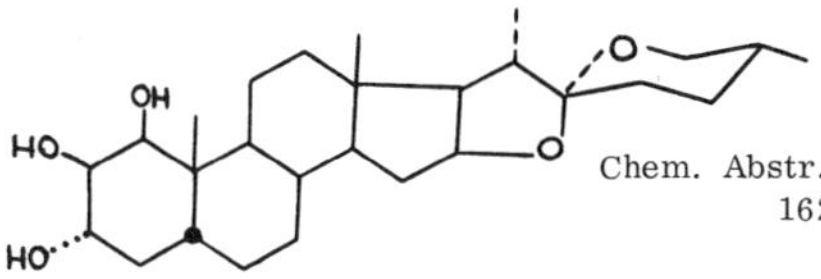

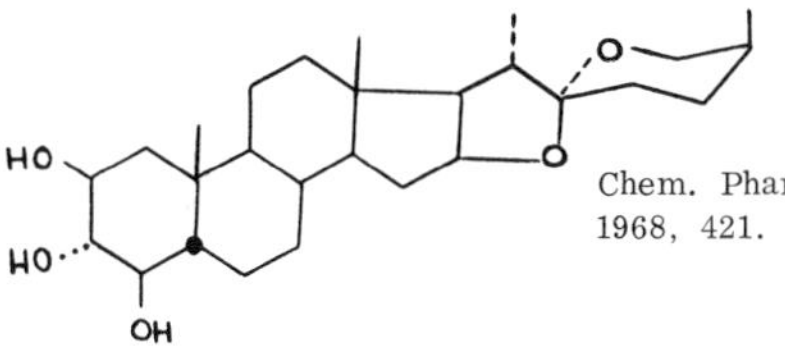

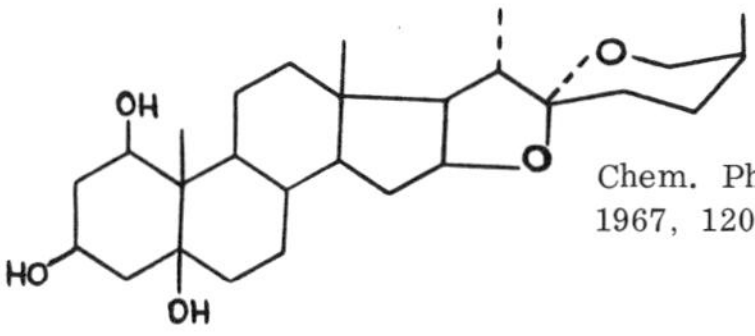

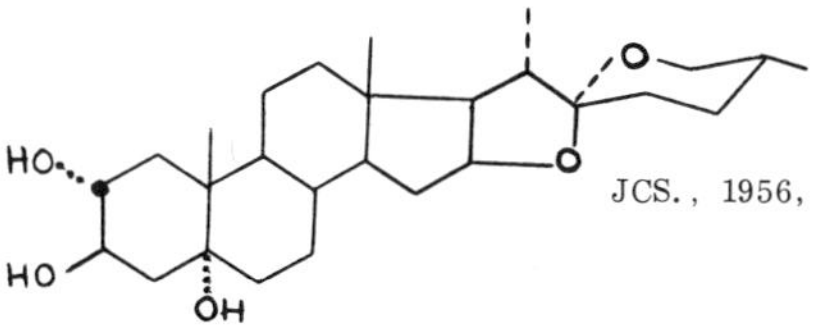

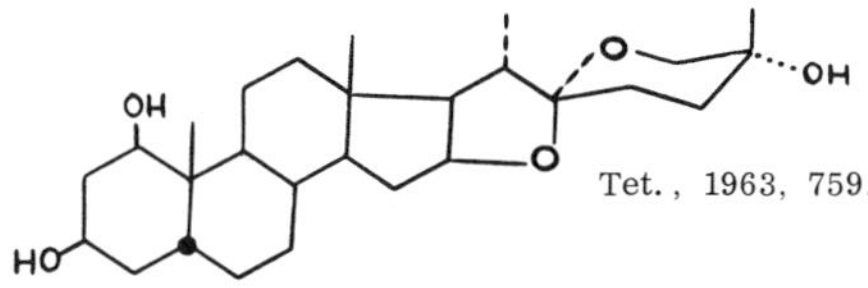

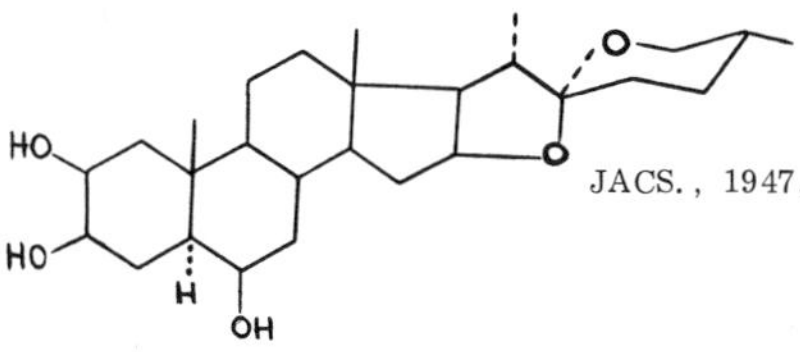

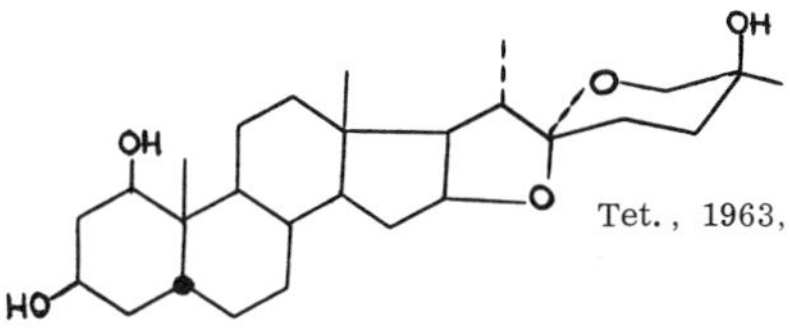

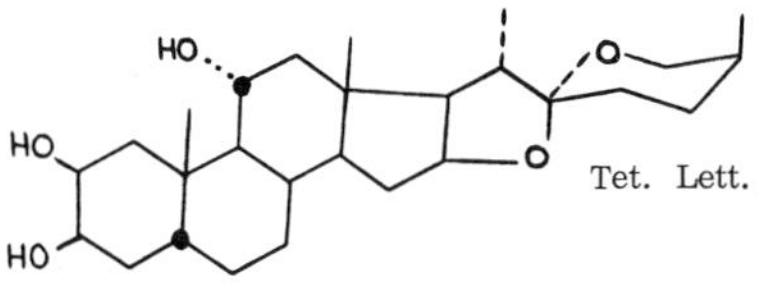

AGAVOGENIN $C_{27}H_{44}O_5$ 448 -70°c 242°C JACS., 1951, 2496. 4504-041	ASPERAGENIN $C_{27}H_{44}O_5$ 448 -136° 266°C Chem. Abstr., 1968 (68) 49884. 4504-046
DIGITOGENIN $C_{27}H_{44}O_5$ 448 -81°c 280°C JACS., 1956, 3166. 4504-042	KOGAGENIN $C_{27}H_{44}O_6$ 464 -27°p 320°C Tet., 1959 (7) 62. 4504-047
ISOPLEXIGENIN C $C_{27}H_{44}O_5$ 448 -62° 273°C Tet., 1970, 3233. 4504-043	KITIGENIN $C_{27}H_{44}O_6$ 464 -35° 298°C Chem. Abstr., 1961 (56) 7390. 4504-048
ISOPLEXIGENIN D $C_{27}H_{44}O_5$ 448 -74° 280°C Tet., 1970, 3233. 4504-044	CONVALLAGENIN B $C_{27}H_{44}O_6$ 464 -43°c/m 277°C Chem. Pharm. Bull., 1967, 1713. 4504-049
IGAGENIN $C_{27}H_{44}O_5$ 448 -44°m 253°C Tet., 1968, 6535. 4504-045	HISPIDOGENIN $C_{27}H_{42}O_4$ 430 -46°c 208°C Chem. Ind., 1965, 1653. 4504-050

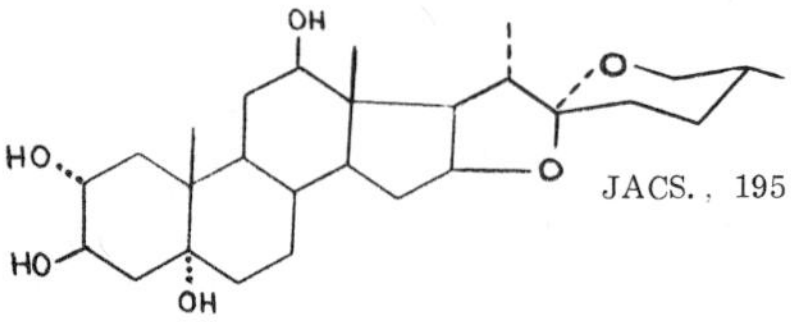

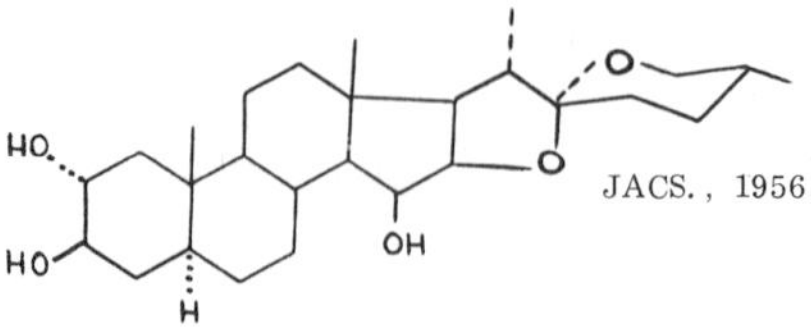

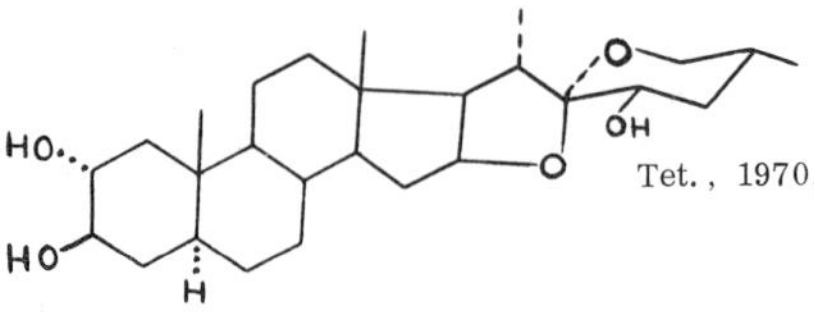

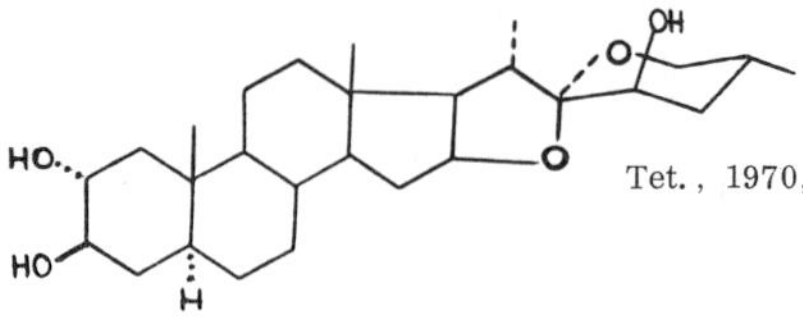

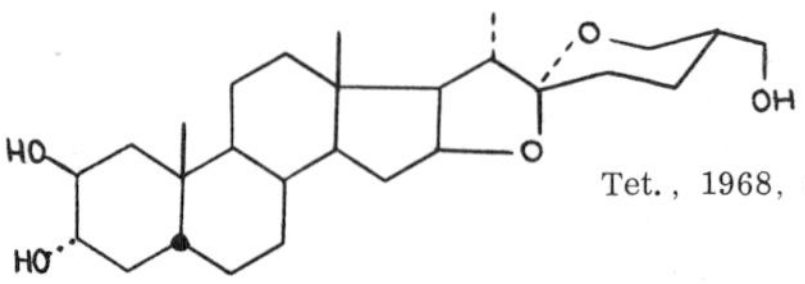

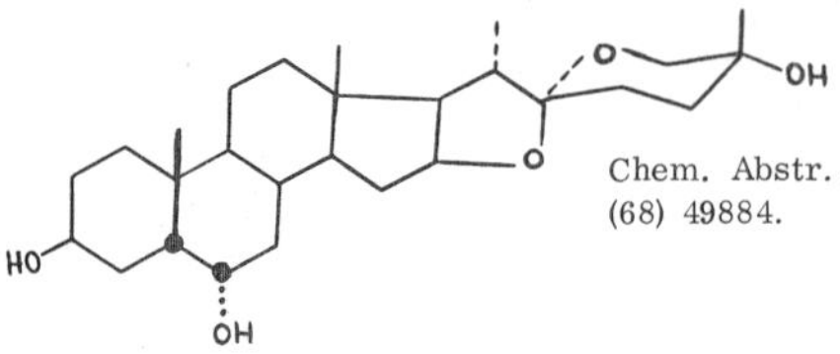

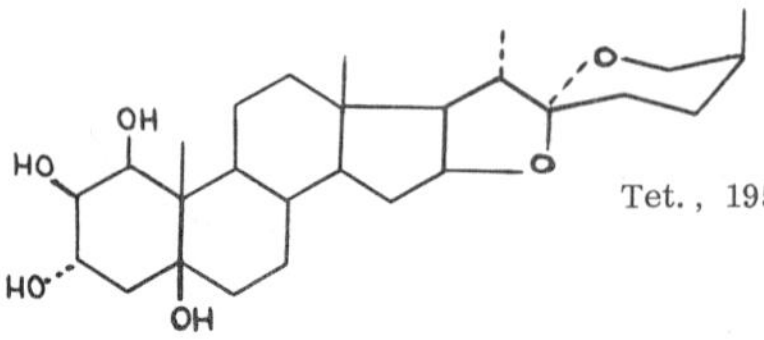

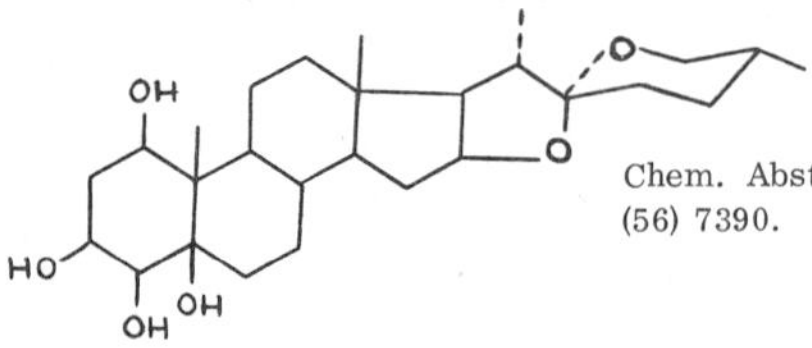

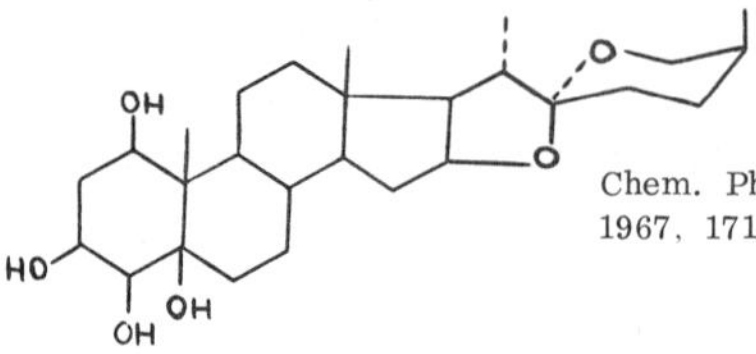

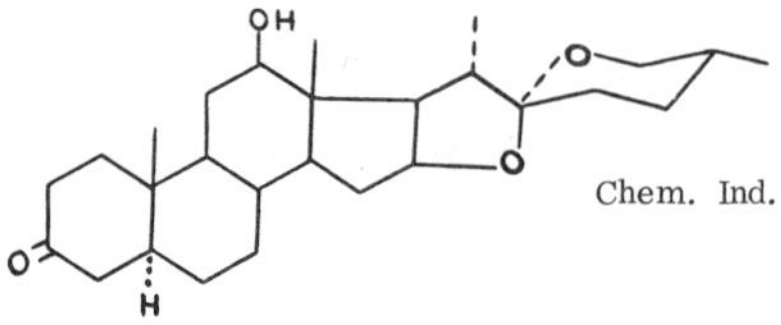

RAXOGENIN $C_{27}H_{42}O_4$ 430 211°C Chem. Abstr., 1966 (64) 5185. 4504-051	WILLAGENIN $C_{27}H_{42}O_4$ 430 5° 167°C JOC., 1957 (22) 468. 4504-056
TAMUSGENIN $C_{27}H_{40}O_4$ 428 -75° 182°C Chem. Abstr., 1969 (70) 44800. 4504-052	JIMOGENIN $C_{27}H_{42}O_4$ 430 225°C Chem. Abstr., 1950, 2000. 4504-057
HECOGENIN $C_{27}H_{42}O_4$ 430 +10°c 264°C Tet. Lett., 1964, 2281. 4504-053	BOTOGENIN $C_{27}H_{40}O_4$ 428 -57°c 215 °C JACS., 1947, 2167. 4504-058
SISALAGENIN $C_{27}H_{42}O_4$ 430 -5°c 244°C Tet. Lett., 1964, 2281. 4504-054	CORRELLOGENIN $C_{27}H_{40}O_4$ 428 -69°c 210°C JOC., 1957, 182. 4504-059
HECOGENIN, 9-dehydro- $C_{27}H_{40}O_4$ 428 -3°c 235°C JACS., 1952, 3201. 4504-055	MEXOGENIN $C_{27}H_{42}O_5$ 446 -6°c 246°C JACS., 1955, 4291. 4504-060

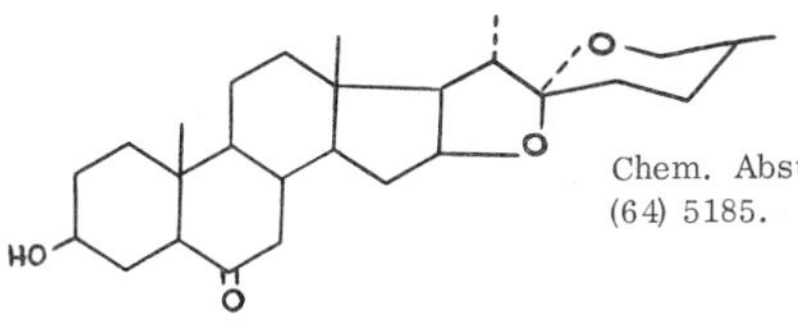

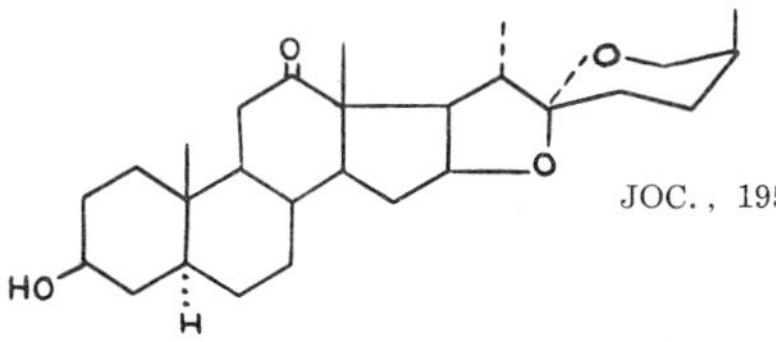

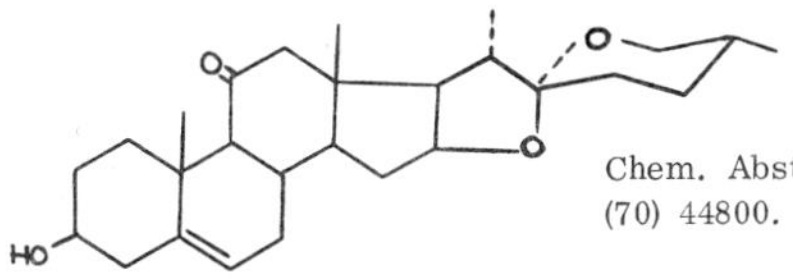

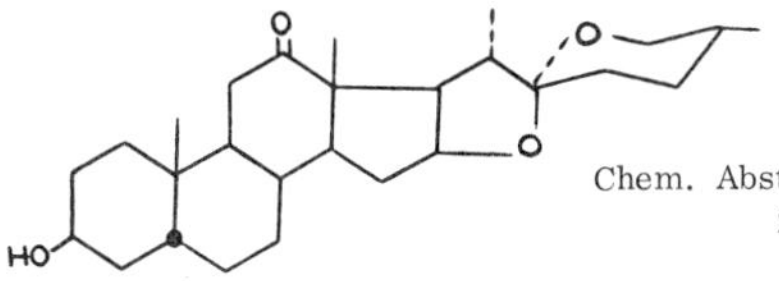

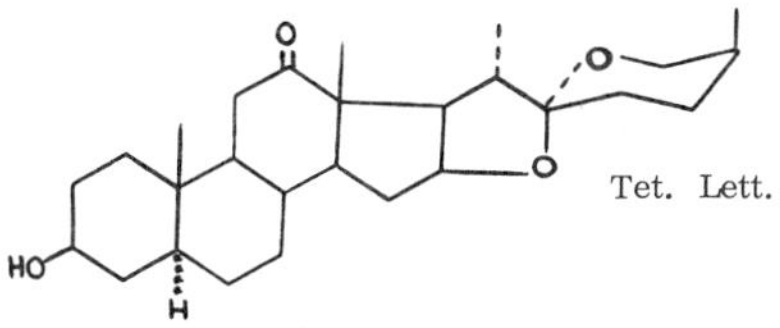

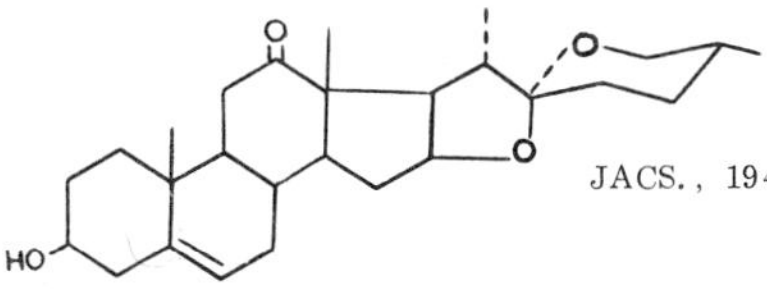

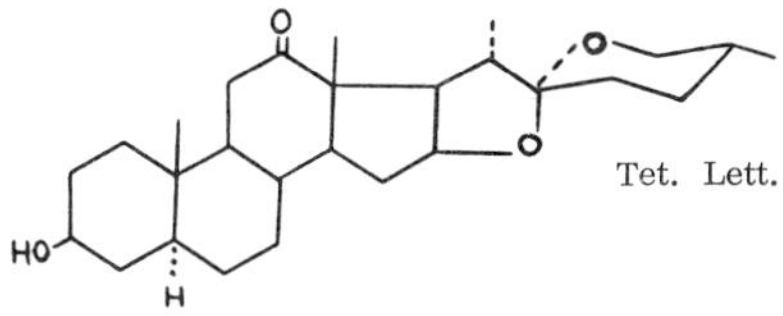

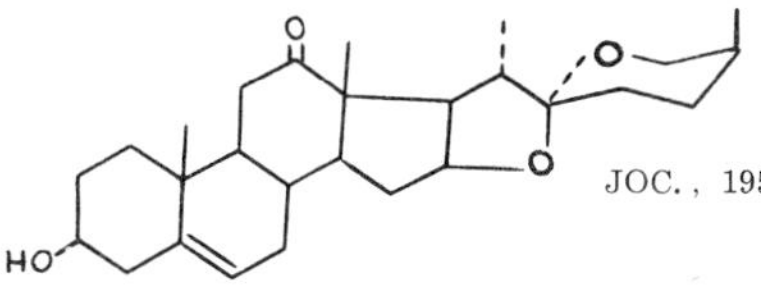

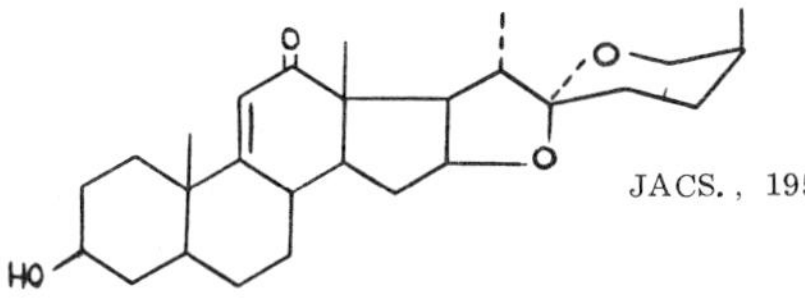

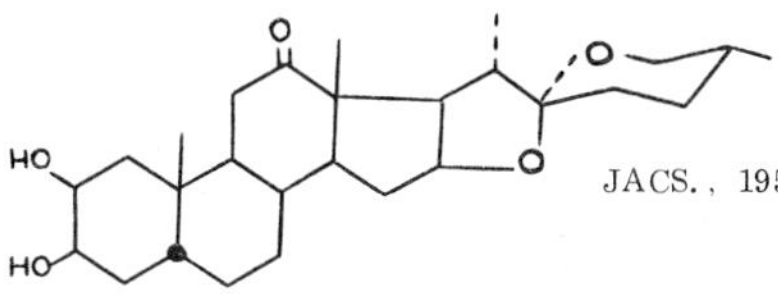

MEXOGENIN, neo- $C_{27}H_{42}O_5$ 446 222^{0}C JACS., 1955, 1232. 4504-061	MANOGENIN, 9, 25(27)-di-dehydro- $C_{27}H_{38}O_5$ 442 -36^{0}c 231^{0}C Tet., 1965, 2089. 4504-066
MANOGENIN $C_{27}H_{42}O_5$ 446 -5^{0}c 244^{0}C JACS., 1956, 3749. 4504-062	KAMMOGENIN $C_{27}H_{40}O_5$ 444 -53^{0}c 244^{0}C JACS., 1951, 5375. 4504-067
MANOGENIN, neo- $C_{27}H_{42}O_5$ 446 242^{0}C JACS., 1947, 2375. 4504-063	KAMMOGENIN, neo- $C_{27}H_{40}O_5$ 444 230^{0}C JACS., 1955, 1230/2. 4504-068
MANOGENIN, 9-dehydro- $C_{27}H_{40}O_5$ 444 -16^{0}d 240^{0}C JACS., 1952, 3201. 4504-064	COLOGENIN $C_{27}H_{42}O_5$ 446 Chem. Abstr., 1950, 2000. 4504-069
MANOGENIN, 25(27)-dehydro- $C_{27}H_{40}O_5$ 444 239^{0}C Tet., 1965, 2089. 4504-065	CACOGENIN $C_{27}H_{42}O_6$ 462 278^{0}C JACS., 1947, 2399. 4504-070

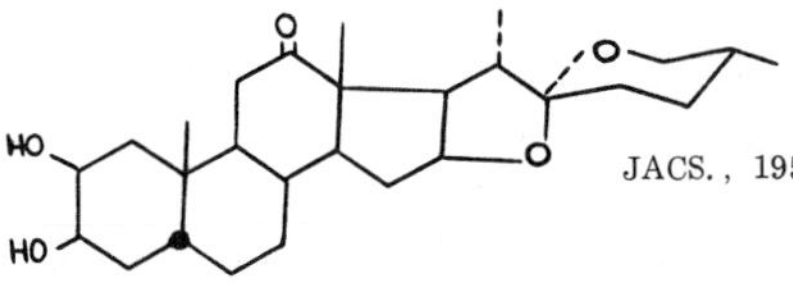

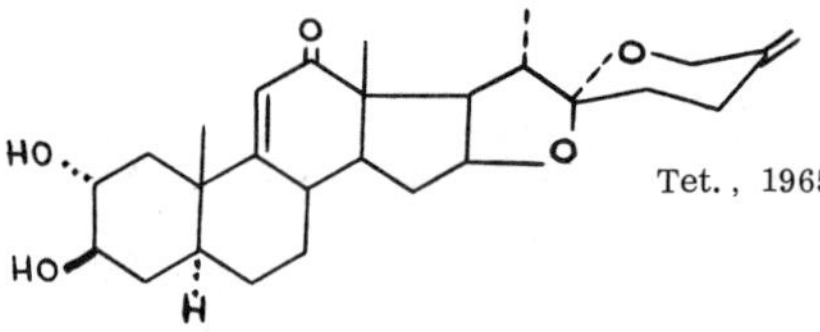

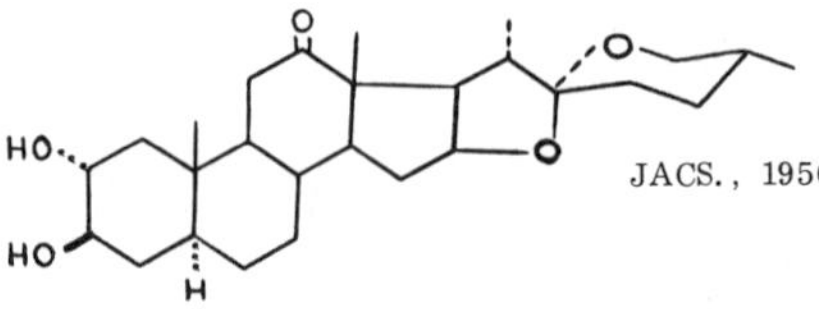

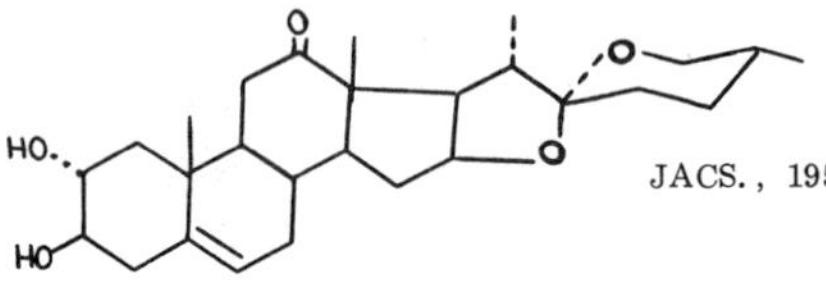

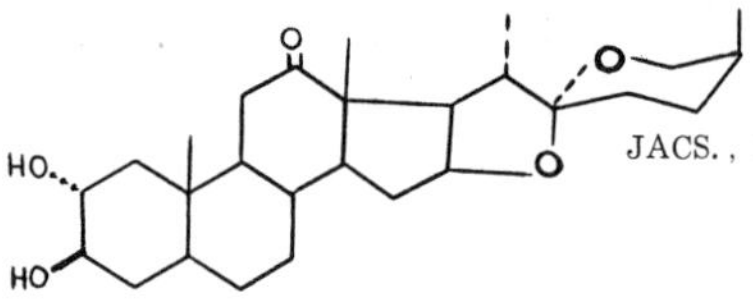

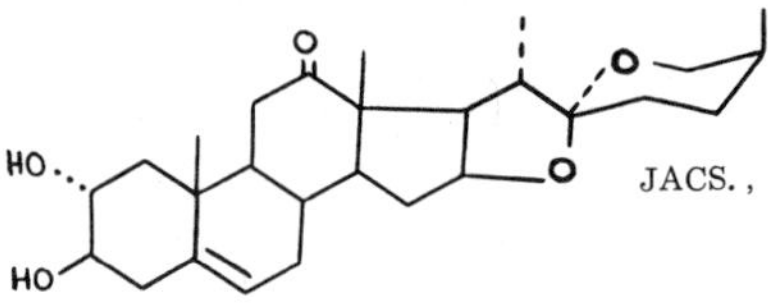

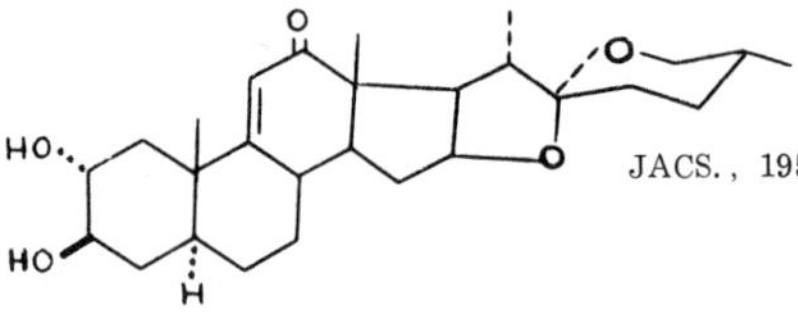

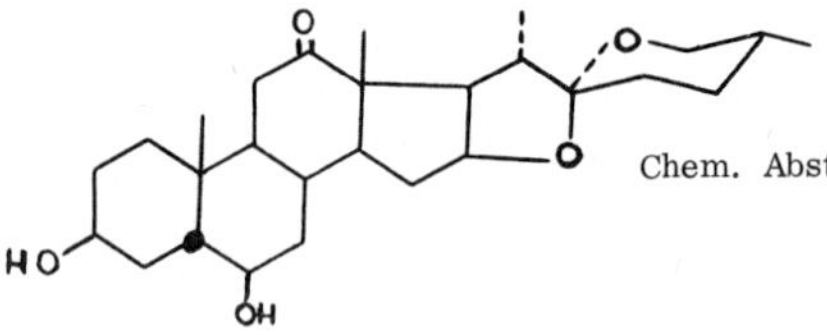

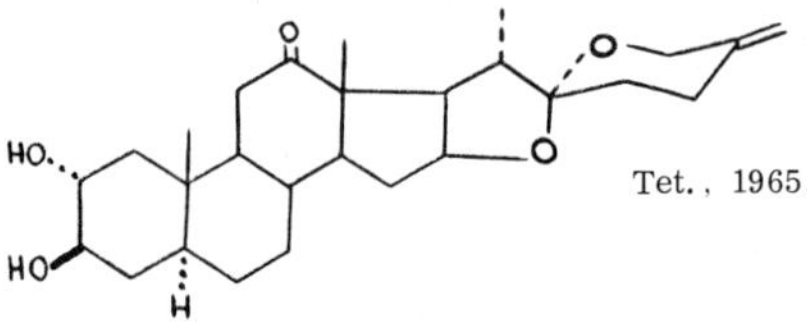

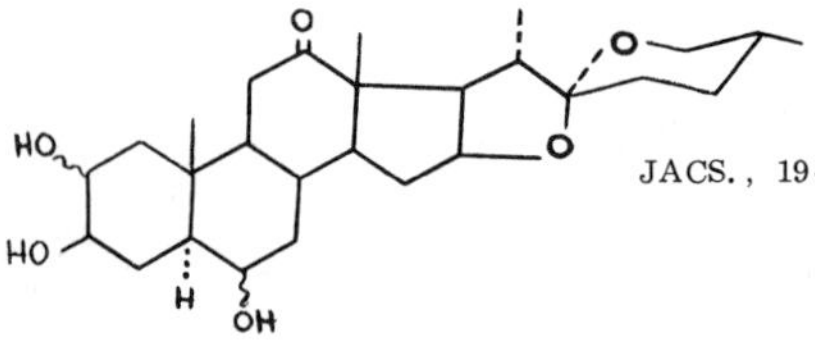

CHOLEGENIN $C_{27}H_{44}O_4$ 432
-25°c
193°C

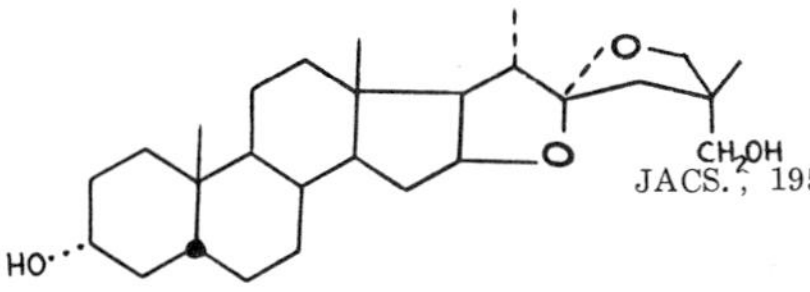

CH₂OH

JACS., 1959, 5225.

4504-071

NUATIGENIN $C_{27}H_{42}O_4$ 430
-93°c
212°C

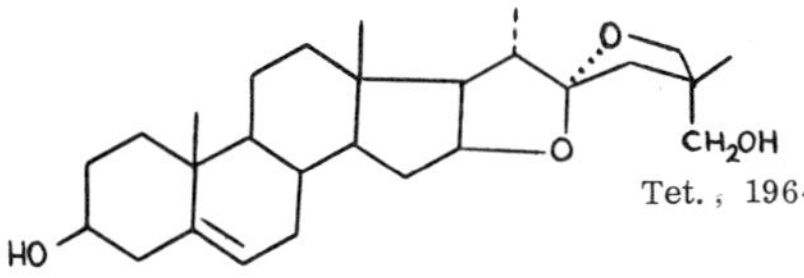

CH₂OH

Tet., 1964, 387.

4504-072

YONONIN $C_{32}H_{52}O_8$ 564
-11°
239°C

Yonogenin (018)
3-0-L-Arabinoside

Tet., 1965, 3633.

4504-073

PARILLIN $C_{51}H_{84}O_{22}$ 1048
-64°e
220°C

Sarsasapogenin (004)
3-glycoside
Gly = 2,6-di-glu-4-rha-glucosyl

Ann., 1966 (699) 212.

4504-074

SARSPARILLOSIDE $C_{57}H_{96}O_{28}$ 1228
-44°
amorph.

26-0-Glucosyl-Parillin

Tet. Lett., 1967, 2785.

4504-075

ADYNERIGENIN — $C_{23}H_{32}O_4$ — 372, -18^0p, 240^0C

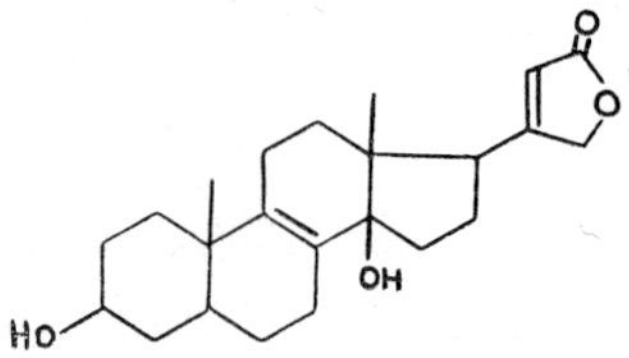

Helv., 1963 (46) 374.

4505-001

ADYNERIN — $C_{30}H_{44}O_7$ — 516, $+7^0$p, 234^0C

Helv., 1963 (46) 374.

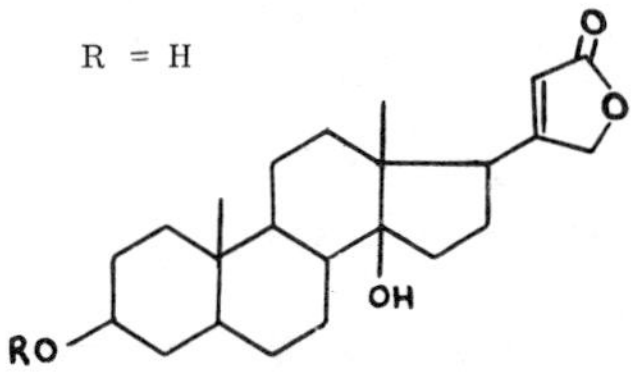

4505-002

ACOFRIOGENIN — $C_{23}H_{32}O_4$ — 372

R = H

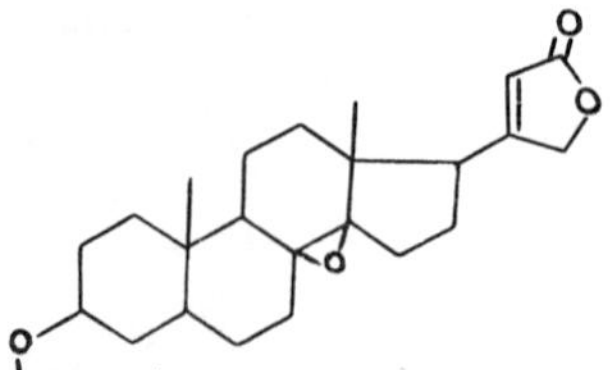

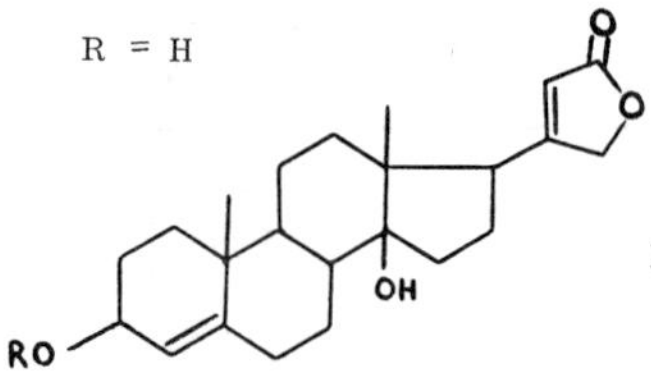

Helv., 1954, 403.

4505-003

ACOFRIOSIDE L — $C_{30}H_{44}O_8$ — 532, -54^0, 248^0C

R = Acofriosyl

Helv., 1954, 403.

4505-004

XYSMALOGENIN — $C_{23}H_{34}O_4$ — 374, $+21^0$e, 230^0C

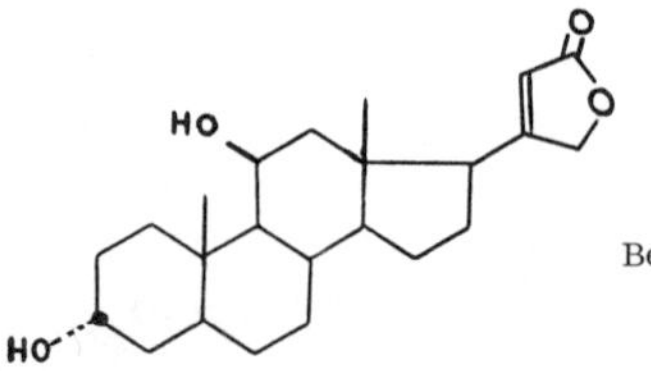

Ber., 1955, 686.

4505-005

UZARIGENIN — $C_{23}H_{34}O_4$ — 374, $+14^0$, 246^0C

R = H

Helv., 1949, 930.

4505-006

ODOROSIDE B — $C_{30}H_{46}O_7$ — 518, -5^0, 160^0C

R = diginosyl

Helv., 1949, 930.

4505-007

CHEIROSIDE A — $C_{35}H_{54}O_{13}$ — 682, -25^0p, 293^0C

R = glucosido-fucosyl-

Helv., 1954, 755.

4505-008

CHEIROSIDE H — $C_{35}H_{54}O_{13}$ — 682, -27^0, 302^0C

R = rhamnosido-glucosyl

Helv., 1949, 930.

4505-009

UZAROSIDE — $C_{41}H_{64}O_{19}$ — 860, -10^0

R = $(glucosyl)_3$

Ber., 1952, 1042.

4505-010

UREZIGENIN $C_{23}H_{34}O_4$ 374 +4°c 272°C R = H Ber., 1952, 1042. 4505-011	EVOMONOSIDE $C_{29}H_{44}O_8$ 520 -31°m 239°C R = rhamnosyl JOC., 1962, 1766. 4505-016
ASCLEPOSIDE $C_{29}H_{44}O_8$ 520 -27°m 260°C R = allomethylosyl Helv., 1963, 8. 4505-012	BEAUMONTOSIDE $C_{30}H_{46}O_7$ 518 -32°c 202°C R = oleandrosyl Helv., 1963, 1691. 4505-017
UREZIN $C_{35}H_{54}O_{14}$ 698 -5°e 185°C R = (glucosyl)$_2$ Ber., 1952, 1042. 4505-013	SOMALIN $C_{30}H_{46}O_7$ 518 +10° 197°C R = D-cymarosyl Helv., 1953, 85. 4505-018
DIGITOXIGENIN $C_{23}H_{34}O_4$ 374 +18°c 250°C R =H JACS., 1965, 3475. 4505-014	WALLICHOSIDE $C_{30}H_{46}O_7$ 518 -66°c 195°C R = L-cymarosyl Pharm. Acta Helv., 1964, 168. 4505-019
DIGIPROSIDE $C_{29}H_{44}O_8$ 520 0° 150°C R = fucosyl Chem. Pharm. Bull., 1956, 284. 4505-015	ODOROSIDE A $C_{30}H_{46}O_7$ 518 -5° 183°C R = diginosyl Helv., 1949 (32) 930. 4505-020

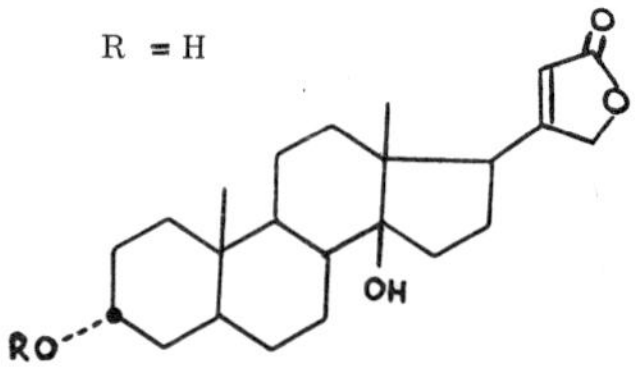

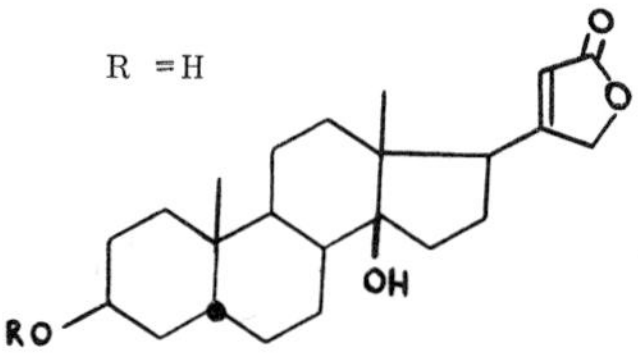

ODOROSIDE H — $C_{30}H_{46}O_8$ — 534, +6°, 235°C R = digitalosyl Helv., 1952, 687. 4505-021	GRACILOSIDE — $C_{36}H_{56}O_{13}$ — 696, 15.3°m, 300°C R = glucosido-digitalosyl Helv., 1953, 787. 4505-026
NERIIFOLIN — $C_{30}H_{46}O_8$ — 534, -50°, 225°C R = thevetosyl Helv., 1948 (31) 2097. 4505-022	ODOROBIOSIDE G — $C_{36}H_{56}O_{13}$ — 696, -7°, 243°C R = glucosido-digitalosyl Helv., 1952, 687. 4505-027
NERIIFOLIN, acetyl- — $C_{32}H_{48}O_9$ — 576, -86°m, 312°C R = thevetosyl 14-O-Acetyl Helv., 1948, 1470. 4505-023	DIGITOXIN — $C_{40}H_{70}O_{14}$ — 774, +5°d, 255°C R = digitoxosyl Bull. Soc. Chim. Fr., 1950, 288. 4505-028
DIGIFUCOSIDE, gluco- — $C_{35}H_{54}O_{13}$ — 682, +1.5°e, 186°C R = glucosido-fucosyl Experimentia, 1965 (21) 575. 4505-024	THEVETIN — $C_{42}H_{66}O_{18}$ — 858, -67°, 193°C R = diglucosido-thevetosyl Helv., 1948, 2097. 4505-029
ECHUBIOSIDE — $C_{36}H_{56}O_{12}$ — 680, +4°, 231°C R = glucosido-cymarosyl Helv., 1952, 2202. 4505-025	ECHUJIN — $C_{42}H_{66}O_{17}$ — 842, -6°m, 165°C R = strophanthotriosyl Helv., 1952, 2202. 4505-030

ODOROSIDE-G-ACETATE $C_{44}H_{68}O_{19}$ 900

R = diglucosido-acetyl-
-digitalosyl

Helv., 1952, 687.

4505-031

LANATOSIDE A, des-Ac- $C_{47}H_{74}O_8$ 926
$+12^0$
268^0C

R = glucosido-didigitoxosido-
-digitoxosyl

Ber. 1959 2258.

4505-032

LANATOSIDE A $C_{49}H_{76}O_{19}$ 968
$+32^0e$
248^0C

R = glucosido-Ac-digitoxosido-
-digitoxido-digitoxosyl

Ber. 1959 2258.

4505-033

deleted

4505-034

deleted

4505-035

HYRCANOGENIN $C_{22}H_{28}O_5$ 372

R = H

RO CHO OH O O

Chem. Abstr., 1967 (66) 483.

4505-036

HYRCANOSIDE $C_{34}H_{48}O_{14}$ 680
$+7^0$
200^0C

R = glucosido-glucosyl

Chem. Abstr., 1967 (66) 483.

4505-037

COROTOXIGENIN $C_{23}H_{32}O_5$ 388
$+43^0m$
216^0C

R = H

RO CHO OH O O

Helv., 1952, 730.

4505-038

BOISTROSIDE $C_{29}H_{42}O_8$ 518
$+5^0$
213^0C

R = digitoxosyl

Helv., 1952, 730.

4505-039

STROBOSIDE $C_{29}H_{42}O_8$ 518
-13^0
204^0C

R = boivinosyl

Helv., 1952, 730.

4505-040

GOFRUSIDE $C_{29}H_{42}O_9$ 534 -5⁰ 250⁰C R = allomethylosyl Helv., 1952, 1073. 4505-041	TANGHININ $C_{32}H_{46}O_{10}$ 590 -81⁰m 128⁰C R = thevetosyl Helv., 1965, 1113. 4505-046
PAULIOSIDE $C_{30}H_{44}O_8$ 532 -10⁰ 203⁰C R = sarmentosyl Helv., 1952, 730. 4505-042	COROGLAUCIGENIN $C_{23}H_{34}O_5$ 390 +26⁰m 246⁰C R = H OH CH₂ OH RO Helv. 1952 1073. 4505-047
MILLOSIDE $C_{30}H_{44}O_8$ 532 -1⁰ 146⁰C R = cymarosyl Helv., 1952, 730. 4505-043	COROGLAUCIGENIN-3-RHAMNOSIDE $C_{29}H_{44}O_9$ 536 -38⁰m 231⁰C R = rhamnosyl Helv., 1963, 2886. 4505-048
CHRISTYOSIDE $C_{30}H_{44}O_9$ 548 +14⁰ 213⁰C R = digitalosyl Helv., 1952, 730. 4505-044	FRUGOSIDE $C_{29}H_{44}O_9$ 536 -17⁰e 170⁰C R = allomethylosyl Helv., 1952, 1073. 4505-049
TANGHINIGENIN $C_{23}H_{32}O_5$ 388 +14⁰c 187⁰C R = H O OH RO Helv., 1965, 1113. 4505-045	DIGOXIGENIN $C_{23}H_{34}O_5$ 390 +23⁰m 221⁰C R = H OH OH RO JCS., 1953, 3325. 4505-050

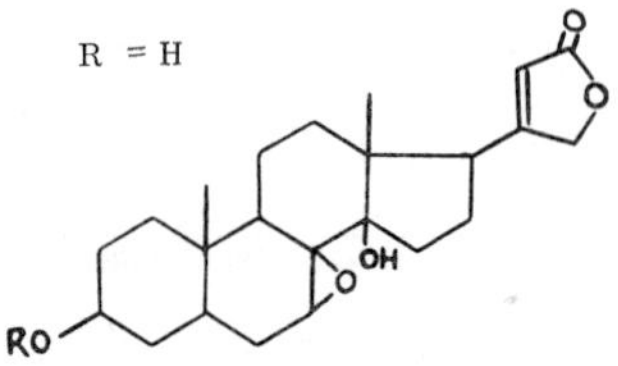

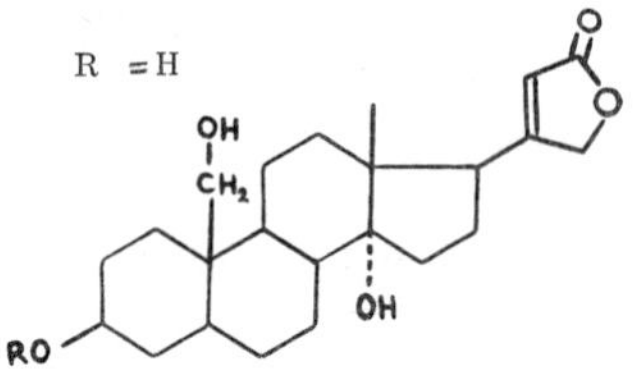

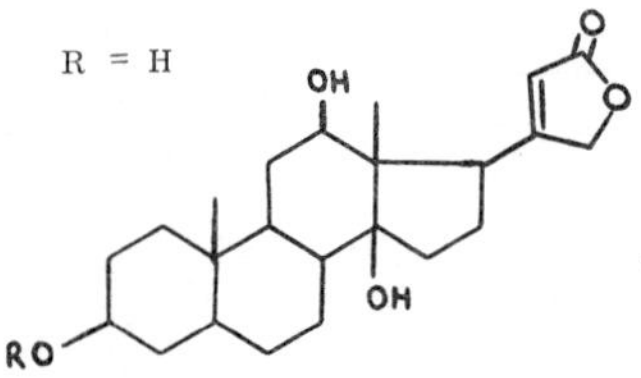

DIGOXIN	$C_{41}H_{64}O_{14}$	780 $+13^0p$ 265^0C
R = (digitoxosyl)$_3$	JCS., 1930, 508.	
4505-051		
GITOXIGENIN	$C_{23}H_{34}O_5$	390 $+35^0m$ 224^0C
R = H	Helv., 1946, 1908.	
4505-056		
DIGORID	$C_{43}H_{66}O_{15}$	822
R = (glucosyl)$_3$	Arch. Pharm., 1941 (279) 223.	
4505-052		
RHODEXIN B	$C_{29}H_{44}O_9$	536 -40^0 262^0C
R = rhamnosyl	Chem. Abstr., 1953 (47) 2190.	
4505-057		
SYRIOGENIN	$C_{23}H_{34}O_5$	390
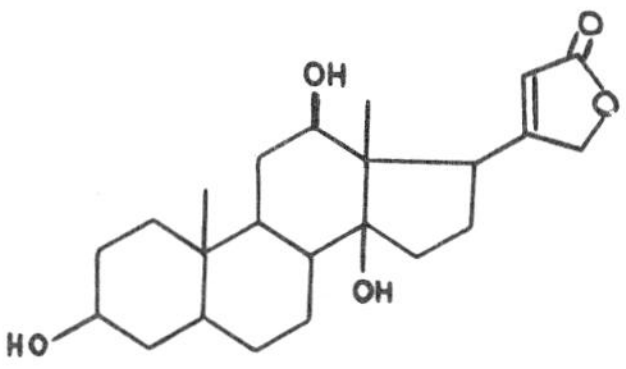	Coll. Czech. Chem. Comm., 1962, 895.	
4505-053		
GITORIN	$C_{29}H_{44}O_{10}$	552 $+7^0m$ 210^0C
R = glucosyl	Chem. Pharm. Bull., 1956 (4) 319.	
4505-058		
GOMPHOGENIN	$C_{23}H_{34}O_5$	390 235^0C
$R_1 = R_2 = H$	Helv., 1969, 1940.	
4505-054		
STROSPESIDE	$C_{30}H_{46}O_9$	550 $+19^0e$ 252^0C
R = digitalosyl	Helv., 1952, 442.	
4505-059		
GOMPHOSIDE	$C_{29}H_{42}O_8$	518 $+16^0m$ 238^0C
R_1, R_2 } =	Aust. J. C., 1964, 573.	
4505-055		
RHODEXIN C	$C_{35}H_{54}O_{14}$	698 -18^0e 75^0C
R = glucosido-rhamnosyl	Chem. Abstr., 1953 (47) 2190.	
4505-060		

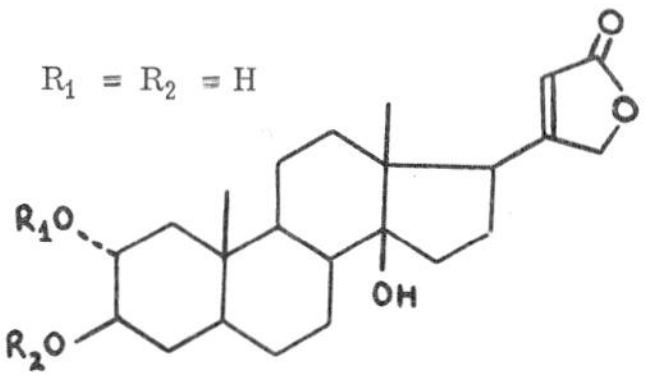

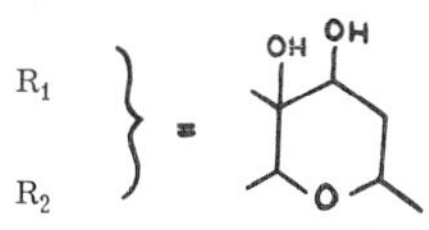

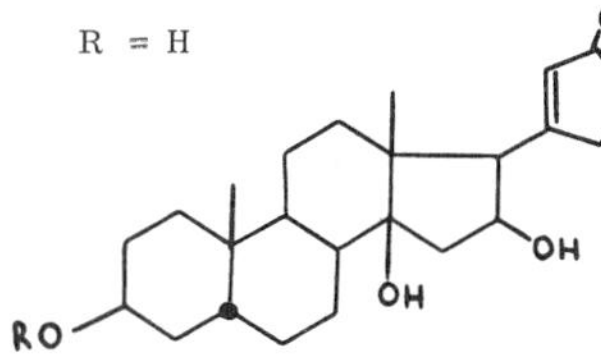

HONGHELOSIDE B $C_{38}H_{58}O_{15}$ 754; -3^0e; 260^0C

R = acetyl-digitalosido-
-glucosyl-

Helv., 1952, 442.

4505-061

GITOXIN $C_{41}H_{64}O_{14}$ 780; $+22^0e$; 285^0C

R = digitoxosyl

Bull. Soc. Chim. Fr., 1950 (17) 288.

4505-062

LANATOSIDE B, de-Ac- $C_{47}H_{74}O_{19}$ 942; $+20^0m$; 240^0C

R = glucosido-tridigitoxosyl

Helv., 1954, 1134.

4505-063

LANATOSIDE B $C_{49}H_{76}O_{20}$ 984; $+35^0m$; 248^0C

R = glucosido-Ac-digitoxosido-
didigitoxosyl

Helv., 1954, 1134.

4505-064

ACOVENOSIGENIN A $C_{23}H_{34}O_5$ 390; $+2^0m$; 298^0C

R = H

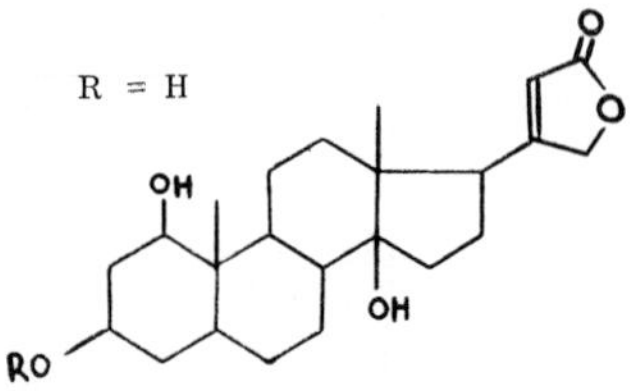

Helv., 1950, 485.

4505-065

ACOVENOSIDE A $C_{30}H_{46}O_9$ 550; -65^0d; 222^0C

R = acovenosyl

Helv., 1962, 2116.

4505-066

ACOVENOSIDE C $C_{42}H_{66}O_{19}$ 874; -63^0; 189^0C

R = gentiobiosido-acovenosyl

Helv., 1962, 2116.

4505-067

PERIPLOGENIN $C_{23}H_{34}O_5$ 390; $+32^0$; 185^0C

R = H

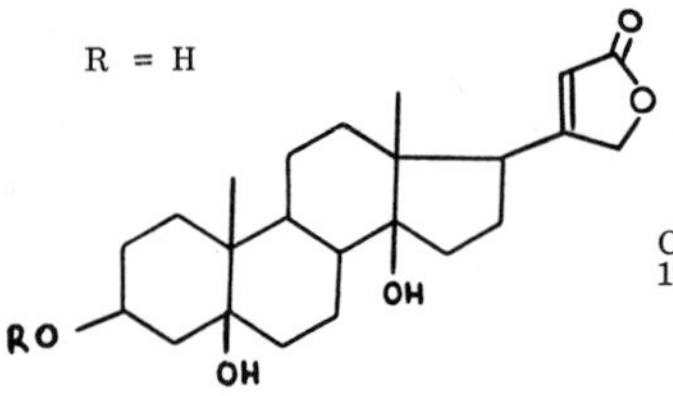

Chem. Pharm. Bull., 1968, 326.

4505-068

PERIPLOCYMARIN $C_{30}H_{46}O_8$ 534; $+29^0m$; 212^0C

R = cymarosyl

Helv., 1948, 883.

4505-069

EMICYMARIN $C_{30}H_{46}O_9$ 550; $+13^0$; 159^0C

R = digitalosyl

Helv., 1950, 639.

4505-070

PERIPLOCIN $C_{36}H_{56}O_{13}$ 696 +23°m 209°C R = glucosido-cymarosyl Helv., 1939, 1193. 4505-071	SARNOVIDE $C_{30}H_{46}O_9$ 550 +7°m 225°C R = digitalosyl Helv., 1951, 1477. 4505-076
SARMENTOGENIN $C_{23}H_{34}O_5$ 390 +21°e 266°C R = H HO, OH, RO, O JCS., 1952, 2299. 4505-072	SARGENOSIDE $C_{36}H_{56}O_{14}$ 712 +2°m 260°C R = glucosido-digitalosyl Helv., 1952, 1560. 4505-077
RHODEXIN A $C_{29}H_{44}O_9$ 536 -20° 265°C R = rhamnosyl Chem. Abstr., 1953 (47) 2190. 4505-073	STROPHANTHIDIN $C_{23}H_{32}O_6$ 404 +40°c R = H CHO, OH, RO, OH, O Helv., 1947, 2024. 4505-078
SARMENTOCYMARIN $C_{30}H_{46}O_8$ 534 -13°m 130°C R = sarmentosyl Helv., 1948, 993. 4505-074	CHEIROTOXIN, degluco- $C_{28}H_{40}O_{10}$ 536 +1° 191°C R = lyxosyl Pharm. Acta Helv., 1949, 113. 4505-079
DIVARICOSIDE $C_{30}H_{46}O_8$ 534 -46°m 222°C R = oleandrosyl Helv., 1954, 667. 4505-075	COCHOROSIDE A $C_{29}H_{42}O_9$ 534 +19°m 189°C R = boivinosyl Helv., 1957, 593. 4505-080

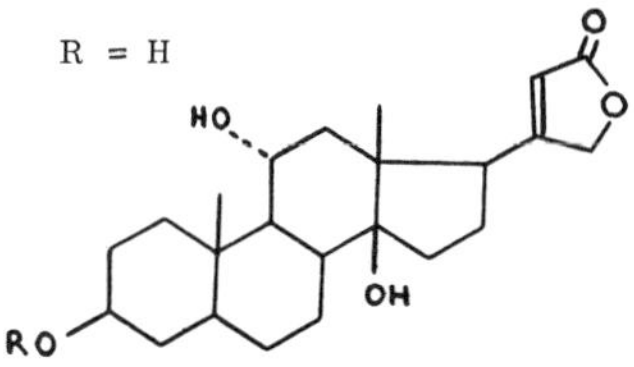

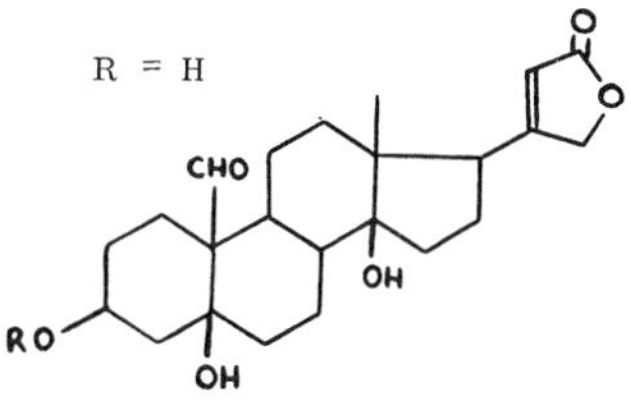

STROPHALLOSIDE $C_{29}H_{42}O_{10}$ 550 R = 6-deoxy-allosyl Helv., 1964, 2320. 4505-081	ERYSCENOSIDE $C_{35}H_{52}O_{15}$ 712 +21⁰m 140⁰ R = eryscenobiosyl Tet. Lett., 1966, 1703. 4505-086
CONVALLATOXIN $C_{29}H_{42}O_{10}$ 550 0⁰ 247⁰C R = rhamnosyl Helv., 1950, 1541. 4505-082	CONVALLOSIDE $C_{35}H_{52}O_{15}$ 712 -10⁰ 204⁰C R = glucosido-rhamnosyl Pharm. Acta Helv., 1947 (22) 359. 4505-087
CYMARIN $C_{30}H_{44}O_{9}$ 548 +38⁰c 139⁰C R = cymarosyl Helv., 1948, 883. 4505-083	CHEIROTOXIN $C_{35}H_{52}O_{15}$ 712 -17⁰ 211⁰C R = glucosido-lyxosyl Pharm. Acta Helv., 1949 (24) 113. 4505-088
ERYSIMINE $C_{29}H_{42}O_{9}$ 534 +44⁰e 170⁰C R = digitoxosyl Planta Med., 1960, 145. 4505-084	ADONITOXIGENIN $C_{23}H_{32}O_{6}$ 404 174⁰C R = H 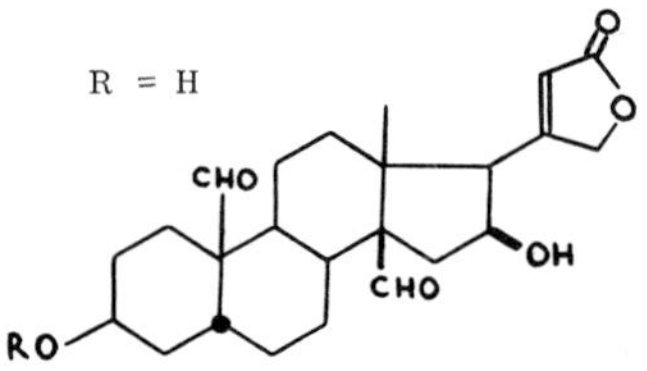 CHO OH CHO RO Ber., 1953, 574. 4505-089
ERYSIMOSIDE $C_{35}H_{52}O_{14}$ 696 +16⁰ R = glucosido-digitoxosyl Planta Med., 1960, 145. 4505-085	ADONITOXIN $C_{29}H_{42}O_{10}$ 550 -27⁰m 263⁰C R = rhamnosyl Coll. Czech. Chem. Comm., 1961, 1551. 4505-090

CALOTROPAGENIN $C_{23}H_{32}O_6$ 404 +43° 240°C

R = H

JCS., 1959, 85.

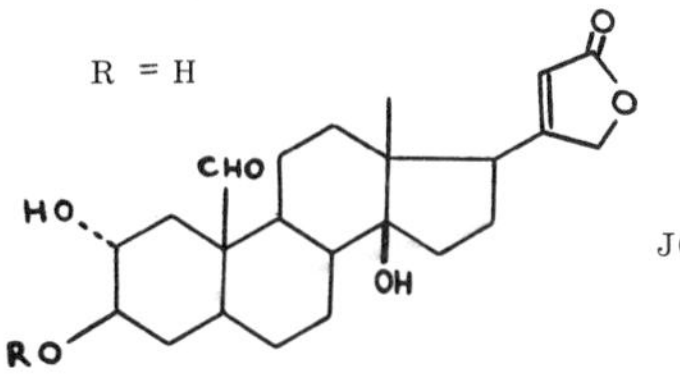

4505-091

USCHARIN $C_{31}H_{41}NO_8S$ 587

R =

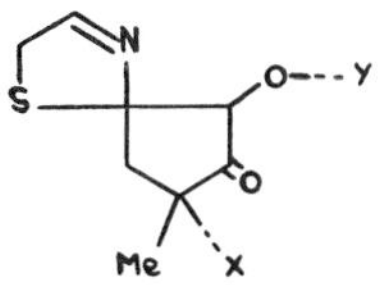

X = C_3-0- Y = C_{19}-lactol

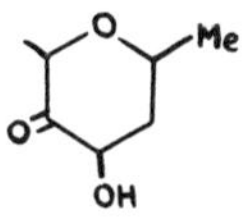

Ann., 1960 (632) 158.

4505-096

CALOTROPIN $C_{29}H_{40}O_9$ 532 +56° 221°C

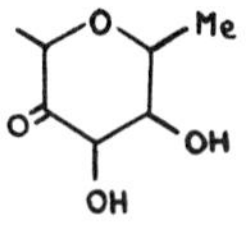

R =

JCS., 1964, 2187.

4505-092

STROPHANTHIDOL $C_{23}H_{34}O_6$ 406 +37°m 141°C

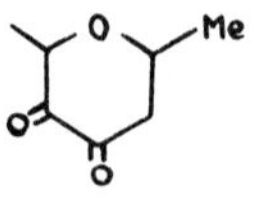

R = H

Helv., 1957, 299.

4505-097

CALOTOXIN $C_{29}H_{40}O_{10}$ 548 +66° 268°C

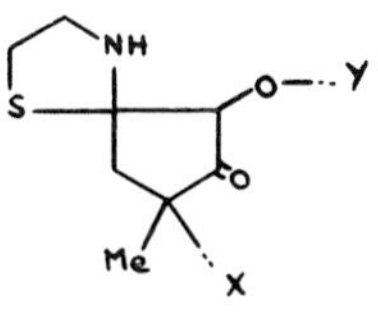

R =

JCS., 1964, 2187.

4505-093

CONVALLATOXOL $C_{29}H_{44}O_{10}$ 552 -10°m 171°C

R = rhamnosyl

Helv., 1952, 635.

4505-098

USCHARIDIN $C_{29}H_{38}O_9$ 530 +38°e 290°C

R =

JCS., 1964, 2187.

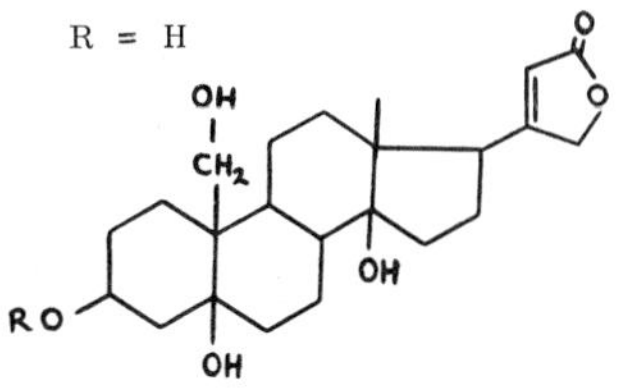

4505-094

CYMAROL $C_{30}H_{46}O_9$ 550 +31°m 242°C

R = cymarosyl

Helv., 1954, 141.

4505-099

VORUSCHARIN $C_{31}H_{43}NO_8S$ 589 -61°e 165°C

R =

X = C_3-0- Y = C_{19}-lactol

Ann., 1960 (632) 158.

4505-095

CYMAROL, gluco- $C_{35}H_{54}O_{14}$ 698 amorph.

R = strophanthobiosyl

Ann., 1961 (643) 192.

4505-100

ERYSIMOSOL $C_{35}H_{54}O_{14}$ 698

R = digilanidobiosyl

Ann., 1961 (643) 192.

4505-101

BEAUWALLOSIDE $C_{30}H_{46}O_8$ 534
-81°c
224°C

R = cymarosyl

Helv., 1963, 1691.

4505-106

ADIGENIN $C_{28}H_{42}O_6$ 474
-7.5°m
amorph.

R = H

O-i-But

OH

RO

Helv., 1966, 1855.

4505-102

CRYPTOGRANDOSIDE A $C_{32}H_{48}O_9$ 576
-33°m
123°C

R = sarmentosyl

Helv., 1950, 1013.

4505-107

ADIGOSIDE $C_{35}H_{54}O_9$ 618
-21°m
143°C

R = diginosyl

Helv., 1966, 1855.

4505-103

CRYPTOGRANDOSIDE B $C_{38}H_{58}O_{14}$ 738
-34°m
ac. 124°C

R = glucosido-sarmentosyl

Helv., 1950, 1013.

4505-108

OLEANDRIGENIN $C_{25}H_{36}O_6$ 432
-9°m
224°C

R = H

OAc

OH

RO

Helv., 1950, 1013.

4505-104

DIGINATIGENIN $C_{23}H_{34}O_6$ 406
+34°m
157°C

R = H

OH

OH

OH

RO

Chem. Pharm. Bull., 1960, 535.

4505-109

OLEANDRIN $C_{32}H_{48}O_9$ 576
-52°
250°C

R = oleandrosyl

Ber., 1937 (70) 2264.

4505-105

DIGINATIN $C_{41}H_{64}O_{15}$ 796
+24°m
234°C

R = digitoxosyl

J. Am. Pharm. Assoc., 1955 (44) 719.

4505-110

SINOGENIN $C_{23}H_{32}O_6$ 404 R = H HO, O, OH, RO Helv., 1959, 182. 4505-111	SARMUTOSIDE $C_{30}H_{44}O_9$ 548 R = L-diginosyl Helv., 1959, 182. 4505-116
SINOSIDE $C_{30}H_{44}O_9$ 548 R = oleandrosyl Helv., 1959, 182. 4505-112	AMBOSIDE $C_{30}H_{44}O_9$ 548 R = D-diginosyl Helv., 1956, 904. 4505-117
SINOSTROSIDE $C_{30}H_{44}O_9$ 548 R = digitosyl Helv., 1959, 182. 4505-113	MUSAROSIDE $C_{30}H_{44}O_{10}$ 564 R = digitalosyl Helv.. 1956, 904. 4505-118
SARMUTOGENIN $C_{23}H_{32}O_6$ 404 $+49^0$m 260^0C R = H OH, O, OH, RO Helv., 1959, 182. 4505-114	GITALOXIGENIN $C_{24}H_{34}O_6$ 418 -7^0p 251^0C R = H O.CHO, OH, RO Ber. 1956 (89) 1353. 4505-119
CAUDOSIDE, pseudo- $C_{30}H_{44}O_9$ 548 $+100^0$m 250^0C R = oleandrosyl Helv., 1959, 182. 4505-115	GITALOXIN $C_{42}H_{64}O_{15}$ 808 -7^0 p 252^0C R = tri-diginosyl Ber. 1956 (89) 1353. 4505-120

LANATOSIDE E $C_{50}H_{76}O_{21}$ 1013 27⁰ m 210⁰C R = glucosido-Ac-digitoxosido--di-digitoxosyl Helv. 1958 479. 4505-121	PANSTROSIDE $C_{30}H_{44}O_{11}$ 580 +30⁰m 125⁰C R = digitalosyl Helv. 1952 2143. 4505-126
SARMENTOSIGENIN-B $C_{25}H_{36}O_{6}$ 432 AcO, HO, OH, O Helv. 1952 1560. 4505-122	STROPHADOGENIN $C_{23}H_{32}O_{7}$ 420 23⁰ p 240⁰C R = H CHO, RO, OH, OH, OH, O Chem. Ind. 1964 1766. 4505-127
SARVEROGENIN $C_{23}H_{30}O_{7}$ 418 +45⁰m 224⁰C R = H HO, O, RO, O, OH Helv. 1969 616. 4505-123	VERNADIGIN $C_{30}H_{44}O_{10}$ 564 +39⁰p 229⁰C R = diginosyl Chem. Ind. 1963 1766. 4505-128
SARVEROSIDE $C_{30}H_{44}O_{10}$ 564 +9⁰m 123⁰C R = diginosyl JACS. 1952 4709. 4505-124	INERTOGENIN $C_{23}H_{30}O_{7}$ 418 R = H OH, O, RO, O, OH Helv. 1955 98. 4505-129
INTERMEDIOSIDE $C_{30}H_{44}O_{10}$ 564 +19⁰m 125⁰C R = sarmentosyl Helv. 1952 152. 4505-125	INERTOSIDE $C_{30}H_{42}O_{10}$ 562 R = diginosyl Helv. 1955 98. 4505-130

LEPTOGENIN $C_{23}H_{30}O_7$ 418

R = H

Helv. 1955 98.

4505-131

SARMENTOSIGENIN-E $C_{23}H_{30}O_7$ 418 +32° 170°C

R = H

Helv. 1960 1570.

4505-136

LEPTOSIDE $C_{30}H_{42}O_{10}$ 562

R = diginosyl

Helv. 1955 98.

4505-132

SARMENTOSIDE-E $C_{29}H_{40}O_{11}$ 564

R = talomethylosyl

Helv. 1960 1570.

4505-137

ANTIARIGENIN $C_{23}H_{32}O_7$ 420 +42°m 244°C

R = H

Helv. 1948 688.

4505-133

NIGRESCIGENIN $C_{23}H_{32}O_7$ 420 + 25°m 152°C

R = H

Helv. 1966 1844.

4505-138

ANTIARIN, alpha- $C_{29}H_{42}O_{11}$ 566 -4°m 240°C

R = antiarosyl

Helv. 1962 2285.

4505-134

SARMENTOSIDE-A $C_{29}H_{42}O_{11}$ 566 -33°m 240°C

R = talomethylosyl

Helv. 1958 736.

4505-139

ANTIARIN, beta- $C_{29}H_{42}O_{11}$ 566 + 2°m 242°C

R = rhamnosyl

Helv. 1962 2285.

4505-135

THOLLOSIDE $C_{29}H_{42}O_{11}$ 566 -16°m 260°C

R = rhamnosyl

Helv. 1958 736.

4505-140

CANESCEIN $C_{29}H_{42}O_{11}$ 566 -23^{0}m 194^{0}C

R = 6-deoxy-gulosyl

Chem. Nat. Comp. (USSR) 1968 (4) 6.

4505-141

CERBERTIN, deacetyl- $C_{30}H_{44}O_9$ 548 -21^{0}m 206^{0}C

R = thevetosyl

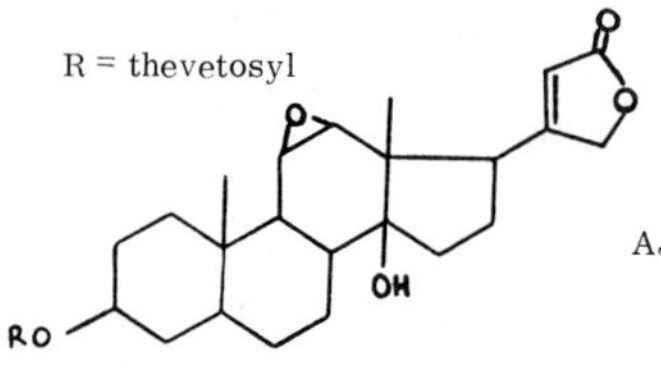

AJC. 1965 1079.

4505-145

OUABAGENIN $C_{23}H_{34}O_8$ 438 + 11^{0}w 255^{0}C

R = H

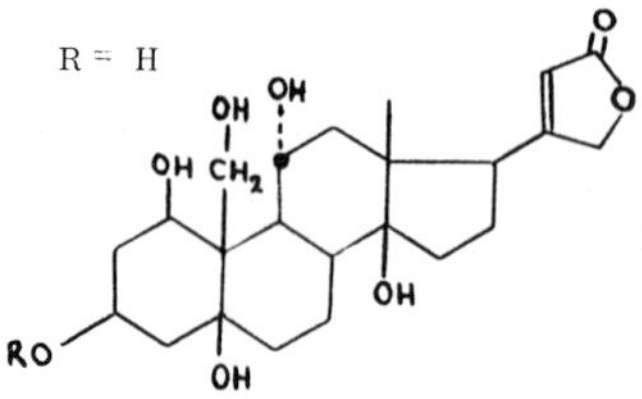

Helv. 1948 2111.

4505-142

CERBERTIN $C_{32}H_{46}O_{10}$ 590 -57^{0}c 249^{0}C

R = acetyl-thevetosyl

AJC 1965 1079.

4505-146

OUABAIN $C_{29}H_{44}O_{12}$ 584 -34^0 200^{0}C

R = rhamnosyl

JACS. 1942 720.

4505-143

DESAROGENIN $C_{23}H_{32}O_5$ 388 +11^{0}m 252^{0}C

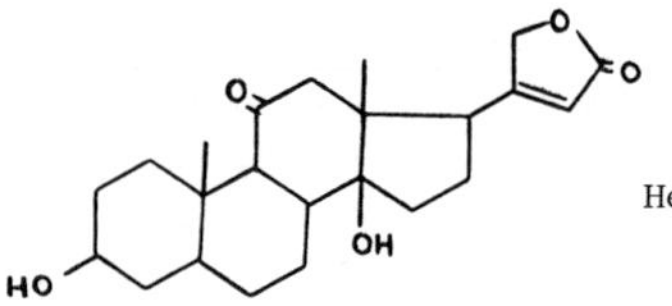

Helv., 1955, 538.

4505-147

ACOLONGIFLOROSIDE K $C_{29}H_{44}O_{12}$ 584 -53^{0}m 224^{0}C

R = talomethylosyl

Helv., 1962, 1244.

4505-144

DIGOXOSIDE $C_{47}H_{74}O_{17}$ 910 +18^{0}p 266^{0}C

Digoxigenin-3-0-(digitoxosyl)$_4$ (050)

Natwiss., 1963 (50) 668.

4505-148

BUFOTALIN $C_{26}H_{36}O_6$ 444 -9^0 168^{0}C

R = H

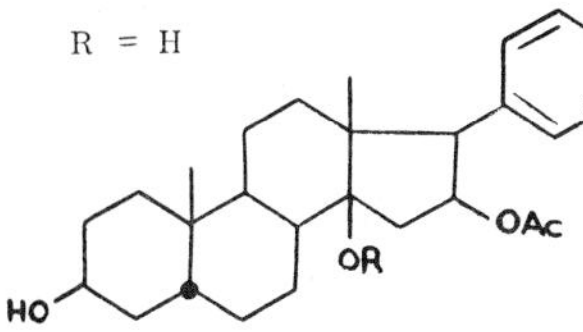

Helv., 1949 (32) 1993.

4506-001

ARTEBUFOGENIN-B $C_{24}H_{32}O_4$ 384 -72^{0}c 127^{0}C

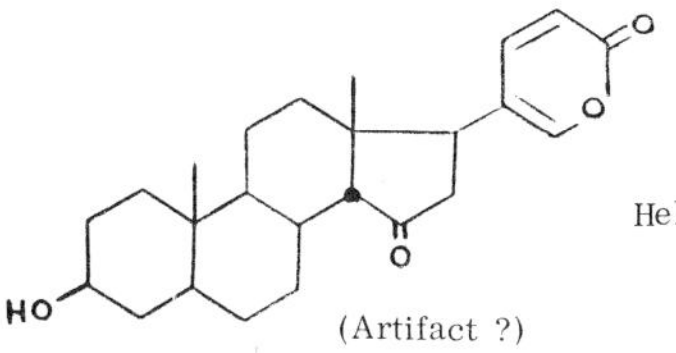

Helv., 1962 (45) 1794.

(Artifact ?)

4506-006

BUFOTOXIN $C_{40}H_{60}N_4O_{10}$ 756 +4^{0}m 205^{0}C

R = suberyl-arginine

Ann., 1941 (549) 209.

4506-002

BUFALIN $C_{24}H_{34}O_4$ 386 236^{0}C

R = H

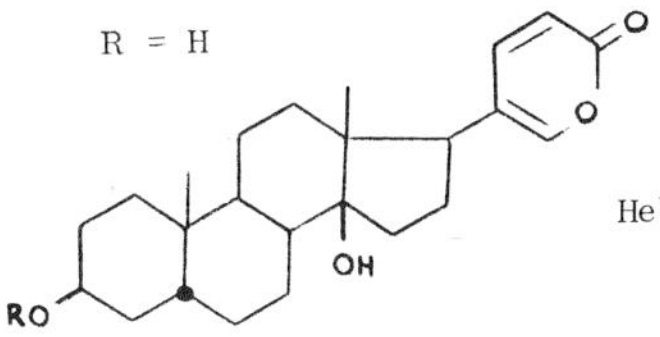

Helv., 1949 (32) 1238.

4506-007

RESIBUFOGENIN $C_{24}H_{32}O_4$ 384 -5^{0}c 155^{0}C

R = H

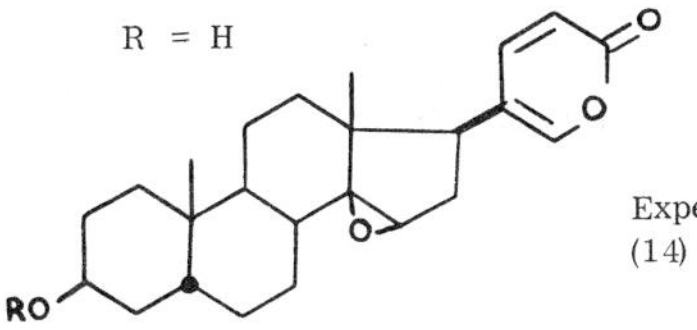

Experimentia. 1958 (14) 238.

4506-003

BUFALIN-3-(H-SUBERATE) $C_{32}H_{46}O_7$ 542

R = suberyl

Tet. Lett., 1968. 5673.

4506-008

RESIBUFOGENIN-3-(H-SUBERATE) $C_{32}H_{44}O_7$ 540

R = suberyl

Tet. Lett., 1968, 5673.

4506-004

CINOBUFAGIN $C_{26}H_{34}O_6$ 442 213^{0}C

R = H

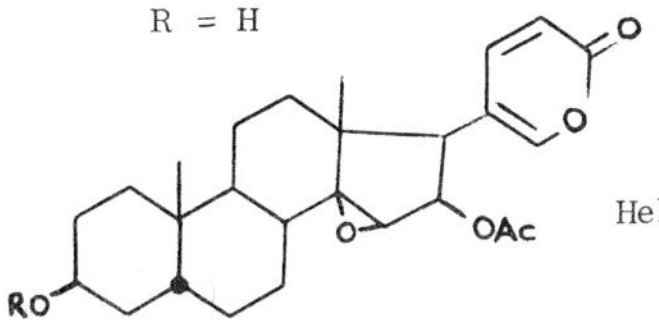

Helv., 1962 240.

4506-009

ARTEBUFOGENIN-A $C_{24}H_{32}O_4$ 384 +29^{0}c 275^{0}C

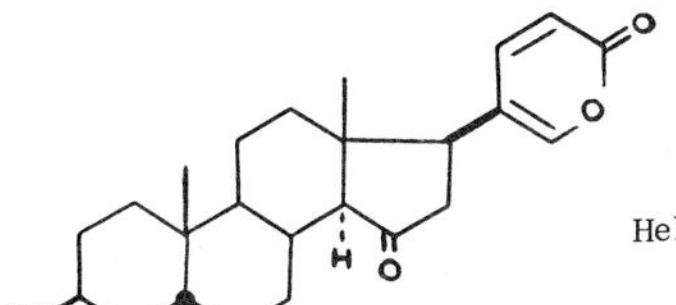

Helv., 1962 1794.

(Artifact ?)

4506-005

CINOBUFAGIN-3-(H-SUBERATE) $C_{34}H_{46}O_9$ 598 139^{0}C

R = suberyl

Tet. Lett., 1968. 5673.

4506-010

CINOBUFOTALIN — $C_{26}H_{34}O_7$ — 458; + 11°; 256°C

Helv., 1962. 240.

4506-011

SCILLARENIN — $C_{24}H_{32}O_4$ — 384; -17°m; 227°C

R = H

Helv., 1964 1228.

4506-012

SCILLARENIN GLUCOSIDE — $C_{30}H_{42}O_9$ — 546; -54°; 210°C

R = glucosyl

Helv., 1964. 1228.

4506-013

SCILLAREN-A — $C_{36}H_{52}O_{13}$ — 692; -73°e; 193°C

R = scillabiosyl

Ber., 1953 (86) 392.

4506-014

SCILLARENIN. DI-RHAMNOSYL- — $C_{36}H_{52}O_{12}$ — 676; -75°m; 160°C

R = Rhamnosido-rhamnosyl

Tet. Lett., 1967 4563.

4506-015

SCILLAREN-A, Gluco- — $C_{42}H_{62}O_{18}$ — 854; -66°m; 230°C

R = scillatriosyl

Helv., 1952 2495.

4506-016

GAMABUFOTALIN — $C_{24}H_{34}O_5$ — 402; 262°C

Helv., 1949. 1599.

4506-017

ARENOBUFAGIN — $C_{24}H_{32}O_6$ — 416; +56°m; 226°C

Tet. Lett., 1959 (7) 8.

4506-018

BUFARENOGENIN — $C_{24}H_{32}O_6$ — 416

Helv., 1960. 1495.

4506-019

BUFARENOGENIN, pseudo- — $C_{24}H_{32}O_6$ — 416

12-epimer

Chem. Abstr., 1968 (68) 3113.

4506-020

SCILLIROSIDINE $C_{26}H_{34}O_7$ 458 -23^0m 174^0C R = H Helv., 1959. 1620. 4506-021	TELOCINOBUFOGENIN $C_{24}H_{34}O_5$ 402 $+4^0$ 165^0C Helv.. 1949. 1593. 4506-026
SCILLIROSIDE $C_{32}H_{44}O_{12}$ 620 -59^0 169^0C R = glucosyl Helv., 1959. 1620. 4506-022	SCILLIGLAUCOSIDINE $C_{24}H_{30}O_5$ 398 $+50^0$m 247^0C R = H Experimentia. 1956 (12) 285. 4506-027
SCILLIROSIDINE. 1. 2-epoxy- $C_{26}H_{32}O_8$ 472 -11^0c 157^0C R = H Tet.. 1966. 3213. 4506-023	SCILLAREN-F $C_{30}H_{40}O_{10}$ 560 $+106^0$ 165^0C R = glucosyl Helv. 1953. 1531. 4506-028
SCILLIROSIDIN. 1. 2-epoxy-12-OH- $C_{26}H_{32}O_9$ 488 R = OH J. S. Afr. Chem. Inst.. 1970 (22) 665. 4506-024	BERSALDEGENIN-3-ACETATE $C_{26}H_{34}O_8$ 474 285^0C R_1 = R_3 = H: R_2 = Ac Tet. Lett.. 1969. 1709. 4506-029
MARINOBUFAGIN $C_{24}H_{32}O_5$ 400 223^0C Tet. Lett.. 1968. 5669. 4506-025	BERSALDEGENIN, 1. 3. 5-ortho-Ac- $C_{26}H_{32}O_7$ 456 -24^0c 292^0C R_1, R_2, R_3 = orthoacetyl Tet. Lett.. 1969. 1709. 4506-030

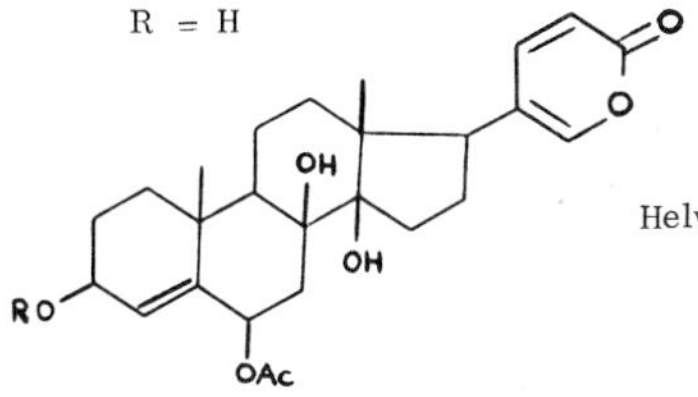

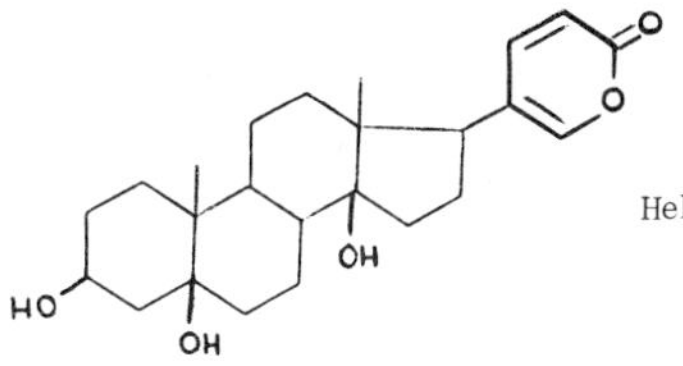

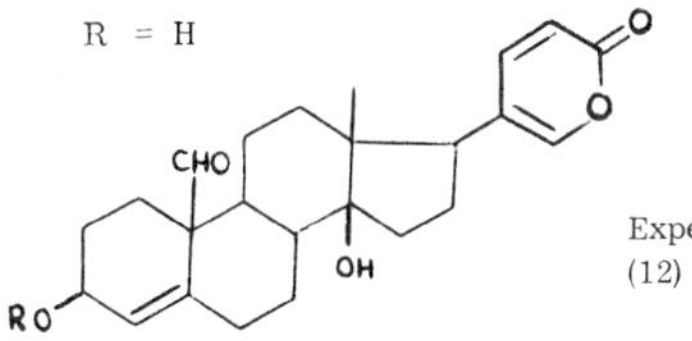

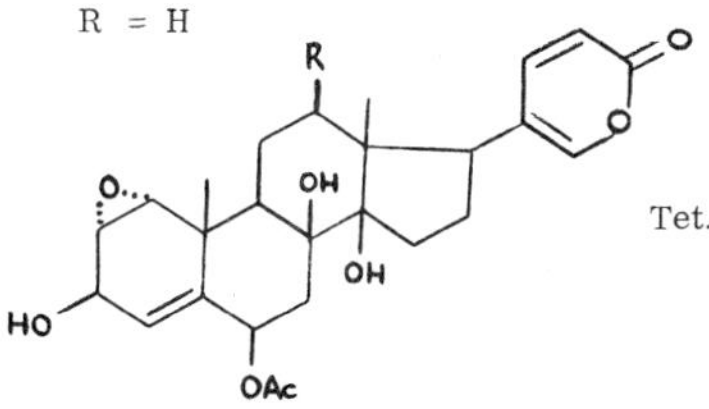

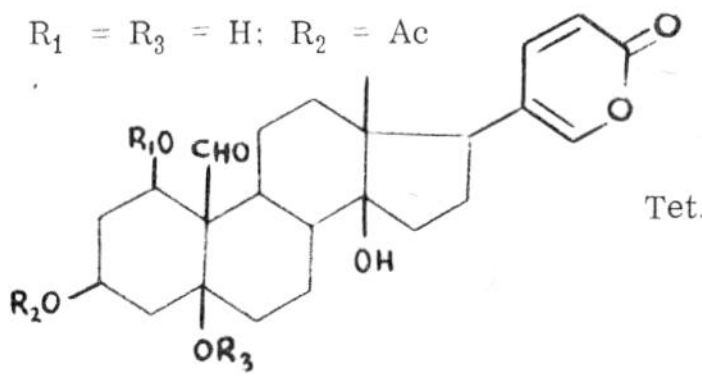

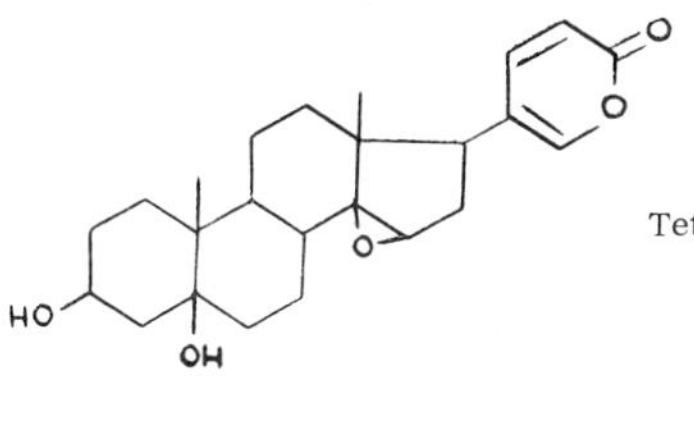

BOVOGENIN A $C_{24}H_{32}O_5$ 400 -74° 244°C

R = H

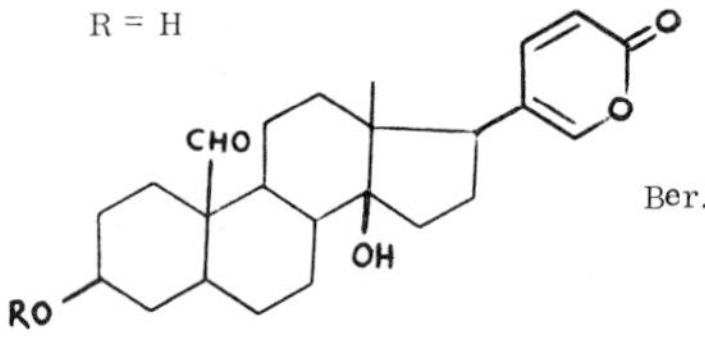

Ber. 1957 2378.

4506-031

RESIBUFAGIN $C_{24}H_{30}O_5$ 398 211°C

R = H

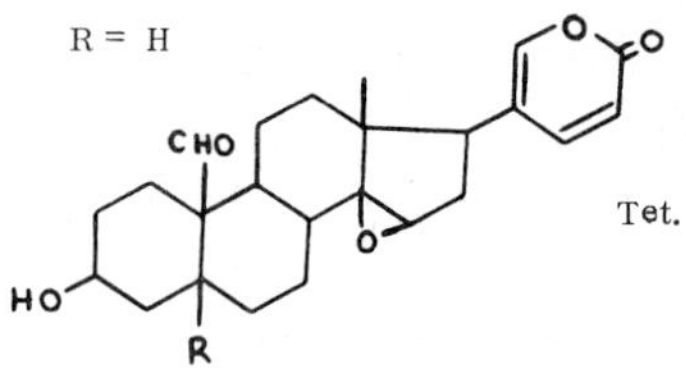

Tet. Lett. 1968 5669.

4506-036

BOVOSIDE A $C_{31}H_{42}O_9$ 560 +73°m 245°C

R = thevetosyl

Helv. 1954 451.

4506-032

BUFOTALININ $C_{24}H_{30}O_6$ 414 +19° 233°C

R = OH

Helv. 1955 883.

4506-037

BOVOGENIN-D $C_{24}H_{32}O_6$ 416 +51°m 244°C

R = H

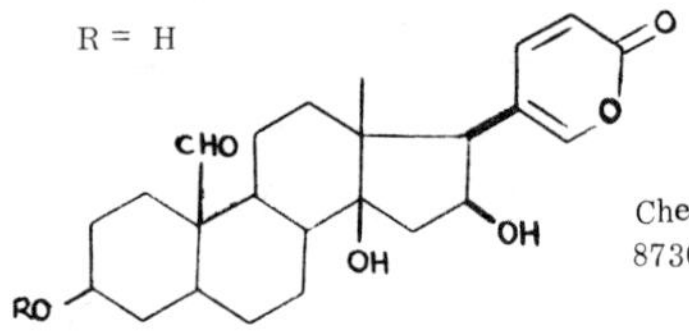

Chem. Abstr. 1968 (69) 87307.

4506-033

HELLEBRIGENIN $C_{24}H_{32}O_6$ 416 +18°a 152°+238°C

$R_1 = R_2 = H$

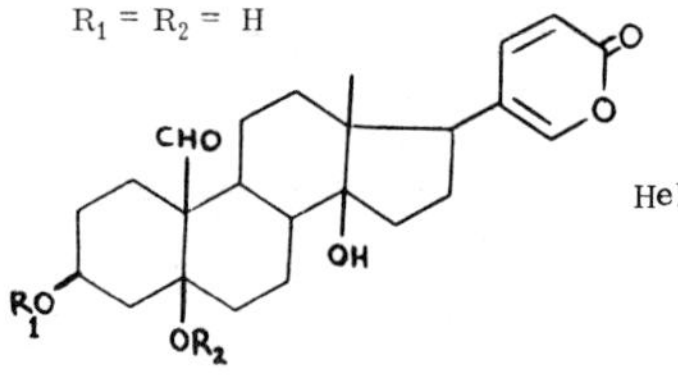

Helv. 1949 1442.

4506-038

BOVOSIDE-D $C_{31}H_{44}O_{10}$ 576 -67° 288°C

R = thevetosyl

Ber. 1953 54.

4506-034

HELLEBRIGENIN, 3-Acetate- $C_{26}H_{34}O_7$ 458 +30° 231°C

R_1 = Ac; R_2 = H

Tet. Lett. 1968 149.

4506-039

BOVOGENIN A, 16-formyloxy- $C_{25}H_{32}O_7$ 444 +17°m 224°C

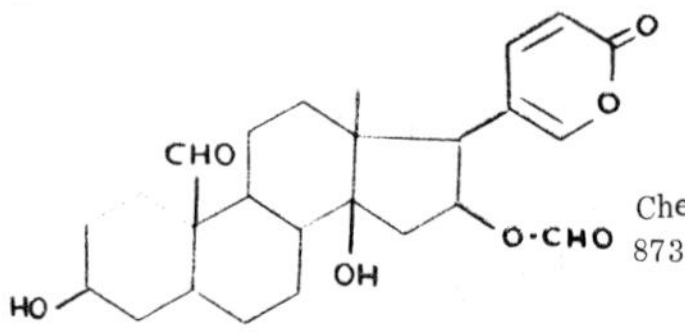

Chem. Abstr. 1968 (69) 87307.

4506-035

HELLEBRIGENIN, 3,5-diAc- $C_{28}H_{36}O_8$ 500 -23°c 218°C

$R_1 = R_2$ = Ac

Tet. Lett. 1968 149.

4506-040

HELLEBRIN $C_{36}H_{52}O_{15}$ 724 -23^{0}m 283^{0}C R_1 = glucosido-rhamnosyl R_2 = H Helv. 1949 (32) 1442. 4506-041	CINOBUFAGIN, 19-oxo- $C_{26}H_{32}O_7$ 456 193^{0}C R = H Helv. 1969 1097. 4506-043
CINOBUFOTALIN, 19-oxo- $C_{26}H_{32}O_8$ 472 +19^0 amorph. R = OH CHO, HO, R, O, OAc Helv. 1969 1097. 4506-042	CINOBUFAGINOL $C_{26}H_{34}O_7$ 458 OH, CH$_2$, HO, O, OAc Helv. 1966 1243. 4506-044

PREGNAN-3α-OL $C_{21}H_{36}O$ 304 148°C JACS., 1938. 2928. 4507-001	PREGNAN-3β, 20β-DIOL. allo- $C_{21}H_{36}O_2$ 320 196°C JACS., 1937. 614. 4507-006
PREGNAN-3α, 20α-DIOL $C_{21}H_{36}O_2$ 320 +27° 244°C JACS., 1940. 518. 4507-002	PREGN-5-EN-3, 20-DIOL $C_{21}H_{34}O_2$ 318 -56°c 182°C Chem. Pharm. Bull., 1968. 326. 4507-007
PREGNAN-3β. 20α-DIOL $C_{21}H_{36}O_2$ 320 182°C JACS., 1937. 2291. 4507-003	PREGN-4-EN-20α-OL-3-ONE $C_{21}H_{32}O_2$ 316 +102°e 161°C Tet., 1970 2061. 4507-008
PREGNAN-3α, 20α-DIOL. allo- $C_{21}H_{36}O_2$ 320 +17° 248°C JACS., 1938. 2931. 4507-004	PREGN-4-EN-20β-OL-3-ONE $C_{21}H_{32}O_2$ 316 +84°c 171°C R = H Bull. Chem. Soc. Jap., 1958 (31) 450. 4507-009
PREGNAN-3β, 20α-DIOL. allo- $C_{21}H_{36}O_2$ 320 +23° 218°C JACS., 1937. 2291. 4507-005	PREGN-4-EN-20β-OL-3-ONE, Ac- $C_{23}H_{34}O_3$ 358 +140°c 159°C R = Ac Chem. Ind., 1967. 1365. 4507-010

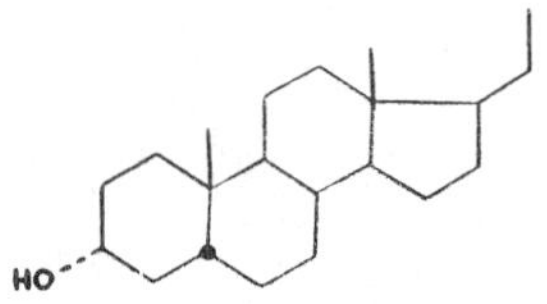

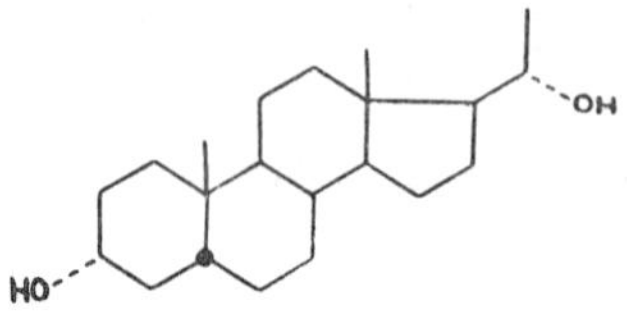

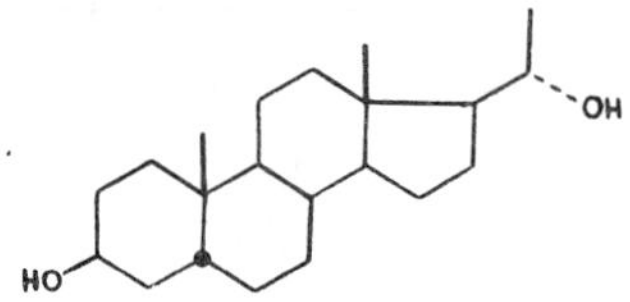

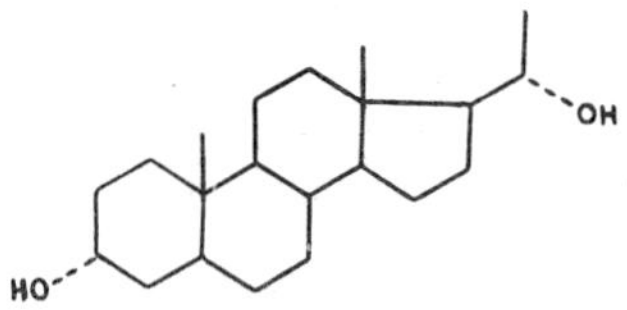

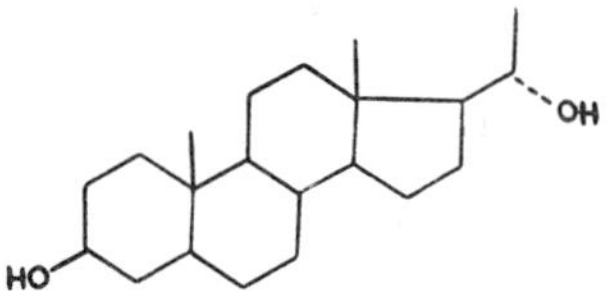

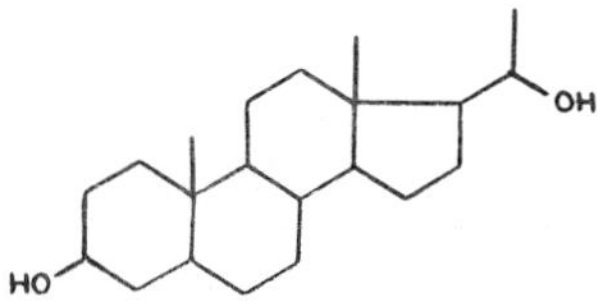

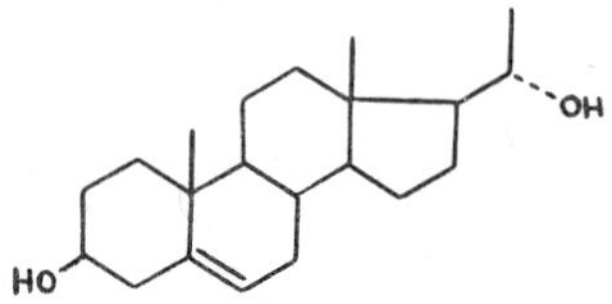

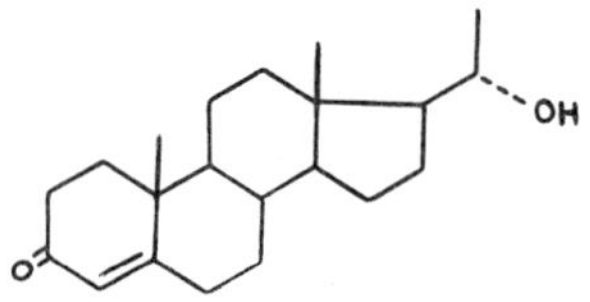

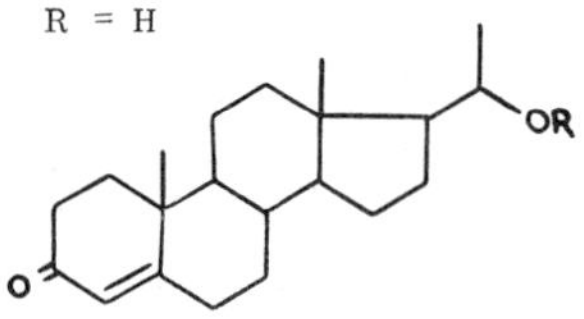

CYBISTERONE $C_{21}H_{30}O_2$ 314
153^{0}C

Ann., 1967 (703) 182.

4507-011

PREGNAN-3α, 11β, 20α-TRIOL $C_{21}H_{36}O_3$ 336
+41^0
205^{0}C

J. Biol. Chem., 1954 (210) 129.

4507-012

PREGNAN-3, 17, 20α-TRIOL, allo- $C_{21}H_{36}O_3$ 336
-13^{0}e
223^{0}C

Helv., 1940, 170.

4507-013

PREGNAN-3, 17, 20β-TRIOL, allo- $C_{21}H_{36}O_3$ 336
-8^0
223^{0}C

20-epimer

Helv., 1940, 170.

4507-014

PREGN-5-EN-3, 16, 20-TRIOL $C_{21}H_{34}O_3$ 334
-65^0
251^{0}C

Chem. Pharm. Bull. 1968 326.

4507-015

PREGN-5-EN-3, 17, 20-TRIOL $C_{21}H_{34}O_3$ 334
-69^{0}e
230^{0}C

Chem. Pharm. Bull. 1968 326.

4507-016

PREGN-4-ENE-15, 20-DIOL-
-3-ONE $C_{21}H_{32}O_3$ 332
133^{0}C

Natwiss., 1969 (56) 37.

4507-017

PREGNAN-3, 20-DIOL-11-ONE $C_{21}H_{34}O_3$ 334
+59^{0}e
217-19^{0}C

J. Biol. Chem. 1948 (172) 783.

4507-018

PREGNAN-3, 17, 20, 21-
-TETROL, allo- $C_{21}H_{36}O_4$ 352
-1^{0}e
200^{0}C

Ber. 1939 391.

4507-019

PREGN-4-ENE-20, 21-DIOL-
- 3, 11-DIONE $C_{21}H_{30}O_4$ 346
+176^0
224^{0}C

Helv. 1942 988.

4507-020

PREGN-4-ENE-17, 20, 21-TRIOL-3, 11-DIONE $C_{21}H_{30}O_5$ 362 208°C Helv. 1942 988. 4507-021	CORTOL, β- $C_{21}H_{36}O_5$ 368 +33°e 266-9°C 20-epimer J. Biol. Chem. 1955 (212) 449. 4507-026
PREGN-4-ENE-11, 17, 20, 21-TETROL-3-ONE $C_{21}H_{32}O_5$ 364 +87° 125°C Helv. 1941 247. 4507-022	CORTOLONE $C_{21}H_{34}O_5$ 366 +44°e 208°C R = H J. Biol. Chem. 1955 (212) 449. 4507-027
UTENDIN $C_{21}H_{34}O_5$ 366 244°C Chem. Pharm. Bull. 1966 717. 4507-023	CORTOLONE, β- $C_{21}H_{34}O_5$ 366 +37° 263°C 20-epimer R = H J. Biol. Chem. 1955 (212) 449. 4507-028
TOMENTOGENIN $C_{21}H_{36}O_5$ 368 + 40°m 258°C Chem. Pharm. Bull. 1968 1634. 4507-024	β-CORTOLONE-3-GLUCURONOSIDE, $C_{27}H_{44}O_{11}$ 544 20-epimer R = glucuronosyl JACS. 1955 4184. 4507-029
CORTOL $C_{21}H_{36}O_5$ 368 +24°e 250°C J. Biol. Chem. 1955 (212) 449. 4507-025	PREGNAN-3, 11, 17, 20, 21-PENTOL, allo- $C_{21}H_{36}O_5$ 368 +16°e 222°C Helv. 1941 247. 4507-030

MARSECTOHEXOL $C_{21}H_{34}O_6$ 382

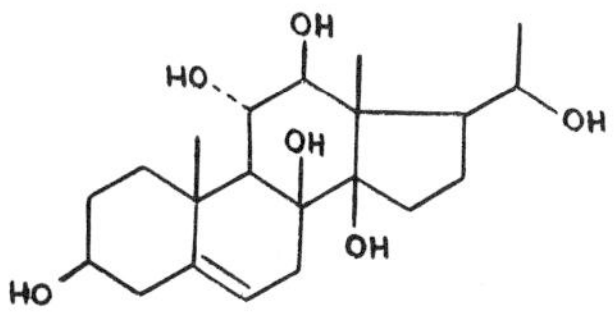

Helv. 1970 221.

4507-031

SARCOSTIN $C_{27}H_{40}O_9$ 508

R = Ac

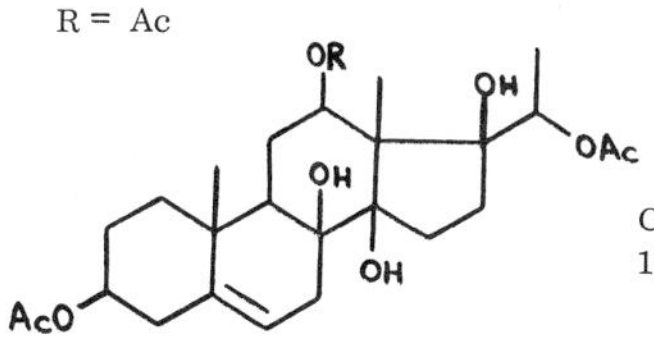

Chem. Pharm. Bull. 1966 717.

4507-032

PENUPOGENIN $C_{36}H_{46}O_{10}$ 638

R = cinnamoyl

Tet. 1968 4147.

4507-033

KONDURANGOGENIN C $C_{32}H_{44}O_7$ 540 $+20^0$

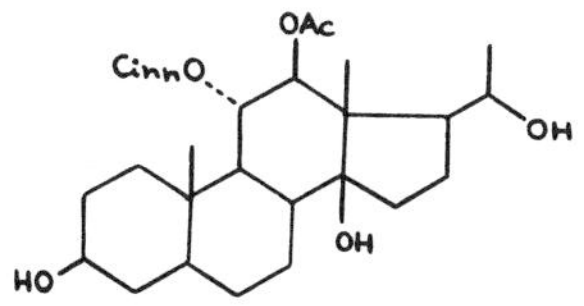

Tet. 1967 1461.

4507-034

STEPHANOL $C_{21}H_{34}O_7$ 398 $+37^0$m 222^0C

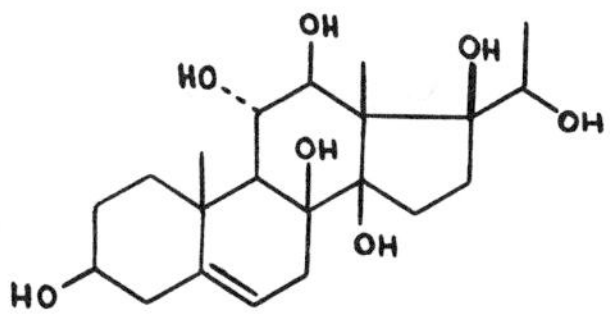

Chem. Pharm. Bull. 1968 553.

4507-035

PREGNAN-3α-OL-20-ONE $C_{21}H_{34}O_2$ 318 $+107^0$e 149^0C

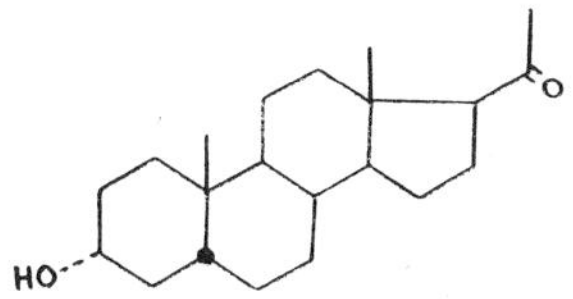

J. Biol. Chem. 1948 (172) 263.

4507-036

PREGNAN-3α-OL-20-ONE, allo- $C_{21}H_{34}O_2$ 318 $+96^0$ 175^0C

5α-H

JACS 1937 616.

4507-037

PREGNAN-3β-OL-20-ONE, allo- $C_{21}H_{34}O_2$ 318 $+91^0$ 194^0C

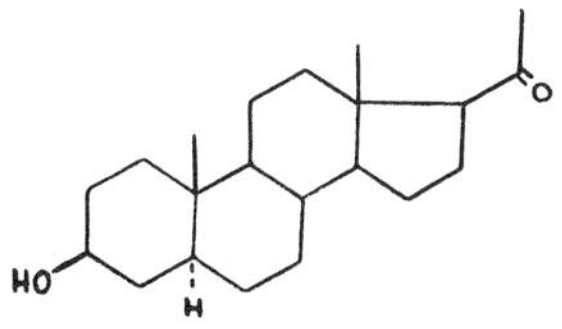

Helv. 1946 33.

4507-038

PREGNAN-3, 20-DIONE $C_{21}H_{32}O_2$ 316 123^0C

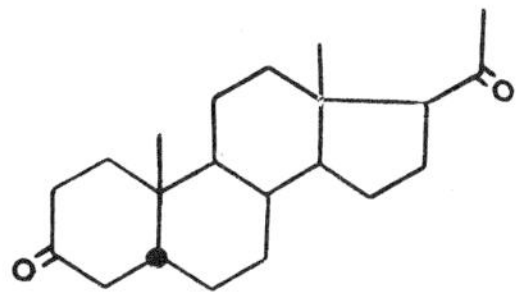

JACS. 1938 1559.

4507-039

PREGNAN-3, 20-DIONE, allo- $C_{21}H_{32}O_2$ 316 +127 200^0C

5α-H

Z. Physiol. Chem. 1934 (227) 84.

4507-040

PROGESTERONE $C_{21}H_{30}O_2$ 314
+ 192°w
122°C

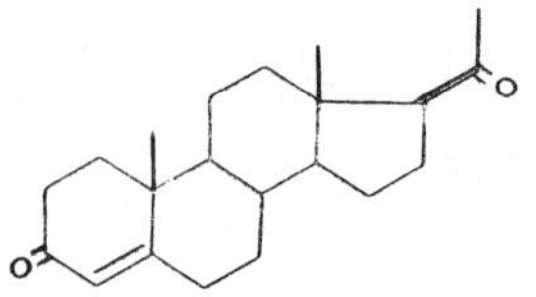

Biochim. Biophys. Acta 1960 (40) 289.

4507-041

CYBISTEROL $C_{21}H_{28}O_3$ 328
196°C

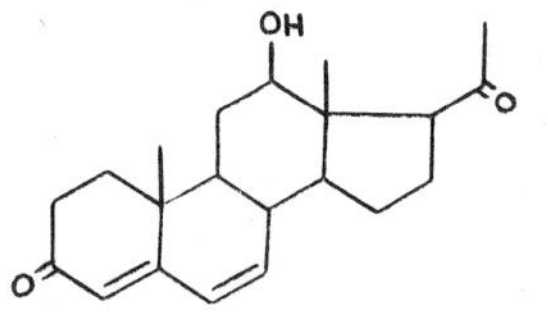

Tet. 1960 2061.

4507-046

PREGN-14-EN-20-ONE. 3-OH-allo- $C_{21}H_{32}O_2$ 316
+53°a
190°C

R = H

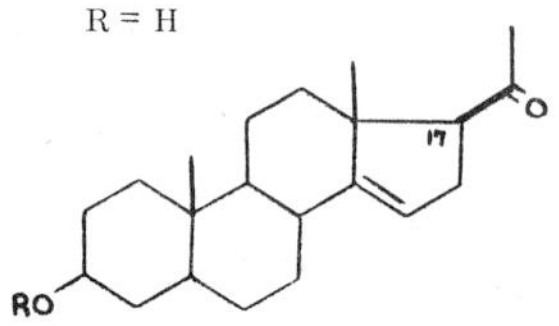

Helv. 1969 2583.

4507-042

HOLACURTENINE $C_{21}H_{34}O_3$ 334
+57°
137°C

R = H

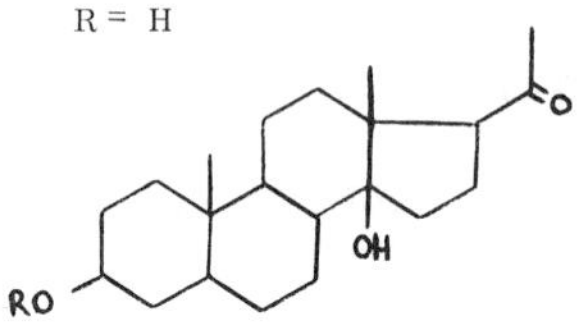

Chem. Abstr. 1968 (68) 87472.

4507-047

HOLADIOLONE $C_{29}H_{47}NO_4$ 473
+ 39°
242°C

R =

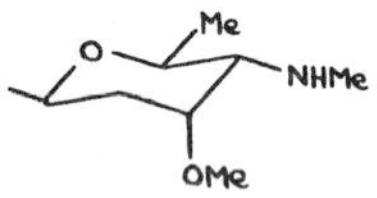

Chem. Abstr. 1968 (68) 87472.

4507-043

HOLACURTINE $C_{29}H_{49}NO_5$ 491
+ 42°
162°C

R =

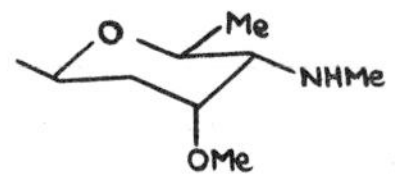

Chem. Abstr. 1968 (68) 87472.

4507-048

PREGN-14-EN-20-ONE, 3-OH-allo-iso- $C_{21}H_{32}O_2$ 316
+117°a
140°C

17-epimer

R = H

Helv., 1969, 2583.

4507-044

PROGESTERONE, 17-OH- $C_{21}H_{30}O_3$ 330
+ 105°
223°C

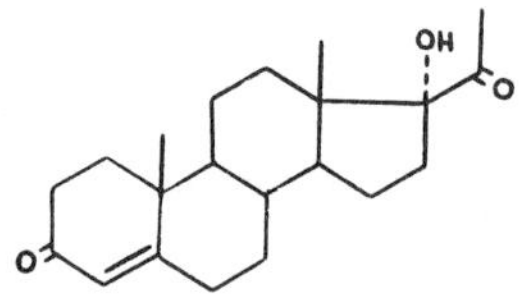

Helv. 1941 879.

4507-049

PREGN-4-EN-12-OL-3, 20-DIONE $C_{21}H_{30}O_3$ 330

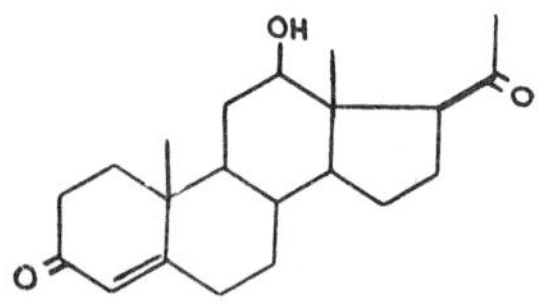

Tet. 1970 2061.

4507-045

CORTICOSTERONE, deoxy- $C_{21}H_{30}O_3$ 330
+ 178°e
142°C

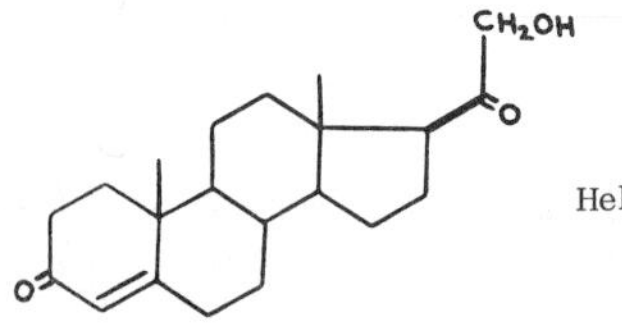

Helv. 1938 1183; 1197.

4507-050

PREGNAN-3β, 17α-DIOL-3-ONE, allo- $C_{21}H_{34}O_3$ 334 + 38⁰ 264⁰C

Helv. 1938 1181.

4507-051

PREGNAN-20-ONE, 3, 8, 14-triOH-allo-iso- $C_{21}H_{34}O_4$ 350 -28⁰m 214⁰C

Helv. 1969 2583.

4507-052

PREGNAN-3β, 11, 21-TRIOL-20-ONE, allo- $C_{21}H_{34}O_4$ 350 + 35⁰ 204⁰C

Helv. 1938 1490.

4507-053

PREGNAN-3α, 11, 21-TRI-OL-20-ONE, allo- $C_{21}H_{34}O_4$ 350 205⁰C

3-epimer

Fed. Proc. 1954 (13) 204.

4507-054

CORTICOSTERONE, tetrahydro- $C_{21}H_{32}O_4$ 350 156⁰C

JCS. 1964 537.

4507-055

CORTICOSTERONE $C_{21}H_{30}O_4$ 346 +223⁰e 181⁰C

JACS. 1956 1414.

4507-056

PREGNAN-3, 21-DIOL-11, 20-DIONE, allo- $C_{21}H_{32}O_4$ 348 + 94⁰ 191⁰C

J. Biol. Chem. 1937 (120) 719.

4507-057

CORTICOSTERONE, 11-dehydro- $C_{21}H_{28}O_4$ 344 239⁰d 184⁰C

J. Biol. Chem. 1951 (188) 287.

4507-058

DIGIPURPUROGENIN-I $C_{21}H_{32}O_4$ 348 + 51⁰c 167-75⁰C

Tet. Lett. 1964 473.

4507-059

RAMANONE, Benzoyl $C_{28}H_{36}O_5$ 452 + 48⁰p 226⁰C

Chem. Pharm. Bull. 1966 1332.

4507-060

PURPNIGENIN $C_{21}H_{32}O_4$ 348 +21° 240°C

R = H

RO, OH, OH

Chem. Pharm. Bull. 1960 657.

4507-061

LINEOLONE $C_{21}H_{32}O_5$ 364 +13°m 242°C

R = H

OR, OH, 17, OH, HO

Tet. 1968 4143.

4507-066

PURPNIN $C_{39}H_{62}O_{13}$ 738 285°C

R = digitoxosyl

Ann. 1957 (606) 160.

4507-062

LINEOLONE, iso- $C_{21}H_{32}O_5$ 364 +53°m 246°C

R = H

17-epimer

Aust. J. C. 1968 1625.

4507-067

PREGNAN-3, 17, 21-TRIOL-20-ONE, allo- $C_{21}H_{34}O_4$ 350 +48°e 249°C

CH₂OH, OH, HO

Helv. 1938 1185.

4507-063

LINEOLONE, benzoyl- $C_{28}H_{36}O_6$ 468 -65°m 256°C

R = benzoyl

Aust. J. C. 1968 1625.

4507-068

CORTICOSTERONE, 17-OH-11-deoxy- $C_{21}H_{30}O_4$ 346 +132° 208°C

CH₂OH, OH

JACS. 1950 5145.

4507-064

LINEOLONE, benzoyl-iso- $C_{28}H_{36}O_6$ 468 -52°m 272°C

R = benzoyl

17-epimer

Aust. J. C. 1968 1625.

4507-069

POSTERONE $C_{21}H_{30}O_5$ 362 235°C

HO, HO, OH

Steroids 1960 (16) 393.

4507-065

CYNANCHOGENIN $C_{28}H_{42}O_6$ 574

R =

Tet. 1968 4143.

4507-070

DREVOGENIN D $C_{21}H_{34}O_5$ 366 -10^0m 228^0C HO, OH, OH, OH, HO Helv. 1966 1632. 4507-071	DIGIPROGENIN. gamma- $C_{21}H_{28}O_5$ 360 -122^0p 238^0C R = H Chem. Pharm. Bull. 1966 552. 4507-076
DREVOGENIN-P $C_{21}H_{32}O_5$ 364 $R_1 = R_2$ = H Helv. 1966 1632. 4507-072	DIGIPRONIN $C_{28}H_{40}O_9$ 520 -73^0p $236-41^0$C R = digitalosyl JCS. 1962 3610. 4507-077
DREVOGENIN-B $C_{23}H_{44}O_6$ 406 R_1 = Ac; R_2 = H Helv. 1966 1632. 4507-073	PREGNAN-3β, 11. 17. 21--TETROL-20-ONE. allo- $C_{21}H_{34}O_5$ 366 $+51^0$e 225^0C Helv. 1942 988. 4507-078
DREVOGENIN A $C_{28}H_{42}O_7$ 590 182^0C R_1 = Ac; R_2 = i-Val Helv. 1966 1655. 4507-074	PREGNAN-3α, 11. 17. 21--TETROL-20-ONE. allo- $C_{21}H_{34}O_5$ 366 $+73^0$e 276^0C 3-epimer Biochem. J. 1957 (67) 689. 4507-079
KONDURANGOGENIN A $C_{32}H_{42}O_7$ 538 Tet. 1967 1461. 4507-075	PREGNAN-3β, 17, 21-TRIOL--11, 20-DIONE. allo- $C_{21}H_{32}O_5$ 364 $+62^0$e 242^0C Helv. 1942 988. 4507-080

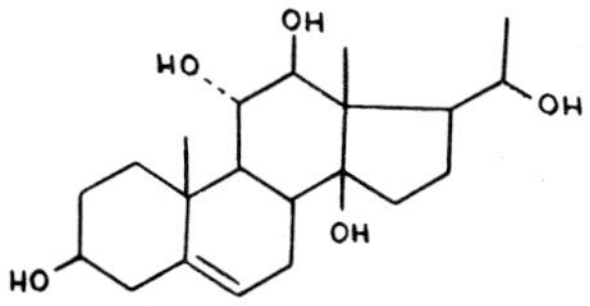

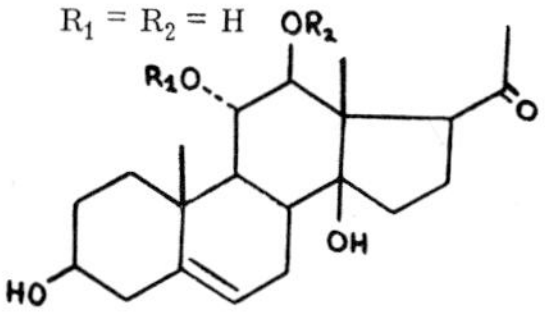

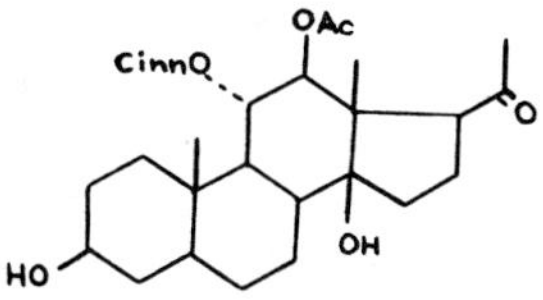

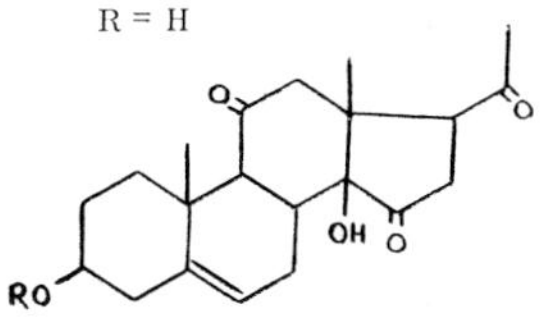

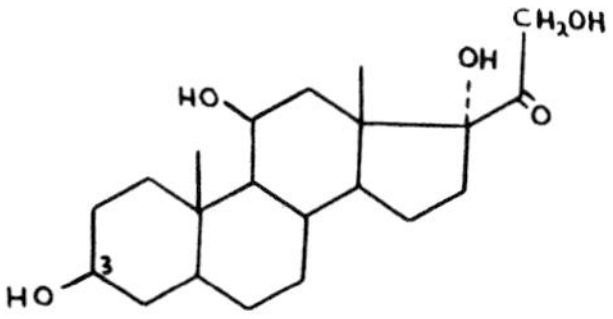

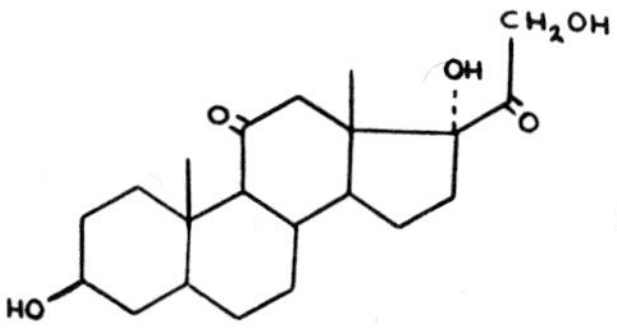

CORTISOL $C_{21}H_{30}O_5$ 362 +167° 220°C HO, OH, CH_2OH, O JACS 1951 3818. 4507-081	PURPROGENIN $C_{21}H_{30}O_5$ 362 O, OH, OH, HO Chem. Pharm. Bull. 1968 178. 4507-086
CORTISONE $C_{21}H_{28}O_5$ 360 +209°e 230°C O, OH, CH_2OH, O JCS. 1954 125. 4507-082	MARSDENIN $C_{21}H_{32}O_6$ 380 -16°m 262°C OH, HO, 17, OH, OH, HO Chem. Pharm. Bull. 1968 2522. 4507-087
ALDOSTERONE $C_{21}H_{28}O_5$ 360 +161°c 166°C CHO, CH_2OH, HO, O JACS. 1954 5035. 4507-083	MARSDENIN, 17β- $C_{21}H_{32}O_6$ 380 17-epimer Helv. 1970 221. 4507-088
PERGULARIN $C_{21}H_{32}O_5$ 364 220°C OH, OH, OH, HO Chem. Pharm. Bull. 1966 717. 4507-084	METAPLEXIGENIN, deacyl- $C_{21}H_{32}O_6$ 380 220°C R = H OR, OH, OH, OH, HO Chem. Pharm. Bull. 1966 717. 4507-089
DIGACETIGENIN $C_{23}H_{32}O_6$ 404 OAc, OH, O, HO Tet. Lett. 1968 3243; 3897. 4507-085	METAPLEXIGENIN $C_{23}H_{34}O_7$ 422 R = Ac Chem. Pharm. Bull. 1966 717. 4507-090

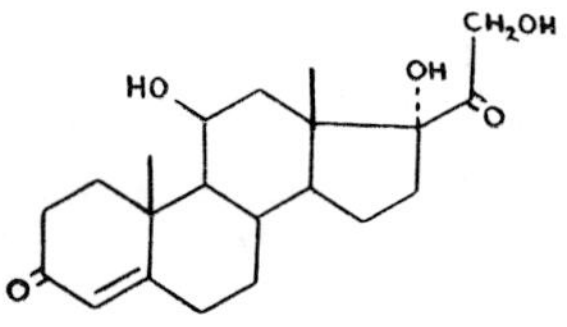

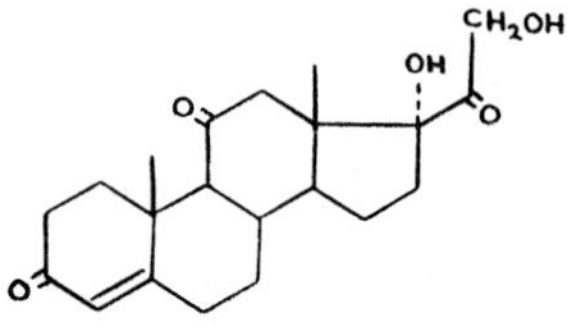

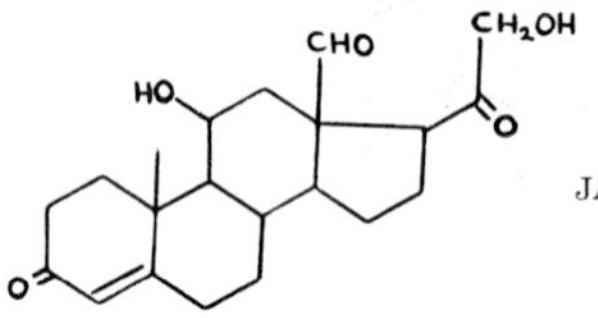

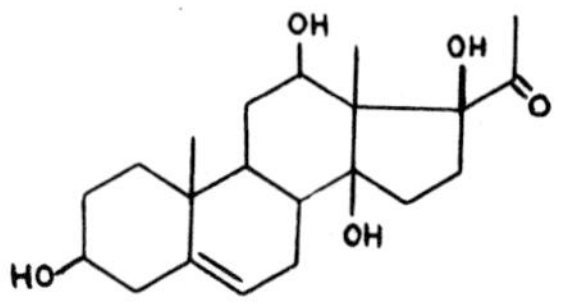

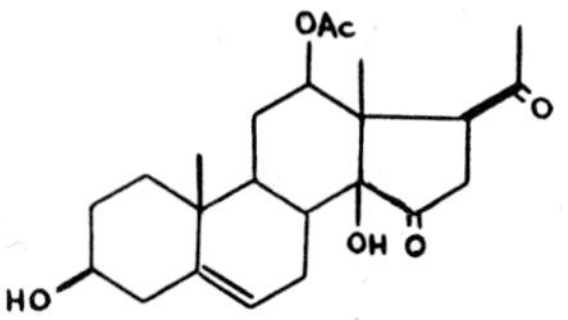

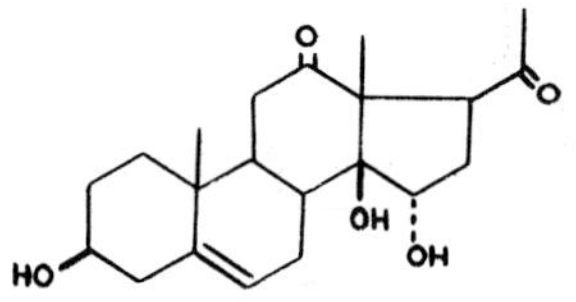

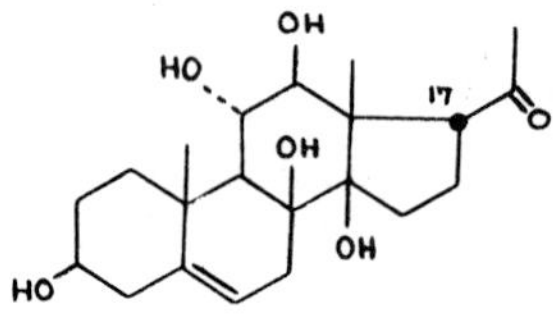

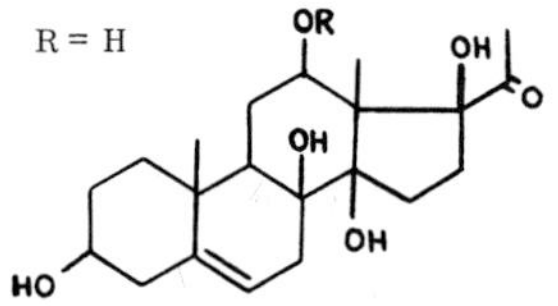

METAPLEXIGENIN, 12-O-
- benzoyl-de-Ac-

$C_{28}H_{36}O_7$ 484

R = benzoyl

Helv. 1968 767.

4507-091

VIMINOLONE, di-OBz-

$C_{35}H_{40}O_9$ 604

R = benzoyl

Helv. 1968 767.

4507-092

SIBIRIGENIN

$C_{28}H_{42}O_6$ 574

Chem. Abstr. 1970 (73) 77477.

(cf 4507-070?)

4507-093

DIGINIGENIN

$C_{21}H_{28}O_4$ 344
-226^0a
115^0C

R = H

JCS. 1962 3610.

4507-094

DIGININ

$C_{28}H_{40}O_7$ 488
-223^0c

R = diginosyl

JCS. 1962 3610.

4507-095

DIGITALONIN

$C_{28}H_{40}O_8$ 504
-167^0m
$208-13^0$C

R = digitalosyl

Chem. Pharm. Bull. 1957 253.

4507-096

DIGIFOLOGENIN

$C_{21}H_{28}O_5$ 360
-270^0a
174^0C

R = H

JCS. 1963 3281.

4507-097

DIGIFOLEIN

$C_{28}H_{40}O_8$ 504
-189^0m
200^0C

R = diginosyl

JCS. 1963 3281.

4507-098

PREGN-4-EN-3-ON-18-COOH-
LACTONE. 16 20-diOH-

$C_{21}H_{28}O_4$ 344
$+44^0$c
242^0C

Tet. Lett. 1964 1671.

4507-099

PREGNA-4. 6-DIENE-18-COOH-
LACTONE 20-OH-3-oxo-

$C_{21}H_{26}O_3$ 326
$+32^0$c
238^0C

Tet. Lett. 1964 1671.

4507-100

HOLANTOSINE A $C_{28}H_{47}NO_6$ 493 -28^{0}c 260^{0}C OH OH O Holosaminosyl Tet., 1970, 1695. 4507-101	STAPELOGENIN $C_{21}H_{30}O_4$ 346 300^{0}C OH O HO Helv. 1966 1505. 4507-103
HOLANTOSINE B $C_{28}H_{45}NO_5$ 475 -29^{0}c 290^{0}C O Holosaminosyl Tet., 1970, 1695. 4507-102	

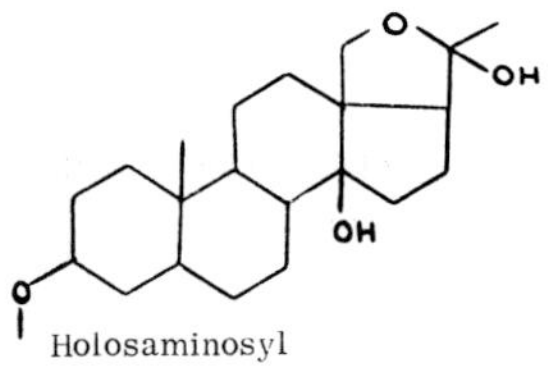

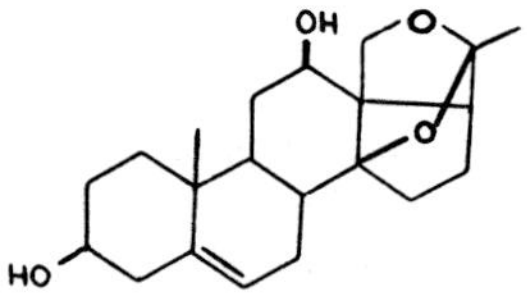

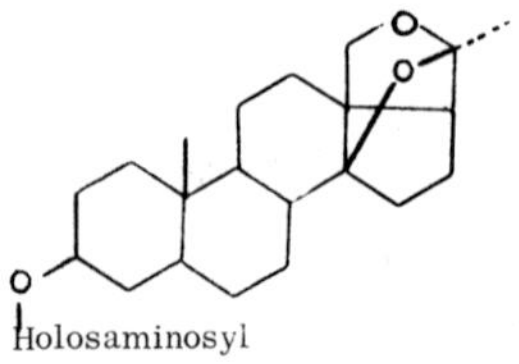

ANDROSTAN-3α, 17β-DIOL, 5β- $C_{19}H_{32}O_2$ 292 +26^{0}e 232^{0}C

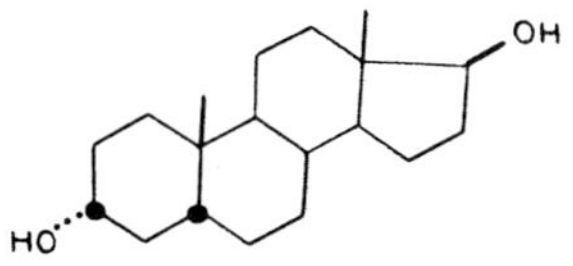

Helv. 1937 (20) 547.

4508-001

ANDROST-5-EN-3, 17-DIOL $C_{19}H_{30}O_2$ 290

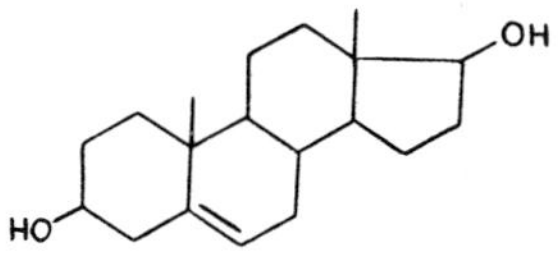

J. Biol. Chem. 1945 (157) 601.

4508-002

ANDROSTAN-3β, 16α, 17α-TRIOL, 5α- $C_{19}H_{32}O_3$ 308 -17^{0}c 265-80^{0}C

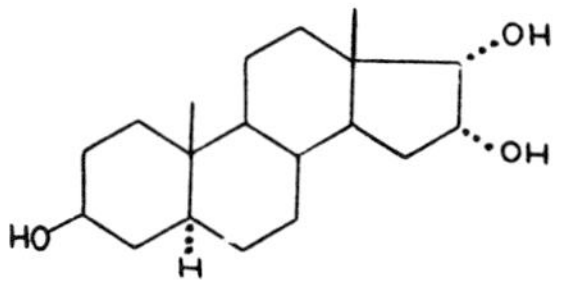

Tet. Lett. 1964 217.

4508-003

ANDROST-5-ENE-3β, 16α, 17β-TRIOL $C_{19}H_{30}O_3$ 306 -74^{0}d 265-70^{0}C

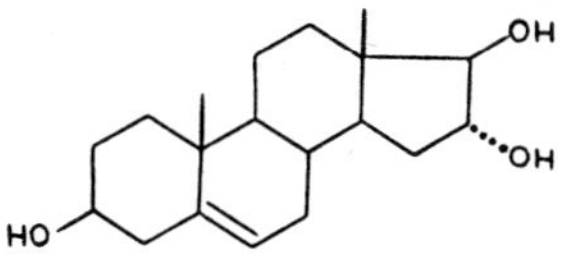

J. Biol. Chem. 1946 (164) 363.

4508-004

ANDROST-5-EN-3β, 16β17β-TRIOL $C_{19}H_{30}O_3$ 306

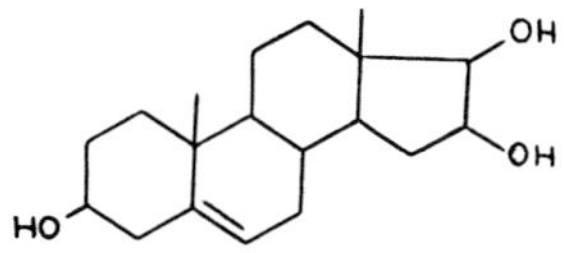

Biochem. J. 1957 (67) 25.

4508-005

ANDROSTA-3, 5-DIEN-17-ONE $C_{19}H_{26}O$ 270 -31^{0}e 88^{0}C

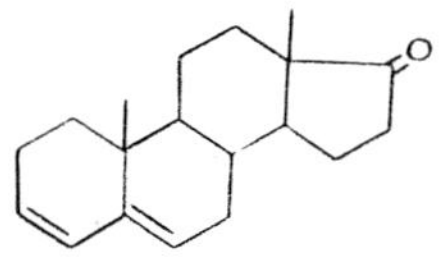

Biochem. J. 1937 (31) 956.

4508-006

ANDROSTERONE $C_{19}H_{30}O_2$ 290 +95^{0}e 183^{0}C

R = H

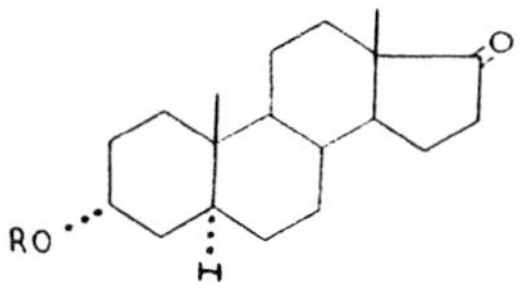

J. Biol. Chem. 1952 (198) 871.

4508-007

ANDROSTERONE-3β-GLUCURONOSIDE $C_{25}H_{38}O_8$ 466

R = Glucuronosyl

J. Biol. Chem. 1963 (238) 907.

4508-008

ANDROSTERONE, 5β- $C_{19}H_{30}O_2$ 290 +105^{0}e 151^{0}C

R = H

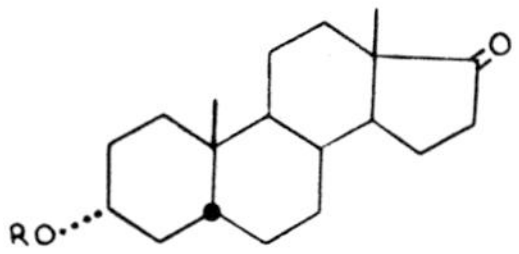

J. Biol. Chem. 1952 (198) 871.

4508-009

CHOLANOLONE-3β-GLUCURONOSIDE, AETIO- $C_{25}H_{38}O_8$ 446 +26^{0}m 221^{0}C

R = Glucuronosyl

J. Biol. Chem. 1963 (238) 907.

4508-010

ANDROSTENOLONE	$C_{19}H_{28}O_2$	288 +11⁰e 153⁰C	JACS. 1935 1504. 4508-011
RUBROSTERONE	$C_{19}H_{26}O_5$	334 +119⁰m	Tet. 1969 3389. 4508-016
ANDROSTAN-17-ONE. 3α, 11β-diOH-	$C_{19}H_{30}O_3$	306 +97⁰e 198⁰C	JCS. 1946 1134. 4508-012
ANDROSTA-1, 4-DIEN-12-OL-3, 17-DIONE	$C_{19}H_{24}O_3$	300 +83⁰c 222⁰C	Compt. Rend., 1970 (271) 153. 4508-017
ANDROST-5-EN-17-ONE. 3. 16-diOH-	$C_{19}H_{28}O_3$	304	Biochem. J. 1957 (66) 664. 4508-013
ANDROSTA-1, 4-DIEN-17-OL-3-ONE	$C_{19}H_{26}O_2$	286 164⁰C	Ber., 1969 (102) 1859. 4508-018
TESTERONE	$C_{19}H_{28}O_2$	288 +109⁰e 154⁰C	JACS. 1955 817. 4508-014
ANDROST-4-EN-17-OL-3-ONE	$C_{19}H_{28}O_2$	288	Ber., 1969 (102) 1859. 4508-019
ANDRENOSTERONE	$C_{19}H_{24}O_3$	300 +262⁰e 222⁰C	JACS. 1946 2478. 4508-015

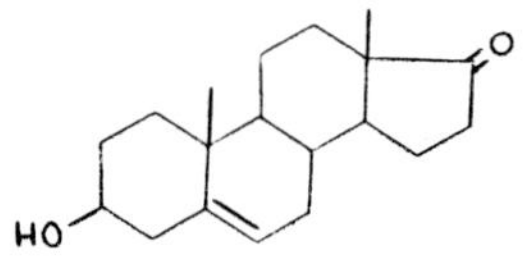

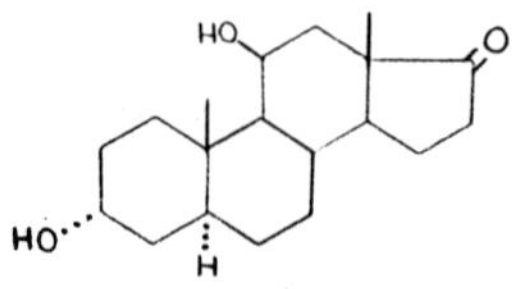

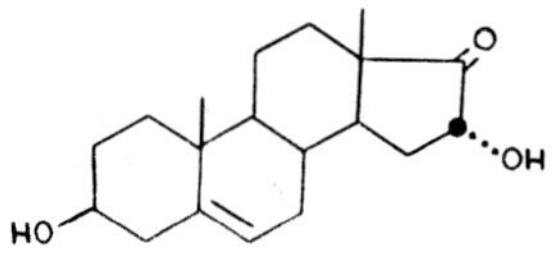

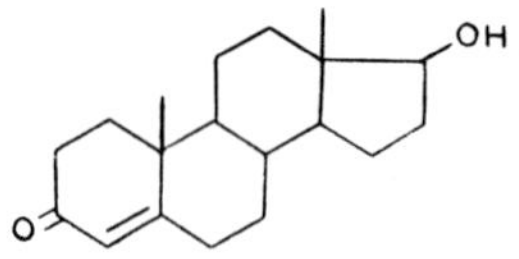

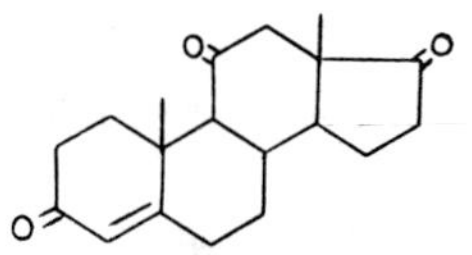

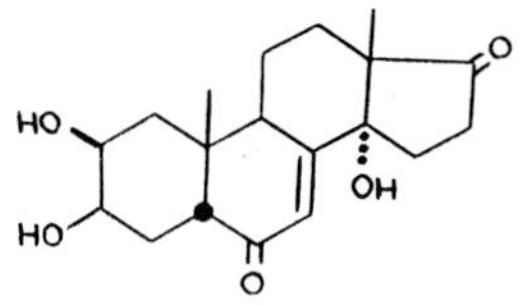

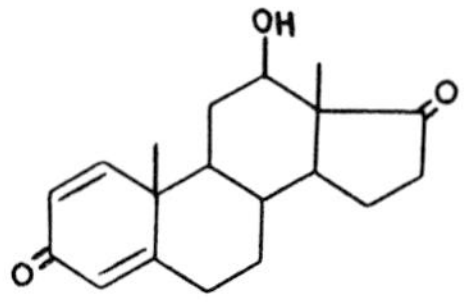

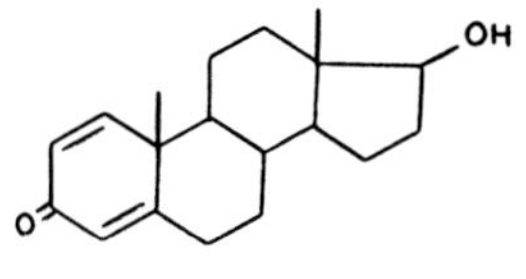

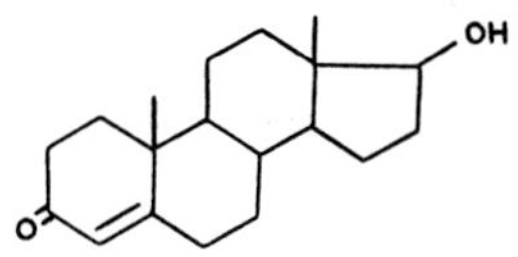

COPROSTEROL $C_{27}H_{48}O$ 388 +28^{0}c 101^{0}C Z. Physiol., 1967 (348) 1688. 4509-001	CHIMAEROL $C_{27}H_{48}O_5$ 452 +42^{0}e 181^{0}C Biochem. J., 1963 (87) 28. 4509-006
COPROCHOLIC ACID, cheno-deoxy- $C_{27}H_{46}O_4$ 434 Biochem. J., 1966 (99) 9P. 4509-002	ECDYSONE, deoxy- $C_{27}H_{44}O_5$ 448 232^{0}C Chem. Comm., 1970, 1217. 4509-007
COPROCHOLIC ACID $C_{27}H_{46}O_5$ 450 +27^{0}e 181^{0}C J. Biol. Chem., 1961 (236) 688. 4509-003	CHEILANTHONE-B $C_{27}H_{46}O_5$ 450 227^{0}C R = H Chem. Comm., 1970, 243. 4509-008
COPROSTAN-3, 7, 12, 25, 26-PENTAOL $C_{27}H_{48}O_5$ 452 +38^0 178^{0}C J. Biochem. (Jap.), 1961 (51) 48. 4509-004	CHEILANTHONE A $C_{27}H_{46}O_6$ 466 237^{0}C R = OH Chem. Comm., 1970, 243. 4509-009
SCYMNOL $C_{27}H_{46}O_5$ 450 +38^{0}e 187^{0}C JACS., 1943, 477. 4509-005	deleted 4509-010

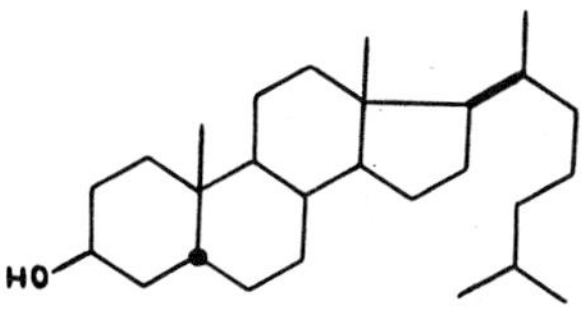

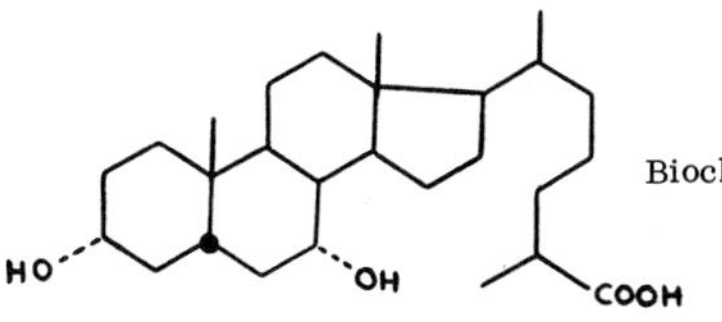

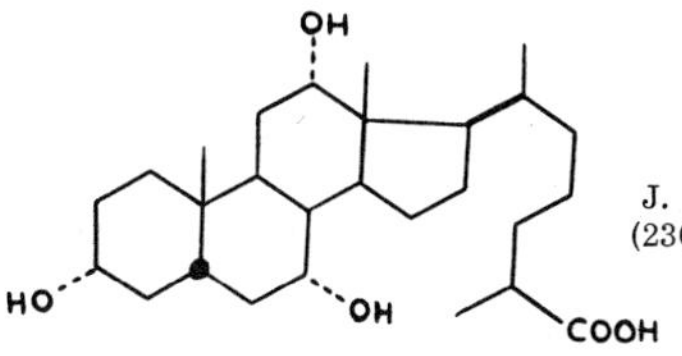

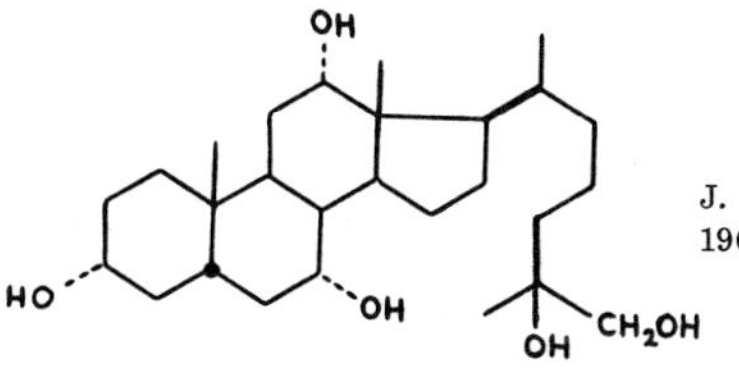

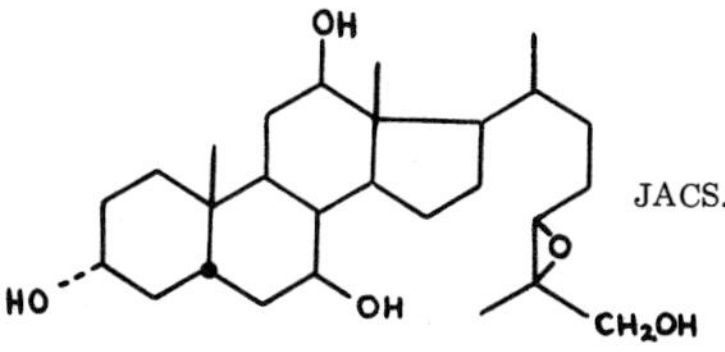

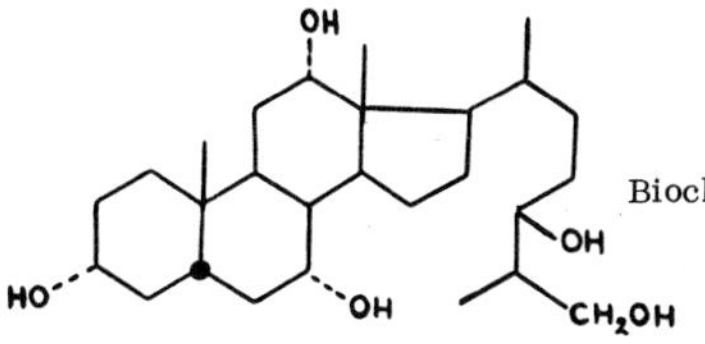

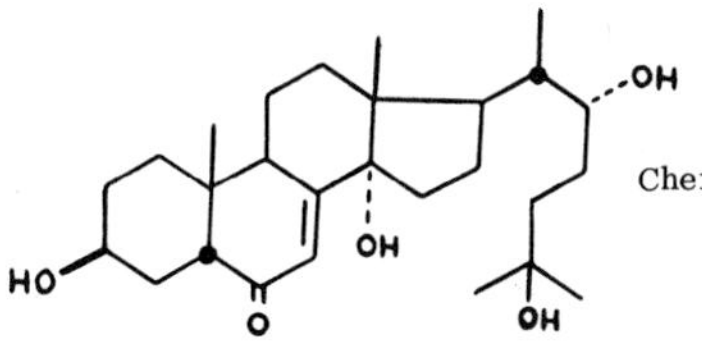

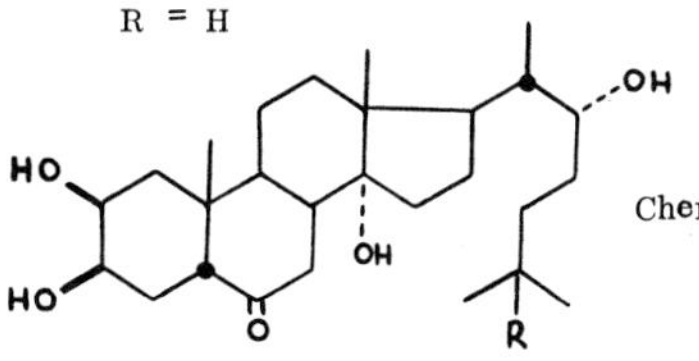

STACHYSTERONE B — $C_{27}H_{42}O_6$ — 462 — amorph.

JACS. 1970. 7510.

4509-011

PONASTERONE B — $C_{27}H_{44}O_6$ — 464 — 129^0C

Tet. Lett., 1968, 1105.

4509-016

STACHYSTERONE C — $C_{27}H_{42}O_6$ — 462 — 238^0C

Chem. Comm., 1970. 352.

4509-012

ECDYSONE — $C_{27}H_{44}O_6$ — 464

Ber., 1965, 2394.

4509-017

PODECDYSONE B — $C_{27}H_{42}O_6$ — 462 — 126^0C

Chem. Comm., 1969. 402.

4509-013

CRUSTECDYSONE, deoxy- — $C_{27}H_{44}O_6$ — 464 — 234^0C

Chem. Comm., 1968. 83.

4509-018

PONASTERONE A — $C_{27}H_{44}O_6$ — 464 — $+90^0m$ — 160

R = H

Chem. Comm., 1966 (24) 915.

4509-014

AJUGASTERONE C — $C_{27}H_{44}O_7$ — 480 — amorph.

Chem. Comm., 1969. 546.

4509-019

PONASTEROSIDE A — $C_{33}H_{54}O_{11}$ — 626 — $+28^0p$ — 278^0C

R = glucosyl

Tet., 1969. 3909.

4509-015

PTEROSTERONE — $C_{27}H_{44}O_7$ — 480 — $+7^0m$ — 230^0C

Tet. Lett., 1968. 375.

4509-020

CRUSTECDYSONE $C_{27}H_{44}O_7$ 480

242°C

R = H

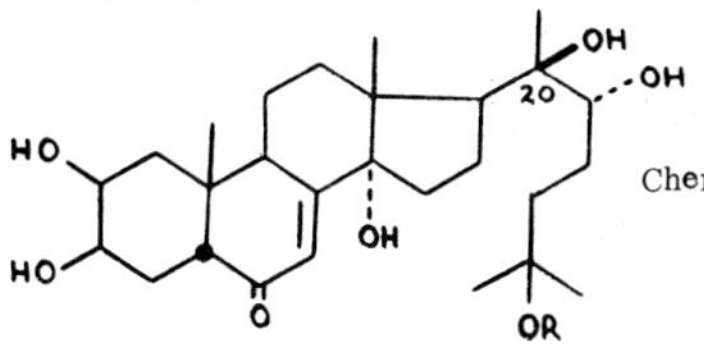

Chem. Comm., 1966, 339.

4509-021

STACHYSTERONE D $C_{27}H_{42}O_6$ 462

248°C

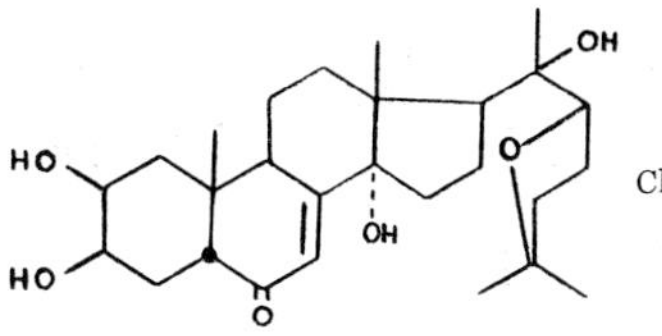

Chem. Comm., 1970, 352.

4509-026

VITICOSTERONE-E $C_{29}H_{46}O_8$ 522

199°C

R = Ac

Tet. Lett., 1969, 329.

4509-022

deleted

4509-027

SHIDASTERONE $C_{27}H_{44}O_7$ 480

258°C

R = H

(21 epimer ?)

Tet. Lett., 1968, 6095.

4509-023

deleted

4509-028

INOKOSTERONE $C_{27}H_{44}O_7$ 480

255°C
A/C

R = H

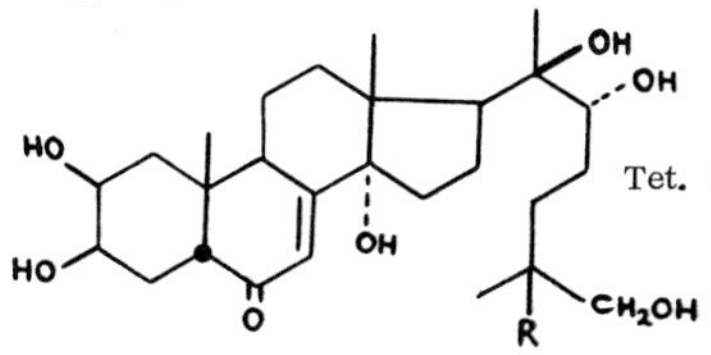

Tet. Lett., 1968, 2475.

4509-024

CONVALLAMARETIN $C_{26}H_{40}O_5$ 432

-86°c
250°C

$R_1 = R_2 = H$

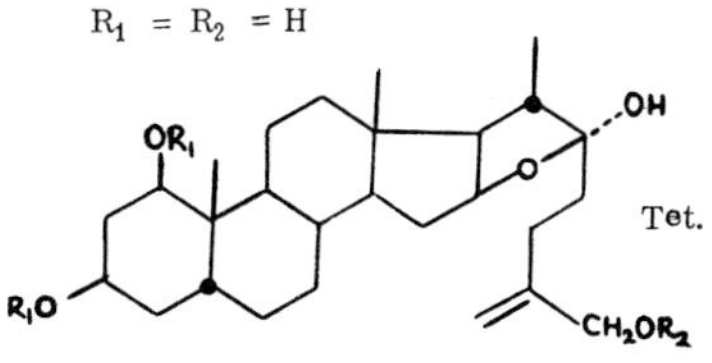

Tet. Lett., 1968, 5141.

4509-029

ECDYSONE, 20, 26-diOH- $C_{27}H_{44}O_8$ 496

150°C

R = OH

Chem. Comm., 1967, 650.

4509-025

CONVALLAMAROSIDE $C_{51}H_{82}O_{23}$ 1062

-66°

R_1 = glucose + (rhamnose)$_2$ (?)
R_2 = glucosyl

Tet. Lett., 1968, 5141.

4509-030

POLYPODINE B $C_{27}H_{44}O_8$ 496, 256°C. Tet. Lett., 1968, 6063. 4509-031	MYXINOL $C_{27}H_{48}O_4$ 436, -15°e, 206°C. Biochem. J., 1966, 238. 4509-033
PONASTERONE C $C_{27}H_{44}O_8$ 496, 271°C. Chem. Comm., 1970, 351. 4509-032	

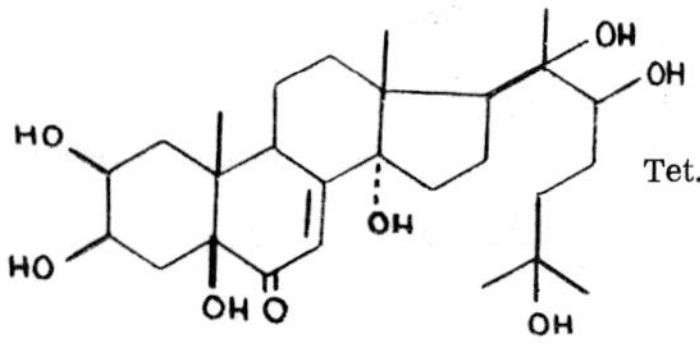

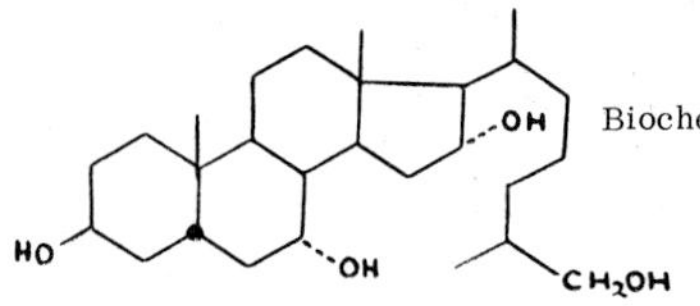

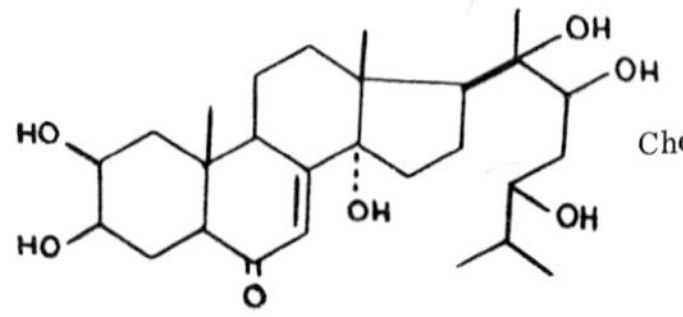

CHOLIC ACID, litho- — $C_{24}H_{40}O_3$ — 376 — $+32^0$e — 186^0C

Chem. Reviews 1934 (15) 311.

4510-001

CHOLIC ACID, glyco-cheno-desoxy- — $C_{26}H_{43}O_5N$ — 449 — $+9^0$e — 181^0C

R = -NH-COOH

Acta Chem. Scand. 1963 (17) 173.

4510-006

CHOLIC ACID, beta-lagodeoxy- — $C_{24}H_{40}O_4$ — 392 — $+42^0$e — 215^0C

J. Biol. Chem. 1963 (238) 3846.

4510-002

CHOLIC ACID, tauro-cheno-desoxy- — $C_{26}H_{45}NO_6S$ — 499

R = . NH. CH_2CH_2. SO_3H

Acta Chem. Scand., 1963 (17) 173.

4510-007

CHOLIC ACID, desoxy- — $C_{24}H_{40}O_4$ — 392 — $+53^0$e — 176^0C

R = OH

Acta Chem. Scand. 1953 (7) 481.

4510-003

CHOLIC ACID, hyo-desoxy- — $C_{24}H_{40}O_4$ — 392 — $+8^0$e — $198\text{-}9^0$C

R = OH

JACS. 1947 1995.

4510-008

CHOLIC ACID, glyco-desoxy- — $C_{26}H_{43}O_5N$ — 349 — $+46^0$e — 188^0C

R = -NH-COOH

J. Biol. Chem. 1952 (195) 263.

4510-004

CHOLIC ACID, glyco-hyo-desoxy- — $C_{26}H_{43}O_5N$ — 349

R = -NH-COOH

Arkiv. Kemi. 1955 (8) 331.

4510-009

CHOLIC ACID, chenodesoxy- — $C_{24}H_{40}O_4$ — 392 — $+11^0$e — 143^0C

R = OH

Acta Chem. Scand. 1963 (17) 173.

4510-005

CHOLIC ACID, β-hyo-desoxy- — $C_{24}H_{40}O_4$ — 392 — $+5^0$e — 190^0C

Z. Physiol. 1937 (248) 280.

4510-010

CHOLIC ACID, ursodesoxy- $C_{24}H_{40}O_4$ 392 +57⁰e 198-200⁰C

R= OH

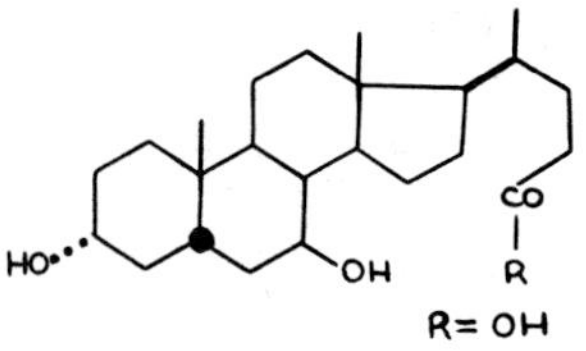

Z. Physiol. 1937 (250) 31.

4510-011

CHOLIC ACID, glyco-urso-desoxy- $C_{26}H_{43}O_5N$ 449 +50⁰e 228⁰C

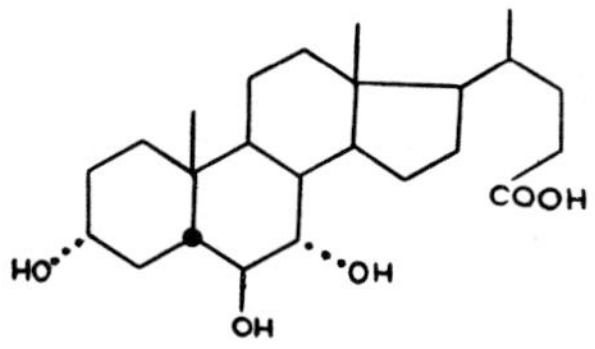

R = . NH. CH_2. COOH

Chem. Abstr. 1955 (49) 14785.

4510-012

CHOLANOIC ACID, 3α-OH-6-oxo-5α-H- $C_{24}H_{38}O_4$ 390 194⁰C

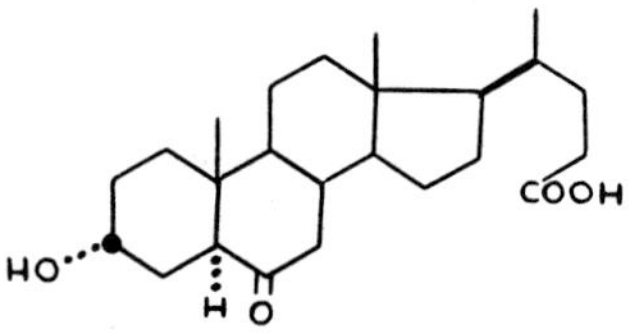

Chem. Abstr. 1957 (51) 14769.

4510-013

CHOLIC ACID, hyo- $C_{24}H_{40}O_5$ 408 +5⁰e 188⁰C

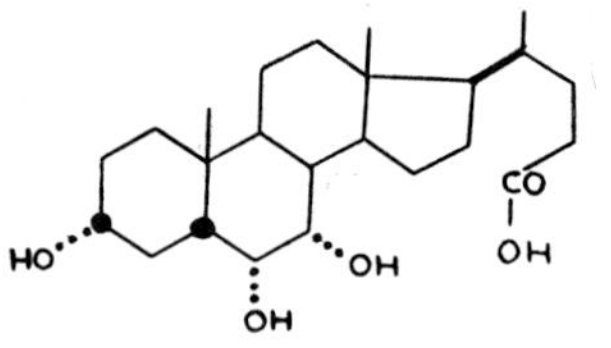

Can. J. C. 1956 (34) 523.

4510-014

CHOLIC ACID I, hyo- $C_{24}H_{40}O_5$ 408 +63⁰e 226⁰C

J. Biol. Chem. 1958 (233) 1337.

4510-015

CHOLIC ACID II. hyo- $C_{24}H_{40}O_5$ 408 +38⁰e 200⁰C

J. Biol. Chem. 1958 (233) 1337.

4510-016

CHOLIC ACID $C_{24}H_{40}O_5$ 408 +37⁰e 197⁰C

R= OH

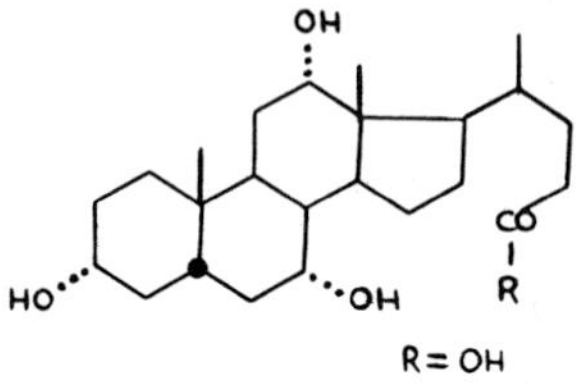

JACS. 1953 6306.

4510-017

CHOLIC ACID, glyco- $C_{26}H_{43}O_6N$ 465 +24⁰w 154⁰C

R = NH COOH

Acta Chem. Scand. 1953 (7) 1126.

4510-018

CHOLIC ACID. Tauro- $C_{26}H_{45}O_7NS$ 515 +39⁰e

R = . NH . CH_2CH_2 . SO_3H

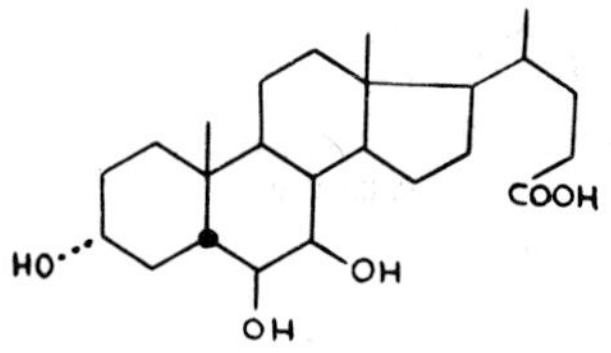

Z. Physiol. 1934 (224) 160.

4510-019

CHOLIC ACID, β-Phocae- $C_{24}H_{40}O_5$ 408 223⁰C

R= OH

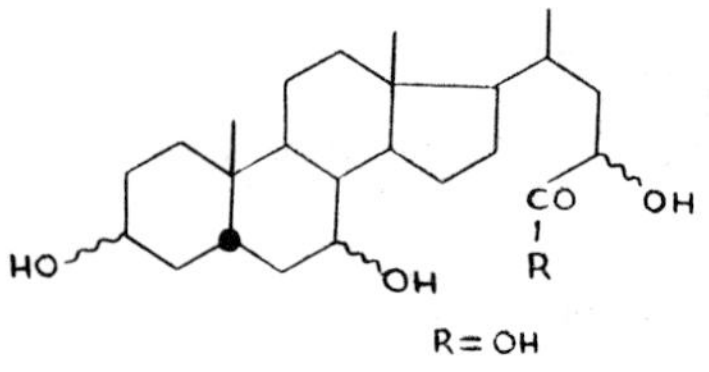

Z. Physiol. 1928 (173) 312.

4510-020

CHOLIC ACID. tauro-β-phocae- $C_{26}H_{45}O_7NS$ 515

R = . NH. CH_2CH_2. SO_3H

Z. Physiol. 1928 (173) 312.

4510-021

CHOLIC ACID. Pytho- $C_{24}H_{40}O_5$ 408

187°C

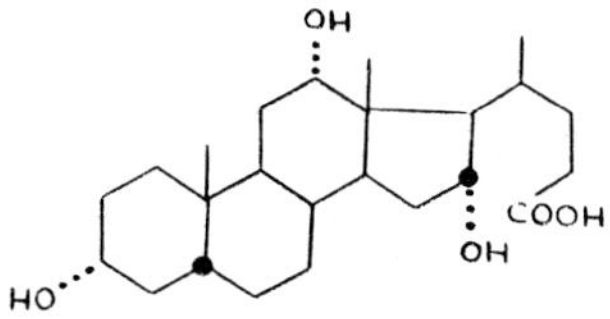

JCS. 1954 1979.

4510-022

FUNTUMINE $C_{21}H_{35}NO$ 317 $+95^0$ 126^0C

Compt. Rend., 1959 (248) 982.

4511-001

FUNTUMIDINE $C_{21}H_{37}NO$ 319 $+10^0$ 182^0C

Compt. Rend., 1959 (248) 982.

4511-002

HOLAMINE $C_{21}H_{33}NO$ 315 $+23^0c$

Bull. Soc. Chim. Fr., 1962, 646.

4511-003

HOLAPHYLLAMINE $C_{21}H_{33}NO$ 315 $+33^0m$ 260^0C

Compt. Rend., 1960 (251) 559.

4511-004

HOLAPHYLLINE $C_{22}H_{35}NO$ 329 $+17^0c$ 128^0C

Bull. Soc. Chim. Fr., 1959, 896.

4511-005

HOLAPHYLLIDINE $C_{22}H_{37}NO$ 331

Compt. Rend., 1964 (259) 3401.

4511-006

HOLADYSAMINE $C_{22}H_{35}ON$ 339 -78^0c 173^0C

Bull. Soc. Chim. Fr. 1966 1212.

4511-007

HOLADYSINE $C_{22}H_{35}NO$ 339 -199^0c 120^0C

Bull. Soc. Chim. Fr. 1966 1212.

4511-008

KURCHILINE $C_{23}H_{37}NO_2$ 369 $+46^0c$ 219^0C

Bull. Soc. Chim. Fr. 1964 2158.

4511-009

KURCHIPHYLLAMINE $C_{22}H_{35}NO_2$ 345 -211^0c 161^0C

Bull. Soc. Chim. Fr. 1966 1212.

4511-010

KURCHIPHYLLINE $C_{23}H_{37}NO_2$ 359, -173°c, 184°C Bull. Soc. Chim. Fr. 1966 1212. 4511-011	PARAVALLARINE, 7β-OH- $C_{22}H_{33}NO_3$ 359 Chem. Abstr. 1967 (67) 108845. 4511-016
KURCHALINE $C_{23}H_{37}NO_2$ 359, -37°c, 185°C Bull. Soc. Chim. Fr. 1966 1212. 4511-012	PARAVALLARINE, 11α-OH- $C_{22}H_{33}NO_3$ 359, -55°, 235°C Chem. Abstr. 1967 (67) 108845. 4511-017
GITINGENSINE $C_{22}H_{31}NO_2$ 341, -65°, 161°C JOC. 1967 2642. 4511-013	PARAVALLARIDINE $C_{22}H_{33}NO_3$ 359, -48°c, 231°C Bull. Soc. Chim. Fr. 1963 597. 4511-018
PARAVALLARINE $C_{22}H_{32}NO_2$ 343, -52°c, 181°C Chem. Abstr. 1967 (67) 116997. 4511-014	KIBATALINE $C_{23}H_{35}NO_2$ 357, -42°c, 171°C Bull. Soc. Chim. Fr. 1964, 2415. 4511-019
PARAVALLARINE, 7α-OH- $C_{22}H_{33}NO_3$ 359 Chem. Abstr. 1967 (67) 108845. 4511-015	PARAVALLARINE, 20-epi-N-Me- $C_{23}H_{35}NO_2$ 357, -33°c, 191°C Bull. Soc. Chim. Fr. 1965 2502. 4511-020

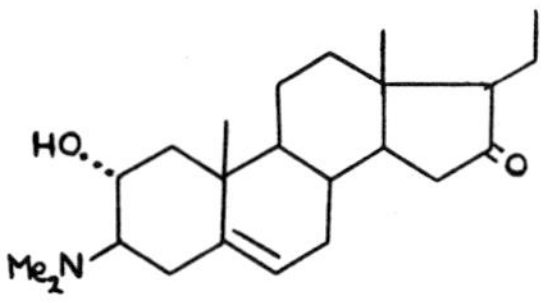

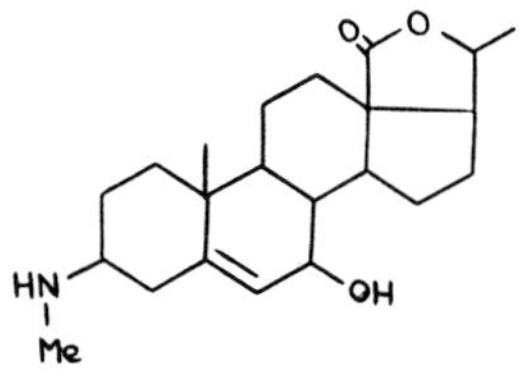

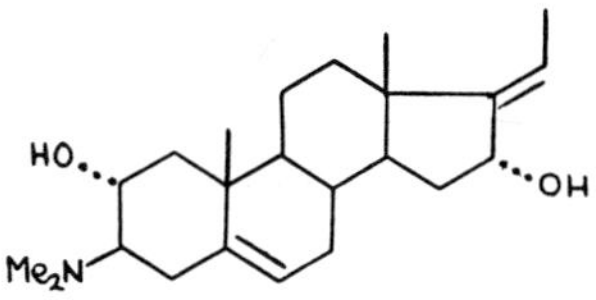

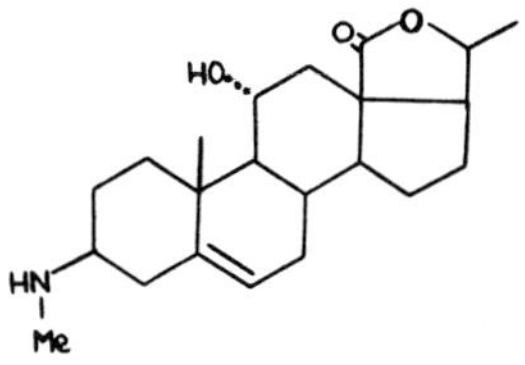

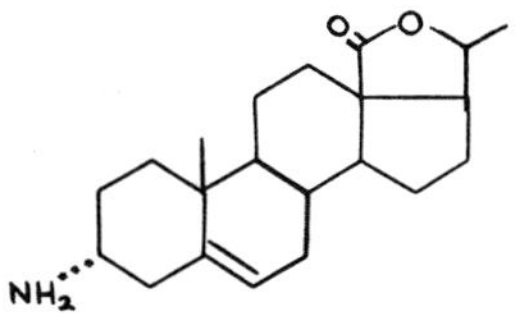

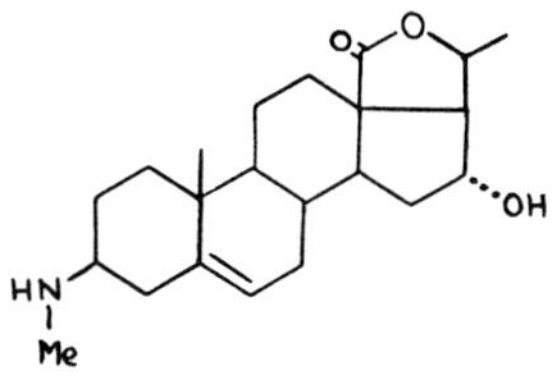

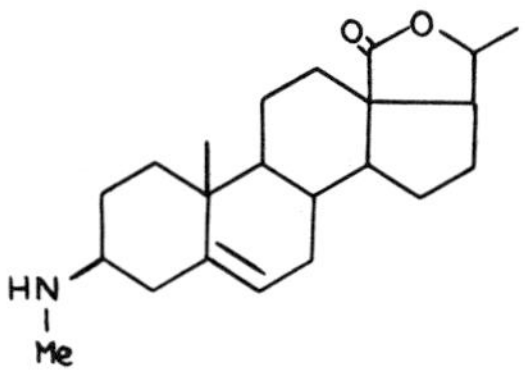

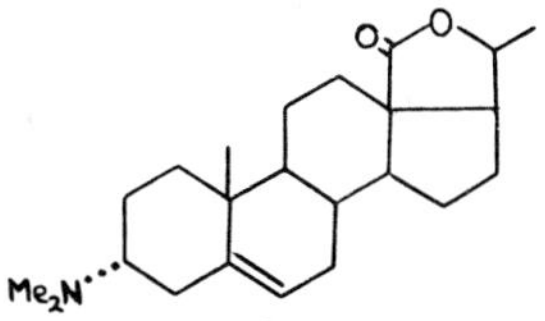

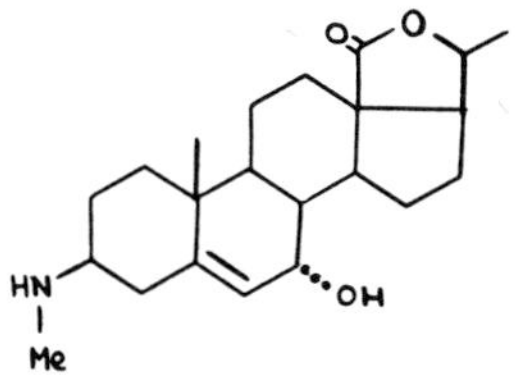

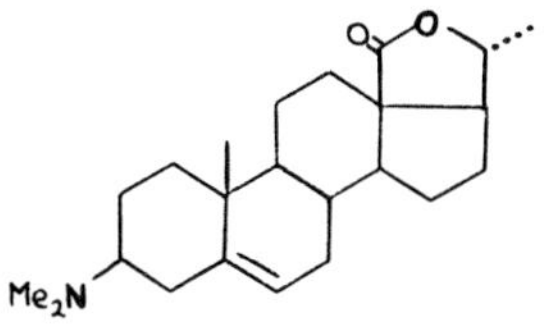

FUNTUPHYLLAMINE A — $C_{21}H_{37}NO$ — 319; $+13^0$; 173^0C

Compt. Rend. 1960 (250) 2445.

4512-001

FUNTUPHYLLAMINE B — $C_{22}H_{39}NO$ — 333; $+24^0c$; 214^0C

Compt. Rend. 1960 (250) 2445.

4512-002

FUNTUPHYLLAMINE C — $C_{23}H_{41}NO$ — 347; $+24^0$; 172^0C

Compt. Rend. 1960 (250) 2445.

4512-003

FUNTUMAFRINE B — $C_{22}H_{37}ON$ — 331; $+43^0$; 160^0C

Compt. Rend. 1960 (250) 2445.

4512-004

FUNTUMAFRINE C — $C_{23}H_{39}ON$ — 345; $+45^0$; 174^0C

Bull. Soc. Chim. Fr. 1962 111.

4512-005

HOLAFEBRINE — $C_{21}H_{35}NO$ — 317; -61^0c; 177^0C

Bull. Soc. Chim. Fr. 1962 285.

4512-006

CONOPHARYNGINE — $C_{27}H_{45}NO_6$ — 479; 288^0C

JACS. 1960 5688.

4512-007

IREHAMINE — $C_{22}H_{37}NO$ — 331; -33^0c; 230^0C

Bull. Soc. Chim. Fr. 1963 2332.

4512-008

IREHINE — $C_{23}H_{39}NO$ — 345; -46^0; 174^0C

Coll. Czech. Chem. Comm. 1965 (30) 348.

4512-009

TERMINALINE — $C_{23}H_{41}O_2N$ — 363; $+29^0$; 234^0C

Chem. Pharm. Bull. 1967 316.

4512-010

DICTYODIAMINE $C_{23}H_{42}N_2$ 346 R = H MeNH; N(Me)R Bull. Soc. Chim. Fr., 1965, 3035. 4513-001	EPIPACHYSAMINE-C $C_{23}H_{42}N_2$ 346 -16^{0}c 242^{0}C Me_2N; NH_2 Chem. Pharm. Bull. 1967 307. 4513-006
DICTYOPHLEBINE $C_{24}H_{44}N_2$ 360 R = Me Bull. Soc. Chim. Fr. 1965 3035. 4513-002	EPIPACHYSAMINE-B $C_{29}H_{45}N_3O$ 451 +38^{0}c 260-3^{0}C R = nicotinoyl R-NH; NMe_2 Chem. Pharm. Bull. 1967 307. 4513-007
PACHYSAMINE-A $C_{24}H_{44}N_2$ 360 +20^{0}c 167^{0}C R = H RN(Me); NMe_2 Chem. Pharm. Bull. 1967 302. 4513-003	EPIPACHYSAMINE-D $C_{30}H_{46}N_2O$ 450 +13^{0}c 248^{0}C R = benzoyl Chem. Pharm. Bull. 1967 307. 4513-008
EPIPACHYSAMINE-A $C_{26}H_{46}N_2O$ 402 -17^{0}c 202^{0}C Me_2N; N(Ac)Me Chem. Pharm. Bull. 1967 307. 4513-004	EPIPACHYSAMINE-E $C_{28}H_{48}N_2O$ 420 +19^{0}c 200-5^{0}C R = senecioyl Chem. Pharm. Bull. 1967 307 4513-009
PACHYSAMINE B $C_{29}H_{50}ON_2$ 442 +67^{0}c 174^{0}C R = senecioyl Chem. Pharm. Bull. 1967 302. 4513-005	EPIPACHYSAMINE-F $C_{23}H_{42}N_2$ 346 +6^{0}c 252^{0}C Me_2N; NH_2 Chem. Pharm. Bull. 1967 307. 4513-010

PACHYSANDRINE A $C_{33}H_{50}O_3N_2$ 522 +80°c 235°C R = benzoyl NMe₂ RN Me OAc Chem. Pharm. Bull. 1967 193. 4513-011	PACHYSANTERMINE A $C_{29}H_{48}O_2N_2$ 456 +43°c 260-3°C NMe₂ HN O O Chem. Pharm. Bull. 1967 571. 4513-016
EPIPACHYSANDRINE-A $C_{30}H_{46}O_2N_2$ 466 +12° 291°C NMe₂ Ph O NH OH Chem. Pharm. Bull. 1967 577. 4513-012	PACHYSTERMINE A $C_{29}H_{48}O_2N_2$ 456 +24° 222°C R = oxo NMe₂ N O R Chem. Pharm. Bull. 1967 449. 4513-017
PACHYSANDRINE B $C_{31}H_{52}O_3N_2$ 500 +93°c 187°C R = senecioyl Chem. Pharm. Bull. 1967 207. 4513-013	PACHYSTERMINE B $C_{29}H_{50}O_2N_2$ 458 -16°c 257° R = β-OH Chem. Pharm. Bull. 1967 549. 4513-018
PACHYSANDRINE C $C_{24}H_{44}ON_2$ 376 -38°c 214°C R = H NMe₂ MeNH OR Chem. Pharm. Bull. 1967 207. 4513-014	PREGNANE, 3α-NMe₂-20α-NMeAc-5α- $C_{26}H_{46}N_2O$ 402 -14°c 245°C N Me Ac Me₂N Tet. 1967 3829. 4513-019
PACHYSANDRINE D $C_{29}H_{50}O_2N_2$ 458 +2°c 185°C R = senecioyl Chem. Pharm. Bull. 1967 207. 4513-015	IREHDIAMINE-A $C_{21}H_{38}N_2$ 316 -47° 148°C R = H NH₂ RNH Bull. Soc. Chim. Fr. 1963 594. 4513-020

IREHDIAMINE-B | $C_{22}H_{38}N_2$ | 330 | -56^0c | 116^0C

R = CH_3

Bull. Soc. Chim. 1963, 594.

4513-021

KURCHESSINE | $C_{25}H_{44}N_2$ | 372 | -37^0c | 140^0C

Tet. Lett. 1965 67.

4513-022

SARACOCINE | $C_{26}H_{44}N_2O$ | 400

Tet. Lett. 1965 67.

4513-023

SARACOCINE, 3-epi- | $C_{26}H_{44}N_2O$ | 400 | -56^0c | 233^0C

3-epimer

Tet. 1967 3829.

4513-024

SARACODINE | $C_{26}H_{46}N_{20}$ | 402 | 191^0C

* dihydro

Tet. Lett. 1965 67.

4513-025

CHONEMORPHINE, N-senecioyl- | $C_{27}H_{48}N_2O$ | 416 | $+46^0c$ | 149^0C

R = senecioyl

Chem. Ind. 1966 769.

4513-026

CHONEMORPHINE | $C_{23}H_{42}N_2$ | 346 | $+24^0c$ | 144^0C

R = H

Chem. Abstr. 67 117068.

4513-027

MALOUPHYLLINE | $C_{25}H_{42}O_2N_2$ | 402 | $+32^0c$ | 280^0C

Bull. Soc. Chim. 648 1962.

4513-028

MALOUETINE | $[C_{27}H_{52}N_2]^{++}$ | 404 | 264^0C

Ann. Pharm. Fr. 18 673 1960.

4513-029

HOLARHIMINE | $C_{21}H_{36}ON_2$ | 332 | -14.2^0 | 183^0

Coll. Czech. Chem. Comm. 1953 1115.

4513-030

HOLARHIDINE $C_{21}H_{36}ON_2$ 332 -23^{0}c 180-1^0 3-epimer Coll. Czech. Chem. Comm. 1959 24 370 4513-031	

CONAN-5-ENE. 3α-amino-

$C_{22}H_{36}N_2$ 328 $+11^0$e 94^0C

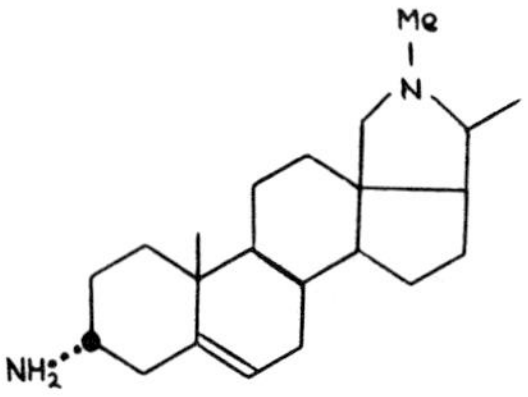

Coll. Czech. Chem. Comm. 1964 1591.

4514-001

CONIMINE

$C_{22}H_{36}N_2$ 328 -30^0 130^0C

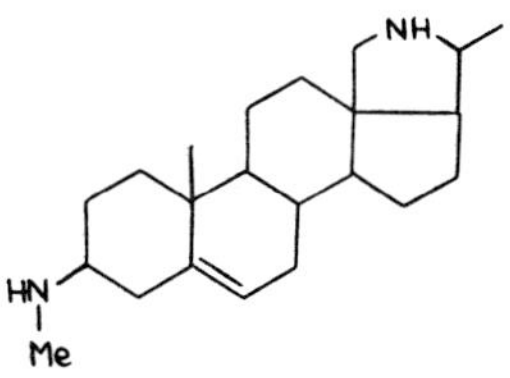

Chem. Abstr. 1968 (68) 87462.

4514-002

CONAMINE

$C_{22}H_{36}N_2$ 328 $+16^0$ 98^0C

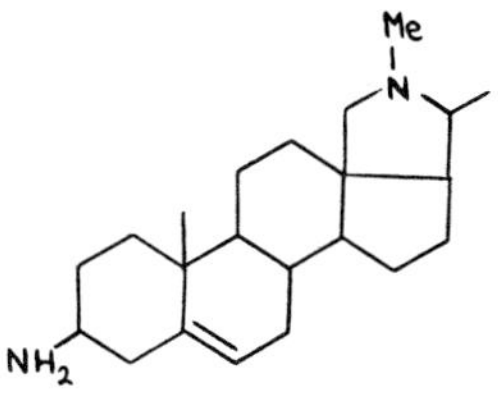

Bull. Soc. Chim. Fr. 1963 1861.

4514-003

CONESSIMINE

$C_{23}H_{38}N_2$ 342 -22^0c 100^0C

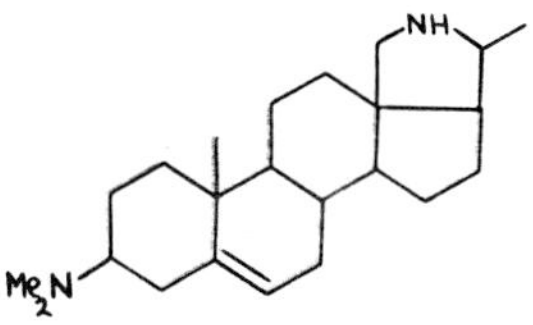

Coll. Czech. Chem. Comm. 1964 1591.

4514-004

CONESSIMINE, iso-

$C_{23}H_{38}N_2$ 342 -22^0 92^0C

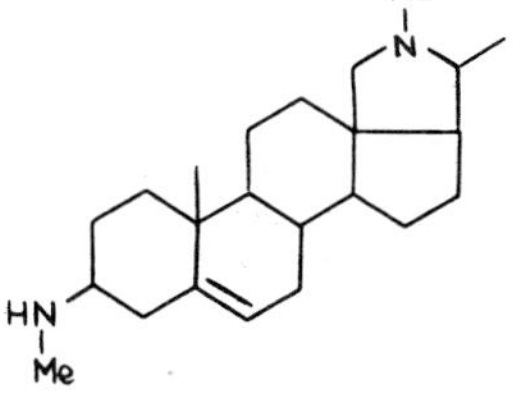

Chem. Abstr. 1968 (68) 87462.

4514-005

CONESSIMINE. dihydro-iso-

$C_{23}H_{40}N_2$ 344 98^0C

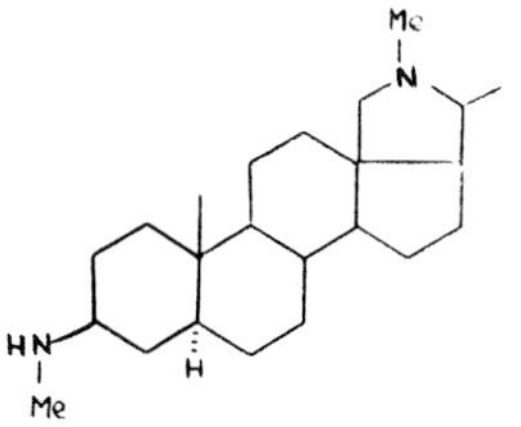

Coll. Czech. Chem. Comm. 1964 1591.

4514-006

MALOUPHYLLAMINE

$C_{24}H_{40}N_2O$ 372 $+42^0$c 220^0C

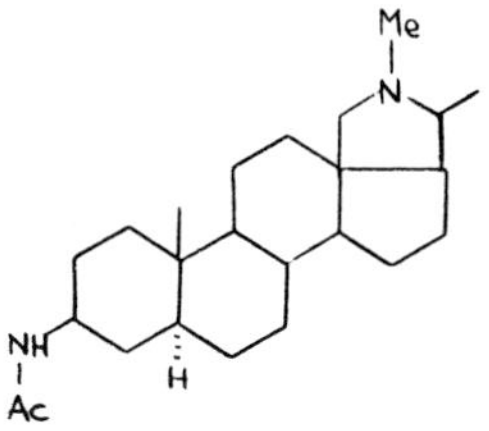

Bull. Soc. Chim. Fr. 1963 641.

4514-007

CONKURESSINE

$C_{24}H_{40}N_2$ 356 $+7^0$c 87^0C

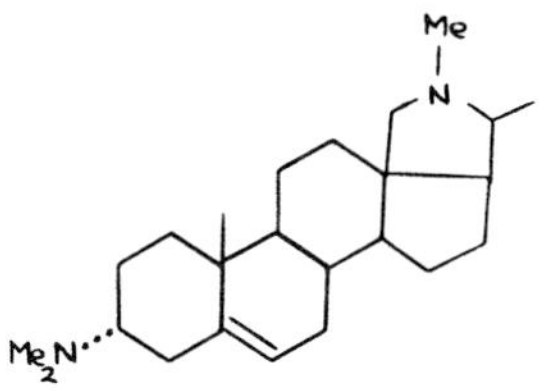

Coll. Czech. Chem. Comm. 1963 2345.

4514-008

CONKURESSINE. dihydro-

$C_{24}H_{42}N_2$ 358 $+48^0$c 94^0C

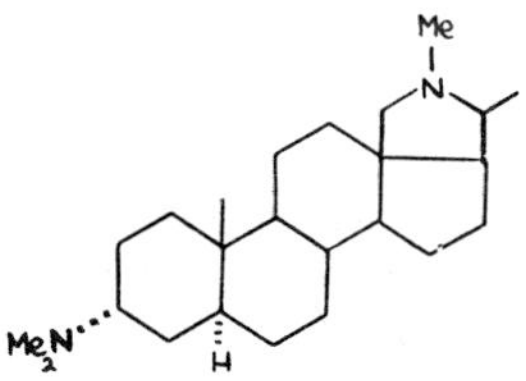

Coll. Czech. Chem. Comm. 1963 2345.

4514-009

CONESSINE

$C_{24}H_{40}N_2$ 356 $+22^0$c 126^0C

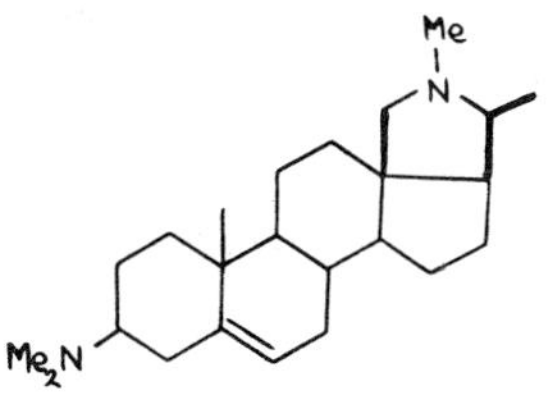

Chem. Ind. 1957 1296.

4514-010

Compound	Formula	Data	Reference	No.
CONESSINE, dihydro-	$C_{24}H_{42}N_2$	358	Coll. Czech. Chem. Comm. 1964 1591.	4514-011
LATIFOLINE, NOR-	$C_{21}H_{33}ON$	315 -26^0c 192^0	Bull. Soc. Chim. 2486 1963.	4514-016
CONESSINE, 7-OH-	$C_{24}H_{40}N_2O$	372 -43^0e 172^0C	Ber. 1964 (97) 2316.	4514-012
HOLALINE	$C_{24}H_{42}N_2O$	374 $+36^0c$ 267^0C	Bull. Soc. Chim. Fr. 1967 4315.	4514-017
CONKURCHINE	$C_{21}H_{32}N_2$	312 -67^0e 153^0C	Coll. Czech. Chem. Comm. 1965 1016.	4514-013
HOLARRHENINE	$C_{24}H_{40}N_2O$	372 -7^0c 198^0C	Tet. 1959 (7) 104.	4514-018
CONESSIDINE	$C_{22}H_{34}N_2$	326 -52^0c 123^0C	Coll. Czech. Chem. Comm. 1965 1016.	4514-014
HOLARRHELINE	$C_{23}H_{38}ON_2$	358 $+18^0c$ 190^0C	Compt. Rend. 1965 (260) 6631.	4514-019
LATIFOLINE	$C_{22}H_{35}ON$	329 -4^0c 129^0C	Tet. Lett. 869 1963.	4514-015
FUNTESSINE	$C_{22}H_{38}N_2O$	346 $+49^0c$ 195^0C	Bull. Soc. Chim. Fr. 1965 1831.	4514-020

HOLAFRINE — $C_{29}H_{46}N_2O_2$ — 454; -19°c; 116°C

Helv. 1958 (41) 11.

4514-021

HOLARRHETINE — $C_{30}H_{48}N_2O_2$ — 468; -15°c; 75°C

Helv. 1958 (41) 11.

4514-022

FUNTULINE — $C_{22}H_{35}NO_2$ — 345; -6°c; 235°C

Bull. Soc. Chim. Fr. 1964 787.

4514-023

KURCHOLESSINE — $C_{25}H_{44}N_2O_2$ — 404; +5°e; 220°C

Tet. Lett. 1964 1659.

4514-024

HOLONAMINE — $C_{21}H_{27}NO_2$ — 325; -15°e; 258°C

Ber. 1964 (97) 2326.

4514-025

FUNTUDIENINE — $C_{22}H_{31}NO$ — 325; -228°c

Bull. Soc. Chim. Fr. 1964 787.

4514-026

MALARBORINE — $C_{21}H_{31}NO$ — 313; 165°C

Tet. Lett. 1967 1437.

4514-027

MALARBOREINE — $C_{21}H_{27}NO$ — 309; +57°e; 174°C

Tet. Lett. 1967 1437.

4514-028

HOLADIENINE — $C_{22}H_{31}ON$ — 325; +80°c; 109°C

Compt. Rendu. 1965 (260) 6631.

4514-029

LATIFOLININE — $C_{22}H_{33}ON$ — 327; 110°C

Bull. Soc. Chim. 2486 1963.

4514-030

MAINGAYINE

$C_{21}H_{27}NO$ 309
+2°c
151°C

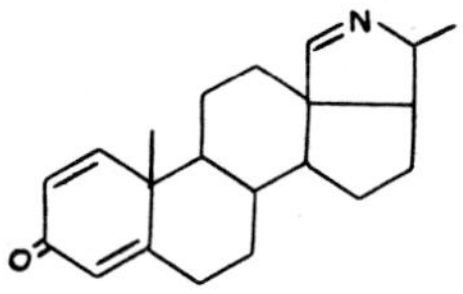

Chem. Ind., 1970, 627.

4514-031

SOLACONGESTIDINE $C_{27}H_{45}NO$ 399 +25°c 167°C JOC. 1969 1577. 4515-001	BAIKEINE $C_{27}H_{45}NO_3$ 431 "Alkaloids," Ed. R. Manske, Vol. X, p. 30. 4515-006
VERAZINE $C_{27}H_{43}NO$ 397 -90°e 177°C Tet. 1967 167. 4515-002	PETILINE $C_{27}H_{43}NO_2$ 413 -52°m 205°C JOC. 1970 2422. 4515-007
SOLAFLORIDINE $C_{27}H_{45}NO_2$ 415 +115°c 169°C JOC. 1969 1577. 4515-003	KORSEVININE $C_{27}H_{41}NO_3$ 427 -16° 224°C Khim. Prir. Soedin. 1969 600. 4515-008
TOMATILLIDINE, 5,6-dihydro- $C_{27}H_{43}NO_2$ 413 +21°c 180°C JOC. 1965 754. 4515-004	SOLAPHYLLIDINE $C_{29}H_{47}NO_5$ 489 -25°m 165-70°C JACS. 1970 700. 4515-009
TOMATILLIDINE $C_{27}H_{41}NO_2$ 411 -18°c 220°C JOC. 1965 754. 4515-005	SOLADULCIDINE $C_{27}H_{45}O_2N$ 415 -30°p 207°C Planta Med. 1958 (6) 93. 4515-010

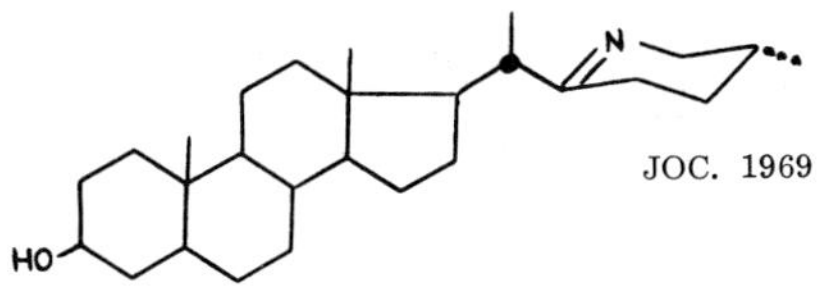

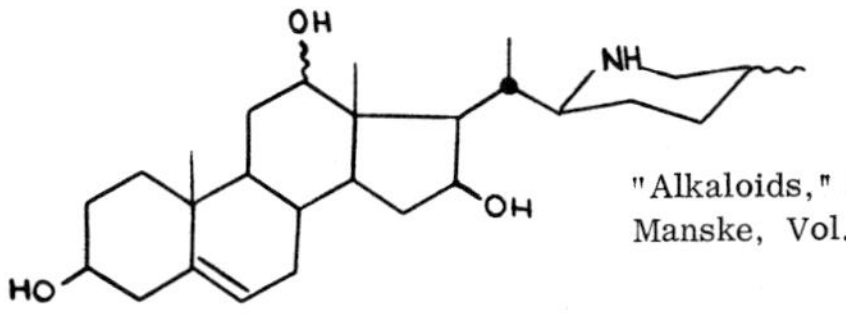

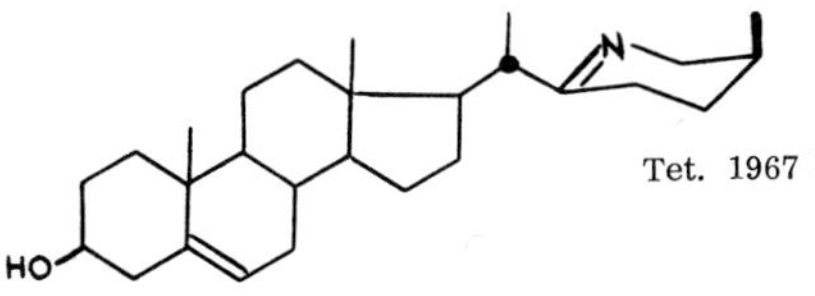

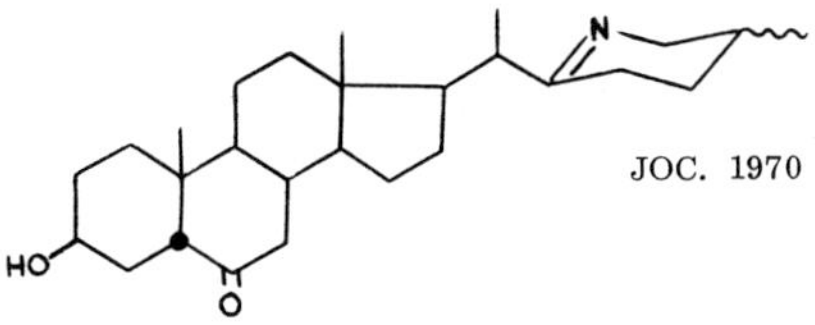

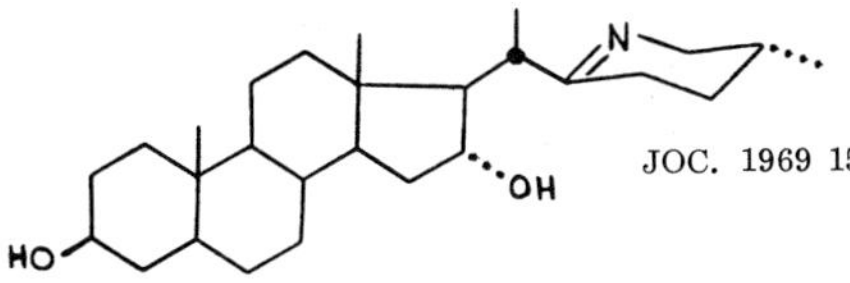

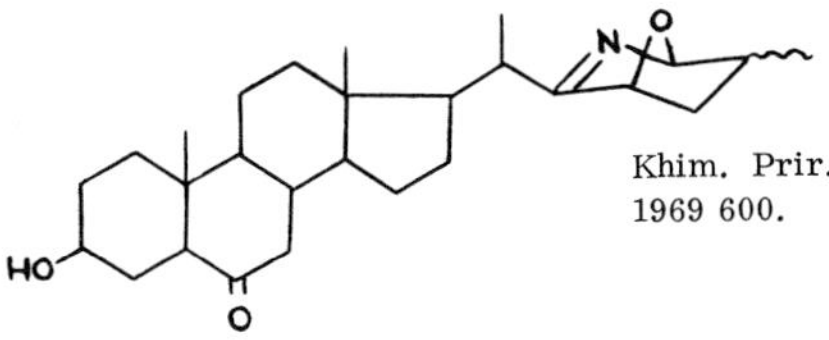

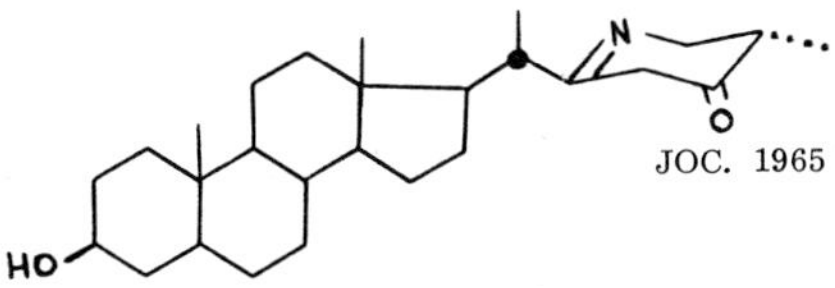

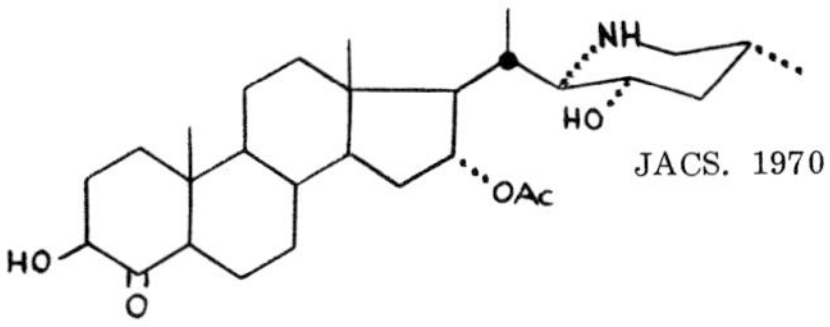

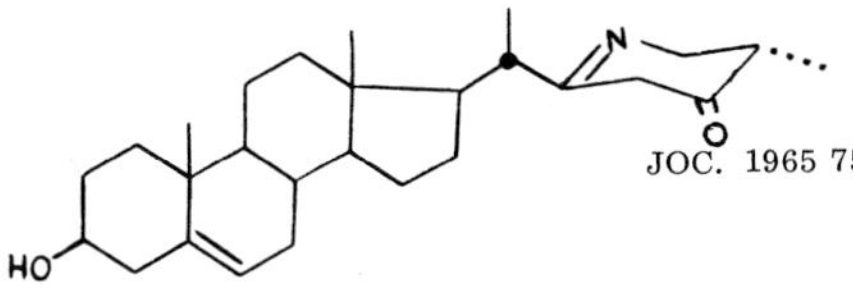

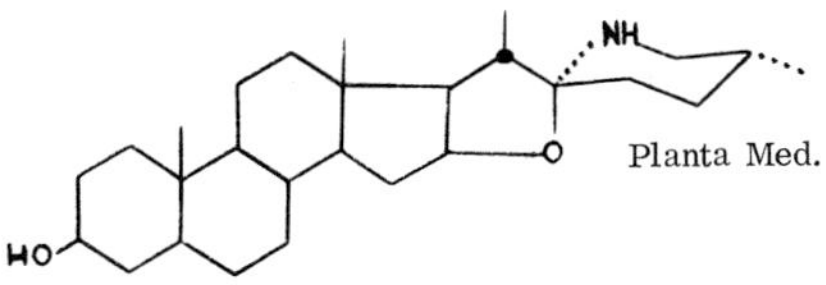

SOLADULCIDINE, 15α-OH- $C_{27}H_{45}NO_3$ 431 -41° 214°C

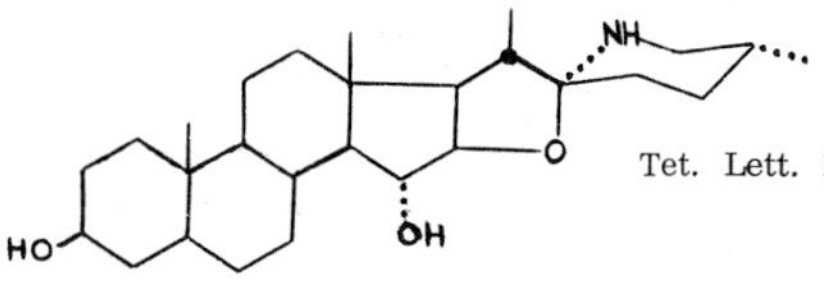

Tet. Lett. 1965 1947.

4515-011

TOMATIDENOL $C_{27}H_{43}O_2N$ 413 -45°c 206°C

R = H

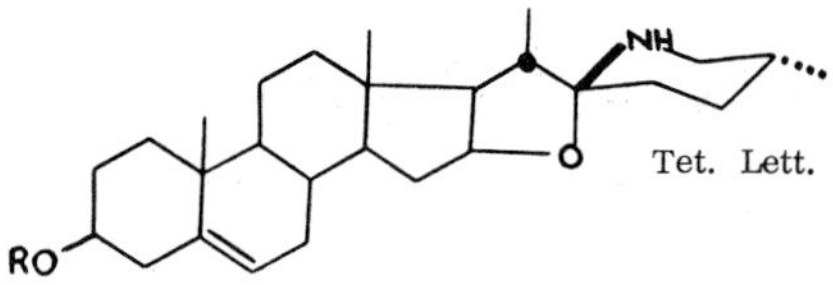

Tet. Lett. 1963 329.

4515-016

SOLADULCIDINE, 15β-OH- $C_{27}H_{45}NO_3$ 431 -66°c 1 205°C

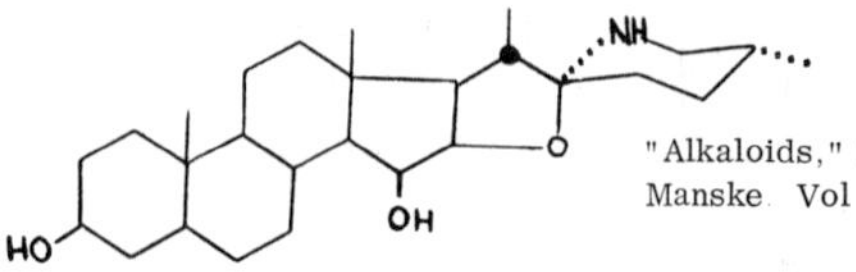

"Alkaloids," Ed. R. Manske Vol X, p. 30.

4515-012

SOLANGUSTINE $C_{33}H_{53}NO_7$ 575 235°C

R = Glucosyl

JCS. 1914 568.

4515-017

TOMATIDINE $C_{27}H_{45}O_2N$ 415 +5°m 206°C

R = H

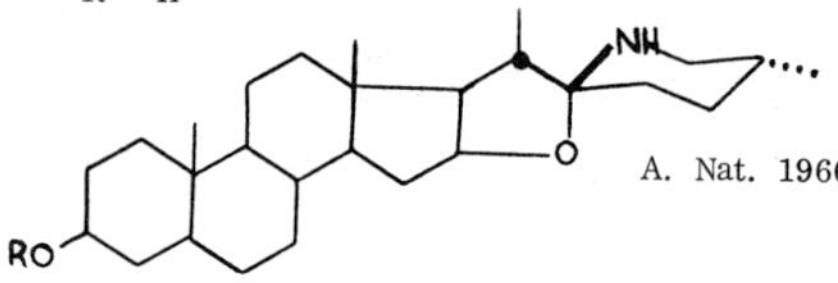

A. Nat. 1966 (21) 893.

4515-013

TOMATIDENOL, 15α-OH- $C_{27}H_{43}NO_3$ 429 -24° 238°C

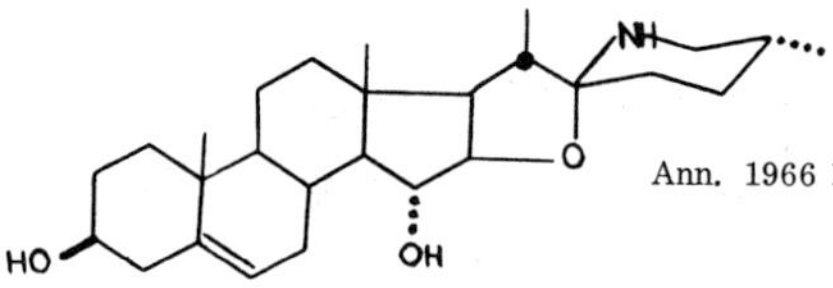

Ann. 1966 169.

4515-018

TOMATINE $C_{50}H_{83}NO_{21}$ 1033 -19°p 270°C

R = Lycotetraosyl

Ber. 1950 (83) 448.

4515-014

SOLASODINE $C_{27}H_{43}O_2N$ 413 -113°c 200°C

R = H

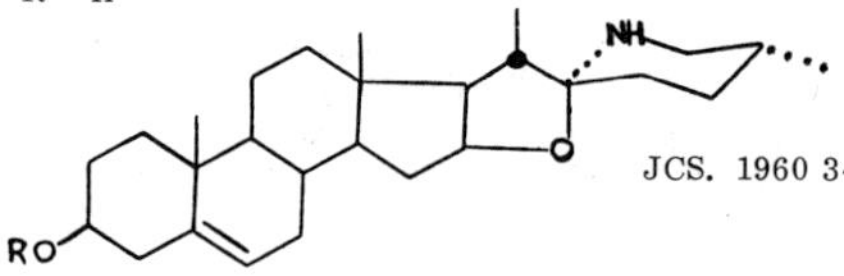

JCS. 1960 3417.

4515-019

TOMATIDINE, 15α-OH- $C_{27}H_{45}NO_3$ 431 +20°c 150-5°C

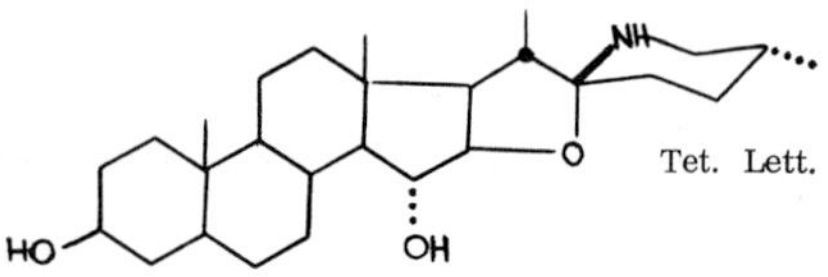

Tet. Lett. 1965 1947.

4515-015

SOLAMARGINE, γ- $C_{33}H_{53}NO_7$ 575

R = Glucosyl

JCS. 1952 3587.

4515-020

SOLAMARGINE, β- $C_{39}H_{63}NO_{11}$ 721 R = Glu-Rhamnosyl- Planta Med. 1964 (12) 541. 4515-021	SOLASODINE, 15β-OH- $C_{27}H_{43}NO_3$ 429 -114°c 214°C HO, NH, O, OH "Alkaloids," Ed. R. Manske, Vol. X. p. 30. 4515-025
SOLAMARGINE, α- $C_{45}H_{73}NO_{15}$ 867 -105°m 296°C R = Chacotriosyl- Ber. 1955 (88) 289. 4515-022	SOLACONGESTIDINE, 23-oxo- $C_{27}H_{43}NO_2$ 413 212°C HO, N, O JOC. 1969 1577. 4515-026
SOLASONINE $C_{45}H_{73}NO_{16}$ 883 -72°m 302°C R = Solatriosyl- JCS. 1961 4645. 4515-023	VERALOSIDINE $C_{27}H_{43}NO_2$ 413 R_1 = R_2 = H R_1O, N, OR_2 Khim. prir. Soedin 1970 6 339. 4515-027
SOLASODINE, 15α-OH- $C_{27}H_{43}O_3N$ 429 -85°c 215°C HO, NH, O, OH Ann. 1966 169. 4515-024	VERALOSINE $C_{33}H_{53}NO_7$ 575 R_1 = glucosyl R_2 = acetyl Khim Prir. Soedin 1970 6 343. 4515-028

VERAMARINE — $C_{27}H_{43}NO_3$ — 429 — -85^0c — amorph

Coll. Czech. Chem. Comm. 1968 4429.

4516-001

ZYGADENINE — $C_{27}H_{43}O_7N$ — 493 — -48^0c — 219^0C

R = H

Tet. 1964 913.

4516-002

ZYGACINE — $C_{29}H_{45}O_8N$ — 535

R = Ac

JACS. 1959 1925.

4516-003

ZYGADENINE, angeloyl- — $C_{32}H_{49}NO_8$ — 575

R = Angeloyl

J. Pharm. Soc. Jap. 1959 (79) 619.

4516-004

ZYGADENINE, 3-0-(2-Me-butyroyl)- — $C_{32}H_{51}NO_8$ — 577 — -8^0c — 175^0C

R = α-Me-butyroyl

J. Am. Pharm. Assoc. 1959 (48) 303.

4516-005

ZYGADENINE, Veratroyl- — $C_{36}H_{51}NO_{10}$ — 657

R = Veratroyl

JACS. 1953 1025.

4516-006

ZYGADENINE, Vanilloyl- — $C_{35}H_{49}NO_{10}$ — 643

R = Vanilloyl

JACS. 1953 1025.

4516-007

SABINE — $C_{27}H_{45}NO_7$ — 495 — -33^0e — 175^0C

R = H

Tet. 1964 913.

4516-008

SABADINE — $C_{29}H_{47}NO_8$ — 537

R = Ac

J. Med. Pharm. Chem. 1962 (5) 690.

4516-009

CEVACINE — $C_{29}H_{45}O_9N$ — 551

Coll Czech. Chem. Comm. 1957 98.

4516-010

VERACEVINE — $C_{27}H_{43}O_8N$ — 509, -30^0c, 182^0C

R = H

Tet. Lett. 1960 24.

4516-011

CEVADINE — $C_{32}H_{49}NO_9$ — 591

R = Angeloyl

J. Am. Pharm. Assoc. 1960 (49) 242.

4516-012

VERATRIDINE — $C_{36}H_{51}NO_{11}$ — 673

R = Veratroyl

Coll. Czech. Chem. Comm. 1957 98.

4516-013

VERACEVINE, Vanilloyl- — $C_{35}H_{49}NO_{11}$ — 659

R = Vanilloyl

J. Am Pharm. Assoc. 1956 (45) 252.

4516-014

GERMINE — $C_{27}H_{43}O_8N$ — 509, $+5^0$e, 220^0C

$R_1 = R_2 = R_3 =$ H

JACS. 1959 1913.

4516-015

GERMIDINE, Neo- — $C_{34}H_{53}NO_{10}$ — 635

R_1 = H

R_2 = Ac

R_3 = α-Me-butyroyl

JACS. 1959 1921.

4516-016

GERMIDINE — $C_{34}H_{53}O_{10}N$ — 635, $+13^0$c, 242^0C

R_1 = Ac

R_2 = H

R_3 = α-Me-butyroyl

JACS 1959 1921.

4516-017

GERMITRINE, Neo- — $C_{36}H_{55}NO_{11}$ — 677

R_1 = Ac

R_2 = AC

R_3 = α-Me-Butyroyl

JACS. 1959 1921.

4516-018

GERMBUDINE — $C_{37}H_{59}NO_{12}$ — 709, -7^0, 163^0C

R_1 = (threo)

R_2 = H

R_3 = α-Me-Butyroyl

J. Am. Pharm. Assoc. 1959, 48, 737.

4516-019

GERMBUDINE, Neo- — $C_{37}H_{59}NO_{12}$ — 709

R_1 = (erythro)

R_2 = H

R_3 = α-Me-Butyroyl

JACS. 1956 1621.

4516-020

GERMERINE — $C_{37}H_{59}NO_{11}$ — 693 — +11°c — 194°C

R_1 = α-Me-Butyroyl

R_2 = [structure]

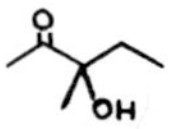

R_3 = H

JACS. 1959 1921.

4516-021

PROTOVERINE — $C_{27}H_{43}O_9N$ — 525 — -12°p — 212°C

R_1 = R_2 = R_3 = R_4 = H

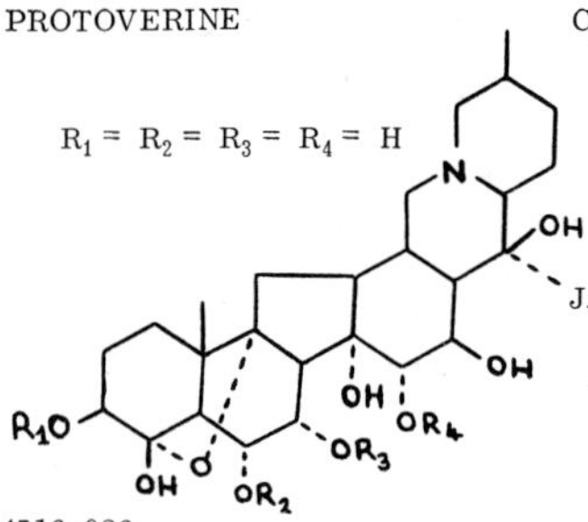

JACS. 1959 1921.

4516-026

GERMITETRINE — $C_{41}H_{63}NO_{14}$ — 793 — -74°p — 230°C

R_1 = [structure] (erythro)

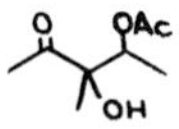

R_2 = Ac

R_3 = α-Me-Butyroyl

J. Am. Pharm. Assoc. 1959 (48) 440.

4516-022

PROTOVERATRINE A — $C_{41}H_{63}NO_{14}$ — 793

R_1 = [structure]

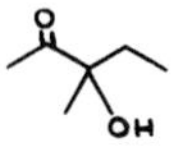

R_2 = Ac

R_3 = Ac

R_4 = α-Me-Butyroyl

JACS. 1960 2252.

4516-027

GERMITRINE — $C_{39}H_{61}NO_{12}$ — 735 — +11° — 198°C

R_1 = α-Me-Butyroyl

R_2 = Ac

R_3 = [structure]

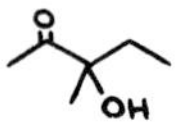

JACS. 1959 (81) 1921.

4516-023

PROTOVERATRINE A, des-Ac- — $C_{39}H_{61}NO_{13}$ — 751

R_1 = [structure]

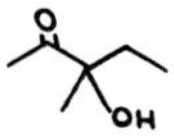

R_2 = Ac

R_3 = H

R_4 = α-Me-Butyroyl

JACS. 1960 2616.

4516-028

GERMANITRINE — $C_{39}H_{59}NO_{11}$ — 717 — -61°p — 228°C

R_1 = Angeloyl

R_2 = Ac

R_3 = α-Me-Betyroyl

J. Am Pharm. Assoc. 1959 (48) 1731.

4516-024

PROTOVERATRINE B — $C_{41}H_{63}NO_{13}$ — 777

R_1 = [structure] (threo)

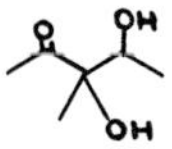

R_2 = Ac

R_3 = Ac

R_4 = α-Me-Butyroyl

JACS. 1960 2252.

4516-029

PROTOVERATRIDINE — $C_{32}H_{51}NO_9$ — 593 — -9°p — 273°C

R_1 = α-Me-Butyroyl

R_2 = R_3 = H

JACS. 1959 1921.

4516-025

PROTOVERATRINE B, des-Ac- — $C_{39}H_{61}NO_{12}$ — 735

R_1 = [structure]

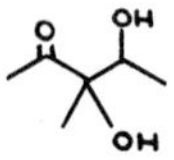

R_2 = Ac

R_3 = H

R_4 = α-Me-Butyroyl

JACS. 1960 2616.

4516-030

ESCHOLERINE $C_{41}H_{61}NO_{13}$ 775

R_1 = Angeloyl

R_2 = Ac

R_3 = Ac

R_4 = α-Me-Butyroyl

J. Am. Pharm. Assoc. 1959 (48) 735.

4516-031

KORSEVERIDINE $C_{27}H_{43}O_2N$ 413
-49°
290-2°C

HO OH N

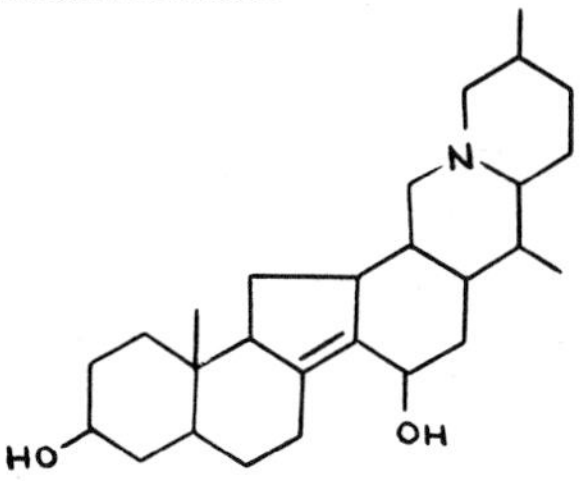

Chem. Abstr. 1968 (69) 77560.

4516-035

VERTICINE $C_{27}H_{45}NO_3$ 431
-20°c
224°C

R = H

N OH RO OH

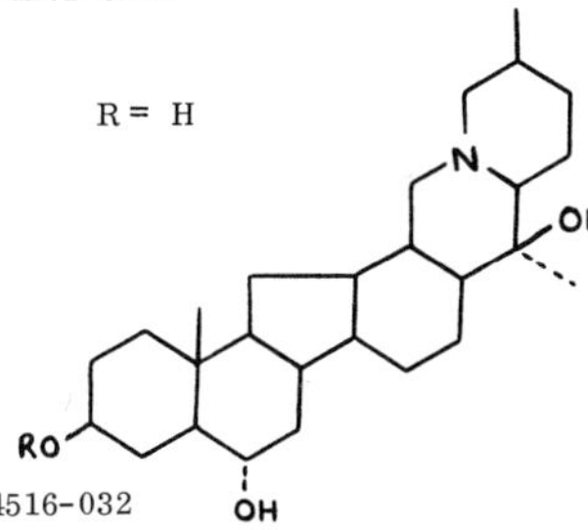

Tet. Lett. 1968 5373.

4516-032

KORSINE $C_{27}H_{43}O_3N$ 429
+88°
302°C

OH N HO HO

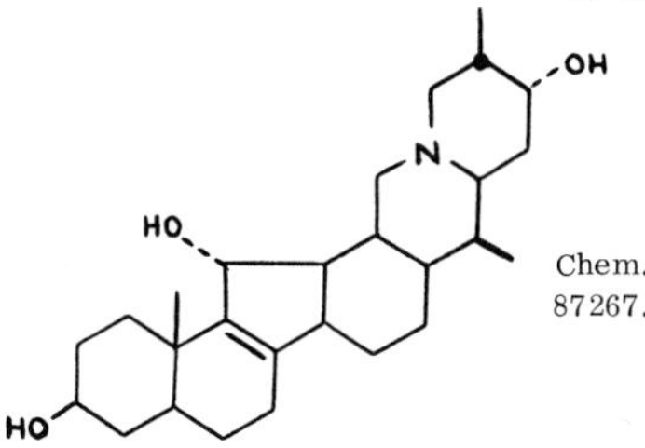

Chem. Abstr. 1968 (69) 87267.

4516-036

PEIMINOSIDE $C_{33}H_{55}NO_8$ 593

R = Glucosyl

Chem. Pharm. Bull. 1960 302.

4516-033

EDUARDINE $C_{27}H_{43}O_2N$ 413
-53°m
250°C

N O OH

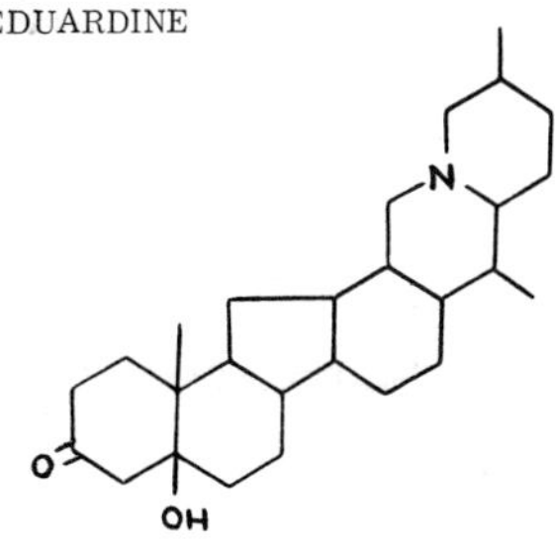

Chem. Abstr. 1968 (68) 69168.

4516-037

VERTICINONE $C_{27}H_{43}O_3N$ 429
-54°c
212°C

N OH HO O

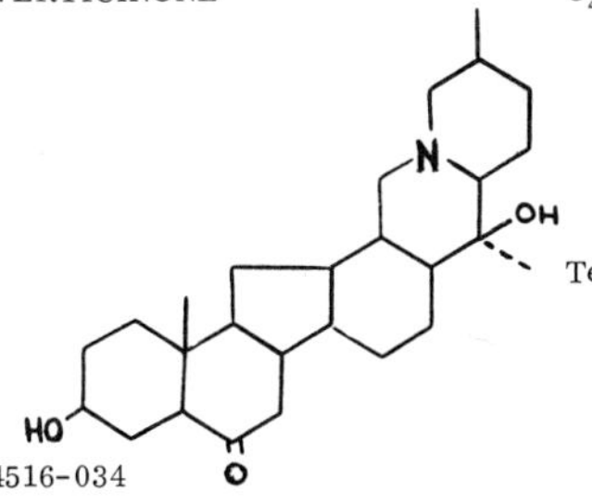

Tet. Lett. 1968 5375.

4516-034

VERATRAMINE | $C_{27}H_{39}O_2N$ | 409 | -70^0m | 206^0C

R = OH

Tet. Lett 1967 2381.

4517-001

VERARINE | $C_{27}H_{39}ON$ | 393 | -68^0e | 175^0C

R = H

Tet. Lett. 1967 2381.

4517-002

JERVINE | $C_{27}H_{39}O_3N$ | 425 | -150^0e | 244^0C

R = oxo

JACS. 1968 2730.

4517-003

JERVINE, 11-deoxo- | $C_{27}H_{41}O_2N$ | 411 | -44^0e | 238^0C

R = H

JACS. 1968 2730.

4517-004

VERATROBASINE | $C_{27}H_{41}O_3N$ | 427 | -77^0e | 285^0C

R = β- OH

JACS. 1968 2730.

4517-005

CORSEVINE | $C_{28}H_{45}O_2N$ | 427 | -84^0m | 170^0C

Chem. Abstr. 1968 68 105400.

4517-006

EDPETINE | $C_{27}H_{41}NO_4$ | 443 | 315^0C

Chem. Abstr. 1970 73 25713.

4517-007

DEMISSIDINE $C_{27}H_{45}ON$ 399 +28°m 220°C R = H 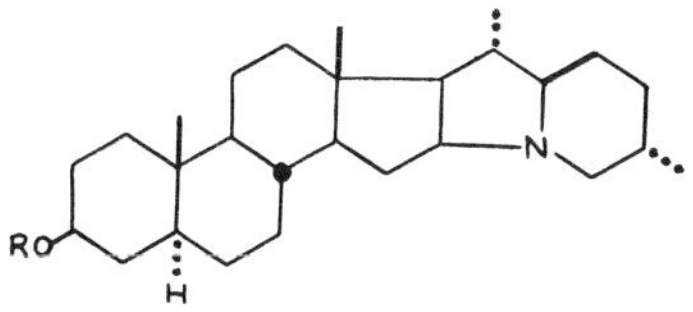 Tet. 1966 673. 4518-001	CHACONINE, β1 $C_{39}H_{63}NO_{10}$ 705 R = 1, 2-Glu-Rhamnosyl Ber. 1955 (88) 1690. 4518-006
DEMISSINE $C_{50}H_{83}NO_{20}$ 1017 -20°p 306°C R = Lycotetraosyl Ber. 1957 (90) 203/217. 4518-002	CHACONINE, β2- $C_{39}H_{63}NO_{10}$ 705 R = 1, 4-Glu-Rhamnosyl Ber. 1955 (88) 1690. 4518-007
SOLANIDINE $C_{27}H_{43}ON$ 397 -27°c 219°C R = H 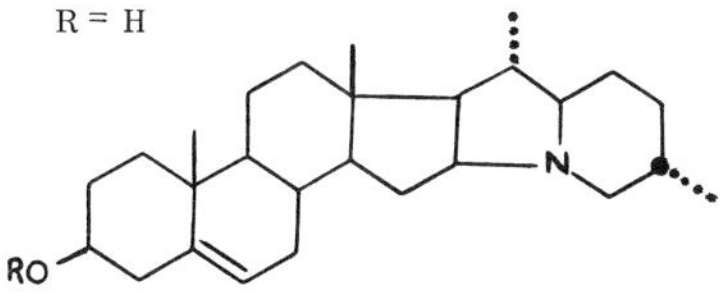 Tet. 1966 673. 4518-003	SOLANINE, β- $C_{39}H_{63}NO_{11}$ 721 -31°m 295°C R = Solabiosyl Ber. 1955 (88) 1492. 4518-008
SOLANINE, $C_{33}H_{53}NO_6$ 559 -26°m 250°C R = Galactosyl Angew. 1955 (67) 127. 4518-004	CHACONINE, α- $C_{45}H_{73}NO_{14}$ 851 -85°p 242°C R = Chacotriosyl Ber. 1955 (88) 1690. 4518-009
CHACONINE, γ- $C_{33}H_{53}NO_6$ 559 -40°p 244°C R = Glucosyl Angew. 1955 (67) 127. 4518-005	RUBIJERVINE $C_{27}H_{43}O_2N$ 413 +19°e 241°C 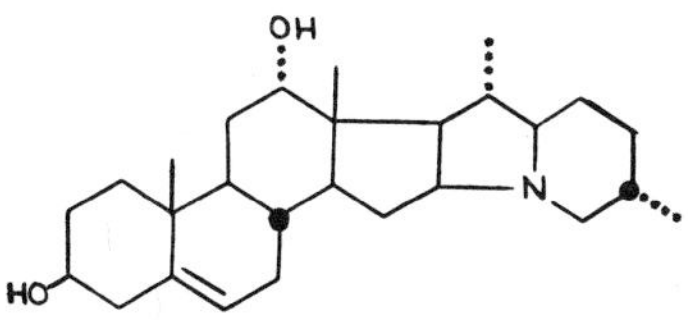 Tet. 1966 673. 4518-010

RUBIJERVINE, Iso- $C_{27}H_{43}O_2N$ 413 +8°e 239°C CH₂OH, RO, N Tet. 1966 673. 4518-011	VERALOBINE $C_{27}H_{41}O_2N$ 411 +67°m 238°C CH₂OH, O, N Tet. 1966 673. 4518-013
RUBIJERVOSINE, iso- $C_{33}H_{53}NO_7$ 575 -20°e 280°C R = Glucosyl JACS. 1953 2133. 4518-012	LEPTINIDINE $C_{27}H_{43}NO_2$ 413 -24°c 248°C OH, HO, N Ber., 1967 (100) 1381. 4518-014

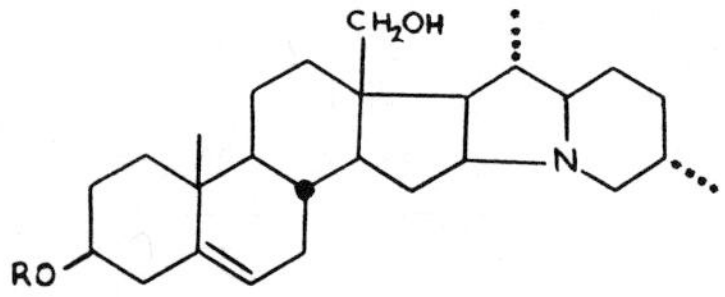

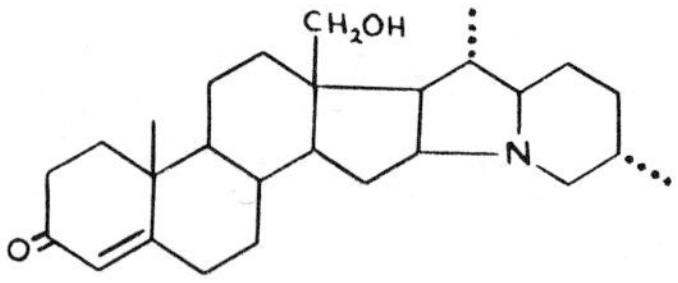

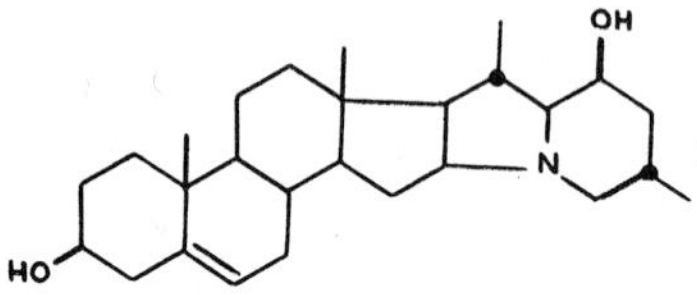

OESTRADIOL, 17α-

$C_{18}H_{24}O_2$ 272
+57°e
223°C

R = α-OH

JACS. 1938 60 2927.

4519-001

OESTRADIOL, 17β-

$C_{18}H_{24}O_2$ 272
+78°e
178°C

R = β - OH

JACS. 1938 60 2927.

4519-002

OESTRONE

$C_{18}H_{22}O_2$ 270
+152°c
255°C

R = oxo

Nature 1947 409.

4519-003

OESTRONE, 18-OH

$C_{18}H_{22}O_3$ 286
+146°e
256°C

JCS. 1968 2283.

4519-004

OESTRONE, 6α-OH-

$C_{18}H_{22}O_3$ 286
+163°e
220°C

Biochem. J. 1966 101 397.

4519-005

OESTRIOL

$C_{18}H_{24}O_3$
+30°p
285°C

Nature 1947 409.

4519-006

OESTRIOL, 16-Epi

$C_{18}H_{24}O_3$ 288
+81°d
277-8°C

Biochem. J. 1956 63 64.

4519-007

OESTRIOL 16-GLUCOSIDURONIC ACID

$C_{24}H_{32}O_9$ 464
+1.5°e
226°C

J. Bio. Chem. 1963 238 1273.

4519-008

EQUILENIN (+)

$C_{18}H_{18}O_2$ 266
+87°e
250-1°

JCS. 1945 582.

4519-009

STACHYSTERONE A — $C_{27}H_{42}O_6$ — 462

amorph.

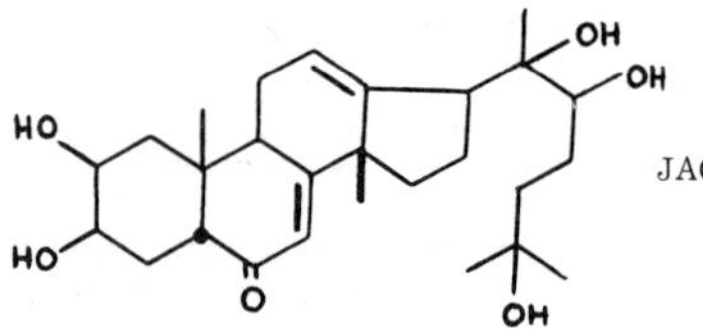

JACS. 1970 7510.

4591-001

DICTYOLUCIDINE — $C_{22}H_{39}NO_2$ — 349

$+30^0$

198^0C

R = H

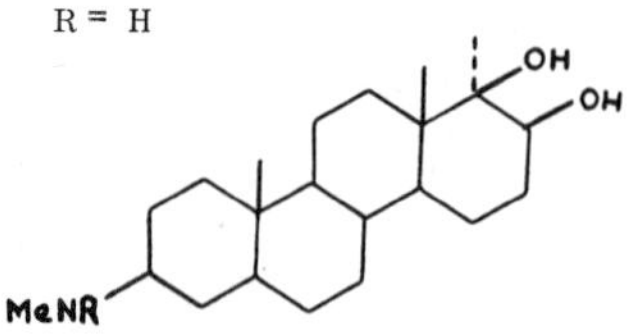

Bull. Soc. Chim. Fr. 1966 3472.

4591-002

DICTYOLUCIDAMINE — $C_{23}H_{41}NO_2$ — 363

$+6^0$

205^0C

R = Me

Bull. Soc. Chim. Fr. 1966 3472.

4591-003

ZYGADENILIC ACID LACTONE, Angeloyl- — $C_{32}H_{47}NO_8$ — 573

-11^0c

235^0C

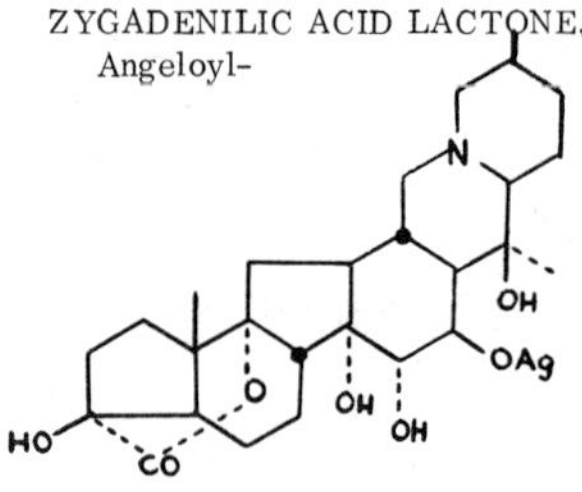

Chem. Pharm. Bull. 1962, 519.

4591-004

VERALININ — $C_{27}H_{43}NO$ — 397

-80^0

125^0C

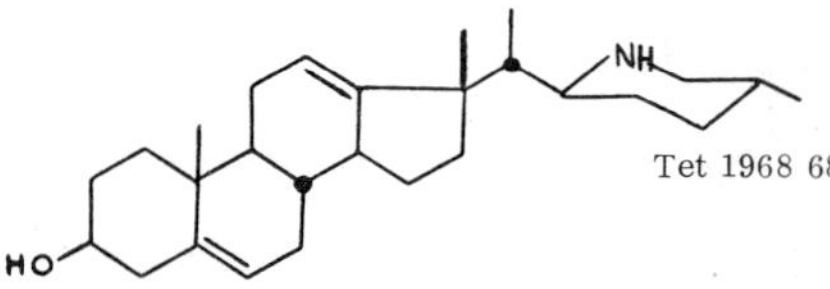

Tet 1968 6839.

4591-005

VERALKAMINE — $C_{27}H_{43}NO_2$ — 413

-88^0e

120^0C

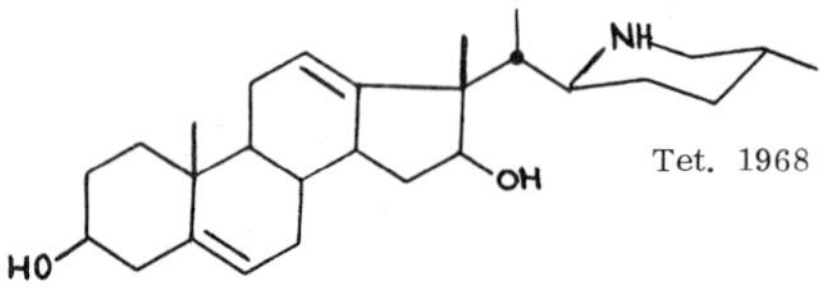

Tet. 1968 4865.

4591-006

VERAMINE — $C_{27}H_{41}NO_2$ — 411

-94^0

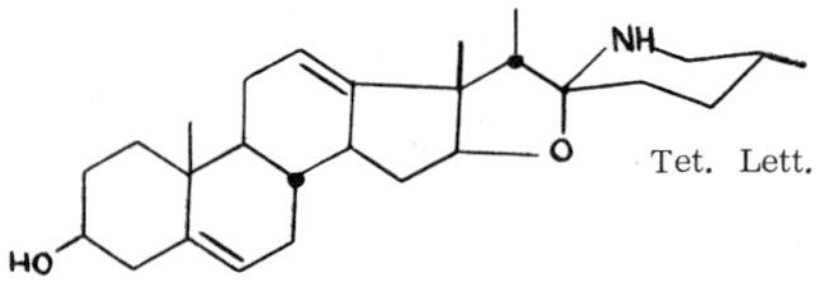

Tet. Lett. 1968 2817.

4591-007

FUKUJUSONORONE $C_{20}H_{26}O_3$ 314 89°C Exp. 1969 25 1129. 4592-001	VIRIDIN $C_{20}H_{16}O_6$ 352 -224° R = oxo JCS. 1969 549. 4592-006
CHIOGRALACTONE $C_{23}H_{34}O_4$ 374 -113° 239°C Tet. Supp. 8 1967 (Pt I) 123. 4592-002	VIRIDIOL $C_{20}H_{18}O_6$ 354 198-201°C R = β - OH Chem. Comm. 1969 839. 4592-007
CHOLECALCIFEROL $C_{27}H_{44}O$ 384 Chem. Comm. 1968 801. 4592-003	HIRUNDIGENIN $C_{21}H_{30}O_5$ 362 -45°c 190-6°C Tet. Lett. 1968 3799. 4592-008
CHOLECALCIFEROL, 25-OH- $C_{27}H_{44}O_2$ 400 Chem. Comm. 1968 801. 4592-004	HIRUNDIGENIN, anhydro- $C_{21}H_{28}O_4$ 344 -27°c 212-17°C Tet. Lett. 1968 3799. 4592-009
WORTMANNIN $C_{23}H_{24}O_8$ 428 240°C Chem. Comm 1968 613. 4592-005	PHYSALIN A $C_{28}H_{30}O_{10}$ 526 266°C R = OH Tet. Lett. 1969 1084. 4592-010

PHYSALIN C	$C_{28}H_{30}O_9$	510 -160°c 273-7°C
R = H	Tet. 1970 1743.	
4592-011		

SAMANDENONE	$C_{22}H_{33}NO_3$	359 191°C
	Ber. 1966 1439.	
4592-016		

PHYSALIN B	$C_{28}H_{30}O_9$	510 269-72°C
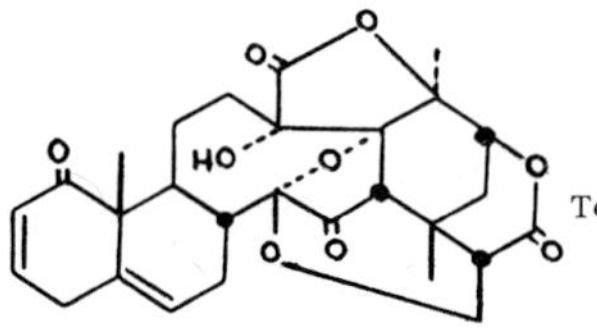	Tet. Lett. 1969 1765.	
4592-012		

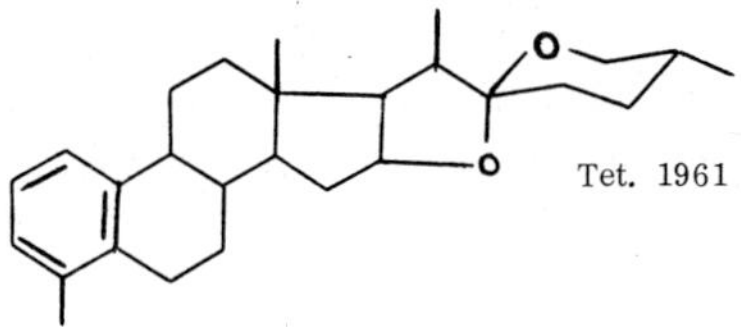

SAMANDARONE	$C_{19}H_{29}O_2N$	303 190°C
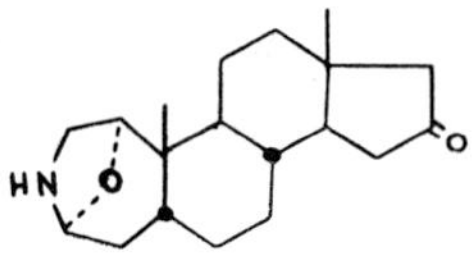	Natwiss. 1966 123.	
4592-017		

deleted

4592-013

RANOL	$C_{26}H_{46}O_5$	438 +21°e amorph.
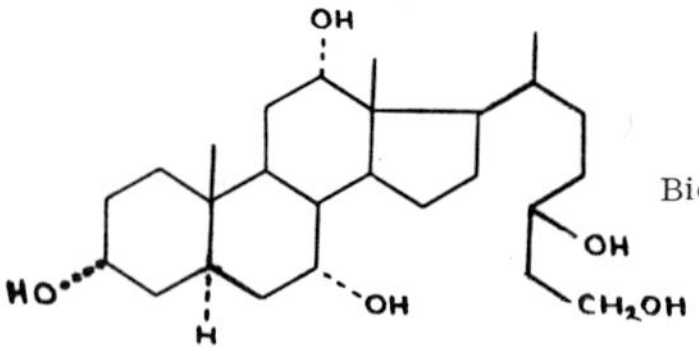	Biochem. J. 1964 90 314.	
4592-018		

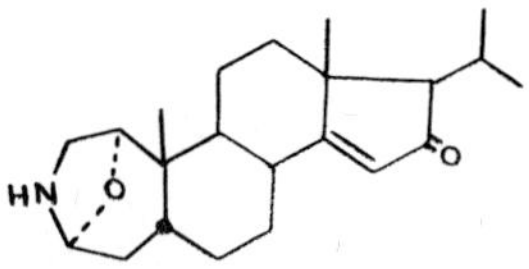

LUVIGENIN	$C_{27}H_{38}O_2$	394 -35° 184°C
	Tet. 1961 183.	
4592-014		

CHOLAN-3, 7, 12, 24, 26--PENTAOL, bis-homo-	$C_{26}H_{46}O_5$	438 186°C
5-β-H epimer	J. Biochem. 1963 53 331.	
4592-019		

HOLAROMINE	$C_{22}H_{31}N$	309 +90°c 198°C
	Chem. Abstr. 1968 68 87462.	
4592-015		

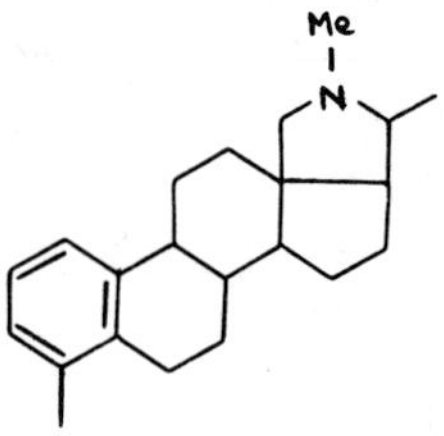

CHOL-24-EN-3, 7, 12-TRIOL, homo-5β-H-	$C_{25}H_{42}O_3$	390 179°C
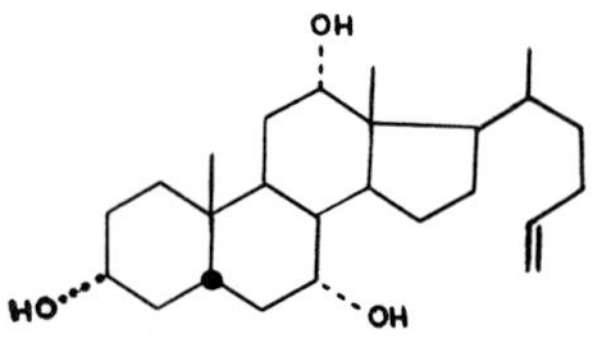	J. Biochem. 1962 51 119.	
4592-020		

CHOL-24-EN-3, 7, 12-TRIOL, homo-5α-H- | $C_{25}H_{42}O_3$ | 390 | +43^0 | 146^{0}C

J. Biochem. 1963 54 41.

4592-021

HIRUNDOSIDE-A | $C_{28}H_{40}O_7$ | 488 | -66^{0}a | 210^{0}C

anhydro-hirundigenin (009)

3-oleandroside

Helv. 1969 1175.

4592-022

SAMANINE | $C_{19}H_{33}NO$ | 291 | 196^{0}C

Tet. Lett. 1969 1193.

4592-023

BATRACHOTOXININ A | $C_{24}H_{35}NO_5$ | 417

R = H

JACS. 1968 1917.

4592-024

BATRACHOTOXIN | $C_{31}H_{42}N_2O_6$ | 538

R =

JACS. 1963 3931.

4592-025

BATRACHOTOXIN, homo- | $C_{32}H_{44}N_2O_6$ | 552

R =

JACS. 1969 3931.

4592-026

POLLINASTANOL $C_{28}H_{48}O$ 400 +35°c 95°C Bull. Soc. Chim. Fr. 1964 2012. 4599-001	JURUBIDINE $C_{27}H_{45}O_2N$ 415 -50°p 184°C Tet. Lett. 1966 5997. 4599-005
GORGOSTEROL $C_{30}H_{50}O$ 426 -45° 187°C JACS. 1970 5281. 4599-002	JURUBINE $C_{33}H_{57}NO_8$ 595 -28°p amorph. Ber. 1967 100 1725. 4599-006
SOLANOCAPSINE $C_{27}H_{46}O_2N_2$ 430 +25°m 222°C Tet. Lett. 1970 2913. 4599-003	SPIROPACHYSINE $C_{31}H_{46}ON_2$ 462 +35° 290-2°C Tet. Lett. 1969 2519. 4599-007
SEVCORINE $C_{34}H_{55}O_6N$ -40°m 238°C R = glucosyl Chem. Abstr. 1968 69 27562. 4599-004	GORGOSTEROL, 23-deMe- $C_{29}H_{48}O$ 412 -35° 163°C JACS. 1970 6073. 4599-008

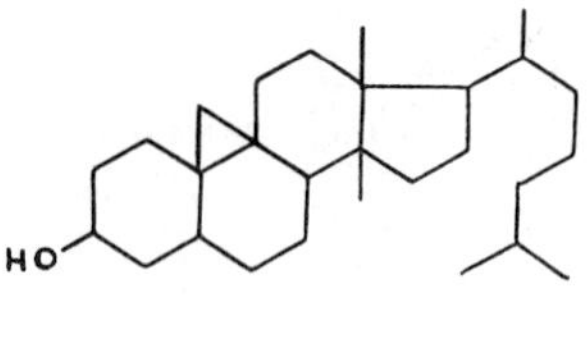

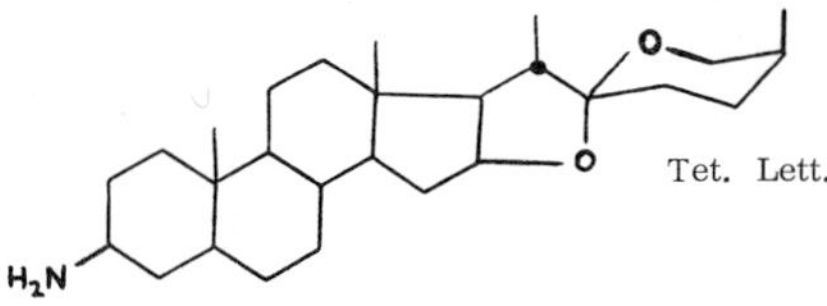

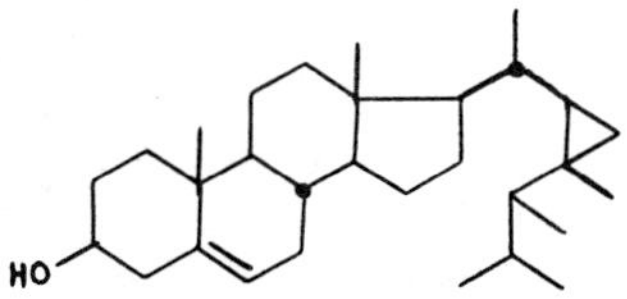

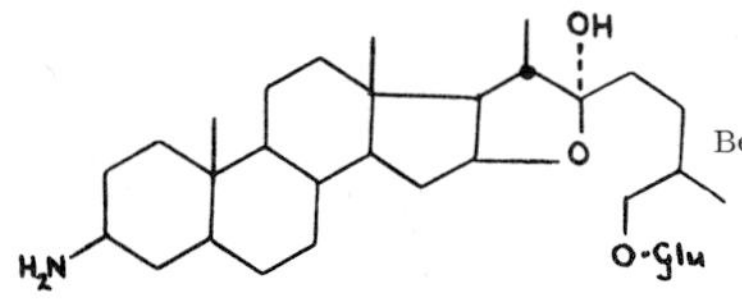

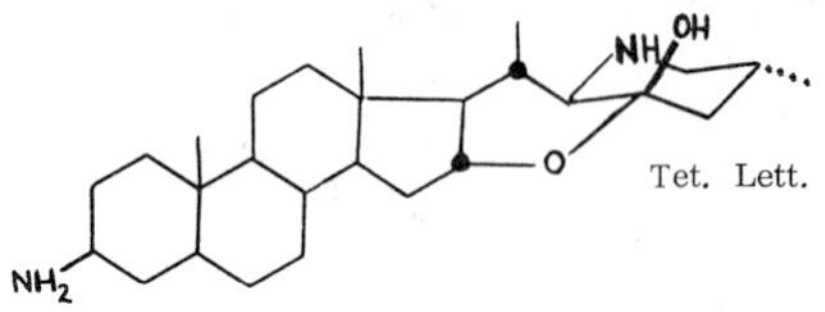

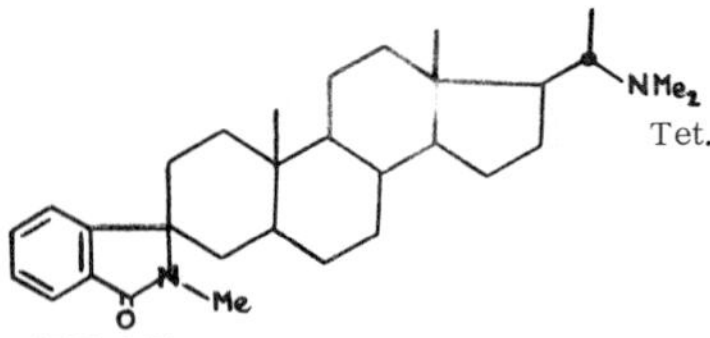

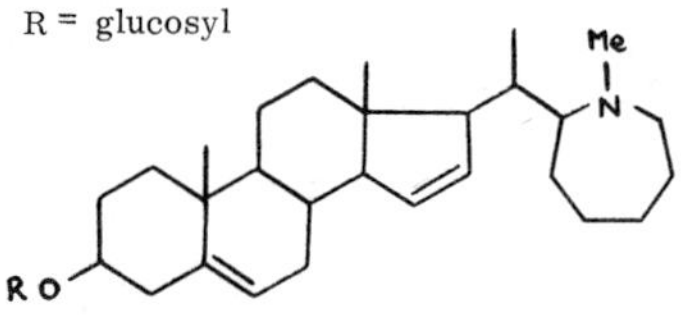

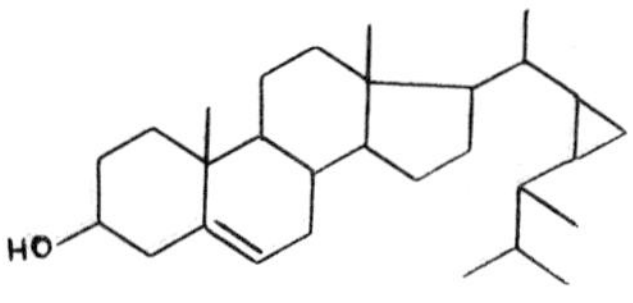

46--   THE CAROTENOIDS

The carotenoids are a relatively small class within the terpene family, there only being some 135 compounds in this present compilation. Also there is very little diversity of structural type among these compounds, six main subclasses serving to accommodate most of the carotenoids (see Skeleton Key). The remainder have been placed in subclass 4699 which includes degraded, rearranged and C_{50} carotenoids.

Carotenoids— Skeleton Classification

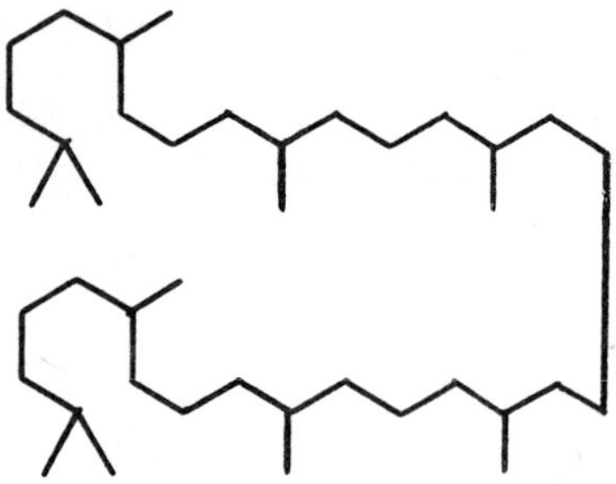 4601 Acyclic	4605 Allenic
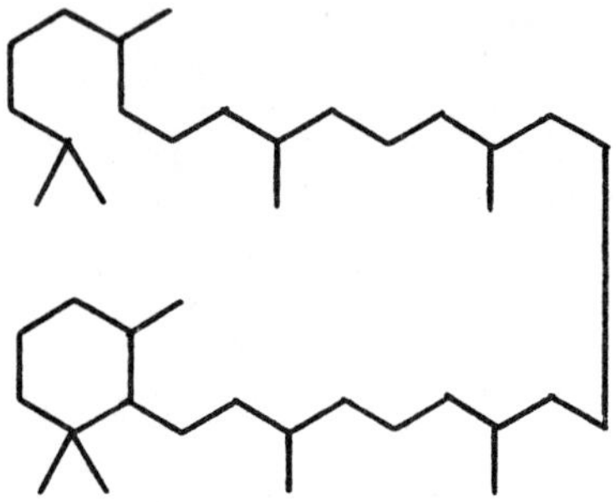 4602 Monocyclic	4606 Cyclopentanoid
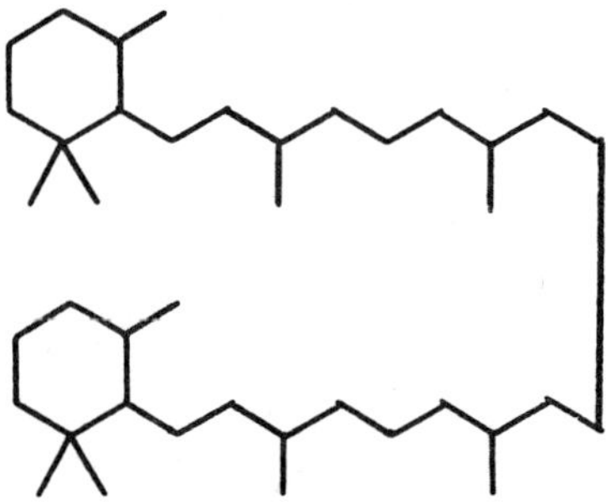 4603 Bicyclic	4699 Miscellaneous
4604 Acetylenic	

PHYTOENE	$C_{40}H_{64}$	544 OI Oil
cis	Biochem. J. 1956 62 269.	
4601-001		
NEUROSPORENE, 1, 2-dihydro-	$C_{40}H_{60}$	540 OI Oil
	Chem. Comm. 1970 127.	
4601-006		
CAROTENE, eta-	$C_{40}H_{64}$	544 Oil
t	Arch. Biochem. Biophys. 1954 53 480.	
4601-002		
LYCOPENE	$C_{40}H_{56}$	536 OI 174^{0}C
2 1	Biochim. Biophys. Acta. 1958 29 477.	
4601-007		
PHYTOFLUENE	$C_{40}H_{62}$	542 IO Oil
	Experimentia 1954 1.	
4601-003		
LYCOPENE, 1, 2-dihydro-	$C_{40}H_{58}$	538 OI
	Chem. Comm. 1970 127.	
4601-008		
CAROTENE, zeta-	$C_{40}H_{60}$	540 OI Oil
	JCS. 1966 2154.	
4601-004		
LYCOPENE, 1, 2-dihydro-3, 4-dehydro-	$C_{40}H_{56}$	536 OI
	Chem. Comm. 1970 127.	
4601-009		
NEUROSPORENE	$C_{40}H_{58}$	538 OI Oil
2 1	JCS. 1966 2154.	
4601-005		
LYCOPENE, 3, 4-dehydro-	$C_{40}H_{54}$	534 OI
	Phytochemistry 1965 925.	
4601-010		

LYCOPENE, 7, 8, 11, 12-tetrahydro- $C_{40}H_{60}$ 540

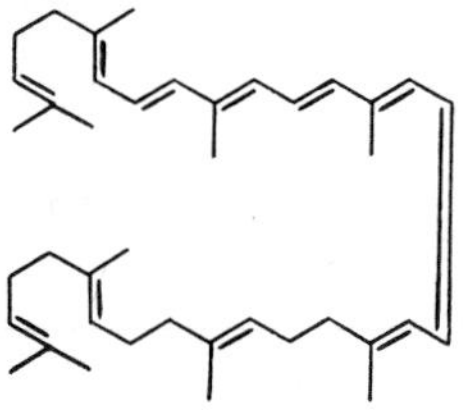

JCS. 1969 1266.

4601-011

LYCOPENE, 3, 4, 3', 4'-bisdehydro- $C_{40}H_{52}$ 532 OI 240°C

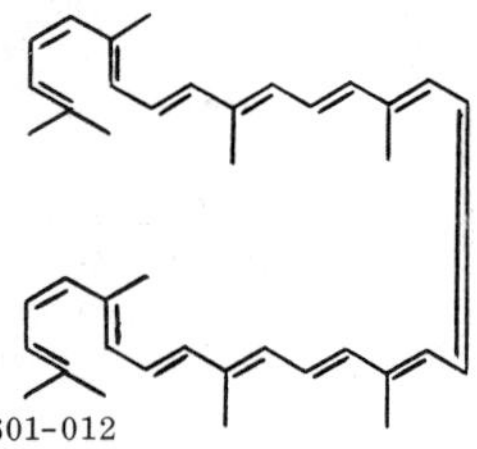

JOC. 1963 28 2735.

4601-012

CHLOROXANTHIN $C_{40}H_{60}O$ 556 OI

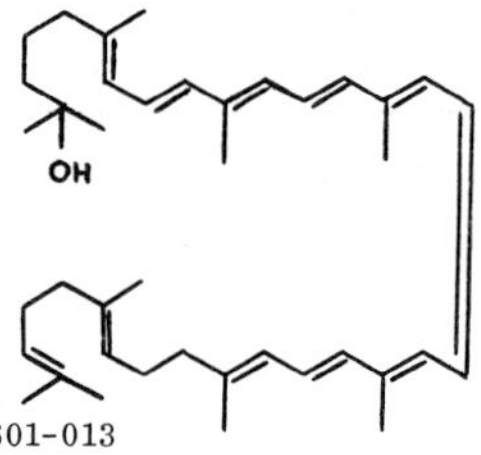

OH

JCS. 1966 2166.

4601-013

RHODOPIN $C_{40}H_{58}O$ 554 OI 168°C

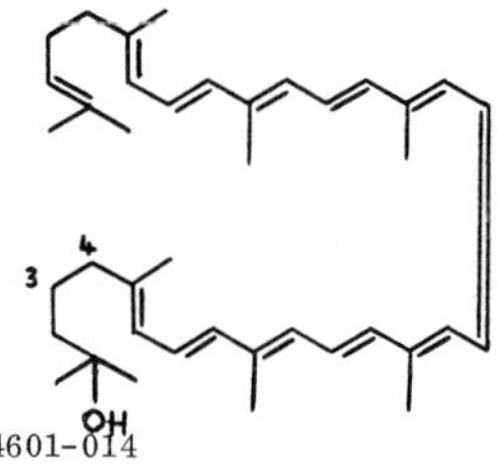

4 3 OH

Acta Chem. Scand. 1959 842.

4601-014

RHODOPIN, 3, 4-dehydro- $C_{40}H_{56}O$ 552 OI 190°C

JCS. 1967 1686.

4601-015

RHODOVIBRIN $C_{41}H_{60}O_2$ 584 OI 191°C

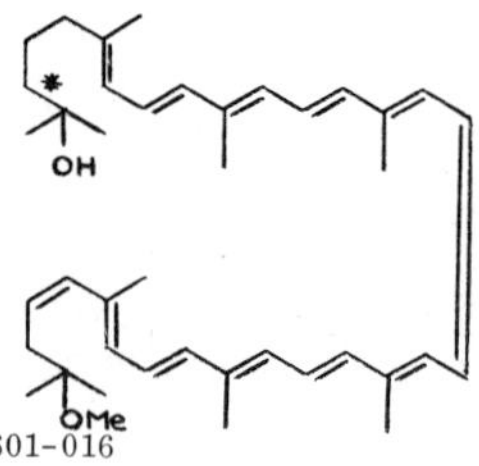

* OH OMe

JOC. 1966 186.

4601-016

RHODOVIBRIN, anhydro- $C_{41}H_{58}O$ 566 OI

* anhydro

JOC. 1966 186.

4601-017

SPHEROIDENE $C_{41}H_{60}O$ 568 OI 145°C

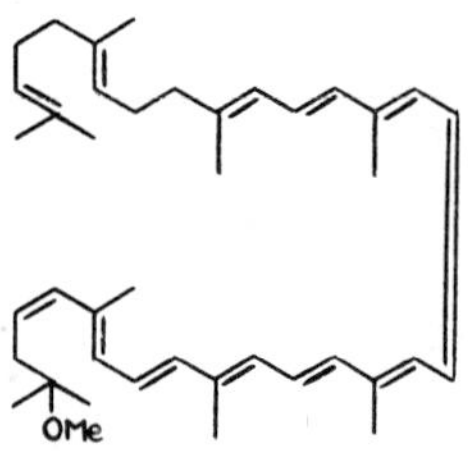

OMe

JCS. 1966 2166.

4601-018

SPHEROIDENE, OH- $C_{41}H_{62}O_2$ 586 OI 157°C

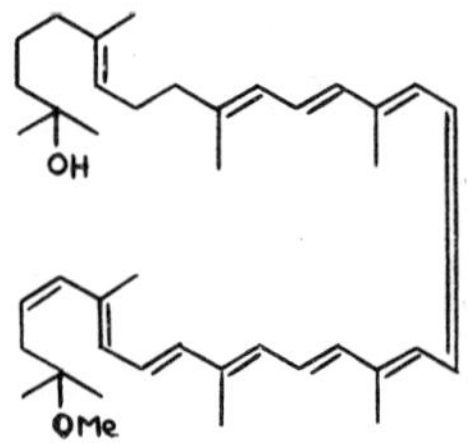

OH OMe

Acta Chem. Scand. 1964 1403.

4601-019

SPHEROIDENONE $C_{41}H_{58}O_2$ 582 OI 163°C

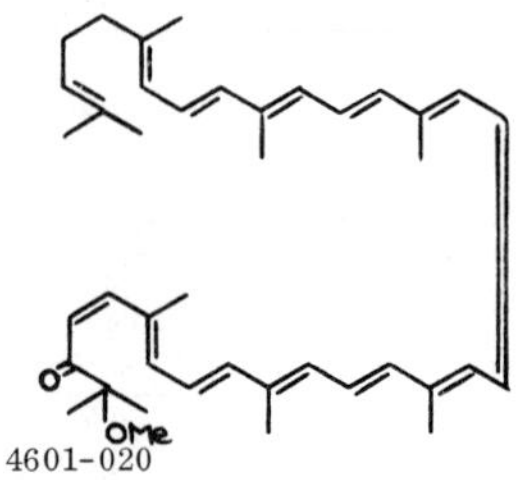

O OMe

JCS. 1966 2166.

4601-020

SPHEROIDENONE, OH- | $C_{41}H_{60}O_3$ | 600 OI 159^0C

OH

O

OMe

Acta Chem. Scand. 1964 1403.

4601-021

BACTERIOERYTHRIN | $C_{41}H_{58}O_2$ | 582 OI 209^0C

3' 4'

R = H

OR

3 4

OMe

Acta Chem. Scand. 1963, 500.

4601-026

SPHEROIDENE, 11', 12'- dihydro- | $C_{41}H_{62}O$ | 570

3 4

OMe

JCS. 1969 1266.

4601-022

SPIRILLOXANTHIN | $C_{42}H_{60}O_2$ | 596 OI 218^0C

R = Me

JOC. 1963 2735.

4601-027

SPHEROIDENE, 3, 4, 11', 12'- tetrahydro- | $C_{41}H_{64}O$ | 572

JCS. 1969 1265.

4601-023

SPIRILLOXANTHIN, 3, 4, 3', 4', - tetrahydro- | $C_{42}H_{64}OZ$ | 600 OI 175^0C

R = Me

Acta Chem. Scand. 1967 371.

4601-028

LYCOXANTHIN | $C_{40}H_{56}O$ | 552 OI 168^0C

R = CH_3

CH_2OH

R

Tet. Lett. 1968 1931.

4601-024

SPIRILLOXANTHIN, 2, 2'-dioxo- | $C_{42}H_{56}O_4$ | 624 OI 222^0C

O

OMe

O

OMe

Tet. Lett. 1966 989.

4601-029

LYCOPHYLL | $C_{40}H_{56}O_2$ | 568 OI 179^0C

R = CH_2OH

Tet. Lett. 1968 1931.

4601-025

LYCOPENAL | $C_{40}H_{54}O$ | 550 OI

CHO

Acta Chem. Scand. 1970 2705.

4601-030

LYCOPENAL, OMe- $C_{41}H_{58}O_2$ 582 OI

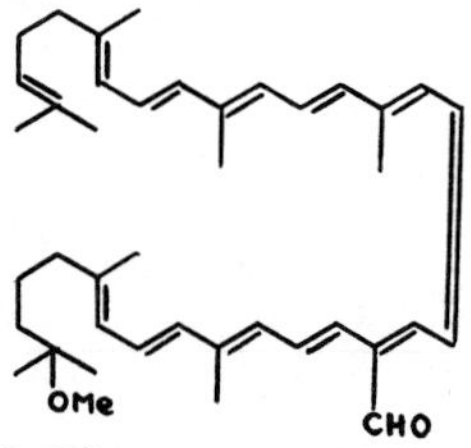

OMe CHO

Acta Chem. Scand. 1970 2705.

4601-031

SPIRILLOXANTHINAL, 3, 4, 3', 4'-tetrahydro- $C_{42}H_{62}O_3$ 614 OI

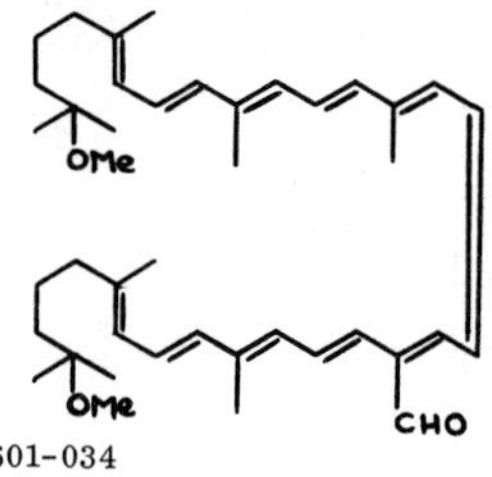

OMe OMe CHO

Acta Chem. Scand. 1970 2705.

4601-034

RHODOPINOL $C_{40}H_{58}O_2$ 570 OI

R = CH_2OH

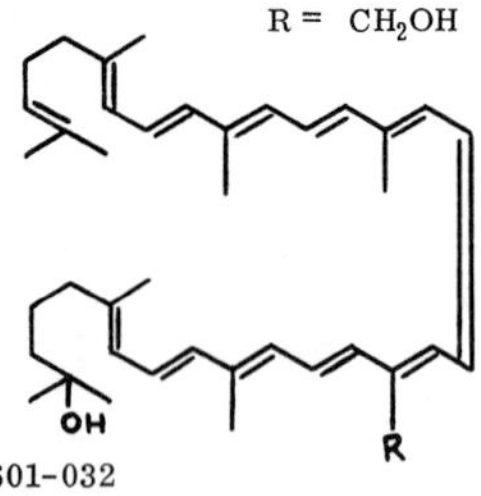

OH R

Acta Chem. Scand. 1970 2705.

4601-032

OSCILLAXANTHIN $C_{52}H_{76}O_{12}$ 892

R = thamnosyl

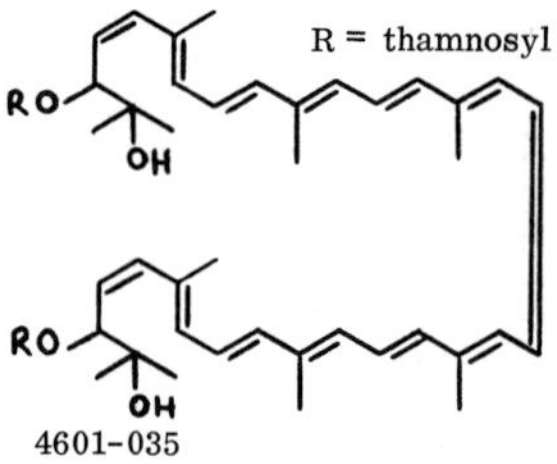

RO OH RO OH

Phytochemistry 1969 1280.

4601-035

RHODOPINAL $C_{40}H_{56}O_2$ 568 OI

R = CHO

Acta Chem. Scand. 1950 2705.

4601-033

CAROTENE, gamma- $C_{40}H_{56}$ 536 OI 153°C

Helv., 1961 (44) 985.

4602-001

ZEACAROTENE, β- $C_{40}H_{58}$ 538 OI 96°C

Phytochem., 1965, 193.

4602-006

CAROTENE, 4-keto-gamma- $C_{40}H_{54}O$ 550 OI 146°C

Acta Chem. Scand., 1966, 1187.

4602-002

CAROTENE, delta- $C_{40}H_{56}$ 536 +317° 140°C

JCS., 1965, 2019.

4602-007

CAROTENE, 1', 2'-dihydro-1'-OH-gamma- $C_{40}H_{58}O$ 554 OI

OH

Acta Chem. Scand., 1966, 1187.

4602-003

TORULENE $C_{40}H_{54}$ 534 OI 185°C

R = CH_3

R

Helv., 1961, 994.

4602-008

CAROTENE, 1', 2'-dihydro-1'-OH-4-keto-gamma- $C_{40}H_{56}O_2$ 568 OI

OH

Acta Chem. Scand., 1966, 1187.

4602-004

TORULENAL $C_{40}H_{52}O$ 548 OI

R = CHO

Ber., 1960, 2951.

4602-009

ZEACAROTENE, α- $C_{40}H_{58}$ 538

Arch. Biochem. Biophys., 1959 (82) 117.

4602-005

TORULARHODIN $C_{40}H_{52}O_2$ 564 OI 211°C

R = COOH

Biochem. J., 1964 (92) 508.

4602-010

CELAXANTHIN $C_{40}H_{54}O$ 550 210°C Arch. Biochem., 1943 (1) 17. HO 4602-011	FLEXIXANTHIN $C_{40}H_{54}O_3$ 582 R = OH OH R Acta Chem. Scand., 1966, 1970. 4602-016
RUBIXANTHIN $C_{40}H_{56}O$ 552 160°C 5' t 6' HO * Chem. Comm., 1968, 382. 4602-012	FLEXIXANTHIN, $C_{40}H_{54}O_2$ 566 OI 155°C R = H Acta Chem. Scand., 1966, 1970. 4602-017
RUBIXANTHIN-5, 6-EPOXIDE $C_{40}H_{56}O_2$ 568 170°C *epoxide Helv., 1947, 531. 4602-013	PHLEIXANTHOPHYLL $C_{46}H_{66}O_8$ 746 HO O·Glu HO 4 Acta Chem. Scand., 1967, 15. 4602-018
GAZANIAXANTHIN $C_{40}H_{56}O$ 552 136°C 5', 6'-cis isomer of Rubixanthin Chem. Comm., 1968, 382. 4602-014	PHLEIXANTHOPHYLL, 4-keto- $C_{46}H_{64}O_9$ 760 Acta Chem. Scand., 1967, 15. 4602-019
SAPROXANTHIN $C_{40}H_{56}O_2$ 568 178°C OH HO Acta Chem. Scand., 1966, 811. 4602-015	MYXOXANTHOPHYLL $C_{46}H_{66}O_7$ 730 173°C R = rhamnosyl RO OH HO Phytochem., 1969, 1259. 4602-020

RUBICHROME

$C_{40}H_{56}O_2$ 568

154^{0}C

HO

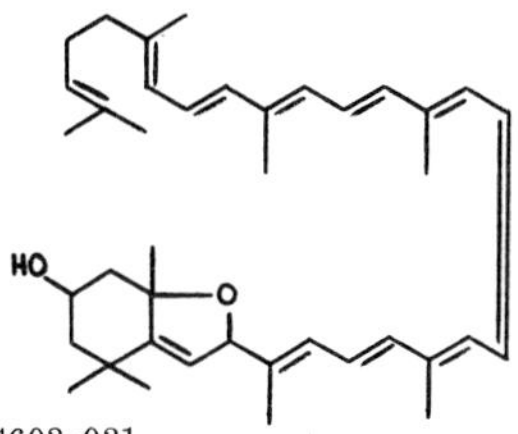

O

Helv., 1947, 531.

4602-021

CAROTENE, α- — $C_{40}H_{56}$ — 536, +385⁰b, 188⁰C

Biochem. J. 1953 (53) 538.

4603-001

CAROTENE EPOXIDE, α- — $C_{40}H_{56}O$ — 552, 175⁰C

*epoxide

Helv. 1945 (28) 1146.

4603-002

CAROTENE, β- — $C_{40}H_{56}$ — 536, OI, 183⁰C

JOC. 1964 187.

4603-003

CAROTENE EPOXIDE, β- — $C_{40}H_{56}O$ — 552, 160⁰C

*epoxide

Acta Chim. Acad. Sci. Hung. 1958 (16) 227.

4603-004

ZEINOXANTHIN — $C_{40}H_{56}O$ — 552, +508⁰a, 175⁰C

Arch. Biochem. Biophys. 1960 (86) 163.

4603-005

CRYPTOXANTHIN, α- — $C_{40}H_{56}O$ — 552, +360⁰b, 182⁰C

Ann. 1958 (616) 207.

4603-006

CRYPTOXANTHIN — $C_{40}H_{56}O$ — 552, 169⁰C

Ind. J. C. 1964 451.

4603-007

CRYPTOXANTHIN EPOXIDE — $C_{40}H_{56}O_2$ — 568

*epoxide

J. Food Sci. 1961 (26) 442.

4603-008

ZEAXANTHIN — $C_{40}H_{56}O_2$ — 568, 215⁰C

R = H

Chem. Comm. 1969 1311.

4603-009

PHYSALIEN — $C_{72}H_{116}O_4$ — 1044, -45⁰c, 99⁰C

R = palmitoyl

Ber. 1939 (72) 1678.

4603-010

ANTHERAXANTHIN — $C_{40}H_{56}O_3$ — 584
205^{0}C

Helv. 1949 (32) 50.

4603-011

VIOLAXANTHIN — $C_{40}H_{56}O_4$ — 600
+35^{0}c
200^{0}C

* epoxide

Ber. 1931 (64) 326.

4603-012

LUTEIN — $C_{40}H_{56}O_2$ — 568
+160^{0}c
193^{0}C
A/C

R = H

Nature 1958 (182) 55.
Chem. Comm., 1970, 1578.

4603-013

HELENEIN — $C_{72}H_{116}O_4$ — 1044
92^0

R = palmitoyl

Natwiss. 1930 (18) 754.

4603-014

LUTEIN EPOXIDE — $C_{40}H_{56}O_3$ — 584
192 — 192^{0}C

R = H

* epoxide

Fortschr. Chem. Org. Nat. 1948 (5) 15.

4603-015

ECHINENONE — $C_{40}H_{54}O$ — 550
OI
192^{0}C

R = H

Phytochemistry 1966 565.

4603-016

ECHINENONE, 3-OH- — $C_{40}H_{54}O_2$ — 566
156^{0}c

R = OH

Chem. Comm. 1967 49.

4603-017

ESCHSCHOLTZXANTHIN — $C_{40}H_{54}O_2$ — 566
+225^{0}c
186^{0}C

Helv. 1958 (41) 402.

4603-018

ESCHSCHOLTZXANTHONE — $C_{40}H_{52}O_2$ — 564

Ber. 1933 (66) 828.

4603-019

RHODOXANTHIN — $C_{40}H_{50}O_2$ — 562
OI
219^{0}C

Fortschr. Chem. Org. Nat. 1960 (18) 183.

4603-020

CANTHAXANTHIN $C_{40}H_{52}O_2$ 564 OI 218^{0}C Natwiss 1961 (48) 581. 4603-021	SIPHONEIN $C_{52}H_{76}O_5$ 780 R = dodecanoyl or dodecenoyl depending on natural source Phytochemistry 1950 2545. 4603-026
ASTAXANTHIN $C_{40}H_{52}O_4$ 596 216^{0}C RO, O; R = H Angew Chem. 1966 (5) 957. 4603-022	CITROXANTHIN $C_{40}H_{56}O$ 552 R; O; R = H Helv. 1947 (30) 536. 4603-027
ASTAXANTHIN DIPALMITATE $C_{72}H_{112}O_6$ 1072 72^{0}C R = Palmitoyl Ber. 1938 (71) 1879. 4603-023	CRYPTOFLAVIN $C_{40}H_{56}O_2$ 568 R = OH Ind. J. C. 1964 451. 4603-028
CRUSTAXANTHIN $C_{40}H_{56}O_4$ 600 172^{0}c OH, HO Chem. Abstr. 1968 (68) 13214. 4603-024	CHRYSANTHEMAXANTHIN $C_{40}H_{56}O_3$ 584 HO; O; *; R = H; HO; R Helv. 1945 (28) 1156. 4603-029
SIPHONAXANTHIN $C_{40}H_{56}O_4$ 600 HO; O; CH_2OR; R = H Phytochemistry 1970 2545. 4603-025	FLAVOXANTHIN $C_{40}H_{56}O_3$ 584 +190^{0}b 184^{0}C *epimer R = H Helv. 1942 (25) 1144. 4603-030

TROLLICHROME	$C_{40}H_{56}O_4$	600 -57^0 206^0C	LOROXANTHIN	$C_{40}H_{56}O_3$	584
R = OH	Helv. 1955 (38) 638.		HO HO CH₂OH	Phytochemistry 1970 2545.	
4603-031			4603-032		

CROCOXANTHIN $C_{40}H_{54}O$ 550 R = H Chem. Comm. 1967 301. 4604-001	HETEROXANTHIN $C_{40}H_{56}O_4$ 600 Chem. Comm. 1970 877. 4604-006
MONADOXANTHIN $C_{40}H_{54}O_2$ 566 R = OH Chem. Comm 1967 301. 4604-002	PECTENOLONE $C_{40}H_{52}O_3$ 580 Chem. Comm. 1967 941. 4604-007
DIATOXANTHIN $C_{40}H_{54}O_2$ 566 OI 201^{0}C Chem. Comm. 1967 301. 4604-003	ALLOXANTHIN $C_{40}H_{52}O_2$ 564 Chem. Comm. 1967 301. 4604-008
DIADINOXANTHIN $C_{40}H_{54}O_3$ 582 158-62^{0}C * epoxide Chem. Comm. 1968 32. 4604-004	ASTAXANTHIN, 7, 8, 7', 8'-tetradehydro- $C_{40}H_{48}O_4$ 592 Acta Chem. Scand. 1970 3050. 4604-009
ASTAXANTHIN, 7, 8-di-dehydro- $C_{40}H_{50}O_4$ 594 Acta Chem. Scand. 1970 3050. 4604-005	

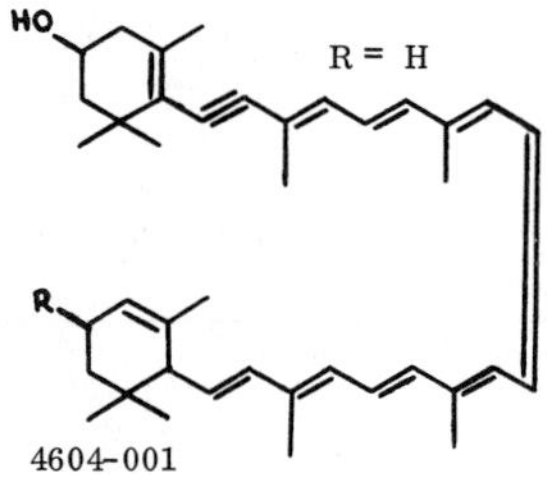

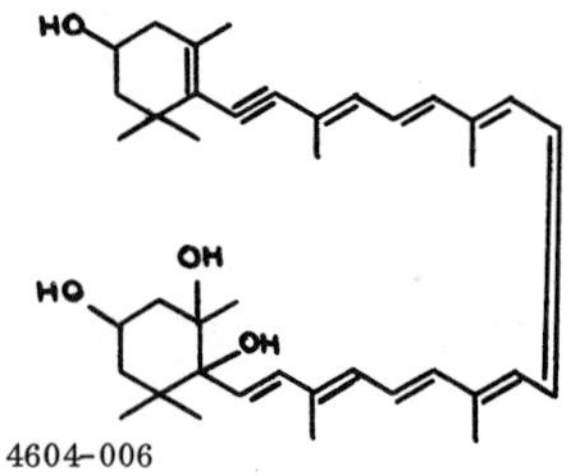

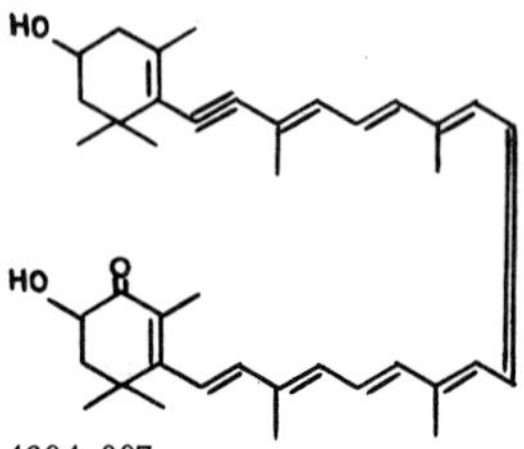

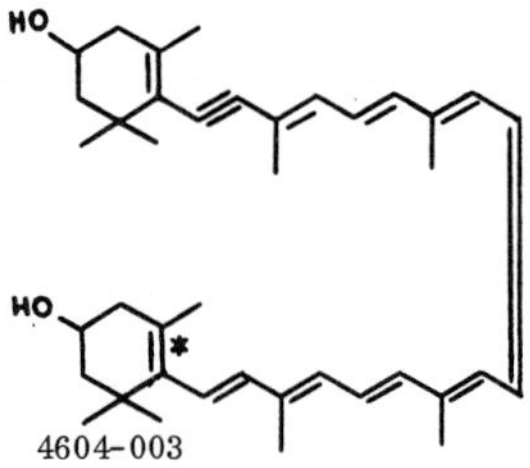

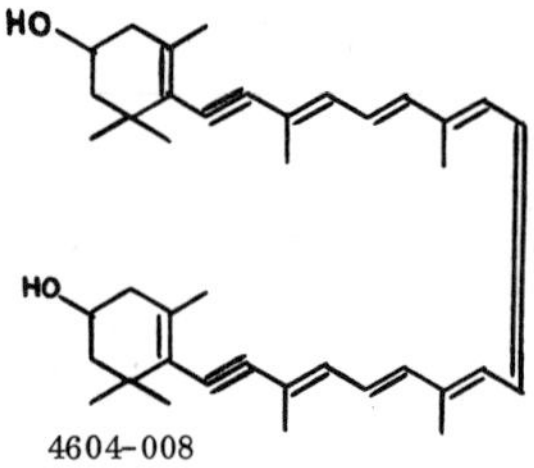

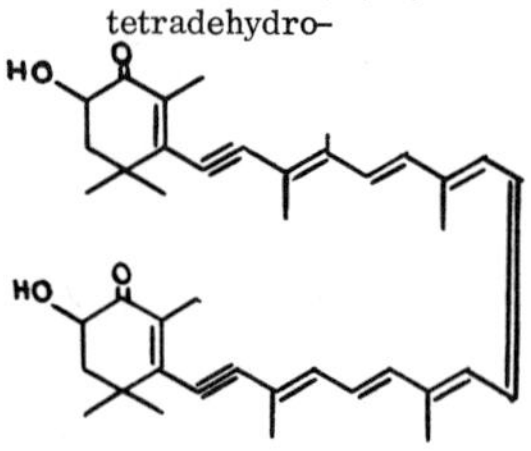

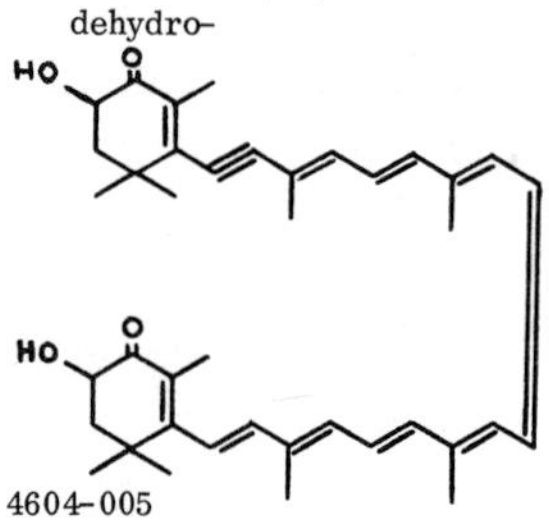

NEOXANTHIN $C_{40}H_{56}O_4$ 600 128°C R = Me HO, OH, R, HO, O JCS. 1969 1256. 4605-001	FUCOXANTHINOL $C_{40}H_{56}O_5$ 616 R = H RO, OH, HO, O, O JCS. 1969 431. 4605-003
VAUCHERIAXANTHIN $C_{40}H_{56}O_5$ 616 R = CH_2OH Tet. Lett. 1435 1970. 4605-002	FUCOXANTHIN $C_{42}H_{58}O_6$ 658 166°C R = Ac JCS. 1969 431. 4605-004

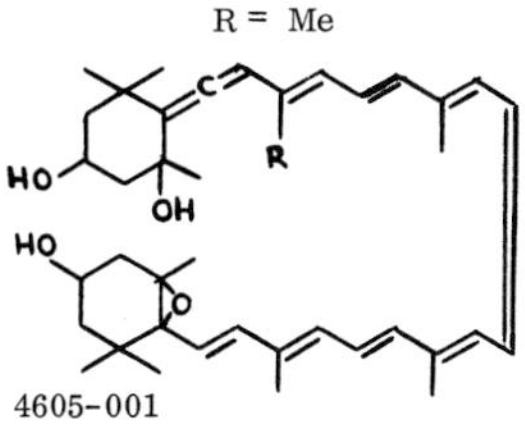

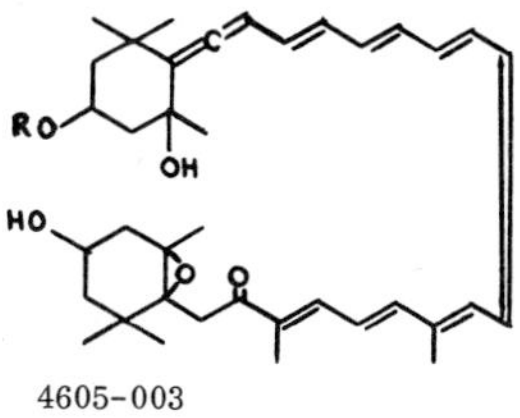

KRYPTOCASIN $C_{40}H_{56}O_2$ 568 160-1°C R = H Tet. 1963 1257. 4606-001	ACTINIOERYTHROL $C_{38}H_{48}O_4$ 568 91°C R = OH Chem. Comm. 1969 128. 4606-004
CAPSANTHIN $C_{40}H_{56}O_3$ 584 175°C R = OH Helv. 1964 47 741. 4606-002	CAPSORUBIN $C_{40}H_{56}O_4$ 600 Helv. 1964 47 741. 4606-005
VIOLERYTHRIN $C_{38}H_{44}O_4$ 564 OI 236-8°C R = oxo Acta Chem. Scand. 1968 1714. 4606-003	

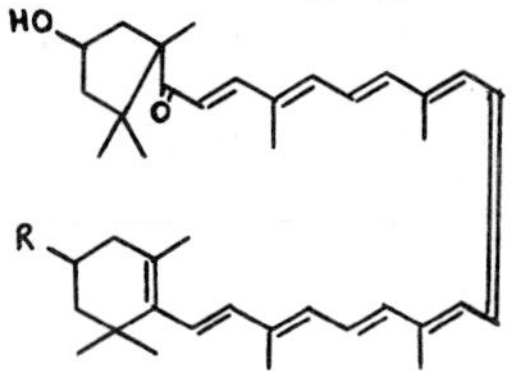

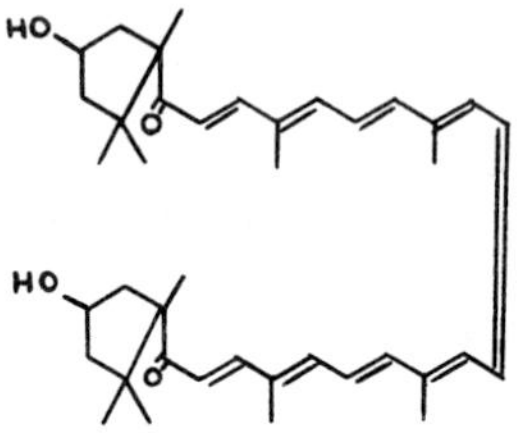

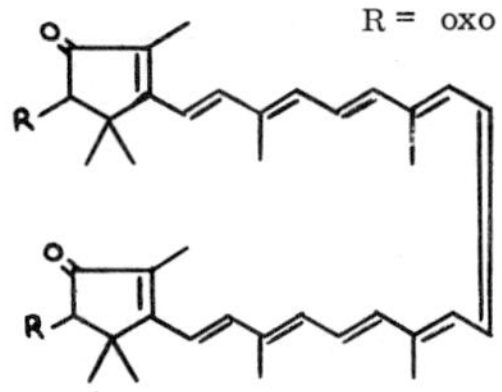

CAROTENONE, semi-alpha-	$C_{40}H_{56}O_2$	568 OI 135°C	Phytochemistry 1970 231.
4699-001			
OKENONE	$C_{42}H_{54}O_2$	578 OI	Acta Chem. Scand. 1967 961.
OMe 4699-006			
CAROTENONE, semi-β- R = H	$C_{40}H_{56}O_2$	568 OI	Phytochemistry 1968 1031.
R 4699-002			
RENIERAPURPURIN	$C_{40}H_{48}$	528 OI 230°C	Bull. Chem. Soc. Jap. 1960 1560.
4699-007			
TRIPHASIAXANTHIN R = OH	$C_{40}H_{56}O_3$	584 OI 96°C	JOC. 1970 2080.
4699-003			
RENIERATENE	$C_{40}H_{48}$	528 OI 185°C	Bull. Chem. Soc. Jap. 1960 1560.
4699-008			
CHLOROBACTENE	$C_{40}H_{52}$	532 OI 156°C	Acta Chem. Scand. 1964 1739.
4699-004			
RENIERATENE, iso- R = R' = H	$C_{40}H_{48}$	528 OI 200°C	Acta Chem. Scand. 1964 1562.
R R' 4699-009			
CHLOROBACTENE, hydroxy-	$C_{40}H_{54}O$	550 OI 131°C	Acta Chem. Scand. 1964 1739.
OH 4699-005			
LEPROTENE, 3-OH- R = H; R' = OH	$C_{40}H_{48}O$	544 OI 184°C	Experimentia 1969 25 241.
4699-010			

LEPROTENE, 3, 3'-di-OH- — $C_{40}H_{48}O_2$ — 560 — OI — 200°C

R = R' = OH

Experimentia 1969 25 241.

4699-011

CITRANAXANTHIN, 8'-OH-8', 9'-dihydro- — $C_{33}H_{44}O_2$ — 472 — 146°C

JOC. 1966 3452.

4699-016

NEUROSPORAXANTHIN — $C_{35}H_{46}O_2$ — 498 — OI — 193°C

Acta Chem. Scand. 1965 1843.

4699-012

SINTAXANTHIN — $C_{31}H_{42}O$ — 430 — OI

JOC. 1965 3394.

4699-017

RETICULATAXANTHIN — $C_{35}H_{48}O_2$ — 500

J. Food Sci. 1962 27 537.

4699-013

CAROTENAL, β-apo-8'- — $C_{30}H_{40}O$ — 416 — OI — 139°C

R = H

J. Food Sci. 1964 30 13.

4699-018

TANGERAXANTHIN — $C_{34}H_{44}O_2$ — 484

J. Food Sci. 1962 27 537.

4699-014

CITRAURIN, β- — $C_{30}H_{40}O_2$ — 432 — 147°C

R = OH

Endeavour 1951 10 175.

4699-019

CITRANAXANTHIN — $C_{33}H_{42}O$ — 454 — OI

JOC. 1965 2481.

4699-015

AZAFRIN — $C_{27}H_{38}O_4$ — 426 — -76°e — 213°C

Biochem. J. 1935 29 2553.

4699-020

BIXIN — $C_{25}H_{30}O_4$ — 394 — OI — 198°C

Proc. Chem. Soc. 1960 23.

4699-021

CROCETIN — $C_{20}H_{24}O_4$ — 328 — OI — 285°C

R = H

Helv. 1932 1399.

4699-022

CROCIN, α- — $C_{44}H_{64}O_{24}$ — 976 — 186°C

R = gentiobiosyl

Helv. 1932 15 1399.

4699-023

CROCETIN ALDEHYDE — $C_{20}H_{24}O_3$ — 312

R = CHO; R = COOH

Helv. 1969 806.

4699-024

CROCETIN DIALDEHYDE — $C_{20}H_{24}O_2$ — 296 — OI — 193°C

R = R' = CHO

Helv. 1969 806.

4699-025

SARCINAXANTHIN — $C_{50}H_{72}O_2$ — 704 — 149°C

R, R' = CH_3, CH_2OH(?)

Acta Chem. Scand. 1970 1460.

4699-026

PARACENTRONE — $C_{31}H_{42}O_3$ — 462

JCS. 1969 1264.

4699-027

BACTERIORUBERIN — $C_{50}H_{76}O_4$ — 730

Acta Chem. Scand. 1970 2169.

4699-028

DECAPRENOXANTHIN — $C_{50}H_{72}O_2$ — 704

Acta Chem. Scand. 1970 1460.

4699-029

RHODOPIN, 2-isopentenyl-3, 4-dehydro- — $C_{45}H_{64}O$ — 620 — 153°C

Acta Chem. Scand. 1969 1463.

4699-030

CAROTENAL, β-apo-10'- $C_{32}H_{42}O$ 442
OI
98^{0}C

Phytochemistry 1966
5 1159.

4699-031

47-- THE POLYPRENOIDS

This small group is comprised of acyclic compounds derived from 6 or more "isoprene" units linked in a "head-to-tail" manner (cf. squalene and the carotenoids which derive from the union of *two chains* in a "tail-to-tail" fashion). There still remains some ambiguity as to the precise configurations of the olefinic bonds in some of these compounds and so at present the isomeric compounds are presumed not to be identical.

BETULAPRENOL-6 $C_{30}H_{50}O$ 426 OI n = 6 $H\left(\ldots\right)_n OH$ Acta Chem. Scand. 1965 (19) 1317. 4701-001	CASTAPRENOL-11 $C_{55}H_{90}O$ 766 OI n = 11 Biochem. J. 1967 (102) 313/325. 4701-006
BETULAPRENOL-7 $C_{35}H_{58}O$ 494 OI n = 7 Acta Chem. Scand. 1965 (19) 1317. 4701-002	CASTAPRENOL-12 $C_{60}H_{98}O$ 834 OI n = 12 Biochem. J. 1967 (102) 313/325. 4701-007
BETULARPRENOL-8 $C_{40}H_{66}O$ 562 OI n = 8 Acta Chem. Scand. 1965 (19) 1317. 4701-003	CASTAPRENOL-13 $C_{65}H_{106}O$ 902 OI n = 13 Biochem. J. 1967 (102) 313/325. 4701-008
BETULAPRENOL-9 $C_{45}H_{74}O$ 630 OI n = 9 Acta Chem. Scand. 1965 (19) 1317. 4701-004	SOLANESOL $C_{45}H_{74}O$ 630 OI 42^{0}C n = 9 (all trans) Helv. 1959 (42) 2252. 4701-009
CASTAPRENOL-10 $C_{50}H_{82}O$ 698 OI n = 10 Biochem. J. 1967 (102) 313/325. 4701-005	SPADICOL $C_{50}H_{82}O$ 698 OI n = 10 (all trans) Proc. Roy. Soc. 1963 (158B) 291. 4701-010

UNDECAPRENOL-1 $C_{55}H_{90}O$ 766 n = 11 Tet. Lett. 1966 6209. 4701-011	BOMBIPRENONE $C_{43}H_{70}O$ 602 OI 46°C Tet. Lett. 1968 3839. 4701-015
DODECAPRENOL-1 $C_{60}H_{98}O$ 834 n = 12 Tet. Lett. 1966 6209. 4701-012	MALLOPRENOL $C_{45}H_{74}O$ 630 OI Oil R = H Bull. Chem. Soc. Jap. 1970 (43) 2174. 4701-016
BACTOPRENOL $C_{55}H_{92}O$ 768 n = 10 Biochem. J. 1966 (99) 123. 4701-013	MALLOPRENOL LINOLENATE $C_{63}H_{102}O_2$ 894 OI Oil R = linolenoyl Bull. Chem. Soc. Jap. 1970 2174. 4701-017
DOLICHOL-20 $C_{100}H_{164}O$ 1380 n = 19 Biochem. J. 1963 (88) 470. 4701-014	

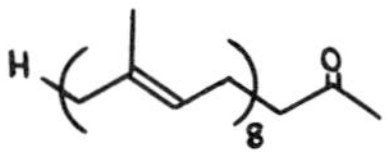

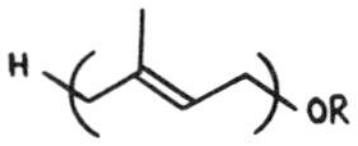

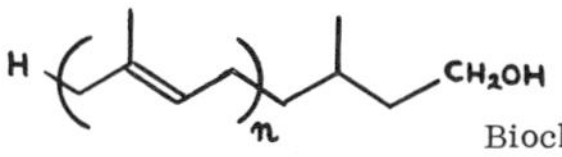

49-- MISCELLANEOUS COMPOUNDS OF TERPENOID ORIGIN

This inevitable final category for miscellaneous compounds of terpenoid origin arises through the somewhat frequent occurrence of oxidative degradation in terpenoid biosynthesis. In general the compounds in this section have degraded structures such that there exists some ambiguity as to their precise biosynthetic precursors. Thus the ionones (C_{13}) could be regarded as built up monoterpenes, degraded sesquiterpenes or degraded carotenoids; whilst the carotenoid origin is favoured currently it is clear that they are better placed in this section than with the carotenoids. Similarly fraxinellone (also C_{13}) would be lost amid the triterpenes (its likely biosynthetic origin). As subclassification by structure or biogenesis cannot be entertained, this section has simply been subclassified by the number of carbon atoms *in the carbon skeleton.* Thus both the ionones and fraxinellone would be found in the class 4913.

MEVALONIC ACID LACTONE	$C_6H_{10}O_3$	130 27^{0}C
JACS. 1957 3294.		
4906-001		

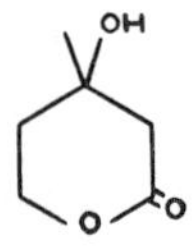

PROPYLISOBUTYL KETONE	$C_8H_{16}O$	128 OI Oil
Chim. Ind. 1956 (38) 1015.		
4908-003		

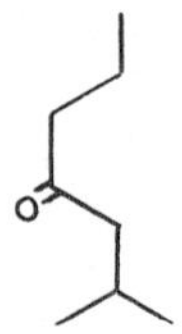

GENTIOCRUCINE	$C_6H_7NO_3$	141 139-40^{0}C
Chem. Abstr. 1970 (73) 15051.		
4906-002		

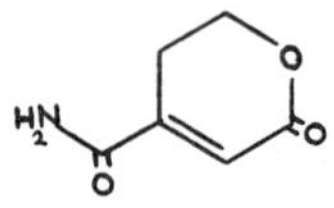

BICYCLO(3, 2, 1)OCTANE, -1, 5-diMe-6, 8-dioxa-	$C_8H_{14}O_2$	142 Oil
Nature 1969 (221) 477.		
4908-004		

GENTIANAINE	$C_6H_7NO_3$	141 OI 149^{0}C
Chem. Abstr. 1969 (71) 13247.		
4906-003		

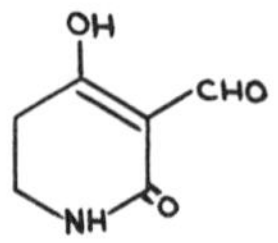

CYCLOPENTANONE, 2-ISOPROPYL-	$C_8H_{14}O$	126 Oil
Chem. Abstr. 1963 (59) 3780.		
4908-005		

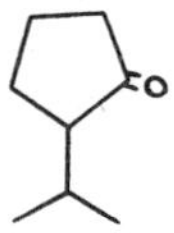

HEPT-2-EN-6-OL, 2-Me-	$C_8H_{16}O$	128 Oil
Biochem. J. 1950 55.		
4908-001		

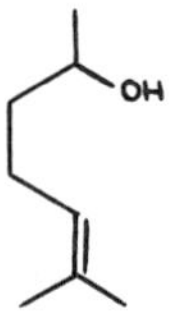

CYCLOHEX-1-ENE, 1-Me-4-Ac-	$C_9H_{14}O$	138 Oil
Helv. 1934 372.		
4909-001		

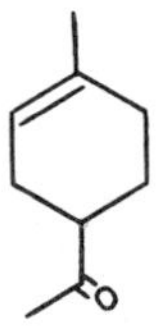

HEPT-2-ENE-6-ONE, 2-Me-	$C_8H_{14}O$	126 OI Oil
Biochem. J. 1950 55.		
4908-002		

BENZALDEHYDE, 3-Ac-6-OMe-	$C_{10}H_{10}O_3$	178 OI 144^{0}C
JACS. 1948 1249.		
4909-002		

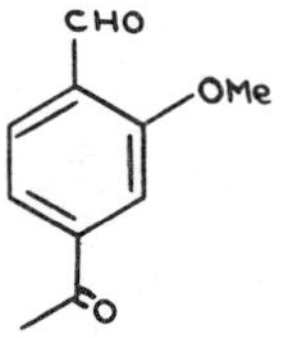

TOLYLMETHYL CARBINOL	$C_9H_{12}O$	136 Oil
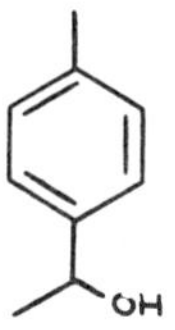	Arch. Pharm. 1933 337.	
4909-003		
CRYPTONE	$C_9H_{14}O$	138 -119^0e Oil
	JACS. 1955 1003.	
4909-008		
ANISOLE, Me-vinyl-	$C_{10}H_{12}O$	148
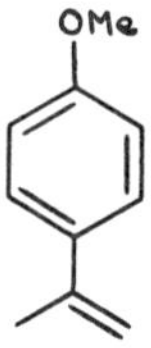	Can. J. C. 1968 3743.	
4909-004		
CYCLOHEXANONE, 2 2.6- -tri-Me-	$C_9H_{16}O$	140 Oil
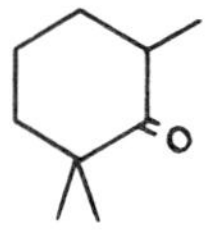	Chem. Abstr. 1951 (45) 3996.	
4909-009		
AUSTRALOL	$C_9H_{12}O$	136 OI 60^0C
	Chem. Abstr. 1926 (20) 2560.	
4909-005		
CYCLOHEXANONE, 2-OH- 2, 6, 6-tri-Me-	$C_9H_{16}O_2$	156
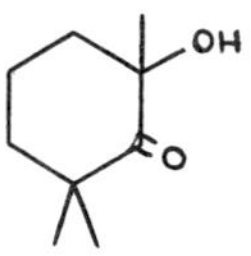	Helv. 1969 549.	
4909-010		
CHAMENOL	$C_{10}H_{14}O_2$	166 OI 42^0C
	Chem. Abstr. 1956 7081.	
4909-006		
PALASONIN	$C_9H_{10}O_4$	182 110^0C
	Tet. Lett. 1968 1971.	
4909-011		
THYMOQUINONE	$C_{10}H_{12}O_2$	164 OI 46^0C
	JOC. 1955 788.	
4909-007		
CANTHARIDIN	$C_{10}H_{12}O_4$	196 218^0C
	JACS 1953 384.	
4910-001		

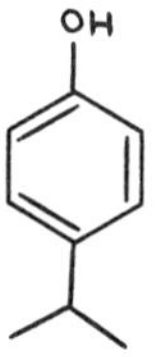

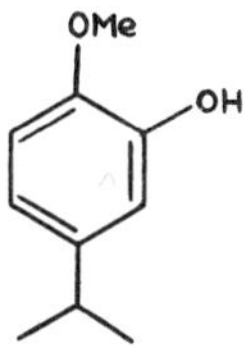

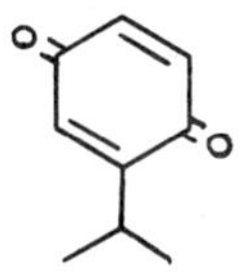

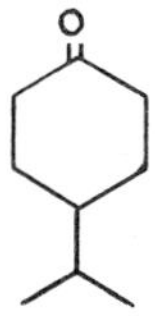

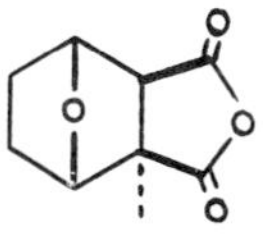

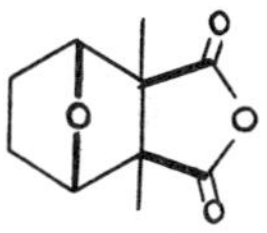

$C_{15}H_{20}O_3$ 248 Oil O.Ang / CHO Ber. 1969 (102) 2211. 4910-002	LOLIOLIDE, deoxy- $C_{11}H_{16}O_2$ JOC 1968 2819. 4911-004
$C_{17}H_{22}O_5$ 306 Oil OAc / CHO Ber. 1969 (102) 2211. 4910-003	$C_{12}H_{20}O_3$ 212 COOH / CH₂ OH Tet. Lett., 1968, 4893. 4912-001
ACTINIDIOLIDE $C_{11}H_{14}O_2$ 178 +77⁰ Tet. Lett. 1968 5325. 4911-001	CURCUMONE $C_{12}H_{16}O$ 176 +81⁰ Oil Helv., 1934, 372. 4912-002
ACTINIDIOLIDE, dihydro- $C_{11}H_{16}O_2$ 189 +7.1⁰ 40⁰C *dihydro- Tet. Lett. 1968 5325. 4911-002	NAPHTHALENE, 1,6-DIMETHYL $C_{12}H_{12}$ 156 Chem. Abstr. 1970 (73) 438925. 4912-003
LOLIOLIDE $C_{11}H_{16}O_3$ 196 -92⁰c 148⁰C HO Tet. 1964 1463. 4911-003	TETRAL-1-ONE, 4,7-DIMETHYL $C_{12}H_{14}O$ 174 Chem. Abstr. 1970 (73) 438925. 4912-004

SOLANONE	$C_{13}H_{24}O$	196	ROMALEA ALLENE	$C_{13}H_{20}O_3$	224 128°C
	JOC. 1966 1797.			Tet. Lett. 1968 2959.	
4913-001			4913-006		
EUXINE	$C_{13}H_{20}$	176 Oil	IONONE, α,	$C_{13}H_{20}O$	192 +379° Oil
	Chim. Ind. 1970 (52) 581.			Chem. Abstr. 1956 896.	
4913-002			4913-007		
ACTINIDOL	$C_{13}H_{20}O_2$	208	IONONE, β-	$C_{13}H_{20}O$	192 OI
	Tet. Lett. 1967 1623.			Chem. Abstr. 1956 896.	
4913-003			4913-008		
THEASPIRONE	$C_{13}H_{20}O_2$	208	IONONE, cis-DIHYDRO-	$C_{13}H_{22}O$	194
	Tet. Lett. 1968 2777.			Helv. 1949 1064.	
4913-004			4913-009		
VOMIFOLIOL	$C_{13}H_{20}O_3$	224 +233°c 115°C	IONENE, 3, 4-dehydro-	$C_{13}H_{16}$	172 OI
	Tet. Lett. 1969 1173.			Phytochemistry 1970 1159.	
4913-005			4913-010		

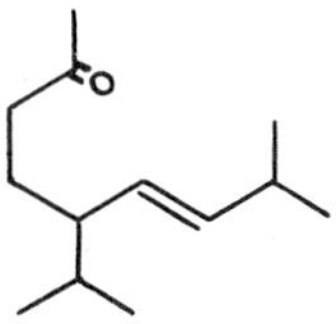

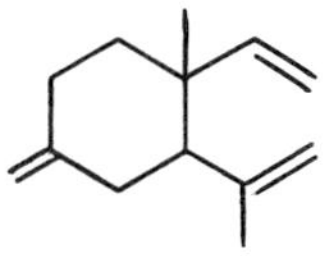

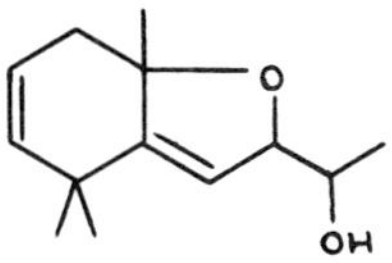

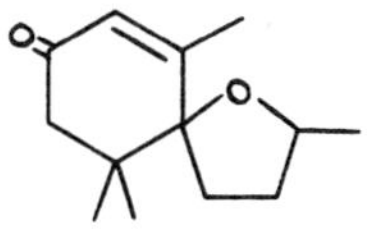

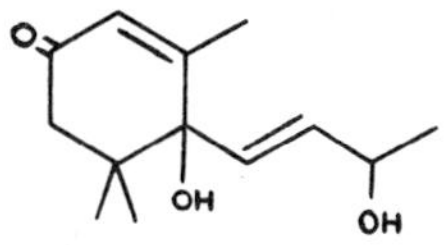

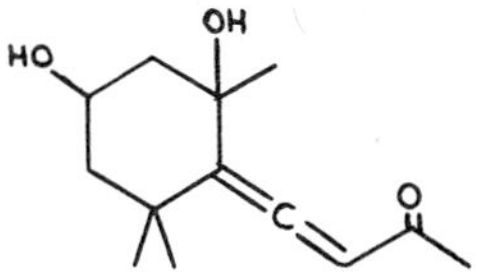

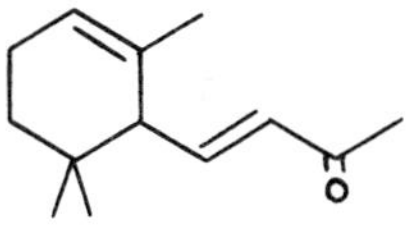

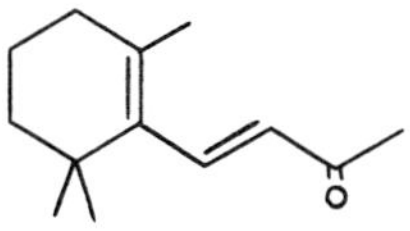

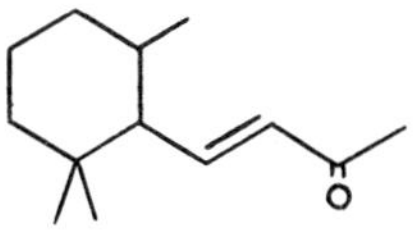

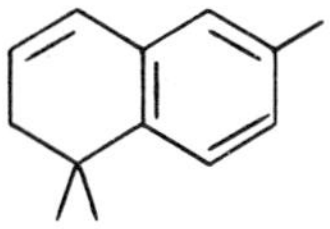

IRONE, α- $C_{14}H_{22}O$ 206 +70°e Oil Helv., 1952, 1820. 4914-001	LATIA LUCIFERIN $C_{15}H_{24}O_2$ 236 O·CHO Tet. Lett. 1969 3951. 4914-006
IRONE, neo-α- $C_{14}H_{22}O$ 206 -8.3° Oil stereoisomer Helv., 1948, 1876. 4914-002	TRISPORONE $C_{15}H_{22}O_3$ 250 +19° O, CH_2OH, CH_2OH Chem. Abstr. 1968 (68) 3004. 4915-001
IRONE, β- $C_{14}H_{22}O$ 206 + 44° Oil Helv., 1948, 257. 4914-003	TRISPORONE, deoxy- $C_{15}H_{22}O_2$ 234 OI O, CH_2 OH Chim. Ind. 1970 (52) 584. 4915-002
IRONE, γ - $C_{14}H_{22}O$ 206 +43° Oil Helv. 1952 1826. 4914-004	TRISPORONE, anhydro- $C_{15}H_{20}O_2$ 232 -84° 110°C Chem. Abstr. 1968 (68) 3004. 4915-003
FRAXINELLONE $C_{14}H_{16}O_3$ 232 -38°c 116°C Chem. Comm. 1969 828. 4914-005	ABSCISIN II $C_{15}H_{20}O_4$ 264 R = CH_3 COOH, OH, R Chem. Comm 1967 114. 4915-004

METABOLITE C $C_{15}H_{20}O_5$ 280 R = CH_2OH Chem. Comm. 1969 966. 4915-005	LL-Z1271-GAMMA $C_{16}H_{18}O_5$ 290 -259⁰m 239⁰C R = H JACS. 1969 2134. 4916-002
PHASEIC ACID $C_{15}H_{20}O_5$ 280 208⁰C COOH, OH Chem. Comm. 1969 966. 4915-006	AMBREINOLIDE, iso- $C_{17}H_{28}O_2$ 264 98⁰C Chem. Abstr. 1967 (67) 54276. 4917-001
XANTHOXIN, trans, trans- $C_{15}H_{22}O_3$ 250 HO, CHO Nature 1970 (227) 302. 4915-007	FARNESYL ACETONE $C_{18}H_{30}O$ 262 OI Oil Chem. Ind. 1969 238. 4918-001
XANTHOXIN, cis, trans- $C_{15}H_{22}O_3$ 250 *cis Nature 1970 (227) 302. 4915-008	TRISPORIC ACID B, 13-keto- $C_{18}H_{24}O_4$ 304 Experimentia 1970 (26) 348. COOH 4918-002
LL-Z1271-ALPHA $C_{17}H_{20}O_5$ 304 -203⁰m 215⁰C R = Me OR, CO—O JACS. 1969 2134. 4916-001	TRISPORIC-C ACID $C_{18}H_{26}O_4$ 306 R = COOH Chem. Comm. 1970 255. OH, R 4918-003

TRISPOROL C — $C_{18}H_{28}O_3$ — 292

R = CH_2OH

Expermentia 1970 (26) 348.

4918-004

VITAMIN A_2 — $C_{20}H_{28}O$ — 284, OI

R = CH_2OH

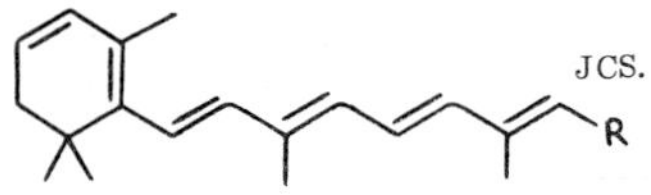

R

JCS. 1955 2765.

4920-003

ZAMENE — $C_{19}H_{38}$ — 266, Oil

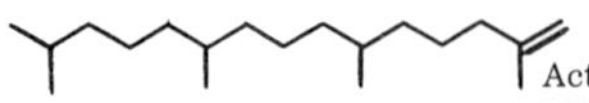

Acta Chem. Scand. 1951 (5) 751.

4919-001

RETINENE-2 — $C_{20}H_{26}O$ — 282, IO, 77-8^{0}C

R = CHO

JCS. 1952 2657.

4920-004

PRISTANE — $C_{19}H_{40}$ — 268, 290^{0}C

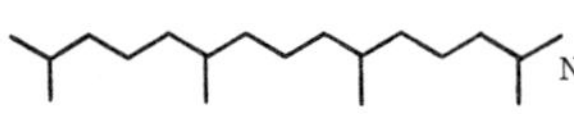

Nature 1963 (140) 974.

4919-002

MOENOCINOL — $C_{25}H_{42}O$ — 358

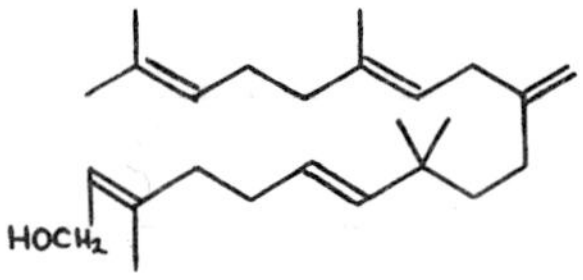

$HOCH_2$

Tet. Lett. 1968 2905.

4925-001

VITAMIN A_1 — $C_{20}H_{30}O$ — 286, OI, 63^{0}C

R = CH_2OH

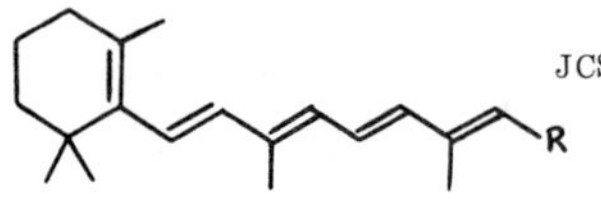

R

JCS. 1955 2765.

4920-001

MOENOCINOL, iso- — $C_{25}H_{42}O$ — 358

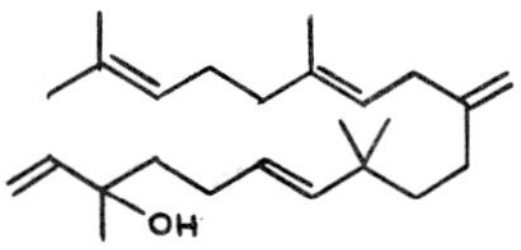

OH

Tet. Lett. 1968 2905.

4925-002

RETINENE-1 — $C_{20}H_{28}O$ — 284, OI, 64-5^{0}C

R = CHO

JCS. 1952 2657.

4920-002

DIUMYCINOL — $C_{25}H_{42}O$ — 358, +6^{0}e

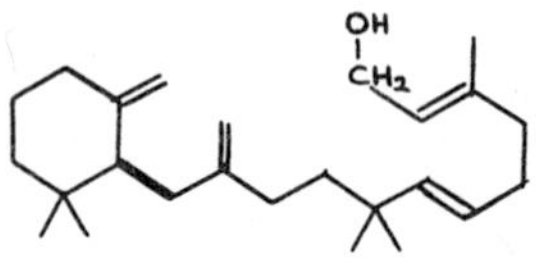

OH
CH_2

JACS 1970 4486.

4925-003

DIUMYCINOL, iso- $C_{25}H_{42}O$ 358 $+8^0e$ JACS 1970 4486. 4925-004	KITOL $C_{40}H_{60}O_2$ 572 -3^0c 136^0C Bull. Soc. Chim. Fr. 1966 3299. 4940-001
DIUMYCENE $C_{25}H_{40}$ 340 JACS 1970 4486. 4925-005	

INDICES GUIDE

In an attempt to cater for maximum access to this handbook three indices have been provided in addition to the structural guides.

The Molecular Weight Index provides a listing of the classification numbers in ascending molecular weight sequence.

The Molecular Formula Index alternatively provides a listing of accession numbers in ascending order of the number of C atoms in the molecular formula; within each carbon number, hydrogen, nitrogen, oxygen, etc., are similarly sequenced.

The Alphabetical Index provides access to the handbook by compound name and also serves to indicate alternative names for compounds. No attempt has been made to apply IUPAC nomenclature for compound names – those used in the index are in general the trivial names found commonly in the literature. In view of the structural organisation of this handbook there should be no special need to provide IUPAC nomenclature. It should be noted that in the alphabetical listing the common prefixes (iso, neo, dihydro, etc.) are placed *after* the parent name of the compound unless the prefix has some more meaningful connotation (e. g. cyclo- in the Buxus alkaloids). Thus:

4204-024	ABIETIC ACID
4204-025	ABIETIC ACID, DEHYDRO-
4204-041	ABIETIC ACID, 4-EPI-
4204-027	ABIETIC ACID, NEO-
4204-039	ABIETIC ACID, 15-OH-

and

4101-003	FARNESOL
4191-047	FARNESOL, CYCLO-

but

4417-001	CYCLOARTANOL (No such compound as ARTANOL)
4419-042	CYCLOBUXINE-D (Refers to the Cyclobuxane skeleton)

ALPHABETICAL INDEX

4508-019 ANDROST-4-EN-17-OL-3-ONE
4508-002 ANDROST-5-EN-3,17-DIOL
4508-005 ANDROST-5-EN-3,16,17-TRIOL
4508-004 ANDROST-5-ENE-3,16,17-TRIOL
4508-013 ANDROST-5-EN-17-ONE, 3,16-DIOH-
4508-018 ANDROSTA-1,4-DIEN-17-OL-3-ONE
4508-017 ANDROSTA-1,4-DIEN-12-OL-3,17-DIONE
4508-006 ANDROSTA-3,5-DIEN-17-ONE
4508-001 ANDROSTAN-3 ALPHA, 17 BETA-DIOL, 5 BETA-
4508-003 ANDROSTAN-3,16,17-TRIOL, 5-ALPHA-H-
4508-012 ANDROSTAN-17-ONE, 3,11-DI-OH-
4508-011 ANDROSTENOLONE
4508-007 ANDROSTERONE
4508-009 ANDROSTERONE, 5-BETA-
4508-008 ANDROSTERONE-3 BETA-GLUCURONOPYRANOSIDE
4106-035 ANGELIKOREANOL
4428-004 ANGOLENSATE,METHYL-
4428-006 ANGOLENSATE,6-OAC-ME-
4428-005 ANGOLENSATE,6-OH-ME-
4424-018 ANGUSTIDIENOLIDE
4424-019 ANGUSTIDIENOLIDE,2-HYDROXY-
4424-017 ANGUSTINOLIDE
4192-074 ANISATIN
4192-075 ANISATIN,NEO-
4909-004 ANISOLE, ME-VINYL-
4008-010 ANISOMORPHAL
4191-006 ANTHEMOIDIN
4002-098 ANTHEMOL
4603-011 ANTHERAXANTHIN
4503-038 ANTHERIDIOL
4426-011 ANTHOTHECOL
4505-133 ANTIARIGENIN
4505-134 ANTIARIN,ALPHA-
4505-135 ANTIARIN,BETA-
4008-066 ANTIRRHIDE
4008-051 ANTIRRHINOSIDE
4106-012 ANYMOL
4493-042 APETALACTONE
4405-023 APHANAMIXIN
4428-003 APHANAMIXININ
4202-020 APHANAMIXOL = LABD-13-EN-8,15-DIOL, ENT-
4603-021 APHANICIN = CANTHAXANTHIN
4306-016 APHANIN = ECHINENONE
4104-010 APITONENE-I
4111-003 APLYSIN
4111-004 APLYSIN,DE-BR-
4111-005 APLYSINOL
4202-025 APLYSIN-20
4203-019 ARAUCARENOLONE
4203-020 ARAUCAROL
4203-017 ARAUCAROLONE
4203-018 ARAUCARONE
4104-069 ARBIGLOVIN
4104-057 ARBORESCIN
4422-001 ARBORINOL
4422-002 ARBORINOL ME ETHER
4422-003 ARBORINOL,ISO-
4422-005 ARBORINONE
4102-062 ARCTIOPICRINE
4506-018 ARENOBUFAGIN
4120-073 ARGLANINE
4102-012 ARISTOLACTONE
4119-004 ARISTOLA-1(10),8-DIEN-2-ONE
4119-006 ARISTOLENE,D- = FERULENE, ALPHA-
4119-005 ARISTOLENE,L-
4119-003 ARISTOLEN-2-ONE, 1(10)-
4119-008 ARISTOLONE
4003-024 ARITASONE
4408-094 ARJUNOLIC ACID
4408-119 ARMILLARIGENIN
4413-006 ARNENEDIOL = ARNIDIOL
4413-006 ARNIDIOL
4125-001 AROMADENDRENE,D-
4125-002 AROMADENDRENE,L-
4114-025 AROMATICIN
4114-024 AROMATIN
4104-056 ARTABSINE
4506-005 ARTEBUFOGENIN A
4506-006 ARTEBUFOGENIN B
4291-015 ARTEMISENE
4099-008 ARTEMISIA ALCOHOL
4099-006 ARTEMISIA KETONE
4099-007 ARTEMISIA KETONE,ISO-
4102-043 ARTEMISIIFOLIN
4120-072 ARTEMISIN
4002-039 ARTEMISOL
4190-001 ARTEMONE
4102-080 ARTEMORIN
4120-075 ARTICALIN
4107-041 ARTICULOL
4107-053 ARTICULONE
4402-007 ARUNDOIN
4402-011 ARUNDOIN,12-OXO-
4002-054 ASCARIDOL
4505-012 ASCLEPOSIDE
4502-012 ASCOSTEROL
4218-001 ASEBOTOXIN I
4218-002 ASEBOTOXIN II
4218-003 ASEBOTOXIN III
4409-017 ASIATIC ACID
4409-023 ASIATIC ACID,11-OXO-
4493-046 ASIATIC ACID ME ESTER, 2,3-SECO-
4504-046 ASPERAGENIN
4120-057 ASPERILIN
4008-043 ASPERULOSIDE
4008-044 ASPERULOSIDE,DES-AC-
4603-022 ASTAXANTHIN
4603-023 ASTAXANTHIN DI-PALMITATE
4604-005 ASTAXANTHIN, 7,8-DI-DEHYDRO-
4604-009 ASTAXANTHIN, 7,8,7',8'-TETRA-DEHYDRO-
ASTERINSAURE = MIXTURE
4212-003 ATISINE
4212-001 ATISIRENE
4212-002 ATISIRENE,ISO-
4106-020 ATLANTONE,ALPHA-
4106-021 ATLANTONE,GAMMA-
4293-045 ATRACTYLIGENIN
4122-006 ATRACTYLOL = HINESOL
4120-041 ATRACTYLONE
4293-046 ATRACTYLOSIDE
4008-047 AUCUBIN
AUROXANTHIN = ARTIFACT
4003-001 AUSTRALENE = PINENE, D-ALPHA-
4909-005 AUSTRALOL
4104-078 AUSTRICINE = MATRICARIN, DES-AC-
4114-056 AUTUMNOLIDE
4213-031 AVADHARIDINE
4503-011 AVENASTEROL, 5-DEHYDRO- = FUCOSTEROL, ISO-
4503-017 AVENASTEROL,7-DEHYDRO-
4193-004 AXIVALIN
4426-008 AZADIRADIONE
4426-009 AZADIRADIONE,EPOXY-
4426-007 AZADIRONE
4699-020 AZAFRIN
4192-007 AZULENE
4192-006 AZULENE,1-COOME-4-ME-
4192-005 AZULENE,1,4-DIME-
4192-001 AZULENE, 4-ME-7-ISOPROPYL-
4215-010 BACCATIN
4215-011 BACCATIN-I
4215-012 BACCATIN-I,5-DE-AC-
4215-013 BACCATIN-I,1-OH-
4215-010 BACCATIN III = BACCATIN
4215-014 BACCATIN-V
4499-008 BACCHARIS OXIDE
4404-018 BACOGENIN
4404-019 BACOSIDE A
4601-026 BACTERIOERYTHRIN
4601-026 BACTERIOPURPURIN = BACTERIOERYTHRIN
4699-028 BACTERIORUBERIN
4701-013 BACTOPRENOL
4104-084 BADKHYSIN
4104-087 BAHIA I
4104-088 BAHIA II
4515-006 BAIKEINE
4014-010 BAKANKOSIDE
4014-010 BAKANKOSIN = BAKANKOSIDE
4192-053 BAKKENOLIDE A = FUKINANOLIDE
4192-056 BAKKENOLIDE C
4192-055 BAKKENOLIDE D = S-FUKINOLIDE
4120-062 BALCHANIN = SANTAMARINE
4102-050 BALCHANOLIDE
4102-051 BALCHANOLIDE,ISO-
4102-053 BALCHANOLIDE,AC-
4102-052 BALCHANOLIDE,OH-
4114-031 BALDUILIN A
4419-034 BALEABUXIDINE
4419-016 BALEABUXINE
4210-025 BAMBOO GIBBERELLIN = GIBBERELLIN A19
4408-026 BARRIGENOL,A1-
4408-020 BARRIGENOL A2 = CAMELLIAGENIN A
4408-036 BARRIGENOL,R1-
4408-027 BARRIGENOL R2 = BARRINGTOGENOL C
4408-035 BARRIGENOL, 7-OH-A1-
4408-109 BARRINGTOGENIC ACID
4408-018 BARRINGTOGENOL
4408-031 BARRINGTOGENOL B
4408-027 BARRINGTOGENOL C
4408-015 BARRINGTOGENOL C,16-DESOXY-
4408-016 BARRINGTOGENOL D
4408-032 BARRINGTOGENOL E
4504-004 BASSEOL = BUTYROSPERMOL
4408-093 BASSIAIC ACID
4001-030 BATATIC ACID
4592-025 BATRACHOTOXIN
4592-026 BATRACHOTOXIN, HOMO-
4592-024 BATRACHOTOXININ-A
4492-013 BAUERENOL
4492-014 BAUERENOL,ISO-
4192-051 BAZZANENE

4505-091 CALOTROPAGENIN
4505-092 CALOTROPIN
4408-020 CAMELLIAGENIN A
4408-021 CAMELLIAGENIN A,16-CINNAMATE
4408-047 CAMELLIAGENIN B
4408-029 CAMELLIAGENIN C
4408-049 CAMELLIAGENIN D
4414-004 CAMPANULIN = DENDROPANOXIDE
4502-019 CAMPEST-4-EN-3-ONE
4502-003 CAMPESTEROL
4099-003 CAMPHENE,D-
4099-022 CAMPHENE,DL-
4099-021 CAMPHENE,L-
4192-057 CAMPHERENOL
4192-058 CAMPHERENONE
4004-004 CAMPHOR,D-
4004-005 CAMPHOR,L-
4004-007 CAMPHOR, (+-)-
4120-020 CAMPHOR,(+)-JUNIPER-
4299-001 CAMPHORENE,ALPHA-
4191-046 CANABRIN
4493-047 CANARIC ACID
4120-026 CANARONE
4206-010 CANDICANDIOL
4505-141 CANESCEIN
4099-017 CANNABICYCLOL
4099-017 CANNABIPINOL = CANNABICYCLOL
4403-010 CANOPHYLLAL
4403-011 CANOPHYLLIC ACID
4403-009 CANOPHYLLOL
4910-001 CANTHARIDIN
4603-021 CANTHAXANTHIN
4101-005 CAPARRAPIDIOL
4101-006 CAPARRATRIOL
4192-067 CAPENICIN
4503-041 CAPITASTERONE
4606-004 CAPSANTHIN
4606-005 CAPSORUBIN
4192-010 CARABRONE
4424-014 CARAPIN
4424-016 CARAPIN,6-OH-
4424-015 CARAPIN ACETATE,3-BETA-DIHYDRO-
4501-003 CARCINOLIPIN
4005-002 CARENE,D-3-
4005-001 CARENE,L-3-
4005-003 CARENE,4-D-
4003-016 CARENE-5,6-EPOXIDE,L-3- = CHRYSANTHENONE
4603-007 CARICAXANTHIN = CRYPTOXANTHIN
4120-031 CARISSONE
4404-005 CARNAUBADIOL
4504-036 CARNEAGENIN,ISO-
4204-032 CARNOSOL
4699-018 CAROTENAL,BETA-APO-8'-
4699-031 CAROTENAL,BETA-APO-10'-
4603-001 CAROTENE,ALPHA-
4603-003 CAROTENE,BETA-
4602-007 CAROTENE,DELTA-
4601-002 CAROTENE, ETA-
4602-001 CAROTENE,GAMMA-
4601-004 CAROTENE,ZETA-
4602-003 CAROTENE,1,2-DIHYDRO-1-OH-GAMMA-
4602-004 CAROTENE,1,2-DIHYDRO-1-OH-4-KETO-GAMMA-
4603-002 CAROTENE EPOXIDE,ALPHA-
4603-004 CAROTENE EPOXIDE, BETA-
4602-002 CAROTENE,4-KETO-GAMMA-
4699-001 CAROTENONE, SEMI-ALPHA-
4699-002 CAROTENONE, SEMI-BETA-
4108-001 CAROTOL
4104-070 CARPESIA LACTONE
4492-016 CARPESTEROL
4099-019 CARQUEJOL ACETATE
4002-044 CARVACROL
4002-045 CARVACROL ME ETHER
4002-046 CARVACROL,P-METHOXY-
4002-023 CARVEOL
4002-014 CARVEOL,D-DIHYDROL
4002-016 CARVEOL,D-NEO-DIHYDRO-
4002-017 CARVEOL,L-NEODIHYDRO-
4002-015 CARVEOL,L-ISODIHYDRO-
4002-074 CARVOMENTHONE,L-
4002-071 CARVONE
4002-072 CARVONE,L-
4002-073 CARVONE,L-DIHYDRO-
4002-075 CARVOTACETONE,D-
4116-001 CARYOPHYLLENE
4116-002 CARYOPHYLLENE, D-BETA-
4116-003 CARYOPHYLLENE, L-BETA-
4116-004 CARYOPHYLLENE,GAMMA- = CARYOPHYLLENE,ISO-
4116-005 CARYOPHYLLENE OXIDE
4116-004 CARYOPHYLLENE,ISO-
4291-021 CASBENE
4217-025 CASCARILLIN
4217-043 CASCARILLIN A
4207-003 CASSAIC ACID
4207-011 CASSAIDIC ACID
4207-012 CASSAIDINE
4207-004 CASSAINE
4207-007 CASSAMIC ACID
4207-008 CASSAMINE
4701-005 CASTAPRENOL-10
4701-006 CASTAPRENOL-11
4701-007 CASTAPRENOL-12
4701-008 CASTAPRENOL-13
4198-010 CASTORAMINE
4008-053 CATALPOL
4008-054 CATALPOL-6-ME ETHER
4008-055 CATALPOSIDE
4202-065 CATIVIC ACID
4202-084 CATIVIC ACID,6-OXO-
4103-008 CAUCALOL,DI-ACETATE
CAUDOGENIN = ARTEFACT
4505-115 CAUDOSIDE, PSEUDO-
4494-002 CEANOTHENIC ACID
4494-005 CEANOTHIC ACID
4101-017 CECROPIA JUVENILE HORMONE
4505-034 CEDILANIDE = LANOTOSIDE C
4105-006 CEDR-8-EN-13-OL-
4105-008 CEDRANE-8,9-DIOL
4105-007 CEDRANEDIOL,8S,14-
4105-011 CEDRANOLIDE,8,14-
4105-012 CEDRANOXIDE,8,14-
4424-006 CEDRELA ODORATA B = MEXICANOLIDE
4107-047 CEDRELANOL = TORREYOL,(-)-
4426-010 CEDRELONE
4105-001 CEDRENE,ALPHA-
4105-002 CEDRENE,BETA-
4105-003 CEDROL
4105-013 CEDROLIC ACID
4493-066 CEDRONINE
4493-065 CEDRONYLINE
4493-027 CELASTROL
4602-011 CELAXANTHIN
4216-004 CEMBRENE
4216-005 CEMBRENE,ISO-
4216-006 CEMBROL
4216-007 CEMBROL,ISO- = THUNBERGOL
4302-004 CEPHALONIC ACID = OPHIOBOLIN D
4429-006 CEPHALOSPORIN P1
4429-008 CEPHALOSPORIN P1,ISO-
4429-007 CEPHALOSPORIN P1,MONODES-AC-
4505-023 CERBERIN = NERIIFOLIN, AC-
4505-145 CERBERTIN
4505-146 CERBERTIN,DEACETYL-
4502-018 CEREVISTEROL
4403-006 CERIN
4302-008 CEROPLASTERIC ACID
4302-006 CEROPLASTOL I
4302-007 CEROPLASTOL II
4516-010 CEVACINE
4516-012 CEVADINE
4516-011 CEVINE = VERACEVINE
4518-009 CHACONINE, ALPHA-
4518-006 CHACONINE, BETA-1-
4518-007 CHACONINE, BETA-2-
4518-005 CHACONINE, GAMMA-
4099-028 CHAKSINE
4011-013 CHAMAECIN
4192-043 CHAMAECYNENAL,DEHYDRO-
4192-040 CHAMAECYNENOL
4192-041 CHAMAECYNENOL ACETATE
4192-042 CHAMAECYNENOL,DEHYDRO-
4192-036 CHAMAECYNONE
4192-037 CHAMAECYNONE,ISO-
4192-038 CHAMAECYNONE,OH-ISO-
4192-039 CHAMAECYNONE,DIHYDRO-ISO-
4192-002 CHAMAZULENE
4104-012 CHAMAZULENE-COOH
4192-003 CHAMAZULENE,3,6-DIHYDRO-
4192-004 CHAMAZULENE,5,6-DIHYDRO-
4104-055 CHAMAZULENOGEN,PRO-
4909-006 CHAMENOL
4130-003 CHAMIGRENAL
4130-002 CHAMIGRENE
4130-001 CHAMIGRENE,ALPHA-
4005-004 CHAMIC ACID
4005-005 CHAMINIC ACID
4102-077 CHAMISSONIN
4192-016 CHANOOTINE
4425-027 CHAPARRIN
4425-015 CHAPARRINONE
4213-013 CHASMACONITINE
4213-006 CHASMANINE
4213-007 CHASMANINE,HOMO-
4217-028 CHASMANTHIN
4213-012 CHASMANTHININE
4509-009 CHEILANTHONE A
4509-008 CHEILANTHONE-B
4505-008 CHEIROSIDE A
4505-009 CHEIROSIDE H
4505-088 CHEIROTOXIN

4505-038	COROTOXIGENIN
4517-006	CORSEVINE
4507-056	CORTICOSTERONE
4507-058	CORTICOSTERONE, 11-DEHYDRO-
4507-050	CORTICOSTERONE, DEOXY-
4507-064	CORTICOSTERONE, 17-OH-11-DEOXY-
4507-055	CORTICOSTERONE, TETRAHYDRO-
4507-081	CORTISOL
4507-082	CORTISONE
4507-025	CORTOL
4507-026	CORTOL,BETA-
4507-027	CORTOLONE
4507-028	CORTOLONE,BETA-
4507-029	CORTOLONE-3-GLUCURONOSIDE,BETA-
4206-014	CORYMBOL
4299-002	CORYMBOSITIN
4001-004	COSMENE
4120-079	COSTAL,ALPHA-
4120-080	COSTAL,BETA-
4120-032	COSTIC ACID
4120-014	COSTOL,ALPHA-
4102-044	COSTUNOLIDE
4102-045	COSTUNOLIDE,DIHYDRO-
4102-046	COSTUNOLIDE,OH-
4104-045	COSTUS LACTONE,DEHYDRO-
4207-006	COUMINGIDINE
4207-005	COUMINGIDINIC ACID
4207-002	COUMINGINE
4207-001	COUMINGINIC ACID
4216-014	CRASSIN ACETATE
4408-078	CRATAEGOLIC ACID
4502-007	CRINOSTEROL
4699-022	CROCETIN
4699-024	CROCETIN ALDEHYDE
4699-025	CROCETIN DI-ALDEHYDE
4699-023	CROCIN
4699-023	CROCIN,ALPHA-
4604-001	CROCOXANTHIN
4110-008	CROTOCIN
4299-004	CROTONIN
4416-003	CRUSGALLIN = SAWAMILLETIN
4603-024	CRUSTAXANTHIN
4509-021	CRUSTECDYSONE
4509-018	CRUSTECDYSONE,DESOXY-
4129-004	CRYPTOFAURONOL
4603-028	CRYPTOFLAVIN
4501-019	CRYPTOGENIN
4505-107	CRYPTOGRANDOSIDE A
4505-108	CRYPTOGRANDOSIDE B
4204-010	CRYPTOJAPONOL
4120-024	CRYPTOMERIDIOL
4106-022	CRYPTOMERIONE
4106-023	CRYPTOMERONE
4909-008	CRYPTONE,L-
4203-022	CRYPTOPINONE
4603-007	CRYPTOXANTHIN
4603-006	CRYPTOXANTHIN, ALPHA-
4603-008	CRYPTOXANTHIN EPOXIDE
4603-007	CRYPTOXANTHOL = CRYPTOXANTHIN
4214-007	CUAUCHICHICINE
4193-020	CUBEBENE,ALPHA-
4193-021	CUBEBENE,BETA-
4193-019	CUBEBOL
4107-038	CUBENOL
4107-039	CUBENOL,EPI-
4412-004	CUCURBITACIN A
4412-005	CUCURBITACIN B
4412-006	CUCURBITACIN B,2-EPI-
4412-007	CUCURBITACIN C
4412-008	CUCURBITACIN D
4412-009	CUCURBITACIN D,22-DE-OXO-
4412-010	CUCURBITACIN D,22-DE-OXO-ISO-
4493-031	CUCURBITACIN D, HEXANOR-
4412-011	CUCURBITACIN E
4412-012	CUCURBITACIN F
4412-013	CUCURBITACIN G
4412-014	CUCURBITACIN H
4412-015	CUCURBITACIN I
4412-017	CUCURBITACIN J
4412-018	CUCURBITACIN K
4412-016	CUCURBITACIN L
4412-019	CUCURBITACIN O
4412-020	CUCURBITACIN P
4412-021	CUCURBITACIN Q
4193-033	CULMORIN
4104-051	CUMAMBRIN A
4104-052	CUMAMBRIN B
4114-040	CUMANIN
4002-083	CUMINALDEHYDE
4002-052	CUMINYL ALCOHOL
4115-003	CUPARENE
4115-001	CUPARENENE,ALPHA-
4115-002	CUPARENENE,BETA-
4115-010	CUPARENIC ACID
4115-004	CUPARENOL,ALPHA-
4115-005	CUPARENOL,ALPHA-ISO-
4115-006	CUPARENOL,BETA-
4115-007	CUPARENOL,GAMMA-
4115-008	CUPARENONE,ALPHA-
4115-009	CUPARENONE,BETA-
4209-002	CUPRESSENE = HIBAENE, (-)-
4202-055	CUPRESSIC ACID,ISO-
4202-054	CUPRESSIC ACID
4120-039	CURCOLONE
4106-006	CURCUMENE,ALPHA-D-
4106-007	CURCUMENE,ALPHA-L-
4106-006	CURCUMENE,AR- = CURCUMENE,ALPHA-D-
4106-009	CURCUMENE,BETA-L-
4106-008	CURCUMENE,BETA-D-
4106-010	CURCUMENE,GAMMA-
4104-025	CURCUMENOL
4104-024	CURCUMENOL,PRO-
4104-083	CURCUMOL
4912-002	CURCUMONE
4102-011	CURDIONE
4121-015	CURZERENE = GERMACRENE,ISO-FURANO-
4121-020	CURZERENONE
4121-021	CURZERENONE,EPI-
4107-073	CURZERENONE, PYRO-
4503-039	CYASTERONE
4503-043	CYASTERONE, PRE-
4507-046	CYBISTEROL
4507-011	CYBISTERONE
4408-053	CYCLAMIGENIN A1
4493-025	CYCLAMIGENIN A2
4408-052	CYCLAMIGENIN B
4408-054	CYCLAMIGENIN C
4408-055	CYCLAMIGENIN D
4408-050	CYCLAMIRETIN A
4408-051	CYCLAMIRETIN D
4417-001	CYCLOARTANOL
4417-016	CYCLOARTANOL,24-METHYLENE-
4417-020	CYCLOARTANOL, 24-ME-24-OME-
4493-015	CYCLOARTANOL, 29-NOR-
4417-003	CYCLOARTANOL, 25-OH-
4417-004	CYCLOARTANOL, 25-OET-
4417-021	CYCLOARTAN-3,27-DIOL, 24-METHYLENE-
4417-023	CYCLOARTANONE,24-METHYLENE-
4417-002	CYCLOARTENOL
4417-008	CYCLOARTENONE
4417-007	CYCLOART-23-ENE, 3-OAC-25-OME-
4417-006	CYCLOART-23-EN-3,25-DIOL
4417-022	CYCLOART-24-EN-27-AL, 3-OH-
4417-005	CYCLOART-25-EN-3,4-DIOL
4417-009	CYCLOART-25-EN-24-ONE, 3-OAC-
4419-024	CYCLOBALEABUXINE = CYCLOMICROPHYLLINE B
4417-017	CYCLOBRANOL
4099-023	CYCLOBUTANE,1-(2-OH-ETHYL)-2-ISOPROPENYL-1-ME-
4419-041	CYCLOBUXAMINE
4419-035	CYCLOBUXIDINE-F,N-BENZOYL-
4419-042	CYCLOBUXINE D
4418-004	CYCLOBUXOMICREINE
4418-010	CYCLOBUXOPHYLLINE
4418-009	CYCLOBUXOPHYLLININE
4418-007	CYCLOBUXOSUFFRINE
4418-017	CYCLOBUXOVIRIDINE
4419-037	CYCLOBUXOXAZINE-A
4419-036	CYCLOBUXOXAZINE-C
4418-001	CYCLOBUXOXINE = BUXTAUINE
4125-009	CYCLOCOLORENONE
4493-014	CYCLOEUCALENOL
4909-009	CYCLOHEXANONE, 2,2,6-TRI-ME-
4909-010	CYCLOHEXANONE, 2-OH-2,6,6-TRI-ME-
4909-001	CYCLOHEX-1-ENE, 1-ME-4-AC-
4211-002	CYCLOKAURANIC ACID = TRACHYLOBANIC ACID
4419-015	CYCLOKOREANINE B
4417-018	CYCLOLAUDENOL
4419-012	CYCLOMALAYANINE B
4419-038	CYCLOMETHOXAZINE B
4418-003	CYCLOMICROBUXEINE
4418-002	CYCLOMICROBUXINE = BUXPIINE
4418-001	CYCLOMICROBUXININE = BUXTAUINE
4419-027	CYCLOMICROPHYLLIDINE A
4419-026	CYCLOMICROPHYLLINE A
4419-023	CYCLOMICROPHYLLINE-A, DIHYDRO-
4419-024	CYCLOMICROPHYLLINE B
4419-025	CYCLOMICROPHYLLINE C
4419-021	CYCLOMICROPHYLLINE-C, DIHYDRO-
4419-020	CYCLOMICROPHYLLINE-F, DIHYDRO-
4419-022	CYCLOMICROPHYLLINE F,N-BENZOYL-DIHYDRO-
4419-028	CYCLOMICROSINE
4418-016	CYCLOMIKURANINE
4499-004	CYCLONEOLITSIN
4908-005	CYCLOPENTANONE, 2-ISOPROPYL-
4099-026	CYCLOPENTENE, 1-AC-4-ISOPROPENYL-1-
4419-004	CYCLOPROTOBUXINE A
4419-003	CYCLOPROTOBUXINE C
4419-001	CYCLOPROTOBUXINE D
4419-002	CYCLOPROTOBUXINE-D, 21-N-AC-

4509-007 ECDYSONE, DEOXY-
4509-025 ECDYSONE,20,26-DIOH-
4603-016 ECHINENONE
4603-017 ECHINENONE, 3-OH-
4493-026 ECHINOCYSTENOLONE,NOR- = ALBIGENIN
4408-081 ECHINOCYSTIC ACID
4408-082 ECHINOCYSTIC ACID,16-EPI- = COCHALIC ACID
4410-031 ECHINODOL
4505-025 ECHUBIOSIDE
4505-030 ECHUJIN
4193-047 ECSANTALAL, NOR-
4517-007 EDPETINE
4516-037 EDUARDINE
4001-033 EGOMAKETONE
4001-048 EGOMAKETONE,ISO-
4193-046 EKASANTALAL,TRICYCLO-
4412-008 ELATERICIN A = CUCURBITACIN D
4412-015 ELATERICIN B = CUCURBITACIN I
4412-011 ELATERIN, ALPHA- = CUCURBITACIN E
4213-039 ELATINE
4213-037 ELDELIDINE = DELTAMINE
4213-038 ELDELINE = DELTALINE
4121-014 ELEMENAL
4121-001 ELEMENE,D-ALPHA-
4121-002 ELEMENE,D-BETA-
4121-003 ELEMENE,L-BETA-
4121-004 ELEMENE,DELTA-
4121-005 ELEMENOL,DELTA-
4121-006 ELEMENOL,DELTA-EPI-
4121-008 ELEMENONE,ALPHA-
4121-009 ELEMENONE,BETA-
4121-010 ELEMENONE,GAMMA-
4121-007 ELEMOL
4405-008 ELEMOLIC ACID,ALPHA-
4405-009 ELEMOLIC ACID, ALPHA-ISO-
4405-010 ELEMOLIC ACID, BETA-
4405-011 ELEMONIC ACID,BETA-
4009-008 ELENOLIDE
4102-041 ELEPHANTIN
4102-040 ELEPHANTOPIN
4102-074 ELEPHANTOPIN,DEOXY-
4202-007 ELLIOTINOL
4217-035 ELONGATOLIDE A
4217-036 ELONGATOLIDE B
4217-037 ELONGATOLIDE C
4217-038 ELONGATOLIDE D
4217-039 ELONGATOLIDE E
4001-049 ELSHOLTZIDIOL
4001-034 ELSHOLTZIONE
4001-035 ELSHOLTZIONE,BETA-DEHYDRO-
4219-004 EMEMODIN
4505-070 EMICYMARIN
EMICYMARIN, 17-ISO- = ARTIFACT
4494-003 EMMOLACTONE
4494-004 EMMOLACTONE, 22-DEOXY-
4409-007 EMPETROSIDE A
4120-046 ENCELIN
4206-028 ENMEDOL
4219-006 ENMEIN
4219-007 ENMEIN-3-ACETATE
4219-008 ENMEIN,16,17-DIHYDRO-
4206-026 ENMELOL
4206-020 ENMENIN = TRICHOKAURIN
4206-025 ENMENOL
4408-095 ENTAGENIC ACID
4427-015 ENTANDROPHRAGMIN
4202-020 EPERUENEDIOL = LABD-13-EN-8,15-DIOL, ENT-
4202-061 EPERUA-7,13-DIEN-15-OIC ACID
4202-042 EPERUAN-14,15,18-TRIOL, 8,13-EPOXY-
4202-018 EPERUANE-8,15-DIOL
4202-027 EPERUANE-8,15,18-TRIOL
4202-078 EPERU-8(20)-EN-15,18-DIOIC ACID
4202-091 EPERU-8(20)-EN-15,18-DIOIC ACID-BUTENOLIDE
4202-060 EPERU-8(20)-EN-18-OIC ACID,15-OH-
4202-058 EPERU-8(20)-EN-18-OIC ACID,15,16-DIOH-
4202-064 EPERU-7-ENOIC ACID
4202-067 EPERUIC ACID
4502-009 EPISTEROL
4519-009 EQUILENIN
4291-011 EREMOLACTONE
4117-007 EREMOLIGENOL
4117-010 EREMOPHILA-1,11-DIEN-9-ONE,7ALPHA-H-
4117-030 EREMOPHILANDIOL,FURANO-
4117-026 EREMOPHILANE,FURANO-
4117-027 EREMOPHILANE,9-OH-FURO-
4117-050 EREMOPHILAN-14BETA,6ALPHA-OLIDE,FURANO
4117-055 EREMOPHILANE,DIOME-DIHYDRO-FURO-
4117-020 EREMOPHILA-1,11-DIEN-9-ONE,8ALPHA-OH
4117-019 EREMOPHILA-10,11-DIEN-9-ONE,8-OH-
4117-001 EREMOPHILENE
4117-047 EREMOPHILENOLIDE
4117-049 EREMOPHILENOLIDE,6-OH-
4117-009 EREMOPHILONE
4117-012 EREMOPHILONE,ALLO-
4117-017 EREMOPHILONE,OXY-
4117-018 EREMOPHILONE,OXY-DIHYDRO-
4117-036 EREMOPHILONE,FURANO-
4502-001 ERGOSTANOL
4502-022 ERGOSTA-4,6,8(14),22-TETRA-EN-3-ONE
4502-015 ERGOSTA-5,8,17(20)-TRIEN-3-OL
4502-013 ERGOSTEROL
4502-016 ERGOSTEROL,14-DEHYDRO-
4502-017 ERGOSTEROL,24(28)-DEHYDRO-
4502-005 ERGOSTEROL, GAMMA- = FUNGISTEROL
4502-023 ERGOSTEROL PEROXIDE
4502-010 ERGOSTEROL,5,6-DIHYDRO-
4502-021 ERGOSTERONE,ISO-
4502-020 ERGOST-7,22-DIENE-3-ONE
4505-086 ERYSCENOSIDE
4505-084 ERYSIMINE
4505-085 ERYSIMOSIDE
4505-101 ERYSIMOSOL
4408-004 ERYTHRODIOL
4408-010 ERYTHRODIOL,22ALPHA-OH-
4408-005 ERYTHRODIOL-3-PALMITATE
4207-009 ERYTHROPHLAMIC ACID
4207-010 ERYTHROPHLAMINE
4207-023 ERYTHROPHLEADIENOLIC ACID
4207-013 ERYTHROPHLEGUINE
4209-008 ERYTHROXYDIOL A
4291-003 ERYTHROXYDIOL X
4292-002 ERYTHROXYDIOL Y
4292-003 ERYTHROXYDIOL Z
4209-004 ERYTHROXYLOL A
4209-010 ERYTHROXYLOL A ACETATE EPOXIDE
4209-009 ERYTHROXYLOL A EPOXIDE
4209-005 ERYTHROXYLOL B
4208-008 ERYTHROXYTRIOL P
4291-004 ERYTHROXYTRIOL Q
4516-031 ESCHOLERINE
4603-018 ESCHSCHOLTZXANTHIN
4603-019 ESCHSCHOLTZXANTHONE
4408-027 ESCINIDIN = BARRINGTOGENOL C
4699-020 ESCOBIDIN = AZAFRIN
4104-058 ESTAFIATINE
4012-001 EUCARVOL = EUCARVONE
4012-001 EUCARVONE
4120-001 EUDESMENE = SELINENE,ALPHA-
4120-007 EUDESMOL,ALPHA-
4120-008 EUDESMOL,BETA-
4120-009 EUDESMOL,GAMMA-
4291-009 EUNICELLIN
4216-015 EUNICIN
4104-065 EUPACHLORIN
4104-066 EUPACHLORIN ACETATE
4104-067 EUPACHLOROXIN
4104-060 EUPAROTIN
4104-061 EUPAROTIN ACETATE
4107-017 EUPATENE,D-
4107-018 EUPATENE,L-
4102-047 EUPATOLIDE
4102-065 EUPATORIOPICRINE
4104-062 EUPATOROXIN
4104-063 EUPATOROXIN,10-EPI-
4104-064 EUPATUNDIN
4405-004 EUPHA-7,24-DIEN-3-OL = BUTYROSPERMOL
4405-001 EUPHOL
4220-012 EUPHORBIA COMPOUND C
4220-013 EUPHORBIA COMPOUND C'
4405-005 EUPHORBOL
4405-006 EUPHORBOL HEXACOSANOATE
4493-071 EUPTELEOGENIN
4493-069 EURYCOMALACTONE
4493-070 EURYCOMALACTONE, 'DIHYDRO-
4117-034 EURYOPSOL
4117-041 EURYOPSONOL
4913-002 EUXINE
4423-002 EVODIN = LIMONIN
4423-003 EVODOL
4002-094 EVODONE
4505-016 EVOMONOSIDE
4413-005 FARADIOL
4120-047 FARINOSIN
4101-007 FARNESAL
4101-001 FARNESENE,ALPHA-
4101-002 FARNESENE,BETA-
4101-023 FARNESENE, C,T-ALLO-
4101-024 FARNESENE, T,T-ALLO-
4126-002 FARNESIFEROL A
4191-040 FARNESIFEROL B
4191-041 FARNESIFEROL C
4101-003 FARNESOL
4191-047 FARNESOL, TRANS-GAMMA-MONOCYCLO-
4918-001 FARNESYL ACETONE
4114-037 FASTIGILIN A
4114-038 FASTIGILIN B
4114-036 FASTIGILIN C
4192-046 FAURINONE
4129-005 FAURONYL ACETATE
4502-011 FECOSTEROL

4210-006	GIBBERELLIN A2
4210-001	GIBBERELLIN A3 = GIBBERELLIC ACID
4210-007	GIBBERELLIN A4
4210-008	GIBBERELLIN A5
4210-009	GIBBERELLIN A6
4210-010	GIBBERELLIN A7
4210-011	GIBBERELLIN A7,DEOXY-ISO-
4210-013	GIBBERELLIN A8
4210-014	GIBBERELLIN A8,3-O-GLUCOSYL-
4210-015	GIBBERELLIN A9
4210-016	GIBBERELLIN A10
4210-017	GIBBERELLIN A11
4210-018	GIBBERELLIN A12
4210-019	GIBBERELLIN A13
4210-020	GIBBERELLIN A14
4210-021	GIBBERELLIN A15
4210-022	GIBBERELLIN A16
4210-023	GIBBERELLIN A17
4210-024	GIBBERELLIN A18
4210-025	GIBBERELLIN A19
4210-026	GIBBERELLIN A20
4210-027	GIBBERELLIN A21
4210-028	GIBBERELLIN A22
4210-029	GIBBERELLIN A23
4210-030	GIBBERELLIN A24
4210-031	GIBBERELLIN A26
4210-032	GIBBERELLIN A26,3-O-GLUCOSYL-
4210-033	GIBBERELLIN A27
4210-034	GIBBERELLIN A27,3-O-GLUCOSYL-
4210-036	GIBBERELLIN A29
4210-037	GIBBERELLIN A29,3-O-GLUCOSYL-
4210-024	GIBBERELLIN I ,LUPINUS- = GIBBERELLIN A18
4293-033	GINKOLIDE A
4293-034	GINKOLIDE B
4293-036	GINKOLIDE C
4293-035	GINKOLIDE M
4404-012	GINSENOSIDE
4505-056	GITALIGENIN,B1- = GITOXIGENIN
4505-119	GITALOXIGENIN
4505-120	GITALOXIN
4511-013	GITINGENSINE
4504-013	GITOGENIN
4504-028	GITOGENIN,25(27)-DEHYDRO-
4504-015	GITOGENIN,NEO-
4504-014	GITONIN,F-
4505-058	GITORIN
4506-056	GITOSTIN,NEO- = GITOXIGENIN
4506-056	GITOXIGENIN
4505-062	GITOXIN
4408-069	GLABROLIDE
4408-068	GLABROLIDE,DEOXY-
4408-070	GLABROLIDE, ISO-
4408-071	GLABROLIDE,21ALPHA-OH-ISO-
4425-026	GLAUCARUBIN
4425-017	GLAUCARUBINONE
4425-016	GLAUCARUBOLONE
	GLAUCOTOXIGENIN, ALLO- = ARTEFACT
4104-050	GLOBICINE
4008-056	GLOBULARIN
4125-004	GLOBULOL,L-
4406-004	GLOCHIDIOL
4406-009	GLOCHIDONE
4406-010	GLOCHIDONOL
4414-001	GLUTINOL
4414-002	GLUTINOL, 3-EPI-
4413-003	GLUTINONE = ALNUSENONE
4408-063	GLYCYRRHETIC ACID
4408-066	GLYCYRRHETIC ACID,18ALPHA-OH-
4408-061	GLYCYRRHETIC ACID,11-DEOXO-
4408-062	GLYCYRRHETIC ACID,24-OH-11-DEOXY-
4408-067	GLYCYRRHETIC ACID,24OH-
4408-014	GLYCYRRHETOL
4408-064	GLYCYRRHINIC ACID
4212-022	GNAVINE, ISO-HYPO-
4505-041	GOFRUSIDE
4505-054	GOMPHOGENIN
4505-055	GOMPHOSIDE
4192-026	GORGONENE,(+)-BETA-
4599-002	GORGOSTEROL
4599-008	GORGOSTEROL, 23-DES-ME-
4107-071	GOSSYPOL
4505-026	GRACILOSIDE
4206-039	GRANDIFLORIC ACID
4492-011	GRANDIFOLIOLENONE
4424-008	GRANDIFOLIOLIN = FISSINOLIDE
4426-017	GRANDIFOLIONE
4191-002	GRAPHINONE = OVALICIN
4412-022	GRATIOGENIN
4412-024	GRATIOGENIN,16-OH-
4406-013	GRATIOLONE = BETULINIC ACID
4412-023	GRATIOSIDE
4218-004	GRAYANOTOXIN-I
4218-005	GRAYANOTOXIN-II
4218-006	GRAYANOYOXIN-III
4218-011	GRAYANOTOXIN-IV
4218-012	GRAYANOTOXIN-V
4191-009	GREENEIN
4191-020	GRIESENIN
4191-021	GRIESENIN,DIHYDRO-
4202-081	GRINDELIC ACID
4202-083	GRINDELIC ACID,6-OXO-
4202-082	GRINDELIC ACID,EPOXY-
4202-110	GRINDELIC ACID, OXY-
4410-026	GRISEOGENIN
4104-080	GROSSHEMIN
4104-006	GUAIA-3,7-DIENE
4104-007	GUAIAZULENE
4104-001	GUAIENE,ALPHA-
4104-003	GUAIENE,DELTA- = BULNESENE,ALPHA-
4104-013	GUAIENE,(+)-EPOXY-
4104-002	GUAIENE,EPSILON-
4104-020	GUAI-1,3,5,9,11-PENTAENE,14-OH-
4104-021	GUAI-1,3,5,9,11-PENTAENE,14-STEARYLOXY-
4104-027	GUAIOXIDE
4104-015	GUAIOL
4198-016	GUAI-PYRIDINE,EPI
4408-044	GUMMOSOGENIN
4125-003	GURJUNENE,ALPHA-L-
4119-007	GURJUNENE, BETA-
4104-089	GURJUNENE,GAMMA-
4408-039	GYMNEMAGENIN
4408-030	GYMNESTROGENIN
4408-118	GYMNOSPOROL
4408-106	GYPSOGENIN
4499-005	HAKONANOL,KETO-
4217-020	HARDWICKIIC ACID,(-)-
4217-021	HARDWICKIIC ACID(+)-
4008-059	HARPAGIDE
4008-060	HARPAGIDE,8-OAC-
4008-061	HARPAGOSIDE
4217-022	HAUTRIWAIC ACID LACTONE
4426-004	HAVANENSIN 1,7-DIAC-
4426-005	HAVANENSIN 3,7-DIAC-
4426-003	HAVANENSIN,3,7-DIACETATE,DEOXY-
4426-006	HAVANENSIN,1,3,7-TRIAC-
4504-053	HECOGENIN
4504-055	HECOGENIN,9-DEHYDRO-
4504-054	HECOGENIN,NEO- = SISALAGENIN
4408-090	HEDERAGENIN
4102-005	HEDYCARYOL
4114-027	HELENALIN
4104-042	HELENALIN,ISO-
4114-024	HELENALIN,6-DEOXY - = AROMATIN
4603-014	HELENEIN
4002-101	HELENIUM LACTOL
4104-039	HELENIUM LACTONE
4102-061	HELIANGINE
4115-012	HELICOBASIDIN,DEOXY-
4115-011	HELICOBASIDINE
4506-038	HELLEBRIGENIN
4506-039	HELLEBRIGENIN 3-ACETATE
4506-040	HELLEBRIGENIN 3,5-DI-ACETATE
4506-041	HELLEBRIN
4192-059	HELMINTHOSPORAL
4192-061	HELMINTHOSPORAL,PRE-
4192-060	HELMINTHSPOROL
4192-062	HELMINTHOSPOROL,PRE-
4192-063	HELMINTHOSPOROL,9-OH-PRE-
4429-003	HELVOLIC ACID
4429-005	HELVOLIC ACID, 6-DES-AC- = HELVOLINIC ACID
4429-005	HELVOLIC ACID, 6-DES-OAC-
4429-004	HELVOLINIC ACID
4908-001	HEPT-2-EN-6-OL,2-ME-
4908-002	HEPT-2-EN-6-ONE, 2-ME-
4293-044	HETERATISINE
4293-043	HETEROPHYLLIDINE
4293-041	HETEROPHYLLINE
4293-042	HETEROPHYLLISINE
4604-006	HETEROXANTHIN
4212-010	HETIDINE
4212-013	HETISINE
4212-014	HETISINONE
4426-015	HEUDELOTTIN C
4426-013	HEUDELOTTIN D
4426-014	HEUDELOTTIN E
4426-016	HEUDELOTTIN F
4209-002	HIBAENE,(-)-
4293-029	HIBAENE,4ALPHA-OH-18-NOR-
4293-028	HIBAENE,4BETA-OH-18-NOR-
4209-003	HIBAENE EPOXIDE,(+)-
4109-001	HIMACHALENE,ALPHA-
4109-002	HIMACHALENE,BETA-
4109-003	HIMACHALOL,(+)-
4192-020	HIMACHALOL,(+)-ALLO-
4122-006	HINESOL
4204-006	HINOKIOL
4204-007	HINOKIONE
4193-052	HIRSUTIC ACID C
4426-020	HIRTIN
4426-019	HIRTIN,DESAC-

4206-015 KAURANE-3,16,17-TRI-OH-(-)-
4206-045 KAURAN-19-OIC ACID, 12,17-DIOH-(-)-
4206-047 KAURAN-17-OIC ACID, 1,19-DIOH-(-)-
4206-044 KAURAN-19-OIC ACID, 16,17-DIOH-(-)-
4206-023 KAURAN-19-AL,16-ALPHA-OH-(-)-
4206-035 KAURANOIC ACID, 16-OH-(-)-
4206-034 KAURANOIC ACID,17-OH-19-
4206-048 KAURANOIC ACID,18-OH-3-OXO-17-
4206-049 KAURANOIC ACID,18-OAC-3-OXO-17-
4206-046 KAURANOIC ACID, 19-OH-17-
4206-005 KAURANOL, (-)-
4206-004 KAURANOL, (+)-
4206-033 KAUR-9(11),16-DIEN-19-OIC ACID,(-)-
4206-042 KAUR-9(11),16-DIEN-19-OIC ACID,15-OAC-
4206-043 KAUR-9(11),16-DIEN-19-OIC ACID,15-OH-
4206-008 KAUR-16-EN-3ALPHA,19-DIOL,(-)-
4206-022 KAUR-15-EN-19-AL
4206-009 KAUR-16-ENE, 3-OH-19-SUCCINOYLOXY-
4206-007 KAUREN-17,19-DIOL, ISO-(-)-
4206-001 KAURENE,(+)-
4206-002 KAURENE,(-)-
4206-003 KAURENE,ISO-(+)-
4206-029 KAURENIC ACID
4206-038 KAURENIC ACID,ACETOXY-
4206-031 KAURENIC ACID,ISO-
4206-032 KAURENIC ACID,19-ISO-
4206-030 KAUR-16-EN-19-OIC ACID,(-)-
4206-040 KAUR-16-EN-19-OIC ACID,15BETA-OH-(-)-
4206-054 KAURENOLIDE, 7,13-DIOH-
4206-052 KAURENOLIDE,7,18-DIOH-
4206-053 KAURENOLIDE,7,16,18-TRIOH-
4206-051 KAURENOLIDE,7-OH-
4104-031 KESSANE
4104-032 KESSANOL
4104-033 KESSANOL,8-EPI-
4104-034 KESSYL ALCOHOL,ALPHA-
4104-035 KESSYL GLYCOL
4426-018 KHAYANTHONE
4424-009 KHAYASIN
4424-011 KHAYASIN B
4424-010 KHAYASIN T
4424-012 KHAYASIN,DIHYDRO-
4427-008 KHIVORIN
4427-009 KHIVORIN,3-DES-AC-
4427-011 KHIVORIN, 3,7-DI-DES-AC-7-OXO-
4427-010 KHIVORIN,7-DESAC-
4427-012 KHIVORIN,7-DES-OAC-7-KETO-
4427-013 KHIVORIN,11BETA-OAC-
4124-001 KHUSENE
4124-004 KHUSENIC ACID
4124-006 KHUSENIC ACID,ISO-
4124-002 KHUSENOL
4192-023 KHUSILAL
4193-051 KHUSIMENE
4124-002 KHUSIMOL = KHUSENOL
4193-048 KHUSIMONE
4107-042 KHUSINOL
4107-043 KHUSINOL,EPI-
4107-044 KHUSINOLOXIDE
4192-024 KHUSITONE
4107-048 KHUSOL
4511-019 KIBATALINE
4107-019 KIGANENE
4107-029 KIGANOL
4009-004 KINGISIDE
4425-017 KIRONDRINE,ALPHA- = GLAUCARUBINONE
4504-048 KITIGENIN
4940-001 KITOL
4425-012 KLAINEANONE
4212-015 KOBUSINE
4212-020 KOBUSINE, PSEUDO-
4410-023 KOELLIKERIGENIN
4504-047 KOGAGENIN
4217-002 KOLAVELOOL
4217-004 KOLAVENIC ACID
4217-003 KOLAVENOL
4217-005 KOLAVIC ACID
4507-075 KONDURANGOGENIN A
4507-034 KONDURANGOGENIN C
4516-035 KORSEVERIDINE
4515-008 KORSEVININE
4516-036 KORSINE
4606-001 KRYPTOCASIN
4405-028 KULACTONE
4405-016 KULINONE
4405-027 KULOLACTONE
4511-012 KURCHALINE
4513-022 KURCHESSINE
4511-009 KURCHILINE
4511-010 KURCHIPHYLLAMINE
4511-011 KURCHIPHYLLINE
4514-024 KURCHOLESSINE
4193-032 KUROMATSUOL = LONGIBORNEOL
4202-069 LABDAN-8BETA-OL-15-OIC ACID,ENANTIO-
4202-118 LABDAN-15-OIC ACID, 6,8-DI-OH-ENT-
4202-117 LABDAN-15-OIC ACID, 7,8-DI-OH-ENT-
4202-026 LABD-8-EN-3,15-DIOL
4293-037 LABD-11-EN-13-ONE,14,15-BIS-NOR-8-OH-
4202-037 LABD-11-EN-13-ONE,14,15-BIS-NOR-8ALPHA-OH-
4202-020 LABD-13-EN-8,15-DIOL,ENT-
4202-068 LABD-13-EN-8-OL-15-OIC ACID, ENT-
4202-063 LABDA-8(20),13-DIENOIC ACID
4202-003 LABDA-8(20),13-DIEN-15-OL
4202-105 LABDA-8(20)-EN-15,18-DIOIC ACID, ENT-
4202-017 LABDA-8(20)-EN-15,18-DIOL, ENT-
4202-071 LABDA-8(20)-EN-15-OIC ACID, 3-BETA-OH-
4202-066 LABDA-8(20)-EN-15-OIC ACID, ENT-
4202-073 LABDA-8(20)-EN-15-OIC ACID, ENT-18-OME-
4202-052 LABDA-8(20)-EN-19-OIC ACID, 12-OH-
4202-072 LABDA-8(20)-EN-18-OL-15-OIC ACID, ENT-
4202-075 LABDA-8(20),13-DIEN-15-OIC ACID, ENT-18-OAC-
4202-111 LABDA-8(20),13-DIEN-15-OIC ACID ME ESTER,11-OAC-
4202-019 LABDANE-8ALPHA,15-DIOL
4202-028 LABDANE-8ALPHA,15,19ALPHA-TRIOL
4202-070 LABDANOLIC ACID
4105-017 LACCIJALARIC ACID
4105-018 LACCISHELLOLIC ACID
4105-019 LACCISHELLOLIC ACID,EPI-
4104-008 LACTARAZULENE
4104-009 LACTAROFULVENE
4104-011 LACTAROVIOLIN
4413-001 LACTUCEROL, ALPHA- = TARAXASTEROL
4104-072 LACTUCIN
4104-073 LACTUCOPICRIN
4120-013 LAEVOJUNEOL = JUNENOL, L-
4202-029 LAGOCHILIN
4105-020 LAKSHOLIC ACID
4105-021 LAKSHOLIC ACID,EPI-
4202-094 LAMBERTIANIC ACID
4192-066 LAMBICIN
4008-039 LAMIOL
4008-040 LAMIOSODE
4193-037 LAMPTEROL = ILLUDIN S
4505-033 LANATOSIDE-A
4505-032 LANATOSIDE-A,DES-AC-
4505-064 LANATOSIDE B
4505-063 LANATOSIDE B, DES-AC-
4505-034 LANATOSIDE C
4505-035 LANATOSIDE D
4505-121 LANATOSIDE E
4106-016 LANCEOL
4410-009 LANOSTANE-7,11-DIONE, 3-OH-
4410-007 LANOST-8-EN-7-ONE, 3-OH-
4410-008 LANOST-8-EN-7,11-DIONE, 3-OH-
4410-010 LANOSTA-8,24-DIEN-7-ONE, 3-OH-
4410-014 LANOSTA-7,9(11),24-TRIEN-21-OIC ACID, 3-BETA-OH
4410-018 LANOSTA-8,24-DIEN-OIC ACID,3-OAC-ME ESTER
4410-013 LANOSTA-8,24-DIEN-3-ONE,21-OH-
4410-011 LANOSTA-7,9,(11),24-TRIENE-3,21-DIOL
4410-012 LANOSTA-7,9,(11),24-TRIENE-3-ONE, 21-OH-
4410-003 LANOSTA-9(11),24-DIEN-3BETA-OL = PARKEOL
4410-001 LANOSTEROL
4410-002 LANOSTEROL,DIHYDRO-
4410-006 LANOSTEROL,23-OH-
4493-032 LANOSTEROL, 3-EPI-HEXANOR-
4493-003 LANOSTEROL, 28,30-BISNOR-
4493-004 LANOSTEROL, 29,30-BISNOR-
4493-061 LANOSTEROL, 29,30-BISNOR-DIHYDRO-
4493-002 LANOSTEROL, 29-NOR-
4493-001 LANOSTEROL, 30-NOR-
4505-051 LANOXIN = DIGOXIN
4493-055 LANSIC ACID
4408-084 LANTADEN A
4408-085 LANTADEN B
4102-076 LANUGINOLIDE
4213-035 LAPPACONINE
4202-015 LARIXOL
4202-016 LARIXYL ACETATE
4108-003 LASEROL
4108-004 LASERPITINE
4108-005 LASERPITINE,ISO-
4291-016 LATHYROL, 6,20-EPOXY-
4291-017 LATHYROL DIACETATE PHENYLACETATE, EPOXY-
4914-006 LATIA LUCIFERIN
4514-015 LATIFOLINE
4514-016 LATIFOLINE, NOR-
4514-030 LATIFOLININE
4111-001 LAURENE
4111-002 LAURENISOL
4193-034 LAURINTEROL
4193-035 LAURINTEROL,DEBROMO-
4111-006 LAURINTEROL, ISO-
4001-044 LAVENDER PYRAN
4001-045 LAVENDER PYRAN,EPI-
4001-046 LAVENDER PYRAN,DEHYDRO-
4001-047 LAVENDER PYRAN,DEHYDRO-EPI-
4099-013 LAVANDULAL,BETA-CYCLO-
4099-014 LAVANDULIC ACID,BETA-CYCLO-

4202-041 MANOYL OXIDE,19-COOH-13-EPI-ENANTIO-
4202-040 MANOYL OXIDE,19-OH-13-EPI-ENANTIO-
4293-003 MANOYL OXIDE,SECO-EPI-
4293-047 MANOYL OXIDE, NOR-SECO-EPI-
4107-056 MANSONONE A
4107-057 MANSONONE B
4107-058 MANSONONE C
4107-059 MANSONONE D
4107-060 MANSONONE E
4107-061 MANSONONE F
4107-062 MANSONONE G
4107-063 MANSONONE H
4107-064 MANSONONE I
4107-065 MANSONONE L
4193-036 MARASMIC ACID
4506-025 MARINOBUFAGIN
4504-017 MARKOGENIN
4202-116 MARRUBENOL
4202-100 MARRUBIIN
4202-099 MARRUBIN,PRE-
4507-087 MARSDENIN
4507-088 MARSDENIN, 17-BETA-
4507-031 MARSECTOHEXOL
4408-078 MASLINIC ACID = CRATAEGOLIC ACID
4405-014 MASTICADIENONIC ACID
4405-015 MASTICADIENONIC ACID,ISO-
MATABILACTONE = IRIDOMYRMECIN+ISO-IRIDOMYRMECIN
4008-012 MATATABIETHER
4008-014 MATATABIOL,ISO-NEO-
4008-013 MATATABIOL,NEO
4104-077 MATRICARINE
4104-078 MATRICARIN,DES-AC-
4104-079 MATRICARIN,DES-OAC-
4104-054 MATRICIN
4192-035 MATURIN
4192-033 MATURININ
4192-034 MATURINONE
4192-032 MATURONE
4204-033 MAYTENONE
4120-084 MAYTINE
4120-085 MAYTOLINE
4193-044 MAYURONE
4408-108 MEDICAGENIC ACID
4515-010 MEGACARPIDINE = SOLADULCIDINE
4406-015 MELALEUCIC ACID
4426-002 MELDENIN
4405-026 MELIANADIOL
4405-021 MELIANOL
4405-024 MELIANONE
4405-025 MELIANTRIOL
4008-050 MELITTOSIDE
4008-049 MELITTOSIDE,MONO-
4128-002 MELLITOXIN = HYENANCIN
4002-011 MENTHA-1,3,8-TRIENE
4002-038 MENTHANDIOL,(-)-2,5-TRANS-
4009-005 MENTHIAFOLIN
4002-029 MENTHA-1(7),8-DIEN-2OL,(+)-CIS-
4002-093 MENTHA-2,8-DIEN-1-OL
4002-030 MENTHA-2,8-DIEN-1-OL,(+)-CIS-
4002-031 MENTHA-2,8-DIEN-1-OL,(+)-TRANS
4002-095 MENTHOFURAN
4002-024 MENTHOL,L-
4002-025 MENTHOL,D-NEO-
4002-061 MENTHONE,D-
4002-062 MENTHONE,L-
4002-064 MENTHONE,D-ISO-
4002-063 MENTHONE,L-ISO-
4002-100 MENTHYL ACETATE,1,2-EPOXY-
4408-060 MERISTOTROPIC ACID
4213-018 MESACONITINE
4408-089 MESEMBRYANTHEMOIDIGENIN
4915-005 METABOLITE C
4504-040 METAGENIN
4507-090 METAPLEXIGENIN
4507-091 METAPLEXIGENIN,12-O-BENZOYL-DE-AC-
4507-089 METAPLEXIGENIN,DEACYL-
4107-012 METROSIDERENE
4906-001 MEVALONIC ACID LACTONE
4114-022 MEXICANIN A
4114-026 MEXICANIN C
4192-011 MEXICANIN D
4192-008 MEXICANIN E
4192-009 MEXICANIN E,DIHYDRO-
4104-043 MEXICANIN H
4114-028 MEXICANIN I
4114-025 MEXICANIN,6-DEOXY- = AROMATICIN
4405-029 MEXICANOL = ODORATOL
4424-006 MEXICANOLIDE
4424-007 MEXICANOLIDE,3-BETA-DIHYDRO-
4504-060 MEXOGENIN
4504-061 MEXOGENIN,NEO-
4293-039 MEZEREIN
4120-064 MIBULACTONE
4120-054 MICROCEPHALIN
4409-010 MICROMERIC ACID
4102-035 MIKANOLIDE
4102-037 MIKANOLIDE, DESOXY-
4102-036 MIKANOLIDE,DIHYDRO-
4505-043 MILLOSIDE
4407-002 MILIACIN
4293-016 MILTIRONE
4121-027 MISCANDENIN
4212-011 MIYACONITINE
4212-012 MIYACONITINONE
4925-001 MOENOCINOL
4925-002 MOENOCINOL, ISO-
4104-044 MOKKU-LACTONE
4401-017 MOLLUGOGENOL A
4408-077 MOMORDIC ACID
4604-002 MONADOXANTHIN
4191-047 MONOCYCLOFARNESOL, TRANS-GAMMA-
4406-007 MONOGYNOL A
4008-032 MONOTROPEIN
4008-062 MONOTROPEIN,BIS-DESOXY-DIHYDRO-
4104-090 MONTANOLIDE
4401-006 MORETENOL
4401-007 MORETENOL,3-EPI-
4401-018 MORETENONE
4407-005 MOROLIC ACID
4407-006 MOROLIC ACID,AC-
4009-003 MORRONISIDE
4402-008 MOTIDIOL
4402-005 MOTIOL
4492-004 MOTIOL,NEO-
4002-037 MULLILAM DIOL
4492-015 MULTIFLORENOL
4116-006 MULTIJUGENOL, ALPHA-
4107-033 MURROLOL,T-
4505-118 MUSCAROSIDE
4408-123 MUSENNIN
4193-022 MUSTAKONE
4603-027 MUTOCHROME = CITROXANTHIN
4107-020 MUUROLENE,(-)-ALPHA-
4107-022 MUUROLENE,EPSILON-
4107-021 MUUROLENE,(+)-GAMMA-
4107-035 MUUROL-3-ENE-2A,9B-DIOL
4107-034 MUUROL-3-ENE-2B,9A-DIOL
4107-055 MUUROL-3-ENE-9-OL-2-ONE
4101-019 MYOPORONE
4001-002 MYRCENE
4191-033 MYRCENE,CYCLO-ISOPROPENYL-
4001-017 MYRCENOL
4001-020 MYRCENOL,ISO-
4001-021 MYRCENONE
4416-007 MYRICADIOL
4416-008 MYRICOLAL
4003-017 MYRTENAL,D-
4003-008 MYRTENOL,D-
4003-009 MYRTENOL,L-
4003-022 MYRTENOL,ISOVALERATE
4003-018 MYRTENOIC ACID
4003-019 MYRTENOIC ACID,DIHYDRO-
4509-033 MYXINOL
4603-016 MYXOXANTHIN = ECHINENONE
4602-020 MYXOXANTHOPHYLL
4293-005 NAGILACTONE A
4293-006 NAGILACTONE B
4293-007 NAGILACTONE C
4293-010 NAGILACTONE D
4214-004 NAPELLINE, ISO-
4912-003 NAPHTHALENE, 1,6-DI-ME-
4104-016 NARDOL
4192-021 NARDOSINONE
4192-022 NARDOSINONE, ISO-
4117-016 NARDOSTACHONE
4504-031 NARTHOGENIN,ISO-
4213-005 NEOLINE
4499-004 NEOLITSIN , CYCLO-
4213-008 NEOPELLINE
4605-001 NEOXANTHIN
4202-115 NEPETAEFOLIN
4202-114 NEPETAEFURAN
4202-113 NEPETAEFURANOL
4008-017 NEPETALACTONE,CIS-TRANS-
4008-018 NEPETALACTONE,TRANS-CIS-
4008-019 NEPETALACTONE,DIHYDRO-
4008-020 NEPETALACTONE,ISO-DIHYDRO
4008-021 NEPETALACTONE,NEO-
4405-003 NERIFOLIOL
4505-022 NERIIFOLIN
4505-023 NERIIFOLIN, AC-
4401-003 NERIIFOLIOL
4001-011 NEROL
4001-012 NEROL GLUCOSIDE
4001-040 NEROLIC ACID
4191-042 NEROLIDIOL,CYCLO-
4101-004 NEROLIDOL
4699-012 NEUROSPORAXANTHIN
4601-005 NEUROSPORENE

ROSMARICINE = ARTIFACT
4208-007 ROSOLOLACTONE
4409-019 ROTUNDIC ACID
4002-068 ROTUNDIFOLONE = PIPERITENONE OXIDE
4104-022 ROTUNDONE,(-)-
4120-027 ROTUNOL,ALPHA-
4120-028 ROTUNOL,BETA-
4403-007 ROXBURGHOLONE
4204-018 ROYLEANONE
4204-019 ROYLEANONE,7-OAC-
4204-020 ROYLEANONE,6-DEHYDRO-
4602-021 RUBICHROME
4518-010 RUBIJERVINE
4518-011 RUBIJERVINE,ISO-
4518-012 RUBIJERVOSINE, ISO-
4602-012 RUBIXANTHIN
4602-013 RUBIXANTHIN-5,6-EPOXIDE
4508-016 RUBROSTERONE
4504-011 RUSCOGENIN
4504-012 RUSCOGENIN,NEO-
4423-004 RUTAEVIN
4013-012 RW 47
4291-010 RYANODINE
4516-009 SABADINE
4516-008 SABADINE,NEO- = SABINE
SABATINE = SABADINE
4006-001 SABINANE = THUJANE
4516-008 SABINE
4006-002 SABINENE,D-
4006-003 SABINENE,L-
4006-012 SABINENE,DL-
4006-005 SABINOL
4010-001 SAFRANAL
4408-022 SAIKOGENIN A
4408-012 SAIKOGENIN E
4408-024 SAIKOGENIN F
4408-025 SAIKOGENIN G
4408-124 SAIKOSAPONIN A
4408-125 SAIKOSAPONIN C
4408-126 SAIKOSAPONIN D
4408-120 SOYASAPOGENOL E
4493-060 SALANNIN
4102-058 SALONITENOLIDE
4102-030 SALONITOLIDE
4006-011 SALVENE
4204-031 SALVIN
4493-067 SAMADERIN B
4493-068 SAMADERIN C
4592-017 SAMANDARONE
4592-016 SAMANDENONE
4592-023 SAMANINE
4504-016 SAMOGENIN
4203-006 SANDARACOPIMAR-15-ENE,8BETA-OH-
4203-034 SANDARACOPIMARA-8(14),15-DIENE
4203-004 SANDARACOPIMARADIEN-3-OL
4203-002 SANDARACOPIMARADIEN-19-OL
4203-010 SANDARACOPIMARADIENE-3,18-DIOL
4203-009 SANDARACOPIMARADIEN-3BETA,19-DIOL
4203-013 SANDARACOPIMARA-8(14),16-DIEN-2A,18,19-TRIOL
4203-014 SANDARACOPIMARA-8(14),16-DIEN-3B,18,19-TRIOL
4203-015 SANDARACOPIMARA-8(14),16-DIEN-2A,3B,18,19-TETROL
4203-021 SANDARACOPIMARADIENE-3-ONE
4203-024 SANDARACOPIMARIC ACID
4203-033 SANDARACOPIMARIC ACID,OAC-
4203-028 SANDARACOPIMARIC ACID,6-OH-
4203-026 SANDARACOPIMARIC ACID,12ALPHA-OH-
4118-006 SANTALAL,(-)-ALPHA-
4118-001 SANTALENE,ALPHA-
4118-002 SANTALENE,BETA-
4118-003 SANTALENE,EPI-BETA-
4118-007 SANTALIC ACID,BETA-
4118-004 SANTALOL,ALPHA-
4118-005 SANTALOL,BETA-
4120-062 SANTAMARINE
4120-091 SANTAMARINE, EPOXY-
4099-001 SANTENE
4004-006 SANTENONE
4004-003 SANTENONE ALCOHOL
4099-012 SANTOLINA TRIENE
4002-060 SANTOLINENONE,ALPHA-
4099-002 SANTOLINENONE, BETA-
4120-065 SANTONIN,ALPHA-
4120-066 SANTONIN,BETA-
4120-067 SANTONIN,PSEUDO-
4120-068 SANTONIN,DESOXY-PSEUDO-
4120-070 SANTONIN,11-OH-
4405-018 SAPELIN A
4405-019 SAPELIN B
4204-028 SAPIETIC ACID,(-)-
4408-076 SAPINDIC ACID
4602-015 SAPROXANTHIN
4513-023 SARACOCINE
4513-024 SARACOCINE, 3-EPI-
4513-025 SARACODINE
4513-022 SARACODININE = KURCHESSINE
4699-026 SARCINAXANTHIN
4507-032 SARCOSTIN
4503-012 SARGASTEROL
4505-077 SARGENOSIDE
4503-026 SARINGOSTEROL
4505-074 SARMENTOCYMARIN
4505-072 SARMENTOGENIN
4505-139 SARMENTOSIDE A
4505-137 SARMENTOSIDE E
4505-122 SARMENTOSIGENIN B
4505-136 SARMENTOSIGENIN E
4505-114 SARMUTOGENIN
4505-116 SARMUTOSIDE
4505-076 SARNOVIDE
4504-004 SARSASAPOGENIN
4504-075 SARSPARILLOSIDE
4505-123 SARVEROGENIN
4505-124 SARVEROSIDE
4193-018 SATIVENE
4194-005 SATIVENE,CYCLO-
4104-081 SAURINE
4121-023 SAUSSUREA LACTONE
4416-003 SAWAMILLETIN
4408-002 SAWAMILLETIN,ISO-
4102-031 SCABIOLIDE
4102-038 SCANDENOLIDE
4102-039 SCANDENOLIDE,DIHYDRO-
4008-033 SCANDOSIDE
4504-008 SCEPTRUMGENIN
4405-007 SCHINOL
4202-097 SCIADINE
4202-096 SCIADINONE
4202-095 SCIADINONATE,DIME-
4202-059 SCIADOPIC ACID,ME ESTER
4506-014 SCILLAREN A
4506-016 SCILLAREN A,GLUCO-
4506-028 SCILLAREN-F
4506-012 SCILLARENIN
4506-013 SCILLARENIN GLUCOSIDE
4506-015 SCILLARENIN, DI-RHAMNOSYL-
4506-028 SCILLIGLAUCOSIDE = SCILLAREN-F
4506-022 SCILLIROSIDE
4506-024 SCILLIROSIDIN, 1,2-EPOXY-12-OH-
4506-021 SCILLIROSIDINE
4506-023 SCILLIROSIDINE,1ALPHA,2ALPHA-EPOXY-
4506-027 SCILLIGLAUCOSIDINE
4110-007 SCIRPENOL,DIOAC-
4110-009 SCIRP-9-EN-3ALPHA-OL, 8ALPHA-ISOVALEROYLOXY-4BETA,15-DIACETOXY-
4110-010 SCIRP-9-ENE,4,8,15-TRIOAC-3ALPHA,7ALPHA-DIOH-
4202-001 SCLARENE
4202-021 SCLAREOL
4509-005 SCYMNOL
4120-004 SELINA-4,11-DIENE
4120-086 SELINA-4(14),7(11)-DIENE
4120-076 SELINADIENE, 3,7(11)-
4120-001 SELINENE,ALPHA-
4120-002 SELINENE,BETA-
4120-003 SELINENE,GAMMA-
4120-007 SELINELOL = EUDESMOL,ALPHA-
4120-019 SELIN-11-EN-4ALPHA-OL
4292-009 SEMPERVIROL
4408-103 SENEGENIC ACID = POLYGALIC ACID
4503-040 SENGOSTERONE
4121-018 SERICEALACTONE
4121-019 SERICEALACTONE,DESOXY-
4102-017 SERICENIC ACID
4102-018 SERICENINE
4121-026 SERICENINE,ISO-
4408-115 SERJANIC ACID
4408-114 SERRATAGENIC ACID
4415-001 SERRATENE
4415-002 SERRATENE,ISO-
4415-001 SERRAT-13-ENE = SERRATENE, ISO-
4415-003 SERRAT-14-ENE, 3-BETA-,21-ALPHA-DIOH- = SERRATENEDIOL
4415-003 SERRATENEDIOL
4415-004 SERRATENEDIOL 3-ME-ETHER
4415-006 SERRATENEDIOL DI-ME-ETHER
4415-007 SERRATENEDIOL,21-EPI-
4415-008 SERRATENEDIOL 3-ME ETHER, 21-EPI-
4415-009 SERRATENEDIOL,DI-EPI-
4415-010 SERRATENEDIOL 3-ME ETHER, 3,21-DI-EPI-
4415-011 SERRATENEDIOL DI-ME ETHER, 3,21-DI-EPI-
4415-012 SERRATENEDIOL 3-ME ETHER, 3,21-DI-EPI-ISO-
4415-019 SERRATENEDIOL,16-OXO-
4415-020 SERRATENEDIOL,16-OXO-EPI-
4415-021 SERRATENEDIOL,16-OXO-DI-EPI-
4415-013 SERRATENE,3BETA-OH-21-OXO-

4192-019	WIDDROL
4504-056	WILLAGENIN
4126-021	WINTERINE
4502-031	WITHAFERIN A
4502-033	WITHAFERIN-A, 27-DEOXY-
4502-034	WITHAFERIN A,27-DEOXY-14-HYDROXY-
4502-032	WITHAFERIN A,2,3-DIHYDRO-3-OME-
4502-035	WITHANOLIDE-D
4502-036	WITHANOLIDE-D, 24,25-DIHYDRO-
4502-037	WITHANOLIDE D,20-DEOXY-24,25-DIHYDRO-
4592-005	WORTMANNIN
4191-016	XANTHANOL
4191-017	XANTHANOL,ISO-
4191-015	XANTHATIN
4191-018	XANTHININ
4204-014	XANTHOPEROL
4204-015	XANTHOPEROL,5-EPI-
4603-013	XANTHOPHYLL = LUTEIN
4915-008	XANTHOXIN, C,T-
4915-007	XANTHOXIN, TRANS,TRANS-
4424-003	XYLOCARPIN
4206-041	XYLOPIC ACID
4505-005	XYSMALOGENIN
4002-091	YABUNIKKEOL,CIS-
4002-092	YABUNIKKEOL,TRANS-
4504-006	YAMOGENIN
4193-025	YLANGENE,ALPHA-D-
4193-026	YLANGENE,ALPHA-L-
4193-027	YLANGENE,BETA-
4099-011	YOMOGI ALCOHOL A
4120-049	YOMOGIN
4504-018	YONOGENIN
4504-073	YONONIN
4504-026	YUCCAGENIN
4504-027	YUCCAGENIN,NEO- = LILAGENIN
4193-001	ZALUZANIN A
4193-002	ZALUZANIN B
4104-046	ZALUZALIN C
4104-047	ZALUZANIN D
4919-001	ZAMENE
4202-053	ZANZIBARIC ACID
4423-007	ZAPOTERIN
4602-005	ZEACAROTENE, ALPHA-
4602-006	ZEACAROTENE, BETA-
4603-009	ZEAXANTHIN
4102-016	ZEDERONE
4121-029	ZEDOARONE
4603-005	ZEINOXANTHIN
4401-009	ZEORIN
4103-007	ZERUMBONE
4191-024	ZERUMBONE,(RACEMIC)-PSI-PHOTO-
4191-025	ZERUMBONE,(RACEMIC)-PSI-PHOTO-DIHYDRO-
4191-026	ZERUMBONE,(RACEMIC)-PSI-PHOTO-DIHYDRO-
4102-042	ZEXBREVIN
4102-026	ZEYLANANE
4102-027	ZEYLANICINE
4102-028	ZEYLANIDINE
4102-025	ZEYLANINE
4192-013	ZIERONE
4106-004	ZINGIBERENE,ALPHA-
4106-005	ZINGIBERENE,BETA-
4107-069	ZIZANENE
4193-049	ZIZAENE, PRE-
4302-003	ZIZANIN A = OPHIOBOLIN C
4302-002	ZIZANIN B = OPHIOBOLIN B
4124-004	ZIZANOIC ACID = KHUSENIC ACID
4124-005	ZIZANOIC ACID,EPI-
4124-003	ZIZANOL
4516-003	ZYGACINE
4591-004	ZYGADENILIC ACID LACTONE, ANGELOYL-
4516-002	ZYGADENINE
4516-004	ZYGADENINE, ANGELOYL-
4516-005	ZYGADENINE, 3-O-(2-ME-BUTYROYL)-
4516-007	ZYGADENINE, VANILLOYL-
4516-006	ZYGADENINE, VERATROYL-
4501-007	ZYMOSTEROL
4493-009	ZYMOSTEROL, 24,25-DIHYDRO-4-ALPHA-ME-24-METHYLENE-
4002-104	
4002-105	
4002-106	
4008-065	
4099-029	
4405-013	
4099-030	
4099-031	
4099-032	
4101-020	
4293-004	
4293-031	
4293-032	
4293-040	
4910-002	
4910-003	
4912-001	

MOLECULAR WEIGHT INDEX

No.	MW	Formula
4099-001	122	C 9 H 14
4908-002	126	C 8 H 14 O 1
4908-005	126	C 8 H 14 O 1
4908-001	128	C 8 H 16 O 1
4908-003	128	C 8 H 16 O 1
4192-007	128	C10 H 8
4906-001	130	C 6 H 10 O 3
4002-013	132	C10 H 12
4013-016	133	C 9 H 11 N
4001-004	134	C10 H 14
4002-012	134	C10 H 14
4002-011	134	C10 H 14
4002-010	136	C10 H 16
4003-021	136	C10 H 16
4003-020	136	C10 H 16
4003-004	136	C10 H 16
4003-003	136	C10 H 16
4003-002	136	C10 H 16
4003-001	136	C10 H 16
4002-089	136	C10 H 16
4002-088	136	C10 H 16
4010-005	136	C10 H 16
4010-004	136	C10 H 16
4006-012	136	C10 H 16
4006-004	136	C10 H 16
4006-003	136	C10 H 16
4006-002	136	C10 H 16
4005-003	136	C10 H 16
4005-002	136	C10 H 16
4005-001	136	C10 H 16
4001-001	136	C10 H 16
4001-003	136	C10 H 16
4001-002	136	C10 H 16
4002-009	136	C10 H 16
4002-008	136	C10 H 16
4002-007	136	C10 H 16
4002-006	136	C10 H 16
4002-005	136	C10 H 16
4002-004	136	C10 H 16
4002-003	136	C10 H 16
4002-002	136	C10 H 16
4002-001	136	C10 H 16
4099-022	136	C10 H 16
4099-021	136	C10 H 16
4099-020	136	C10 H 16
4099-012	136	C10 H 16
4099-003	136	C10 H 16
4909-005	136	C 9 H 12 O 1
4909-003	136	C 9 H 12 O 1
4909-001	138	C 9 H 14 O 1
4909-008	138	C 9 H 14 O 1
4006-001	138	C10 H 18
4006-011	138	C10 H 18
4004-006	138	C 9 H 14 O 1
4004-003	140	C 9 H 16 O 1
4008-001	140	C10 H 20
4909-009	140	C 9 H 16 O 1
4906-002	141	C 6 H 7 N 1 O 3
4908-004	142	C 8 H 14 O 2
4013-015	147	C10 H 13 N 1
4001-032	148	C10 H 12 O 1
4011-004	148	C10 H 12 O 1
4002-083	148	C10 H 12 O 1
4909-004	148	C10 H 12 O 1
4013-013	149	C 9 H 11 N 1 O 1
4013-012	149	C 9 H 11 N 1 O 1
4014-003	149	C 9 H 11 N 1 O 1
4014-002	149	C 8 H 7 N 1 O 2
4099-018	150	C10 H 14 O 1
4099-025	150	C10 H 14 O 1
4099-024	150	C10 H 14 O 1
4099-026	150	C10 H 14 O 1
4001-031	150	C10 H 14 O 1
4001-036	150	C10 H 14 O 1
4001-022	150	C10 H 14 O 1
4001-021	150	C10 H 14 O 1
4002-082	150	C10 H 14 O 1
4002-081	150	C10 H 14 O 1
4002-095	150	C10 H 14 O 1
4003-017	150	C10 H 14 O 1
4003-016	150	C10 H 14 O 1
4003-014	150	C10 H 14 O 1
4003-013	150	C10 H 14 O 1
4002-072	150	C10 H 14 O 1
4002-071	150	C10 H 14 O 1
4002-070	150	C10 H 14 O 1
4002-067	150	C10 H 14 O 1
4002-052	150	C10 H 14 O 1
4002-049	150	C10 H 14 O 1
4002-044	150	C10 H 14 O 1
4011-005	150	C 9 H 10 O 2
4010-001	150	C10 H 14 O 1
4012-001	150	C10 H 14 O 1
4006-010	150	C10 H 14 O 1
4906-003	151	C 6 H 7 N 1 O 3
4006-009	152	C10 H 16 O 1
4006-008	152	C10 H 16 O 1
4006-005	152	C10 H 16 O 1
4004-007	152	C10 H 16 O 1
4007-002	152	C10 H 16 O 1
4008-011	152	C10 H 16 O 1
4008-012	152	C10 H 16 O 1
4012-002	152	C10 H 16 O 1
4010-006	152	C10 H 16 O 1
4002-066	152	C10 H 16 O 1
4002-073	152	C10 H 16 O 1
4002-065	152	C10 H 16 O 1
4002-060	152	C10 H 16 O 1
4002-059	152	C10 H 16 O 1
4002-058	152	C10 H 16 O 1
4002-023	152	C10 H 16 O 1
4002-035	152	C10 H 16 O 1
4002-034	152	C10 H 16 O 1
4002-031	152	C10 H 16 O 1
4002-030	152	C10 H 16 O 1
4002-029	152	C10 H 16 O 1
4003-011	152	C10 H 16 O 1
4003-010	152	C10 H 16 O 1
4003-009	152	C10 H 16 O 1
4003-008	152	C10 H 16 O 1
4003-005	152	C10 H 16 O 1
4003-015	152	C10 H 16 O 1
4004-004	152	C10 H 16 O 1
4004-005	152	C10 H 16 O 1
4002-093	152	C10 H 16 O 1
4002-092	152	C10 H 16 O 1
4002-091	152	C10 H 16 O 1
4002-090	152	C10 H 16 O 1
4002-098	152	C10 H 16 O 1
4002-080	152	C10 H 16 O 1
4002-079	152	C10 H 16 O 1
4002-075	152	C10 H 16 O 1
4001-020	152	C10 H 16 O 1
4001-027	152	C10 H 16 O 1
4001-026	152	C10 H 16 O 1
4001-019	152	C10 H 16 O 1
4001-018	152	C10 H 16 O 1
4001-017	152	C10 H 16 O 1
4099-002	152	C10 H 16 O 1
4099-006	152	C10 H 16 O 1
4099-013	152	C10 H 16 O 1
4099-010	152	C10 H 16 O 1
4099-015	152	C10 H 16 O 1
4013-009	153	C10 H 19 N
4099-011	154	C10 H 18 O 1
4099-009	154	C10 H 18 O 1
4099-008	154	C10 H 18 O 1
4099-023	154	C10 H 18 O 1
4001-016	154	C10 H 18 O 1
4001-014	154	C10 H 18 O 1
4001-013	154	C10 H 18 O 1
4001-025	154	C10 H 18 O 1
4001-024	154	C10 H 18 O 1
4001-023	154	C10 H 18 O 1
4001-011	154	C10 H 18 O 1
4001-008	154	C10 H 18 O 1
4001-043	154	C10 H 18 O 1
4001-042	154	C10 H 18 O 1
4001-047	154	C10 H 18 O 1
4001-046	154	C10 H 18 O 1
4002-074	154	C10 H 18 O 1
4002-099	154	C10 H 18 O 1
4004-002	154	C10 H 18 O 1
4004-001	154	C10 H 18 O 1

4003-006	154	C10	H 18	O 1	
4003-007	154	C10	H 18	O 1	
4003-012	154	C10	H 18	O 1	
4002-028	154	C10	H 18	O 1	
4002-027	154	C10	H 18	O 1	
4002-026	154	C10	H 18	O 1	
4002-039	154	C10	H 18	O 1	
4002-022	154	C10	H 18	O 1	
4002-020	154	C10	H 18	O 1	
4002-019	154	C10	H 18	O 1	
4002-018	154	C10	H 18	O 1	
4002-017	154	C10	H 18	O 1	
4002-016	154	C10	H 18	O 1	
4002-015	154	C10	H 18	O 1	
4002-014	154	C10	H 18	O 1	
4002-061	154	C10	H 18	O 1	
4002-064	154	C10	H 18	O 1	
4002-063	154	C10	H 18	O 1	
4002-062	154	C10	H 18	O 1	
4002-043	154	C10	H 18	O 1	
4002-042	154	C10	H 18	O 1	
4002-057	154	C10	H 18	O 1	
4002-056	154	C10	H 18	O 1	
4002-055	154	C10	H 18	O 1	
4007-001	154	C10	H 18	O 1	
4006-006	154	C10	H 18	O 1	
4006-007	154	C10	H 18	O 1	
4002-021	156	C10	H 20	O 1	
4002-025	156	C10	H 20	O 1	
4002-024	156	C10	H 20	O 1	
4001-045	156	C10	H 20	O 1	
4001-044	156	C10	H 20	O 1	
4001-005	156	C10	H 20	O 1	
4001-006	156	C10	H 20	O 1	
4001-015	156	C10	H 20	O 1	
4912-003	156	C12	H 12		
4909-010	156	C 9	H 16	O 2	
4192-005	156	C12	H 12		
4013-017	161	C10	H 11	N 1	O 1
4002-101	162	C10	H 10	O 2	
4011-007	162	C10	H 10	O 2	
4191-043	162	C12	H 18		
4191-045	162	C12	H 18		
4192-027	162	C12	H 18		
4013-014	163	C10	H 13	N 1	O 1
4099-007	164	C11	H 16	O 1	
4001-033	164	C10	H 12	O 2	
4001-035	164	C10	H 12	O 2	
4001-048	164	C10	H 12	O 2	
4012-003	164	C10	H 12	O 2	
4011-001	164	C10	H 12	O 2	
4011-003	164	C10	H 12	O 2	
4011-002	164	C10	H 12	O 2	
4002-094	164	C10	H 12	O 2	
4002-097	164	C10	H 12	O 2	
4002-050	164	C11	H 16	O 1	
4002-045	164	C11	H 16	O 1	
4193-047	164	C11	H 16	O 1	
4909-007	164	C10	H 12	O 2	
4014-004	165	C 9	H 11	N 1	O 2
4099-016	166	C10	H 14	O 2	
4099-027	166	C10	H 14	O 2	
4001-034	166	C10	H 14	O 2	
4001-029	166	C10	H 14	O 2	
4001-037	166	C 9	H 10	O 3	
4002-047	166	C10	H 14	O 2	
4002-068	166	C10	H 14	O 2	
4002-077	166	C10	H 14	O 2	
4002-076	166	C10	H 14	O 2	
4003-018	166	C10	H 14	O 2	
4005-005	166	C10	H 14	O 2	
4005-004	166	C10	H 14	O 2	
4008-010	166	C10	H 14	O 2	
4008-009	166	C10	H 14	O 2	
4008-018	166	C10	H 14	O 2	
4008-017	166	C10	H 14	O 2	
4008-021	166	C10	H 14	O 2	
4909-006	166	C10	H 14	O 2	
4013-002	167	C11	H 21	N	
4013-001	167	C11	H 21	N	
4013-003	167	C11	H 21	N	
4010-002	168	C10	H 16	O 2	
4008-020	168	C10	H 16	O 2	
4008-016	168	C10	H 16	O 2	
4008-019	168	C10	H 16	O 2	
4008-015	168	C10	H 16	O 2	
4008-008	168	C10	H 16	O 2	
4003-019	168	C10	H 16	O 2	
4002-078	168	C10	H 16	O 2	
4002-086	168	C10	H 16	O 2	
4002-069	168	C10	H 16	O 2	
4002-054	168	C10	H 16	O 2	
4001-039	168	C10	H 16	O 2	
4001-040	168	C10	H 16	O 2	
4099-014	168	C10	H 16	O 2	
4099-004	168	C10	H 16	O 2	
4001-038	170	C10	H 18	O 2	
4001-028	170	C10	H 18	O 2	
4001-051	170	C10	H 18	O 2	
4001-050	170	C10	H 18	O 2	
4002-036	170	C10	H 18	O 2	
4002-033	170	C10	H 18	O 2	
4002-032	170	C10	H 18	O 2	
4008-007	170	C10	H 18	O 2	
4008-014	170	C10	H 18	O 2	
4008-013	170	C10	H 18	O 2	
4008-006	172	C10	H 20	O 2	
4008-005	172	C10	H 20	O 2	
4008-004	172	C10	H 20	O 2	
4008-003	172	C10	H 20	O 2	
4008-002	172	C10	H 20	O 2	
4002-038	172	C10	H 20	O 2	
4002-041	172	C10	H 20	O 2	
4002-040	172	C10	H 20	O 2	
4913-010	172	C13	H 16		
4912-004	174	C12	H 14	O 1	
4014-001	175	C10	H 9	N 1	O 2
4912-002	176	C12	H 16	O 1	
4913-002	176	C13	H 20		
4013-018	177	C10	H 11	N 1	O 2
4011-006	178	C10	H 10	O 3	
4911-001	178	C11	H 14	O 2	
4909-002	178	C10	H 10	O 3	
4193-046	178	C12	H 18	O 1	
4013-011	179	C11	H 17	N 1	O 1
4014-005	179	C10	H 13	N 1	O 2
4011-011	180	C10	H 12	O 3	
4011-010	180	C10	H 12	O 3	
4002-051	180	C11	H 16	O 2	
4002-046	180	C11	H 16	O 2	
4911-002	180	C11	H 16	O 2	
4911-004	180	C11	H 16	O 2	
4909-011	182	C 9	H 10	O 4	
4002-087	182	C10	H 14	O 3	
4002-096	182	C10	H 14	O 3	
4192-029	182	C12	H 22	O 1	
4013-004	183	C11	H 21	N 1	O 1
4013-006	183	C11	H 21	N 1	O 1
4013-005	183	C11	H 21	N 1	O 1
4013-010	183	C11	H 21	N 1	O 1
4013-008	183	C11	H 21	N 1	O 1
4013-007	183	C11	H 21	N 1	O 1
4013-021	183	C11	H 21	N 1	O 1
4001-049	184	C10	H 16	O 3	
4008-022	184	C 9	H 12	O 4	
4002-084	184	C10	H 16	O 3	
4192-002	184	C14	H 16		
4192-001	184	C14	H 16		
4192-004	186	C14	H 18		
4192-003	186	C14	H 18		
4002-037	186	C10	H 18	O 3	
4099-019	192	C12	H 16	O 2	
4913-008	192	C13	H 20	O 1	
4913-007	192	C13	H 20	O 1	
4014-006	193	C10	H 11	N 1	O 3
4002-048	194	C12	H 18	O 2	
4011-012	194	C11	H 14	O 3	
4011-008	194	C11	H 14	O 3	
4913-009	194	C13	H 22	O 1	
4911-003	196	C11	H 16	O 3	
4910-001	196	C10	H 12	O 4	
4913-001	196	C13	H 24	O 1	
4104-008	196	C15	H 16		
4104-009	196	C15	H 16		
4001-030	196	C10	H 12	O 4	
4001-009	196	C12	H 20	O 2	
4104-007	198	C15	H 18		
4107-027	198	C15	H 18		
4107-026	200	C15	H 20		
4111-001	200	C15	H 20		
4107-025	200	C15	H 20		
4192-043	200	C14	H 16	O 1	
4192-006	200	C13	H 12	O 2	
4115-003	202	C15	H 22		
4122-001	202	C15	H 22		
4192-042	202	C14	H 18	O 1	
4193-023	202	C15	H 22		
4122-002	202	C15	H 22		
4113-002	202	C15	H 22		
4192-037	202	C14	H 18	O 1	
4117-056	202	C15	H 22		
4122-003	202	C15	H 22		
4192-036	202	C14	H 18	O 1	
4192-023	202	C14	H 18	O 1	
4122-004	202	C15	H 22		
4117-004	202	C15	H 22		
4117-003	202	C15	H 22		
4107-024	202	C15	H 22		
4106-007	202	C15	H 22		
4106-006	202	C15	H 22		

4106-009	204	C15	H 24		
4112-001	204	C15	H 24		
4112-002	204	C15	H 24		
4106-008	204	C15	H 24		
4106-002	204	C15	H 24		
4112-005	204	C15	H 24		
4112-006	204	C15	H 24		
4112-007	204	C15	H 24		
4106-003	204	C15	H 24		
4106-005	204	C15	H 24		
4106-004	204	C15	H 24		
4106-001	204	C15	H 24		
4107-017	204	C15	H 24		
4112-013	204	C15	H 24		
4113-001	204	C15	H 24		
4107-010	204	C15	H 24		
4107-009	204	C15	H 24		
4107-018	204	C15	H 24		
4106-010	204	C15	H 24		
4107-020	204	C15	H 24		
4107-019	204	C15	H 24		
4107-008	204	C15	H 24		
4107-021	204	C15	H 24		
4104-002	204	C15	H 24		
4104-003	204	C15	H 24		
4104-004	204	C15	H 24		
4104-001	204	C15	H 24		
4107-023	204	C15	H 24		
4107-022	204	C15	H 24		
4107-006	204	C15	H 24		
4107-005	204	C15	H 24		
4107-004	204	C15	H 24		
4109-001	204	C15	H 24		
4107-003	204	C15	H 24		
4105-001	204	C15	H 24		
4105-002	204	C15	H 24		
4107-015	204	C15	H 24		
4107-001	204	C15	H 24		
4104-006	204	C15	H 24		
4107-016	204	C15	H 24		
4107-002	204	C15	H 24		
4103-002	204	C15	H 24		
4103-001	204	C15	H 24		
4106-033	204	C15	H 24		
4107-069	204	C15	H 24		
4110-001	204	C15	H 24		
4109-002	204	C15	H 24		
4104-005	204	C15	H 24		
4107-011	204	C15	H 24		
4107-013	204	C15	H 24		
4101-001	204	C15	H 24		
4104-089	204	C15	H 24		
4107-012	204	C15	H 24		
4102-004	204	C15	H 24		
4102-003	204	C15	H 24		
4102-002	204	C15	H 24		
4102-001	204	C15	H 24		
4101-024	204	C15	H 24		
4101-023	204	C15	H 24		
4101-002	204	C15	H 24		
4121-004	204	C15	H 24		
4117-002	204	C15	H 24		
4117-001	204	C15	H 24		
4118-003	204	C15	H 24		
4191-033	204	C15	H 24		
4118-002	204	C15	H 24		
4116-002	204	C15	H 24		
4116-003	204	C15	H 24		
4130-002	204	C15	H 24		
4192-024	204	C14	H 20	O 1	
4127-001	204	C15	H 24		
4192-039	204	C14	H 20	O 1	
4116-001	204	C15	H 24		
4193-049	204	C15	H 24		
4124-001	204	C15	H 24		
4127-002	204	C15	H 24		
4123-003	204	C15	H 24		
4123-002	204	C15	H 24		
4130-001	204	C15	H 24		
4120-003	204	C15	H 24		
4123-001	204	C15	H 24		
4120-004	204	C15	H 24		
4120-002	204	C15	H 24		
4127-003	204	C15	H 24		
4120-001	204	C15	H 24		
4192-040	204	C14	H 20	O 1	
4116-004	204	C15	H 24		
4121-003	204	C15	H 24		
4118-001	204	C15	H 24		
4193-027	204	C15	H 24		
4192-051	204	C15	H 24		
4193-010	204	C15	H 24		
4193-021	204	C15	H 24		
4193-020	204	C15	H 24		
4193-007	204	C15	H 24		
4193-048	204	C14	H 20	O 1	
4193-031	204	C15	H 24		
4192-072	204	C15	H 24		
4192-026	204	C15	H 24		
4193-008	204	C15	H 24		
4192-050	204	C15	H 24		
4120-076	204	C15	H 24		
4193-030	204	C15	H 24		
4193-051	204	C15	H 24		
4120-006	204	C15	H 24		
4120-086	204	C15	H 24		
4193-018	204	C15	H 24		
4119-005	204	C15	H 24		
4194-006	204	C15	H 24		
4121-002	204	C15	H 24		
4120-005	204	C15	H 24		
4125-001	204	C15	H 24		
4115-002	204	C15	H 24		
4121-001	204	C15	H 24		
4115-001	204	C15	H 24		
4119-006	204	C15	H 24		
4125-003	204	C15	H 24		
4125-002	204	C15	H 24		
4192-048	204	C15	H 24		
4193-026	204	C15	H 24		
4119-007	204	C15	H 24		
4193-025	204	C15	H 24		
4193-024	204	C15	H 24		
4193-044	204	C14	H 20	O 1	
4192-047	204	C15	H 24		
4194-005	204	C15	H 24		
4193-005	204	C15	H 24		
4014-007	205	C11	H 11	N 1	O 3
4104-010	206	C15	H 26		
4101-022	206	C15	H 26		
4101-021	206	C15	H 26		
4107-014	206	C15	H 26		
4193-009	206	C14	H 22	O 1	
4914-002	206	C14	H 22	O 1	
4914-001	206	C14	H 22	O 1	
4914-004	206	C14	H 22	O 1	
4914-003	206	C14	H 22	O 1	
4913-003	208	C13	H 20	O 2	
4913-004	208	C13	H 20	O 2	
4107-007	208	C15	H 28		
4104-011	210	C15	H 14	O 1	
4128-002	210	C15	H 18	O 7	
4099-005	212	C11	H 16	O 4	
4002-100	212	C12	H 20	O 3	
4107-073	212	C15	H 16	O 1	
4912-001	212	C12	H 20	O 3	
4107-050	214	C15	H 18	O 1	
4104-020	214	C15	H 18	O 1	
4120-040	214	C15	H 18	O 1	
4191-032	214	C15	H 18	O 1	
4193-012	214	C15	H 18	O 1	
4198-016	215	C15	H 21	N 1	
4198-017	215	C15	H 21	N 1	
4119-004	216	C15	H 20	O 1	
4193-035	216	C15	H 20	O 1	
4115-008	216	C15	H 20	O 1	
4115-009	216	C15	H 20	O 1	
4121-015	216	C15	H 20	O 1	
4120-041	216	C15	H 20	O 1	
4117-016	216	C15	H 20	O 1	
4106-025	216	C15	H 20	O 1	
4102-013	216	C15	H 20	O 1	
4106-019	216	C15	H 20	O 1	
4111-004	216	C15	H 20	O 1	
4106-021	218	C15	H 22	O 1	
4106-020	218	C15	H 22	O 1	
4107-072	218	C15	H 22	O 1	
4106-018	218	C15	H 22	O 1	
4115-004	218	C15	H 22	O 1	
4115-005	218	C15	H 22	O 1	
4115-006	218	C15	H 22	O 1	
4115-007	218	C15	H 22	O 1	
4106-015	218	C15	H 22	O 1	
4107-053	218	C15	H 22	O 1	
4106-022	218	C15	H 22	O 1	
4104-022	218	C15	H 22	O 1	
4102-009	218	C15	H 22	O 1	
4101-008	218	C15	H 22	O 1	
4101-014	218	C15	H 22	O 1	
4103-007	218	C15	H 22	O 1	
4104-014	218	C15	H 22	O 1	
4107-051	218	C15	H 22	O 1	
4117-012	218	C15	H 22	O 1	
4117-013	218	C15	H 22	O 1	
4117-015	218	C15	H 22	O 1	
4192-038	218	C14	H 18	O 2	
4118-006	218	C15	H 22	O 1	
4194-007	218	C15	H 22	O 1	
4121-014	218	C15	H 22	O 1	
4121-010	218	C15	H 22	O 1	

4117-010	218	C15	H 22	O 1	
4192-013	218	C15	H 22	O 1	
4117-009	218	C15	H 22	O 1	
4121-009	218	C15	H 22	O 1	
4121-008	218	C15	H 22	O 1	
4120-078	218	C15	H 22	O 1	
4117-026	218	C15	H 22	O 1	
4127-008	218	C15	H 22	O 1	
4120-079	218	C15	H 22	O 1	
4125-009	218	C15	H 22	O 1	
4120-030	218	C15	H 22	O 1	
4127-009	218	C15	H 22	O 1	
4120-080	218	C15	H 22	O 1	
4130-003	218	C15	H 22	O 1	
4119-003	218	C15	H 22	O 1	
4120-029	218	C15	H 22	O 1	
4191-024	218	C15	H 22	O 1	
4193-022	218	C15	H 22	O 1	
4192-028	218	C15	H 22	O 1	
4127-007	218	C15	H 22	O 1	
4122-007	218	C15	H 22	O 1	
4119-008	220	C15	H 24	O 1	
4192-058	220	C15	H 24	O 1	
4120-022	220	C15	H 24	O 1	
4120-034	220	C15	H 24	O 1	
4120-021	220	C15	H 24	O 1	
4194-003	220	C15	H 24	O 1	
4191-026	220	C15	H 24	O 1	
4120-077	220	C15	H 24	O 1	
4125-008	220	C15	H 24	O 1	
4192-052	220	C15	H 24	O 1	
4191-025	220	C15	H 24	O 1	
4120-026	220	C15	H 24	O 1	
4194-004	220	C15	H 24	O 1	
4124-007	220	C15	H 24	O 1	
4127-004	220	C15	H 24	O 1	
4121-012	220	C15	H 24	O 1	
4120-015	220	C15	H 24	O 1	
4121-011	220	C15	H 24	O 1	
4121-013	220	C15	H 24	O 1	
4127-005	220	C15	H 24	O 1	
4119-001	220	C15	H 24	O 1	
4121-006	220	C15	H 24	O 1	
4118-004	220	C15	H 24	O 1	
4118-005	220	C15	H 24	O 1	
4121-005	220	C15	H 24	O 1	
4120-010	220	C15	H 24	O 1	
4120-035	220	C15	H 24	O 1	
4120-014	220	C15	H 24	O 1	
4191-037	220	C15	H 24	O 1	
4191-031	220	C15	H 24	O 1	
4124-002	220	C15	H 24	O 1	
4124-003	220	C15	H 24	O 1	
4116-006	220	C15	H 24	O 1	
4116-007	220	C15	H 24	O 1	
4116-008	220	C15	H 24	O 1	
4103-005	220	C15	H 24	O 1	
4103-004	220	C15	H 24	O 1	
4103-003	220	C15	H 24	O 1	
4107-048	220	C15	H 24	O 1	
4117-005	220	C15	H 24	O 1	
4105-005	220	C15	H 24	O 1	
4104-019	220	C15	H 24	O 1	
4105-004	220	C15	H 24	O 1	
4105-006	220	C15	H 24	O 1	
4116-005	220	C15	H 24	O 1	
4117-011	220	C15	H 24	O 1	
4104-013	220	C15	H 24	O 1	
4103-010	220	C15	H 24	O 1	
4117-014	220	C15	H 24	O 1	
4014-008	220	C11	H 12	N 2	O 3
4101-007	220	C15	H 24	O 1	
4102-010	220	C15	H 24	O 1	
4105-012	220	C15	H 24	O 1	
4107-041	220	C15	H 24	O 1	
4112-011	220	C15	H 24	O 1	
4112-012	220	C15	H 24	O 1	
4107-052	220	C15	H 24	O 1	
4107-043	220	C15	H 24	O 1	
4106-016	220	C15	H 24	O 1	
4107-042	220	C15	H 24	O 1	
4103-011	220	C15	H 24	O 1	
4107-028	220	C15	H 24	O 1	
4107-038	222	C15	H 26	O 1	
4107-031	222	C15	H 26	O 1	
4107-030	222	C15	H 26	O 1	
4107-032	222	C15	H 26	O 1	
4107-033	222	C15	H 26	O 1	
4104-031	222	C15	H 26	O 1	
4104-028	222	C15	H 26	O 1	
4106-013	222	C15	H 26	O 1	
4106-012	222	C15	H 26	O 1	
4106-011	222	C15	H 26	O 1	
4106-014	222	C15	H 26	O 1	
4108-001	222	C15	H 26	O 1	
4112-004	222	C15	H 26	O 1	
4112-003	222	C15	H 26	O 1	
4107-040	222	C15	H 26	O 1	
4107-039	222	C15	H 26	O 1	
4104-027	222	C15	H 26	O 1	
4104-026	222	C15	H 26	O 1	
4102-005	222	C15	H 26	O 1	
4101-003	222	C15	H 26	O 1	
4101-004	222	C15	H 26	O 1	
4109-003	222	C15	H 26	O 1	
4107-047	222	C15	H 26	O 1	
4107-046	222	C15	H 26	O 1	
4103-009	222	C15	H 26	O 1	
4104-015	222	C15	H 26	O 1	
4104-016	222	C15	H 26	O 1	
4105-003	222	C15	H 26	O 1	
4117-007	222	C15	H 26	O 1	
4117-006	222	C15	H 26	O 1	
4104-018	222	C15	H 26	O 1	
4107-045	222	C15	H 26	O 1	
4107-049	222	C15	H 26	O 1	
4104-017	222	C15	H 26	O 1	
4120-012	222	C15	H 26	O 1	
4192-045	222	C14	H 22	O 2	
4120-013	222	C15	H 26	O 1	
4192-044	222	C14	H 22	O 2	
4120-020	222	C15	H 26	O 1	
4120-036	222	C15	H 26	O 1	
4120-019	222	C15	H 26	O 1	
4120-011	222	C15	H 26	O 1	
4123-004	222	C15	H 26	O 1	
4192-046	222	C15	H 26	O 1	
4120-018	222	C15	H 26	O 1	
4121-007	222	C15	H 26	O 1	
4120-016	222	C15	H 26	O 1	
4120-017	222	C15	H 26	O 1	
4125-006	222	C15	H 26	O 1	
4125-005	222	C15	H 26	O 1	
4125-004	222	C15	H 26	O 1	
4193-006	222	C15	H 26	O 1	
4125-007	222	C15	H 26	O 1	
4120-007	222	C15	H 26	O 1	
4122-005	222	C15	H 26	O 1	
4122-006	222	C15	H 26	O 1	
4126-001	222	C15	H 26	O 1	
4191-047	222	C15	H 26	O 1	
4192-057	222	C15	H 26	O 1	
4129-001	222	C15	H 26	O 1	
4120-009	222	C15	H 26	O 1	
4193-029	222	C15	H 26	O 1	
4120-008	222	C15	H 26	O 1	
4193-019	222	C15	H 26	O 1	
4193-032	222	C15	H 26	O 1	
4192-025	222	C15	H 26	O 1	
4192-020	222	C15	H 26	O 1	
4192-019	222	C15	H 26	O 1	
4002-103	224	C15	H 28	O 1	
4002-102	224	C15	H 28	O 1	
4913-006	224	C13	H 20	O 3	
4913-005	224	C13	H 20	O 3	
4008-041	226	C11	H 14	O 5	
4008-036	226	C11	H 14	O 5	
4192-034	226	C14	H 10	O 3	
4193-013	228	C15	H 16	O 2	
4107-058	228	C15	H 16	O 2	
4009-008	228	C10	H 12	O 6	
4008-065	228	C11	H 16	O 5	
4107-067	228	C15	H 16	O 2	
4104-012	228	C15	H 16	O 2	
4102-014	230	C15	H 18	O 2	
4102-015	230	C15	H 18	O 2	
4104-045	230	C15	H 18	O 2	
4117-025	230	C15	H 18	O 2	
4192-030	230	C15	H 18	O 2	
4121-021	230	C15	H 18	O 2	
4120-045	230	C15	H 18	O 2	
4121-020	230	C15	H 18	O 2	
4117-039	230	C15	H 18	O 2	
4117-038	230	C15	H 18	O 2	
4117-046	230	C15	H 18	O 2	
4193-014	230	C15	H 18	O 2	
4121-029	230	C15	H 18	O 2	
4191-036	230	C15	H 18	O 2	
4198-009	231	C15	H 21	N 1	O 1
4192-076	232	C15	H 20	O 2	
4117-037	232	C15	H 20	O 2	
4117-036	232	C15	H 20	O 2	
4191-030	232	C15	H 20	O 2	
4193-028	232	C15	H 20	O 2	
4191-029	232	C15	H 20	O 2	
4121-017	232	C15	H 20	O 2	
4191-034	232	C15	H 20	O 2	
4192-008	232	C14	H 16	O 3	
4126-008	232	C15	H 20	O 2	
4120-043	232	C15	H 20	O 2	

4120-042	232	C15	H 20	O 2		
4115-010	232	C15	H 20	O 2		
4107-056	232	C15	H 20	O 2		
4104-044	232	C15	H 20	O 2		
4102-012	232	C15	H 20	O 2		
4101-011	232	C15	H 20	O 2		
4107-066	232	C15	H 20	O 2		
4099-028	232	C11	H 20	N 3	O 2	+
4106-034	232	C15	H 20	O 2		
4102-044	232	C15	H 20	O 2		
4914-005	232	C14	H 16	O 3		
4915-003	232	C15	H 20	O 2		
4198-006	233	C15	H 23	N 1	O 1	
4198-007	233	C15	H 23	N 1	O 1	
4198-002	233	C15	H 23	N 1	O 1	
4198-004	233	C15	H 23	N 1	O 1	
4192-053	234	C15	H 22	O 2		
4192-078	234	C15	H 22	O 2		
4121-023	234	C15	H 22	O 2		
4126-006	234	C15	H 22	O 2		
4126-007	234	C15	H 22	O 2		
4120-044	234	C15	H 22	O 2		
4191-027	234	C15	H 22	O 2		
4126-003	234	C15	H 22	O 2		
4120-032	234	C15	H 22	O 2		
4119-002	234	C15	H 22	O 2		
4117-047	234	C15	H 22	O 2		
4120-027	234	C15	H 22	O 2		
4120-028	234	C15	H 22	O 2		
4194-001	234	C15	H 22	O 2		
4194-002	234	C15	H 22	O 2		
4126-016	234	C15	H 22	O 2		
4126-017	234	C15	H 22	O 2		
4191-023	234	C15	H 22	O 2		
4192-014	234	C15	H 22	O 2		
4191-028	234	C15	H 22	O 2		
4192-059	234	C15	H 22	O 2		
4124-004	234	C15	H 22	O 2		
4118-007	234	C15	H 22	O 2		
4124-005	234	C15	H 22	O 2		
4192-009	234	C14	H 18	O 3		
4124-006	234	C15	H 22	O 2		
4915-002	234	C15	H 22	O 2		
4102-045	234	C15	H 22	O 2		
4117-008	234	C15	H 22	O 2		
4105-011	234	C15	H 22	O 2		
4101-010	234	C15	H 22	O 2		
4117-017	234	C15	H 22	O 2		
4117-027	234	C15	H 22	O 2		
4117-028	234	C15	H 22	O 2		
4117-021	234	C15	H 22	O 2		
4113-003	234	C15	H 22	O 2		
4117-019	234	C15	H 22	O 2		
4117-020	234	C15	H 22	O 2		
4104-024	234	C15	H 22	O 2		
4104-025	234	C15	H 22	O 2		
4003-022	236	C15	H 24	O 2		
4112-010	236	C15	H 24	O 2		
4112-009	236	C15	H 24	O 2		
4112-008	236	C15	H 24	O 2		
4107-054	236	C15	H 24	O 2		
4107-055	236	C15	H 24	O 2		
4101-009	236	C15	H 24	O 2		
4101-016	236	C15	H 24	O 2		
4117-018	236	C15	H 24	O 2		
4102-011	236	C15	H 24	O 2		
4104-083	236	C15	H 24	O 2		
4103-006	236	C15	H 24	O 2		
4107-044	236	C15	H 24	O 2		
4914-006	236	C15	H 24	O 2		
4192-060	236	C15	H 24	O 2		
4192-017	236	C15	H 24	O 2		
4121-028	236	C15	H 24	O 2		
4192-049	236	C15	H 24	O 2		
4190-001	236	C15	H 24	O 2		
4120-031	236	C15	H 24	O 2		
4120-023	238	C15	H 26	O 2		
4192-077	238	C15	H 26	O 2		
4193-033	238	C15	H 26	O 2		
4192-018	238	C15	H 26	O 2		
4120-025	238	C15	H 26	O 2		
4193-011	238	C15	H 26	O 2		
4129-002	238	C15	H 26	O 2		
4120-037	238	C15	H 26	O 2		
4193-045	238	C15	H 26	O 2		
4129-004	238	C15	H 26	O 2		
4105-008	238	C15	H 26	O 2		
4105-007	238	C15	H 26	O 2		
4102-072	238	C15	H 26	O 2		
4107-034	238	C15	H 26	O 2		
4107-037	238	C15	H 26	O 2		
4107-036	238	C15	H 26	O 2		
4107-035	238	C15	H 26	O 2		
4104-034	238	C15	H 26	O 2		
4104-033	238	C15	H 26	O 2		

4106-017	238	C15	H 26	O 2	
4104-032	238	C15	H 26	O 2	
4104-029	238	C15	H 26	O 2	
4108-002	238	C15	H 26	O 2	
4107-061	240	C15	H 12	O 3	
4101-015	240	C15	H 12	O 3	
4101-005	240	C15	H 28	O 2	
4191-042	240	C15	H 28	O 2	
4120-024	240	C15	H 28	O 2	
4192-032	242	C14	H 10	O 4	
4107-060	242	C15	H 14	O 3	
4107-059	242	C15	H 14	O 3	
4008-038	242	C11	H 14	O 6	
4107-062	244	C15	H 16	O 3	
4104-040	244	C15	H 16	O 3	
4102-019	244	C15	H 16	O 3	
4104-086	244	C15	H 16	O 3	
4015-002	244	C14	H 12	O 4	
4102-078	244	C15	H 16	O 3	
4120-049	244	C15	H 16	O 3	
4121-025	244	C15	H 16	O 3	
4193-016	244	C16	H 20	O 2	
4193-017	244	C16	H 20	O 2	
4120-046	244	C15	H 16	O 3	
4192-031	244	C15	H 16	O 3	
4192-041	246	C16	H 22	O 2	
4120-048	246	C15	H 18	O 3	
4120-050	246	C15	H 18	O 3	
4120-081	246	C15	H 18	O 3	
4120-066	246	C15	H 18	O 3	
4192-016	246	C15	H 18	O 3	
4191-015	246	C15	H 18	O 3	
4120-071	246	C15	H 18	O 3	
4117-050	246	C15	H 18	O 3	
4120-065	246	C15	H 18	O 3	
4120-039	246	C15	H 18	O 3	
4104-079	246	C15	H 18	O 3	
4114-024	246	C15	H 18	O 3	
4114-025	246	C15	H 18	O 3	
4102-083	246	C15	H 18	O 3	
4102-016	246	C15	H 18	O 3	
4102-017	246	C15	H 18	O 3	
4114-057	246	C15	H 18	O 3	
4104-046	246	C15	H 18	O 3	
4104-074	246	C15	H 18	O 3	
4104-069	246	C15	H 18	O 3	
4114-009	246	C15	H 18	O 3	
4114-002	246	C15	H 18	O 3	
4104-053	246	C15	H 18	O 3	
4104-058	246	C15	H 18	O 3	
4104-057	248	C15	H 20	O 3	
4104-056	248	C15	H 20	O 3	
4104-055	248	C15	H 20	O 3	
4106-024	248	C15	H 20	O 3	
4102-057	248	C15	H 20	O 3	
4102-054	248	C15	H 20	O 3	
4104-049	248	C15	H 20	O 3	
4104-070	248	C15	H 20	O 3	
4115-012	248	C15	H 20	O 3	
4114-004	248	C15	H 20	O 3	
4104-037	248	C15	H 20	O 3	
4104-039	248	C15	H 20	O 3	
4107-057	248	C15	H 20	O 3	
4099-031	248	C15	H 20	O 3	
4102-080	248	C15	H 20	O 3	
4105-010	248	C15	H 20	O 3	
4105-009	248	C15	H 20	O 3	
4106-035	248	C15	H 20	O 3	
4114-019	248	C15	H 20	O 3	
4106-029	248	C15	H 20	O 3	
4102-046	248	C15	H 20	O 3	
4102-047	248	C15	H 20	O 3	
4102-056	248	C15	H 20	O 3	
4120-057	248	C15	H 20	O 3	
4121-016	248	C15	H 20	O 3	
4120-056	248	C15	H 20	O 3	
4126-021	248	C15	H 20	O 3	
4120-063	248	C15	H 20	O 3	
4126-011	248	C15	H 20	O 3	
4120-069	248	C15	H 20	O 3	
4120-068	248	C15	H 20	O 3	
4117-041	248	C15	H 20	O 3	
4120-090	248	C15	H 20	O 3	
4120-060	248	C15	H 20	O 3	
4120-061	248	C15	H 20	O 3	
4192-010	248	C15	H 20	O 3	
4120-059	248	C15	H 20	O 3	
4120-058	248	C15	H 20	O 3	
4120-062	248	C15	H 20	O 3	
4910-002	248	C15	H 20	O 3	
4198-010	249	C15	H 23	N 1	O 2
4198-003	249	C15	H 23	N 1	O 2
4198-008	249	C15	H 23	N 1	O 2
4198-005	249	C15	H 23	N 1	O 2
4121-022	250	C15	H 22	O 3	

4192-021	250	C15	H 22	O 3		
4191-035	250	C15	H 22	O 3		
4126-009	250	C15	H 22	O 3		
4192-022	250	C15	H 22	O 3		
4192-015	250	C15	H 22	O 3		
4117-049	250	C15	H 22	O 3		
4126-019	250	C15	H 22	O 3		
4915-008	250	C15	H 22	O 3		
4915-007	250	C15	H 22	O 3		
4915-001	250	C15	H 22	O 3		
4102-050	250	C15	H 22	O 3		
4102-051	250	C15	H 22	O 3		
4110-003	250	C15	H 22	O 3		
4101-013	250	C15	H 22	O 3		
4101-012	250	C15	H 22	O 3		
4101-019	250	C15	H 22	O 3		
4104-038	250	C15	H 22	O 3		
4102-075	250	C15	H 22	O 3		
4117-030	250	C15	H 22	O 3		
4117-032	250	C15	H 22	O 3		
4102-068	250	C15	H 22	O 3		
4102-067	250	C15	H 22	O 3		
4114-001	250	C15	H 22	O 3		
4198-001	251	C15	H 25	N 1	O 2	
4193-040	252	C15	H 24	O 3		
4127-006	252	C15	H 24	O 3		
4192-063	252	C15	H 24	O 3		
4106-023	252	C15	H 24	O 3		
4105-013	252	C15	H 24	O 3		
4110-002	252	C15	H 24	O 3		
4102-055	252	C15	H 24	O 3		
4106-026	252	C15	H 24	O 3		
4104-035	254	C15	H 26	O 3		
4120-038	254	C15	H 26	O 3		
4192-033	254	C16	H 14	O 3		
4107-065	256	C15	H 12	O 4		
4107-070	256	C15	H 28	O 3		
4293-015	256	C19	H 28			
4101-006	258	C15	H 30	O 3		
4107-064	258	C15	H 14	O 4		
4107-063	258	C15	H 14	O 4		
4107-068	258	C16	H 18	O 3		
4102-018	260	C16	H 20	O 3		
4102-020	260	C15	H 16	O 4		
4102-022	260	C15	H 16	O 4		
4102-021	260	C15	H 16	O 4		
4102-034	260	C15	H 16	O 4		
4121-026	260	C16	H 20	O 3		
4191-020	260	C15	H 16	O 4		
4120-073	262	C15	H 18	O 4		
4120-047	262	C15	H 18	O 4		
4193-036	262	C15	H 18	O 4		
4192-073	262	C15	H 18	O 4		
4191-021	262	C15	H 18	O 4		
4192-011	262	C15	H 18	O 4		
4120-089	262	C15	H 18	O 4		
4120-072	262	C15	H 18	O 4		
4193-003	262	C15	H 18	O 4		
4191-011	262	C15	H 18	O 4		
4120-070	262	C15	H 18	O 4		
4104-087	262	C15	H 18	O 4		
4114-028	262	C15	H 18	O 4		
4114-022	262	C15	H 18	O 4		
4114-027	262	C15	H 18	O 4		
4104-080	262	C15	H 18	O 4		
4104-078	262	C15	H 18	O 4		
4104-042	262	C15	H 18	O 4		
4104-043	262	C15	H 18	O 4		
4114-011	262	C15	H 18	O 4		
4104-071	262	C15	H 18	O 4		
4104-075	262	C15	H 18	O 4		
4293-026	262	C19	H 34			
4918-001	262	C18	H 30	O 1		
4198-022	263	C16	H 25	N 1	O 2	
4120-075	264	C15	H 20	O 4		
4120-088	264	C15	H 20	O 4		
4191-003	264	C15	H 20	O 4		
4120-052	264	C15	H 20	O 4		
4120-087	264	C15	H 20	O 4		
4191-004	264	C15	H 20	O 4		
4120-053	264	C15	H 20	O 4		
4193-052	264	C15	H 20	O 4		
4191-009	264	C15	H 20	O 4		
4120-074	264	C15	H 20	O 4		
4120-091	264	C15	H 20	O 4		
4193-037	264	C15	H 20	O 4		
4193-001	264	C15	H 20	O 4		
4120-067	264	C15	H 20	O 4		
4192-062	264	C17	H 28	O 2		
4191-013	264	C15	H 20	O 4		
4191-012	264	C15	H 20	O 4		
4915-004	264	C15	H 20	O 4		
4917-001	264	C17	H 28	O 2		
4114-005	264	C15	H 20	O 4		
4114-007	264	C15	H 20	O 4		
4105-017	264	C15	H 20	O 4		
4115-011	264	C15	H 20	O 4		
4102-059	264	C15	H 20	O 4		
4102-058	264	C15	H 20	O 4		
4104-052	264	C15	H 20	O 4		
4114-058	264	C15	H 20	O 4		
4114-012	264	C15	H 20	O 4		
4104-041	264	C15	H 20	O 4		
4102-077	264	C15	H 20	O 4		
4102-043	264	C15	H 20	O 4		
4106-028	264	C16	H 24	O 3		
4114-026	264	C15	H 20	O 4		
4102-081	264	C15	H 20	O 4		
4102-082	264	C15	H 20	O 4		
4114-021	264	C15	H 20	O 4		
4106-030	264	C15	H 20	O 4		
4099-029	264	C15	H 20	O 4		
4114-035	264	C15	H 20	O 4		
4114-050	264	C15	H 20	O 4		
4104-082	264	C15	H 20	O 4		
4114-042	264	C15	H 20	O 4		
4114-041	266	C15	H 22	O 4		
4114-040	266	C15	H 22	O 4		
4114-039	266	C15	H 22	O 4		
4102-030	266	C15	H 22	O 4		
4114-043	266	C15	H 22	O 4		
4106-027	266	C16	H 26	O 3		
4102-052	266	C15	H 22	O 4		
4114-010	266	C15	H 22	O 4		
4117-034	266	C15	H 22	O 4		
4114-003	266	C15	H 22	O 4		
4110-006	266	C15	H 22	O 4		
4519-009	266	C18	H 18	O 2		
4919-001	266	C19	H 38			
4126-020	266	C15	H 22	O 4		
4126-015	266	C15	H 22	O 4		
4120-064	266	C15	H 22	O 4		
4191-014	266	C15	H 22	O 4		
4126-013	266	C15	H 22	O 4		
4126-014	266	C15	H 22	O 4		
4191-006	266	C15	H 22	O 4		
4191-005	266	C15	H 22	O 4		
4120-054	266	C15	H 22	O 4		
4193-038	266	C15	H 22	O 4		
4191-044	268	C15	H 24	O 4		
4919-002	268	C19	H 40			
4013-019	268	C18	H 22	N 1	O 1	+
4014-009	269	C16	H 15	N 1	O 3	
4508-006	270	C19	H 26	O 1		
4519-003	270	C18	H 22	O 2		
4204-001	270	C20	H 30			
4192-035	270	C16	H 14	O 4		
4291-015	270	C20	H 30			
4198-018	271	C16	H 17	N 1	O 3	
4206-056	272	C20	H 32			
4208-001	272	C20	H 32			
4202-001	272	C20	H 32			
4209-002	272	C20	H 32			
4202-002	272	C20	H 32			
4193-015	272	C17	H 20	O 3		
4206-055	272	C20	H 32			
4117-040	272	C16	H 16	O 4		
4209-001	272	C20	H 32			
4212-002	272	C20	H 32			
4212-001	272	C20	H 32			
4203-034	272	C20	H 32			
4203-035	272	C20	H 32			
4206-002	272	C20	H 32			
4206-003	272	C20	H 32			
4206-001	272	C20	H 32			
4293-012	272	C18	H 24	O 2		
4291-002	272	C20	H 32			
4291-021	272	C20	H 32			
4216-005	272	C20	H 32			
4216-004	272	C20	H 32			
4292-004	272	C20	H 32			
4299-001	272	C20	H 32			
4519-002	272	C18	H 24	O 2		
4519-001	272	C18	H 24	O 2		
4293-013	274	C17	H 22	O 3		
4293-014	274	C19	H 30	O 1		
4293-029	274	C19	H 30	O 1		
4293-028	274	C19	H 30	O 1		
4121-027	274	C15	H 14	O 5		
4121-024	276	C15	H 16	O 5		
4121-019	276	C16	H 20	O 4		
4192-068	276	C15	H 16	O 5		
4121-030	276	C15	H 16	O 5		
4127-010	276	C17	H 24	O 3		
4192-064	276	C15	H 16	O 5		
4293-024	276	C18	H 12	O 3		
4293-025	276	C18	H 12	O 3		
4102-037	276	C15	H 16	O 5		
4102-023	276	C15	H 16	O 5		
4104-072	276	C15	H 16	O 5		

4102-069	278	C15	H 18	O 5		
4106-032	278	C15	H 18	O 5		
4102-070	278	C15	H 18	O 5		
4114-013	278	C15	H 18	O 5		
4015-006	278	C14	H 14	O 6		
4104-091	278	C15	H 18	O 5		
4104-092	278	C15	H 18	O 5		
4104-081	278	C15	H 18	O 5		
4293-037	278	C18	H 30	O 2		
4191-022	278	C16	H 22	O 4		
4201-007	278	C20	H 38			
4201-008	278	C20	H 38			
4128-005	278	C15	H 18	O 5		
4192-079	278	C15	H 18	O 5		
4201-005	278	C20	H 38			
4201-006	278	C20	H 38			
4198-023	279	C16	H 25	N 1	O 3	
4198-021	279	C16	H 25	N 1	O 3	
4117-055	280	C17	H 28	O 3		
4129-003	280	C17	H 28	O 3		
4191-010	280	C15	H 20	O 5		
4129-005	280	C17	H 28	O 3		
4193-039	280	C15	H 20	O 5		
4191-046	280	C15	H 20	O 5		
4193-041	280	C15	H 20	O 5		
4114-056	280	C15	H 20	O 5		
4101-018	280	C17	H 28	O 3		
4106-031	280	C15	H 20	O 5		
4105-018	280	C15	H 20	O 5		
4105-014	280	C15	H 20	O 5		
4105-019	280	C15	H 20	O 5		
4104-059	280	C15	H 20	O 5		
4104-030	280	C17	H 28	O 3		
4915-006	280	C15	H 20	O 5		
4915-005	280	C15	H 20	O 5		
4920-004	282	C20	H 26	O 1		
4105-021	282	C15	H 22	O 5		
4105-020	282	C15	H 22	O 5		
4293-016	282	C19	H 22	O 2		
4013-020	284	C18	H 22	N 1	O 2	+
4920-002	284	C20	H 28	O 1		
4920-003	284	C20	H 28	O 1		
4204-017	284	C20	H 28	O 1		
4204-005	284	C20	H 28	O 1		
4204-004	286	C20	H 30	O 1		
4204-003	286	C20	H 30	O 1		
4209-013	286	C20	H 30	O 1		
4206-022	286	C20	H 30	O 1		
4203-036	286	C20	H 30	O 1		
4203-022	286	C20	H 30	O 1		
4203-021	286	C20	H 30	O 1		
4203-023	286	C20	H 30	O 1		
4204-036	286	C20	H 30	O 1		
4205-001	286	C20	H 30	O 1		
4508-018	286	C19	H 26	O 2		
4519-004	286	C18	H 22	O 3		
4519-005	286	C18	H 22	O 3		
4920-001	286	C20	H 30	O 1		
4108-003	286	C15	H 26	O 5		
4292-009	286	C20	H 30	O 1		
4293-031	288	C20	H 32	O 1		
4104-047	288	C17	H 20	O 4		
4508-014	288	C19	H 28	O 2		
4519-006	288	C18	H 24	O 3		
4508-011	288	C19	H 28	O 2		
4519-007	288	C18	H 24	O 3		
4508-019	288	C19	H 28	O 2		
4203-005	288	C20	H 32	O 1		
4203-004	288	C20	H 32	O 1		
4209-003	288	C20	H 32	O 1		
4203-007	288	C20	H 32	O 1		
4204-035	288	C20	H 32	O 1		
4203-003	288	C20	H 32	O 1		
4211-001	288	C20	H 32	O 1		
4117-042	288	C17	H 20	O 4		
4202-006	288	C20	H 32	O 1		
4203-002	288	C20	H 32	O 1		
4203-001	288	C20	H 32	O 1		
4202-007	288	C20	H 32	O 1		
4209-005	288	C20	H 32	O 1		
4209-004	288	C20	H 32	O 1		
4202-008	288	C20	H 32	O 1		
4203-040	288	C20	H 32	O 1		
4206-005	290	C20	H 34	O 1		
4202-009	290	C20	H 34	O 1		
4202-011	290	C20	H 34	O 1		
4202-031	290	C20	H 34	O 1		
4202-032	290	C20	H 34	O 1		
4202-034	290	C20	H 34	O 1		
4202-033	290	C20	H 34	O 1		
4202-030	290	C20	H 34	O 1		
4202-010	290	C20	H 34	O 1		
4206-004	290	C20	H 34	O 1		
4204-016	290	C20	H 34	O 1		
4202-005	290	C20	H 34	O 1		
4206-057	290	C20	H 34	O 1		
4202-004	290	C20	H 34	O 1		
4202-003	290	C20	H 34	O 1		
4201-003	290	C20	H 34	O 1		
4203-006	290	C20	H 34	O 1		
4201-001	290	C20	H 34	O 1		
4201-002	290	C20	H 34	O 1		
4508-007	290	C19	H 30	O 2		
4508-002	290	C19	H 30	O 2		
4508-009	290	C19	H 30	O 2		
4916-002	290	C65	H 18	O 5		
4015-005	290	C15	H 14	O 6		
4015-003	290	C15	H 14	O 6		
4102-035	290	C15	H 14	O 6		
4102-049	290	C17	H 22	O 4		
4102-048	290	C17	H 22	O 4		
4217-002	290	C20	H 34	O 1		
4217-003	290	C20	H 34	O 1		
4292-012	290	C19	H 30	O 2		
4216-007	290	C20	H 34	O 1		
4216-006	290	C20	H 34	O 1		
4291-006	290	C20	H 34	O 1		
4217-001	290	C20	H 34	O 1		
4592-023	291	C19	H 33	N 1	O 1	
4198-024	291	C17	H 25	N 1	O 3	
4128-006	292	C15	H 16	O 6		
4126-010	292	C17	H 24	O 4		
4121-018	292	C16	H 20	O 5		
4204-002	292	C20	H 36	O 1		
4508-001	292	C19	H 32	O 2		
4918-004	292	C18	H 28	O 3		
4102-036	292	C15	H 16	O 6		
4015-004	292	C15	H 16	O 6		
4102-053	292	C17	H 24	O 4		
4110-004	292	C17	H 24	O 4		
4198-020	293	C17	H 27	N 1	O 3	
4193-034	294	C15	H 19	O 1	BR 1	
4128-001	294	C15	H 18	O 6		
4111-002	294	C15	H 19	O 1	BR 1	
4111-003	294	C15	H 19	O 1	BR 1	
4111-006	294	C15	H 19	O 1	BR 1	
4101-017	294	C18	H 30	O 3		
4293-017	294	C19	H 18	O 3		
4293-020	294	C19	H 18	O 3		
4293-021	296	C19	H 20	O 3		
4293-019	296	C19	H 20	O 3		
4102-006	296	C17	H 28	O 4		
4105-015	296	C15	H 20	O 6		
4105-016	296	C15	H 20	O 6		
4201-004	296	C20	H 40	O 1		
4191-002	296	C16	H 24	O 5		
4699-025	296	C20	H 24	O 2		
4192-071	298	C15	H 23	O 1	BR 1	
4204-009	298	C20	H 26	O 2		
4291-011	298	C20	H 26	O 2		
4217-012	300	C20	H 28	O 2		
4206-033	300	C20	H 28	O 2		
4204-008	300	C20	H 28	O 2		
4204-007	300	C20	H 28	O 2		
4205-005	300	C20	H 28	O 2		
4205-003	300	C20	H 28	O 2		
4209-015	300	C20	H 28	O 2		
4204-026	300	C20	H 28	O 2		
4204-025	300	C20	H 28	O 2		
4508-017	300	C19	H 24	O 3		
4508-015	300	C19	H 24	O 3		
4003-024	300	C20	H 28	O 2		
4007-003	300	C19	H 24	O 3		
4292-006	301	C20	H 31	N 1	O 1	
4217-015	302	C20	H 30	O 2		
4217-010	302	C20	H 30	O 2		
4102-024	302	C17	H 18	O 5		
4102-025	302	C17	H 18	O 5		
4204-024	302	C20	H 30	O 2		
4204-027	302	C20	H 30	O 2		
4208-004	302	C20	H 30	O 2		
4204-029	302	C20	H 30	O 2		
4205-002	302	C20	H 30	O 2		
4204-028	302	C20	H 30	O 2		
4211-002	302	C20	H 30	O 2		
4117-043	302	C19	H 26	O 3		
4211-005	302	C20	H 30	O 2		
4209-016	302	C20	H 30	O 2		
4202-047	302	C20	H 30	O 2		
4202-048	302	C20	H 30	O 2		
4202-049	302	C20	H 30	O 2		
4202-050	302	C20	H 30	O 2		
4204-040	302	C20	H 30	O 2		
4117-044	302	C19	H 26	O 3		
4203-037	302	C20	H 30	O 2		
4203-039	302	C20	H 30	O 2		
4204-041	302	C20	H 30	O 2		
4204-042	302	C20	H 30	O 2		
4203-042	302	C20	H 30	O 2		
4206-029	302	C20	H 30	O 2		

4204-006	302	C20	H 30	O 2	
4205-006	302	C20	H 30	O 2	
4206-032	302	C20	H 30	O 2	
4206-030	302	C20	H 30	O 2	
4206-031	302	C20	H 30	O 2	
4203-027	302	C20	H 30	O 2	
4203-029	302	C20	H 30	O 2	
4203-024	302	C20	H 30	O 2	
4203-025	302	C20	H 30	O 2	
4592-017	303	C19	H 29	N 1	O 2
4916-001	304	C17	H 20	O 5	
4507-001	304	C21	H 36	O 1	
4918-002	304	C18	H 24	O 4	
4508-013	304	C19	H 28	O 3	
4209-006	304	C20	H 32	O 2	
4209-007	304	C20	H 32	O 2	
4202-035	304	C20	H 32	O 2	
4209-008	304	C20	H 32	O 2	
4209-009	304	C20	H 32	O 2	
4203-041	304	C20	H 32	O 2	
4202-038	304	C20	H 32	O 2	
4206-023	304	C20	H 32	O 2	
4202-045	304	C20	H 32	O 2	
4202-046	304	C20	H 32	O 2	
4202-109	304	C20	H 32	O 2	
4204-034	304	C20	H 32	O 2	
4206-011	304	C20	H 32	O 2	
4206-008	304	C20	H 32	O 2	
4203-008	304	C20	H 32	O 2	
4206-007	304	C20	H 32	O 2	
4206-010	304	C20	H 32	O 2	
4203-010	304	C20	H 32	O 2	
4203-011	304	C20	H 32	O 2	
4202-062	304	C20	H 32	O 2	
4202-061	304	C20	H 32	O 2	
4202-063	304	C20	H 32	O 2	
4203-009	304	C20	H 32	O 2	
4114-029	304	C17	H 20	O 5	
4114-031	304	C17	H 20	O 5	
4114-030	304	C17	H 20	O 5	
4104-077	304	C17	H 20	O 5	
4104-076	304	C17	H 20	O 5	
4114-023	304	C17	H 20	O 5	
4217-004	304	C20	H 32	O 2	
4216-010	304	C20	H 32	O 2	
4216-011	304	C20	H 32	O 2	
4292-003	306	C20	H 34	O 2	
4216-001	306	C20	H 34	O 2	
4217-008	306	C20	H 34	O 2	
4292-002	306	C20	H 34	O 2	
4216-009	306	C20	H 34	O 2	
4216-008	306	C20	H 34	O 2	
4216-012	306	C20	H 34	O 2	
4216-013	306	C20	H 34	O 2	
4291-003	306	C20	H 34	O 2	
4293-047	306	C19	H 30	O 3	
4114-032	306	C17	H 22	O 5	
4099-030	306	C17	H 22	O 5	
4099-032	306	C17	H 22	O 5	
4104-085	306	C17	H 22	O 5	
4102-073	306	C17	H 22	O 5	
4104-036	306	C17	H 22	O 5	
4114-006	306	C17	H 22	O 5	
4002-104	306	C18	H 26	O 4	
4104-054	306	C17	H 22	O 5	
4104-048	306	C17	H 22	O 5	
4104-051	306	C17	H 22	O 5	
4104-050	306	C17	H 22	O 5	
4191-018	306	C17	H 22	O 5	
4202-065	306	C20	H 34	O 2	
4202-064	306	C20	H 34	O 2	
4193-004	306	C17	H 22	O 5	
4202-066	306	C20	H 34	O 2	
4202-067	306	C20	H 34	O 2	
4120-055	306	C17	H 22	O 5	
4120-051	306	C17	H 22	O 5	
4192-012	306	C17	H 22	O 5	
4120-082	306	C17	H 22	O 5	
4206-006	306	C20	H 34	O 2	
4202-037	306	C20	H 34	O 2	
4202-043	306	C20	H 34	O 2	
4202-040	306	C20	H 34	O 2	
4202-023	306	C20	H 34	O 2	
4202-024	306	C20	H 34	O 2	
4202-012	306	C20	H 34	O 2	
4202-036	306	C20	H 34	O 2	
4193-002	306	C17	H 22	O 5	
4202-107	306	C20	H 34	O 2	
4202-014	306	C20	H 34	O 2	
4202-013	306	C20	H 34	O 2	
4202-015	306	C20	H 34	O 2	
4202-017	306	C20	H 34	O 2	
4910-003	306	C17	H 22	O 5	
4918-003	306	C18	H 26	O 4	
4508-012	306	C19	H 30	O 3	
4508-005	306	C19	H 30	O 3	
4508-004	306	C19	H 30	O 3	
4198-025	307	C17	H 25	N 1	O 4
4192-061	308	C19	H 32	O 3	
4126-012	308	C17	H 24	O 5	
4202-020	308	C20	H 36	O 2	
4202-021	308	C20	H 36	O 2	
4202-022	308	C20	H 36	O 2	
4202-026	308	C20	H 36	O 2	
4126-018	308	C17	H 24	O 5	
4191-016	308	C17	H 24	O 5	
4191-017	308	C17	H 24	O 5	
4126-005	308	C17	H 24	O 5	
4191-019	308	C17	H 24	O 5	
4508-003	308	C19	H 32	O 3	
4114-008	308	C17	H 24	O 5	
4102-076	308	C17	H 24	O 5	
4114-044	308	C17	H 24	O 5	
4217-006	308	C20	H 36	O 2	
4592-015	309	C22	H 31	N 1	
4514-031	309	C21	H 27	N 1	O 1
4514-028	309	C21	H 27	N 1	O 1
4293-018	310	C19	H 18	O 4	
4293-022	310	C19	H 18	O 4	
4291-001	310	C20	H 22	O 3	
4111-005	310	C15	H 19	O 2	BR 1
4128-007	310	C15	H 18	O 7	
4191-007	310	C17	H 26	O 5	
4202-019	310	C20	H 38	O 2	
4202-018	310	C20	H 38	O 2	
4212-016	311	C20	H 25	N 1	O 2
4192-075	312	C15	H 20	O 7	
4128-003	312	C15	H 20	O 7	
4201-009	312	C20	H 40	O 2	
4002-107	312	C19	H 20	O 4	
4110-011	312	C15	H 20	O 7	
4291-019	312	C20	H 24	O 3	
4699-024	312	C20	H 24	O 3	
4514-013	312	C21	H 32	N 2	
4514-027	313	C21	H 31	N 1	O 1
4212-015	313	C20	H 27	N 1	O 2
4212-017	313	C20	H 27	N 1	O 2
4210-011	314	C19	H 22	O 4	
4204-023	314	C20	H 26	O 3	
4204-020	314	C20	H 26	O 3	
4204-014	314	C20	H 26	O 3	
4204-015	314	C20	H 26	O 3	
4204-013	314	C20	H 26	O 3	
4592-001	314	C20	H 26	O 3	
4507-011	314	C21	H 30	O 2	
4507-041	314	C21	H 30	O 2	
4292-014	314	C20	H 26	O 3	
4217-017	314	C20	H 26	O 3	
4217-022	314	C20	H 26	O 3	
4291-018	314	C20	H 26	O 3	
4293-004	314	C20	H 26	O 3	
4099-017	314	C21	H 30	O 2	
4493-054	314	C19	H 22	O 4	
4514-016	315	C21	H 33	N 1	O 1
4511-004	315	C21	H 33	N 1	O 1
4511-003	315	C21	H 33	N 1	O 1
4507-009	316	C21	H 32	O 2	
4513-020	316	C21	H 38	N 2	
4507-008	316	C21	H 32	O 2	
4507-044	316	C21	H 32	O 2	
4507-042	316	C21	H 32	O 2	
4507-039	316	C21	H 32	O 2	
4507-040	316	C21	H 32	O 2	
4001-012	316	C16	H 28	O 6	
4001-010	316	C16	H 28	O 6	
4117-029	316	C20	H 28	O 3	
4117-023	316	C20	H 28	O 3	
4117-022	316	C20	H 28	O 3	
4217-018	316	C20	H 28	O 3	
4292-005	316	C20	H 28	O 3	
4217-020	316	C20	H 28	O 3	
4217-021	316	C20	H 28	O 3	
4217-013	316	C20	H 28	O 3	
4292-013	316	C20	H 28	O 3	
4299-004	316	C19	H 24	O 4	
4206-051	316	C20	H 28	O 3	
4204-038	316	C20	H 28	O 3	
4205-004	316	C20	H 28	O 3	
4202-092	316	C20	H 28	O 3	
4204-018	316	C20	H 28	O 3	
4208-003	316	C20	H 28	O 3	
4204-022	316	C20	H 28	O 3	
4202-087	316	C20	H 28	O 3	
4202-093	316	C20	H 28	O 3	
4202-094	316	C20	H 28	O 3	
4202-085	316	C20	H 28	O 3	
4207-018	316	C20	H 28	O 3	
4203-032	316	C20	H 28	O 3	
4512-006	317	C21	H 35	N 1	O 1
4511-001	317	C21	H 35	N 1	O 1

4507-007	318	C21	H 34	O 2	
4507-038	318	C21	H 34	O 2	
4507-036	318	C21	H 34	O 2	
4507-037	318	C21	H 34	O 2	
4206-036	318	C20	H 30	O 3	
4203-030	318	C20	H 30	O 3	
4203-028	318	C20	H 30	O 3	
4203-031	318	C20	H 30	O 3	
4202-103	318	C20	H 30	O 3	
4203-026	318	C20	H 30	O 3	
4202-102	318	C20	H 30	O 3	
4202-076	318	C20	H 30	O 3	
4208-007	318	C20	H 30	O 3	
4204-030	318	C21	H 34	O 2	
4208-002	318	C20	H 30	O 3	
4202-090	318	C20	H 30	O 3	
4206-040	318	C20	H 30	O 3	
4202-053	318	C20	H 30	O 3	
4211-003	318	C20	H 30	O 3	
4203-018	318	C20	H 30	O 3	
4202-051	318	C20	H 30	O 3	
4206-039	318	C20	H 30	O 3	
4204-039	318	C20	H 30	O 3	
4206-043	318	C20	H 30	O 3	
4217-011	318	C20	H 30	O 3	
4217-016	318	C20	H 30	O 3	
4293-001	318	C20	H 30	O 3	
4217-035	318	C20	H 30	O 3	
4001-007	318	C16	H 30	O 6	
4102-026	318	C17	H 18	O 6	
4102-029	318	C17	H 18	O 6	
4102-027	318	C17	H 18	O 6	
4101-020	318	C20	H 30	O 3	
4512-001	319	C21	H 37	N 1	O 1
4511-002	319	C21	H 37	N 1	O 1
4507-006	320	C21	H 36	O 2	
4507-005	320	C21	H 36	O 2	
4507-004	320	C21	H 36	O 2	
4507-003	320	C21	H 36	O 2	
4507-002	320	C21	H 36	O 2	
4002-105	320	C18	H 24	O 5	
4114-014	320	C17	H 20	O 6	
4293-045	320	C19	H 28	O 4	
4206-021	320	C20	H 32	O 3	
4203-014	320	C20	H 32	O 3	
4203-013	320	C20	H 32	O 3	
4203-012	320	C20	H 32	O 3	
4206-018	320	C20	H 32	O 3	
4202-054	320	C20	H 32	O 3	
4202-055	320	C20	H 32	O 3	
4202-112	320	C20	H 32	O 3	
4206-046	320	C20	H 32	O 3	
4202-074	320	C20	H 32	O 3	
4209-011	320	C20	H 32	O 3	
4202-081	320	C20	H 32	O 3	
4202-104	320	C20	H 32	O 3	
4203-020	320	C20	H 32	O 3	
4206-034	320	C20	H 32	O 3	
4206-035	320	C20	H 32	O 3	
4202-039	320	C20	H 32	O 3	
4202-041	320	C20	H 32	O 3	
4202-084	320	C20	H 32	O 3	
4203-038	322	C20	H 34	O 3	
4202-108	322	C20	H 34	O 3	
4202-071	322	C20	H 34	O 3	
4202-072	322	C20	H 34	O 3	
4206-015	322	C20	H 34	O 3	
4206-013	322	C20	H 34	O 3	
4206-014	322	C20	H 34	O 3	
4202-068	322	C20	H 34	O 3	
4206-016	322	C20	H 34	O 3	
4206-017	322	C20	H 34	O 3	
4202-060	322	C20	H 34	O 3	
4202-056	322	C20	H 34	O 3	
4202-052	322	C20	H 34	O 3	
4291-005	322	C20	H 34	O 3	
4291-004	322	C20	H 34	O 3	
4216-002	322	C20	H 34	O 3	
4114-015	322	C17	H 22	O 6	
4114-016	322	C17	H 22	O 6	
4114-020	322	C17	H 22	O 6	
4114-046	324	C17	H 24	O 6	
4114-045	324	C17	H 24	O 6	
4114-018	324	C17	H 24	O 6	
4114-051	324	C17	H 24	O 6	
4128-004	324	C16	H 20	O 7	
4208-008	324	C20	H 36	O 3	
4202-069	324	C20	H 36	O 3	
4202-070	324	C20	H 36	O 3	
4514-029	325	C22	H 31	N 1	O 1
4514-025	325	C21	H 27	N 1	O 2
4514-026	325	C22	H 31	N 1	O 1
4507-100	326	C21	H 26	O 3	
4514-014	326	C22	H 34	N 2	
4191-038	326	C22	H 30	O 2	

4202-027	326	C20	H 38	O 3	
4202-028	326	C20	H 38	O 3	
4212-014	327	C20	H 25	N 1	O 3
4514-030	327	C22	H 33	N 1	O 1
4514-001	328	C22	H 36	N 2	
4699-022	328	C20	H 24	O 4	
4514-003	328	C22	H 36	N 2	
4514-002	328	C22	H 36	N 2	
4507-046	328	C21	H 28	O 3	
4205-007	328	C21	H 28	O 3	
4202-096	328	C20	H 24	O 4	
4202-097	328	C20	H 24	O 4	
4192-074	328	C15	H 20	O 8	
4208-006	328	C20	H 24	O 4	
4204-012	328	C21	H 28	O 3	
4117-045	328	C20	H 24	O 4	
4114-053	328	C20	H 24	O 4	
4216-003	328	C20	H 24	O 4	
4212-013	329	C20	H 27	N 1	O 3
4212-020	329	C20	H 27	N 1	O 3
4514-015	329	C22	H 35	N 1	O 1
4511-005	329	C22	H 35	N 1	O 1
4507-050	330	C21	H 30	O 3	
4507-049	330	C21	H 30	O 3	
4513-021	330	C22	H 38	N 2	
4507-045	330	C21	H 30	O 3	
4204-010	330	C21	H 30	O 3	
4207-021	330	C21	H 30	O 3	
4207-019	330	C21	H 30	O 3	
4210-010	330	C19	H 22	O 5	
4210-008	330	C19	H 22	O 5	
4204-037	330	C22	H 34	O 2	
4207-022	330	C20	H 26	O 4	
4210-017	330	C19	H 22	O 5	
4204-032	330	C20	H 26	O 4	
4210-021	330	C20	H 26	O 4	
4219-001	330	C20	H 26	O 4	
4010-003	330	C16	H 26	O 7	
4008-066	330	C15	H 22	O 8	
4512-004	331	C22	H 37	N 1	O 1
4511-006	331	C22	H 37	N 1	O 1
4512-008	331	C22	H 37	N 1	O 1
4513-031	332	C21	H 36	N 2	O 1
4507-017	332	C21	H 32	O 3	
4513-030	332	C21	H 36	N 2	O 1
4008-058	332	C14	H 20	O 9	
4117-031	332	C20	H 28	O 4	
4110-005	332	C19	H 24	O 5	
4117-024	332	C20	H 28	O 4	
4110-008	332	C19	H 24	O 5	
4293-048	332	C20	H 28	O 4	
4217-019	332	C20	H 28	O 4	
4293-010	332	C18	H 20	O 6	
4208-005	332	C20	H 28	O 4	
4210-026	332	C19	H 24	O 5	
4204-021	332	C20	H 28	O 4	
4202-091	332	C20	H 28	O 4	
4206-054	332	C20	H 28	O 4	
4206-052	332	C20	H 28	O 4	
4210-018	332	C20	H 28	O 4	
4203-019	332	C20	H 28	O 4	
4204-011	332	C20	H 28	O 4	
4210-007	332	C19	H 24	O 5	
4202-099	332	C20	H 28	O 4	
4117-051	332	C20	H 28	O 4	
4204-031	332	C20	H 28	O 4	
4117-052	332	C20	H 28	O 4	
4202-100	332	C20	H 28	O 4	
4512-002	333	C22	H 39	N 1	O 1
4508-016	334	C19	H 26	O 5	
4507-015	334	C21	H 34	O 3	
4507-018	334	C21	H 34	O 3	
4507-016	334	C21	H 34	O 3	
4507-051	334	C21	H 34	O 3	
4507-047	334	C21	H 34	O 3	
4206-050	334	C20	H 30	O 4	
4206-048	334	C20	H 30	O 4	
4202-083	334	C20	H 30	O 4	
4203-017	334	C20	H 30	O 4	
4210-016	334	C19	H 26	O 5	
4202-077	334	C20	H 30	O 4	
4207-003	334	C20	H 30	O 4	
4216-015	334	C20	H 30	O 4	
4217-005	334	C20	H 30	O 4	
4102-028	334	C17	H 18	O 7	
4102-038	334	C17	H 18	O 7	
4002-106	334	C19	H 26	O 5	
4198-019	335	C19	H 29	N 1	O 4
4202-073	336	C21	H 36	O 2	
4202-080	336	C20	H 32	O 4	
4202-079	336	C20	H 32	O 4	
4202-078	336	C20	H 32	O 4	
4202-116	336	C20	H 32	O 4	
4202-110	336	C20	H 32	O 4	
4203-015	336	C20	H 32	O 4	

ID	MW	C	H	N/O	O/S
4206-045	336	C20	H 32	O 4	
4202-082	336	C20	H 32	O 4	
4206-044	336	C20	H 32	O 4	
4207-011	336	C20	H 32	O 4	
4206-047	336	C20	H 32	O 4	
4202-105	336	C20	H 32	O 4	
4102-039	336	C17	H 20	O 7	
4217-007	336	C20	H 32	O 4	
4292-007	336	C20	H 32	O 4	
4292-011	336	C20	H 32	O 4	
4293-002	336	C20	H 32	O 4	
4215-001	336	C20	H 32	O 4	
4507-013	336	C21	H 36	O 3	
4507-014	336	C21	H 36	O 3	
4507-012	336	C21	H 36	O 3	
4293-023	338	C20	H 18	O 5	
4102-007	338	C19	H 30	O 5	
4103-008	338	C19	H 30	O 5	
4206-019	338	C20	H 34	O 4	
4203-016	338	C20	H 34	O 4	
4202-058	338	C20	H 34	O 4	
4212-019	339	C22	H 29	N 1	O 2
4511-008	339	C22	H 35	N 1	O 1
4511-007	339	C22	H 35	N 1	O 1
4925-005	340	C25	H 40		
4202-042	340	C20	H 36	O 4	
4202-118	340	C20	H 36	O 4	
4202-117	340	C20	H 36	O 4	
4011-009	340	C20	H 20	O 5	
4511-013	341	C22	H 31	N 1	O 2
4514-005	342	C23	H 38	N 2	
4514-004	342	C23	H 38	N 2	
4191-039	342	C22	H 30	O 3	
4212-009	343	C22	H 33	N 1	O 2
4212-003	343	C22	H 33	N 1	O 2
4511-014	343	C22	H 33	N 1	O 2
4214-005	343	C22	H 33	N 1	O 2
4214-009	343	C22	H 33	N 1	O 2
4214-007	343	C22	H 33	N 1	O 2
4214-006	343	C22	H 33	N 1	O 2
4293-030	344	C19	H 20	O 6	
4299-005	344	C19	H 20	O 6	
4507-058	344	C21	H 28	O 4	
4507-094	344	C21	H 28	O 4	
4514-006	344	C23	H 40	N 2	
4592-009	344	C21	H 28	O 4	
4507-099	344	C21	H 28	O 4	
4207-020	344	C22	H 32	O 3	
4102-074	344	C19	H 20	O 6	
4104-084	344	C20	H 24	O 5	
4514-023	345	C22	H 35	N 1	O 2
4512-005	345	C23	H 39	N 1	O 1
4512-009	345	C23	H 39	N 1	O 1
4511-010	345	C22	H 35	N 1	O 2
4513-001	346	C23	H 42	N 2	
4513-006	346	C23	H 42	N 2	
4507-020	346	C21	H 30	O 4	
4513-027	346	C23	H 42	N 2	
4507-103	346	C21	H 30	O 4	
4514-020	346	C22	H 38	N 2	O 1
4507-064	346	C21	H 30	O 4	
4507-056	346	C21	H 30	O 4	
4513-010	346	C23	H 42	N 2	
4114-034	346	C20	H 26	O 5	
4114-033	346	C20	H 26	O 5	
4102-042	346	C19	H 22	O 6	
4008-047	346	C15	H 22	O 9	
4008-067	346	C15	H 22	O 9	
4102-060	346	C20	H 26	O 5	
4104-068	346	C19	H 22	O 6	
4202-101	346	C20	H 26	O 5	
4210-028	346	C19	H 22	O 6	
4210-009	346	C19	H 22	O 6	
4210-030	346	C20	H 26	O 5	
4206-012	346	C22	H 34	O 3	
4210-001	346	C19	H 22	O 6	
4209-010	346	C22	H 34	O 3	
4210-004	346	C19	H 22	O 6	
4120-083	346	C20	H 26	O 5	
4219-002	346	C20	H 26	O 5	
4219-009	346	C20	H 26	O 5	
4217-023	346	C20	H 26	O 5	
4512-003	347	C23	H 41	N 1	O 1
4507-061	348	C21	H 32	O 4	
4507-057	348	C21	H 32	O 4	
4507-059	348	C21	H 32	O 4	
4217-043	348	C20	H 28	O 5	
4293-005	348	C19	H 24	O 6	
4219-003	348	C20	H 28	O 5	
4292-010	348	C21	H 32	O 4	
4210-005	348	C19	H 24	O 6	
4210-020	348	C20	H 28	O 5	
4210-022	348	C19	H 24	O 6	
4192-056	348	C20	H 28	O 5	
4202-044	348	C22	H 36	O 3	
4202-098	348	C20	H 28	O 5	
4202-086	348	C20	H 28	O 5	
4202-016	348	C22	H 36	O 3	
4210-036	348	C19	H 24	O 6	
4102-064	348	C19	H 24	O 6	
4493-069	348	C19	H 24	O 6	
4591-002	349	C22	H 39	N 1	O 2
4507-055	350	C21	H 34	O 4	
4507-054	350	C21	H 34	O 4	
4507-053	350	C21	H 34	O 4	
4507-052	350	C21	H 34	O 4	
4507-063	350	C21	H 34	O 4	
4493-070	350	C19	H 26	O 6	
4102-062	350	C19	H 26	O 6	
4202-106	350	C20	H 30	O 5	
4117-053	350	C19	H 26	O 4	S 1
4117-054	350	C19	H 26	O 4	S 1
4206-026	350	C20	H 30	O 5	
4202-088	350	C20	H 30	O 5	
4202-059	350	C21	H 34	O 4	
4210-006	350	C19	H 26	O 6	
4206-053	350	C20	H 30	O 5	
4291-016	350	C20	H 30	O 5	
4198-015	351	C20	H 33	N 1	O 4
4192-066	352	C18	H 24	O 7	
4120-033	352	C15	H 24	O 3	
4293-003	352	C20	H 32	O 5	
4218-005	352	C20	H 32	O 5	
4592-006	352	C20	H 16	O 6	
4507-019	352	C21	H 36	O 4	
4418-005	353	C24	H 35	N 1	O 1
4592-007	354	C20	H 18	O 6	
4293-032	354	C20	H 34	O 5	
4009-002	354	C16	H 18	O 9	
4110-012	354	C17	H 22	O 8	
4214-008	355	C22	H 29	N 1	O 3
4212-018	355	C22	H 29	N 1	O 3
4202-029	356	C20	H 36	O 5	
4302-007	356	C25	H 40	O 1	
4302-006	356	C25	H 40	O 1	
4217-032	356	C20	H 20	O 6	
4514-010	356	C24	H 40	N 2	
4514-008	356	C24	H 40	N 2	
4511-019	357	C23	H 35	N 1	O 2
4511-020	357	C23	H 35	N 1	O 2
4214-001	357	C22	H 31	N 1	O 3
4212-010	357	C21	H 27	N 1	O 4
4212-007	357	C22	H 31	N 1	O 3
4212-005	357	C22	H 33	N 1	O 3
4212-004	357	C22	H 33	N 1	O 3
4014-010	357	C16	H 23	N 1	O 8
4009-012	358	C16	H 22	O 9	
4002-053	358	C22	H 30	O 4	
4126-022	358	C22	H 30	O 4	
4301-001	358	C25	H 42	O 1	
4301-002	358	C25	H 42	O 1	
4217-027	358	C20	H 22	O 6	
4302-005	358	C25	H 42	O 1	
4514-011	358	C24	H 42	N 2	
4925-001	358	C25	H 42	O 1	
4514-009	358	C24	H 42	N 2	
4514-017	358	C24	H 42	N 2	O 1
4514-019	358	C23	H 38	N 2	O 1
4507-010	358	C23	H 34	O 3	
4925-002	358	C25	H 42	O 1	
4925-003	358	C25	H 42	O 1	
4925-004	358	C25	H 42	O 1	
4493-032	358	C24	H 38	O 2	
4592-016	359	C22	H 33	N 1	O 3
4511-011	359	C23	H 37	N 1	O 2
4511-012	359	C23	H 37	N 1	O 2
4511-015	359	C22	H 33	N 1	O 3
4511-018	359	C22	H 33	N 1	O 3
4511-017	359	C22	H 33	N 1	O 3
4511-016	359	C22	H 33	N 1	O 3
4214-004	359	C22	H 33	N 1	O 3
4214-002	359	C22	H 33	N 1	O 3
4217-036	360	C22	H 32	O 4	
4507-082	360	C21	H 28	O 5	
4507-083	360	C21	H 28	O 5	
4513-002	360	C24	H 44	N 2	
4507-076	360	C21	H 28	O 5	
4513-003	360	C24	H 44	N 2	
4507-097	360	C21	H 28	O 5	
4203-033	360	C22	H 32	O 4	
4209-014	360	C22	H 32	O 4	
4206-041	360	C22	H 32	O 4	
4206-042	360	C22	H 32	O 4	
4207-023	360	C21	H 28	O 5	
4206-038	360	C22	H 32	O 4	
4211-004	360	C22	H 32	O 4	
4008-062	360	C16	H 24	O 9	
4102-040	360	C19	H 20	O 7	
4114-036	360	C20	H 24	O 6	
4114-055	360	C20	H 24	O 6	

4114-054	360	C20	H 24	O 6	
4293-041	361	C21	H 31	N 1	O 4
4293-007	362	C19	H 22	O 7	
4219-011	362	C20	H 26	O 6	
4299-007	362	C19	H 22	O 7	
4217-009	362	C22	H 34	O 4	
4219-004	362	C20	H 26	O 6	
4219-006	362	C20	H 26	O 6	
4219-010	362	C20	H 26	O 6	
4114-038	362	C20	H 26	O 6	
4114-037	362	C20	H 26	O 6	
4008-051	362	C15	H 22	O10	
4008-049	362	C15	H 22	O10	
4008-053	362	C15	H 22	O10	
4102-071	362	C19	H 22	O 7	
4102-065	362	C20	H 26	O 6	
4102-061	362	C20	H 26	O 6	
4210-025	362	C20	H 26	O 6	
4202-075	362	C22	H 34	O 4	
4210-031	362	C19	H 22	O 7	
4210-033	362	C20	H 26	O 6	
4207-007	362	C21	H 30	O 5	
4210-027	362	C19	H 22	O 7	
4507-065	362	C21	H 30	O 5	
4592-008	362	C21	H 30	O 5	
4507-021	362	C21	H 30	O 5	
4507-081	362	C21	H 30	O 5	
4507-086	362	C21	H 30	O 5	
4493-067	362	C19	H 22	O 7	
4512-010	363	C23	H 41	N 1	O 2
4591-003	363	C23	H 41	N 1	O 2
4507-066	364	C21	H 32	O 5	
4507-067	364	C21	H 32	O 5	
4507-072	364	C21	H 32	O 5	
4507-080	364	C21	H 32	O 5	
4507-022	364	C21	H 31	O 5	
4507-084	364	C21	H 32	O 5	
4425-012	364	C20	H 28	O 6	
4493-066	364	C19	H 24	O 7	
4425-029	364	C20	H 28	O 6	
4493-068	364	C19	H 24	O 7	
4206-024	364	C20	H 28	O 6	
4207-015	364	C20	H 28	O 6	
4210-024	364	C20	H 28	O 6	
4210-013	364	C19	H 24	O 7	
4202-057	364	C22	H 36	O 4	
4210-015	364	C19	H 24	O 7	
4008-059	364	C15	H 24	O10	
4114-047	364	C20	H 28	O 6	
4219-008	364	C20	H 28	O 6	
4219-012	364	C20	H 28	O 6	
4301-003	364	C25	H 48	O 1	
4293-011	364	C18	H 20	O 8	
4293-006	364	C19	H 24	O 7	
4220-001	364	C20	H 28	O 6	
4110-007	366	C19	H 26	O 7	
4114-017	366	C19	H 26	O 7	
4114-052	366	C19	H 26	O 7	
4207-016	366	C20	H 30	O 6	
4206-025	366	C20	H 30	O 6	
4493-065	366	C19	H 26	O 7	
4507-023	366	C21	H 34	O 5	
4507-071	366	C21	H 34	O 5	
4507-078	366	C21	H 34	O 5	
4507-079	366	C21	H 34	O 5	
4507-028	366	C21	H 34	O 5	
4507-027	366	C21	H 34	O 5	
4418-003	367	C25	H 37	N 1	O 1
4418-006	367	C25	H 37	N 1	O 1
4507-030	368	C21	H 36	O 5	
4507-026	368	C21	H 36	O 5	
4507-025	368	C21	H 36	O 5	
4507-024	368	C21	H 36	O 5	
4218-009	368	C20	H 32	O 6	
4511-009	369	C23	H 37	N 1	O 2
4418-007	369	C25	H 39	N 1	O 1
4418-004	369	C25	H 39	N 1	O 1
4418-009	369	C25	H 39	N 1	O 1
4418-008	369	C25	H 39	N 1	O 1
4493-034	370	C25	H 42	N 2	
4501-001	370	C27	H 46		
4218-006	370	C20	H 34	O 6	
4302-009	370	C25	H 38	O 2	
4303-001	370	C25	H 38	O 2	
4302-008	370	C25	H 38	O 2	
4418-001	371	C24	H 37	N 1	O 2
4505-003	372	C23	H 32	O 4	
4505-036	372	C22	H 28	O 5	
4505-001	372	C23	H 32	O 4	
4217-026	372	C21	H 24	O 6	
4293-040	372	C22	H 28	O 5	
4217-033	372	C20	H 20	O 7	
4514-007	372	C24	H 40	N 2	O 1
4514-012	372	C24	H 40	N 2	O 1
4513-022	372	C25	H 44	N 2	
4514-018	372	C24	H 40	N 2	O 1
4008-044	372	C16	H 20	O10	
4009-010	374	C16	H 22	O10	
4008-030	374	C17	H 26	O 9	
4102-041	374	C20	H 22	O 7	
4592-002	374	C23	H 34	O 4	
4217-031	374	C20	H 22	O 7	
4292-001	374	C21	H 26	O 6	
4217-030	374	C20	H 22	O 7	
4217-028	374	C20	H 22	O 7	
4217-029	374	C20	H 22	O 7	
4505-011	374	C23	H 34	O 4	
4419-041	374	C24	H 42	N 2	O 1
4505-014	374	C23	H 34	O 4	
4505-006	374	C23	H 34	O 4	
4505-005	374	C23	H 34	O 4	
4204-019	374	C22	H 30	O 5	
4209-017	374	C22	H 30	O 5	
4293-042	375	C22	H 33	N 1	O 4
4217-038	376	C22	H 32	O 5	
4299-006	376	C20	H 24	O 7	
4207-005	376	C22	H 32	O 5	
4202-111	376	C23	H 36	O 4	
4206-049	376	C22	H 32	O 5	
4425-028	376	C21	H 28	O 6	
4425-013	376	C20	H 24	O 7	
4510-001	376	C24	H 40	O 3	
4513-014	376	C24	H 44	N 2	O 1
4104-088	376	C20	H 24	O 7	
4008-031	376	C16	H 24	O10	
4008-054	376	C16	H 24	O10	
4104-064	376	C20	H 24	O 7	
4104-060	376	C20	H 24	O 7	
4104-023	376	C22	H 32	O 5	
4293-043	377	C21	H 31	N 1	O 5
4216-014	378	C22	H 34	O 5	
4293-008	378	C19	H 22	O 8	
4292-008	378	C22	H 34	O 5	
4102-066	378	C20	H 26	O 7	
4102-063	378	C20	H 26	O 7	
4008-039	378	C16	H 26	O10	
4102-079	378	C19	H 22	O 8	
4425-015	378	C20	H 26	O 7	
4425-004	378	C21	H 30	O 6	
4207-009	378	C21	H 30	O 6	
4210-029	378	C20	H 26	O 7	
4210-019	378	C20	H 26	O 7	
4210-023	378	C20	H 26	O 7	
4192-065	380	C19	H 24	O 8	
4425-011	380	C21	H 32	O 6	
4425-027	380	C20	H 28	O 7	
4008-048	380	C15	H 24	O11	
4507-087	380	C21	H 32	O 6	
4507-089	380	C21	H 32	O 6	
4507-088	380	C21	H 32	O 6	
4493-033	381	C26	H 39	N 1	O 1
4501-006	382	C27	H 42	O 1	
4507-031	382	C21	H 34	O 6	
4114-048	382	C19	H 26	O 8	
4191-041	382	C24	H 30	O 4	
4191-040	382	C24	H 30	O 4	
4126-002	382	C24	H 30	O 4	
4418-017	383	C26	H 41	N 1	O 1
4418-010	383	C26	H 41	N 1	O 1
4506-012	384	C24	H 32	O 4	
4506-005	384	C24	H 32	O 4	
4493-035	384	C26	H 44	N 2	
4506-006	384	C24	H 32	O 4	
4501-004	384	C27	H 44	O 1	
4506-003	384	C24	H 32	O 4	
4501-005	384	C27	H 44	O 1	
4501-008	384	C27	H 44	O 1	
4501-007	384	C27	H 44	O 1	
4592-003	384	C27	H 44	O 1	
4418-015	385	C26	H 43	N 1	O 1
4418-002	385	C25	H 39	N 1	O 2
4419-042	386	C25	H 42	N 2	O 1
4501-002	386	C27	H 46	O 1	
4506-007	386	C24	H 34	O 4	
4419-001	386	C26	H 46	N 2	
4202-025	386	C20	H 35	O 2	BR 1
4302-003	386	C25	H 38	O 3	
4418-014	387	C25	H 41	N 1	O 2
4425-001	388	C22	H 28	O 6	
4505-045	388	C23	H 32	O 5	
4505-038	388	C23	H 32	O 5	
4505-147	388	C23	H 32	O 5	
4219-013	388	C22	H 28	O 6	
4217-034	388	C20	H 20	O 8	
4202-095	388	C22	H 28	O 6	
4210-003	388	C21	H 24	O 7	
4509-001	388	C27	H 48	O 1	
4009-001	388	C17	H 24	O10	
4008-042	388	C17	H 24	O10	
4008-029	390	C17	H 26	O10	

4008-033	390	C16	H 22	O11	
4008-032	390	C16	H 22	O11	
4592-020	390	C25	H 42	O 3	
4592-021	390	C25	H 42	O 3	
4510-013	390	C24	H 38	O 4	
4192-054	390	C22	H 30	O 6	
4206-028	390	C22	H 30	O 6	
4217-024	390	C22	H 30	O 6	
4505-068	390	C23	H 34	O 5	
4505-053	390	C23	H 34	O 5	
4505-054	390	C23	H 34	O 5	
4505-047	390	C23	H 34	O 5	
4505-065	390	C23	H 34	O 5	
4505-050	390	C23	H 34	O 5	
4425-003	390	C22	H 30	O 6	
4505-072	390	C23	H 34	O 5	
4505-056	390	C23	H 34	O 5	
4293-044	391	C22	H 33	N 1	O 5
4502-022	392	C28	H 40	O 1	
4425-005	392	C22	H 32	O 6	
4192-067	392	C20	H 24	O 8	
4514-024	392	C24	H 44	N 2	O 2
4510-011	392	C24	H 40	O 4	
4510-010	392	C24	H 40	O 4	
4510-008	392	C24	H 40	O 4	
4510-005	392	C24	H 40	O 4	
4510-002	392	C24	H 40	O 4	
4510-003	392	C24	H 40	O 4	
4104-063	392	C20	H 24	O 8	
4104-062	392	C20	H 24	O 8	
4517-002	393	C27	H 39	N 1	O 1
4592-014	394	C27	H 38	O 2	
4699-021	394	C25	H 30	O 4	
4502-021	394	C28	H 42	O 1	
4425-016	394	C20	H 26	O 8	
4425-019	394	C20	H 26	O 8	
4502-016	394	C28	H 42	O 1	
4293-009	394	C19	H 22	O 9	
4218-011	394	C22	H 34	O 6	
4219-005	394	C21	H 30	O 7	
4502-020	396	C28	H 44	O 1	
4502-014	396	C28	H 44	O 1	
4502-013	396	C28	H 44	O 1	
4502-015	396	C28	H 44	O 1	
4110-019	396	C19	H 24	O 9	
4518-003	397	C27	H 43	N 1	O 1
4591-005	397	C27	H 43	N 1	O 1
4515-002	397	C27	H 43	N 1	O 1
4507-035	398	C21	H 34	O 7	
4502-006	398	C28	H 46	O 1	
4502-012	398	C28	H 46	O 1	
4502-007	398	C28	H 46	O 1	
4502-011	398	C28	H 46	O 1	
4502-009	398	C28	H 46	O 1	
4502-008	398	C28	H 46	O 1	
4502-010	398	C28	H 46	O 1	
4493-030	398	C27	H 42	O 2	
4502-019	398	C28	H 46	O 1	
4493-004	398	C28	H 46	O 1	
4493-003	398	C28	H 46	O 1	
4506-036	398	C24	H 30	O 5	
4493-006	398	C28	H 46	O 1	
4506-027	398	C24	H 30	O 5	
4518-001	399	C27	H 45	N 1	O 1
4515-001	399	C27	H 45	N 1	O 1
4212-006	399	C24	H 33	N 1	O 4
4513-024	400	C26	H 44	N 2	O 1
4513-023	400	C26	H 44	N 2	O 1
4592-004	400	C27	H 44	O 2	
4599-001	400	C28	H 48	O 1	
4506-031	400	C24	H 32	O 5	
4419-003	400	C27	H 48	N 2	
4493-005	400	C28	H 48	O 1	
4501-020	400	C27	H 44	O 1	
4502-002	400	C28	H 48	O 1	
4493-061	400	C28	H 48	O 1	
4419-040	400	C27	H 48	N 2	
4493-036	400	C26	H 44	N 2	O 1
4506-025	400	C24	H 32	O 5	
4502-004	400	C28	H 48	O 1	
4502-005	400	C28	H 48	O 1	
4502-003	400	C28	H 48	O 1	
4302-001	400	C25	H 36	O 4	
4217-037	400	C25	H 36	O 4	
4302-004	400	C25	H 36	O 4	
4214-003	401	C24	H 35	N 1	O 4
4418-016	401	C26	H 43	N 1	O 2
4502-032	402	C29	H 42	O 7	
4419-008	402	C26	H 46	N 2	O 1
4501-010	402	C27	H 46	O 2	
4501-009	402	C27	H 46	O 2	
4506-017	402	C24	H 34	O 5	
4506-026	402	C24	H 34	O 5	
4502-001	402	C28	H 50	O 1	
4493-031	402	C24	H 34	O 5	
4302-002	402	C25	H 38	O 4	
4513-028	402	C25	H 42	O 2	N 2
4513-025	402	C26	H 46	N 2	O 1
4513-004	402	C26	H 46	N 2	O 1
4513-019	402	C26	H 46	N 2	O 1
4513-029	404	C27	H 52	N 2	++
4507-085	404	C23	H 32	O 6	
4219-007	404	C22	H 28	O 7	
4219-015	404	C22	H 28	O 7	
4219-014	404	C22	H 28	O 7	
4505-111	404	C23	H 32	O 6	
4505-089	404	C23	H 32	O 6	
4505-114	404	C23	H 32	O 6	
4505-091	404	C23	H 32	O 6	
4425-002	404	C22	H 28	O 7	
4505-078	404	C23	H 32	O 6	
4206-009	404	C24	H 36	O 5	
4202-114	404	C22	H 28	O 7	
4202-115	404	C22	H 28	O 7	
4008-035	404	C17	H 24	O11	
4009-004	404	C17	H 24	O11	
4207-004	405	C24	H 39	N 1	O 4
4009-003	406	C17	H 26	O11	
4008-060	406	C17	H 26	O11	
4104-090	406	C22	H 30	O 7	
4505-097	406	C23	H 34	O 6	
4494-004	406	C28	H 38	O 2	
4505-109	406	C23	H 34	O 6	
4425-030	406	C22	H 30	O 7	
4507-073	406	C23	H 44	O 6	
4213-001	407	C23	H 37	N 1	O 5
4207-012	407	C24	H 41	N 1	O 4
4213-002	407	C23	H 37	N 1	O 5
4192-055	408	C21	H 28	O 6	S 1
4206-027	408	C22	H 32	O 7	
4193-043	408	C23	H 36	O 6	
4510-020	408	C24	H 40	O 5	
4510-017	408	C24	H 40	O 5	
4510-016	408	C24	H 40	O 5	
4510-022	408	C24	H 40	O 5	
4510-015	408	C24	H 40	O 5	
4510-014	408	C24	H 40	O 5	
4492-001	408	C30	H 48		
4425-008	408	C22	H 32	O 7	
4102-031	408	C21	H 28	O 8	
4008-027	408	C22	H 32	O 7	
4402-004	408	C30	H 48		
4217-025	408	C22	H 32	O 7	
4218-002	408	C23	H 36	O 6	
4293-033	408	C20	H 24	O 9	
4517-001	409	C27	H 39	N 1	O 2
4218-010	410	C22	H 34	O 7	
4218-008	410	C22	H 34	O 7	
4402-003	410	C30	H 50		
4402-001	410	C30	H 50		
4402-003	410	C30	H 50		
4402-002	410	C30	H 50		
4401-001	410	C30	H 50		
4104-073	410	C23	H 22	O 7	
4503-030	410	C29	H 46	O 1	
4492-008	410	C30	H 50		
4503-022	410	C29	H 46	O 1	
4503-044	410	C29	H 46	O 1	
4492-002	410	C30	H 50		
4415-002	410	C30	H 50		
4503-024	410	C29	H 46	O 1	
4415-001	410	C30	H 50		
4492-003	410	C30	H 50		
4502-017	410	C28	H 42	O 2	
4503-025	410	C29	H 46	O 1	
4493-024	410	C29	H 46	O 1	
4492-005	410	C30	H 50		
4416-001	410	C30	H 50		
4499-001	410	C30	H 50		
4425-018	410	C20	H 26	O 9	
4503-029	410	C29	H 46	O 1	
4425-010	410	C22	H 34	O 7	
4503-023	410	C29	H 46	O 1	
4591-007	411	C27	H 41	N 1	O 2
4515-005	411	C27	H 41	N 1	O 2
4518-013	411	C27	H 41	N 1	O 2
4517-004	411	C27	H 41	N 1	O 2
4599-008	412	C29	H 48	O 1	
4503-017	412	C29	H 48	O 1	
4503-027	412	C29	H 48	O 1	
4503-015	412	C29	H 48	O 1	
4503-028	412	C29	H 48	O 1	
4503-011	412	C29	H 48	O 1	
4503-014	412	C29	H 48	O 1	
4425-024	412	C20	H 28	O 9	
4493-020	412	C29	H 48	O 1	
4503-012	412	C29	H 48	O 1	
4503-013	412	C29	H 48	O 1	
4503-009	412	C29	H 48	O 1	
4503-021	412	C29	H 48	O 1	

4504-008	412	C27	H 40	O 3	
4503-010	412	C29	H 48	O 1	
4503-018	412	C29	H 48	O 1	
4503-019	412	C29	H 48	O 1	
4493-002	412	C29	H 48	O 1	
4493-001	412	C29	H 48	O 1	
4503-020	412	C29	H 48	O 1	
4493-009	412	C29	H 48	O 1	
4493-008	412	C29	H 48	O 1	
4503-032	412	C29	H 48	O 1	
4503-008	412	C29	H 48	O 1	
4104-065	412	C20	H 25	O 7	CL 1
4218-004	412	C22	H 36	O 7	
4299-002	412	C23	H 40	O 6	
4299-003	412	C23	H 40	O 6	
4299-002	412	C23	H 40	O 6	
4505-123	413	C28	H 30	O 7	
4516-037	413	C27	H 43	N 1	O 2
4516-035	413	C27	H 43	N 1	O 2
4515-016	413	C27	H 43	N 1	O 2
4591-006	413	C27	H 43	N 1	O 2
4515-019	413	C27	H 43	N 1	O 2
4515-026	413	C27	H 43	N 1	O 2
4518-010	413	C27	H 43	N 1	O 2
4518-011	413	C27	H 43	N 1	O 2
4515-004	413	C27	H 43	N 1	O 2
4518-014	413	C27	H 43	N 1	O 2
4515-007	413	C27	H 43	N 1	O 2
4515-027	413	C27	H 43	N 1	O 2
4212-012	413	C23	H 27	N 1	O 6
4503-005	414	C29	H 50	O 1	
4506-037	414	C24	H 30	O 6	
4503-007	414	C29	H 50	O 1	
4419-004	414	C28	H 50	N 2	
4493-015	414	C29	H 50	O 1	
4419-010	414	C27	H 46	N 2	O 1
4419-015	414	C27	H 46	N 2	O 1
4504-005	414	C27	H 42	O 3	
4504-006	414	C27	H 42	O 3	
4503-004	414	C29	H 50	O 1	
4504-007	414	C27	H 42	O 3	
4503-003	414	C29	H 50	O 1	
4008-043	414	C18	H 22	O11	
4418-018	415	C26	H 41	N 1	O 3
4212-011	415	C23	H 29	N 1	O 6
4515-013	415	C27	H 45	N 1	O 2
4599-005	415	C27	H 45	N 1	O 2
4515-003	415	C27	H 45	N 1	O 2
4515-010	415	C27	H 45	N 1	O 2
4699-018	416	C30	H 40	O 1	
4513-026	416	C27	H 48	N 2	O 1
4493-013	416	C28	H 48	O 2	
4503-002	416	C29	H 52	O 1	
4506-038	416	C24	H 32	O 6	
4506-020	416	C24	H 32	O 6	
4503-001	416	C29	H 52	O 1	
4504-002	416	C27	H 44	O 3	
4504-001	416	C27	H 44	O 3	
4506-019	416	C24	H 32	O 6	
4504-004	416	C27	H 44	O 3	
4506-018	416	C24	H 32	O 6	
4504-003	416	C27	H 44	O 3	
4506-033	416	C24	H 32	O 6	
4217-014	416	C25	H 36	O 5	
4217-039	416	C25	H 36	O 5	
4592-024	417	C24	H 35	N 1	O 5
4217-040	418	C22	H 26	O 8	
4217-042	418	C22	H 26	O 8	
4220-009	418	C24	H 34	O 6	
4419-020	418	C26	H 46	N 2	O 2
4505-136	418	C23	H 30	O 7	
4505-119	418	C24	H 34	O 6	
4505-131	418	C23	H 30	O 7	
4501-011	418	C27	H 46	O 3	
4505-129	418	C23	H 30	O 7	
4104-061	418	C22	H 26	O 8	
4008-040	420	C18	H 28	O11	
4505-127	420	C23	H 32	O 7	
4505-133	420	C23	H 32	O 7	
4505-138	420	C23	H 32	O 7	
4215-002	420	C24	H 36	O 6	
4213-003	421	C24	H 39	N 1	O 5
4202-113	422	C22	H 30	O 8	
4193-042	422	C23	H 34	O 7	
4406-009	422	C30	H 46	O 1	
4426-010	422	C26	H 30	O 5	
4507-090	422	C23	H 34	O 7	
4494-003	422	C28	H 38	O 3	
4008-028	422	C23	H 34	O 7	
4008-023	422	C22	H 30	O 8	
4213-035	423	C23	H 37	N 1	O 6
4404-013	424	C30	H 48	O 1	
4401-018	424	C30	H 48	O 1	
4293-035	424	C20	H 24	O10	
4409-004	424	C30	H 48	O 1	
4414-003	424	C30	H 48	O 1	
4402-010	424	C30	H 48	O 1	
4293-034	424	C20	H 24	O10	
4410-004	424	C30	H 48	O 1	
4406-020	424	C30	H 48	O 1	
4008-025	424	C22	H 32	O 8	
4492-009	424	C30	H 48	O 1	
4422-005	424	C30	H 48	O 1	
4417-008	424	C30	H 48	O 1	
4416-005	424	C30	H 48	O 1	
4425-009	424	C22	H 32	O 8	
4517-003	425	C27	H 39	N 1	O 3
4701-001	426	C30	H 50	O 1	
4599-002	426	C30	H 50	O 1	
4699-020	426	C27	H 38	O 4	
4499-002	426	C30	H 50	O 1	
4491-006	426	C30	H 50	O 1	
4422-001	426	C30	H 50	O 1	
4422-003	426	C30	H 50	O 1	
4499-006	426	C30	H 50	O 1	
4491-001	426	C30	H 50	O 1	
4493-017	426	C29	H 46	O 2	
4492-007	426	C30	H 50	O 1	
4492-004	426	C30	H 50	O 1	
4493-026	426	C29	H 46	O 2	
4492-010	426	C30	H 50	O 1	
4493-012	426	C30	H 50	O 1	
4493-014	426	C30	H 50	O 1	
4421-001	426	C30	H 50	O 1	
4493-010	426	C30	H 50	O 1	
4421-002	426	C30	H 50	O 1	
4499-008	426	C30	H 50	O 1	
4493-011	426	C30	H 50	O 1	
4493-007	426	C30	H 50	O 1	
4492-012	426	C30	H 50	O 1	
4492-013	426	C30	H 50	O 1	
4492-019	426	C30	H 50	O 1	
4492-014	426	C30	H 50	O 1	
4492-015	426	C30	H 50	O 1	
4413-001	426	C30	H 50	O 1	
4409-035	426	C30	H 50	O 1	
4413-002	426	C30	H 50	O 1	
4410-001	426	C30	H 50	O 1	
4406-001	426	C30	H 50	O 1	
4408-001	426	C30	H 50	O 1	
4406-019	426	C30	H 50	O 1	
4401-006	426	C30	H 50	O 1	
4410-005	426	C30	H 50	O 1	
4402-006	426	C30	H 50	O 1	
4401-007	426	C30	H 50	O 1	
4402-005	426	C30	H 50	O 1	
4401-020	426	C30	H 50	O 1	
4410-003	426	C30	H 50	O 1	
4403-017	426	C30	H 50	O 1	
4405-001	426	C30	H 50	O 1	
4407-001	426	C30	H 50	O 1	
4406-003	426	C30	H 50	O 1	
4404-001	426	C30	H 50	O 1	
4405-002	426	C30	H 50	O 1	
4218-001	426	C23	H 38	O 7	
4409-001	426	C30	H 50	O 1	
4401-004	426	C30	H 50	O 1	
4414-004	426	C30	H 50	O 1	
4416-002	426	C30	H 50	O 1	
4414-002	426	C30	H 50	O 1	
4416-004	426	C30	H 50	O 1	
4218-012	426	C23	H 38	O 7	
4414-001	426	C30	H 50	O 1	
4403-004	426	C30	H 50	O 1	
4406-012	426	C30	H 50	O 1	
4405-004	426	C30	H 50	O 1	
4417-002	426	C30	H 50	O 1	
4405-003	426	C30	H 50	O 1	
4418-012	427	C27	H 41	N 1	O 3
4517-006	427	C28	H 45	N 1	O 2
4515-008	427	C27	H 41	N 1	O 3
4517-005	427	C27	H 41	N 1	O 3
4513-009	428	C28	H 48	N 2	O 1
4592-005	428	C23	H 24	O 8	
4503-031	428	C29	H 48	O 2	
4502-023	428	C28	H 44	O 3	
4419-002	428	C28	H 48	N 2	O 1
4493-016	428	C29	H 48	O 2	
4504-058	428	C27	H 40	O 4	
4425-025	428	C20	H 28	O10	
4493-021	428	C29	H 48	O 2	
4504-059	428	C27	H 40	O 4	
4493-022	428	C29	H 48	O 2	
4503-026	428	C29	H 48	O 2	
4419-013	428	C28	H 48	N 2	O 1
4491-005	428	C30	H 52	O 1	
4504-052	428	C27	H 40	O 4	
4504-055	428	C27	H 40	O 4	
4491-007	428	C30	H 52	O 1	
4499-005	428	C29	H 48	O 2	

4417-001	428	C30	H 52	O 1	
4401-002	428	C30	H 52	O 1	
4403-003	428	C30	H 52	O 1	
4403-001	428	C30	H 52	O 1	
4410-002	428	C30	H 52	O 1	
4401-003	428	C30	H 52	O 1	
4104-067	428	C20	H 25	O 8	CL 1
4515-018	429	C27	H 43	N 1	O 3
4516-001	429	C27	H 43	N 1	O 3
4516-034	429	C27	H 43	N 1	O 3
4516-036	429	C27	H 43	N 1	O 3
4515-025	429	C27	H 43	N 1	O 3
4515-024	429	C27	H 43	N 1	O 3
4599-003	430	C27	H 46	N 2	O 2
4699-017	430	C31	H 42	O 1	
4117-035	430	C25	H 34	O 6	
4008-046	430	C18	H 22	O10	S 1
4220-011	430	C25	H 34	O 6	
4504-054	430	C27	H 42	O 4	
4504-056	430	C27	H 42	O 4	
4419-036	430	C27	H 46	N 2	O 2
4504-057	430	C27	H 42	O 4	
4419-030	430	C27	H 46	N 2	O 2
4504-051	430	C27	H 42	O 4	
4504-031	430	C27	H 42	O 4	
4504-053	430	C27	H 42	O 4	
4419-025	430	C27	H 46	N 2	O 2
4419-024	430	C27	H 46	N 2	O 2
4504-026	430	C27	H 42	O 4	
4504-030	430	C27	H 42	O 4	
4501-019	430	C27	H 42	O 4	
4504-050	430	C27	H 42	O 4	
4501-015	430	C27	H 42	O 4	
4501-016	430	C27	H 42	O 4	
4504-027	430	C27	H 42	O 4	
4502-018	430	C28	H 46	O 3	
4504-072	430	C27	H 42	O 4	
4504-028	430	C27	H 42	O 4	
4504-022	430	C27	H 42	O 4	
4419-009	430	C28	H 50	N 2	O 1
4504-025	430	C27	H 42	O 4	
4504-012	430	C27	H 42	O 4	
4504-011	430	C27	H 42	O 4	
4191-008	430	C21	H 34	O 9	
4516-032	431	C27	H 45	N 1	O 3
4515-015	431	C27	H 45	N 1	O 3
4515-006	431	C27	H 45	N 1	O 3
4515-011	431	C27	H 45	N 1	O 3
4515-012	431	C27	H 45	N 1	O 3
4699-019	432	C30	H 40	O 2	
4504-024	432	C27	H 44	O 4	
4504-010	432	C27	H 44	O 4	
4504-018	432	C27	H 44	O 4	
4505-122	432	C25	H 36	O 6	
4504-023	432	C27	H 44	O 4	
4504-019	432	C27	H 44	O 4	
4504-017	432	C27	H 44	O 4	
4504-016	432	C27	H 44	O 4	
4504-015	432	C27	H 44	O 4	
4504-013	432	C27	H 44	O 4	
4504-020	432	C27	H 44	O 4	
4504-009	432	C27	H 44	U 4	
4504-021	432	C27	H 44	O 4	
4419-021	432	C27	H 48	N 2	O 2
4504-071	432	C27	H 44	O 4	
4509-029	432	C26	H 40	O 5	
4505-104	432	C25	H 36	O 6	
4220-010	432	C25	H 36	O 6	
4212-022	433	C27	H 31	N 1	O 4
4207-006	433	C25	H 39	N 1	O 5
4207-008	433	C25	H 39	N 1	O 5
4207-017	434	C24	H 34	O 7	
4206-020	434	C24	H 34	O 7	
4207-001	434	C25	H 38	O 6	
4217-044	434	C24	H 34	O 7	
4425-006	434	C24	H 34	O 7	
4509-002	434	C27	H 46	O 4	
4509-033	436	C27	H 48	O 4	
4426-007	436	C28	H 36	O 4	
4213-005	437	C24	H 39	N 1	O 6
4491-004	438	C30	H 46	O 2	
4417-023	438	C31	H 50	O 1	
4505-142	438	C23	H 34	O 8	
4410-012	438	C30	H 46	O 2	
4592-018	438	C26	H 46	O 5	
4592-019	438	C26	H 46	O 5	
4008-026	438	C23	H 34	O 8	
4110-010	440	C21	H 28	O10	
4410-013	440	C30	H 48	O 2	
4416-003	440	C31	H 52	O 1	
4417-016	440	C31	H 52	O 1	
4417-017	440	C31	H 52	O 1	
4417-018	440	C31	H 52	O 1	
4410-011	440	C30	H 48	O 2	
4410-010	440	C30	H 48	O 2	

4403-010	440	C30	H 48	O 2	
4403-008	440	C30	H 48	O 2	
4408-007	440	C30	H 48	O 2	
4411-001	440	C31	H 52	O 1	
4408-042	440	C30	H 48	O 2	
4407-002	440	C31	H 52	O 1	
4405-005	440	C31	H 52	O 1	
4405-016	440	C30	H 48	O 2	
4218-003	440	C23	H 36	O 8	
4406-010	440	C30	H 48	O 2	
4408-006	440	C30	H 48	O 2	
4406-002	440	C31	H 52	O 1	
4402-007	440	C31	H 52	O 1	
4408-002	440	C31	H 52	O 1	
4415-013	440	C30	H 48	O 2	
4409-002	440	C31	H 52	O 1	
4403-014	440	C30	H 48	O 2	
4403-018	440	C30	H 48	O 2	
4427-003	440	C26	H 32	O 6	
4427-001	440	C26	H 32	O 6	
4422-002	440	C31	H 52	O 1	
4493-044	440	C30	H 48	O 2	
4422-004	440	C31	H 52	O 1	
4418-013	441	C28	H 43	N 1	O 3
4415-007	442	C30	H 50	O 2	
4415-009	442	C30	H 50	O 2	
4403-012	442	C30	H 50	O 2	
4402-008	442	C30	H 50	O 2	
4410-033	442	C30	H 50	O 2	
4403-013	442	C30	H 50	O 2	
4413-005	442	C30	H 50	O 2	
4415-003	442	C30	H 50	O 2	
4413-006	442	C30	H 50	O 2	
4409-003	442	C30	H 50	O 2	
4410-006	442	C30	H 50	O 2	
4403-007	442	C30	H 50	O 2	
4406-006	442	C30	H 50	O 2	
4401-019	442	C30	H 50	O 2	
4410-007	442	C30	H 50	O 2	
4406-005	442	C30	H 50	O 2	
4408-004	442	C30	H 50	O 2	
4403-006	442	C30	H 50	O 2	
4403-015	442	C30	H 50	O 2	
4402-009	442	C30	H 50	O 2	
4403-009	442	C30	H 50	O 2	
4406-023	442	C30	H 50	O 2	
4417-006	442	C30	H 50	O 2	
4417-005	442	C30	H 50	O 2	
4404-014	442	C30	H 50	O 2	
4413-007	442	C30	H 50	O 2	
4406-018	442	C30	H 50	O 2	
4407-003	442	C30	H 50	O 2	
4416-007	442	C30	H 50	O 2	
4408-003	442	C30	H 50	O 2	
4404-004	442	C30	H 50	O 2	
4407-004	442	C30	H 50	O 2	
4408-116	442	C30	H 50	O 2	
4406-004	442	C30	H 50	O 2	
4491-002	442	C30	H 50	O 2	
4491-003	442	C30	H 50	O 2	
4506-009	442	C26	H 34	O 6	
4493-043	442	C30	H 50	O 2	
4493-040	442	C30	H 50	O 2	
4513-005	442	C29	H 50	N 2	O 1
4493-025	442	C29	H 46	O 3	
4492-006	442	C30	H 50	O 2	
4427-002	442	C26	H 34	O 6	
4504-066	442	C27	H 38	O 5	
4421-003	442	C30	H 50	O 2	
4699-031	442	C32	H 42	O 1	
4517-007	443	C27	H 41	N 1	O 4
4493-047	444	C30	H 52	O 2	
4504-067	444	C27	H 40	O 5	
4504-065	444	C27	H 40	O 5	
4504-068	444	C27	H 40	O 5	
4506-035	444	C25	H 32	O 7	
4504-064	444	C27	H 40	O 5	
4501-018	444	C27	H 40	O 5	
4504-029	444	C28	H 44	O 4	
4506-001	444	C26	H 36	O 6	
4416-006	444	C30	H 52	O 2	
4404-003	444	C30	H 52	O 2	
4419-037	444	C28	H 48	N 2	O 2
4419-038	444	C28	H 48	N 2	O 2
4419-026	444	C28	H 48	N 2	O 2
4417-003	444	C30	H 52	O 2	
4403-005	444	C30	H 52	O 2	
4401-012	444	C30	H 52	O 2	
4401-008	444	C30	H 52	O 2	
4413-004	444	C30	H 52	O 2	
4401-009	444	C30	H 52	O 2	
4404-002	444	C30	H 52	O 2	
4413-003	444	C30	H 52	O 2	
4406-007	444	C30	H 52	O 2	
4401-011	444	C30	H 52	O 2	

4419-023	446	C28	H 50	N 2	O 2	
4419-039	446	C27	H 46	N 2	O 3	
4504-069	446	C27	H 42	O 5		
4504-063	446	C27	H 42	O 5		
4504-062	446	C27	H 42	O 5		
4504-060	446	C27	H 42	O 5		
4504-061	446	C27	H 42	O 5		
4008-034	446	C19	H 26	O12		
4008-045	448	C18	H 24	O11	S 1	
4102-032	448	C23	H 28	O 9		
4504-035	448	C27	H 44	O 5		
4504-036	448	C27	H 44	O 5		
4504-033	448	C27	H 44	O 5		
4504-037	448	C27	H 44	O 5		
4504-034	448	C27	H 44	O 5		
4504-039	448	C27	H 44	O 5		
4504-038	448	C27	H 44	O 5		
4504-032	448	C27	H 44	O 5		
4504-046	448	C27	H 44	O 5		
4504-040	448	C27	H 44	O 5		
4501-017	448	C27	H 44	O 5		
4509-007	448	C27	H 44	O 5		
4504-045	448	C27	H 44	O 5		
4504-044	448	C27	H 44	O 5		
4504-041	448	C27	H 44	O 5		
4504-042	448	C27	H 44	O 5		
4504-043	448	C27	H 44	O 5		
4207-014	448	C24	H 32	O 8		
4213-004	449	C25	H 39	N 1	O 6	
4207-010	449	C25	H 39	N 1	O 6	
4207-013	449	C25	H 39	N 1	O 6	
4212-023	449	C27	H 31	N 1	O 5	
4212-021	449	C27	H 31	N 1	O 5	
4510-006	449	C26	H 43	N 1	O 5	
4510-004	449	C26	H 43	N 1	O 5	
4510-009	449	C26	H 43	N 1	O 5	
4510-012	449	C26	H 43	N 1	O 5	
4213-036	449	C25	H 39	N 1	O 6	
4425-007	450	C24	H 34	O 8		
4501-013	450	C27	H 46	O 5		
4513-008	450	C30	H 46	N 2	O 1	
4509-008	450	C27	H 46	O 5		
4493-027	450	C29	H 38	O 4		
4493-029	450	C28	H 50	O 4		
4426-008	450	C28	H 34	O 5		
4509-005	450	C27	H 46	O 5		
4509-003	450	C27	H 46	O 5		
4209-012	450	C29	H 38	O 4		
4108-005	450	C25	H 38	O 7		
4108-004	450	C25	H 38	O 7		
4213-006	451	C25	H 41	N 1	O 6	
4513-007	451	C29	H 45	N 3	O 1	
4213-022	451	C24	H 37	N 1	O 7	
4417-024	452	C31	H 48	O 3		
4410-020	452	C30	H 44	O 3		
4218-007	452	C24	H 36	O 8		
4405-028	452	C30	H 44	O 3		
4507-060	452	C28	H 36	O 5		
4501-012	452	C27	H 48	O 5		
4501-014	452	C27	H 48	O 5		
4509-006	452	C27	H 48	O 5		
4509-004	452	C27	H 48	O 5		
4426-001	452	C28	H 36	O 5		
4428-001	452	C27	H 32	O 6		
4494-001	452	C29	H 40	O 4		
4213-021	453	C24	H 39	N 1	O 7	
4213-009	453	C24	H 39	N 1	O 7	
4410-014	454	C30	H 56	O 3		
4423-001	454	C26	H 30	O 7		
4405-015	454	C30	H 46	O 3		
4409-032	454	C30	H 46	O 3		
4408-068	454	C30	H 46	O 3		
4403-020	454	C30	H 46	O 3		
4423-006	454	C26	H 30	O 7		
4409-033	454	C30	H 46	O 3		
4409-011	454	C30	H 46	O 3		
4409-010	454	C30	H 46	O 3		
4408-118	454	C30	H 46	O 3		
4417-022	454	C31	H 50	O 2		
4402-011	454	C31	H 50	O 2		
4410-019	454	C30	H 46	O 3		
4406-021	454	C30	H 46	O 3		
4405-011	454	C30	H 46	O 3		
4408-076	454	C30	H 46	O 3		
4417-015	454	C30	H 46	O 3		
4405-014	454	C30	H 46	O 3		
4405-027	454	C30	H 46	O 3		
4406-016	454	C30	H 46	O 3		
4410-022	454	C30	H 46	O 3		
4408-075	454	C30	H 46	O 3		
4409-008	454	C30	H 46	O 3		
4415-014	454	C31	H 50	O 2		
4415-015	454	C31	H 50	O 2		
4494-002	454	C29	H 42	O 4		
4426-002	454	C28	H 38	O 5		
4503-016	454	C31	H 50	O 2		
4493-071	454	C29	H 42	O 4		
4502-033	454	C28	H 38	O 5		
4514-021	454	C29	H 46	N 2	O 2	
4493-051	454	C30	H 46	O 3		
4502-027	454	C28	H 38	O 5		
4502-026	454	C28	H 38	O 5		
4493-052	454	C30	H 46	O 3		
4104-066	454	C22	H 27	O 8	CL 1	
4699-015	454	C33	H 42	O 1		
4499-007	456	C31	H 52	O 2		
4513-017	456	C29	H 48	N 2	O 2	
4513-016	456	C29	H 48	N 2	O 2	
4506-030	456	C26	H 32	O 7		
4503-006	456	C31	H 52	O 2		
4492-018	456	C30	H 48	O 3		
4493-048	456	C30	H 48	O 3		
4502-037	456	C28	H 40	O 5		
4506-043	456	C26	H 32	O 7		
4493-019	456	C29	H 44	O 4		
4415-012	456	C31	H 52	O 2		
4408-120	456	C30	H 48	O 3		
4415-010	456	C31	H 52	O 2		
4407-005	456	C30	H 48	O 3		
4403-019	456	C30	H 48	O 3		
4409-024	456	C30	H 48	O 3		
4405-012	456	C30	H 48	O 3		
4408-013	456	C30	H 48	O 3		
4408-056	456	C30	H 48	O 3		
4415-008	456	C31	H 52	O 2		
4408-074	456	C30	H 48	O 3		
4415-004	456	C31	H 52	O 2		
4405-007	456	C30	H 48	O 3		
4408-012	456	C30	H 48	O 3		
4417-014	456	C30	H 48	O 3		
4405-010	456	C30	H 48	O 3		
4408-127	456	C30	H 48	O 3		
4409-005	456	C30	H 48	O 3		
4408-072	456	C30	H 48	O 3		
4405-020	456	C30	H 48	O 3		
4405-008	456	C30	H 48	O 3		
4406-011	456	C30	H 48	O 3		
4405-009	456	C30	H 48	O 3		
4406-013	456	C30	H 48	O 3		
4403-016	456	C30	H 48	O 3		
4409-034	456	C30	H 48	O 3		
4408-014	456	C30	H 48	O 3		
4408-061	456	C30	H 48	O 3		
4409-027	456	C30	H 48	O 3		
4423-005	456	C26	H 32	O 7		
4410-015	456	C30	H 48	O 3		
4412-001	456	C30	H 48	O 3		
4410-008	456	C30	H 48	O 3		
4408-044	456	C30	H 48	O 3		
4415-021	456	C30	H 48	O 3		
4415-020	456	C30	H 48	O 3		
4415-019	456	C30	H 48	O 3		
4415-018	458	C30	H 50	O 3		
4415-017	458	C30	H 50	O 3		
4404-017	458	C30	H 50	O 3		
4404-015	458	C30	H 50	O 3		
4408-011	458	C30	H 50	O 3		
4404-016	458	C30	H 50	O 3		
4408-010	458	C30	H 50	O 3		
4401-026	458	C30	H 50	O 3		
4403-011	458	C30	H 50	O 3		
4408-009	458	C30	H 50	O 3		
4410-009	458	C30	H 50	O 3		
4408-008	458	C30	H 50	O 3		
4406-022	458	C30	H 50	O 3		
4404-027	458	C30	H 50	O 3		
4404-025	458	C30	H 50	O 3		
4420-002	458	C30	H 50	O 3		
4415-016	458	C30	H 50	O 3		
4404-024	458	C30	H 50	O 3		
4404-005	458	C31	H 54	O 2		
4506-011	458	C26	H 34	O 7		
4493-018	458	C29	H 46	O 4		
4506-044	458	C26	H 34	O 7		
4493-042	458	C30	H 50	O 3		
4493-050	458	C30	H 50	O 3		
4513-018	458	C29	H 50	N 2	O 2	
4506-039	458	C26	H 34	O 7		
4513-015	458	C29	H 50	O 2	N 2	
4506-021	458	C26	H 34	O 7		
4191-001	458	C26	H 34	O 7		
4493-041	460	C30	H 52	O 3		
4404-023	460	C30	H 52	O 3		
4404-006	460	C30	H 52	O 3		
4404-007	460	C30	H 52	O 3		
4404-009	460	C30	H 52	O 3		
4404-026	460	C30	H 52	O 3		
4404-021	460	C30	H 52	O 3		
4217-041	460	C24	H 28	O 9		
4406-008	460	C30	H 52	O 3		

4220-006	460	C26	H 36	O 7	
4415-022	460	C30	H 52	O 3	
4420-001	460	C30	H 52	O 3	
4401-014	460	C30	H 52	O 3	
4509-026	462	C27	H 42	O 6	
4504-070	462	C27	H 42	O 6	
4509-011	462	C27	H 42	O 6	
4509-012	462	C27	H 42	O 6	
4509-013	462	C27	H 42	O 6	
4591-001	462	C27	H 42	O 6	
4699-027	462	C31	H 42	O 3	
4599-007	462	C31	H 46	N 2	O 1
4519-008	464	C24	H 32	O 9	
4504-049	464	C27	H 44	O 6	
4493-028	464	C30	H 40	O 4	
4509-014	464	C27	H 44	O 6	
4504-047	464	C27	H 44	O 6	
4509-018	464	C27	H 44	O 6	
4509-016	464	C27	H 44	O 6	
4509-017	464	C27	H 44	O 6	
4504-048	464	C27	H 44	O 6	
4102-033	464	C23	H 28	O10	
4510-018	465	C26	H 43	N 1	O 6
4213-037	465	C25	H 39	N 1	O 7
4213-026	465	C25	H 39	N 1	O 7
4213-007	465	C26	H 43	N 1	O 6
4411-005	466	C31	H 46	O 3	
4426-012	466	C28	H 34	O 6	
4426-009	466	C28	H 34	O 6	
4513-012	466	C30	H 46	N 2	O 2
4508-008	466	C25	H 38	O 8	
4508-010	466	C25	H 38	O 8	
4509-009	466	C27	H 46	O 6	
4493-058	466	C27	H 30	O 7	
4110-009	466	C24	H 34	O 9	
4008-052	466	C22	H 26	O11	
4213-027	467	C25	H 41	N 1	O 7
4213-025	467	C25	H 41	N 1	O 7
4213-024	467	C25	H 41	N 1	O 7
4424-006	468	C27	H 32	O 7	
4428-002	468	C27	H 32	O 7	
4410-021	468	C31	H 48	O 3	
4424-014	468	C27	H 32	O 7	
4410-027	468	C30	H 44	O 4	
4411-004	468	C31	H 48	O 3	
4410-029	468	C30	H 44	O 4	
4401-005	468	C32	H 52	O 2	
4408-087	468	C30	H 44	O 4	
4408-069	468	C30	H 44	O 4	
4408-070	468	C30	H 44	O 4	
4408-060	468	C30	H 44	O 4	
4114-049	468	C23	H 32	O10	
4507-068	468	C28	H 36	O 6	
4507-069	468	C28	H 36	O 6	
4499-004	468	C33	H 56	O 1	
4514-022	468	C30	H 48	N 2	O 2
4502-034	470	C28	H 38	O 6	
4502-035	470	C28	H 38	O 6	
4503-038	470	C29	H 42	O 5	
4502-031	470	C28	H 38	O 6	
4493-055	470	C30	H 46	O 4	
4015-001	470	C21	H 26	O12	
4403-002	470	C32	H 54	O 2	
4424-001	470	C27	H 34	O 7	
4408-065	470	C30	H 46	O 4	
4408-117	470	C31	H 50	O 3	
4408-100	470	C30	H 46	O 4	
4419-016	470	C30	H 50	N 2	O 2
4409-016	470	C30	H 46	O 4	
4423-002	470	C26	H 30	O 8	
4408-106	470	C30	H 46	O 4	
4408-063	470	C30	H 46	O 4	
4408-077	470	C30	H 46	O 4	
4423-007	470	C26	H 30	O 8	
4408-099	470	C30	H 46	O 4	
4410-016	470	C31	H 50	O 3	
4408-052	470	C30	H 46	O 4	
4405-024	470	C30	H 46	O 4	
4415-011	470	C32	H 54	O 2	
4411-002	470	C31	H 50	O 3	
4410-023	470	C30	H 46	O 4	
4415-006	470	C32	H 54	O 2	
4428-004	470	C27	H 34	O 7	
4424-007	470	C27	H 34	O 7	
4408-122	470	C30	H 46	O 4	
4428-003	470	C27	H 34	O 7	
4410-017	472	C30	H 48	O 4	
4408-016	472	C30	H 48	O 4	
4409-029	472	C30	H 48	O 4	
4408-019	472	C30	H 48	O 4	
4409-025	472	C30	H 48	O 4	
4408-024	472	C30	H 48	O 4	
4406-017	472	C30	H 48	O 4	
4412-003	472	C30	H 48	O 4	
4409-012	472	C30	H 48	O 4	
4404-020	472	C30	H 48	O 4	
4409-014	472	C30	H 48	O 4	
4220-008	472	C27	H 36	O 7	
4408-051	472	C30	H 48	O 4	
4408-119	472	C30	H 48	O 4	
4409-013	472	C30	H 48	O 4	
4408-022	472	C30	H 48	O 4	
4408-079	472	C30	H 48	O 4	
4423-008	472	C26	H 32	O 8	
4408-082	472	C30	H 48	O 4	
4405-031	472	C30	H 48	O 4	
4408-078	472	C30	H 48	O 4	
4415-024	472	C30	H 48	O 4	
4405-017	472	C30	H 48	O 4	
4423-011	472	C26	H 32	O 8	
4415-025	472	C30	H 48	O 4	
4408-057	472	C30	H 48	O 4	
4408-062	472	C30	H 48	O 4	
4408-050	472	C30	H 48	O 4	
4408-086	472	C30	H 48	O 4	
4408-081	472	C30	H 48	O 4	
4408-083	472	C30	H 48	O 4	
4408-088	472	C30	H 48	O 4	
4408-089	472	C30	H 48	O 4	
4417-021	472	C31	H 52	O 3	
4408-080	472	C30	H 48	O 4	
4417-020	472	C32	H 56	O 2	
4408-090	472	C30	H 48	O 4	
4406-014	472	C30	H 48	O 4	
4412-022	472	C30	H 48	O 4	
4421-007	472	C30	H 48	O 4	
4408-025	472	C30	H 48	O 4	
4413-008	472	C30	H 48	O 4	
4417-004	472	C32	H 56	O 2	
4405-021	472	C30	H 48	O 4	
4506-042	472	C26	H 32	O 8	
4506-023	472	C26	H 32	O 8	
4502-036	472	C28	H 40	O 6	
4499-003	472	C30	H 46	O 4	
4502-029	472	C28	H 40	O 6	
4493-049	472	C30	H 48	O 4	
4699-016	472	C33	H 44	O 2	
4507-043	473	C29	H 47	N 1	O 4
4418-011	473	C32	H 43	N 1	O 2
4207-024	473	C28	H 43	N 1	O 5
4408-020	474	C30	H 50	O 4	
4220-007	474	C27	H 38	O 7	
4408-017	474	C30	H 50	O 4	
4408-018	474	C30	H 50	O 4	
4420-003	474	C30	H 50	O 4	
4412-002	474	C30	H 50	O 4	
4404-018	474	C30	H 50	O 4	
4408-015	474	C30	H 50	O 4	
4401-023	474	C30	H 50	O 4	
4401-021	474	C30	H 50	O 4	
4405-018	474	C30	H 50	O 4	
4405-019	474	C30	H 50	O 4	
4405-029	474	C30	H 50	O 4	
4405-030	474	C30	H 50	O 4	
4493-063	474	C26	H 34	O 8	
4506-029	474	C26	H 34	O 8	
4505-102	474	C28	H 42	O 6	
4507-102	475	C28	H 45	N 1	O 5
4415-023	476	C30	H 52	O 4	
4404-029	476	C30	H 52	O 4	
4401-017	476	C30	H 52	O 4	
4404-011	476	C30	H 52	O 4	
4215-017	476	C26	H 36	O 8	
4404-022	476	C30	H 52	O 4	
4404-010	476	C30	H 52	O 4	
4291-007	476	C26	H 36	O 8	
4512-007	479	C27	H 45	N 1	O 6
4509-020	480	C27	H 44	O 7	
4509-019	480	C27	H 44	O 7	
4509-021	480	C27	H 44	O 7	
4509-024	480	C27	H 44	O 7	
4509-023	480	C27	H 44	O 7	
4215-015	480	C29	H 36	O 6	
4426-011	480	C28	H 32	O 7	
4425-022	480	C23	H 28	O11	
4202-089	480	C26	H 40	O 8	
4003-023	480	C23	H 28	O11	
4104-021	480	C33	H 52	O 2	
4213-028	481	C26	H 43	N 1	O 7
4411-011	482	C31	H 46	O 4	
4427-004	482	C28	H 34	O 7	
4417-009	482	C32	H 50	O 3	
4417-019	482	C33	H 54	O 2	
4408-043	482	C32	H 50	O 3	
4293-038	482	C27	H 30	O 8	
4008-055	482	C22	H 26	O12	
4008-024	482	C24	H 34	O10	
4110-016	484	C27	H 32	O 8	
4408-071	484	C30	H 44	O 5	
4423-003	484	C26	H 28	O 9	

4427-005	484	C28	H 36	O 7		
4410-024	484	C31	H 48	O 4		
4424-016	484	C27	H 32	O 8		
4410-028	484	C30	H 44	O 5		
4411-008	484	C31	H 48	O 4		
4507-091	484	C28	H 36	O 7		
4699-014	484	C34	H 44	O 2		
4410-026	486	C30	H 46	O 5		
4424-013	486	C27	H 34	O 8		
4401-010	486	C32	H 54	O 3		
4409-020	486	C30	H 46	O 5		
4411-009	486	C31	H 50	O 4		
4408-107	486	C30	H 46	O 5		
4494-005	486	C30	H 46	O 5		
4409-028	486	C30	H 46	O 5		
4411-010	486	C31	H 50	O 4		
4410-030	486	C30	H 46	O 5		
4428-005	486	C27	H 34	O 8		
4410-032	486	C30	H 46	O 5		
4406-015	486	C30	H 46	O 5		
4423-004	486	C26	H 30	O 9		
4408-114	486	C30	H 46	O 5		
4401-013	486	C32	H 54	O 3		
4408-067	486	C30	H 46	O 5		
4423-009	486	C27	H 34	O 8		
4408-113	486	C30	H 46	O 5		
4401-024	486	C32	H 54	O 3		
4411-006	486	C31	H 50	O 4		
4408-066	486	C30	H 46	O 5		
4408-093	486	C30	H 46	O 5		
4409-018	486	C30	H 46	O 5		
4493-037	486	C30	H 50	N 2	O 3	
4493-053	486	C27	H 34	O 8		
4409-026	488	C30	H 48	O 5		
4423-013	488	C26	H 39	O 9		
4408-033	488	C30	H 48	O 5		
4409-017	488	C30	H 48	O 5		
4419-033	488	C29	H 48	N 2	O 4	
4408-094	488	C30	H 48	O 5		
4408-058	488	C30	H 48	O 5		
4493-023	488	C29	H 44	O 6		
4417-010	488	C30	H 48	O 5		
4415-026	488	C30	H 48	O 5		
4409-015	488	C30	H 48	O 5		
4412-024	488	C30	H 48	O 5		
4419-005	488	C34	H 52	N 2		
4409-030	488	C30	H 48	O 5		
4405-026	488	C30	H 48	O 5		
4409+019	488	C30	H 48	O 5		
4408-095	488	C30	H 48	O 5		
4408-047	488	C30	H 48	O 5		
4408-098	488	C30	H 48	O 5		
4409-031	488	C30	H 48	O 5		
4408-034	488	C30	H 48	O 5		
4592-022	488	C28	H 40	O 7		
4506-024	488	C26	H 32	O 9		
4507-095	488	C28	H 40	O 7		
4515-009	489	C29	H 47	N 1	O 5	
4502-028	490	C28	H 39	O 5	CL 1	
4502-030	490	C28	H 39	O 5	CL 1	
4405-025	490	C30	H 50	O 5		
4408-027	490	C30	H 50	O 5		
4421-004	490	C30	H 50	O 5		
4421-005	490	C30	H 50	O 5		
4408-026	490	C30	H 50	O 5		
4408-029	490	C30	H 50	O 5		
4408-030	490	C30	H 50	O 5		
4507-101	491	C28	H 47	N 1	O 6	
4507-048	491	C29	H 49	N 1	O 5	
4405-013	492	C30	H 52	O 5		
4008-057	492	C24	H 28	O11		
4008-056	492	C24	H 28	O11		
4291-010	493	C25	H 35	N 1	O 9	
4516-002	493	C27	H 43	N 1	O 7	
4425-017	494	C25	H 34	O10		
4502-025	494	C28	H 46	O 7		
4502-024	494	C28	H 46	O 7		
4008-061	494	C24	H 30	O11		
4008-063	494	C24	H 30	O11		
4701-002	494	C35	H 58	O 1		
4198-011	494	C30	H 42	N 2	O 2	S 1
4198-012	494	C30	H 42	N 2	O 2	S 1
4213-023	495	C26	H 41	N 1	O 8	
4516-008	495	C27	H 45	N 1	O 7	
4509-025	496	C27	H 44	O 8		
4509-031	496	C27	H 44	O 8		
4509-032	496	C27	H 44	O 8		
4409-009	496	C32	H 48	O 4		
4429-002	496	C31	H 44	O 5		
4419-011	496	C32	H 52	N 2	O 2	
4215-019	496	C29	H 36	O 7		
4425-026	496	C25	H 36	O10		
4426-003	498	C30	H 42	O 6		
4408-073	498	C32	H 50	O 4		
4408-059	498	C31	H 46	O 5		
4417-007	498	C33	H 54	O 3		
4411-012	498	C31	H 46	O 5		
4407-006	498	C32	H 50	O 4		
4410-031	498	C32	H 50	O 4		
4699-012	498	C35	H 46	O 2		
4510-007	499	C26	H 45	N 1	O 6	S 1
4513-013	500	C31	H 52	N 2	O 3	
4506-040	500	C28	H 36	O 8		
4699-013	500	C35	H 48	O 2		
4408-054	500	C32	H 52	O 4		
4408-053	500	C32	H 52	O 4		
4410-025	500	C31	H 48	O 5		
4408-115	500	C31	H 48	O 5		
4427-011	500	C28	H 36	O 8		
4401-025	500	C32	H 52	O 4		
4110-014	500	C27	H 32	O 9		
4110-015	500	C27	H 32	O 9		
4110-013	502	C27	H 34	O 9		
4409-023	502	C30	H 46	O 6		
4419-034	502	C30	H 50	N 2	O 4	
4401-015	502	C32	H 54	O 4		
4493-045	502	C30	H 46	O 6		
4417-011	502	C31	H 50	O 5		
4412-010	502	C30	H 46	O 6		
4412-009	502	C30	H 46	O 6		
4408-108	502	C30	H 46	O 6		
4408-109	502	C30	H 46	O 6		
4409-021	504	C30	H 48	O 6		
4215-003	504	C28	H 40	O 8		
4408-048	504	C30	H 48	O 6		
4408-103	504	C30	H 48	O 6		
4408-049	504	C30	H 48	O 6		
4408-104	504	C30	H 48	O 6		
4493-039	504	C30	H 52	N 2	O 4	
4503-041	504	C29	H 44	O 7		
4409-022	504	C30	H 48	O 6		
4419-017	504	C33	H 48	N 2	O 2	
4493-064	504	C27	H 36	O 9		
4408-101	504	C30	H 48	O 6		
4408-102	504	C30	H 48	O 6		
4507-098	504	C28	H 40	O 8		
4507-096	504	C28	H 40	O 8		
4207-002	505	C29	H 47	N 1	O 6	
4408-039	506	C30	H 50	O 6		
4419-019	506	C33	H 50	N 2	O 2	
+408-037	506	C30	H 50	O 6		
4408-038	506	C30	H 50	O 6		
4417-012	506	C30	H 46	O 6		
4408-035	506	C30	H 50	O 6		
4408-036	506	C30	H 50	O 6		
4419-007	506	C33	H 50	N 2	O 2	
4408-040	506	C30	H 50	O 6		
4503-035	506	C29	H 46	O 7		
4213-038	507	C27	H 41	N 1	O 8	
4503-034	508	C29	H 48	O 7		
4503-033	508	C29	H 48	O 7		
4503-036	508	C29	H 48	O 7		
4503-037	508	C29	H 48	O 7		
4210-002	508	C25	H 32	O11		
4507-032	508	C27	H 40	O 9		
4002-085	508	C22	H 36	O13		
4516-015	509	C27	H 43	N 1	O 8	
4516-011	509	C27	H 43	N 1	O 8	
4592-011	510	C28	H 30	O 9		
4592-012	510	C28	H 30	O 9		
4210-037	510	C25	H 34	O11		
4424-018	510	C29	H 34	O 8		
4424-015	510	C29	H 34	O 8		
4429-005	510	C31	H 42	O 6		
4411-003	512	C33	H 52	O 4		
4410-018	512	C33	H 52	O 4		
4424-017	512	C29	H 36	O 8		
4424-008	512	C29	H 36	O 8		
4424-002	512	C29	H 36	O 8		
4409-006	512	C33	H 52	O 4		
4423-010	514	C28	H 34	O 9		
4426-004	514	C30	H 42	O 7		
4426-005	514	C30	H 42	O 7		
4421-008	514	C32	H 50	O 5		
4405-022	514	C32	H 50	O 5		
4412-015	514	C30	H 42	O 7		
4427-014	514	C28	H 34	O 9		
4405-023	514	C32	H 50	O 5		
4510-021	515	C26	H 45	N 1	O 7	S 1
4510-019	515	C26	H 45	N 1	O 7	S 1
4505-002	516	C30	H 44	O 7		
4429-001	516	C31	H 48	O 6		
4412-008	516	C30	H 44	O 7		
4408-055	516	C32	H 52	O 5		
4412-016	516	C30	H 44	O 7		
4220-012	516	C29	H 40	O 8		
4503-042	516	C29	H 40	O 8		
4408-023	516	C32	H 52	O 5		
4215-018	518	C28	H 38	O 9		
4408-110	518	C30	H 46	O 7		

4412-012	518	C30	H 46	O 7		
4412-019	518	C30	H 46	O 7		
4291-008	518	C28	H 38	O 9		
4505-020	518	C30	H 46	O 7		
4505-007	518	C30	H 46	O 7		
4505-055	518	C29	H 42	O 8		
4505-040	518	C29	H 42	O 8		
4505-017	518	C30	H 46	O 7		
4505-039	518	C29	H 42	O 8		
4505-018	518	C30	H 46	O 7		
4505-019	518	C30	H 46	O 7		
4107-071	518	C30	H 30	O 8		
4505-012	520	C29	H 44	O 8		
4505-016	520	C29	H 44	O 8		
4507-077	520	C28	H 40	O 9		
4505-015	520	C29	H 44	O 8		
4419-006	520	C34	H 52	N 2	O 2	
4425-020	520	C26	H 32	O11		
4419-018	520	C33	H 48	N 2	O 3	
4412-020	520	C30	H 48	O 7		
4503-039	520	C29	H 44	O 8		
4408-105	520	C30	H 48	O 7		
4493-038	520	C33	H 48	N 2	O 3	
4419-029	520	C33	H 48	N 2	O 3	
4503-043	520	C29	H 44	O 8		
4419-031	520	C33	H 28	N 2	O 3	
4425-021	522	C26	H 34	O11		
4419-022	522	C33	H 50	N 2	O 3	
4291-009	522	C28	H 42	O 9		
4509-022	522	C29	H 46	O 8		
4513-011	522	C33	H 50	N 2	O 3	
4008-050	524	C21	H 32	O15		
4210-034	524	C26	H 36	O11		
4210-032	524	C25	H 32	O12		
4516-026	525	C27	H 43	N 1	O 9	
4592-010	526	C28	H 30	O10		
4198-013	526	C30	H 42	N 2	O 4	S 1
4198-014	526	C30	H 42	N 2	O 4	S 1
4210-014	526	C25	H 34	O12		
4424-019	526	C29	H 34	O 9		
4429-004	526	C31	H 42	O 7		
4492-011	528	C32	H 48	O 6		
4426-017	528	C30	H 40	O 8		
4417-013	528	C32	H 48	O 6		
4493-046	528	C32	H 48	O 6		
4428-006	528	C29	H 36	O 9		
4421-009	528	C32	H 48	O 6		
4411-007	528	C33	H 52	O 5		
4424-003	528	C29	H 36	O 9		
4401-022	530	C33	H 54	O 5		
4505-094	530	C29	H 38	O 9		
4110-018	530	C29	H 38	O 9		
4110-017	532	C29	H 40	O 9		
4601-012	532	C40	H 52			
4505-043	532	C30	H 44	O 8		
4505-004	532	C30	H 44	O 8		
4505-092	532	C29	H 40	O 9		
4505-042	532	C30	H 44	O 8		
4429-007	532	C31	H 48	O 7		
4421-006	532	C32	H 52	O 6		
4412-018	532	C30	H 44	O 8		
4412-017	532	C30	H 44	O 8		
4408-121	532	C31	H 48	O 7		
4699-004	532	C40	H 52			
4602-008	534	C40	H 54			
4412-014	534	C30	H 46	O 8		
4419-028	534	C34	H 50	N 2	O 3	
4412-013	534	C30	H 46	O 8		
4408-111	534	C30	H 46	O 8		
4408-112	534	C30	H 46	O 8		
4505-074	534	C30	H 46	O 8		
4505-080	534	C29	H 42	O 9		
4505-022	534	C30	H 46	O 8		
4505-084	534	C29	H 42	O 9		
4505-106	534	C30	H 46	O 8		
4505-021	534	C30	H 46	O 8		
4601-010	534	C40	H 54			
4505-041	534	C29	H 42	O 9		
4505-069	534	C30	H 46	O 8		
4505-075	534	C30	H 46	O 8		
4516-003	535	C29	H 45	N 1	O 8	
4601-009	536	C40	H 56			
4601-007	536	C40	H 56			
4505-048	536	C29	H 44	O 9		
4602-007	536	C40	H 56			
4505-049	536	C29	H 44	O 9		
4505-073	536	C29	H 44	O 9		
4505-079	536	C28	H 40	O10		
4505-057	536	C29	H 44	O 9		
4602-001	536	C40	H 56			
4419-035	536	C33	H 48	N 2	O 4	
4503-040	536	C29	H 44	O 9		
4603-001	536	C40	H 56			
4603-003	536	C40	H 56			
4516-009	537	C29	H 47	N 1	O 8	
4601-005	538	C40	H 58			
4601-008	538	C40	H 58			
4602-006	538	C40	H 58			
4602-005	538	C40	H 58			
4592-025	538	C31	H 42	N 2	O 6	
4507-075	538	C32	H 42	O 7		
4699-008	538	C40	H 58			
4699-007	538	C40	H 58			
4699-009	538	C40	H 58			
4506-004	540	C32	H 44	O 7		
4601-006	540	C40	H 60			
4601-004	540	C40	H 60			
4601-011	540	C40	H 60			
4507-034	540	C32	H 44	O 7		
4493-059	540	C30	H 36	O 9		
4424-009	540	C31	H 40	O 8		
4427-006	540	C30	H 36	O 9		
4009-009	540	C25	H 32	O13		
4009-006	540	C26	H 36	O12		
4009-005	540	C26	H 36	O12		
4009-007	542	C26	H 38	O12		
4008-064	542	C26	H 38	O12		
4424-012	542	C31	H 42	O 8		
4427-012	542	C30	H 38	O 9		
4423-012	542	C29	H 34	O10		
4506-008	542	C32	H 46	O 7		
4601-003	542	C40	H 62			
4601-002	544	C40	H 64			
4601-001	544	C40	H 64			
4507-029	544	C27	H 44	O11		
4401-016	544	C34	H 56	O 5		
4427-010	544	C30	H 40	O 9		
4427-009	544	C30	H 40	O 9		
4699-010	544	C40	H 48	O 1		
4215-004	546	C31	H 46	O 8		
4506-013	546	C30	H 42	O 9		
4505-145	548	C30	H 44	O 9		
4505-083	548	C30	H 44	O 9		
4505-112	548	C30	H 44	O 9		
4505-115	548	C30	H 44	O 9		
4505-113	548	C30	H 44	O 9		
4602-009	548	C40	H 52	O 1		
4505-117	548	C30	H 44	O 9		
4505-093	548	C29	H 40	O10		
4505-044	548	C30	H 44	O 9		
4505-116	548	C30	H 44	O 9		
4215-005	548	C30	H 44	O 9		
4419-027	548	C35	H 52	N 2	O 3	
4505-066	550	C30	H 46	O 9		
4505-090	550	C29	H 42	O10		
4602-011	550	C40	H 54	O 1		
4602-002	550	C40	H 54	O 1		
4505-082	550	C29	H 42	O10		
4601-030	550	C40	H 54	O 1		
4505-070	550	C30	H 46	O 9		
4603-016	550	C40	H 54	O 1		
4505-081	550	C29	H 42	O10		
4505-076	550	C30	H 46	O 9		
4505-099	550	C30	H 46	O 9		
4699-005	550	C40	H 54	O 1		
4604-001	550	C40	H 54	O 1		
4008-037	550	C23	H 34	O15		
4516-010	551	C29	H 45	N 1	O 9	
4603-002	552	C40	H 56	O 1		
4603-027	552	C40	H 56	O 1		
4603-007	552	C40	H 56	O 1		
4592-026	552	C32	H 44	N 2	O 6	
4602-014	552	C40	H 56	O 1		
4602-012	552	C40	H 56	O 1		
4603-005	552	C40	H 56	O 1		
4603-004	552	C40	H 56	O 1		
4601-024	552	C40	H 56	O 1		
4505-058	552	C29	H 44	O10		
4603-006	552	C40	H 56	O 1		
4601-015	552	C40	H 56	O 1		
4505-098	552	C29	H 44	O10		
4408-084	552	C35	H 52	O 5		
4408-085	552	C35	H 52	O 5		
4291-017	552	C32	H 40	O 8		
4424-010	552	C32	H 40	O 8		
4426-019	554	C30	H 34	O10		
4602-003	554	C40	H 58	O 1		
4601-014	554	C40	H 58	O 1		
4601-013	556	C40	H 60	O 1		
4426-006	556	C32	H 44	O 8		
4412-011	556	C32	H 44	O 8		
4412-006	558	C32	H 46	O 8		
4412-005	558	C32	H 46	O 8		
4220-013	558	C31	H 42	O 9		
4518-005	559	C33	H 53	N 1	O 6	
4518-004	559	C33	H 53	N 1	O 6	
4506-032	560	C31	H 44	O 9		
4506-028	560	C30	H 40	O10		
4419-012	560	C36	H 52	N 2	O 3	
4412-007	560	C32	H 48	O 8		

4412-021	560	C32	H 48	O 8		
4404-028	560	C34	H 56	O 6		
4699-011	560	C40	H 48	O 2		
4701-003	562	C40	H 66	O 1		
4419-032	562	C35	H 50	N 2	O 4	
4419-014	562	C35	H 50	N 2	O 4	
4215-007	562	C30	H 42	O10		
4215-008	562	C30	H 42	O10		
4492-016	562	C37	H 54	O 4		
4603-020	562	C40	H 50	O 2		
4505-130	562	C30	H 42	O10		
4505-132	562	C30	H 42	O10		
4606-003	564	C38	H 44	O 4		
4604-008	564	C40	H 52	O 2		
4603-019	564	C40	H 52	O 2		
4602-010	564	C40	H 52	O 2		
4603-021	564	C40	H 52	O 2		
4505-124	564	C30	H 44	O10		
4505-128	564	C30	H 44	O10		
4505-125	564	C30	H 44	O10		
4505-137	564	C29	H 40	O11		
4505-118	564	C30	H 44	O10		
4425-023	564	C28	H 36	O12		
4504-073	564	C32	H 52	O 8		
4602-017	566	C40	H 54	O 2		
4604-002	566	C40	H 54	O 2		
4603-017	566	C40	H 54	O 2		
4604-003	566	C40	H 54	O 2		
4505-139	566	C29	H 42	O11		
4505-140	566	C29	H 42	O11		
4505-141	566	C29	H 42	O11		
4601-017	566	C41	H 58	O 1		
4603-018	566	C40	H 54	O 2		
4505-135	566	C29	H 42	O11		
4505-134	566	C29	H 42	O11		
4606-004	568	C38	H 48	O 4		
4602-015	568	C40	H 56	O 2		
4602-013	568	C40	H 56	O 2		
4606-001	568	C40	H 56	O 2		
4699-002	568	C40	H 56	O 2		
4699-001	568	C40	H 56	O 2		
4602-004	568	C40	H 56	O 2		
4603-009	568	C40	H 56	O 2		
4603-013	568	C40	H 56	O 2		
4603-028	568	C40	H 56	O 2		
4601-018	568	C41	H 60	O 1		
4601-025	568	C40	H 56	O 2		
4603-008	568	C40	H 56	O 2		
4601-033	568	C40	H 56	O 2		
4429-003	568	C33	H 44	O 8		
4426-018	570	C32	H 42	O 9		
4424-004	570	C31	H 38	O10		
4424-005	570	C32	H 42	O 9		
4601-022	570	C41	H 62	O 1		
4601-032	570	C40	H 58	O 2		
4205-008	570	C40	H 58	O 2		
4601-023	572	C41	H 64	O 1		
4408-028	572	C35	H 56	O 6		
4411-013	572	C34	H 52	O 7		
4408-031	572	C35	H 56	O 6		
4940-001	572	C40	H 60	O 2		
4599-004	573	C34	H 55	N 1	O 6	
4591-004	573	C32	H 47	N 1	O 8	
4507-070	574	C28	H 42	O 6		
4507-093	574	C28	H 42	O 6		
4213-029	574	C31	H 46	N 2	O 8	
4424-011	574	C34	H 38	O 8		
4412-004	574	C32	H 46	O 9		
4429-006	574	C33	H 50	O 8		
4429-008	574	C33	H 50	O 8		
4516-004	575	C32	H 49	N 1	O 8	
4515-017	575	C33	H 53	N 1	O 7	
4515-028	575	C33	H 53	N 1	O 7	
4515-020	575	C33	H 53	N 1	O 7	
4518-012	575	C33	H 53	N 1	O 7	
4505-105	576	C32	H 48	O 9		
4506-034	576	C31	H 44	O10		
4505-107	576	C32	H 48	O 9		
4505-023	576	C32	H 48	O 9		
4516-005	577	C32	H 51	N 1	O 8	
4699-006	578	C41	H 54	O 2		
4604-007	580	C40	H 52	O 3		
4505-126	580	C30	H 44	O11		
4505-059	582	C30	H 46	O 9		
4602-016	582	C40	H 54	O 3		
4601-026	582	C41	H 58	O 2		
4601-031	582	C41	H 58	O 2		
4604-004	582	C40	H 54	O 3		
4601-020	582	C41	H 58	O 2		
4213-008	583	C33	H 45	N 1	O 8	
4606-002	584	C40	H 56	O 3		
4699-003	584	C40	H 56	O 3		
4505-143	584	C29	H 44	O12		
4603-030	584	C40	H 56	O 3		
4603-029	584	C40	H 56	O 3		
4603-015	584	C40	H 56	O 3		
4505-144	584	C29	H 44	O12		
4603-032	584	C40	H 56	O 3		
4603-011	584	C40	H 56	O 3		
4601-016	584	C41	H 60	O 2		
4601-019	586	C41	H 62	O 2		
4291-020	586	C36	H 58	O 6		
4427-008	586	C32	H 42	O10		
4427-016	586	C32	H 42	O12		
4009-013	586	C29	H 30	O13		
4505-096	587	C31	H 41	N 1	O 8	S 1
4215-014	588	C31	H 40	O11		
4505-095	589	C31	H 43	N 1	O 8	S 1
4505-046	590	C32	H 46	O10		
4507-074	590	C28	H 42	O 7		
4505-146	590	C32	H 46	O10		
4215-006	590	C32	H 46	O10		
4516-012	591	C32	H 49	N 1	O 9	
4120-084	591	C29	H 37	N 1	O12	
4604-009	592	C40	H 48	O 4		
4516-025	593	C32	H 51	N 1	O 9	
4516-033	593	C33	H 55	N 1	O 8	
4604-005	594	C40	H 50	O 4		
4215-012	594	C30	H 42	O12		
4599-006	595	C33	H 57	N 1	O 8	
4603-022	596	C40	H 52	O 4		
4601-027	596	C42	H 60	O 2		
4493-060	596	C34	H 44	O 9		
4213-033	596	C30	H 48	N 2	O10	
4426-020	596	C32	H 36	O11		
4291-014	596	C32	H 52	O10		
4427-007	598	C32	H 38	O11		
4506-010	598	C34	H 46	O 9		
4213-010	599	C33	H 45	N 1	O 9	
4601-028	600	C42	H 64	O 2		
4603-012	600	C40	H 56	O 4		
4603-031	600	C40	H 56	O 4		
4606-005	600	C40	H 56	O 4		
4601-021	600	C41	H 60	O 3		
4603-024	600	C40	H 56	O 4		
4605-001	600	C40	H 56	O 4		
4603-025	600	C40	H 56	O 4		
4604-006	600	C40	H 56	O 4		
4701-015	602	C43	H 70	O 1		
4220-005	602	C35	H 54	O 8		
4009-011	602	C29	H 30	O14		
4408-021	604	C39	H 56	O 5		
4507-092	604	C35	H 40	O 9		
4204-033	604	C40	H 60	O 4		
4215-016	606	C35	H 42	O 9		
4415-005	606	C39	H 58	O 5		
4120-085	607	C29	H 37	N 1	O13	
4213-013	613	C34	H 47	N 1	O 9	
4601-034	614	C42	H 62	O 3		
4213-019	615	C33	H 45	N 1	O10	
4220-002	616	C36	H 56	O 8		
4605-002	616	C40	H 56	O 5		
4605-003	616	C40	H 56	O 5		
4505-103	618	C35	H 54	O 9		
4409-007	618	C36	H 58	O 8		
4215-009	620	C32	H 44	O12		
4506-022	620	C32	H 44	O12		
4699-030	620	C45	H 64	O 1		
4601-029	624	C42	H 56	O 4		
4509-015	626	C33	H 54	O11		
4213-030	628	C34	H 48	N 2	O 9	
4213-016	629	C34	H 47	N 1	O10	
4213-020	629	C34	H 47	N 1	O10	
4408-041	630	C37	H 58	O 8		
4220-004	630	C37	H 58	O 8		
4701-009	630	C45	H 74	O 1		
4701-004	630	C45	H 74	O 1		
4701-016	630	C45	H 74	O 1		
4205-009	630	C40	H 54	O 6		
4213-018	631	C33	H 45	N 1	O11	
4412-023	634	C36	H 58	O 9		
4516-016	635	C34	H 53	N 1	O10	
4516-017	635	C34	H 53	N 1	O10	
4215-011	636	C32	H 44	O13		
4291-013	638	C34	H 54	O11		
4501-003	638	C44	H 78	O 2		
4507-033	638	C36	H 46	O10		
4213-012	639	C36	H 49	N 1	O 9	
4426-013	642	C35	H 46	O11		
4426-021	642	C39	H 46	O 8		
4516-007	643	C35	H 49	N 1	O10	
4220-003	644	C38	H 60	O 8		
4427-013	644	C34	H 44	O12		
4213-014	645	C34	H 47	N 1	O11	
4215-010	646	C33	H 42	O13		
4215-013	652	C32	H 44	O14		
4293-039	654	C38	H 38	O10		
4516-006	657	C36	H 51	N 1	O10	
4605-004	658	C42	H 58	O 6		
4516-014	659	C35	H 49	N 1	O11	

4408-097	666	C41	H 62	O 7		
4408-096	666	C41	H 62	O 7		
4213-011	673	C36	H 51	N 1	O11	
4516-013	673	C36	H 51	N 1	O11	
4493-057	674	C39	H 46	O10		
4213-015	675	C35	H 49	N 1	O12	
4506-015	676	C36	H 52	O12		
4516-018	677	C36	H 55	N 1	O11	
4505-037	680	C34	H 48	O14		
4505-025	680	C36	H 56	O12		
4408-005	680	C46	H 80	O 3		
4291-012	680	C36	H 56	O12		
4213-034	682	C37	H 50	N 2	O10	
4505-008	682	C35	H 54	O13		
4505-009	682	C35	H 54	O13		
4505-024	682	C35	H 54	O13		
4426-014	684	C38	H 52	O11		
4213-031	685	C36	H 51	N 3	O10	
4506-014	692	C36	H 52	O13		
4516-021	693	C37	H 59	N 1	O11	
4213-039	694	C38	H 50	N 2	O10	
4505-071	696	C36	H 56	O13		
4505-027	696	C36	H 56	O 3		
4505-026	696	C36	H 56	O13		
4505-085	696	C35	H 52	O14		
4505-100	698	C35	H 54	O14		
4701-005	698	C50	H 82	O 1		
4701-010	698	C50	H 82	O 1		
4505-060	698	C35	H 54	O14		
4505-101	698	C35	H 54	O14		
4505-013	698	C35	H 54	O14		
4426-015	698	C39	H 54	O11		
4408-032	698	C44	H 58	O 7		
4213-032	699	C37	H 53	N 3	O10	
4213-017	703	C37	H 53	N 1	O12	
4699-026	704	C50	H 72	O 2		
4699-029	704	C50	H 72	O 2		
4518-006	705	C39	H 63	N 1	O10	
4518-007	705	C39	H 63	N 1	O10	
4516-020	709	C37	H 59	N 1	O12	
4516-019	709	C37	H 59	N 1	O12	
4505-086	712	C35	H 52	O15		
4505-077	712	C36	H 56	O144		
4505-087	712	C35	H 52	O15		
4505-088	712	C35	H 52	O15		
4516-024	717	C39	H 59	N 1	O11	
4515-021	721	C39	H 63	N 1	O11	
4518-008	721	C39	H 63	N 1	O11	
4506-041	724	C36	H 52	O15		
4426-016	726	C41	H 58	O11		
4602-020	730	C46	H 66	O 7		
4699-028	730	C50	H 76	O 4		
4516-030	735	C39	H 61	N 1	O12	
4516-023	735	C39	H 61	N 1	O12	
4505-108	738	C38	H 58	O14		
4507-062	738	C39	H 62	O13		
4602-018	746	C46	H 66	O 8		
4516-028	751	C39	H 61	N 1	O13	
4505-061	754	C36	H 58	O15		
4506-002	756	C40	H 60	N 4	O10	
4602-019	760	C46	H 64	O 9		
4701-011	766	C55	H 90	O 1		
4701-006	766	C55	H 90	O 1		
4701-013	768	C55	H 92	O 1		
4404-019	768	C41	H 68	O13		
4505-028	774	C40	H 70	O14		
4516-031	775	C41	H 61	N 1	O13	
4516-029	777	C41	H 63	N 1	O13	
4603-026	780	C52	H 76	O 5		
4505-051	780	C41	H 64	O14		
4505-062	780	C41	H 64	O14		
4408-124	780	C42	H 68	O13		
4408-126	780	C42	H 68	O13		
4516-022	793	C41	H 63	N 1	O14	
4516-027	793	C41	H 63	N 1	O14	
4505-110	796	C41	H 64	O15		
4404-012	800	C42	H 72	O14		
4293-046	802	C30	H 44	K 2	O16	S 2
4206-037	804	C38	H 60	O18		
4505-120	808	C42	H 64	O15		
4499-009	816	C41	H 52	O17		
4405-006	818	C57	H102	O 2		
4408-064	822	C42	H 62	O16		
4505-052	822	C43	H 66	O15		
4701-012	834	C60	H 98	O 1		
4701-007	834	C60	H 98	O 1		
4505-030	842	C42	H 66	O17		
4427-015	844	C43	H 56	O17		
4518-009	851	C45	H 73	N 1	O14	
4506-016	854	C42	H 62	O18		
4505-029	858	C42	H 66	O18		
4505-010	860	C41	H 64	O19		
4515-022	867	C45	H 73	N 1	O15	
4505-067	874	C42	H 66	O19		
4515-023	883	C45	H 73	N 1	O16	
4601-035	892	C52	H 76	O12		
4701-017	894	C63	H102	O 2		
4505-031	900	C44	H 68	O19		
4701-008	902	C65	H106	O 1		
4505-148	910	C47	H 74	O17		
4408-125	924	C48	H 76	O17		
4505-032	926	C47	H 74	O18		
4505-063	942	C47	H 74	O19		
4505-033	968	C49	H 76	O19		
4699-023	976	C44	H 64	O24		
4505-064	984	C49	H 76	O20		
4505-121	1013	C50	H 76	O21		
4518-002	1017	C50	H 83	N 1	O20	
4408-123	1030	C51	H 82	O21		
4515-014	1033	C50	H 83	N 1	O21	
4603-014	1044	C72	H116	O 4		
4603-010	1044	C72	H116	O 4		
4504-074	1048	C51	H 84	O22		
4504-014	1050	C50	H 82	O23		
4509-030	1062	C51	H 82	O23		
4603-023	1072	C72	H112	O 6		
4504-075	1228	C57	H 96	O28		

MOLECULAR FORMULA INDEX

4002-030	C10	H 16		O 1	
4002-029	C10	H 16		O 1	
4002-066	C10	H 16		O 1	
4002-073	C10	H 16		O 1	
4002-065	C10	H 16		O 1	
4002-060	C10	H 16		O 1	
4002-059	C10	H 16		O 1	
4002-058	C10	H 16		O 1	
4003-005	C10	H 16		O 1	
4003-011	C10	H 16		O 1	
4003-010	C10	H 16		O 1	
4003-009	C10	H 16		O 1	
4003-008	C10	H 16		O 1	
4003-015	C10	H 16		O 1	
4004-004	C10	H 16		O 1	
4004-005	C10	H 16		O 1	
4002-075	C10	H 16		O 1	
4002-080	C10	H 16		O 1	
4002-079	C10	H 16		O 1	
4002-093	C10	H 16		O 1	
4002-092	C10	H 16		O 1	
4002-091	C10	H 16		O 1	
4002-090	C10	H 16		O 1	
4002-098	C10	H 16		O 1	
4002-078	C10	H 16		O 2	
4002-086	C10	H 16		O 2	
4003-019	C10	H 16		O 2	
4002-069	C10	H 16		O 2	
4002-054	C10	H 16		O 2	
4008-015	C10	H 16		O 2	
4008-016	C10	H 16		O 2	
4008-019	C10	H 16		O 2	
4008-020	C10	H 16		O 2	
4008-008	C10	H 16		O 2	
4010-002	C10	H 16		O 2	
4001-039	C10	H 16		O 2	
4001-040	C10	H 16		O 2	
4099-014	C10	H 16		O 2	
4099-004	C10	H 16		O 2	
4001-049	C10	H 16		O 3	
4002-084	C10	H 16		O 3	
4006-011	C10	H 18			
4006-001	C10	H 18			
4006-007	C10	H 18		O 1	
4006-006	C10	H 18		O 1	
4007-001	C10	H 18		O 1	
4002-074	C10	H 18		O 1	
4002-099	C10	H 18		O 1	
4004-002	C10	H 18		O 1	
4004-001	C10	H 18		O 1	
4003-007	C10	H 18		O 1	
4003-012	C10	H 18		O 1	
4003-006	C10	H 18		O 1	
4002-055	C10	H 18		O 1	
4002-057	C10	H 18		O 1	
4002-056	C10	H 18		O 1	
4002-043	C10	H 18		O 1	
4002-042	C10	H 18		O 1	
4002-061	C10	H 18		O 1	
4002-064	C10	H 18		O 1	
4002-063	C10	H 18		O 1	
4002-062	C10	H 18		O 1	
4002-028	C10	H 18		O 1	
4002-027	C10	H 18		O 1	
4002-026	C10	H 18		O 1	
4002-039	C10	H 18		O 1	
4002-022	C10	H 18		O 1	
4002-020	C10	H 18		O 1	
4002-019	C10	H 18		O 1	
4002-018	C10	H 18		O 1	
4002-017	C10	H 18		O 1	
4002-016	C10	H 18		O 1	
4002-015	C10	H 18		O 1	
4002-014	C10	H 18		O 1	
4001-047	C10	H 18		O 1	
4001-046	C10	H 18		O 1	
4001-042	C10	H 18		O 1	
4001-043	C10	H 18		O 1	
4001-016	C10	H 18		O 1	
4001-014	C10	H 18		O 1	
4001-013	C10	H 18		O 1	
4001-025	C10	H 18		O 1	
4001-024	C10	H 18		O 1	
4001-023	C10	H 18		O 1	
4001-011	C10	H 18		O 1	
4001-008	C10	H 18		O 1	
4099-009	C10	H 18		O 1	
4099-008	C10	H 18		O 1	
4099-011	C10	H 18		O 1	
4099-023	C10	H 18		O 1	
4001-038	C10	H 18		O 2	
4001-028	C10	H 18		O 2	
4001-050	C10	H 18		O 2	
4001-051	C10	H 18		O 2	
4002-036	C10	H 18		O 2	
4002-033	C10	H 18		O 2	
4002-032	C10	H 18		O 2	
4008-007	C10	H 18		O 2	
4008-014	C10	H 18		O 2	
4008-013	C10	H 18		O 2	
4002-037	C10	H 18		O 3	
4013-009	C10	H 19	N		
4008-001	C10	H 20			
4002-021	C10	H 20		O 1	
4002-025	C10	H 20		O 1	
4002-024	C10	H 20		O 1	
4001-045	C10	H 20		O 1	
4001-044	C10	H 20		O 1	
4001-005	C10	H 20		O 1	
4001-006	C10	H 20		O 1	
4001-015	C10	H 20		O 1	
4002-038	C10	H 20		O 2	
4002-041	C10	H 20		O 2	
4002-040	C10	H 20		O 2	
4008-003	C10	H 20		O 2	
4008-002	C10	H 20		O 2	
4008-006	C10	H 20		O 2	
4008-005	C10	H 20		O 2	
4008-004	C10	H 20		O 2	
4192-007	C10	H 8			
4014-007	C11	H 11	N 1	O 3	
4014-008	C11	H 12	N 2	O 3	
4911-001	C11	H 14		O 2	
4011-012	C11	H 14		O 3	
4011-008	C11	H 14		O 3	
4008-041	C11	H 14		O 5	
4008-036	C11	H 14		O 5	
4008-038	C11	H 14		O 6	
4002-045	C11	H 16		O 1	
4002-050	C11	H 16		O 1	
4099-007	C11	H 16		O 1	
4193-047	C11	H 16		O 1	
4002-051	C11	H 16		O 2	
4002-046	C11	H 16		O 2	
4911-002	C11	H 16		O 2	
4911-004	C11	H 16		O 2	
4911-003	C11	H 16		O 3	
4099-005	C11	H 16		O 4	
4008-065	C11	H 16		O 5	
4013-011	C11	H 17	N 1	O 1	
4099-028	C11	H 20	N 3	O 2	+
4013-002	C11	H 21	N		
4013-001	C11	H 21	N		
4013-003	C11	H 2[illegible]	N		
4013-004	C11	H 21	N 1	O 1	
4013-006	C11	H 21	N 1	O 1	
4013-005	C11	H 21	N 1	O 1	
4013-010	C11	H 21	N 1	O 1	
4013-008	C11	H 21	N 1	O 1	
4013-007	C11	H 21	N 1	O 1	
4013-021	C11	H 21	N 1	O 1	
4912-003	C12	H 12			
4192-005	C12	H 12			
4912-004	C12	H 14		O 1	
4912-002	C12	H 16		O 1	
4099-019	C12	H 16		O 2	
4191-045	C12	H 18			
4192-027	C12	H 18			
4191-043	C12	H 18			
4193-046	C12	H 18		O 1	
4002-048	C12	H 18		O 2	
4001-009	C12	H 20		O 2	
4002-100	C12	H 20		O 3	
4912-001	C12	H 20		O 3	
4192-029	C12	H 22		O 1	
4192-006	C13	H 12		O 2	
4913-010	C13	H 16			
4913-002	C13	H 20			
4913-007	C13	H 20		O 1	
4913-008	C13	H 20		O 1	
4913-003	C13	H 20		O 2	
4913-004	C13	H 20		O 2	
4913-005	C13	H 20		O 3	
4913-006	C13	H 20		O 3	
4913-009	C13	H 22		O 1	
4913-001	C13	H 24		O 1	
4192-034	C14	H 10		O 3	
4192-032	C14	H 10		O 4	
4015-002	C14	H 12		O 4	
4015-006	C14	H 14		O 6	
4192-002	C14	H 16			
4192-001	C14	H 16			
4192-043	C14	H 16		O 1	
4192-008	C14	H 16		O 3	
4914-005	C14	H 16		O 3	
4192-004	C14	H 18			
4192-003	C14	H 18			
4192-036	C14	H 18		O 1	
4192-037	C14	H 18		O 1	
4192-023	C14	H 18		O 1	

4102-080	C15	H 20	O 3
4102-054	C15	H 20	O 3
4115-012	C15	H 20	O 3
4107-057	C15	H 20	O 3
4102-057	C15	H 20	O 3
4102-056	C15	H 20	O 3
4099-031	C15	H 20	O 3
4102-046	C15	H 20	O 3
4102-047	C15	H 20	O 3
4104-070	C15	H 20	O 3
4106-029	C15	H 20	O 3
4126-011	C15	H 20	O 3
4120-090	C15	H 20	O 3
4120-057	C15	H 20	O 3
4120-056	C15	H 20	O 3
4120-059	C15	H 20	O 3
4120-058	C15	H 20	O 3
4120-061	C15	H 20	O 3
4120-060	C15	H 20	O 3
4120-062	C15	H 20	O 3
4120-063	C15	H 20	O 3
4120-069	C15	H 20	O 3
4120-068	C15	H 20	O 3
4117-041	C15	H 20	O 3
4192-010	C15	H 20	O 3
4121-016	C15	H 20	O 3
4126-021	C15	H 20	O 3
4191-013	C15	H 20	O 4
4191-004	C15	H 20	O 4
4191-003	C15	H 20	O 4
4120-053	C15	H 20	O 4
4120-052	C15	H 20	O 4
4193-001	C15	H 20	O 4
4193-037	C15	H 20	O 4
4120-067	C15	H 20	O 4
4193-052	C15	H 20	O 4
4120-091	C15	H 20	O 4
4120-088	C15	H 20	O 4
4120-087	C15	H 20	O 4
4191-012	C15	H 20	O 4
4120-074	C15	H 20	O 4
4120-075	C15	H 20	O 4
4191-009	C15	H 20	O 4
4114-007	C15	H 20	O 4
4114-035	C15	H 20	O 4
4102-077	C15	H 20	O 4
4102-043	C15	H 20	O 4
4114-058	C15	H 20	O 4
4099-029	C15	H 20	O 4
4114-042	C15	H 20	O 4
4104-082	C15	H 20	O 4
4102-058	C15	H 20	O 4
4102-059	C15	H 20	O 4
4115-011	C15	H 20	O 4
4114-012	C15	H 20	O 4
4102-081	C15	H 20	O 4
4114-050	C15	H 20	O 4
4102-082	C15	H 20	O 4
4114-005	C15	H 20	O 4
4105-017	C15	H 20	O 4
4104-041	C15	H 20	O 4
4114-021	C15	H 20	O 4
4104-052	C15	H 20	O 4
4114-026	C15	H 20	O 4
4106-030	C15	H 20	O 4
4915-004	C15	H 20	O 4
4915-005	C15	H 20	O 5
4915-006	C15	H 20	O 5
4105-019	C15	H 20	O 5
4105-014	C15	H 20	O 5
4105-018	C15	H 20	O 5
4106-031	C15	H 20	O 5
4104-059	C15	H 20	O 5
4114-056	C15	H 20	O 5
4191-010	C15	H 20	O 5
4193-041	C15	H 20	O 5
4193-039	C15	H 20	O 5
4191-046	C15	H 20	O 5
4105-016	C15	H 20	O 6
4105-015	C15	H 20	O 6
4110-011	C15	H 20	O 7
4128-003	C15	H 20	O 7
4192-075	C15	H 20	O 7
4192-074	C15	H 20	O 8
4198-016	C15	H 21	N 1
4198-017	C15	H 21	N 1
4198-009	C15	H 21	N 1 O 1
4117-056	C15	H 22	
4122-002	C15	H 22	
4193-023	C15	H 22	
4122-004	C15	H 22	
4122-003	C15	H 22	
4122-001	C15	H 22	
4117-004	C15	H 22	
4117-003	C15	H 22	
4115-003	C15	H 22	
4107-024	C15	H 22	
4106-006	C15	H 22	
4106-007	C15	H 22	
4113-002	C15	H 22	
4106-015	C15	H 22	O 1
4104-014	C15	H 22	O 1
4107-051	C15	H 22	O 1
4115-004	C15	H 22	O 1
4106-018	C15	H 22	O 1
4107-072	C15	H 22	O 1
4107-053	C15	H 22	O 1
4115-005	C15	H 22	O 1
4106-021	C15	H 22	O 1
4106-020	C15	H 22	O 1
4106-022	C15	H 22	O 1
4117-009	C15	H 22	O 1
4104-022	C15	H 22	O 1
4117-012	C15	H 22	O 1
4103-007	C15	H 22	O 1
4117-013	C15	H 22	O 1
4115-006	C15	H 22	O 1
4117-010	C15	H 22	O 1
4101-014	C15	H 22	O 1
4101-008	C15	H 22	O 1
4102-009	C15	H 22	O 1
4117-015	C15	H 22	O 1
4115-007	C15	H 22	O 1
4191-024	C15	H 22	O 1
4121-014	C15	H 22	O 1
4121-008	C15	H 22	O 1
4192-013	C15	H 22	O 1
4193-022	C15	H 22	O 1
4117-026	C15	H 22	O 1
4121-009	C15	H 22	O 1
4121-010	C15	H 22	O 1
4127-009	C15	H 22	O 1
4120-029	C15	H 22	O 1
4127-008	C15	H 22	O 1
4127-007	C15	H 22	O 1
4120-030	C15	H 22	O 1
4192-028	C15	H 22	O 1
4118-006	C15	H 22	O 1
4122-007	C15	H 22	O 1
4125-009	C15	H 22	O 1
4120-080	C15	H 22	O 1
4120-079	C15	H 22	O 1
4120-078	C15	H 22	O 1
4194-007	C15	H [illegible]	O 1
4130-003	C15	H 22	O 1
4119-003	C15	H 22	O 1
4124-006	C15	H 22	O 2
4194-002	C15	H 22	O 2
4194-001	C15	H 22	O 2
4119-002	C15	H 22	O 2
4120-044	C15	H 22	O 2
4191-028	C15	H 22	O 2
4118-007	C15	H 22	O 2
4126-016	C15	H 22	O 2
4126-017	C15	H 22	O 2
4192-059	C15	H 22	O 2
4191-023	C15	H 22	O 2
4124-004	C15	H 22	O 2
4124-005	C15	H 22	O 2
4117-021	C15	H 22	O 2
4117-019	C15	H 22	O 2
4117-020	C15	H 22	O 2
4120-027	C15	H 22	O 2
4120-028	C15	H 22	O 2
4121-023	C15	H 22	O 2
4191-027	C15	H 22	O 2
4117-028	C15	H 22	O 2
4117-027	C15	H 22	O 2
4192-053	C15	H 22	O 2
4126-003	C15	H 22	O 2
4192-078	C15	H 22	O 2
4117-047	C15	H 22	O 2
4126-006	C15	H 22	O 2
4126-007	C15	H 22	O 2
4192-014	C15	H 22	O 2
4120-032	C15	H 22	O 2
4117-017	C15	H 22	O 2
4102-045	C15	H 22	O 2
4101-010	C15	H 22	O 2
4104-025	C15	H 22	O 2
4104-024	C15	H 22	O 2
4117-008	C15	H 22	O 2
4105-011	C15	H 22	O 2
4113-003	C15	H 22	O 2
4915-002	C15	H 22	O 2
4915-007	C15	H 22	O 3
4915-008	C15	H 22	O 3
4915-001	C15	H 22	O 3
4104-038	C15	H 22	O 3
4102-068	C15	H 22	O 3

4107-028	C15	H 24	O 1	
4103-011	C15	H 24	O 1	
4103-010	C15	H 24	O 1	
4116-008	C15	H 24	O 1	
4116-007	C15	H 24	O 1	
4105-012	C15	H 24	O 1	
4107-052	C15	H 24	O 1	
4116-005	C15	H 24	O 1	
4106-016	C15	H 24	O 1	
4104-013	C15	H 24	O 1	
4116-006	C15	H 24	O 1	
4105-006	C15	H 24	O 1	
4117-005	C15	H 24	O 1	
4105-005	C15	H 24	O 1	
4105-004	C15	H 24	O 1	
4103-003	C15	H 24	O 1	
4103-005	C15	H 24	O 1	
4127-005	C15	H 24	O 1	
4127-004	C15	H 24	O 1	
4120-010	C15	H 24	O 1	
4120-034	C15	H 24	O 1	
4191-026	C15	H 24	O 1	
4120-026	C15	H 24	O 1	
4192-052	C15	H 24	O 1	
4121-011	C15	H 24	O 1	
4121-013	C15	H 24	O 1	
4121-012	C15	H 24	O 1	
4191-037	C15	H 24	O 1	
4120-035	C15	H 24	O 1	
4120-015	C15	H 24	O 1	
4120-014	C15	H 24	O 1	
4119-008	C15	H 24	O 1	
4125-008	C15	H 24	O 1	
4118-005	C15	H 24	O 1	
4120-077	C15	H 24	O 1	
4121-006	C15	H 24	O 1	
4121-005	C15	H 24	O 1	
4124-002	C15	H 24	O 1	
4124-003	C15	H 24	O 1	
4118-004	C15	H 24	O 1	
4120-021	C15	H 24	O 1	
4192-058	C15	H 24	O 1	
4191-031	C15	H 24	O 1	
4194-004	C15	H 24	O 1	
4194-003	C15	H 24	O 1	
4124-007	C15	H 24	O 1	
4119-001	C15	H 24	O 1	
4120-022	C15	H 24	O 1	
4191-025	C15	H 24	O 1	
4190-001	C15	H 24	O 2	
4192-060	C15	H 24	O 2	
4192-049	C15	H 24	O 2	
4121-028	C15	H 24	O 2	
4117-018	C15	H 24	O 2	
4120-031	C15	H 24	O 2	
4192-017	C15	H 24	O 2	
4107-055	C15	H 24	O 2	
4107-054	C15	H 24	O 2	
4003-022	C15	H 24	O 2	
4112-008	C15	H 24	O 2	
4112-009	C15	H 24	O 2	
4107-044	C15	H 24	O 2	
4104-083	C15	H 24	O 2	
4103-006	C15	H 24	O 2	
4102-011	C15	H 24	O 2	
4101-009	C15	H 24	O 2	
4101-016	C15	H 24	O 2	
4112-010	C15	H 24	O 2	
4914-006	C15	H 24	O 2	
4106-026	C15	H 24	O 3	
4102-055	C15	H 24	O 3	
4105-013	C15	H 24	O 3	
4110-002	C15	H 24	O 3	
4106-023	C15	H 24	O 3	
4193-040	C15	H 24	O 3	
4127-006	C15	H 24	O 3	
4120-033	C15	H 24	O 3	
4192-063	C15	H 24	O 3	
4191-044	C15	H 24	O 4	
4008-059	C15	H 24	O10	
4008-048	C15	H 24	O11	
4198-001	C15	H 25	N 1	O 2
4107-014	C15	H 26		
4104-010	C15	H 26		
4101-022	C15	H 26		
4101-021	C15	H 26		
4102-005	C15	H 26	O 1	
4101-003	C15	H 26	O 1	
4101-004	C15	H 26	O 1	
4104-027	C15	H 26	O 1	
4104-026	C15	H 26	O 1	
4107-045	C15	H 26	O 1	
4107-046	C15	H 26	O 1	
4107-032	C15	H 26	O 1	
4107-033	C15	H 26	O 1	
4107-047	C15	H 26	O 1	
4104-028	C15	H 26	O 1	
4107-049	C15	H 26	O 1	
4107-031	C15	H 26	O 1	
4107-030	C15	H 26	O 1	
4103-009	C15	H 26	O 1	
4107-040	C15	H 26	O 1	
4107-039	C15	H 26	O 1	
4104-018	C15	H 26	O 1	
4107-038	C15	H 26	O 1	
4106-014	C15	H 26	O 1	
4106-013	C15	H 26	O 1	
4106-012	C15	H 26	O 1	
4106-011	C15	H 26	O 1	
4112-004	C15	H 26	O 1	
4112-003	C15	H 26	O 1	
4104-015	C15	H 26	O 1	
4117-007	C15	H 26	O 1	
4104-016	C15	H 26	O 1	
4108-001	C15	H 26	O 1	
4117-006	C15	H 26	O 1	
4105-003	C15	H 26	O 1	
4104-031	C15	H 26	O 1	
4104-017	C15	H 26	O 1	
4109-003	C15	H 26	O 1	
4125-004	C15	H 26	O 1	
4120-016	C15	H 26	O 1	
4120-020	C15	H 26	O 1	
4193-006	C15	H 26	O 1	
4192-046	C15	H 26	O 1	
4125-005	C15	H 26	O 1	
4125-006	C15	H 26	O 1	
4192-025	C15	H 26	O 1	
4125-007	C15	H 26	O 1	
4193-029	C15	H 26	O 1	
4192-057	C15	H 26	O 1	
4129-001	C15	H 26	O 1	
4120-018	C15	H 26	O 1	
4120-019	C15	H 26	O 1	
4121-007	C15	H 26	O 1	
4120-017	C15	H 26	O 1	
4123-004	C15	H 26	O 1	
4122-006	C15	H 26	O 1	
4120-012	C15	H 26	O 1	
4120-009	C15	H 26	O 1	
4120-011	C15	H 26	O 1	
4193-019	C15	H 26	O 1	
4192-020	C15	H 26	O 1	
4191-047	C15	H 26	O 1	
4192-019	C15	H 26	O 1	
4120-008	C15	H 26	O 1	
4120-007	C15	H 26	O 1	
4120-013	C15	H 26	O 1	
4122-005	C15	H 26	O 1	
4120-036	C15	H 26	O 1	
4126-001	C15	H 26	O 1	
4193-032	C15	H 26	O 1	
4192-077	C15	H 26	O 2	
4193-011	C15	H 26	O 2	
4120-037	C15	H 26	O 2	
4120-023	C15	H 26	O 2	
4120-025	C15	H 26	O 2	
4193-033	C15	H 26	O 2	
4192-018	C15	H 26	O 2	
4129-004	C15	H 26	O 2	
4129-002	C15	H 26	O 2	
4193-045	C15	H 26	O 2	
4104-034	C15	H 26	O 2	
4104-033	C15	H 26	O 2	
4104-032	C15	H 26	O 2	
4105-007	C15	H 26	O 2	
4102-072	C15	H 26	O 2	
4105-008	C15	H 26	O 2	
4106-017	C15	H 26	O 2	
4107-037	C15	H 26	O 2	
4107-034	C15	H 26	O 2	
4107-035	C15	H 26	O 2	
4107-036	C15	H 26	O 2	
4104-029	C15	H 26	O 2	
4108-002	C15	H 26	O 2	
4104-035	C15	H 26	O 3	
4120-038	C15	H 26	O 3	
4108-003	C15	H 26	O 5	
4107-007	C15	H 28		
4002-103	C15	H 28	O 1	
4002-102	C15	H 28	O 1	
4101-005	C15	H 28	O 2	
4191-042	C15	H 28	O 2	
4120-024	C15	H 28	O 2	
4107-070	C15	H 28	O 3	
4101-006	C15	H 30	O 3	
4192-033	C16	H 14	O 3	
4192-035	C16	H 14	O 4	
4014-009	C16	H 15	N 1	O 3
4117-040	C16	H 16	O 4	

4293-006	C19	H 24	O 7		
4210-015	C19	H 24	O 7		
4210-013	C19	H 24	O 7		
4192-065	C19	H 24	O 8		
4110-019	C19	H 24	O 9		
4508-006	C19	H 26	O 1		
4508-018	C19	H 26	O 2		
4117-043	C19	H 26	O 3		
4117-044	C19	H 26	O 3		
4117-054	C19	H 26	O 4	S 1	
4117-053	C19	H 26	O 4	S 1	
4210-016	C19	H 26	O 5		
4508-016	C19	H 26	O 5		
4002-106	C19	H 26	O 5		
4102-062	C19	H 26	O 6		
4210-006	C19	H 26	O 6		
4493-070	C19	H 26	O 6		
4493-065	C19	H 26	O 7		
4114-052	C19	H 26	O 7		
4114-017	C19	H 26	O 7		
4110-007	C19	H 26	O 7		
4114-048	C19	H 26	O 8		
4008-034	C19	H 26	O12		
4293-015	C19	H 28			
4508-014	C19	H 28	O 2		
4508-019	C19	H 28	O 2		
4508-011	C19	H 28	O 2		
4508-013	C19	H 28	O 3		
4293-045	C19	H 28	O 4		
4592-017	C19	H 29	N 1	O 2	
4198-019	C19	H 29	N 1	O 4	
4293-028	C19	H 30	O 1		
4293-029	C19	H 30	O 1		
4293-014	C19	H 30	O 1		
4292-012	C19	H 30	O 2		
4508-007	C19	H 30	O 2		
4508-009	C19	H 30	O 2		
4508-002	C19	H 30	O 2		
4508-012	C19	H 30	O 3		
4508-004	C19	H 30	O 3		
4508-005	C19	H 30	O 3		
4293-047	C19	H 30	O 3		
4102-007	C19	H 30	O 5		
4103-008	C19	H 30	O 5		
4508-001	C19	H 32	O 2		
4508-003	C19	H 32	O 3		
4192-061	C19	H 32	O 3		
4592-023	C19	H 33	N 1	O 1	
4293-026	C19	H 34			
4919-001	C19	H 38			
4919-002	C19	H 40			
4592-006	C20	H 16	O 6		
4293-023	C20	H 18	O 5		
4592-007	C20	H 18	O 6		
4011-009	C20	H 20	O 5		
4217-032	C20	H 20	O 6		
4217-033	C20	H 20	O 7		
4217-034	C20	H 20	O 8		
4291-001	C20	H 22	O 3		
4217-027	C20	H 22	O 6		
4217-029	C20	H 22	O 7		
4217-028	C20	H 22	O 7		
4217-030	C20	H 22	O 7		
4217-031	C20	H 22	O 7		
4102-041	C20	H 22	O 7		
4699-025	C20	H 24	O 2		
4699-024	C20	H 24	O 3		
4291-019	C20	H 24	O 3		
4216-003	C20	H 24	O 4		
4699-022	C20	H 24	O 4		
4114-053	C20	H 24	O 4		
4117-045	C20	H 24	O 4		
4208-006	C20	H 24	O 4		
4202-097	C20	H 24	O 4		
4202-096	C20	H 24	O 4		
4104-084	C20	H 24	O 5		
4114-036	C20	H 24	O 6		
4114-054	C20	H 24	O 6		
4114-055	C20	H 24	O 6		
4104-088	C20	H 24	O 7		
4104-064	C20	H 24	O 7		
4104-060	C20	H 24	O 7		
4299-006	C20	H 24	O 7		
4425-013	C20	H 24	O 7		
4104-063	C20	H 24	O 8		
4104-062	C20	H 24	O 8		
4192-067	C20	H 24	O 8		
4293-033	C20	H 24	O 9		
4293-034	C20	H 24	O10		
4293-035	C20	H 24	O10		
4293-036	C20	H 24	O11		
4212-016	C20	H 25	N 1	O 2	
4212-014	C20	H 25	N 1	O 3	
4104-065	C20	H 25	O 7	CL 1	
4104-067	C20	H 25	O 8	CL 1	
4920-004	C20	H 26	O 1		
4204-009	C20	H 26	O 2		
4291-011	C20	H 26	O 2		
4292-014	C20	H 26	O 3		
4217-017	C20	H 26	O 3		
4293-004	C20	H 26	O 3		
4217-022	C20	H 26	O 3		
4291-018	C20	H 26	O 3		
4204-013	C20	H 26	O 3		
4204-015	C20	H 26	O 3		
4204-014	C20	H 26	O 3		
4204-020	C20	H 26	O 3		
4204-023	C20	H 26	O 3		
4592-001	C20	H 26	O 3		
4204-032	C20	H 26	O 4		
4210-021	C20	H 26	O 4		
4207-022	C20	H 26	O 4		
4219-001	C20	H 26	O 4		
4219-009	C20	H 26	O 5		
4219-002	C20	H 26	O 5		
4217-023	C20	H 26	O 5		
4120-083	C20	H 26	O 5		
4202-101	C20	H 26	O 5		
4210-030	C20	H 26	O 5		
4102-060	C20	H 26	O 5		
4114-034	C20	H 26	O 5		
4114-033	C20	H 26	O 5		
4102-065	C20	H 26	O 6		
4114-037	C20	H 26	O 6		
4114-038	C20	H 26	O 6		
4102-061	C20	H 26	O 6		
4210-033	C20	H 26	O 6		
4210-025	C20	H 26	O 6		
4219-004	C20	H 26	O 6		
4219-006	C20	H 26	O 6		
4219-010	C20	H 26	O 6		
4219-011	C20	H 26	O 6		
4210-023	C20	H 26	O 7		
4210-029	C20	H 26	O 7		
4210-019	C20	H 26	O 7		
4102-063	C20	H 26	O 7		
4102-066	C20	H 26	O 7		
4425-015	C20	H 26	O 7		
4425-016	C20	H 26	O 8		
4425-019	C20	H 26	O 8		
4425-018	C20	H 26	O 9		
4212-017	C20	H 27	N 1	O 2	
4212-015	C20	H 27	N 1	O 2	
4212-013	C20	H 27	N 1	O 3	
4212-020	C20	H 27	N 1	O 3	
4204-005	C20	H 28	O 1		
4204-017	C20	H 28	O 1		
4920-003	C20	H 28	O 1		
4920-002	C20	H 28	O 1		
4209-015	C20	H 28	O 2		
4204-025	C20	H 28	O 2		
4204-026	C20	H 28	O 2		
4205-003	C20	H 28	O 2		
4205-005	C20	H 28	O 2		
4204-007	C20	H 28	O 2		
4206-033	C20	H 28	O 2		
4204-008	C20	H 28	O 2		
4003-024	C20	H 28	O 2		
4217-012	C20	H 28	O 2		
4217-013	C20	H 28	O 3		
4292-013	C20	H 28	O 3		
4217-020	C20	H 28	O 3		
4217-018	C20	H 28	O 3		
4217-021	C20	H 28	O 3		
4292-005	C20	H 28	O 3		
4204-038	C20	H 28	O 3		
4202-087	C20	H 28	O 3		
4202-093	C20	H 28	O 3		
4202-092	C20	H 28	O 3		
4208-003	C20	H 28	O 3		
4202-094	C20	H 28	O 3		
4205-004	C20	H 28	O 3		
4202-085	C20	H 28	O 3		
4207-018	C20	H 28	O 3		
4204-022	C20	H 28	O 3		
4206-051	C20	H 28	O 3		
4117-029	C20	H 28	O 3		
4203-032	C20	H 28	O 3		
4117-022	C20	H 28	O 3		
4204-018	C20	H 28	O 3		
4117-023	C20	H 28	O 3		
4204-031	C20	H 28	O 4		
4203-019	C20	H 28	O 4		
4206-052	C20	H 28	O 4		
4206-054	C20	H 28	O 4		
4204-021	C20	H 28	O 4		
4117-024	C20	H 28	O 4		
4117-051	C20	H 28	O 4		
4117-031	C20	H 28	O 4		
4117-052	C20	H 28	O 4		

4206-035	C20	H 32	O 3		
4206-034	C20	H 32	O 3		
4202-116	C20	H 32	O 4		
4202-082	C20	H 32	O 4		
4202-110	C20	H 32	O 4		
4202-079	C20	H 32	O 4		
4202-105	C20	H 32	O 4		
4207-011	C20	H 32	O 4		
4202-080	C20	H 32	O 4		
4206-047	C20	H 32	O 4		
4203-015	C20	H 32	O 4		
4206-044	C20	H 32	O 4		
4206-045	C20	H 32	O 4		
4202-078	C20	H 32	O 4		
4292-011	C20	H 32	O 4		
4293-002	C20	H 32	O 4		
4217-007	C20	H 32	O 4		
4292-007	C20	H 32	O 4		
4215-001	C20	H 32	O 4		
4218-005	C20	H 32	O 5		
4293-003	C20	H 32	O 5		
4218-009	C20	H 32	O 6		
4198-015	C20	H 33	N 1	O 4	
4202-031	C20	H 34	O 1		
4206-004	C20	H 34	O 1		
4206-005	C20	H 34	O 1		
4203-006	C20	H 34	O 1		
4201-001	C20	H 34	O 1		
4201-002	C20	H 34	O 1		
4202-010	C20	H 34	O 1		
4202-009	C20	H 34	O 1		
4202-034	C20	H 34	O 1		
4202-033	C20	H 34	O 1		
4201-003	C20	H 34	O 1		
4202-011	C20	H 34	O 1		
4202-032	C20	H 34	O 1		
4202-030	C20	H 34	O 1		
4202-003	C20	H 34	O 1		
4206-057	C20	H 34	O 1		
4202-005	C20	H 34	O 1		
4204-016	C20	H 34	O 1		
4202-004	C20	H 34	O 1		
4291-006	C20	H 34	O 1		
4216-006	C20	H 34	O 1		
4217-003	C20	H 34	O 1		
4217-001	C20	H 34	O 1		
4217-002	C20	H 34	O 1		
4216-007	C20	H 34	O 1		
4216-013	C20	H 34	O 2		
4216-012	C20	H 34	O 2		
4216-008	C20	H 34	O 2		
4292-002	C20	H 34	O 2		
4292-003	C20	H 34	O 2		
4216-009	C20	H 34	O 2		
4291-003	C20	H 34	O 2		
4216-001	C20	H 34	O 2		
4217-008	C20	H 34	O 2		
4202-043	C20	H 34	O 2		
4202-014	C20	H 34	O 2		
4202-067	C20	H 34	O 2		
4202-015	C20	H 34	O 2		
4202-013	C20	H 34	O 2		
4202-066	C20	H 34	O 2		
4202-064	C20	H 34	O 2		
4202-040	C20	H 34	O 2		
4202-065	C20	H 34	O 2		
4202-037	C20	H 34	O 2		
4202-024	C20	H 34	O 2		
4202-012	C20	H 34	O 2		
4202-036	C20	H 34	O 2		
4202-023	C20	H 34	O 2		
4202-107	C20	H 34	O 2		
4202-017	C20	H 34	O 2		
4206-006	C20	H 34	O 2		
4206-017	C20	H 34	O 3		
4202-072	C20	H 34	O 3		
4202-071	C20	H 34	O 3		
4202-108	C20	H 34	O 3		
4206-015	C20	H 34	O 3		
4206-013	C20	H 34	O 3		
4206-014	C20	H 34	O 3		
4206-016	C20	H 34	O 3		
4202-056	C20	H 34	O 3		
4202-052	C20	H 34	O 3		
4203-038	C20	H 34	O 3		
4202-060	C20	H 34	O 3		
4202-068	C20	H 34	O 3		
4216-002	C20	H 34	O 3		
4291-004	C20	H 34	O 3		
4291-005	C20	H 34	O 3		
4203-016	C20	H 34	O 4		
4202-058	C20	H 34	O 4		
4206-019	C20	H 34	O 4		
4293-032	C20	H 34	O 5		
4218-006	C20	H 34	O 6		
4202-025	C20	H 35	O 2	BR 1	
4204-002	C20	H 36	O 1		
4202-022	C20	H 36	O 2		
4202-026	C20	H 36	O 2		
4202-020	C20	H 36	O 2		
4202-021	C20	H 36	O 2		
4217-006	C20	H 36	O 2		
4202-069	C20	H 36	O 3		
4208-008	C20	H 36	O 3		
4202-070	C20	H 36	O 3		
4202-117	C20	H 36	O 4		
4202-118	C20	H 36	O 4		
4202-042	C20	H 36	O 4		
4202-029	C20	H 36	O 5		
4201-008	C20	H 38			
4201-007	C20	H 38			
4201-006	C20	H 38			
4201-005	C20	H 38			
4202-019	C20	H 38	O 2		
4202-018	C20	H 38	O 2		
4202-027	C20	H 38	O 3		
4202-028	C20	H 38	O 3		
4201-004	C20	H 40	O 1		
4201-009	C20	H 40	O 2		
4217-026	C21	H 24	O 6		
4210-003	C21	H 24	O 7		
4507-100	C21	H 26	O 3		
4292-001	C21	H 26	O 6		
4015-001	C21	H 26	O12		
4514-028	C21	H 27	N 1	O 1	
4514-031	C21	H 27	N 1	O 1	
4514-025	C21	H 27	N 1	O 2	
4212-010	C21	H 27	N 1	O 4	
4204-012	C21	H 28	O 3		
4205-007	C21	H 28	O 3		
4507-046	C21	H 28	O 3		
4507-058	C21	H 28	O 4		
4592-009	C21	H 28	O 4		
4507-094	C21	H 28	O 4		
4507-099	C21	H 28	O 4		
4507-097	C21	H 28	O 5		
4507-082	C21	H 28	O 5		
4507-083	C21	H 28	O 5		
4507-076	C21	H 28	O 5		
4207-023	C21	H 28	O 5		
4425-028	C21	H 28	O 6		
4192-055	C21	H 28	O 6	S 1	
4102-031	C21	H 28	O 8		
4110-010	C21	H 28	O10		
4099-017	C21	H 30	O 2		
4507-011	C21	H 30	O 2		
4507-041	C21	H 30	O 2		
4507-045	C21	H 30	O 3		
4507-049	C21	H 30	O 3		
4507-050	C21	H 30	O 3		
4204-010	C21	H 30	O 3		
4207-019	C21	H 30	O 3		
4207-021	C21	H 30	O 3		
4507-103	C21	H 30	O 4		
4507-020	C21	H 30	O 4		
4507-056	C21	H 30	O 4		
4507-064	C21	H 30	O 4		
4507-086	C21	H 30	O 5		
4507-081	C21	H 30	O 5		
4507-021	C21	H 30	O 5		
4507-065	C21	H 30	O 5		
4592-008	C21	H 30	O 5		
4207-007	C21	H 30	O 5		
4207-009	C21	H 30	O 6		
4425-004	C21	H 30	O 6		
4219-005	C21	H 30	O 7		
4514-027	C21	H 31	N 1	O 1	
4293-041	C21	H 31	N 1	O 4	
4293-043	C21	H 31	N 1	O 5	
4507-022	C21	H 31	O 5		
4514-013	C21	H 32	N 2		
4507-040	C21	H 32	O 2		
4507-008	C21	H 32	O 2		
4507-009	C21	H 32	O 2		
4507-044	C21	H 32	O 2		
4507-039	C21	H 32	O 2		
4507-042	C21	H 32	O 2		
4507-017	C21	H 32	O 3		
4507-057	C21	H 32	O 4		
4507-061	C21	H 32	O 4		
4507-059	C21	H 32	O 4		
4292-010	C21	H 32	O 4		
4507-072	C21	H 32	O 5		
4507-084	C21	H 32	O 5		
4507-080	C21	H 32	O 5		
4507-067	C21	H 32	O 5		
4507-066	C21	H 32	O 5		
4507-088	C21	H 32	O 6		
4507-089	C21	H 32	O 6		
4507-087	C21	H 32	O 6		

No.	C	H		
4425-011	C21	H 32	O 6	
4008-050	C21	H 32	O15	
4514-016	C21	H 33	N 1	O 1
4511-004	C21	H 33	N 1	O 1
4511-003	C21	H 33	N 1	O 1
4507-038	C21	H 34	O 2	
4507-037	C21	H 34	O 2	
4507-007	C21	H 34	O 2	
4507-036	C21	H 34	O 2	
4204-030	C21	H 34	O 2	
4507-047	C21	H 34	O 3	
4507-051	C21	H 34	O 3	
4507-018	C21	H 34	O 3	
4507-016	C21	H 34	O 3	
4507-015	C21	H 34	O 3	
4507-053	C21	H 34	O 4	
4507-052	C21	H 34	O 4	
4507-063	C21	H 34	O 4	
4507-054	C21	H 34	O 4	
4507-055	C21	H 34	O 4	
4202-059	C21	H 34	O 4	
4507-078	C21	H 34	O 5	
4507-079	C21	H 34	O 5	
4507-071	C21	H 34	O 5	
4507-028	C21	H 34	O 5	
4507-027	C21	H 34	O 5	
4507-023	C21	H 34	O 5	
4507-031	C21	H 34	O 6	
4507-035	C21	H 34	O 7	
4191-008	C21	H 34	O 9	
4512-006	C21	H 35	N 1	O 1
4511-001	C21	H 35	N 1	O 1
4513-031	C21	H 36	N 2	O 1
4513-030	C21	H 36	N 2	O 1
4507-001	C21	H 36	O 1	
4507-002	C21	H 36	O 2	
4507-006	C21	H 36	O 2	
4507-003	C21	H 36	O 2	
4507-004	C21	H 36	O 2	
4507-005	C21	H 36	O 2	
4202-073	C21	H 36	O 2	
4507-013	C21	H 36	O 3	
4507-012	C21	H 36	O 3	
4507-014	C21	H 36	O 3	
4507-019	C21	H 36	O 4	
4507-024	C21	H 36	O 5	
4507-025	C21	H 36	O 5	
4507-030	C21	H 36	O 5	
4507-026	C21	H 36	O 5	
4511-002	C21	H 37	N 1	O 1
4512-001	C21	H 37	N 1	O 1
4513-020	C21	H 38	N 2	
4104-061	C22	H 26	O 8	
4217-040	C22	H 26	O 8	
4217-042	C22	H 26	O 8	
4008-052	C22	H 26	O11	
4008-055	C22	H 26	O12	
4104-066	C22	H 27	O 8	CL 1
4293-040	C22	H 28	O 5	
4505-036	C22	H 28	O 5	
4425-001	C22	H 28	O 6	
4219-013	C22	H 28	O 6	
4202-095	C22	H 28	O 6	
4202-114	C22	H 28	O 7	
4202-115	C22	H 28	O 7	
4219-007	C22	H 28	O 7	
4219-015	C22	H 28	O 7	
4219-014	C22	H 28	O 7	
4425-002	C22	H 28	O 7	
4212-019	C22	H 29	N 1	O 2
4212-018	C22	H 29	N 1	O 3
4214-008	C22	H 29	N 1	O 3
4191-038	C22	H 30	O 2	
4191-039	C22	H 30	O 3	
4126-022	C22	H 30	O 4	
4002-053	C22	H 30	O 4	
4204-019	C22	H 30	O 5	
4209-017	C22	H 30	O 5	
4206-028	C22	H 30	O 6	
4192-054	C22	H 30	O 6	
4217-024	C22	H 30	O 6	
4425-003	C22	H 30	O 6	
4425-030	C22	H 30	O 7	
4104-090	C22	H 30	O 7	
4008-023	C22	H 30	O 8	
4202-113	C22	H 30	O 8	
4592-015	C22	H 31	N 1	
4514-029	C22	H 31	N 1	O 1
4514-026	C22	H 31	N 1	O 1
4511-013	C22	H 31	N 1	O 2
4212-007	C22	H 31	N 1	O 3
4214-001	C22	H 31	N 1	O 3
4207-020	C22	H 32	O 3	
4206-042	C22	H 32	O 4	
4211-004	C22	H 32	O 4	
4209-014	C22	H 32	O 4	
4203-033	C22	H 32	O 4	
4206-038	C22	H 32	O 4	
4206-041	C22	H 32	O 4	
4217-036	C22	H 32	O 4	
4217-038	C22	H 32	O 5	
4207-005	C22	H 32	O 5	
4206-049	C22	H 32	O 5	
4104-023	C22	H 32	O 5	
4425-005	C22	H 32	O 6	
4425-008	C22	H 32	O 7	
4008-027	C22	H 32	O 7	
4206-027	C22	H 32	O 7	
4217-025	C22	H 32	O 7	
4008-025	C22	H 32	O 8	
4425-009	C22	H 32	O 8	
4514-030	C22	H 33	N 1	O 1
4511-014	C22	H 33	N 1	O 2
4214-006	C22	H 33	N 1	O 2
4214-005	C22	H 33	N 1	O 2
4214-007	C22	H 33	N 1	O 2
4214-009	C22	H 33	N 1	O 2
4212-003	C22	H 33	N 1	O 2
4212-009	C22	H 33	N 1	O 2
4212-005	C22	H 33	N 1	O 3
4212-004	C22	H 33	N 1	O 3
4214-004	C22	H 33	N 1	O 3
4214-002	C22	H 33	N 1	O 3
4511-016	C22	H 33	N 1	O 3
4511-017	C22	H 33	N 1	O 3
4511-015	C22	H 33	N 1	O 3
4592-016	C22	H 33	N 1	O 3
4511-018	C22	H 33	N 1	O 3
4293-042	C22	H 33	N 1	O 4
4293-044	C22	H 33	N 1	O 5
4514-014	C22	H 34	N 2	
4204-037	C22	H 34	O 2	
4209-010	C22	H 34	O 3	
4206-012	C22	H 34	O 3	
4202-075	C22	H 34	O 4	
4217-009	C22	H 34	O 4	
4292-008	C22	H 34	O 5	
4216-014	C22	H 34	O 5	
4218-011	C22	H 34	O 6	
4218-010	C22	H 34	O 7	
4218-008	C22	H 34	O 7	
4425-010	C22	H 34	O 7	
4511-007	C22	H 35	N 1	O 1
4514-015	C22	H 35	N 1	O 1
4511-005	C22	H 35	N 1	O 1
4511-008	C22	H 35	N 1	O 1
4514-023	C22	H 35	N 1	O 2
4511-010	C22	H 35	N 1	O 2
4514-003	C22	H 36	N 2	
4514-002	C22	H 36	N 2	
4514-001	C22	H 36	N 2	
4202-044	C22	H 36	O 3	
4202-016	C22	H 36	O 3	
4202-057	C22	H 36	O 4	
4218-004	C22	H 36	O 7	
4002-085	C22	H 36	O13	
4511-006	C22	H 37	N 1	O 1
4512-008	C22	H 37	N 1	O 1
4512-004	C22	H 37	N 1	O 1
4513-021	C22	H 38	N 2	
4514-020	C22	H 38	N 2	O 1
4512-002	C22	H 39	N 1	O 1
4591-002	C22	H 39	N 1	O 2
4104-073	C23	H 22	O 7	
4592-005	C23	H 24	O 8	
4212-012	C23	H 27	N 1	O 6
4102-032	C23	H 28	O 9	
4102-033	C23	H 28	O10	
4003-023	C23	H 28	O11	
4425-022	C23	H 28	O11	
4212-011	C23	H 29	N 1	O 6
4505-131	C23	H 30	O 7	
4505-136	C23	H 30	O 7	
4505-129	C23	H 30	O 7	
4505-001	C23	H 32	O 4	
4505-003	C23	H 32	O 4	
4505-045	C23	H 32	O 5	
4505-038	C23	H 32	O 5	
4505-147	C23	H 32	O 5	
4507-085	C23	H 32	O 6	
4505-111	C23	H 32	O 6	
4505-091	C23	H 32	O 6	
4505-114	C23	H 32	O 6	
4505-089	C23	H 32	O 6	
4505-078	C23	H 32	O 6	
4505-133	C23	H 32	O 7	
4505-138	C23	H 32	O 7	
4505-127	C23	H 32	O 7	
4114-049	C23	H 32	O10	
4507-010	C23	H 34	O 3	

4592-002	C23	H 34	O 4	
4505-011	C23	H 34	O 4	
4505-006	C23	H 34	O 4	
4505-005	C23	H 34	O 4	
4505-014	C23	H 34	O 4	
4505-065	C23	H 34	O 5	
4505-053	C23	H 34	O 5	
4505-068	C23	H 34	O 5	
4505-054	C23	H 34	O 5	
4505-056	C23	H 34	O 5	
4505-050	C23	H 34	O 5	
4505-072	C23	H 34	O 5	
4505-047	C23	H 34	O 5	
4505-109	C23	H 34	O 6	
4505-097	C23	H 34	O 6	
4507-090	C23	H 34	O 7	
4008-028	C23	H 34	O 7	
4193-042	C23	H 34	O 7	
4008-026	C23	H 34	O 8	
4505-142	C23	H 34	O 8	
4008-037	C23	H 34	O15	
4511-019	C23	H 35	N 1	O 2
4511-020	C23	H 35	N 1	O 2
4202-111	C23	H 36	O 4	
4193-043	C23	H 36	O 6	
4218-002	C23	H 36	O 6	
4218-003	C23	H 36	O 8	
4511-009	C23	H 37	N 1	O 2
4511-011	C23	H 37	N 1	O 2
4511-012	C23	H 37	N 1	O 2
4213-002	C23	H 37	N 1	O 5
4213-001	C23	H 37	N 1	O 5
4213-035	C23	H 37	N 1	O 6
4514-004	C23	H 38	N 2	
4514-005	C23	H 38	N 2	
4514-019	C23	H 38	N 2	O 1
4218-012	C23	H 38	O 7	
4218-001	C23	H 38	O 7	
4512-005	C23	H 39	N 1	O 1
4512-009	C23	H 39	N 1	O 1
4514-006	C23	H 40	N 2	
4299-002	C23	H 40	O 6	
4299-003	C23	H 40	O 6	
4299-002	C23	H 40	O 6	
4512-003	C23	H 41	N 1	O 1
4591-003	C23	H 41	N 1	O 2
4512-010	C23	H 41	N 1	O 2
4513-006	C23	H 42	N 2	
4513-001	C23	H 42	N 2	
4513-027	C23	H 42	N 2	
4513-010	C23	H 42	N 2	
4507-073	C23	H 44	O 6	
4217-041	C24	H 28	O 9	
4008-057	C24	H 28	O11	
4008-056	C24	H 28	O11	
4191-041	C24	H 30	O 4	
4126-002	C24	H 30	O 4	
4191-040	C24	H 30	O 4	
4506-027	C24	H 30	O 5	
4506-036	C24	H 30	O 5	
4506-037	C24	H 30	O 6	
4008-061	C24	H 30	O11	
4008-063	C24	H 30	O11	
4506-012	C24	H 32	O 4	
4506-003	C24	H 32	O 4	
4506-006	C24	H 32	O 4	
4506-005	C24	H 32	O 4	
4506-025	C24	H 32	O 5	
4506-031	C24	H 32	O 5	
4506-033	C24	H 32	O 6	
4506-038	C24	H 32	O 6	
4506-020	C24	H 32	O 6	
4506-019	C24	H 32	O 6	
4506-018	C24	H 32	O 6	
4207-014	C24	H 32	O 8	
4519-008	C24	H 32	O 9	
4212-006	C24	H 33	N 1	O 4
4506-007	C24	H 34	O 4	
4493-031	C24	H 34	O 5	
4506-026	C24	H 34	O 5	
4506-017	C24	H 34	O 5	
4505-119	C24	H 34	O 6	
4220-009	C24	H 34	O 6	
4217-044	C24	H 34	O 7	
4425-006	C24	H 34	O 7	
4206-020	C24	H 34	O 7	
4207-017	C24	H 34	O 7	
4425-007	C24	H 34	O 8	
4110-009	C24	H 34	O 9	
4008-024	C24	H 34	O10	
4418-005	C24	H 35	N 1	O 1
4214-003	C24	H 35	N 1	O 4
4592-024	C24	H 35	N 1	O 5
4206-009	C24	H 36	O 5	
4215-002	C24	H 36	O 6	

4218-007	C24	H 36	O 8	
4418-001	C24	H 37	N 1	O 2
4213-022	C24	H 37	N 1	O 7
4493-032	C24	H 38	O 2	
4510-013	C24	H 38	O 4	
4207-004	C24	H 39	N 1	O 4
4213-003	C24	H 39	N 1	O 5
4213-005	C24	H 39	N 1	O 6
4213-009	C24	H 39	N 1	O 7
4213-021	C24	H 39	N 1	O 7
4514-008	C24	H 40	N 2	
4514-010	C24	H 40	N 2	
4514-012	C24	H 40	N 2	O 1
4514-007	C24	H 40	N 2	O 1
4514-018	C24	H 40	N 2	O 1
4510-001	C24	H 40	O 3	
4510-008	C24	H 40	O 4	
4510-003	C24	H 40	O 4	
4510-002	C24	H 40	O 4	
4510-005	C24	H 40	O 4	
4510-010	C24	H 40	O 4	
4510-011	C24	H 40	O 4	
4510-020	C24	H 40	O 5	
4510-015	C24	H 40	O 5	
4510-022	C24	H 40	O 5	
4510-017	C24	H 40	O 5	
4510-014	C24	H 40	O 5	
4510-016	C24	H 40	O 5	
4207-012	C24	H 41	N 1	O 4
4514-011	C24	H 42	N 2	
4514-009	C24	H 42	N 2	
4514-017	C24	H 42	N 2	O 1
4419-041	C24	H 42	N 2	O 1
4513-003	C24	H 44	N 2	
4513-002	C24	H 44	N 2	
4513-014	C24	H 44	N 2	O 1
4514-024	C24	H 44	N 2	O 2
4699-021	C25	H 30	O 4	
4506-035	C25	H 32	O 7	
4210-002	C25	H 32	O11	
4210-032	C25	H 32	O12	
4009-009	C25	H 32	O13	
4117-035	C25	H 34	O 6	
4220-011	C25	H 34	O 6	
4425-017	C25	H 34	O10	
4210-037	C25	H 34	O11	
4210-014	C25	H 34	O12	
4291-010	C25	H 35	N 1	O 9
4217-037	C25	H 36	O 4	
4302-004	C25	H 36	O 4	
4302-001	C25	H 36	O 4	
4217-014	C25	H 36	O 5	
4217-039	C25	H 36	O 5	
4220-010	C25	H 36	O 6	
4505-122	C25	H 36	O 6	
4505-104	C25	H 36	O 6	
4425-026	C25	H 36	O10	
4418-006	C25	H 37	N 1	O 1
4418-003	C25	H 37	N 1	O 1
4303-001	C25	H 38	O 2	
4302-009	C25	H 38	O 2	
4302-008	C25	H 38	O 2	
4302-003	C25	H 38	O 3	
4302-002	C25	H 38	O 4	
4207-001	C25	H 38	O 6	
4108-005	C25	H 38	O 7	
4108-004	C25	H 38	O 7	
4508-010	C25	H 38	O 8	
4508-008	C25	H 38	O 8	
4418-004	C25	H 39	N 1	O 1
4418-008	C25	H 39	N 1	O 1
4418-007	C25	H 39	N 1	O 1
4418-009	C25	H 39	N 1	O 1
4418-002	C25	H 39	N 1	O 2
4207-006	C25	H 39	N 1	O 5
4207-008	C25	H 39	N 1	O 5
4207-010	C25	H 39	N 1	O 6
4207-013	C25	H 39	N 1	O 6
4213-004	C25	H 39	N 1	O 6
4213-036	C25	H 39	N 1	O 6
4213-037	C25	H 39	N 1	O 7
4213-026	C25	H 39	N 1	O 7
4925-005	C25	H 40		
4302-006	C25	H 40	O 1	
4302-007	C25	H 40	O 1	
4418-014	C25	H 41	N 1	O 2
4213-006	C25	H 41	N 1	O 6
4213-024	C25	H 41	N 1	O 7
4213-025	C25	H 41	N 1	O 7
4213-027	C25	H 41	N 1	O 7
4493-034	C25	H 42	N 2	
4419-042	C25	H 42	N 2	O 1
4301-002	C25	H 42	O 1	
4301-001	C25	H 42	O 1	
4302-005	C25	H 42	O 1	

4501-020	C27	H 44	O 1	
4501-004	C27	H 44	O 1	
4501-005	C27	H 44	O 1	
4501-007	C27	H 44	O 1	
4501-008	C27	H 44	O 1	
4592-004	C27	H 44	O 2	
4504-003	C27	H 44	O 3	
4504-002	C27	H 44	O 3	
4504-001	C27	H 44	O 3	
4504-004	C27	H 44	O 3	
4504-071	C27	H 44	O 4	
4504-016	C27	H 44	O 4	
4504-009	C27	H 44	O 4	
4504-018	C27	H 44	O 4	
4504-013	C27	H 44	O 4	
4504-024	C27	H 44	O 4	
4504-019	C27	H 44	O 4	
4504-015	C27	H 44	O 4	
4504-017	C27	H 44	O 4	
4504-021	C27	H 44	O 4	
4504-020	C27	H 44	O 4	
4504-023	C27	H 44	O 4	
4504-010	C27	H 44	O 4	
4504-042	C27	H 44	O 5	
4504-038	C27	H 44	O 5	
4504-037	C27	H 44	O 5	
4509-007	C27	H 44	O 5	
4504-041	C27	H 44	O 5	
4504-033	C27	H 44	O 5	
4504-034	C27	H 44	O 5	
4501-017	C27	H 44	O 5	
4504-035	C27	H 44	O 5	
4504-045	C27	H 44	O 5	
4504-043	C27	H 44	O 5	
4504-032	C27	H 44	O 5	
4504-044	C27	H 44	O 5	
4504-039	C27	H 44	O 5	
4504-040	C27	H 44	O 5	
4504-036	C27	H 44	O 5	
4504-046	C27	H 44	O 5	
4509-018	C27	H 44	O 6	
4504-049	C27	H 44	O 6	
4509-017	C27	H 44	O 6	
4509-016	C27	H 44	O 6	
4509-014	C27	H 44	O 6	
4504-047	C27	H 44	O 6	
4504-048	C27	H 44	O 6	
4509-024	C27	H 44	O 7	
4509-023	C27	H 44	O 7	
4509-021	C27	H 44	O 7	
4509-020	C27	H 44	O 7	
4509-019	C27	H 44	O 7	
4509-032	C27	H 44	O 8	
4509-025	C27	H 44	O 8	
4509-031	C27	H 44	O 8	
4507-029	C27	H 44	O11	
4515-001	C27	H 45	N 1	O 1
4518-001	C27	H 45	N 1	O 1
4515-003	C27	H 45	N 1	O 2
4515-010	C27	H 45	N 1	O 2
4599-005	C27	H 45	N 1	O 2
4515-013	C27	H 45	N 1	O 2
4515-012	C27	H 45	N 1	O 3
4516-032	C27	H 45	N 1	O 3
4515-006	C27	H 45	N 1	O 3
4515-011	C27	H 45	N 1	O 3
4515-015	C27	H 45	N 1	O 3
4512-007	C27	H 45	N 1	O 6
4516-008	C27	H 45	N 1	O 7
4501-001	C27	H 46		
4419-015	C27	H 46	N 2	O 1
4419-010	C27	H 46	N 2	O 1
4419-036	C27	H 46	N 2	O 2
4419-024	C27	H 46	N 2	O 2
4419-025	C27	H 46	N 2	O 2
4419-030	C27	H 46	N 2	O 2
4599-003	C27	H 46	N 2	O 2
4419-039	C27	H 46	N 2	O 3
4501-002	C27	H 46	O 1	
4501-009	C27	H 46	O 2	
4501-010	C27	H 46	O 2	
4501-011	C27	H 46	O 3	
4509-002	C27	H 46	O 4	
4509-005	C27	H 46	O 5	
4509-003	C27	H 46	O 5	
4509-008	C27	H 46	O 5	
4501-013	C27	H 46	O 5	
4509-009	C27	H 46	O 6	
4419-040	C27	H 48	N 2	
4419-003	C27	H 48	N 2	
4513-026	C27	H 48	N 2	O 1
4419-021	C27	H 48	N 2	O 2
4509-001	C27	H 48	O 1	
4509-033	C27	H 48	O 4	
4501-014	C27	H 48	O 5	

4509-006	C27	H 48	O 5	
4509-004	C27	H 48	O 5	
4501-012	C27	H 48	O 5	
4513-029	C27	H 52	N 2	++
4505-123	C28	H 30	O 7	
4592-012	C28	H 30	O 9	
4592-011	C28	H 30	O 9	
4592-010	C28	H 30	O10	
4426-011	C28	H 32	O 7	
4426-008	C28	H 34	O 5	
4426-012	C28	H 34	O 6	
4426-009	C28	H 34	O 6	
4427-004	C28	H 34	O 7	
4427-014	C28	H 34	O 9	
4423-010	C28	H 34	O 9	
4426-007	C28	H 36	O 4	
4426-001	C28	H 36	O 5	
4507-060	C28	H 36	O 5	
4507-069	C28	H 36	O 6	
4507-068	C28	H 36	O 6	
4507-091	C28	H 36	O 7	
4427-005	C28	H 36	O 7	
4427-011	C28	H 36	O 8	
4506-040	C28	H 36	O 8	
4425-023	C28	H 36	O12	
4494-004	C28	H 38	O 2	
4494-003	C28	H 38	O 3	
4502-026	C28	H 38	O 5	
4502-033	C28	H 38	O 5	
4502-027	C28	H 38	O 5	
4426-002	C28	H 38	O 5	
4502-035	C28	H 38	O 6	
4502-034	C28	H 38	O 6	
4502-031	C28	H 38	O 6	
4215-018	C28	H 38	O 9	
4291-008	C28	H 38	O 9	
4502-028	C28	H 39	O 5	CL 1
4502-030	C28	H 39	O 5	CL 1
4502-022	C28	H 40	O 1	
4502-037	C28	H 40	O 5	
4502-029	C28	H 40	O 6	
4502-036	C28	H 40	O 6	
4507-095	C28	H 40	O 7	
4592-022	C28	H 40	O 7	
4507-098	C28	H 40	O 8	
4507-096	C28	H 40	O 8	
4215-003	C28	H 40	O 8	
4507-077	C28	H 40	O 9	
4505-079	C28	H 40	O10	
4502-021	C28	H 42	O 1	
4502-016	C28	H 42	O 1	
4502-017	C28	H 42	O 2	
4505-102	C28	H 42	O 6	
4507-070	C28	H 42	O 6	
4507-093	C28	H 42	O 6	
4507-074	C28	H 42	O 7	
4291-009	C28	H 42	O 9	
4418-013	C28	H 43	N 1	O 3
4207-024	C28	H 43	N 1	O 5
4502-013	C28	H 44	O 1	
4502-015	C28	H 44	O 1	
4502-014	C28	H 44	O 1	
4502-020	C28	H 44	O 1	
4502-023	C28	H 44	O 3	
4504-029	C28	H 44	O 4	
4517-006	C28	H 45	N 1	O 2
4507-102	C28	H 45	N 1	O 5
4502-019	C28	H 46	O 1	
4502-012	C28	H 46	O 1	
4502-006	C28	H 46	O 1	
4493-006	C28	H 46	O 1	
4493-003	C28	H 46	O 1	
4502-008	C28	H 46	O 1	
4493-004	C28	H 46	O 1	
4502-007	C28	H 46	O 1	
4502-011	C28	H 46	O 1	
4502-010	C28	H 46	O 1	
4502-009	C28	H 46	O 1	
4502-018	C28	H 46	O 3	
4502-025	C28	H 46	O 7	
4502-024	C28	H 46	O 7	
4507-101	C28	H 47	N 1	O 6
4513-009	C28	H 48	N 2	O 1
4419-013	C28	H 48	N 2	O 1
4419-002	C28	H 48	N 2	O 1
4419-026	C28	H 48	N 2	O 2
4419-038	C28	H 48	N 2	O 2
4419-037	C28	H 48	N 2	O 2
4502-003	C28	H 48	O 1	
4502-004	C28	H 48	O 1	
4502-005	C28	H 48	O 1	
4493-005	C28	H 48	O 1	
4502-002	C28	H 48	O 1	
4493-061	C28	H 48	O 1	
4599-001	C28	H 48	O 1	

No.	C	H			
4493-013	C28	H 48	O 2		
4419-004	C28	H 50	N 2		
4419-009	C28	H 50	N 2	O 1	
4419-023	C28	H 50	N 2	O 2	
4502-001	C28	H 50	O 1		
4493-029	C28	H 50	O 4		
4009-013	C29	H 30	O13		
4009-011	C29	H 30	O14		
4424-015	C29	H 34	O 8		
4424-018	C29	H 34	O 8		
4424-019	C29	H 34	O 9		
4423-012	C29	H 34	O10		
4215-015	C29	H 36	O 6		
4215-019	C29	H 36	O 7		
4424-002	C29	H 36	O 8		
4424-017	C29	H 36	O 8		
4424-008	C29	H 36	O 8		
4424-003	C29	H 36	O 9		
4428-006	C29	H 36	O 9		
4120-084	C29	H 37	N 1	O12	
4120-085	C29	H 37	N 1	O13	
4209-012	C29	H 38	O 4		
4493-027	C29	H 38	O 4		
4505-094	C29	H 38	O 9		
4110-018	C29	H 38	O 9		
4494-001	C29	H 40	O 4		
4503-042	C29	H 40	O 8		
4220-012	C29	H 40	O 8		
4505-092	C29	H 40	O 9		
4110-017	C29	H 40	O 9		
4505-093	C29	H 40	O10		
4505-137	C29	H 40	O11		
4494-002	C29	H 42	O 4		
4493-071	C29	H 42	O 4		
4503-038	C29	H 42	O 5		
4502-032	C29	H 42	O 7		
4505-055	C29	H 42	O 8		
4505-039	C29	H 42	O 8		
4505-040	C29	H 42	O 8		
4505-041	C29	H 42	O 9		
4505-084	C29	H 42	O 9		
4505-080	C29	H 42	O 9		
4505-081	C29	H 42	O10		
4505-090	C29	H 42	O10		
4505-082	C29	H 42	O10		
4505-141	C29	H 42	O11		
4505-134	C29	H 42	O11		
4505-135	C29	H 42	O11		
4505-139	C29	H 42	O11		
4493-019	C29	H 44	O 4		
4493-023	C29	H 44	O 6		
4503-041	C29	H 44	O 7		
4503-039	C29	H 44	O 8		
4505-015	C29	H 44	O 8		
4503-043	C29	H 44	O 8		
4505-016	C29	H 44	O 8		
4505-012	C29	H 44	O 8		
4505-049	C29	H 44	O 9		
4505-057	C29	H 44	O 9		
4505-048	C29	H 44	O 9		
4503-040	C29	H 44	O 9		
4505-073	C29	H 44	O 9		
4505-098	C29	H 44	O10		
4505-058	C29	H 44	O10		
4505-144	C29	H 44	O12		
4505-143	C29	H 44	O12		
4516-003	C29	H 45	N 1	O 8	
4516-010	C29	H 45	N 1	O 9	
4513-007	C29	H 45	N 3	O 1	
4514-021	C29	H 46	N 2	O 2	
4503-024	C29	H 46	O 1		
4493-024	C29	H 46	O 1		
4503-025	C29	H 46	O 1		
4503-029	C29	H 46	O 1		
4503-023	C29	H 46	O 1		
4503-044	C29	H 46	O 1		
4503-022	C29	H 46	O 1		
4503-030	C29	H 46	O 1		
4493-026	C29	H 46	O 2		
4493-017	C29	H 46	O 2		
4493-025	C29	H 46	O 3		
4493-018	C29	H 46	O 4		
4503-035	C29	H 46	O 7		
4509-022	C29	H 46	O 8		
4507-043	C29	H 47	N 1	O 4	
4515-009	C29	H 47	N 1	O 5	
4207-002	C29	H 47	N 1	O 6	
4516-009	C29	H 47	N 1	O 8	
4513-016	C29	H 48	N 2	O 2	
4513-017	C29	H 48	N 2	O 2	
4419-033	C29	H 48	N 2	O 4	
4503-019	C29	H 48	O 1		
4503-018	C29	H 48	O 1		
4503-014	C29	H 48	O 1		
4503-013	C29	H 48	O 1		
4503-011	C29	H 48	O 1		
4503-015	C29	H 48	O 1		
4503-010	C29	H 48	O 1		
4503-012	C29	H 48	O 1		
4503-008	C29	H 48	O 1		
4503-020	C29	H 48	O 1		
4503-009	C29	H 48	O 1		
4503-028	C29	H 48	O 1		
4503-021	C29	H 48	O 1		
4503-017	C29	H 48	O 1		
4503-027	C29	H 48	O 1		
4493-020	C29	H 48	O 1		
4493-001	C29	H 48	O 1		
4503-032	C29	H 48	O 1		
4493-009	C29	H 48	O 1		
4493-008	C29	H 48	O 1		
4493-002	C29	H 48	O 1		
4599-008	C29	H 48	O 1		
4493-016	C29	H 48	O 2		
4503-031	C29	H 48	O 2		
4503-026	C29	H 48	O 2		
4493-021	C29	H 48	O 2		
4493-022	C29	H 48	O 2		
4499-005	C29	H 48	O 2		
4503-036	C29	H 48	O 7		
4503-034	C29	H 48	O 7		
4503-037	C29	H 48	O 7		
4503-033	C29	H 48	O 7		
4507-048	C29	H 49	N 1	O 5	
4513-005	C29	H 50	N 2	O 1	
4513-018	C29	H 50	N 2	O 2	
4493-015	C29	H 50	O 1		
4503-003	C29	H 50	O 1		
4503-007	C29	H 50	O 1		
4503-005	C29	H 50	O 1		
4503-004	C29	H 50	O 1		
4513-015	C29	H 50	O 2	N 2	
4503-001	C29	H 52	O 1		
4503-002	C29	H 52	O 1		
4107-071	C30	H 30	O 8		
4426-019	C30	H 34	O10		
4427-006	C30	H 36	O 9		
4493-059	C30	H 36	O 9		
4427-012	C30	H 38	O 9		
4699-018	C30	H 40	O 1		
4699-019	C30	H 40	O 2		
4493-028	C30	H 40	O 4		
4426-017	C30	H 40	O 8		
4427-009	C30	H 40	O 9		
4427-010	C30	H 40	O 9		
4506-028	C30	H 40	O10		
4198-011	C30	H 42	N 2	O 2	S 1
4198-012	C30	H 42	N 2	O 2	S 1
4198-013	C30	H 42	N 2	O 4	S 1
4198-014	C30	H 42	N 2	O 4	S 1
4426-003	C30	H 42	O 6		
4426-004	C30	H 42	O 7		
4426-005	C30	H 42	O 7		
4412-015	C30	H 42	O 7		
4506-013	C30	H 42	O 9		
4505-132	C30	H 42	O10		
4505-130	C30	H 42	O10		
4215-007	C30	H 42	O10		
4215-008	C30	H 42	O10		
4215-012	C30	H 42	O12		
4293-046	C30	H 44	K 2	O16	S 2
4410-020	C30	H 44	O 3		
4405-028	C30	H 44	O 3		
4408-070	C30	H 44	O 4		
4408-060	C30	H 44	O 4		
4410-027	C30	H 44	O 4		
4410-029	C30	H 44	O 4		
4408-069	C30	H 44	O 4		
4408-087	C30	H 44	O 4		
4410-028	C30	H 44	O 5		
4408-071	C30	H 44	O 5		
4412-008	C30	H 44	O 7		
4412-016	C30	H 44	O 7		
4505-002	C30	H 44	O 7		
4505-042	C30	H 44	O 8		
4505-004	C30	H 44	O 8		
4505-043	C30	H 44	O 8		
4412-017	C30	H 44	O 8		
4412-018	C30	H 44	O 8		
4215-005	C30	H 44	O 9		
4505-083	C30	H 44	O 9		
4505-116	C30	H 44	O 9		
4505-117	C30	H 44	O 9		
4505-113	C30	H 44	O 9		
4505-145	C30	H 44	O 9		
4505-115	C30	H 44	O 9		
4505-112	C30	H 44	O 9		
4505-044	C30	H 44	O 9		
4505-118	C30	H 44	O10		
4505-125	C30	H 44	O10		

4505-128	C30	H 44	O10	
4505-124	C30	H 44	O10	
4505-126	C30	H 44	O11	
4513-008	C30	H 46	N 2	O 1
4513-012	C30	H 46	N 2	O 2
4406-009	C30	H 46	O 1	
4410-012	C30	H 46	O 2	
4491-004	C30	H 46	O 2	
4493-052	C30	H 46	O 3	
4493-051	C30	H 46	O 3	
4409-008	C30	H 46	O 3	
4410-019	C30	H 46	O 3	
4406-021	C30	H 46	O 3	
4417-015	C30	H 46	O 3	
4403-020	C30	H 46	O 3	
4410-022	C30	H 46	O 3	
4408-075	C30	H 46	O 3	
4409-033	C30	H 46	O 3	
4408-076	C30	H 46	O 3	
4409-011	C30	H 46	O 3	
4405-027	C30	H 46	O 3	
4405-014	C30	H 46	O 3	
4406-016	C30	H 46	O 3	
4409-010	C30	H 46	O 3	
4405-015	C30	H 46	O 3	
4408-118	C30	H 46	O 3	
4405-011	C30	H 46	O 3	
4408-068	C30	H 46	O 3	
4409-032	C30	H 46	O 3	
4408-106	C30	H 46	O 4	
4408-065	C30	H 46	O 4	
4408-077	C30	H 46	O 4	
4408-122	C30	H 46	O 4	
4408-052	C30	H 46	O 4	
4408-063	C30	H 46	O 4	
4408-100	C30	H 46	O 4	
4408-099	C30	H 46	O 4	
4410-023	C30	H 46	O 4	
4409-016	C30	H 46	O 4	
4405-024	C30	H 46	O 4	
4493-055	C30	H 46	O 4	
4499-003	C30	H 46	O 4	
4494-005	C30	H 46	O 5	
4409-018	C30	H 46	O 5	
4408-066	C30	H 46	O 5	
4408-067	C30	H 46	O 5	
4406-015	C30	H 46	O 5	
4408-114	C30	H 46	O 5	
4410-032	C30	H 46	O 5	
4408-113	C30	H 46	O 5	
4408-093	C30	H 46	O 5	
4409-028	C30	H 46	O 5	
4410-030	C30	H 46	O 5	
4408-107	C30	H 46	O 5	
4409-020	C30	H 46	O 5	
4410-026	C30	H 46	O 5	
4408-109	C30	H 46	O 6	
4412-009	C30	H 46	O 6	
4408-108	C30	H 46	O 6	
4412-010	C30	H 46	O 6	
4409-023	C30	H 46	O 6	
4417-012	C30	H 46	O 6	
4493-045	C30	H 46	O 6	
4505-017	C30	H 46	O 7	
4505-020	C30	H 46	O 7	
4505-007	C30	H 46	O 7	
4505-019	C30	H 46	O 7	
4505-018	C30	H 46	O 7	
4412-019	C30	H 46	O 7	
4412-012	C30	H 46	O 7	
4408-110	C30	H 46	O 7	
4408-111	C30	H 46	O 8	
4408-112	C30	H 46	O 8	
4412-013	C30	H 46	O 8	
4412-014	C30	H 46	O 8	
4505-069	C30	H 46	O 8	
4505-021	C30	H 46	O 8	
4505-075	C30	H 46	O 8	
4505-022	C30	H 46	O 8	
4505-106	C30	H 46	O 8	
4505-074	C30	H 46	O 8	
4505-066	C30	H 46	O 9	
4505-070	C30	H 46	O 9	
4505-059	C30	H 46	O 9	
4505-099	C30	H 46	O 9	
4505-076	C30	H 46	O 9	
4492-001	C30	H 48		
4402-004	C30	H 48		
4514-022	C30	H 48	N 2	O 2
4213-033	C30	H 48	N 2	O10
4416-005	C30	H 48	O 1	
4401-018	C30	H 48	O 1	
4414-003	C30	H 48	O 1	
4404-013	C30	H 48	O 1	
4417-008	C30	H 48	O 1	

4409-004	C30	H 48	O 1
4410-004	C30	H 48	O 1
4402-010	C30	H 48	O 1
4406-020	C30	H 48	O 1
4422-005	C30	H 48	O 1
4492-009	C30	H 48	O 1
4493-044	C30	H 48	O 2
4402-012	C30	H 48	O 2
4408-007	C30	H 48	O 2
4403-008	C30	H 48	O 2
4406-010	C30	H 48	O 2
4408-042	C30	H 48	O 2
4405-016	C30	H 48	O 2
4410-010	C30	H 48	O 2
4416-008	C30	H 48	O 2
4415-013	C30	H 48	O 2
4403-010	C30	H 48	O 2
4410-011	C30	H 48	O 2
4403-018	C30	H 48	O 2
4410-013	C30	H 48	O 2
4403-014	C30	H 48	O 2
4408-006	C30	H 48	O 2
4415-020	C30	H 48	O 3
4403-016	C30	H 48	O 3
4407-005	C30	H 48	O 3
4408-014	C30	H 48	O 3
4410-015	C30	H 48	O 3
4405-009	C30	H 48	O 3
4403-019	C30	H 48	O 3
4409-027	C30	H 48	O 3
4408-061	C30	H 48	O 3
4408-056	C30	H 48	O 3
4417-014	C30	H 48	O 3
4405-020	C30	H 48	O 3
4409-005	C30	H 48	O 3
4405-010	C30	H 48	O 3
4415-021	C30	H 48	O 3
4406-011	C30	H 48	O 3
4415-019	C30	H 48	O 3
4405-007	C30	H 48	O 3
4410-008	C30	H 48	O 3
4405-008	C30	H 48	O 3
4408-074	C30	H 48	O 3
4409-024	C30	H 48	O 3
4408-044	C30	H 48	O 3
4408-013	C30	H 48	O 3
4408-120	C30	H 48	O 3
4408-012	C30	H 48	O 3
4406-013	C30	H 48	O 3
4405-012	C30	H 48	O 3
4408-072	C30	H 48	O 3
4412-001	C30	H 48	O 3
4408-127	C30	H 48	O 3
4409-034	C30	H 48	O 3
4493-048	C30	H 48	O 3
4492-018	C30	H 48	O 3
4493-049	C30	H 48	O 4
4406-017	C30	H 48	O 4
4415-025	C30	H 48	O 4
4410-017	C30	H 48	O 4
4409-013	C30	H 48	O 4
4408-089	C30	H 48	O 4
4409-012	C30	H 48	O 4
4408-088	C30	H 48	O 4
4409-014	C30	H 48	O 4
4408-090	C30	H 48	O 4
4408-057	C30	H 48	O 4
4408-119	C30	H 48	O 4
4413-008	C30	H 48	O 4
4408-086	C30	H 48	O 4
4408-022	C30	H 48	O 4
4408-019	C30	H 48	O 4
4404-020	C30	H 48	O 4
4408-016	C30	H 48	O 4
4405-021	C30	H 48	O 4
4408-079	C30	H 48	O 4
4408-050	C30	H 48	O 4
4408-051	C30	H 48	O 4
4409-029	C30	H 48	O 4
4405-031	C30	H 48	O 4
4412-022	C30	H 48	O 4
4408-062	C30	H 48	O 4
4406-014	C30	H 48	O 4
4408-024	C30	H 48	O 4
4405-017	C30	H 48	O 4
4408-083	C30	H 48	O 4
4415-024	C30	H 48	O 4
4408-025	C30	H 48	O 4
4408-078	C30	H 48	O 4
4412-003	C30	H 48	O 4
4408-080	C30	H 48	O 4
4409-025	C30	H 48	O 4
4408-082	C30	H 48	O 4
4408-081	C30	H 48	O 4
4421-007	C30	H 48	O 4

4401-002	C30	H 52	O 1		
4410-002	C30	H 52	O 1		
4403-003	C30	H 52	O 1		
4403-001	C30	H 52	O 1		
4401-009	C30	H 52	O 2		
4417-003	C30	H 52	O 2		
4401-008	C30	H 52	O 2		
4401-012	C30	H 52	O 2		
4403-005	C30	H 52	O 2		
4401-011	C30	H 52	O 2		
4406-007	C30	H 52	O 2		
4493-047	C30	H 52	O 2		
4404-003	C30	H 52	O 2		
4413-004	C30	H 52	O 2		
4404-002	C30	H 52	O 2		
4413-003	C30	H 52	O 2		
4416-006	C30	H 52	O 2		
4493-041	C30	H 52	O 3		
4404-009	C30	H 52	O 3		
4420-001	C30	H 52	O 3		
4406-008	C30	H 52	O 3		
4404-021	C30	H 52	O 3		
4404-007	C30	H 52	O 3		
4401-014	C30	H 52	O 3		
4404-026	C30	H 52	O 3		
4404-006	C30	H 52	O 3		
4415-022	C30	H 52	O 3		
4404-023	C30	H 52	O 3		
4404-029	C30	H 52	O 4		
4404-022	C30	H 52	O 4		
4415-023	C30	H 52	O 4		
4404-011	C30	H 52	O 4		
4404-010	C30	H 52	O 4		
4401-017	C30	H 52	O 4		
4405-013	C30	H 52	O 5		
4410-014	C30	H 56	O 3		
4424-004	C31	H 38	O10		
4424-009	C31	H 40	O 8		
4215-014	C31	H 40	O11		
4505-096	C31	H 41	N 1	O 8	S 1
4592-025	C31	H 42	N 2	O 6	
4699-017	C31	H 42	O 1		
4699-027	C31	H 42	O 3		
4429-005	C31	H 42	O 6		
4429-004	C31	H 42	O 7		
4424-012	C31	H 42	O 8		
4220-013	C31	H 42	O 9		
4505-095	C31	H 43	N 1	O 8	S 1
4429-002	C31	H 44	O 5		
4506-032	C31	H 44	O 9		
4506-034	C31	H 44	O10		
4599-007	C31	H 46	N 2	O 1	
4213-029	C31	H 46	N 2	O 8	
4411-005	C31	H 46	O 3		
4411-011	C31	H 46	O 4		
4411-012	C31	H 46	O 5		
4408-059	C31	H 46	O 5		
4215-004	C31	H 46	O 8		
4411-004	C31	H 48	O 3		
4417-024	C31	H 48	O 3		
4410-021	C31	H 48	O 3		
4411-008	C31	H 48	O 4		
4410-024	C31	H 48	O 4		
4408-115	C31	H 48	O 5		
4410-025	C31	H 48	O 5		
4429-001	C31	H 48	O 6		
4429-007	C31	H 48	O 7		
4408-121	C31	H 48	O 7		
4417-023	C31	H 50	O 1		
4417-022	C31	H 50	O 2		
4402-011	C31	H 50	O 2		
4503-016	C31	H 50	O 2		
4415-014	C31	H 50	O 2		
4415-015	C31	H 50	O 2		
4408-117	C31	H 50	O 3		
4411-002	C31	H 50	O 3		
4410-016	C31	H 50	O 3		
4411-006	C31	H 50	O 4		
4411-010	C31	H 50	O 4		
4411-009	C31	H 50	O 4		
4417-011	C31	H 50	O 5		
4513-013	C31	H 52	N 2	O 3	
4411-001	C31	H 52	O 1		
4402-007	C31	H 52	O 1		
4408-002	C31	H 52	O 1		
4406-002	C31	H 52	O 1		
4417-016	C31	H 52	O 1		
4416-003	C31	H 52	O 1		
4417-017	C31	H 52	O 1		
4407-002	C31	H 52	O 1		
4422-002	C31	H 52	O 1		
4422-004	C31	H 52	O 1		
4417-018	C31	H 52	O 1		
4409-002	C31	H 52	O 1		
4405-005	C31	H 52	O 1		
4415-012	C31	H 52	O 2		
4415-010	C31	H 52	O 2		
4499-007	C31	H 52	O 2		
4415-008	C31	H 52	O 2		
4415-004	C31	H 52	O 2		
4503-006	C31	H 52	O 2		
4417-021	C31	H 52	O 3		
4404-005	C31	H 54	O 2		
4426-020	C32	H 36	O11		
4427-007	C32	H 38	O11		
4291-017	C32	H 40	O 8		
4424-010	C32	H 40	O 8		
4699-031	C32	H 42	O 1		
4507-075	C32	H 42	O 7		
4426-018	C32	H 42	O 9		
4424-005	C32	H 42	O 9		
4427-008	C32	H 42	O10		
4427-016	C32	H 42	O12		
4418-011	C32	H 43	N 1	O 2	
4592-026	C32	H 44	N 2	O 6	
4506-004	C32	H 44	O 7		
4507-034	C32	H 44	O 7		
4412-011	C32	H 44	O 8		
4426-006	C32	H 44	O 8		
4215-009	C32	H 44	O12		
4506-022	C32	H 44	O12		
4215-011	C32	H 44	O13		
4215-013	C32	H 44	O14		
4506-008	C32	H 46	O 7		
4412-006	C32	H 46	O 8		
4412-005	C32	H 46	O 8		
4412-004	C32	H 46	O 9		
4215-006	C32	H 46	O10		
4505-046	C32	H 46	O10		
4505-146	C32	H 46	O10		
4591-004	C32	H 47	N 1	O 8	
4409-009	C32	H 48	O 4		
4417-013	C32	H 48	O 6		
4421-009	C32	H 48	O 6		
4493-046	C32	H 48	O 6		
4492-011	C32	H 48	O 6		
4412-007	C32	H 48	O 8		
4412-021	C32	H 48	O 8		
4505-107	C32	H 48	O 9		
4505-023	C32	H 48	O 9		
4505-105	C32	H 48	O 9		
4516-004	C32	H 49	N 1	O 8	
4516-012	C32	H 49	N 1	O 9	
4408-043	C32	H 50	O 3		
4417-009	C32	H 50	O 3		
4407-006	C32	H 50	O 4		
4410-031	C32	H 50	O 4		
4408-073	C32	H 50	O 4		
4421-008	C32	H 50	O 5		
4405-022	C32	H 50	O 5		
4405-023	C32	H 50	O 5		
4516-005	C32	H 51	N 1	O 8	
4516-025	C32	H 51	N 1	O 9	
4419-011	C32	H 52	N 2	O 2	
4401-005	C32	H 52	O 2		
4401-025	C32	H 52	O 4		
4408-053	C32	H 52	O 4		
4408-054	C32	H 52	O 4		
4408-023	C32	H 52	O 5		
4408-055	C32	H 52	O 5		
4421-006	C32	H 52	O 6		
4504-073	C32	H 52	O 8		
4291-014	C32	H 52	O10		
4403-002	C32	H 54	O 2		
4415-011	C32	H 54	O 2		
4415-006	C32	H 54	O 2		
4401-010	C32	H 54	O 3		
4401-013	C32	H 54	O 3		
4401-024	C32	H 54	O 3		
4401-015	C32	H 54	O 4		
4417-004	C32	H 56	O 2		
4417-020	C32	H 56	O 2		
4419-031	C33	H 28	N 2	O 3	
4699-015	C33	H 42	O 1		
4215-010	C33	H 42	O13		
4699-016	C33	H 44	O 2		
4429-003	C33	H 44	O 8		
4213-008	C33	H 45	N 1	O 8	
4213-010	C33	H 45	N 1	O 9	
4213-019	C33	H 45	N 1	O10	
4213-018	C33	H 45	N 1	O11	
4419-017	C33	H 48	N 2	O 2	
4419-018	C33	H 48	N 2	O 3	
4493-038	C33	H 48	N 2	O 3	
4419-029	C33	H 48	N 2	O 3	
4419-035	C33	H 48	N 2	O 4	
4419-019	C33	H 50	N 2	O 2	
4419-007	C33	H 50	N 2	O 2	
4419-022	C33	H 50	N 2	O 3	
4513-011	C33	H 50	N 2	O 3	

4429-008	C33	H 50	O 8	
4429-006	C33	H 50	O 8	
4104-021	C33	H 52	O 2	
4409-006	C33	H 52	O 4	
4411-003	C33	H 52	O 4	
4410-018	C33	H 52	O 4	
4411-007	C33	H 52	O 5	
4518-004	C33	H 53	N 1	O 6
4518-005	C33	H 53	N 1	O 6
4515-017	C33	H 53	N 1	O 7
4515-020	C33	H 53	N 1	O 7
4518-012	C33	H 53	N 1	O 7
4515-028	C33	H 53	N 1	O 7
4417-019	C33	H 54	O 2	
4417-007	C33	H 54	O 3	
4401-022	C33	H 54	O 5	
4509-015	C33	H 54	O11	
4516-033	C33	H 55	N 1	O 8
4499-004	C33	H 56	O 1	
4599-006	C33	H 57	N 1	O 8
4424-011	C34	H 38	O 8	
4699-014	C34	H 44	O 2	
4493-060	C34	H 44	O 9	
4427-013	C34	H 44	O12	
4506-010	C34	H 46	O 9	
4213-013	C34	H 47	N 1	O 9
4213-016	C34	H 47	N 1	O10
4213-020	C34	H 47	N 1	O10
4213-014	C34	H 47	N 1	O11
4213-030	C34	H 48	N 2	O 9
4505-037	C34	H 48	O14	
4419-028	C34	H 50	N 2	O 3
4419-005	C34	H 52	N 2	
4419-006	C34	H 52	N 2	O 2
4411-013	C34	H 52	O 7	
4516-016	C34	H 53	N 1	O10
4516-017	C34	H 53	N 1	O10
4291-013	C34	H 54	O11	
4599-004	C34	H 55	N 1	O 6
4401-016	C34	H 56	O 5	
4404-028	C34	H 56	O 6	
4507-092	C35	H 40	O 9	
4215-016	C35	H 42	O 9	
4699-012	C35	H 46	O 2	
4426-013	C35	H 46	O11	
4699-013	C35	H 48	O 2	
4516-007	C35	H 49	N 1	O10
4516-014	C35	H 49	N 1	O11
4213-015	C35	H 49	N 1	O12
4419-014	C35	H 50	N 2	O 4
4419-032	C35	H 50	N 2	O 4
4419-027	C35	H 52	N 2	O 3
4408-085	C35	H 52	O 5	
4408-084	C35	H 52	O 5	
4505-085	C35	H 52	O14	
4505-087	C35	H 52	O15	
4505-086	C35	H 52	O15	
4505-088	C35	H 52	O15	
4220-005	C35	H 54	O 8	
4505-103	C35	H 54	O 9	
4505-024	C35	H 54	O13	
4505-008	C35	H 54	O13	
4505-009	C35	H 54	O13	
4505-013	C35	H 54	O14	
4505-060	C35	H 54	O14	
4505-101	C35	H 54	O14	
4505-100	C35	H 54	O14	
4408-031	C35	H 56	O 6	
4408-028	C35	H 56	O 6	
4701-002	C35	H 58	O 1	
4507-033	C36	H 46	O10	
4213-012	C36	H 49	N 1	O 9
4516-006	C36	H 51	N 1	O10
4516-013	C36	H 51	N 1	O11
4213-011	C36	H 51	N 1	O11
4213-031	C36	H 51	N 3	O10
4419-012	C36	H 52	N 2	O 3
4506-015	C36	H 52	O12	
4506-014	C36	H 52	O13	
4506-041	C36	H 52	O15	
4516-018	C36	H 55	N 1	O11
4505-027	C36	H 56	O 3	
4220-002	C36	H 56	O 8	
4291-012	C36	H 56	O12	
4505-025	C36	H 56	O12	
4505-026	C36	H 56	O13	
4505-071	C36	H 56	O13	
4505-077	C36	H 56	O144	
4291-020	C36	H 58	O 6	
4409-007	C36	H 58	O 8	
4412-023	C36	H 58	O 9	
4505-061	C36	H 58	O15	
4213-034	C37	H 50	N 2	O10
4213-017	C37	H 53	N 1	O12
4213-032	C37	H 53	N 3	O10
4492-016	C37	H 54	O 4	
4408-041	C37	H 58	O 8	
4220-004	C37	H 58	O 8	
4516-021	C37	H 59	N 1	O11
4516-020	C37	H 59	N 1	O12
4516-019	C37	H 59	N 1	O12
4293-039	C38	H 38	O10	
4606-003	C38	H 44	O 4	
4606-004	C38	H 48	O 4	
4213-039	C38	H 50	N 2	O10
4426-014	C38	H 52	O11	
4505-108	C38	H 58	O14	
4220-003	C38	H 60	O 8	
4206-037	C38	H 60	O18	
4426-021	C39	H 46	O 8	
4493-057	C39	H 46	O10	
4426-015	C39	H 54	O11	
4408-021	C39	H 56	O 5	
4415-005	C39	H 58	O 5	
4516-024	C39	H 59	N 1	O11
4516-023	C39	H 61	N 1	O12
4516-030	C39	H 61	N 1	O12
4516-028	C39	H 61	N 1	O13
4507-062	C39	H 62	O13	
4518-006	C39	H 63	N 1	O10
4518-007	C39	H 63	N 1	O10
4515-021	C39	H 63	N 1	O11
4518-008	C39	H 63	N 1	O11
4699-010	C40	H 48	O 1	
4699-011	C40	H 48	O 2	
4604-009	C40	H 48	O 4	
4603-020	C40	H 50	O 2	
4604-005	C40	H 50	O 4	
4699-004	C40	H 52		
4601-012	C40	H 52		
4602-009	C40	H 52	O 1	
4603-019	C40	H 52	O 2	
4603-021	C40	H 52	O 2	
4602-010	C40	H 52	O 2	
4604-008	C40	H 52	O 2	
4604-007	C40	H 52	O 3	
4603-022	C40	H 52	O 4	
4601-010	C40	H 54		
4602-008	C40	H 54		
4604-001	C40	H 54	O 1	
4601-030	C40	H 54	O 1	
4602-002	C40	H 54	O 1	
4602-011	C40	H 54	O 1	
4699-005	C40	H 54	O 1	
4603-016	C40	H 54	O 1	
4603-018	C40	H 54	O 2	
4603-017	C40	H 54	O 2	
4602-017	C40	H 54	O 2	
4604-003	C40	H 54	O 2	
4604-002	C40	H 54	O 2	
4602-016	C40	H 54	O 3	
4604-004	C40	H 54	O 3	
4205-009	C40	H 54	O 6	
4601-009	C40	H 56		
4602-001	C40	H 56		
4603-003	C40	H 56		
4601-007	C40	H 56		
4602-007	C40	H 56		
4603-001	C40	H 56		
4603-002	C40	H 56	O 1	
4603-006	C40	H 56	O 1	
4603-007	C40	H 56	O 1	
4603-004	C40	H 56	O 1	
4602-014	C40	H 56	O 1	
4602-012	C40	H 56	O 1	
4603-005	C40	H 56	O 1	
4603-027	C40	H 56	O 1	
4601-015	C40	H 56	O 1	
4601-024	C40	H 56	O 1	
4601-025	C40	H 56	O 2	
4602-015	C40	H 56	O 2	
4606-001	C40	H 56	O 2	
4603-013	C40	H 56	O 2	
4602-004	C40	H 56	O 2	
4699-002	C40	H 56	O 2	
4603-028	C40	H 56	O 2	
4603-008	C40	H 56	O 2	
4603-009	C40	H 56	O 2	
4602-013	C40	H 56	O 2	
4699-001	C40	H 56	O 2	
4601-033	C40	H 56	O 2	
4699-003	C40	H 56	O 3	
4606-002	C40	H 56	O 3	
4603-029	C40	H 56	O 3	
4603-030	C40	H 56	O 3	
4603-011	C40	H 56	O 3	
4603-015	C40	H 56	O 3	
4603-032	C40	H 56	O 3	
4603-025	C40	H 56	O 4	
4604-006	C40	H 56	O 4	

4605-001	C40	H 56	O 4	
4603-031	C40	H 56	O 4	
4606-005	C40	H 56	O 4	
4603-012	C40	H 56	O 4	
4603-024	C40	H 56	O 4	
4605-003	C40	H 56	O 5	
4605-002	C40	H 56	O 5	
4699-009	C40	H 58		
4601-008	C40	H 58		
4602-006	C40	H 58		
4699-007	C40	H 58		
4699-008	C40	H 58		
4602-005	C40	H 58		
4601-005	C40	H 58		
4601-014	C40	H 58	O 1	
4602-003	C40	H 58	O 1	
4601-032	C40	H 58	O 2	
4205-008	C40	H 58	O 2	
4601-006	C40	H 60		
4601-011	C40	H 60		
4601-004	C40	H 60		
4506-002	C40	H 60	N 4	O10
4601-013	C40	H 60	O 1	
4940-001	C40	H 60	O 2	
4204-033	C40	H 60	O 4	
4601-003	C40	H 62		
4601-002	C40	H 64		
4601-001	C40	H 64		
4701-003	C40	H 66	O 1	
4505-028	C40	H 70	O14	
4499-009	C41	H 52	O17	
4699-006	C41	H 54	O 2	
4601-017	C41	H 58	O 1	
4601-026	C41	H 58	O 2	
4601-020	C41	H 58	O 2	
4601-031	C41	H 58	O 2	
4426-016	C41	H 58	O11	
4601-018	C41	H 60	O 1	
4601-016	C41	H 60	O 2	
4601-021	C41	H 60	O 3	
4516-031	C41	H 61	N 1	O13
4601-022	C41	H 62	O 1	
4601-019	C41	H 62	O 2	
4408-097	C41	H 62	O 7	
4408-096	C41	H 62	O 7	
4516-029	C41	H 63	N 1	O13
4516-022	C41	H 63	N 1	O14
4516-027	C41	H 63	N 1	O14
4601-023	C41	H 64	O 1	
4505-051	C41	H 64	O14	
4505-062	C41	H 64	O14	
4505-110	C41	H 64	O15	
4505-010	C41	H 64	O19	
4404-019	C41	H 68	O13	
4601-029	C42	H 56	O 4	
4605-004	C42	H 58	O 6	
4601-027	C42	H 60	O 2	
4601-034	C42	H 62	O 3	
4408-064	C42	H 62	O16	
4506-016	C42	H 62	O18	
4601-028	C42	H 64	O 2	
4505-120	C42	H 64	O15	
4505-030	C42	H 66	O17	
4505-029	C42	H 66	O18	
4505-067	C42	H 66	O19	
4408-126	C42	H 68	O13	
4408-124	C42	H 68	O13	
4404-012	C42	H 72	O14	
4427-015	C43	H 56	O17	
4505-052	C43	H 66	O15	
4701-015	C43	H 70	O 1	
4408-032	C44	H 58	O 7	
4699-023	C44	H 64	O24	
4505-031	C44	H 68	O19	
4501-003	C44	H 78	O 2	
4699-030	C45	H 64	O 1	
4518-009	C45	H 73	N 1	O14
4515-022	C45	H 73	N 1	O15
4515-023	C45	H 73	N 1	O16
4701-004	C45	H 74	O 1	
4701-016	C45	H 74	O 1	
4701-009	C45	H 74	O 1	
4602-019	C46	H 64	O 9	
4602-020	C46	H 66	O 7	
4602-018	C46	H 66	O 8	
4408-005	C46	H 80	O 3	
4505-148	C47	H 74	O17	
4505-032	C47	H 74	O18	
4505-063	C47	H 74	O19	
4408-125	C48	H 76	O17	
4505-033	C49	H 76	O19	
4505-064	C49	H 76	O20	
4699-029	C50	H 72	O 2	
4699-026	C50	H 72	O 2	
4699-028	C50	H 76	O 4	
4505-121	C50	H 76	O21	
4701-005	C50	H 82	O 1	
4701-010	C50	H 82	O 1	
4504-014	C50	H 82	O23	
4518-002	C50	H 83	N 1	O20
4515-014	C50	H 83	N 1	O21
4408-123	C51	H 82	O21	
4509-030	C51	H 82	O23	
4504-074	C51	H 84	O22	
4603-026	C52	H 76	O 5	
4601-035	C52	H 76	O12	
4701-011	C55	H 90	O 1	
4701-006	C55	H 90	O 1	
4701-013	C55	H 92	O 1	
4504-075	C57	H 96	O28	
4405-006	C57	H102	O 2	
4701-007	C60	H 98	O 1	
4701-012	C60	H 98	O 1	
4701-017	C63	H102	O 2	
4916-002	C65	H 18	O 5	
4701-008	C65	H106	O 1	
4603-023	C72	H112	O 6	
4603-014	C72	H116	O 4	
4603-010	C72	H116	O 4	

Devon, T K
Handbook of naturally occurring compounds [by] T. K. Devon [and] A. I. Scott. New York, Academic Press, 1972-1975.
2v. illus. 29cm. index.

Contents: v.1. Acetogenins, Shikimates, and carbohydrates.-v.2. Terpenes.

1.Natural products-Handbooks, manuals, etc. I.Scott, Alastair I., joint author II.Title. III.Title: Naturally occurring compounds.